Laser/Optoelektronik in der Technik
Laser/Optoelectronics in Engineering

Vorträge des 7. Internationalen Kongresses
Proceedings of the 7th International Congress

Laser 85 Optoelektronik

Herausgegeben von/Edited by W. Waidelich

W0037803

Mit 682 Abbildungen/With 682 Figures

Springer-Verlag
Berlin Heidelberg NewYork Tokyo 1986

Dr. rer. nat. Wilhelm Waidelich

o. Professor, Vorstand des Instituts für Medizinische Optik der Universität München
Leiter der Abteilung Angewandte Optik der Gesellschaft für Strahlen- und
Umweltforschung, Neuherberg

ISBN-13: 978-3-540-16017-5   e-ISBN-13: 978-3-642-82638-2
DOI: 10.1007/978-3-642-82638-2

CIP-Kurztitelaufnahme der Deutschen Bibliothek.
Optoelektronik in der Technik : Vorträge d. ...
Internat. Kongresses Laser ... Optoelektronik = Optoelectronics in engineering / Laser. – Berlin ;
Heidelberg ; New York ; Tokyo : Springer
Teilw. mit d. Verlagsorten Berlin, Heidelberg, New York
NE: Laser; PT
7. 1985 (1986).

2362/3020-543210

Vorwort

Die Anwendungen des Lasers erweitern sich in rasantem Tempo. Die von der Münchner Messe- und Ausstellungsgesellschaft alle zwei Jahre durchgeführte Kongreß-Messe LASER - OPTOELEKTRONIK vereinigte erneut die internationale Fachwelt in München, an ihrer Spitze Th. Maiman, den Erfinder des Lasers. Die Vorträge des 7. Internationalen Kongresses LASER 85 OPTOELEKTRONIK zeigen den neuesten Stand und einen Ausblick auf zukünftige Entwicklungen. Der vorliegende Band umfaßt den technischen Bereich, die LASER - Medizin wird in einem gesonderten Band publiziert.

Allen Autoren und dem Springer-Verlag sei für die gute Kooperation herzlich gedankt. Möge das inhaltsreiche Werk fachliche Information und Anregungen vermitteln.

München, im November 1985   W. Waidelich

Preface

The applications of laser extend very rapidly. The biennial LASER-OPTOELECTRONICS fair organized by the MMG attracted again the international experts headed by Mr. Maiman, the inventor of the laser. The papers given at the 7th international congress LASER 85 OPTOELECTRONICS present an overview of the state of the art and future developments. This volume contains the contributions from the technical field. Medical contributions are published in a separate volume.

With all due thanks to the authors and the Springer Verlag this book is dedicated to a wide circle of readers. It is intended for anyone who is looking for expert information and ideas.

Munich, November 1985   W. Waidelich

Inhaltsverzeichnis - Contents

LASER - SYSTEME

Laser Systems

LASER - KOMPONENTEN

Laser Components

LASER- UND OPTOELEKTRONISCHE MESSTECHNIK

Lasers and Optoelectronics in Measuring

X

HOLOGRAPHISCHE INTERFEROMETRIE
Holographic Interferometry

LASER IN DER MATERIALBEARBEITUNG
Lasers in Material Processing

LASER UND OPTOELEKTRONIK IN DER UMWELTMESSTECHNIK

Lasers and Optoelectronics in Environmental Measurement Techniques

LASER - ANWENDUNG IN DER CHEMIE
Laser Application in Chemistry

OPTOELEKTRONISCHE SENSORSYSTEME

Optoelectronic Sensor Systems

OPTOELEKTRONISCHE BILDANALYSE,
-VERARBEITUNG UND -SPEICHERUNG

Optoelectronic Image Analysis,
Processing and Storage

EINZELTHEMEN

Individual Papers

LASER ALS WIRTSCHAFTSFAKTOR

Laser - an Economic Factor

VDI-Technologiezentrum Düsseldorf

Sitzungsleiter - Session Chairmen

Dr. D. Basting   Laser-Systeme

Dr. V. Bödecker   Laser als Wirtschaftsfaktor

Prof. Dr. W.L. Bohn   Laser in der Materialbearbeitung III

Prof. Dr. F. Dörr   Laser - Anwendung in der Chemie

Prof. Dr. K. Gürs   Laser und Optoelektronik
                              in der Weltraumtechnik

Prof. Dr. J. Hesse   Laser-Komponenten

Prof. Dr. F. Lanzl   Holographische Interferometrie

Prof. Dr. S. Maslowski   Optoelektronische Signalübertragung

H.-G. Rosen   Laser in der Materialbearbeitung I

Dr. R. Seitner   Einzelthemen

Dr.-Ing. G. Sepold   Laser in der Materialbearbeitung II

Prof. Dr. H. Tiziani   Laser und optoelektronische Meßtechnik

Dr. Chr. Werner   Laser und Optoelektronik
                              in der Umweltmeßtechnik

Dr.R. Wollermann-Windgasse   Laser in der Materialbearbeitung IV

Referenten - Contributors

Laser – Systeme
Laser – Systems

The Status of Commercial Excimer Laser Development – Performance, Reliability, Applications

H. Pummer

Lambda Physik, 289 Great Road, Acton, MA 01720, USA

Introduction

Excimer lasers form a group of pulsed high-pressure gas lasers which emit on a number of lines in the ultra-violet spectral region. Since its comparatively recent discovery (1) and commercialisation in 1975 and 1976 respectively the excimer laser has undergone a rapid commercial and scientific development resulting in products which now are beginning to find acceptance in the industrial market. Applications of excimer lasers cover a wide range of disciplines from photolithography, laser induced chemical vapor deposition (LCVD), laser annealing, photochemical punching, marking, medical applications, LIDAR, and dye laser pumping to applications in laser fusion. Although traditionally a scientific research tool, this laser's potential especially in semiconductor processing (2) together with its strongly enhanced performance, reliability, and user friendliness has attracted considerable interest from industrial users.

The quite unique applications which are accessible to excimer lasers stem largely from their emission wavelengths which, for the four most commonly used types of excimers, are 193 nm (ArF*), 248 nm (KrF*), 308 nm (XeCl*), and 351 nm (XeF*). The corresponding photon energies are 6.4, 5, 4 and 3.5 eV. This leads to three basic features which distinguish excimer lasers when compared to their infrared counterparts like the Nd:YAG or the CO_2 laser.

The high photon energies of ultra-violet light can directly break chemical bonds without the need to expose the material in question to elevated temperatures. This means that ultra-violet radiaton can drastically alter the chemical behaviour of a system in a "cold" environment, a feature which is of great interest for many photochemical processes including laser induced chemical vapor deposition. It further has led to techniques for the well controlled cold removal of many organic materials including living tissue in a process sometimes called chemical punching.

Excimer lasers not only excel in the way their radiation interacts with materials but also in the high degree of spatial control with which this radiation can be applied. The ultimate resolution of imaging systems is limited by diffraction effects so that the smallest structures which can be resolved or produced are approximately equal to the wavelength of the radiation in use. Excimer lasers are therefore of interest for photolithographic applications where their substantially higher photon fluxes compete with conventional light sources. There is, of course, a large number of potential applications like direct writing, mask and chip repair, highly localized LCVD or micromachining of plastics. Standing structures of 0.2 μm dimensions have been produced in photoresists using excimer lasers.

The third feature is associated with the fact that the absorption coefficients of most materials tend to increase as one approaches shorter wavelengths. In many semiconductor materials this leads to penetration depths of ~10 nm and therefore to the possiblity to flash-heat and anneal very thin layers without heating of the bulk material. In plastics, absorption lengths tend to be of the order of ~1 μm so that films can be photo-ablated with this accuracy. The fact that many metals have UV absorption coefficients which are more than a factor of 10 larger than in the IR spectral region may make excimer lasers of interest for certain marking applications as well.

Laser Performance

Nearly all laser applications which involve material processing require the availability of high average powers from systems which are reliable and easy to operate. We will therefore first give a brief discussion of the parameters which determine the average power from excimer lasers (a somewhat more detailed discussion is given in ref. (3).

All commercial excimer lasers are rare gas halide lasers whose upper electronically excited laser level is a diatomic molecule consisting of a rare gas atom and a halogen atom (e.g. ArF*, KrF*, XeCl*, XeF*, KrCl*). The lower laser level is either not or only weakly bound. Such molecules which exist only in excited states are called excimers or exciplexes. The excimers are formed via reaction in a pulsed high voltage discharge whose kinetics is quite complex (4). It is, however possible to define the main reaction channels as

a) excitation or ionization of the rare gas via electron collision

$$Rg + e^- \nearrow\; Rg^* + e^- \atop \searrow\; Rg^+ + 2e^-$$

b) disassociative attachment of the halogen (X_2) molecule

$$X_2 + e^- \longrightarrow X^- + X$$

c) chemical "harpooning" reaction and three body recombination

$$Rg^* + X_2 \longrightarrow RgX^* + X$$

$$Rg^+ + X^- + M \longrightarrow RgX^* + M$$

where M is a collision partner, and

d) stimulated emission

$$RgX^* + \gamma \longrightarrow Rg + X + 2\gamma$$

Because the excited excimer has a spontaneous lifetime of only several ns, its generation has to occur on a fast timescale in order to achieve a sizeable inversion density (e.g. 10^{15} cm^{-3}). This requires high pressures to make the chemical reactions c fast enough as well as very high power density discharges which can prepare high enough concentrations of the excited or ionized rare gas species.

A typical excimer laser gas mixture will consist of e.g. 4 mbar F_2 , 120 mbar Kr and 2400 mbar He where the F_2 and Kr are necessary for the excimer formation and the He is essential for stabilization of the discharge and serves as the third body in the three body recombination. The discharges typically have electron densities of 10^{15} cm^{-3}, electron temperatures of ~1 eV, current densities of 10^3 Acm^{-2}, power densities of 10^6 Wcm^{-3}, and breakdown fields of 10 - 15 kV/cm. These discharges usually work within a parameter space of +/- 25 % with respect to total pressure and discharge voltage but tend to form arcs outside this region. Because broken down excimer gas mixtures have low impedance ($\sim 10^{-1}\Omega$), efficient excitation requires fast low inductance high voltage circuitry. Such free running discharges usually terminate after 30 - 50 ns due to their time constants and the formation of instabilities. It appears that the overall kinetic properties of such discharge pumped systems allow an energy extraction of 1 - 3 mJcm^{-3}. Naturally e-beam pumped systems can operate in a much wider range of parameters, however, due to design problems which are mainly related to the foil no commercial supplier uses this mode of excitation.

Volume Scalability

To generate higher average powers from a discharge pumped excimer laser one has to increase either the excited volume or the pulse repetition rate. In principle the volume can be increased by increasing the electrode spacing, the width of the discharge and its length. There are, however, practical limitations in doing so. When increasing the electrode spacing the charging voltage of the discharge circuitry has to be increased accordingly. To avoid corona problems and discharge instabilities, manufacturers tend to settle for charging voltages of 20 - 30 kV and corresponding electrode spacings of 2 - 3 cm. In order to break down homogeneously, excimer laser discharges need a start-up electron density of 10^7 - 10^8 cm^{-3} which is provided by a preionization mechanism. Most manufacturers achieve this by spark discharges in the laser gas immediately before the fireing of the main discharge. Ultraviolet radiation from the sparks then provides a simple means for photoionization. However, the short absorption length of the ultraviolet radiation in the laser gas limits the usable width of the preionized gas layer to 10 - 20 mm. Somewhat larger preionization widths are possible with X-ray preionization and this technique will probably be used when industrial applications requiring larger single pulse energy justify the increase in complexity.

The length of usable discharge regions can be limited by nonsaturable losses and by the optical roundtrip time in the resonator. Although excimer lasers are high gain lasers, they need two to three resonator round trips in order to efficiently extract their energy. Because the kinetic properties of discharge pumped excimer lasers limit their pulse duration to 10 - 25 ns, resonator lengths tend to be around 1 m or less.

In conclusion, the volumes of excimer laser discharges tend to be several 100 cm^3 and single pulse energies are of the order of 100 - 1000 mJ.

Repetition Rates

At present, excimer lasers with repetition rates of up to 500 Hz and average power in excess of 100 W are available. All these lasers employ gas circulation which is driven by an internal fan because the gas in the discharge region has to be exchanged between shots. It can be expected that with a careful design of the gas flow these repetition rates can be increased substantially.

Component lifetimes and reliability

It appears that the pioneering days of excimer laser development when gas lifetimes were measured in thousands of shots and high voltage components like the thyratron switch were operated close to their maximum specifications are over. Todays generation of excimer lasers has performed nonstop test runs of several 10^8 shots without component replacement. With the arrival of magnetic switching techniques which reduce current rise times and prevent reverse currents in the high voltage circuit, the thyratron switch has ceased to be a critical high voltage component. Indeed with more than 200 Magnetic Switch Control (MSC) lasers in the field, no thyratron failure has ever occurred. At present the limiting high voltage component seems to be the pair of discharge electrodes which are - as i's the case in all high voltage discharges - subject to slow electrode burn-off. It is not straight foreward to define the lifetime of an electrode because as the electrode shape changes due to burn-off the laser will continue to operate although parameters like output energy, beam homogeneity, pulse to pulse stability, and jitter may change. However, good laser performance with XeCl* for greater than 3×10^8 shots has been achieved. As component and gas fill lifetimes increase it becomes more and more time consuming to evaluate such parameters. It takes six months to generate 3×10^8 shots with a 100 Hz repetition rate laser during working hours and it can be doubted that many companies have the capacity to run such tests with statistical significance. In this situation customer feedback from a large number of lasers in the field is certainly most helpful. Electrode burn-off not only changes the shape of the electrodes and thereby limits their lifetime but also introduces gaseous or particulate contaminants into the laser gas and leads to slow consumption of the halogen component. Besides causing absorption or scattering losses in the gas phase, the contaminants can be mechanically or photochemically deposited on the laser windows where they form an absorbing layer. Other contaminants can originate from dark reactions of the halogen and parts of the gas handling system and the laser itself. The first step in building a well performing excimer laser is therefore a judicious choice of halogen compatible materials and good quality control during the assembly and final tests. Furthermore, in systems which are designed for very large numbers of shots, some active devices for contaminant reduction need to be integrated into the laser system. Particles, for instance can be removed from the laser gas by using electrostatic or mechanical dustfilters while many molecular species can be trapped cryogenically. Most modern excimer lasers can be fitted with an external gas processor which continually removes a fraction of the laser gas, cleans it and reinjects it at the laser windows, thereby providing a steady flow of clean gas away from the windows. When operating with XeCl*, periods between window cleaning (the exchange of windows typically takes less than 15 min) should be in excess of $>10^7$ shots although periods of $>10^8$ shots have been observed in well passivated systems.

Summarizing, it can be stated that at present the lifetime of an excimer laser is limited by the predictable wear of exchangeable parts rather than by catastrophic failure of components.

Gas Lifetimes and Computer Control

With the use of better halogen compatible materials and gas processors, the lifetime of excimer laser gases has increased dramatically to several 10^7 shots for XeCl*, such that at this point the laser gases for a continuously running high repetition rate excimer laser have become a small part in the overall operation budget. Even so, the average power from an excimer laser tends to decrease in time due to slow consumption of the halogen component as well as to an increase in impurity level which cannot be completely removed by the gas processing equipment. This is obviously a severe hindrance in many industrial applications which have to rely on a constant power level.

The answer to this problem is computer controlled excimer laser systems which can keep the output power constant in a number of ways and which can be run semi-automatically via operator activated push buttons, automatically under their own computer control, or connected to a main computer through an RS 232 interface. In a representative system, the output of the laser is monitored with a beamsplitter-photodiode package with individual calibrations for the various excimer wavelengths. The operator has the choice to fill the laser per push button with any of a number of preprogrammed "menu" gas mixtures or he can input his individual gas mixture. After the microprocessor has gone through the filling procedure the operator can program a charging voltage and start the laser, which will then run at a fixed voltage while ignoring the energy monitor. Alternatively the operator can program - within a certain range - a desired single shot energy, choose the repetition rate, and start the laser. From then on the laser can run at a constant pulse energy without operator assistance as long as there is a gas supply and no standard maintenance procedure becomes necessary. Initially, the microprocessor will step by step adjust the charging voltage until the programmed pulse energy is reached. As the pulse energy tends to decline the charging voltage is gradually increased until its maximum value is reached. When this occurs, the microprocessor injects a small amount of halogen into the laser which usually increases the energy. It will then readjust the charging voltage and repeat this cycle until injection of halogen ceases to affect the laser output. Upon this the control will proceed in one of two possible ways. In the "total gas replacement" mode the microprocessor will stop the laser, exchange the gas completely and restart the laser. In the "partial gas replacement" mode the laser will continue to operate while part of its gas is removed and replaced by fresh laser gas. At the Conference on Lasers and Electro-Optics 1985 in Baltimore, an excimer laser was operating in this mode during the entire exhibition at a pulse repetition rate of 300 Hz and 45 W average power. For integration into a production line all these functions including the exact temporal sequence of laser pulses can be controlled externally through an RS 232 interface.

Applications

As briefly outlined above excimer lasers offer the advantage of cold photochemical bond breaking, the ability to create submicron structures, a strong absorptive coupling to many materials, and of course any combination of these three features. An example for the use of the excimer radiations photochemical activity is photochemical ablation or punching which is observed in many organic substances, including most plastics (which in turn include the photoresists used in photolithography) as well as living tissue. When a plastic is machined with an infrared laser, the absorbed energy excites molecular vibrations which means that the material will heat up and melt, evaporate or thermally decompose. The process will be strongly influenced by thermal conduction which in turn will lead to strong edge effects which reach into the unilluminated zones. In contrast to this, material removal with an excimer laser can occur under nearly complete absence of thermal effects. Above a certain energy density which, dependent on the type of polymer and the excimer wavelength, can be several 10 to several 100 mJ/cm^2, the excimer laser radiation is obviously capable of simultaneously breaking so many polymer bonds that the remaining small fragments can ablate as a relatively cold high vapor pressure gas. (5) Because thermal effects seem to play only a minor role in this process, extremely sharp and untraumatized edges can be achieved, a finding which is of considerable importance for photolithographic and medical applications. This process can in principle eliminate the development step in photolithography techniques, like direct writing and mask photolithography. With the latter technique self-developing structures of 0,5 μm dimensions have been created in photoresists. (6) When machining larger structures in plastics a typical ablative removal of 1 μm thick layers per laser shot is observed. With the repetition rates of presently available excimer lasers this corresponds to an exposure time of 2 - 10 s per mm of removed material. The quantitative removal of plastic layers from fragile metallic structures has been demonstrated by laser stripping the insulation from micro wires of 50 μm diameter. It is further possible to machine or cut very fragile organic stuctures like certain biological specimens which otherwise could only be handled after embedding them in a supporting material.

The fact that in excimer laser ablation of living tissue the thickness of the thermally influenced edge zone can usually be kept to less than 1 μm, means that regions of charred and denatured tissue are virtually absent. The healing process should therefore be facilitated and the formation of scar tissue should be greatly reduced.

In photolithography the three main steps are the addition, selection, and subtraction of materials. Again, the photochemical properties of excimer laser radiation together with good focusability and high photon fluxes can represent certain advantages. In addition - techniques like chemical vapor deposition (CVD) and subtraction techniques like etching where part of the reactants are present in the gas phase, the chemical reactions involved are often initiated by heating and/or plasma discharges. However, heating may lead to warpage of wafers. In the case of large diameter wafers, it may lead to dissociation in compound semiconductors, and it may cause undesirable diffusion of dopants as well as chemical surface reactions. Techniques which allow production at lower temperatures therefore offer considerable potential advantages. In certain cases laser techniques may offer advantages over plasma processes in that they can provide highly localized processing, may be more specific with respect to the radicals which are created and lead to less damage through bombardment with charged particles.

In laser assisted etching, the substrate is placed in a gaseous environment which often contains a chlorine or fluorine compound. The laser is directed either parallel or perpendicular to the substrate surface. Because excimer wavelengths can readily disassociate molecular species highly reactive gas phase radicals are formed which then diffuse to the surface and initiate the etching process. With perpendicular incidence, more complicated processes like surface heating, photochemical ablation or surface reactions, as well as the formation of photoelectrons and electron-hole pairs, can come into play. Laser assisted etching has been investigated in a variety of materials like Si, SiO_2, GaAs, W, Mo, Al and Cr with etchants like F_2 , Cl_2, NF_3, COF_2 and various Freons. These experiments have resulted in the direct production of submicron structures. Etch rates can vary widely from 100 nm per pulse in Al to 1 atom per 10^3 photons in Si. At higher laser energies ablative processes occur which can be used to remove metal films from dielectric substrates with micron to submicron resolution.

Laser assisted CVD (LCVD) can be used to deposit dielectric, semiconductor or metallic films on various substrates at reduced temperatures. The high average photon fluxes from excimer lasers lead to substantially higher deposition rates then those obtained with ultraviolet lamps. Semiconductor films are mostly deposited by CVD, insulators by CVD and plasma enhanced CVD, and metal films by physical vapor deposition like sputtering or evaporation. A problem with the latter two techniques is poor step coverage which can be overcome by laser depositing metals from their gas phase hexacarbonyls or hexafluorides. Temperatures for growth of Si from SiH_4 , SiH_2Cl_2 and $SiHCl_3$ which is 1050 -1200°C using CVD have been lowered to 650°C in LCVD. Of further importance is the laser recrystallization of silicon films on insulators. Although it may be too early to judge whether LCVD will become an economic technique for certain large area depositions of films, the possibility of direct writing bears considerable potential.

Photolithography using either contact or projection techniques has become the first industrial application of excimer lasers. As driven by the need for higher and higher resolution these techniques move towards an increasing use of the deep UV spectral range, lasers become more and more competitive with respect to the classic lamp sources. The excimer laser with average powers of 50 - 100 W in the deep UV compared to $\sim 10^{-2}$ W from conventional lamps is especially useful for this application because due to its extremely low spatial and temporal coherence it produces virtually no speckle. Not only can the laser reduce exposure times by two

14

orders of magnitude, it can generate structures with 0.2 µm dimension in contact aligners. As the exposure time becomes a less critical factor, the resist sensitivity becomes less important and can be traded for other resist properties like swelling, development time and contrast. In addition, the complexity of structures which optical pattern generators can accomodate are limited by the time such a system can operate without maintenance or failure and are therefore strongly dependant on exposure time.

Conclusion

Excimer lasers whose average powers crossed the 100 W mark in 1984 are well on their way to average powers of several 100 W and repetition rates beyond 10^3 Hz. The unique properties of their high-photon-energy, short-wavelength radiation predestine them for a large variety of possible applications with some very interesting ones in semiconductor processing and medical applications. The requirements of industrial environments have been answered with the development of highly reliable and computer controlled systems. The focus is now on the development of systems which are tailored for certain larger scale applications with respect to their output parameters as well as to the engineering and packaging requirements of OEM companies.

References

(1) S.K. SEARLES, G.A. HART, Appl. Phys. Lett. 27, 243 (1975)

(2) T.F. DEUTSCH in "Laser Processing and Diagnostics"
ed. by D.Bauerle (Springer Verlag, Berlin/Heidelberg/
New York/Tokyo, 1984)

(3) H. PUMMER, Photonics Spectra Vol.19, 5 ,73 (1985)

(4) CH.A. BRAU in "Excimer Lasers" ed. by Ch.K.Rhodes (Springer
Verlag, Berlin/Heidelberg/New York/Tokyo, 2nd edition 1984)

(5) R. SRINIVASAN, W.J. LEIGH, J. Amer. Chem. Soc., 104: 6784 (1982)

(6) K. JAIN, Lasers and Applications, Sept. 1983

Longitudinal gepumpte Excimerlaser

H. J. Eichler, H. Hamisch, B. Nagel und W. Schmid
Optisches Institut, Technische Universität Berlin
Straße des 17. Juni 135, D-1000 Berlin 12

Excimerlaser lassen sich nach der Art der Energiezufuhr grob unterteilen in solche mit Elektronenstrahlanregung und solche mit transversaler oder longitudinaler Gasentladungsanregung. Beim longitudinal
gepumpten Laser erfolgt die Gasentladung parallel zur optischen Achse
des Laserresonators. Eine derartige Anordnung wurde bisher beim XeF-
Laser (λ= 351 nm)/1-3/ und beim XeCl-Laser (λ= 308 nm)/4/ untersucht.
Im folgenden werden zusätzlich Untersuchungen am KrF-Laser (λ= 248 nm)
vorgestellt.
Der prinzipielle Aufbau der Anordnung ist im folgenden Bild gezeigt:

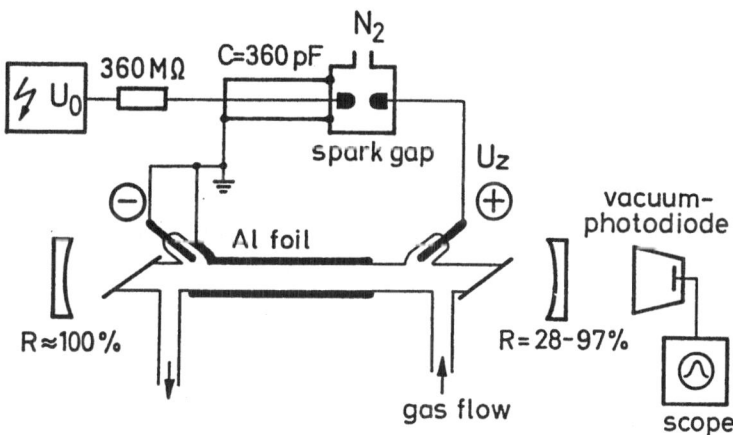

Zwischen den beiden Resonatorspiegeln befindet sich ein einfaches
Entladungsrohr mit zwei Elektroden und zwei Gasanschlüssen und Brew-
ster-Fenstern an den Enden. Im Laserrohr ist ein Gasgemisch aus Krypton (beim KrF-Laser), Fluor und Helium, das während des Betriebes
laufend ausgetauscht wird. Ein koaxialer Kondensator wird mit Span-

nungen um 60 kV geladen und über das Laserrohr entladen. Als schneller
Schalter dient dabei eine Hochdruckfunkenstrecke, deren Zündspannung
über ihren N_2-Innendruck regelbar ist.

Das Laserrohr ist mit einer geerdeten Aluminiumfolie ummantelt. Diese
Ummantelung ist mit der Kathode verbunden; sie stellt eine zusätzliche
kapazitive Kathode dar. Beim XeF-Laser wurden hierdurch die Zündeigen-
schaften wesentlich verbessert, insbesondere bei großen Elektrodenab-
ständen. Außerdem konnte durch die kapazitive Kathode die Laserleistung
gesteigert werden, wie die folgende Tabelle zeigt:

		ohne Ummantelung		mit Ummantelung	
l	[cm]	10 ⎫	Optimum	10	35 ⎫ Optimum
d_i	[mm]	1,5 ⎭		1,5	6,5 ⎭
ϕ	[W]	72		540	12000
$t_{1/2}$	[ns]	30		17	30
E	[μJ]	2,2		9	350

Tabelle 1: XeF-Laser ohne und mit Ummantelung (für U= 60 kV)

Die Laserleistung für das ohne Ummantelung optimale Rohr (10 cm Länge,
1,5 mm Innendurchmesser) kann durch Zufügen der Ummantelung von 72 W
auf 540 W gesteigert werden. Eine weitere Steigerung auf 12 kW erfolgt
durch den Übergang zu den optimalen Rohrmaßen für ummantelte Rohre
(35 cm Länge, 6,5 mm Innendurchmesser). - Die kapazitive Kathode
wurde wegen dieser beim XeF-Laser positiven Eigenschaften auch beim
KrF-Laser beibehalten.

Die Laserleistung des KrF-Lasers wurde in Abhängigkeit von der Ent-
ladungsrohrgeometrie, von der am Laserrohr angelegten Spannung, von
der Reflexion des Auskoppelspiegels und von den Partialdrücken der
Füllgase gemessen. Die Pulsfolgefrequenz betrug dabei 2 Hz.

Die Laserdaten, die nach einer Optimierung der genannten Parameter
erreicht wurden, zeigt die linke Spalte in Tabelle 2. Die Spannung
U= 60 kV ist dabei allerdings nicht optimiert, sondern durch die
Spannungsfestigkeit des Kondensators bedingt. Kurzzeitig darüber
hinausgehende Spannungen bis U= 66 kV ergaben um den Faktor 2,5 höhere

Leistungen. - Der optimale Elektrodenabstand ist 35 cm, wie beim
XeF-Laser. Der optimale Rohrdurchmesser ist jedoch nur 2 mm. Bei
größeren Durchmessern beginnt die Entladung, sich an der Rohrwand
zu konzentrieren, verbunden mit einer Leistungsabnahme. Dieser Effekt
trat bei XeF erst bei größeren Durchmessern auf. - Der Gesamtdruck
besteht zum größten Teil aus dem He-Anteil, der Fluor-Druck beträgt
einige mbar, der Krypton-Druck ist hier 15-mal höher.

l	[cm]	35	2 x 35	2 x 35
d_i	[mm]	2	2	6,5
ϕ	[kW]	0,5	3,3	3,4
$t_{1/2}$	[ns]	16	10	15
E	[μJ]	8	33	51
P_{F_2}	[mbar]	5	5	4
P_{Kr}	[mbar]	72	52	45
P_{He}	[mbar]	660	740	450

Tabelle 2: KrF-Laser mit einer und mit zwei Entladungsstrecken
          (für U= 60 kV)

Eine weitere Leistungssteigerung wurde durch ein segmentiertes Entladungsrohr erreicht, wie die beiden rechten Spalten in Tabelle 2 zeigen.
Bei der Segmentierung wird die gesamte Länge des Lasermediums in mehrere Entladungsstrecken aufgeteilt, die optisch hintereinander liegen
und elektrisch parallel geschaltet sind. D.h. im Entladungsrohr wech-
seln sich Anoden und Kathoden ab. In Tabelle 2 sind die Meßergebnisse
mit optimierten Parametern für das einfache Beispiel einer Zweiteilung
der Gesamtlänge von 70 cm angegeben, d.h. für ein Rohr mit 3 Elektroden im Abstand von jeweils 35 cm. Die übrige elektrische Be-
schaltung wurde dabei nicht geändert. Die rechte Spalte in Tabelle 2
zeigt, daß mit der segmentierten Anordnung ohne Leistungsverschlech-
terung der Übergang zu größeren Rohrdurchmessern möglich ist: Man erhält eine Laserpulsenergie von ca. 50 μJ.
Durch segmentierte Entladungsrohre wird die im Kondensator gespeicherte
Energie, die gegenüber dem einfachen Rohr nicht geändert wurde, offensichtlich besser für die Umwandlung in Laserenergie ausgenutzt. Eine
weitere Steigerung der Laserenergie scheint uns möglich durch die
systematische Untersuchung segmentierter Entladungsanordnungen, d.h.

durch Variation der Anzahl und der Abmessungen der einzelnen Entladungsstrecken.

Zusammenfassung: Der longitudinal gepumpte Excimerlaser liefert zwar
nur geringe Ausgangsleistungen, zeichnet sich aber durch seinen einfachen Aufbau sowie durch vielfältige Variationsmöglichkeiten aus. Die
bisher erreichten Pulsenergien sind 50 µJ für den KrF-Laser und 350 µJ
für den XeF-Laser.

Literatur
/1/ I.M. ISAKOV, A.G. LEONOV, V.E. OGLUZDIN: Sov. Tech. Phys. Lett. $\underline{3}$,
     397 (1977)
/2/ D. CLESCHINSKY, D. DAMMASCH, H. J. EICHLER, J. HAMISCH; Opt.
     Commun. $\underline{39}$, 79 (1981)
/3/ P. BURKHARD, T. GERBER, W. LÜTHY: Appl. Phys. Lett. $\underline{39}$, 19 (1981)
/4/ ZHENZHUO ZHOU, YONGJIAN ZENG, MINGXIN QIU: Appl. Phys. Lett. $\underline{43}$,
     347 (1983)

Multigas-Hochleistungsexcimerlaser hoher Repetitionsrate

P. Oesterlin und U. Rebhan

Lambda Physik Forschungs- und Entwicklungs-GmbH

Hans-Böckler-Str. 12, D-3400 Göttingen

Die mittlere Ausgangsleistung von gepulsten Excimerlasern ist in den letzten Jahren in einem beachtenswerten Tempo gestiegen. Stellten noch vor 5 Jahren kommerzielle Laser mit Leistungen von mehr als 10 W eine Ausnahme dar, so sind solche Laser heutzutage das "Low End" der Produktpalette der Excimerlaserhersteller. Das "High End" liegt momentan bei etwa 150 W, wenn man kommerzielle Laser betrachtet, und in den Entwicklungslaboratorien wesentlich darüber (1). Excimerlaser im kW-Bereich kommen bereits in Sicht, und niemand kann im Augenblick sagen, wo die technologisch gesetzte Grenze liegen wird.

Die enorme Leistungssteigerung wurde dadurch erreicht, daß sowohl die Energie der einzelnen Laserpulse wie auch die Pulswiederholfrequenz gesteigert wurden, deren Produkt ja die mittlere Leistung ergibt. Für einen kommerziellen Laser steht natürlich nicht allein die Lichtleistung im Vordergrund, sondern die Zuverlässigkeit und die Service-Intervalle sind ebenso wichtige Kriterien. Excimerlaser stehen kurz vor ihrem Einsatz in der industriellen Produktion. Die Entwicklungsteams der Excimerlaser-Hersteller sind daher mit der Aufgabe konfrontiert, nicht nur die Leistung der Laser, sondern in gleichem Maße auch die Zuverlässigkeit zu steigern.

Zwei technische Teilbereiche eines Excimerlasers bedürfen besonderer Aufmerksamkeit, wenn ein Hochleistungslaser entwickelt werden soll. Da ist als erstes der elektrische Entladungskreis mit den Speicherkondensatoren und einem Leistungsschalter. Der zweite Teil ist die Gasumwälzung und Gaskühlung, woran sich häufig noch eine Gasreinigung anschließt. Diese werden im folgenden näher dargestellt:

a) Entladungskreis

Alle kommerziellen Excimerlaser werden durch eine gepulste Gasentladung gepumpt, die sich dadurch charakterisieren läßt, daß für 50 - 100 nsec eine elektrische Leistung von einigen hundert Megawatt an das Lasergas abgegeben wird. Dazu sind Spannungen von 20 - 40 kV und Ströme von 20 - 50 kA nötig, um eine Gasentladung mit der notwendigen Energiedichte zu erzeugen.

Diese Zahlen zeigen, daß ein solcher Entladungskreis sorgfältig entwickelt werden muß. Seine Komponenten sind großen Belastungen unterworfen, besonders der Schalter, der die Entladung einleitet. Als Standardschalter dienen heute wasserstoffgefüllte Thyratrons. Um einen Excimerlaser mit sehr hoher Ausgangsleistung zu bauen, genügt es jedoch nicht, das stärkste verfügbare Thyratron einzubauen. Neue Technologien mußten entwickelt werden, um die Probleme zu lösen. Dabei leisten amorphe Metalle wertvolle Dienste. Mit diesen Materialien kann man sättigbare Induktivitäten bauen, die innerhalb weniger Nanosekunden ihre Induktivität um einen Faktor 1000 verringern. Auf diese Weise wird eine Schaltfunktion erzeugt, mit der das Thyratron entlastet werden kann.

Bei Lambda Physik führte der Einsatz dieser Technologie zur Entwicklung der Excimerlaser der MSC-Serie (MSC: Magnetic Switch Control). (2) Inzwischen sind etwa 200 solcher Laser weltweit im Einsatz, ohne daß bisher ein Defekt wesentlicher Komponenten bekannt geworden wäre.

Labortests, die die Lebensdauer der Thyratrons in diesen Lasern ermitteln sollen, sind bisher nicht abgeschlossen. Im Vergleich zu herkömmlichen Lasern wurde jedoch eine Verlängerung auf mindestens das Zehnfache beobachtet, und das bei gleichzeitiger Erhöhung der Laserleistung. Alle anderen Komponenten des elektrischen Systems haben noch längere Standzeiten, so daß die MSC-Technologie neben den hohen Leistungen auch eine wesentlich höhere Zuverlässigkeit der Excimerlaser zur Folge hat. 200 Millionen Laserpulse ohne Ausfall von wesentlichen Komponenten ist bei diesen Lasern bereits Standard.

b) Gasströmung

Ein Excimerlaser kann nur dann befriedigend funktionieren, wenn das Gas zwischen den Entladungselektroden nach jedem Schuß ausgetauscht wird. Für hochrepetierende Excimerlaser bedeutet das, daß Gasströmungen erzeugt werden müssen, die einerseits sehr schnell sind und andererseits möglichst wirbelfrei sein sollen. Wirbel behindern den Gasaustausch, da das im Wirbel "gefangene" Gas nur sehr langsam mit vorbeiströmendem vermischt wird.

Um die Strömung quantitativ zu beschreiben, kann das "Clearing Ratio" benutzt werden. Es ist eine Zahl, die angibt, wie oft das Gas in der Entladungszone zwischen zwei Laserpulsen ausgewechselt wird. Da die Strömungsgeschwindigkeit nicht überall gleich ist, beschreibt das Clearing Ratio nur die mittlere Gasaustauschrate. Ein Laser mit wirbelnder Strömung benötigt daher ein Clearing Ratio über 2, während für einen Laser mit optimierter Gasführung ein Clearing Ratio von 1 noch leicht unterschritten werden kann. So zeigten Experimente bei

EMG 204 MSC

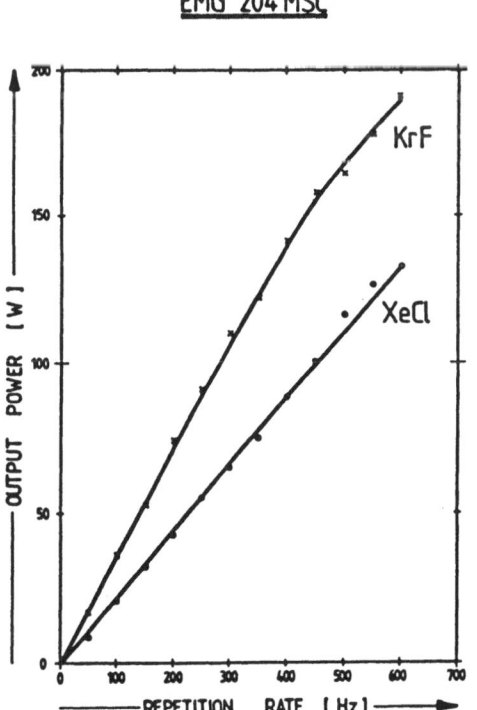

Ausgangsleistung als Funktion der Repetitionsrate bei einem Labormuster des EMG 204 MSC

Lambda Physik, daß mit einer sorgfältig optimierten Strömung, bei der auf Homogenität und Wirbelfreiheit geachtet wurde, bei XeCl ein Clearing Ratio zwischen 0,5 und 1 ausreicht, um einen stabilen Laserbetrieb zu erreichen.

Mit dieser Entwicklung ist es Lambda Physik gelungen, Excimerlaser kommerziell anzubieten, die mehr als 100 W Ausgangsleistung bei 500 Hz Repetitionsrate erreichen (EMG 204 MSC). Im Labor konnten mit solchen Lasern mit geringen Modifikationen fast 200 W bzw. 1 kHz Wiederholfrequenz demonstriert werden (siehe Abbildung).

Dieser Lasertyp kann, wie andere Laser von Lambda Physik auch, mit einer µP-Steuerung ausgerüstet werden. Sie gestattet nicht nur eine unkomplizierte Bedienung, sondern erlaubt auch eine Leistungsstbilisierung für lange Betriebsdauern, was die Anwendungsmöglichkeiten solcher Laser noch erweitert.

Parallel zu solchen Entwicklungen arbeitet Lambda Physik an Forschungsprojekten, deren Ziel der Bau von Excimerlasern mit noch höheren Ausgangsleistungen ist. Die technologischen Erfahrungen aus diesen Projekten werden natürlich in die kommerziellen Laser einfließen, so daß Leistung und Zuverlässigkeit von Excimerlasern auch in Zukunft noch steigen werden.

Literatur

(1) H.PUMMER, U.SOWADA, P.OESTERLIN, U.REBHAN und D.BASTING,
    Laser und Optoelektronik 17, 2, 141 (1985)
(2) D.BASTING, K.HOHLA, E.ALBERS und H.v.BERGMANN,
    Laser und Optoelektronik 16, 2, 128 (1984)

Pulsed High Power CO_2 Lasers

H.Schülke, V.Sturm

TH Darmstadt, Institut für Angewandte Physik

Schloßgartenstr.7, D-6100 Darmstadt

1. Introduction, motivation

With increasing knowledge of the physical processes occuring in material processing appears the interest in controlled laser power modulation. The aim is the creation of processing diagrams for optimum processing-efficiency and quality /1/. The representative frequencies range from kHz (melting and capillary dynamics) to MHz (plasma propagation) /2/.

One modulation technique for the high power range is pulse amplitude modulation of pump power of a radio frequency excited CO_2 laser. Compared with electro-optical techniques the main advantage is, in addition to the higher stability of the rf discharge, that no additional optical components are used in the laser beam, hence there are no limits in high-power application and no disturbances of beam quality. The limiting factors of modulation frequency are the modulation capability of rf power (depending on exciting frequency, /7/) and laser kinetic processes. In this paper the temporal behaviour of the laser emission, the relevant time constants and the modulation behaviour for modulation frequencies up to 100kHz will be discussed. The results reported here in the laser power region of 100W can be carried over into the high-power region of 1-10kW.

2. Experimental set-up

A CO_2 laser system with a fast axial gas flow (v≈300m/s) is used for the experiments (fig.1). A capacitively coupled rf discharge (f=27MHz) excites the laser gas. The discharge length is 30cm with a circular cross-sectional diameter of 24mm ($V=136cm^3$). The typical pulse power density lies in the range of $50W/cm^3$. The cw power of the cavity oscillator is 8kW. The rf power can be modulated up to 100kHz by modulation of the grid voltage of the generator tube. The frequency limit depends on the cavity Q value (Q≈200). The temporal behaviour of the laser power is measured with a pyroelectrical detector (f ≥ 1.5MHz).

3. Fundamental-mode relaxation oscillation

For a better understanding of laser kinetics and relavant time constants, the temporal behaviour of laser emission for fundamental mode operation (aperture A1=10mm) has been examined and described by a computer model.

3.1 Experiments

Fig.2 shows a typical laser pulse. After a time delay t_v the laser emission starts with a narrow peak (typical half-width 1µs) and damped relaxation oscillations. The strong damping is caused by the special excitation and relaxation processes in the CO_2-N_2-He molecular system. The dependence of damping on gas mixture is shown in fig.4a. The peak power is two to five times greater than the cw-power with the same excitation power input. After relaxation oscillations the laser power increases with a typical time constant t_d of 30-100µs. This is caused by two processes: the energy transfer from excited N_2 to the upper laser level (001) of CO_2 with a time constant of 10µs (70mbar,He:N_2:CO_2 8:1:1) and diffusion of excited molecules from the outer discharge volume into the mode volume ($\approx$30µs). At the end of the rf pulse the laser power decreases with a time constant t_a. The dependence of t_a on N_2 fraction is shown in fig.4b. t_a decreases with increasing pressure: t_a(40mbar)=130µs and t_a(100mbar)=25µs for a 8:1:1 gas mixture.

The time constants t_v, t_d and t_a determine the modulation behaviour of laser power in the kHz frequency range. An upper frequency limit (with fixed duty cycle and pump power) results from the delay time t_v (fig.3,fig.8).

3.2 Computer model

The rate equations are formulated in a five-temperature model of the CO_2:N_2:He system /3/, whereby the rotation and intramode relaxation processes have been included /4,5/. The nonlinear differential equation system has been solved numerically (fig.5,fig.2). The vibrational excitation rates for the 27MHz-rf excitation are calculated by averaging the dc values over a rf period /6/. The effects of diffusion ($t_d \approx$30µs) and gasdynamics (t_g=1ms) are neglected in the time region considered here.

4. Higher-order mode laser operation

Increasing the aperture (A1=14mm) higher-order transversal modes can oscillate. This not only influences the spatial behaviour of the laser power, but the temporal behaviour as well. Depending on the rf power coupled into the gas discharge, additional relaxation processes can occur which subject the laser power to large deviations. The time integrated spatial distribution changes, with the ocurrance of these relaxation processes, from a modified fundamental-mode distribution into a donut-mode. The intensity decreases in the center of the beam and increases in outer regions (fig.6). Upon testing the donut-mode point for point, one sees that a large elevation of the first peak only occurs in the center (fig.7).

Fundamental-mode beam quality seems to be necessary for controlled modulation.

5. Resonant modulation

At low power densities the relaxation frequency is in the same range as the max.

modulation frequency and a resonant modulation is possible. With increasing frequency the modulation depth decreases (fig.8).

6. Conclusions

Modulation of the electrical excitation of CO_2 lasers as presented here, subdivides the laser power modulation into two ranges. The time constants t_v,t_d and ta determine the modulation behaviour up to 10kHz. Resonant modulation is possible for frequencies $f \gtrsim 50kHz$. Fine adjustment of frequency and duty cycle is necessary. The frequency limit depends on electrical power density. The necessary modulation frequencies and power densities are only attainable with rf excitation.

Literature
(1) E.Beyer: Dissertation, TH Darmstadt (1985)
(2) S.Wagner: Diplomarbeit, TH Darmstadt (1983)
(3) K.R.Manes, H.J.Seguin: J.Appl.Phys.43,5073 (1972)
(4) K.Smith, R.M.Thomson: Computer modeling of gas lasers, plenum press N.Y. (1978)
(5) P.Loosen: Dissertation, Darmstadt (1985)
(6) R.Wester, K.Schmitt, H.Schülke, H.Schwede: these proceedings
(7) K.Schmitt, H.Schülke, R.Wester: these proceedings

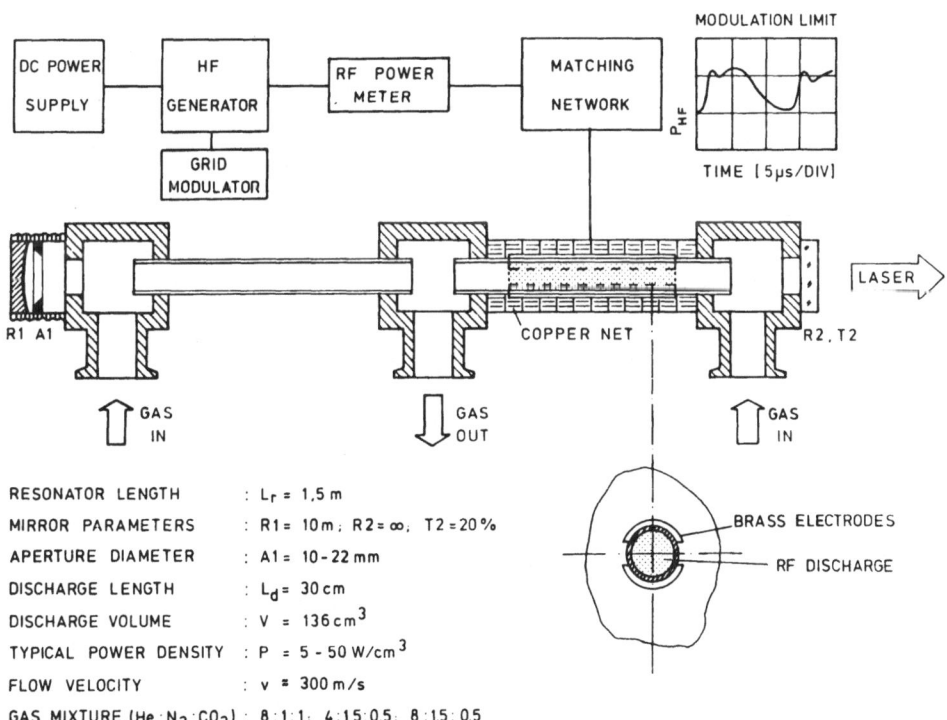

RESONATOR LENGTH   : L_r = 1,5 m
MIRROR PARAMETERS   : R1 = 10 m; R2 = ∞, T2 = 20 %
APERTURE DIAMETER   : A1 = 10 - 22 mm
DISCHARGE LENGTH   : L_d = 30 cm
DISCHARGE VOLUME   : V = 136 cm^3
TYPICAL POWER DENSITY : P = 5 - 50 W/cm^3
FLOW VELOCITY   : v = 300 m/s
GAS MIXTURE (He:N_2:CO_2) : 8:1:1; 4:1,5:0,5; 8:1,5:0,5

fig.1. Experimental set-up

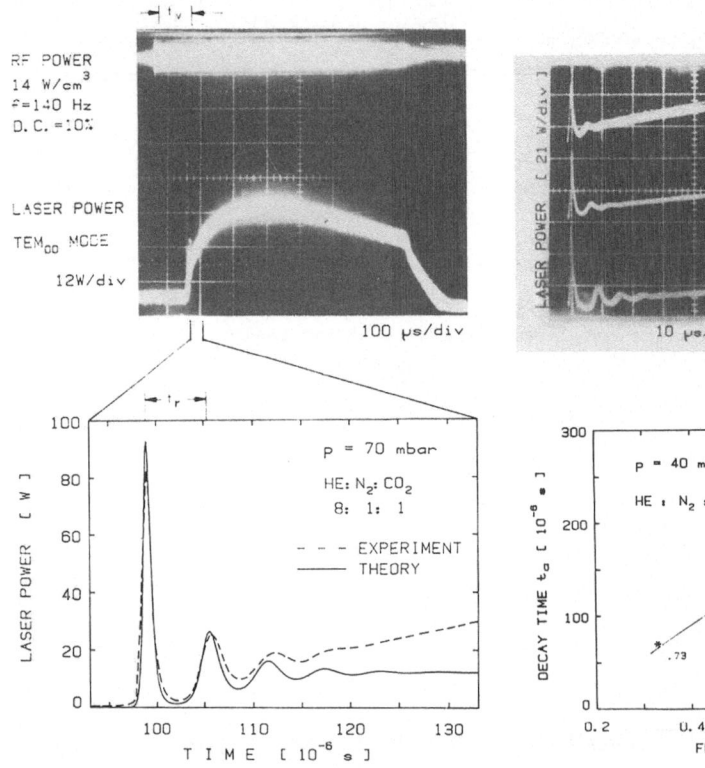

RF POWER
14 W/cm^3
f=140 Hz
D.C.=10%

LASER POWER
TEM$_{00}$ MODE
12W/div

100 μs/div

t_v

t_r

p = 70 mbar
HE: N$_2$: CO$_2$
8: 1: 1

- - - EXPERIMENT
——— THEORY

LASER POWER [W]

T I M E [10^{-6} s]

fig.2. Typical laser pulse

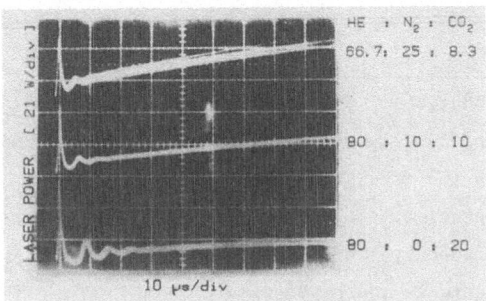

LASER POWER [21 W/div]

HE : N$_2$: CO$_2$

66.7: 25 : 8.3

80 : 10 : 10

80 : 0 : 20

10 μs/div

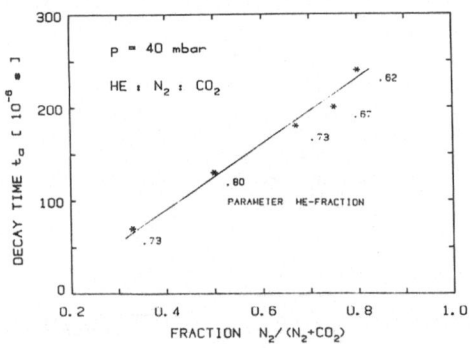

DECAY TIME t_a [10^{-6} s]

p = 40 mbar

HE : N$_2$: CO$_2$

.62
.67
.73
.80
.73

PARAMETER HE-FRACTION

FRACTION N$_2$/(N$_2$+CO$_2$)

fig.4. Variation of gas mixture

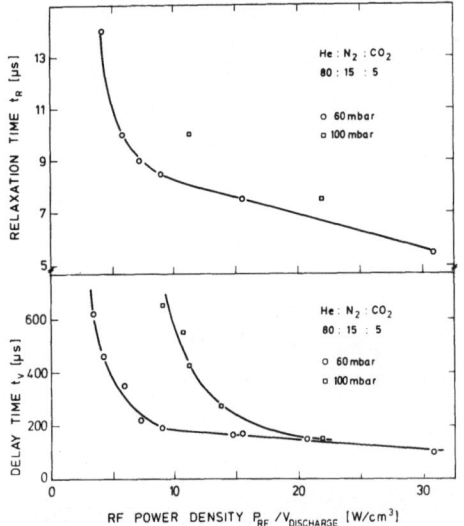

RELAXATION TIME t_R [μs]

He : N$_2$: CO$_2$
80 : 15 : 5

o 60 mbar
□ 100 mbar

DELAY TIME t_v [μs]

He : N$_2$: CO$_2$
80 : 15 : 5

o 60 mbar
□ 100 mbar

RF POWER DENSITY $P_{RF}/V_{DISCHARGE}$ [W/cm^3]

fig.3. Dependence of delay time and
relaxation time on pump power

COMPUTER MODEL OF THE CO$_2$ LASER (SCHEMATIC)

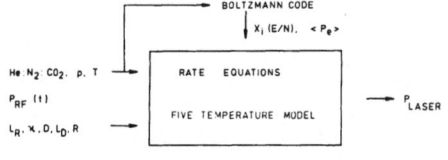

BOLTZMANN CODE

X_i (E/N), < P_e >

He N$_2$ CO$_2$, p, T

P_{RF} (t)

L_R, x, D, L_D, R

RATE EQUATIONS

FIVE TEMPERATURE MODEL

P_{LASER}

P_{RF} (t)   RF POWER ABSORBED IN THE DISCHARGE

P_{LASER} (t) OUTPUT LASER POWER

p   GAS PRESSURE

T   GAS TEMPERATURE

X_i EFFECTIVE ELECTRON VIBRATIONAL EXCITATION RATES

L_R RESONATOR LENGTH

x   RESONATOR LOSSES

D   APERTURE DIAMETER

L_D DISCHARGE LENGTH

R   OUTPUT MIRROR REFLECTIVITY

fig.5. Computer model of CO$_2$ laser
(schematic)

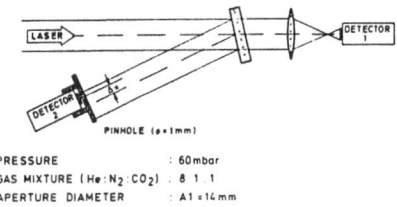

PRESSURE : 60 mbar
GAS MIXTURE (He:N$_2$:CO$_2$) : 8 1 1
APERTURE DIAMETER : A1 = 14 mm

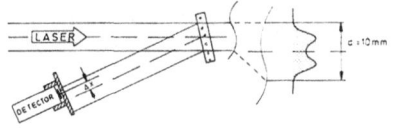

DETECTOR SIGNAL

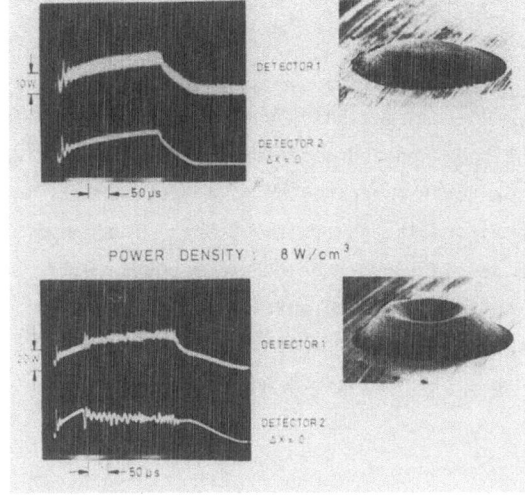

POWER DENSITY : 8 W/cm^3

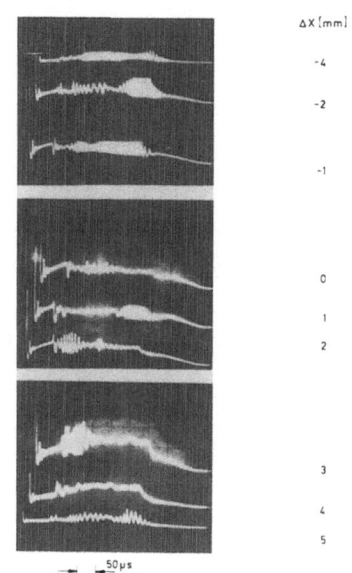

fig.6. Dependence of mode structure on pump power

fig.7. Spatially resolved measurement of laser power

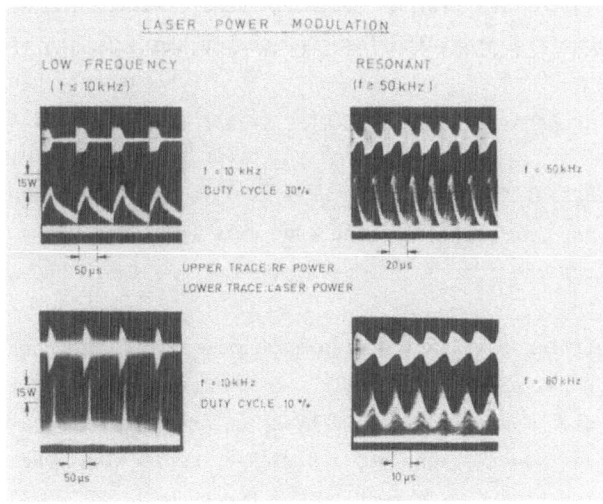

fig.8. Observed laser power modulation

UHF Excitation of an Axial Flow CO$_2$ Laser

K. Schmitt, H. Schülke, R. Wester

Institut für Angewandte Physik, Technische Hochschule Darmstadt, Schloßgartenstr. 7, D-6100 Darmstadt

in cooperation with Siemens Röhrenwerk München

1. Introduction

Radio frequency excitation of CO$_2$ laser discharges has a number of advantages in comparison with DC excitation /1/. It has been applied successfully for frequencies f < 100 MHz /2,3/. In this paper we report on the excitation of a fast axial flow CO$_2$ laser using UHF (ultra-high frequency) radiation at f=500 MHz. The temporal control of the laser power by amplitude modulation of the UHF pump power will be investigated. This temporal control is of interest to particular applications in material processing. Advantage of UHF excitation versus excitation in the lower frequency range (f < 100 MHz) is the higher possible amplitude modulation bandwidth of the pump power.

2. Experimental Set Up

The schematic diagram of the laser system is shown in Fig. 1. It consists of two discharge segments. One segment is provided for HF (high frequency) excitation at f=27 MHz, the second segment for UHF excitation. The laser resonator has a length of 150 cm, the discharge length is about 30 cm for each segment. The discharge tube has an inside diameter of 2.5 cm. The HF generator consists of a cavity resonator and a Siemens RS 3010 triode. The maximum output power is P=8 kW which can be modulated up to 100 kHz by driving the grid voltage of the tube. The HF system is coupled to the discharge tube by two plate electrodes parallel to the tube axis (see Fig. 2.1). Impedance matching of the laser discharge to the HF generator is accomplished by a circuit of lumped components.

The UHF system is an oscillator-amplifier-combination with a Siemens RS 1054 tetrode in the output stage. The maximum output power is P=2 kW. The maximum frequency of the amplitude modulation is f$_{AM}$=5 MHz. Several different types of coupling structures have been tested to accomplish power transfer for laser excitation (see Fig. 2-4). The discharge impedance is matched by a double stub tuner.

3. Experimental Results

The coupling structures shown in Fig. 2 have a number of disadvantages with respect to an effective laser excitation in the UHF frequency range. Due to the build-up of standing waves along the transmission line formed by the parallel plate electrodes (Fig. 2.1) local variations of the discharge intensity are observed. The radial distribution of the electric field inside the helical line (Fig. 2.2) produces a discharge of annular cross section. Excitation using the cylindrical cavity (Fig. 2.3) oscillating in the E_{010}-mode is unstable for pressures p > 10 mbar. For pressures p < 10 mbar, the excitation of surface waves initiating discharges outside the cavity volume is observed.

The most favorable coupling structure tested is a capacitively-loaded cavity oscillating in the H_{101}-mode at a frequency f=500 MHz (Fig. 3,4). The electric field amplitude along the discharge tube inside the cavity can be varied by the use of movable ridge segments. In this way it is possible to adjust for a proper value of the parameter E/n (E: electrical field amplitude, n: gas density). One side wall of the cavity is covered with two copper-mesh windows allowing observation of the discharge inside the cavity.

The maximum efficiency for laser excitation achieved with the capacitively-loaded cavity is about 15 %. The discharge is stable up to a pressure of p=30 mbar. By comparison the laser discharge excited at 27 MHz is stable up to a pressure of p=150 mbar.

Fig. 5a shows the configuration of a discharge observed in the cavity when all ridge segments are equidistant to the resonator axis. Only a portion of the gas volume inside the tube is excited. Proper adjustment of the ridge segments results in the discharge shown in Fig. 5b. By this ridge configuration the electric field amplitude is matched to the gas density decreasing in direction of the flow. The decrease of density is caused by heating and friction at the wall of the discharge tube.

Fig. 6 shows the laser power as a function of the average UHF pump power having an AM frequency of 10 kHz and a duty cycle of 50 %. The laser radiation is in a low order annular mode distribution. By changing the AM frequency in the range from 1 kHz to 100 kHz and the duty cycle in the range from 10 % to 50 % the measured values differ from the curve shown in Fig. 6 by less than 20 %. Fig. 7 shows the relationship between the output power of the laser for a HF excited discharge and a UHF excited discharge. The results are attained from the experimental set up as depicted in Fig. 1 running one of the discharge segments respectively. Saturation of the laser output power occurs at a lower pump power level for HF excitation.

The oscillograms Fig. 8a,b show signals of the UHF pump power modulated at 50 kHz and 100 kHz, respectively, and the corresponding signals of the laser power. By increasing the frequency of the AM modulation the modulation depth of the laser power decreases. The modulation depth observed at an AM frequency of f_{AM}=200 kHz is only 10 %.

4. Conclusions

Modulation of the laser power is limited by the time constants of the relaxation processes in the laser gas /4/. An increase in the modulation depth and modulation frequency of the laser power is possible by increasing the pressure of the laser gas due to a reduction of the relaxation times. Future investigations in UHF laser excitation therefore will concentrate on the problem of exciting stable discharges in a pressure range above the present limit of about p=30 mbar.

References

(1) V.I. MYSHENKOV, N.A. YATSENKO: Sov. J. Quantum. Electron. 11(10), 1297 (1981)

(2) D. HE, D.R. HALL: J. Appl. Phys. 54(8), 4367 (1983)

(3) R.L. SINCLAIR, J. TULIP: Rev. Sci. Instrum. 55(10),1539 (1984)

(4) H. SCHÜLKE, V. STURM: these proceedings

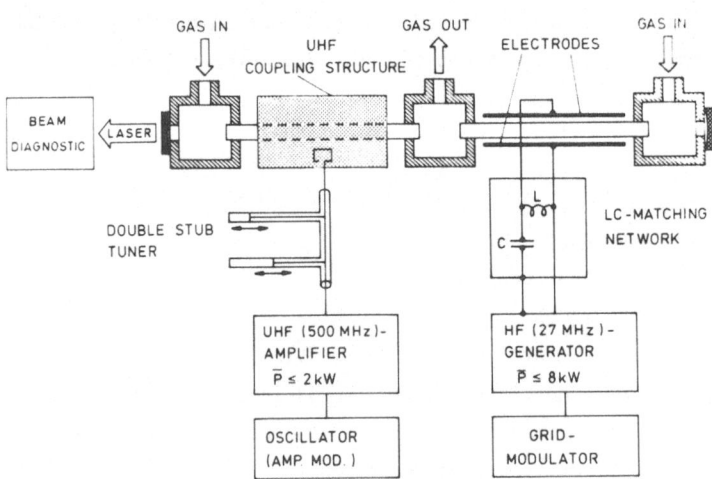

Fig. 1. Experimental set-up for HF- and UHF-excitation

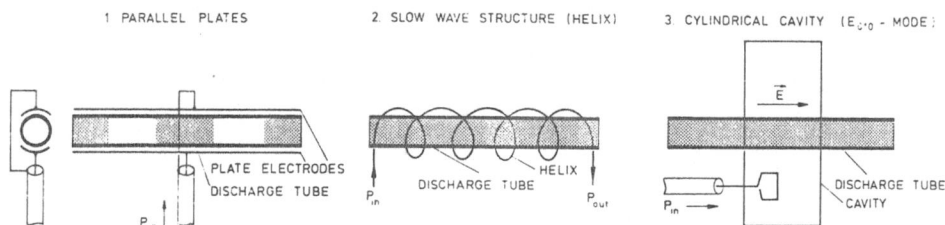

Fig. 2. Coupling structures for laser excitation

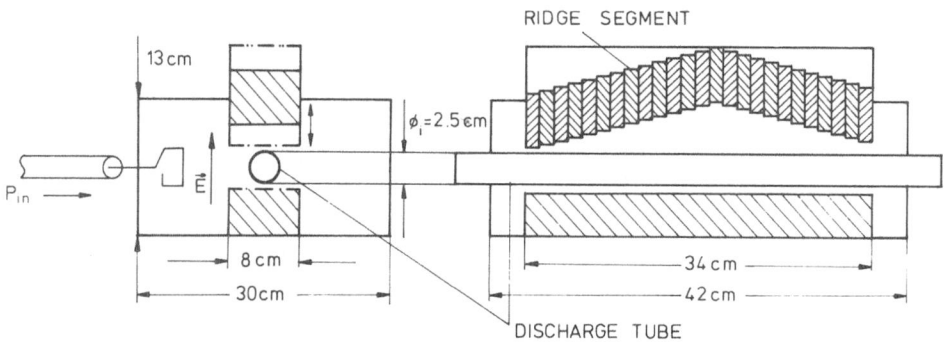

Fig. 3. Capacitively-loaded cavity (H_{101}-mode)

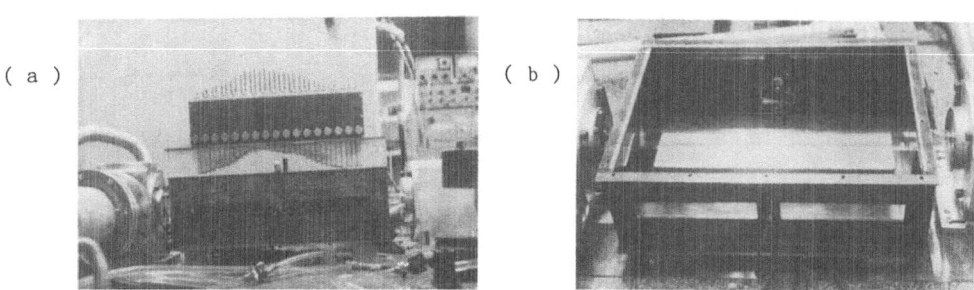

Fig. 4. Side-view of the cavity (a, cavity open) and view into the cavity (b)

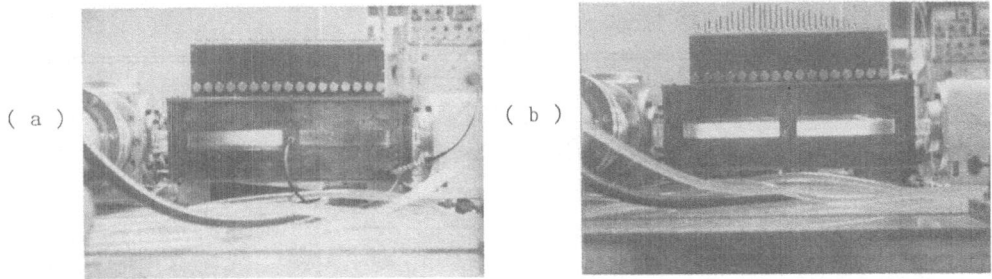

Fig. 5. Discharge observed with parallel ridges (a) and with adjusted ridges (b)

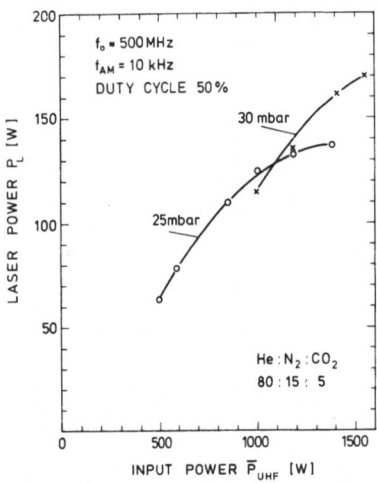

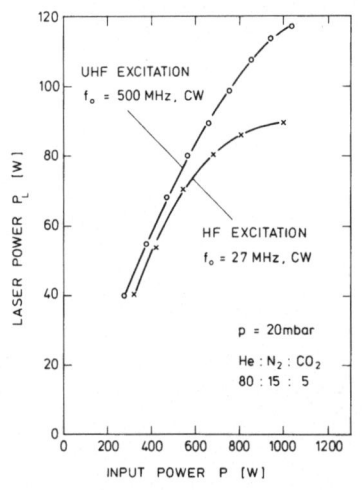

Fig. 6. Output characteristic of the UHF-excited laser segment

Fig. 7. Comparison of HF- and UHF-excited discharge outputs

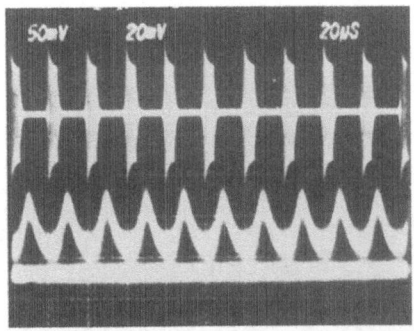

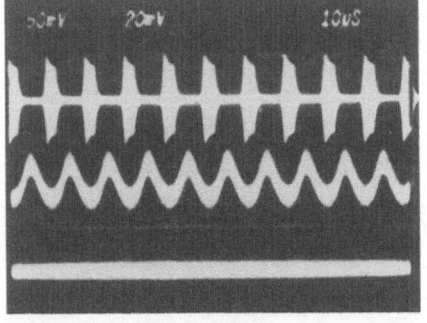

(a) $f_{AM} = 50$ kHz

(b) $f_{AM} = 100$ kHz

Fig. 8. Signal of the UHF pump power (upper trace) and corresponding laser output (lower trace), p=30 mbar, He : N_2 : CO_2 = 8 : 1 : 1

High Frequency Excitation of CO$_2$ Lasers

High Frequency Excitation of CO$_2$ Lasers

R.Wester,H.Schülke,K.Schmitt,H.Schwede
TH-Darmstadt, Institut für Angewandte Physik
Schlossgartenstr. 7, D-6100 Darmstadt

1. Introduction

In the field of material processing there is an increasing need for compact high power CO$_2$ lasers which are pulsable in the 10- or even 100 kHz range. In d.c.-excited systems compactness is limited by the achievable power density of about 10 Wcm^{-3}, whereas the maximal pulse frequency does not exceed the kHz region.

High frequency excitation offers the possibility of attaining a multiple increase in power density and significantly higher pulse frequencies from the generator and gas discharge. The gas discharge parameters, such as average electron energy, rate coefficients and electron density depend on the excitation frequency. Furthermore the discharge stability also proves to depend on the frequency.

2. Plasma Parameters

The properties of a gas discharge are determined by the electron distribution function, the cross sections of the occuring collisional processes and the boundary conditions.

If the energy relaxation frequency of the electrons ν_u greatly exceeds the excitation frequency ω, the isotropic part of the electron distribution function is quasi statically adjusted to the actual value of the reduced field strength:

$$f_o(u,E(t),N,\omega,t) = f_o(u,(E/N)(t))$$

N – neutral gas density , u – electron energy , E – electric field strength
With $\nu_u \ll \omega$ f_o is constant in time with /1/:

$$f_o(u,E,N,\omega,t) = f_o(u,E_o/\sqrt{2}/N)$$

In both cases it is sufficient to solve the time independent Boltzmann equation, whereas in the intermediate range $\nu_u \simeq \omega$, the time dependent Boltzmann equation has to be solved /2/.

The rate coefficient of an electron collisional process is:

$$k = \sqrt{2e/m} \int u\, Q(u)\, f_o(u)\, du \quad , \quad \int \sqrt{u}\, f_o(u)\, du = 1$$

m – electron mass , e – electron charge

Q – collisional cross section

For $\nu_u \ll \omega$ the rate coefficients are the same as in the case of d.c. excitation with $E/N = E_o/\sqrt{2}/N$.

For $\nu_u \gg \omega$ it is useful to average the rate coefficients over one period, if the time constants of the respective processes are small compared to ω^{-1}.

For typical CO_2 lasergas plasmas we have $\nu_u \approx (1-5)10^9\ s^{-1}$.

Fig 1.a) shows the d.c.-excitation rate coefficients for ionization ($k_{CO_2} + 3k_{N_2}$) electron attachment ($CO_2 + e \longrightarrow O^- + CO$) and excitation of the $CO_2(001)$ and the $N_2(v=1-8)$ vibrational levels for a gas mixture ratio of $He:N_2:CO_2 = 40:15:5$. Fig 1.b) shows the averaged rate coefficients for a.c.-excitation and $\nu_u \gg \omega$.

Provided one step ionization and the CO_2 attachment process are the only relevant occuring electron production and loss processes, the points a and b correspond to the stationary states of the d.c.-excited and the a.c.-excited discharges, respectively. (Numerical method and crosssections see /3/).

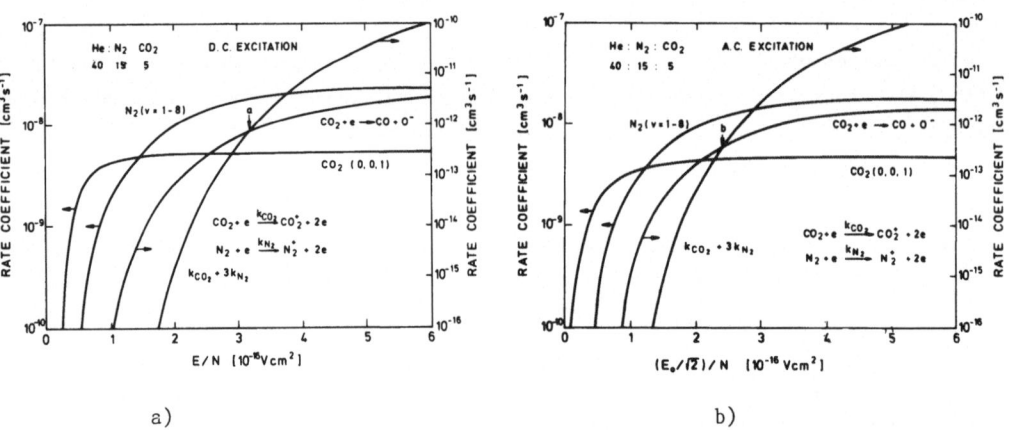

a)   b)

Fig. 1. Electron impact excitation rates for the CO_2 (001) and the combined N_2 (v=1-8) vibrational levels and rate coefficients for the combined ionization of CO_2 and N_2 and the electron attachment to the CO_2 as function of the reduced field streng (effective field strength in the case of a.c. excitation)

a):d.c.-excitation , a denotes the stationary state of the discharge

b):a.c.-excitation , b denotes the stationary state of the discharge

The time averaged absorbed power density is

$$p = 1/2 \text{ Re}(j_0 \text{ } E_0) \text{ , } j_0 \text{ exp}(i\omega t) = \sigma \text{ } E_0 \text{ exp}(i\omega t)$$

with the complex conductivity /4/:

$$\sigma = - \frac{2e^2 n}{3m} \int \frac{u^{3/2}}{(\nu_c(u)+i\omega)} \text{ } df_0/du \text{ } du$$

ν_c - elastic collision frequency , n - electron density.

The conductivity is constant in time in the case of f_0 being constant in time (d.c. or a.c. with $\nu_u << \omega$) or if ν_c does not depend on the electron energy. Otherwise the current density is a nonlinear function of the electric field strength.

Fig 2. shows the average absorbed power per electron p/n_e for d.c.-excitation and a.c.-excitation with $\nu_c, \nu_u >> \omega$. At the stationary state a (d.c.-excitation) the value is about 1.6 times the corresponding value at point b (a.c.-excitation). That means an electron density of about 1.6 times larger corresponding to equal power density. The fraction of power going into the $N_2(v=1-8)$ and the $CO_2(001)$ levels is found to be 67% in the case of d.c.-excitation and 74% in the case of a.c.-excitation. That means a somewhat larger efficiency in the case of a.c.-excitation.

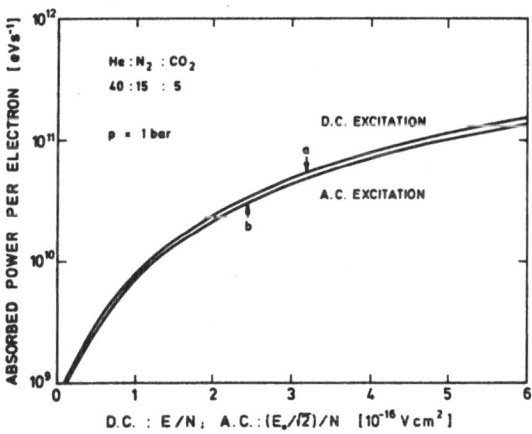

Fig. 2. Absorbed power per electron versus (effective) field strength for d.c.(a.c.)-excitation. a and b denote the stationary states of the d.c. and a.c. excited discharges respectively

3. Coupling of High Frequency Power to a Gas Discharge

Investigations were performed with a capacitive coupling structure at 27 MHz /5/ and a cavity with variable capacitances at 500 MHz /6/. The size of the 27 MHz structure is small compared to the wavelength, whereas for the 500 MHz structure $l \approx c/\nu$ (l-length of the cavity, c-velocity of light, ν=500 MHz). The gas flow is in the direction of the optical axis, whereas the excitation is transverse to it.

In the 27 MHz excitation the capacitance between the metallic electrodes and the plasma acts as a capacitive series resistance. Fig. 3. shows a simplified equivalent circuit. The division of voltage between the capacitive series resistance and the plasma resistance leads to an adjustment of the electric field strength in the plasma depending on electron and gas density. This leads to a homogenisation of the power density in direction of the gas flow at increasing electron density and decreasing neutral gas density. In the case of 500 MHz excitation this voltage division is not sufficient to homogenize the power input due to the larger displacement current. It is possible to attain a homogeneous discharge over a certain distance by adjusting the electric field strength with the variable capacitances. The most favorable adjustment hereby, depends on the gas pressure and the input power /6/.

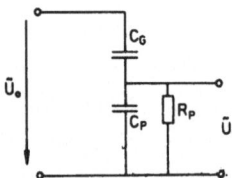

Fig. 3. Simplified equivalent circuit
of the high frequency discharge

For an optimal power input it is necessary to match the complex gas discharge impedance to the generator impedance of 50 Ohm. For the 500 MHz excitation this is done by adjustment of the double stub tuner and the coupling loop. To be able to design the matching network for the 27 MHz discharge the complex discharge impedance must be known. Its value was determined experimentally by measuring the current and the voltage of the discharge, as well as theoretically with a 1-dimensional model of the discharge utilizing gas electronic, gas dynamic and electrodynamic considerations. Fig 4.a) and b) show the absolute value and the phase of the discharge impedance as a function of the voltage.

The 27 MHz excited system presented in /5/, with two discharge tubes, attaines multimode output power of up to 1.5 kW, corresponding to a specific power of 2.3 kW/m, a power density of 40 Wcm^{-3} and an efficiency of 19%.

4. Gas Discharge Stability

Diffuse, homogeneous glow discharges tend to form instabilities which lead to stratification or contraction of the discharge /7,8,9/. Stratification perpendicular to the direction of current flow appears with attachment induced instabilities /9/. This is associated with constant current density /10/. In high frequency discharges the total current density consists of the electron current density and the displacement current density. At sufficiently large frequencies the attachment induced

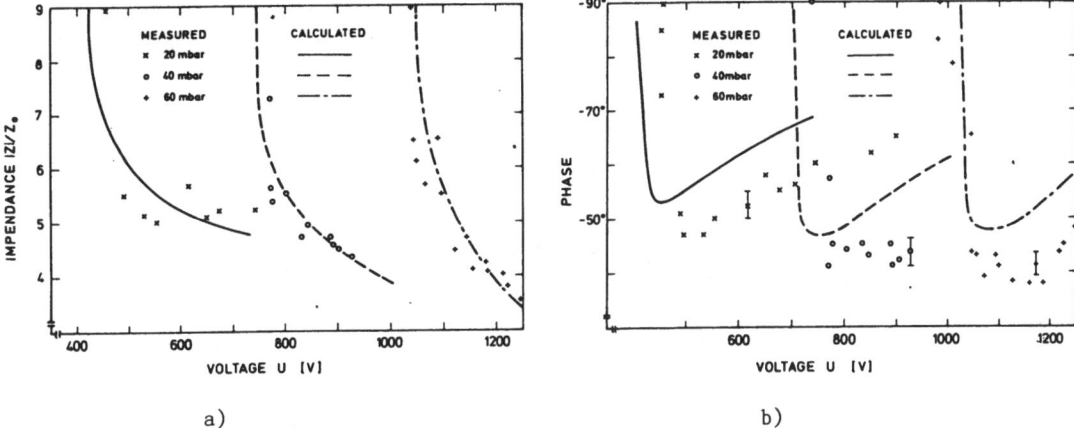

Fig. 4. a) absolute value and b) phase of the complex impedance of the discharge
excited at 27 MHz as function of the discharge voltage

instability is suppressed due to the field stabilizing effect of the displacement
current /10/. Most severe for laser excitation are the thermal instabilities, which
lead lo filamentation in the direction of current flow /9/, which is associated with
none constant current density. Stabilization in that case is achieved by decreasing
the electric field strength with increasing conductivity. If the wavelength is large
compared to the plasma dimensions the electrical field is determined by the equations:

$$\vec{\nabla} (\varepsilon \vec{E}) = 0 \quad , \quad \vec{\nabla} \times \vec{E} \simeq 0$$
$$\varepsilon = 1 + i\sigma/\omega\varepsilon_0$$

With constant elastic collision frequency ν_c and $\nu_u \gg \omega$ a perturbation ansatz and
subsequent fourier transformation lead to:

$$\vec{E}_o = E_o \, \vec{e}_x$$

$$\frac{E_1}{E_o} = - \frac{k_x^2}{k^2} \left(A \frac{n_1}{n_o} - B \frac{N_1}{N_o} \right) \quad , \qquad \begin{array}{l} o - \text{stationary state} \\[4pt] 1 - \text{perturbed state} \end{array}$$

$$A/B/ = \frac{(\omega_{po}^2/\omega \, \nu_{co})^2 - /2/ \; \omega_{po}^2/\nu_{co}^2 \, (1-\omega_{po}^2/\nu_{co}^2)}{(1-\omega_{po}^2/\nu_{co}^2) + (\omega_{po}^2/\omega \, \nu_{co})^2}$$

$$\omega_p^2 = \frac{e^2 n}{\varepsilon_o m} \qquad - \text{ plasma frequency}$$

The variation of the electrical field strength depends on the orientation of the perturbation with regard to the electrical field /9/ and on the frequency. For increasing frequency the field stabilizing displacement current leads to an amplification of the thermal instabilities.

In the 27 MHz and the 500 MHz arrangements the occurence of filaments in the discharge were observed. Fig. 5 shows the electrical power density of the gas discharge at the onset of instability as a function of the gas pressure. At 27 MHz filamentation can only be observed in the pulsed discharge. The critical power density is a linear function of the pressure. The 500 MHz discharge exhibits quit a different behaviour. The critical power density is decreasing with increasing pressure and filamentation also occurs in the cw discharge. This indicates that the excitation frequency has a great impact on the discharge stability.

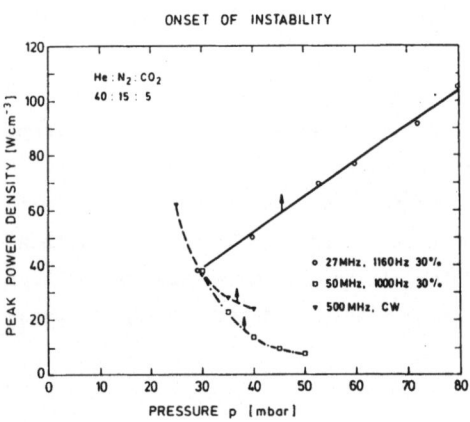

Fig. 5. Critical power density for instability onset as function of pressure. 27 MHz: pulse mode (30% duty cycle). 500 MHz: pulse mode (30 % duty cycle) and cw mode

5. Conclusions

With 27 MHz excitation a maximum power density of about 40 Wcm^{-3} has been obtained, which exceeds the corresponding values of the d.c.-excitation by a factor of 3 or 4. With puls operation even peak power densities of more than 100 Wcm^{-3} are achievable. The maximum pulse frequency is about 100 kHz /5/. At the present the 500 MHz -excitation has a smaller power density capability, but it is pulsable up to the MHz range /6/. This means steeper raising and falling slopes.

High frequency excitation enables the realization of compact CO_2 high power lasers with maximum pulse frequencies which substantially are determined by laser kinetic processes /5/. Modification of the coupling structure should lead to higher power densities also for excitation frequencies in the UHF range.

Literature

/1/ T.Holstein Phys. Rev. Vol.70 No 5 and 6 ,367 (1946)

/2/ R.Winkler,H.Deutsch,J.Wilhelm,Ch.Wilke, Beitr. Plasmaphys. 24 (1984) 3, 285

/3/ K.Smith,R.M. Thompson , "Computer Modelling of Gas Lasers" Plenum Press ,New York (1978)

/4/ C.M.Ferreira,J.Loureiro, J. Phys. D.,17 (1984), 1175

/5/ H. Schülke,V. Sturm , these proceedings

/6/ K. Schmitt,H.Schülke,R.Wester , these proceedings

/7/ W.P. Allis, Physica 82c ,43 (1976)

/8/ W.L.Nighan in "Principals of Laser Plasmas", ed. by G.Bekefi, J.Wiley, New York (1976)

/9/ R.A. Haas , Phys. Rev. A Vol.8, No.2, (1973) 1017

/10/ A.S. Kovalev,A.T. Rakhimov,N.V. Suetin,V.A. Feoktistov in "Gas flow and Chemical Lasers" , ed. by M. Onorato, Plenum Press, New York (1984)

CO_2-Hochleistungslaser mit transversaler RF-Anregung

H. Hügel, W. Schock, A. Giesen, T. Hall und W. Wittwer

DFVLR Institut für Technische Physik

Pfaffenwaldring 38-40, D - 7000 Stuttgart 80

1. Einleitung

Die Effizienz und Qualität der Fertigungsprozesse mit Lasern werden durch die zeitliche und räumliche Intensitätsverteilung des Laserstrahls wesentlich bestimmt. Aus diesem Grunde werden von modernen CO_2-Hochleistungslasern Eigenschaften wie variable Betriebsweise (cw, gepulst, moduliert) und gute Strahlqualität zunehmend gefordert. Diese Merkmale, zusammen mit der erzielbaren Leistungsdichte und dem Wirkungsgrad des Lasers, hängen entscheidend von der elektrischen Anregungstechnik ab.

Die im Hinblick auf die genannten Anforderungen besondere Eignung der Hochfrequenz (RF)-Entladung wird am Institut für Technische Physik der DFVLR schon seit Jahren an mehreren Laserkonzepten demonstriert. Während zunächst Untersuchungen an RF-Entladungen in der Überschallströmung eines CO-Lasers im Vordergrund standen /1/, gewann in Fortführung dieser Arbeiten die Frage nach der einkoppelbaren Leistungsdichte und Homogenität in quergeströmten CO_2-Lasern zunehmendes Interesse /2,3/. Untersuchungen zur variablen Betriebsweise eines längsgeströmten CO_2-Rohr-Lasers mit transversaler RF-Einkopplung /4/ und zur Skalierbarkeit und Strahlqualität des quergeströmten Konzepts /5/ folgten. – In diesem Beitrag wird über experimentelle und theoretische Arbeiten an einem quergeströmten Multi-kW-Modul und an einem längsgeströmten 1 kW-Laser berichtet.

2. Eigenschaften der RF-Entladung

Die physikalischen Eigenschaften von Hochfrequenzentladungen sind schon mehrfach ausführlich behandelt worden /3,5/. Deshalb wird hier nur auf Aspekte der kapazitiven Energieeinkopplung und der variablen Betriebsweise eingegangen.

Die kapazitive Einkopplung der RF-Leistung in das Lasergas erfolgt durch ein Dielektrikum. Über dessen Dicke wird die Entladung homogenisiert und stabilisiert. Dadurch sind höhere elektrische Leistungsdichten im Vergleich zu metallischen Elektroden einkoppelbar. Ebenso entfällt die Segmentierung der Elektroden in quergeströmten Lasern. Zudem befindet sich das Laserplasma nicht in Kontakt mit metallischen Elektroden, was die Gasdegradation gering hält und sich positiv auf den Langzeitbetrieb des Lasers auswirkt.

Der Stromtransport in den Elektrodengrenzschichten ist bei RF-Entladungen mit relativ geringen Verlustspannungen von typischerweise 30-50 Volt verknüpft. Da überdies die Schichtdicken nur sehr gering sind (< 1 mm) bestehen im gesamten Entladungsraum bis unmittelbar zu den Elektroden hin die gleichen elektrophysikalischen Entladungsbedingungen. Dies ist mit ein Grund für die hohe Homogenität des RF-erzeugten laseraktiven Mediums und, insbesondere bei transversalen Entladungen, für den im Vergleich zu DC-Entladungen höheren Wirkungsgrad.

Die bekannt einfache Modulierbarkeit der RF-Leistung erlaubt eine zeitliche Steuerung der Entladung. Damit läßt sich die Laserintensität dem Bearbeitungsprozeß anpassen. Neben dem cw-Betrieb sind Puls- und pulsüberlagerter cw-Betrieb möglich. Beim Pulsbetrieb sind Pulsform, -frequenz und Tastverhältnis frei wählbar.

3. Quergeströmtes Multi-kW-Modul

Nachdem erste Untersuchungen /3/ an einem einfachen aus Plexiglas gefertigten Laserkanal mit stabilem Resonator das hohe Leistungspotential der RF-Entladung experimentell bestätigt hatten (elektr. Leistungsdichte bis 35 W/cm³ bei Drücken bis 160 mbar und Wirkungsgraden von 20 %), wurde ein Gerät mit einem Entladungsvolumen von 4 x 4 x 95 cm³ gebaut und in einen geschlossenen Gaskreislauf integriert. Bild 1 zeigt das mit einem instabilen konfokalen Resonator (single-pass, M = 1.4) ausgestattete Lasermodul.

Bild 1. Multi-kw-Modul

Die im folgenden vorgestellten Messungen wurden in einem Druckbereich zwischen 60 und 80 mbar bei typischen Geschwindigkeiten um 90 m/s durchgeführt. Die in ein homogenes und kontraktionsfreies Plasma eingekoppelte elektrische Leistungsdichte betrug bis zu 26 W/cm³ im kontinuierlichen und bis zu 33 W/cm³ im gepulsten Betrieb. Dabei wurden entsprechende Laserleistungen von 4.2 bzw. 6 kW ausgekoppelt. Da bei diesem kurzen Lasermodul ein Kompromiß zwischen guter Strahlqualität und hoher Leistungsauskopplung angestrebt worden war, überstieg der Wirkungsgrad nicht Werte um 13 %. - In größeren Systemen erhöht sich der Wirkungsgrad bei gleichzeitig besserer Strahlqualität.

42

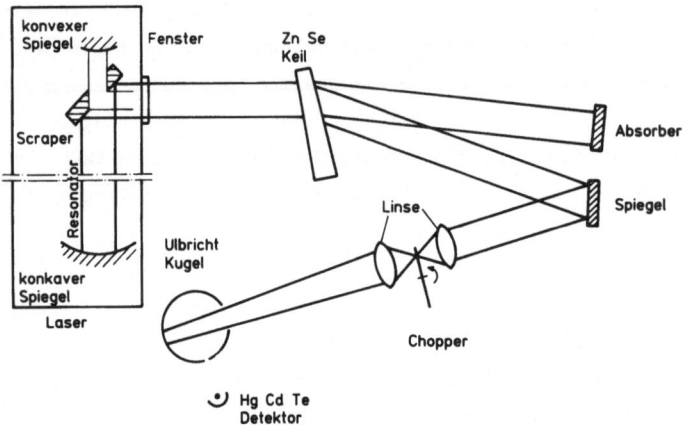

Bild 2. Versuchsaufbau zur Leistungsmessung

Die Leistungsmessungen erfolgten mit einem HgCdTe-Detektor in einem Versuchsaufbau,
wie er in Bild 2 gezeigt ist. Die ZnSe-Keilplatte dient der Leistungsabschwächung und
der Chopper erlaubt Messungen auch im kontinuierlichen Betrieb. Ein Teilstrahl geringer Intensität gelangt nach räumlicher Integration in einer Ulbrichtkugel in den Detektor. Die Bandbreite des verwendeten Meßsystems einschließlich des digitalen Transient-
rekorders beträgt 1 MHz. Die Kalibrierung der Leistungsmessung erfolgte kalorimetrisch.

Im cw-Betrieb wurden Laserleistungen bis 4 kW ausgekoppelt, wobei die Fluktuationen
im gesamten Frequenzbereich bis 1 MHz kleiner als 3 % (RMS) waren. Im Pulsbetrieb dagegen wurden Leistungen bis zu 6 kW erreicht.
Die Frequenz konnte zwischen 100 Hz und 12 kHz
bei veränderlichem Tastverhältnis (1:1 bis
1:5) variiert werden. Ein Beispiel für den
zeitlichen Verlauf der Laserleistung während
des Pulsbetriebes mit der maximal verfügbaren
Generatorleistung von 50 kW zeigt Bild 3. Der
Leistungsabfall ist temperaturbedingt und im
wesentlichen bestimmt durch die Gasgeschwindigkeit in der Entladung. Abschätzungen zeigen, daß die "Abklingdauer" von 0,5 ms jener
Zeit entspricht, nach der sich unter den ge-

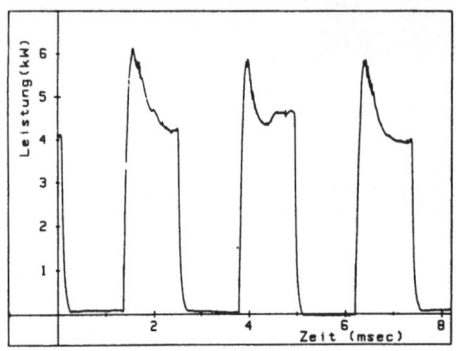

Bild 3. Laserleistung bei 400 Hz
Pulsfrequenz

gebenen Parametern ein stationärer Zustand einstellt. Diese Tatsache verdeutlicht ferner, daß durch Erhöhen der Gasgeschwindigkeit die Laserleistung während der gesamten
Pulslänge wie auch im kontinuierlichen Betrieb auf 6 kW zu steigern wäre.

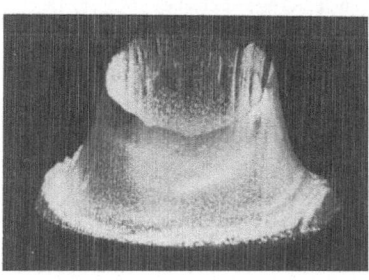 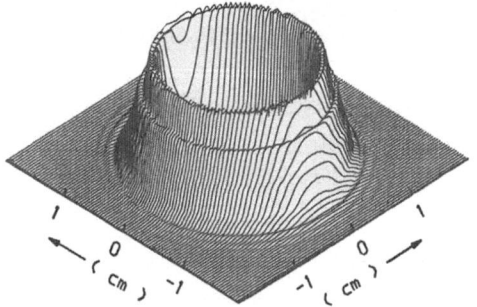

Bild 4. Intensitätsverteilung im Nahfeld

Die Intensitätsverteilungen im Nah- und Fernfeld des Laserstrahls wurden anhand von Einbränden in Plexiglas bestimmt. Abschätzungen aus einer Reihe von Einbränden mit unterschiedlichen Belichtungszeiten und verschiedenen Laserleistungen lassen qualitativ eine gute Übereinstimmung mit den theoretisch berechneten Verteilungen erkennen. Diese erhält man durch iterative Lösung der Kirchhoff-Fresnel-Integralgleichung. Aus dem zunächst ermittelten Resonatormode, bzw. der Nahfeldverteilung, wird dann die Fernfeldverteilung berechnet. In Bild 4 ist der Nahfeldeinbrand der entsprechenden theoretischen Verteilung gegenübergestellt. Den Vergleich der Fernfeldstrukturen zeigt Bild 5. Die gemessene Divergenz, wie auch die Intensitätsverhältnisse im eingebrannten Ringsystem verdeutlichen, daß die Laserleistung nahezu beugungsbegrenzt ausgekoppelt wird.

 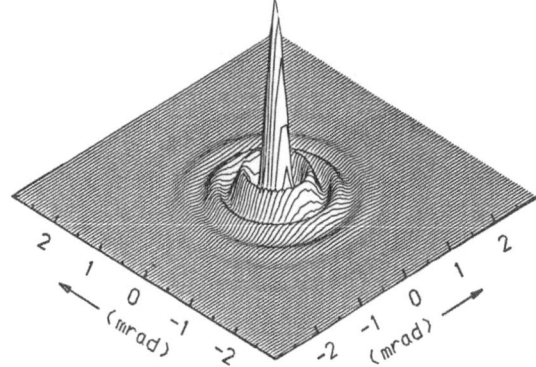

Bild 5. Intensitätsverteilung im Fernfeld

4. Längsgeströmter 1 kW-Laser

Ausgangspunkt der hier vorgestellten Entwicklung war ein längsgeströmtes Gerät im Leistungsbereich um 30 W, an welchem vor allem die variable Betriebsweise eines mit Hochfrequenz angeregten Lasers demonstriert werden konnte /4/. Die Einkopplung der RF-Leistung erfolgte dabei transversal zur Strömungsrichtung durch auf ein Quarzrohr aufgeklebte Metallstreifen.

Diese Technik wird auch bei dem in Bild 6 schematisch gezeigten Experimentiergerät angewandt. Die in den einzelnen Rohrsegmenten (Innendurchmesser 1.5 cm, Entladungslänge 40 cm) anfallende Verlustwärme wird ausschließlich durch die axiale Gasströmung abgeführt. In einem Druckbereich zwischen 100 und 200 mbar werden RF-Leistungen bis zu 10 kW eingekoppelt.

Die Ergebnisse verschiedener Meßreihen sind in Bild 7 zusammengefaßt. Diese Daten sind mit einem stabilen Resonator erzielt worden, dessen Fresnelzahl etwa 1,3 beträgt. Etwa 80 % der Laserstrahlung werden im TEM_{oo}-Mode ausgekoppelt.

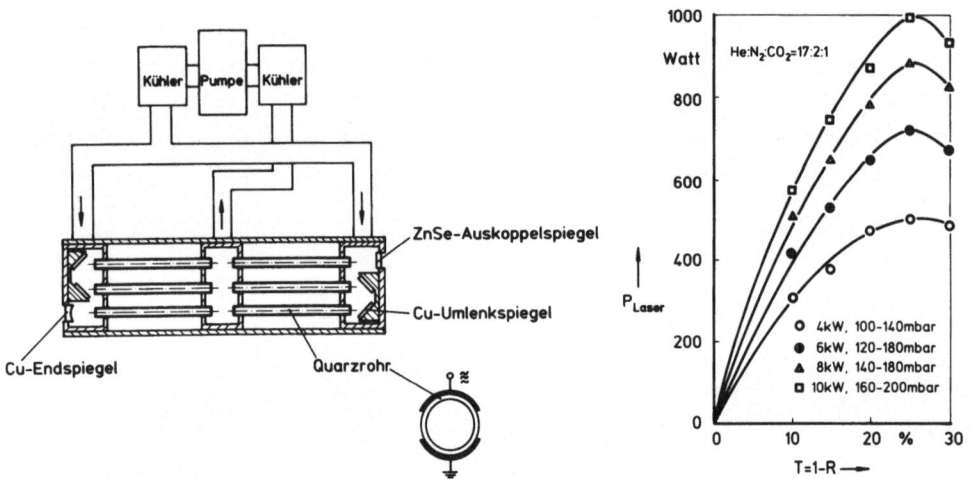

Bild 6. Schema des längsgeströmten
         1 kW-Lasers

Bild 7. Laserleistung

5. Zusammenfassung

Es werden experimentelle Ergebnisse vorgestellt, welche die Eignung der Hochfrequenz-
entladung als effiziente Anregungstechnik für längs- und quergeströmte Laser im kontinuierlichen und gepulsten Betrieb verdeutlichen. Die erzielten elektrischen Leistungsdichten um 30 W/cm³ stellen dabei keine prinzipiellen Grenzen dar, sondern
sind im wesentlichen durch die geometrischen Gegebenheiten der derzeit untersuchten
Geräte bedingt. In gleicher Weise gilt dies für die Werte des Wirkungsgrades, die
z.B. beim quergeströmten Konzept durch eine Doppelpass-Anordnung und beim längsgeströmten durch eine Optimierung des Entladungsvolumens gesteigert werden können.

Eine Zusammenstellung der mit den verschiedenen Geräten erreichten optischen Leistungsdichten ist in Bild 8 gegeben. Es sei darauf hingewiesen, daß diese hohen Werte beim quergeströmten Laser mit instabilem Resonator und beim längsgeströmten bei
sehr guter Strahlqualität erzielt wurden.

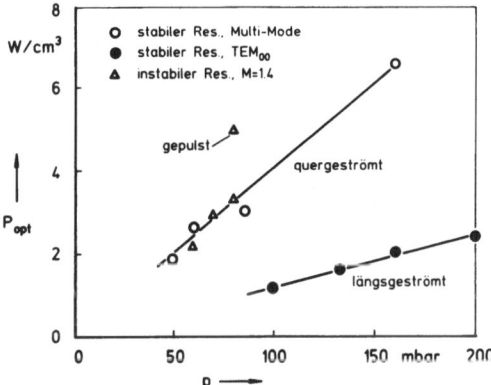

Bild 8. Optische Leistungsdichte

Literatur
/1/ W. SCHOCK, W. SCHALL, H. HÜGEL, P. HOFFMANN: Appl. Phys. Lett. 36, 793 (1980)
/2/ W. SCHOCK, H. HÜGEL, P. HOFFMANN: LASER + Elektro-Optik 2, 76 (1981)
/3/ W. SCHOCK, H. HÜGEL: Proc. 4. Int. Symp. Gas Flow and Chemical Lasers, Stresa 1982
    Plenum Press, New York, 435 (1984)
/4/ W. WITTWER: Verhandl. DPG (VI) 18, 369 (1983)
/5/ W. SCHOCK, A. GIESEN, H. JACOBY, H. HÜGEL: VDI-Berichte 535, 25 (1984)

TEA-CO$_2$-Laser mit Vorionisierung mittels Halbleiterelektroden

B. Walter, D. Schuöcker
Institut für Nachrichtentechnik, Technische Universität Wien
Gußhausstraße 25, A-1040 Wien

Zur Erzeugung einer homogenen Glimmentladung bei Atmosphärendruck benötigt man Elektroden mit homogener Feldstärkeverteilung und eine geeignete Vorionisierung. Diese Vorionisierung ermöglicht nicht nur die Ausbildung der gepulsten Glimmentladung, sondern sie beeinflußt auch die Homogenität der Entladung, die maximal umsetzbare Energie und den Wirkungsgrad des Systems.

Eine besonders einfache Methode zur Vorionisierung des Lasergases ist die Verwendung von UV-Strahlung, die die in einem He-Gasgemisch stets vorhandenen organischen Verunreinigungen direkt ionisiert. Die Erzeugung der Strahlung kann mittels Korona- oder Bogenentladung erfolgen.

Die Koronaentladung /1/ wird meist in Form einer Oberflächenentladung entlang eines Dielektrikums realisiert. Das Dielektrikum schließt dabei jedoch den Entladungsraum seitlich ab und verhindert so einen schnellen transversalen Gasaustausch, der für die Realisierung hoher Wiederholraten notwendig ist. Zur Erzeugung einer Bogenentladung dient im allgemeinen eine Funkenstrecke, die aus zwei Pinelektroden im Abstand von einigen Millimetern besteht. Werden mehrere Funkenstrecken parallel geschaltet, so muß durch Serienschaltung von Kondensatoren /2/ oder Widerständen /3/ eine gleichmäßige Stromaufteilung zwischen den einzelnen Funkenstrecken erzwungen werden. Um zu verhindern, daß zwischen den Pin- und einer der Hauptelektroden Lichtbögen entstehen, muß ein ausreichender Abstand zwischen Vorionisierung und Hauptentladung eingehalten werden. Dadurch wird jedoch die Wirkung der Vorionisierung wegen der höheren Dämpfung der UV-Strahlung verringert.

Werden anstelle der parallel geschalteten Funkenstrecken durchgehende Elektroden verwendet, kommt es zur Ausbildung eines einzelnen Lichtbogens. Sind diese Elektroden jedoch aus einem Material mit hohem spezifischen Widerstand gefertigt, so wird das Auftreten zahlreicher Lichtbögen begünstigt. RICKWOOD /4/, RICKWOOD und McINNES /5/ sowie STARK, CROCKER und LOWDE /6/ verwendeten Halbleiter erfolgreich zur

Vorionisierung von TEA-CO$_2$-Lasern mit kleinem Entladungsquerschnitt. Ziel der vorliegenden Arbeit war es, diese Methode bei TEA-CO$_2$-Lasern mit hohem Elektrodenabstand (12,9mm) jedoch relativ kleinem Entladevolumen (8cm^3) zu erproben.

Für den Aufbau der Vorionisierung standen Siliziumplättchen der Abmessung 17mm x 27mm x 0,5mm mit einem spezifischen Widerstand von $\rho = 22\Omega cm$ sowie Platten mit den Maßen 80mm x 30mm x 2mm mit $\rho = 65\Omega cm$ zur Verfügung. Je fünf der Plättchen wurden zu einer Elektrode (85mm x 27mm x 0,5mm) angeordnet, der Widerstand ergab sich zu $R_q = 140\Omega$. Mit dieser Elektrode oder mit einer der Siliziumplatten ($R_q = 120\Omega$) wurde gemäß Abb. 1 die Vorionisierungseinheit aufgebaut. Um die Bildung von Lichtbögen zwischen der Kontaktierung und der Elektrode der Hauptentladung zu verhindern, mußte die Kontaktierung durch Eingießen in Kunstharz isoliert werden. Zur Ionisierung des Lasergases wurde je eine Halbleiterelektrode zu beiden Seiten der Anode der Hauptentladung angeordnet, elektrisch wurde die Vorionisierung mit der Kathode verbunden (Abb. 2).

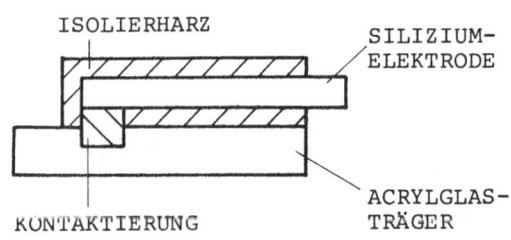

ISOLIERHARZ SILIZIUM-ELEKTRODE

KONTAKTIERUNG ACRYLGLAS-TRÄGER

Abb. 1.
Aufbau der Vorionisierung

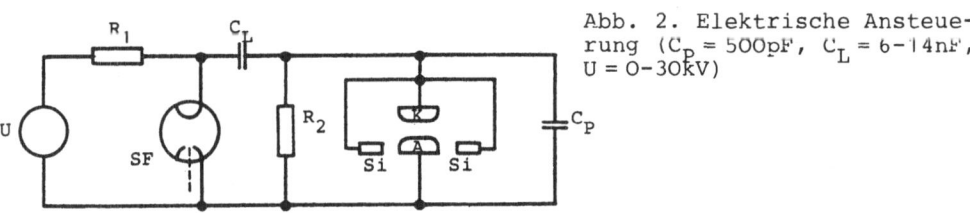

Abb. 2. Elektrische Ansteuerung ($C_P = 500pF$, $C_L = 6-14nF$, $U = 0-30kV$)

Bei Anlegen eines Spannungsimpulses an den Entladungskreis zündet wegen des geringeren Abstandes zwischen Halbleiter und Anode die Entladung der Vorionisierung. Der verteilte Widerstand erzwingt dabei die Ausbildung zahlreicher Lichtbögen, deren UV-Strahlung die Kathode beleuchtet und das Lasergas ionisiert. Ist der Entladungsraum ausreichend ionisiert, zündet die Hauptentladung und schließt die Vorionisierung kurz. Die ersten Ergebnisse zeigten, daß der Abstand zwischen Halbleiterelektrode und Anode die maximale Energieumsetzung bestimmt. Ist der Abstand zu klein, fließt ein Großteil der Energie über die Vorionisierung, zusätzlich bilden sich Lichtbögen zwischen Kathode und Halb-

leiterelektrode aus. Bei zu großem Abstand sinkt die Zahl der sich
ausbildenden Lichtbögen und somit der Anteil der UV-Strahlung. Abb. 3
zeigt als Ergebnis dieser Messungen den Einfluß des gewählten Abstandes bei einseitiger Vorionisierung. Der so ermittelte optimale Wert
für d wurde für die weiteren Untersuchungen beibehalten.

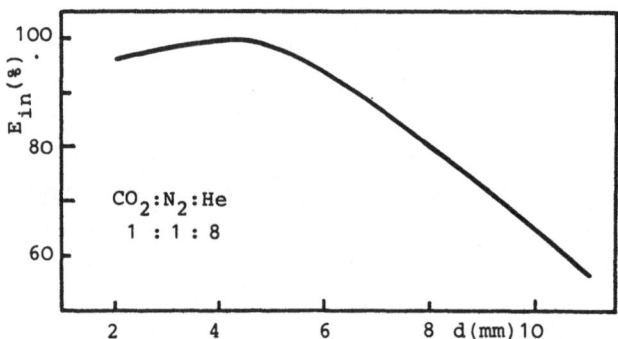

Abb. 3. Eingangsenergie
in Abhängigkeit vom Ab-
stand Siliziumelektrode
zu Rogowski-Elektrode

Wurde das Entladungssystem an einen Gaskreislauf angeschlossen, so
konnten, begrenzt durch die Verwendung einer Schaltfunkenstrecke,
Wiederholraten von 100Hz erzielt werden. Weitere Meßserien ließen
folgende Eigenschaften der Halbleitervorionisierung erkennen:

Vorteile: - geringer Abstand Vorionisierung - Hauptentladung

           - keine eigene elektrische Ansteuerung notwendig

           - schneller transversaler Gasstrom möglich

           - hohe Energieumsetzung möglich

           - Induktivitäten der Zuleitungen stören nicht, da kein
             schneller Spannungsanstieg erforderlich

Nachteile: - wegen Widerstand des Halbleiters geringe Intensität der
             UV-Strahlung

           - I^2R Verluste des Halbleiters verringern Wirkungsgrad

           - nur für kurze Elektrodenlängen geeignet, da sonst der
             Widerstand der Halbleiterelektrode zu gering wird

Die Eigenschaften des Entladekreises als aktives Medium wurden in einem
Resonator, bestehend aus einem ZnSe Endspiegel mit 99% Reflexion und
20m Krümmungsradius und einem im Abstand von 70cm angeordneten planen
ZnSe Auskoppelspiegel mit 95% Reflexion, untersucht. In Abb. 4 ist die
Laserenergie pro Puls bzw. die auf das Volumen der Entladung ($8cm^3$) be-
zogene Laserenergie über der Eingangsenergie darstellt.

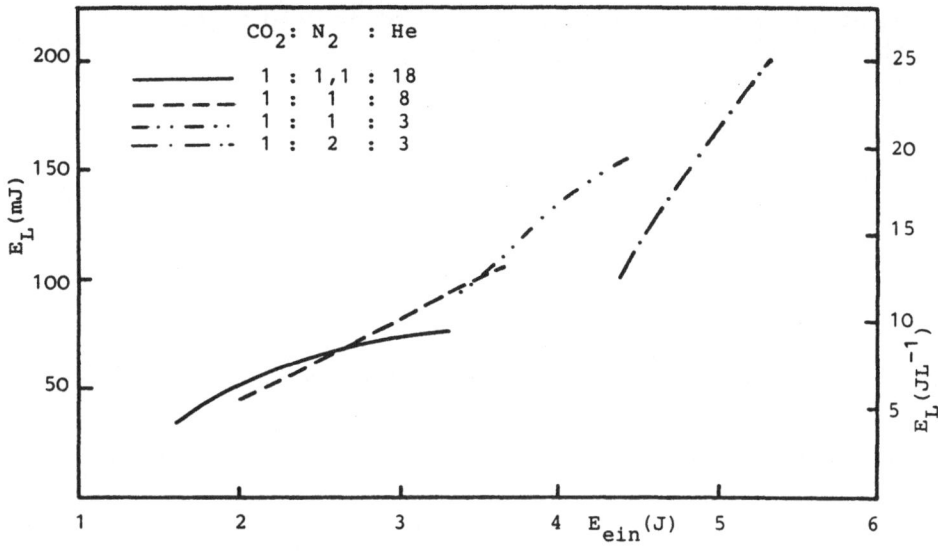

Abb. 4. Ausgangsenergie pro Puls

Der Maximalwert betrug 200mJ pro Puls oder bezogen auf das Volumen der Entladung $25JL^{-1}$. Dieser hohe Wert stellt die Leistungsfähigkeit dieser Art der Vorionisierung unter Beweis.

Literatur

/1/ G.J. ERNST: Rev.Sci.Instrum. <u>48</u>, 1281 (1977).
/2/ N.H. BURNETT, A.A. OFFENBERGER: J.Appl.Phys. <u>44</u>, 3617 (1973).
/3/ O.P. JUDD: Appl.Phys.Lett. <u>22</u>, 95 (1973).
/4/ K.R. RICKWOOD: J.Appl.Phys. <u>53</u>, 2840 (1982).
/5/ K.R. RICKWOOD, J.McINNES: Rev.Sci.Instrum. <u>53</u>, 1667 (1982).
/6/ D.S. STARK, A. CROCKER, N.A. LOWDE: J.Phys.E: Sci.Instrum. <u>16</u>, 1069 (1983).

Untersuchung des thermischen Verhaltens und der Strahlqualität eines Nd-Glas Slab Lasers

J. Eicher
Fachbereich Physik
Universität Kaiserslautern
6750 Kaiserslautern

Einleitung

Festkörperlaser mit hohen Ausgangsleistungen benötigen Entladungslampen als Pumplicht-
quellen. Bedingt durch den niedrigen Wirkungsgrad von einigen Prozent mit dem das
Pumplicht die laseraktiven Atome im Lasermedium anregt, wird der größte Teil der
Pumpenergie als Wärme im gesamten Laserkopf deponiert. Als Folge davon heizt sich
auch das aktive Medium stark auf und muß gekühlt werden. Die entstehenden Temperaturgradienten verursachen Brechungsindexgradienten und mechanische Spannungen. Bei herkömmlichen Festkörperlasern deren aktives Medium die Form eines zylindrischen Stabes
hat, führt dies zu einer Beeinträchtigung der Laserstrahl-Qualität sowie zu einer Begrenzung der Laserausgangsleistung /1-3/.

Vorteile des Slab Lasers /4-6/

Bei einem Slab Laser hat das aktive Medium, der sog. Slab, die Form eines Quaders
(Abb. 1). Gepumpt wird der Slab mit zwei Lampen, die sich in polierten Reflektoren
befinden. Das Laserlicht tritt durch die polierten Brewsterflächen in das aktive Medium ein und durchläuft es unter Ausnutzung der Totalreflexion an den optisch polierten Seitenflächen.

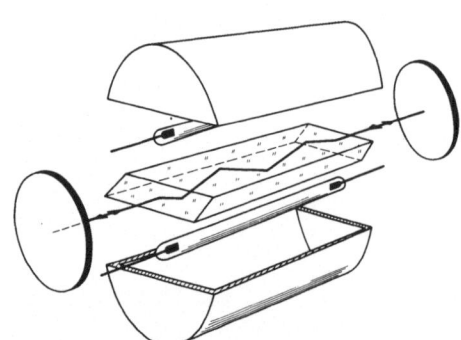

Abb. 1. Prinzip eines Slab
Laser Resonators

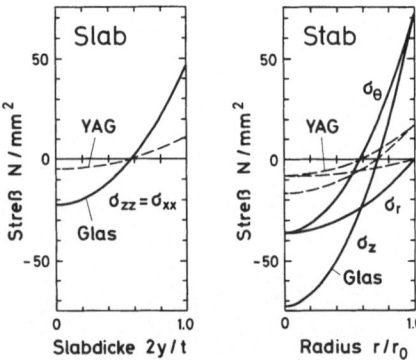

Abb. 2. Vergleich der Streßverteilungen
von Slab und Stab

Durch ein homogenes Pumpen und Kühlen dieser Seitenflächen kann man erreichen, daß über dem größten Teil des Slabs nur Temperaturgradienten in einer Richtung, senkrecht zu diesen Oberflächen auftreten. Als Folge davon stellt sich ein parabolisches Temperaturprofil über der Slab Dicke ein. Durch den Zick-Zack Strahlverlauf im Slab durchläuft das Strahlungsfeld zwischen zwei Reflexionen das gesamte Temperaturprofil und kann so die entstehenden thermischen Störungen in erster Näherung kompensieren. Somit ergibt sich eine hohe Strahlqualität des Slab Lasers bzgl. örtlicher Struktur und Strahldivergenz, unabhängig von der Pumpleistung.

Ausgehend von diesem parabolischen Temperaturprofil unter Vernachlässigung der Randeffekte im Bereich der Brewsterfenster sowie der beiden unpolierten Seitenflächen kann man die mechanischen Spannungen im Slab berechnen. Abb. 2 zeigt die auftretenden Spannungen in einem zylindrischen Stab und in einem Slab bei gleicher Pumpleistungsdichte. Bei diesen Berechnungen für Nd-Glas (durchgezogene Linien) und für Nd-YAG (gestrichelte Linien) als aktives Medium wurde eine Laserausgangsleistung von 40 W angenommen, bei einem Querschnitt des Slabs von 20 x 6 mm. Für beide Wirtsmaterialien ergeben sich die gleichen Verteilungen. Bei Nd-YAG treten jedoch aufgrund der überlegenen Materialeigenschaften wesentlich niedrigere Spannungen auf als bei Glas. Vergleicht man den Streß von Slab und Stab, so ergibt sich ein deutlich geringerer Streß auf der Oberfläche des Slabs. Entsprechend höher liegt somit auch die Bruchgrenze eines Slabs. Die drei Streßkurven beim Stab für radialen (σ_r), tagentialen (σ_θ) und axialen (σ_z) Streß reduzieren sich beim Slab auf eine Streßkurve für die Komponenten in x und y- Richtung. Als Folge dieser geringen mechanischen Spannungen kann ein Slab Laser bei hohen mittleren Pumpleistungen betrieben werden.

Da das Temperaturprofil näherungsweise nur eine Funktion der Slab Dicke ist, hat eine Veränderung der Slab Höhe keinen Einfluß auf die thermischen Eigenschaften des Lasers. Somit kann man das Volumen des aktiven Mediums vergrößern indem man ein größeres Querschnittsverhältnis wählt, ohne die Strahlqualität des Lasers zu beeinträchtigen. Damit ist eine höhere mittlere Laserausgangsleistung bei Slab Lasern möglich.

Die genannten Vorteile gelten sowohl für Glas als Wirtsmaterial, als auch für Kristalle. Eine Verbesserung des Wirkungsgrades scheint möglich durch Verwendung von GGG und GSGG als aktives Medium. Im Vergleich zu YAG haben diese Kristalle jedoch größere thermooptische Konstanten. In der Slab-Geometrie sollten diese thermischen Effekte stark reduziert werden. Diese Kristalle sind daher besonders für den Einsatz in Slab Lasern geeignet, da hier bei hohen Wirkungsgraden große Ausgangsleistungen bei guter Strahlqualität zu erwarten sind.

Sowohl Slab Laser, als auch Laser, die einen zylindrischen Stab als aktives Medium benützen, können durch die Slope Efficiency η_s und den thermooptischen Koeffizienten α unabhängig von der Pumpleistung charakterisiert werden. Beide Parameter sind entscheidend für hohe Ausgangsleistung und Strahlqualität. Der thermooptische Koeffizient α ist geometrieabhängig und eine Funktion der thermooptischen Konstanten.

α wird über die Brechkraft D, die Pumpleistung P_P und die Querschnittsfläche F des aktiven Mediums definiert:

$$D = \frac{\alpha\, P_P}{F}$$

Tab. 1 zeigt den an Hand experimenteller Daten ermittelten Quotienten η_s/α für verschiedene Lasermaterialien in Slab und Stab-Geometrie.

Stäbe	Lasermaterial	η_s/α [kW/mm]
	LG 706 /8/	0.4
	LG 760 /8/	1.0
	YAG /7,8/	1.1
	Alexandrit /9/	10.0
Slabs		
	LHG 5 /4/	0.6/ 1.0
	LHG 8 /5/	9.5/ 9.5
	YAG /7/	6.3/ 18.2
	GGG /7/, estim.	14.0/ 25.0
	Alexandrit, estim.	62 /170

Tab. 1.Quotient η_s/α für verschiedene Lasermaterialien in Slab und Stab-Geometrie, wobei η_s = Slope Efficiency, α = thermooptischer Koeffizient.

Bei Slabs ergeben sich jeweils verschiedene Werte η_s/α senkrecht und parallel zur Strahlebene. Es zeigt sich, daß Slab Laser aufgrund der deutlich größeren Quotienten η_s/α entscheidende Vorteile hinsichtlich der Strahlqualität bei hohen Ausgangsleistungen haben.

Technische Realisierung des Glas Slab Lasers

Der Slab ist in einer Halterung spannungsfrei aufgehängt (Abb. 3). Er hat eine Länge von ca. 300 mm und einen Querschnitt von 20 x 6 mm. Daraus ergeben sich 20 interne

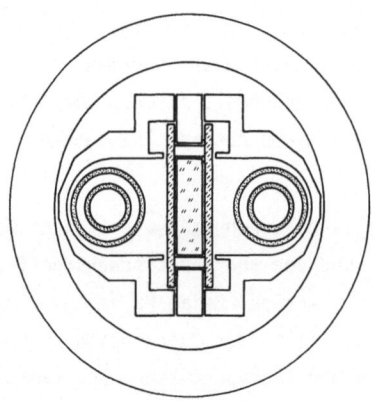

Abb. 3.Querschnitt durch den
Glas Slab Laserkopf

Reflexionen im Laserglas. Der Slab wird von zwei Xenon-Blitzlampen von je 12 Zoll Bogenlänge gepumpt, die sich in Flow Tubes befinden. Die Reflektoren sind poliert und vergoldet. Zur Vermeidung von Temperaturgradienten in vertikaler Richtung sind die Ober- und Unterseite des Slabs thermisch isoliert. Der Slab wird durch eine kombinierte Gas-Flüssigkeitskühlung gekühlt, um eine möglichst homogene Kühlung der Seitenflächen zu erreichen. Dabei ist der Slab ganz von einer Gasatmosphäre umgeben. Bei niedrigen Pumpleistungen genügt Luft als Kühlgas, bei höheren Leistungen wird Helium verwendet. Nur der äußere Teil der Slabhalterung und die Blitzlampen werden mit Wasser gekühlt.

Messungen am Glas Slab Laser

Abb. 4 zeigt die aufgenommenen Feldverteilungen des Slab Lasers bei verschiedenen Blendenquerschnitten im Resonator. Diese Aufnahmen wurden an einem semikonfokalen Resonator mit 10% Auskopplung gemacht. Die Daten geben jeweils die Strahlabmessungen im Resonator in Millimetern an. Bei diesen Aufnahmen waren die Enden des Slabs im Bereich der Brewsterfenster abgeschattet und wurden nicht gepumpt, um mögliche thermische Störungen in diesem Bereich zu vermeiden /7/.

Die aufgenommenen transversalen Moden zeigen ein regelmäßiges reproduzierbares Anschwingen der Feldverteilung während der gesamten Pulsdauer. Dies deutet auf einen homogenen Abbau der Inversion hin, bedingt durch den sich überschneidenden Zick-Zack Weg des Strahls im Slab.

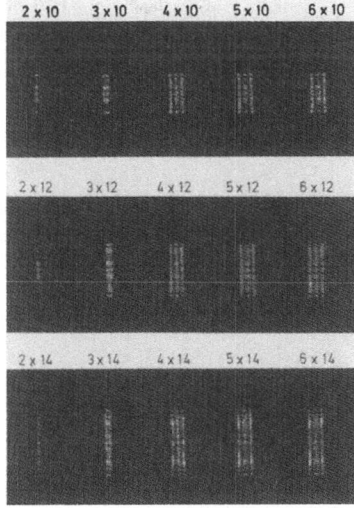

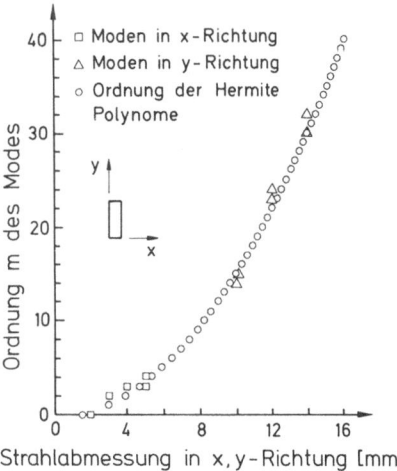

Abb. 4. Transversale Modenstruktur bei verschiedenen Blenden

Abb. 5. Vergleich der berechneten Modenverteilungen mit den gemessenen Feldverteilungen

54

Dieses Verhalten wurde auch durch Messungen bestätigt, bei denen Pumppuls und Laseremission parallel aufgezeichnet wurden. Ein Vergleich beider Intensitäten ergibt eine
lineare Abhängigkeit. Der Laser emittiert somit nahezu während der gesamten Pumppuls-
dauer in einem stabilen Mode.

Berechnet man die Feldverteilungen des Slab Lasers mit den Daten des leeren Resonators, so ergibt sich eine gute Übereinstimmung mit den experimentellen Ergebnissen
(Abb. 5). Es machen sich also in diesem Bereich der Pumpleistung von ca. 1.5 kW keine
störenden thermischen Effekte bemerkbar.

Bei verschiedenen Auskoppelgraden wurden Laserausgangsenergie und Pumpenergie gemessen (Abb. 6). Bei diesen Messungen ergab sich eine maximale Slope Efficiency von 0.6%.
Der Resonator war jedoch diesbezüglich noch nicht optimiert. Eine Verbesserung auf
ca. 1.2% sollte allein durch eine bessere Anpassung des Auskoppelspiegels sowie einer
Optimierung des Modenvolumens möglich sein.

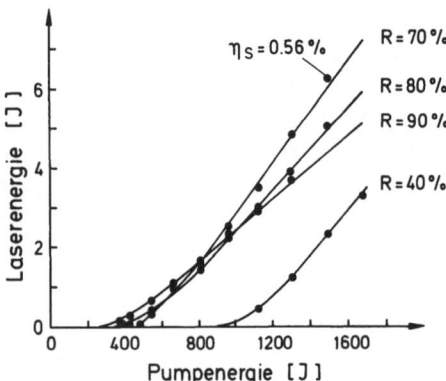

Abb. 6. Laserenergie als Funktion
der Pumpenergie bei ver-
schiedenen Auskoppelgraden

Zusammenfassung

An dem von uns entwickelten Glas Slab Laserkopf konnte gezeigt werden, daß in einem
Bereich bis 1.5 kW Pumpleistung keine störenden thermischen Effekte die Laserstrahl-
qualität beeinträchtigen. Bei einem noch nicht optimierten Resonator wurde eine Slope
Efficiency von 0.6% gemessen.

Literatur

/1/ W. KOECHNER
    Solid State Laser Engineering (Springer, New York, 1976)
/2/ W. KOECHNER, Appl. Opt. 9, 2548 (1970)
/3/ FOSTER, OSTERINK, J. Appl. Phys. 41, 3656 (1970)
/4/ EGGLESTON, KANE, UNTERNÄHRER, BYER, Optics Lett. 7, 405 (1982)
/5/ EGGLESTON, KANE, KUHN, BYER, SPIE 335, 104 (1982)
/6/ EGGLESTON, KANE, KUHN, UNTERNÄHRER, BYER, IEEE J. Quantum Electron. 20, 289 (1984)
/7/ KANE, ECKHARDT, BYER, IEEE J. Quantum Electron. 19, 1351 (1983)
/8/ R. IFFLÄNDER, K.P. DRIEDGER, Priv. Mitteilung, to be publ.
/9/ J.C. WALLING,
    Properties of Alexandrit Lasers, Nato-summer school, Italy 1984

Pumping Efficiency of Flashlamp Excited Slab Lasers

P. Mazzinghi and V. Rivano
Istituto di elettronica Quantistica del CNR
Via Panciatichi 56/30, 50127 Firenze, ITALY

Slab-geometry, flashlamp-pumped, lasers can provide significant performance improvements relative to conventional cylindrical-geometry lasers. For a solid state laser this approach has the advantage of the elimination of the optical distortion of the laser medium induced by the thermal loading, including thermal focusing, stress-induced birefrangence, and stress-induced biaxial focusing (1). For a dye laser the slab geometry is also advantageous at high repetition rates for the possibility to use a transverse flow , and the possibility to obtain an efficient narrowband tuning at high energies (2).

Another potential advantage of the slab laser is the possibility to achieve a higher efficiency (3) because of the better coupling efficiency of the laser medium with two or more flashlamps.

To obtain a good coupling efficiency between the cylindrical surface of a linear flashlamp and a flat surface the pumping reflector may be different from that used for a cylindrical laser. For example several kinds of non imaging, or diffusive reflector can be used.

We have tested 3 different kinds of reflectors for slab lasers: a conventional imaging elliptic reflector, a close-coupled diffusive reflector and a non imaging involute related reflector. The comparison is made in terms of output intensity distribution and total output energy. Results of operation of a slab dye laser are also reported.

The double elliptic reflector has the lamp tangent to one focus and the slab tangent to the image of the lamp, so that the pumping light is not tightly focused in the middle of the slab. The other axis of the ellipse is determined by taking the reflector as close as possible to the lamp water jacket (fig 1b). In this way we have obtained a close coupled configuration, which usually maximizes the transfer efficiency.

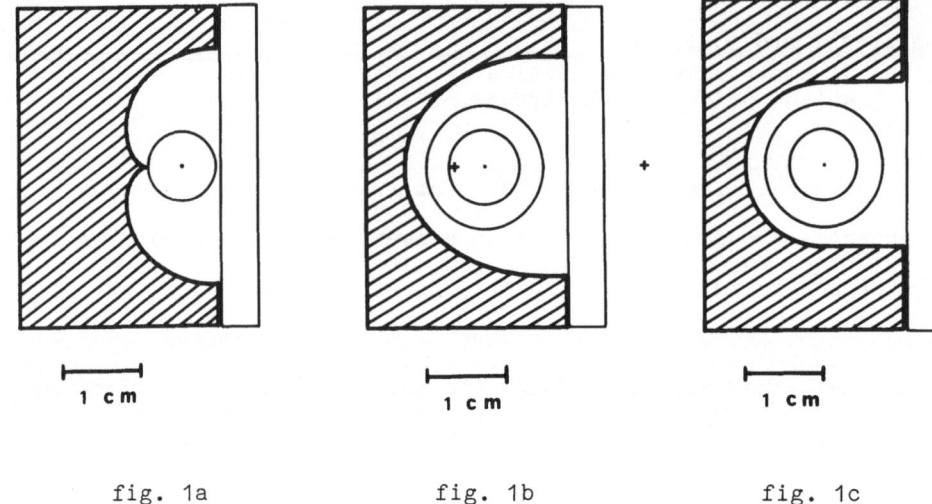

fig. 1a   fig. 1b   fig. 1c

Cross section of the 3 pumping reflectors: a) involute reflector; b) elliptic reflector; c) diffusive reflector. Crosses in fig. 1b indicates the position of the foci of the ellipse.

The non-imaging reflector is involute to the external circumference of the lamp (fig 1a). This curve can be constructed unwrapping a string wrapped around the cylinder (4), so that the theoretical transfer efficiency is 100% and its aperture is equal to the side surface of the generating cylinder. According to the Louville's theorem these facts imply that if the lamp is considered a uniform emitter of diffuse light, the illumination of the slab is uniform, diffuse and with the maximum energy density permitted by the Second Law of thermodynamics. This kind of reflectors were originally proposed to collect the diffuse light from a rectangular aperture on a cylindrical absorber, for solar energy applications (5), but, reversing the optical path, they can also be used for laser pumping by linear flashlamps (6). Both the elliptic and the involute reflectors where made by machining brass blocks. The reflecting surface was then chrome plated and polished and covered by an evaporated aluminum film, protected from scratches and moisture by SiO2.

The diffusive reflectors were made with the MACOR machineable ceramics in order to have a solid reflector which can be easily exchanged with the previous ones without modifying the laser head.The profile, which usually has little importance in this kind of reflectors, was choosen as tight as possible to the lamp water jacket, to obtain a close coupled geometry (fig 1c).

For all the 3 kinds of reflectors the output energy distribution and the total output energy were measured. The energy distribution was measured placing an UV grade optical fibre in the same position of the slab, and scanning it in the direction transverse to the lamp axis. The light transmitted by the fibre was detected by a photovoltaic, UV enhanced Si photodiode. The total output energy was measured with a Gen-Tech ED 500 piroelectric joulemeter. The total energy was then compared with the integral of the energy distribution in order to scale the intensity in energy density units, thus permitting comparisons between the different reflectors.

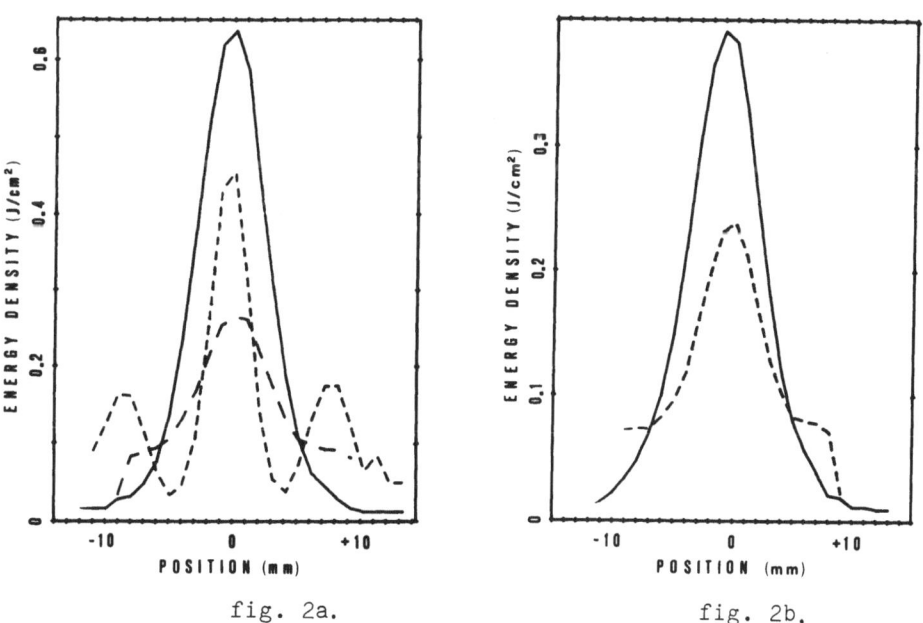

fig. 2a. fig. 2b.

Reflectors output energy density as a function of the distance from the centre: a) lamp without water jacket: full line elliptic reflector, short dashed line involute reflector, long dashed line diffusive reflector. b) lamp with water jacket: full line elliptic reflector, dashed line diffusive reflector.

Fig. 2a shows the output energy profile of the 3 reflectors. The lamp were used without the cooling water jacket, because the involute reflector is tangent to the lamp itself. The elliptic reflector gives clearly the highest energy density in the centre, but it pumps only about half of the slab. The involute reflector shows a three peaks distribution, which can be due to the radiation diagram of the lamp which is not lambertian, but more directional. The total energy is little lower (10%) to that of the elliptic reflector. This means that the average number of reflections is higher, and the light is attenuated by the finite reflectivity of the surface (less than 80% in our case). The diffusive reflector has the smoothest distribution, but the total energy is about half of that of the previous ones. This is due to the reabsorption of part of the radiation by the lamp itself and to the poor reflectivity of the MACOR ceramics in the UV, below 350 nm.

The same comparisons were made between the elliptic and the diffusive reflector with the water jacket installed (fig 2b), with essentially the same results except the attenuation of the water layer.

The operation of a slab dye laser have confirmed the results of these measurements: the elliptic and the involute reflectors have shown similar output energies, but the elliptic gives a lower threshold because lasing action starts in the centre of the slab. The diffusive reflectors give a lower energy (about 25%) with a very high threshold.

REFERENCES

1) Conference on Laser and Electro-Optics, slab laser session, Anaheim, CA, (1984)
2) P. Mazzinghi et al.: IEEE J. Quantum Electron., QE-17, 2245, (1981)
3) P. Mazzinghi, V. Rivano and P. Burlamacchi: Appl. Opt. 22, 3335, (1983)
4) I. M. Basset and G. H. Derrik, Optical and Quantum Electronics, 10, 61, (1978)
5) A. Rabl: Appl.Opt., 15, 1871, (1976)
6) R. J. Rosser: Optica Acta, 25, 727, (1978)

A Tentative Study of Polished Rod Laser

Ying Chengren, Wang Shiaojing and Wang Shaomin

Department of Physics, Hangzhou University

Hangzhou, China

A face polished, face pumped and face cooled slab laser in which the optical path
inside the gain medium undergoes total internal reflection, the effect of the ther-
mal-optical distortions can be reduced[1-5]. Then, a viewpoint on slab lasers having
intrinsic pseudo-phase-conjugate properties was proposed and some demonstrations
were given[6].

This concept may be expanded to a polished rod laser. The principle of transformation for internal reflection of rays or beams is shown in Fig.1. The transformation
could be described by transfer matrix

$$\begin{vmatrix} 1 & 0 \\ 0 & -1 \end{vmatrix}, \quad \text{or} \quad \begin{vmatrix} 1 & 0 \\ -2/\rho_i & 1 \end{vmatrix} . \tag{1}$$

They are identical with that of phase conjugator formed by degenerate four-wave mixing[7,8]. Some demonstrations are shown in Fig.2. Where, if the image distance is
not equal to the object distance, we can get a ring; if the image distance equals
the object distance, then we get a point. Which means that it has intrinsic pseudo-
phase-conjugate property and the energy can be used.

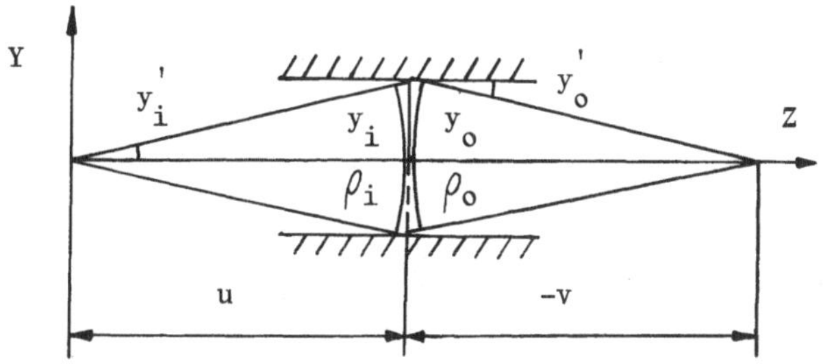

Fig. 1. Principle

Fig. 2. Demonstrations

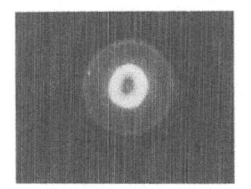

$u = 190$

$v = -120$

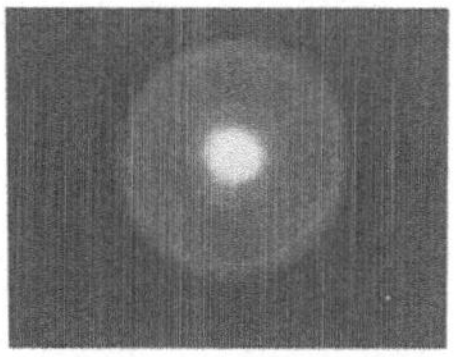

$u = 190$

$v = -190$

It is well-known that the backward-going (pseudo-)conjugators have got ability to compensate wave-front distortions caused by inhomogeneous media, i.e.,

$$
\begin{pmatrix} A_b & B_b \\ C_b & D_b \end{pmatrix} = \begin{pmatrix} d & b \\ c & a \end{pmatrix} \begin{pmatrix} 1 & 0 \\ 0 & -1 \end{pmatrix} \begin{pmatrix} a & b \\ c & d \end{pmatrix} = \begin{pmatrix} 1 & 0 \\ 0 & -1 \end{pmatrix} . \tag{2}
$$

But, this is a forward-going system. For a forward-going conjugator, we have

$$
\begin{pmatrix} A_f & B_f \\ C_f & D_f \end{pmatrix} = \begin{pmatrix} a_2 & b_2 \\ c_2 & d_2 \end{pmatrix} \begin{pmatrix} 1 & 0 \\ 0 & -1 \end{pmatrix} \begin{pmatrix} a_1 & b_1 \\ c_1 & d_1 \end{pmatrix} = \begin{pmatrix} a_1 a_2 - b_1 c_2 & a_1 b_2 - b_1 d_2 \\ c_1 a_2 - d_1 c_2 & c_1 b_2 - d_1 d_2 \end{pmatrix} . \tag{3}
$$

Let

$$
\begin{pmatrix} A_f & B_f \\ C_f & D_f \end{pmatrix} = \begin{pmatrix} 1 & 0 \\ 0 & -1 \end{pmatrix} ;
$$

get

$$
a_1 a_2 - b_1 c_2 = 1, \qquad a_1 b_2 - b_1 d_2 = 0,
$$
$$
c_1 a_2 - d_1 c_2 = 0, \qquad c_1 b_2 - d_1 d_2 = -1. \tag{4}
$$

Then the compensating conditions for a forward-going conjugator can be obtained

$$
b_2 = b_1, \qquad c_2 = c_1,
$$
$$
a_2 = d_1, \qquad d_2 = a_1. \tag{5}
$$

It means that the inhomogeneous medium profiles must have the axis of symmetry. If the rod with length L is pumped symmetrically, the index distributions in Y direction could be expressed by

$$
n(y) = n(0) (1 \pm \beta_0 y^2) ; \tag{6}
$$

the corresponding ray transfer matrix is

$$
\begin{pmatrix} a & b \\ c & d \end{pmatrix} = \begin{pmatrix} Ch(L\sqrt{2\beta_0}) & \frac{1}{\sqrt{2\beta_0}}Sh(L\sqrt{2\beta_0}) \\ -\sqrt{2\beta_0}Sh(L\sqrt{2\beta_0}) & Ch(L\sqrt{2\beta_0}) \end{pmatrix} , \quad \text{for } "+\beta_0" ; \tag{7}
$$

or

$$
\begin{pmatrix} a & b \\ c & d \end{pmatrix} = \begin{pmatrix} Cos(L\sqrt{2\beta_0}) & \frac{1}{\sqrt{2\beta_0}}Sin(L\sqrt{2\beta_0}) \\ -\sqrt{2\beta_0}Sin(L\sqrt{2\beta_0}) & Cos(L\sqrt{2\beta_0}) \end{pmatrix} , \quad \text{for } "-\beta_0" . \tag{8}
$$

The both (7) and (8), they are satisfied the conditions (5).

Then, a set of polished rod laser and a set of conventional roughned rod laser made of the same Nd glass for comparing the output properties are set up. The rod parameters, beam divergences and misalignment sensitivities defined by 9 of these two lasers are shown in the Table. For comparing the output energies and dynamic stabilities (15 Sec. per time) between polished rod laser and roughened rod laser, an illustrating is shown in Fig.3.

Table

Type	Diameter	Length	Beam Divergence	Misalignment Sensitivity
Polished rod	6mm	81mm	1.2×10^{-3}	4.1×10^{4}
Roughened rod	6mm	110mm	3.9×10^{-4}	1.6×10^{5}

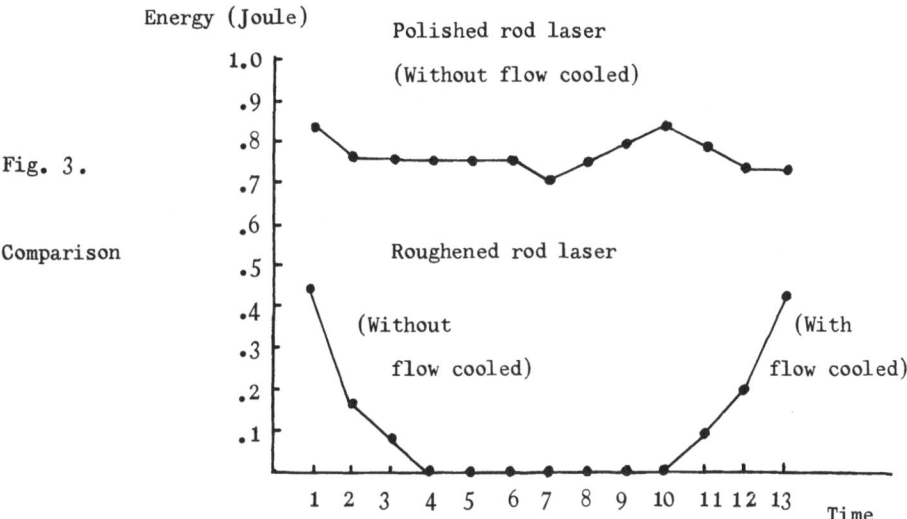

Fig. 3.

Comparison

The conclutions may be that the output energy, dynamic stability and misalignment sensitivity of polished rod laser are better than that of conventional roughened rod laser. But the beam divergence is not so good and the mode structures are rather complicated.

Even though so, the polished rod laser is an expanded slab laser, it is similar to waveguide gas laser, it is one of the simple ways to improve solid state laser in some industrial applications, it is a new recognition of an old problem and it is valuable to study it further.

Some new design and result will be reported.

One of the authers (Wang Shaomin) is grateful to Prof. Dr. H. Weber and Prof. Dr. L. Ronchi for their valuable discussions and grateful to ICTP Programme (Trieste, Italy) and Scientific Foundation (Peking, China) for their support.

Literature

(1) J. M. EGGLESTON, T. J. KANE, J. UNTERNAHRER, R. L. BYER: Opt. Lett. 9, 405 (1982)

(2) T. J. KANE, R. C. ECKARDT, R. L. BYER: IEEE QE-19, 1351 (1983)

(3) J. M. EGGLESTON, T. J. KANE, K. KUHN, J. UNTERNAHRER, R. L. BYER: IEEE QE-20, 289 (1984)

(4) J. EICHER: "Investigation of Slab Laser" Laser 85 Opto-Elektronik, I.1.8, Munich, Germany, July, 1985.

(5) P. MAZZINGHI, V. RIVANO: "Pumping Efficiency of Flashlamp Excited Slab Lasers" Laser 85 Opto-Elektronik, I.1.9, Munich, Germany, July, 1985.

(6) WANG SHAOMIN: Appl. Lasers (in Chinese) 4, 109 (1984)

(7) J. AU YEUNG, D. FEKETE, D. M. PEPPER, A. YARIV: IEEE QE-15, 1180 (1979)

(8) WANG SHAOMIN, H. WEBER: Opt. Comm. 41, 360 (1982)

(9) P.HAUCK, H. P. KORTZ, H. WEBER: Appl. Opt. 19, 598 (1980)

Performance Evaluation of a Q-switched Nd:YAG-Laser Operating at the Fundamental Mode

E.C. Munk - C.J. Nonhof

Nederlandse Philips Bedrijven B.V.

Centre For Manufacturing Technology

Introduction

A performance test for a Q-switched Nd:YAG laser is proposed using a thin film of a substance that is readily attacked by the laser beam on a substrate that is not. The focused laser beam writes a track along the test plate which is tilted such that also burn marks are obtained from the defocused laser beam, see figure 1.

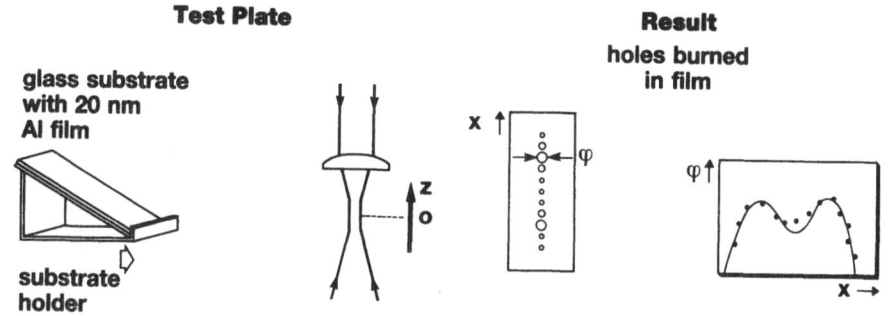

Figure 1

A quick check on the quality of the holes in the film gives an indication of the alignment of the laser. From plots of hole diameter versus defocusing distances we may obtain the beam waist w_{oo} and the ratio E_o/E_t, where E_o is the peak energy density of the laser beam in focus, and E_t is the threshold energy density at which the film is just affected. We employed a 20-30 nm thick aluminium film on a glass substrate.

Qualitative performance evaluation

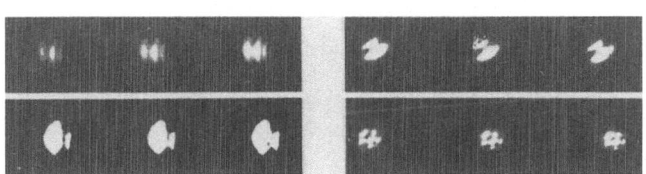

Figure 2. Four lasers that were not properly aligned (defocused).

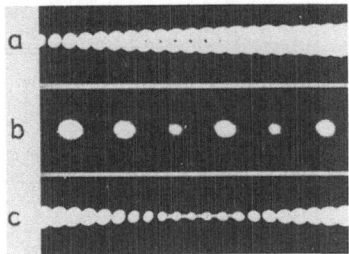

Figure 3.
a) A laser operating at the donut mode. Sometimes this is unavoidable at high output power levels. But on two occasions we found a burned optical component in the laser resonator.

b) Pulse drop outs. Sometimes this is unavoidable at high Q-switch repetition rates. Sometimes the Q-switch is not properly aligned.

c) A leaking Q-switch (focus region). Sometimes this is unavoidable at high output powers. Most of the time the Q-switch is not properly aligned.

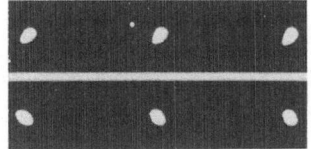

Figure 4.
a) Above focus.

b) Below focus. An indication of an imaging error from the focusing lens.

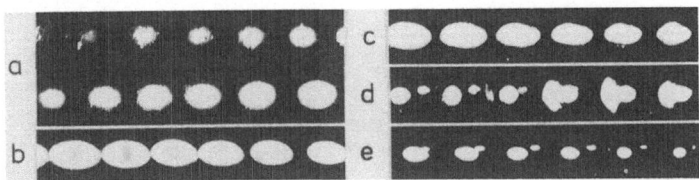

Figure 5.
a) Far above focus for two lamp currents showing a properly aligned laser.

b) Above focus near maximum hole dimensions.

c) Closer to focus. The hole towards the right shows interference from a plasma: a blue flame appears above the substrate.

d) In focus, showing the interference of the plasma plume.

e) In focus at reduced power levels. The plasma vanishes. The satelites are from reflections from laser light from the back surface of the glass substrate, showing that we are slightly above focus.

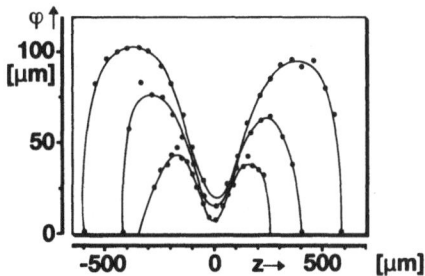

Figure 6.
Plot of hole diameter versus defocusing distance. The left shoulders are significantly higher than the right ones. For a laser operating at the fundamental mode the plot should be symmetrical, indicating that we have here an imaging error in the focusing lens.

Quantitative performance evaluation

In order to describe a laserbeam one needs to know two of the following properties:

- Beam divergence angle θ_{nm} (half angle).
- Beam radius at the waist of the laser beam w_{nm}.
- The mode structure factor C_{nm}, which has a minimum for the fundamental mode.

The relation between the three parameters is given by the formula:

$$\theta_{nm} \, w_{nm} = C_{nm}^2 \frac{\lambda}{\pi} \quad , \quad C_{nm} \geqslant C_{oo} = 1 \tag{1}$$

Since we tried very hard to operate the laser at the fundamental mode we might as well assume that we reached our goal. Not only do we know then that $C_{nm} = 1$, but we also know the detailed energy density profile of a cross-section across the laser beam:

$$E(r,z) = E_o(z) \, \exp \{ -2(r^2/w_{oo}^2(z)) \} \tag{2}$$

Further we know that from the theory of Gaussian Optics

$$w_{oo}(z) = w_{oo} \sqrt{1 + (\frac{\theta_{oo} z}{w_{oo}})^2} \tag{3}$$

and from energy conservation considerations

$$E_o(z) / E_o(o) = (w_{oo}/w_{oo}(z))^2 \qquad (4)$$

When we introduce a damage threshold E_t for the aluminium film we may calculate the diameter of the burned holes.

$$\emptyset(z) = w(z) \sqrt{2 \ln \frac{E_o(o)}{E_t} (\frac{w_{oo}}{w_{oo}(z)})^2} \qquad (5)$$

When the maximum of the hole diameter occurs at defocusing distances where we may approximate (3) by

$$w_{oo}(z) \approx \theta_{oo} z \qquad (3a)$$

we obtain a very simple relation for finding w_{oo}, the second parameter we need to know, as shown in figure 7.

Evaluation

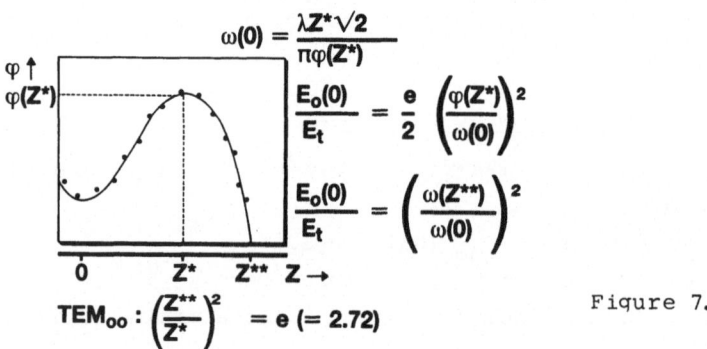

$$\omega(0) = \frac{\lambda Z^* \sqrt{2}}{\pi \varphi(Z^*)}$$

$$\frac{E_0(0)}{E_t} = \frac{e}{2} \left(\frac{\varphi(Z^*)}{\omega(0)}\right)^2$$

$$\frac{E_0(0)}{E_t} = \left(\frac{\omega(Z^{**})}{\omega(0)}\right)^2$$

$$TEM_{oo} : \left(\frac{Z^{**}}{Z^*}\right)^2 = e \ (= 2.72)$$

Figure 7.

Also shown in figure 7 is how we may calculate $E_o(0)/E_t$ and how to check whether the laser really operates at the fundamental mode. In figure 8 the laser beam has some higher order modes mixed in (presumably the donut mode).

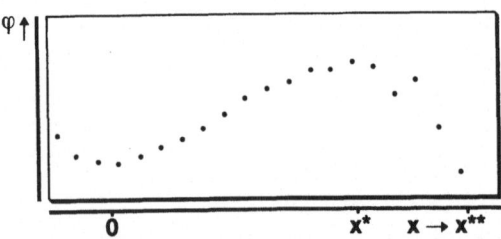

$$(Z^{**}/Z^*)^2 = (x^{**}/x^*)^2 = 2.1 \ (\hat{=} 2.7)$$

Figure 8.

Finally in figure 9 we observe a perfect laser operating at the fundamental mode.

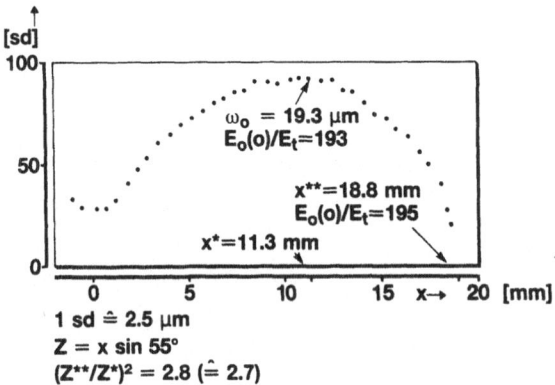

1 sd ≏ 2.5 µm

$Z = x \sin 55°$

$(Z^{**}/Z^{*})^2 = 2.8 \ (\hat{=} 2.7)$

Figure 9.

References

Nakayama, S. and Kashiwabara, M.
Review of the E.C.L. 20 (1972) 145.

Takamoto K and Makayama, S.
Review of the E.C.L. 23 (1975) 353.

Developments in Solid State Lasers

D C Hanna
Department of Physics
University of Southampton
Highfield Southampton SO9 5NH

The past 2-3 years has seen an accelerated pace of research and development in
solid state lasers. Particularly active is the research into tunable solid state
lasers and it is interesting to note how, in the laser's 25th anniversary year,
a great deal of attention is currently being paid to close relations of ruby,
the first laser medium. Another spur to activity is the realisation that solid
state lasers are potentially capable of high efficiency, in the region of 10%,
and very high average powers should also be attainable (1). A further area of
rapid growth involves Nd lasers (NdYAG in particular) pumped by GaAℓAs laser diodes.
This provides a very compact, efficient and long lived source. With the further
prospect of power increases and price decreases the scope for application of
such lasers is clearly considerable.

Tunable Solid State Lasers

A demonstration of tunable operation of a solid state laser was made as early as
1963, by Johnson et al (2), using $Ni:MgF_2$. A detailed understanding of the principles of such 'phonon-terminated' or 'vibronic' lasers was developed at that time
(3,4). However as the laser involved the awkward combination of liquid nitrogen
cooling and flash lamp pumping, its development was limited. Renewed interest
in vibronic lasers was stimulated by two separate developments in the late
1970s: the discovery that Alexandrite (Chromium doped chrysoberyl) displayed
tunable vibronic laser operation at room temperature (5), and the demonstration
of laser-pumped operation of $Ni:MgF_2$ and $Co:MgF_2$ (6,7).

The $Ni:MgF_2$ and $Co:MgF_2$ lasers still suffer from the disadvantage of low temperature operation. Progress towards operation at temperatures above 77K has been
made, for example an arrangement involving a two stage thermoelectric cooler,
allowing operation at 225K has been demonstrated (10), and eventually it is
expected that room temperature operation of vibronic lasers covering the range
$0.7-2.0\mu m$ will be achieved (8). Meanwhile, an impressive range of performance has
been demonstrated at low temperatures, including the following :-

> Pulsed operation of $Co:MgF_2$ (pumped by a pulsed $1.3\mu m$ $NdYAlO_3$ laser)
> giving 150mJ, 1msec pulses at 50Hz, an average output power of ~7w. (9)
> Q-switched operation of $Co:MgF_2$ giving a TEM_{oo} mode output of up to

60mJ in 150ns pulses. [10]

Tuning of $Co:MgF_2$ over the range 1.5 to 2.3μm [10]

Mode-locked operation of $Co:MgF_2$ and $Ni:MgF_2$ yielding pulse durations of 20-30psec [11].

Operation of a $Co:KZnF_3$ laser [12, 8], with narrow linewidth performance, less than 1.5MHz, recently reported.

The three main drawbacks of these Co and Ni lasers are the low temperature requirement, the need for laser pumping and the fact that the emission cross sections are small ($\sim 10^{-21} cm^2$). The small cross section implies a high intensity when Q-switched so that optical damage puts a limit on the pulse energy that can be extracted. Two of these disadvantages are removed in the case of Alexandrite, since room temperature flash lamp pumped operation is possible (pulsed or CW), but the cross section is smaller than ideal. Alexandrite continues to be the subject of research and development [13, 14] but one of its main attributes has been the way it has stimulated the search for other Chromium doped vibronic laser materials. As a result, the past few years has seen the demonstration of room temperature tunable vibronic laser action in a number of Chromium doped materials, such as emerald [15, 16], $Cr:KZnF_3$ [17], $Cr:GSAG$ [18], $Cr:GSGG$ [19] and $Cr:ZnWO_4$ [19]. The essential feature of these materials is that the Chromium ion must reside at a site with a weak crystal field [20-22]. Fig. 1 shows schematically the dependence of the $^4T_2$ and ^2E energy levels of Cr^{3+} on the crystal field strength.

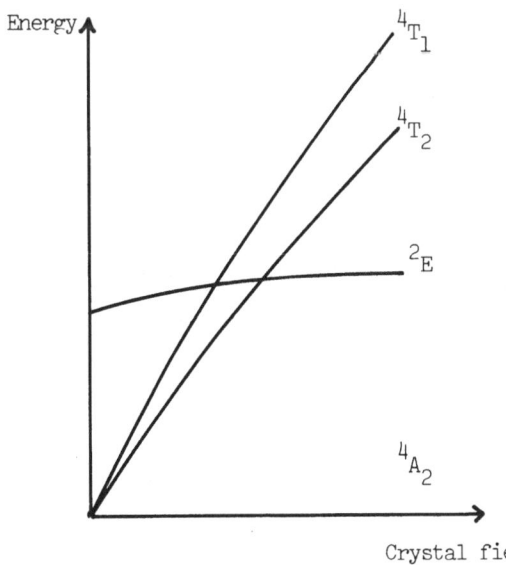

Crystal field

Pumping is via the $^4A_2 \rightarrow {}^4T_2$ and $^4A_2 \rightarrow {}^4T_1$ transitions, with subsequent relaxation of the excited population to an equilibrium distribution amongst the ^2E, $^4T_2$, $^4T_1$ levels. For ruby, having a strong crystal field, the energy difference $E(^4T_2) - E(^2E)$ corresponds to $\sim 2300\text{cm}^{-1}$; hence essentially all the population is in the long lived (3msec) ^2E level. 3 level laser action on the narrowline $^2E \rightarrow {}^4A_2$ transition is observed. In Alexandrite the weaker crystal field leads to a $E(^4T_2) - E(^2E)$ value of $\sim 800\text{cm}^{-1}$. Thus, a significant population resides in $^4T_2$ level, allowing four level, vibronic laser action on the $^4T_2 \rightarrow {}^4A_2$ transition. Since the $^4T_2$ level has a much shorter lifetime (~ 7 µsec) than the ^2E level (1.5msec in Alexandrite), the overall lifetime of the $^4T_2 \rightarrow {}^2E$ populations is ~ 250 µsec at room temperature. Weaker crystal fields lead to the following trends: the emission wavelength for $^4T_2 \rightarrow {}^2E$ gets longer, the $^4T_2$ population is a greater fraction of the total excited state population, and the fluorescence lifetime decreases since more population is in the shorter lived $^4T_2$ level.

Of the various Cr doped materials, particularly interesting results have been obtained from $Cr:KZnF_3$ and Cr:GSAG. Room temperature CW operation, using a Kr laser pump, has been achieved in both of these materials and these lasers are now available commercially. A four mirror astigmatically compensated resonator is used (very similar to CW dye laser designs), with a birefringent filter and undirectional device. A 3mm thick crystal, with ~ 0.5mole % Cr, yields around 50mW output @ 815nm for 1w of absorbed power. Tuning covers the range 780-865nm (17). Cr^{3+}: GSAG shows similar performance but shorter wavelength operation, 735-820nm (23). Cr^{3+}:GSAG has also shown promising performance when flashlamp pumped (100µsec flash), giving 80mJ output when pumped at 120J (23). Clearly a great deal of research and development is still required to assess and perfect these new materials, with improvements sought in crystal quality and losses. Extension to longer wavelength operation is a goal, with initial results of lasing between 950nm and 1090nm from $Cr:ZnWO_4$ (19) and around 925nm from $Cr:SrAlF_5$ reported by Lai and Jenssen at the 1985 topical meeting on Tunable Solid State Lasers.

While Chromium doped laser materials have a relatively long history, a recent discovery by Moulton, of tunable laser action in $Ti:Al_2O_3$ opens the prospect of a new class of laser materials (24, 25, 10). The Ti^{3+} ion derives its lowest energy states from a single d electron. The crystal field splitting yields two low lying states between which the vibronic laser action occurs, with excited state absorption (ESA) eliminated since all other states are very much higher in energy (26). The absence of ESA allows a very wide tuning range, 660-986nm having been demonstrated (10).

The broad pump bands of $Ti:Al_2O_3$ offer a number of pumping schemes, those demonstrated so far include pumping by the second harmonic of a Q-switched NdYAG laser, pumping with a flash-lamp pumped dye laser, pumping by a CW Argon laser and pumping directly by flash lamp. Representative results include 8mJ output when pumped by 35mJ of NdYAG second harmonic (27) and 350mJ output when pumped by 80J of flash lamp input (28). The flash lamp pumping scheme involved a dye fluorescence converter surrounding the 6mm x 100mm rod, thus enhancing the conversion efficiency from ~0.01% (without the converter), to around 0.5%. The flash lamp pulse was kept short (2μsec) since the fluorescence lifetime of the Ti is only ~ 3μsec. Under these pumping conditions lamp life was found to be only 10^3-10^4 shots (28). Clearly Ti:Al$_2$O$_3$ presents an extremely attractive laser medium, provided lamp life problems can be solved, since it provides room temperature operation, reasonable efficiency, very wide tuning (0.66-1μm) and direct pumping with flash lamps.

In addition to the advances being made with tunable solid state lasers progress is being made in the area of high efficiency solid state lasers. NdYAG lasers currently have an efficiency (when flash lamp pumped) of 2-3%. Recent results from Nd:GSGG(29) and Nd:GSAG(30) co-doped with Cr impurity indicate efficiencies around twice that of NdYAG. The principle of co-doping is that the Cr ion, with its broad absorption bands provides an efficient means of absorbing the pump radiation and that this Cr ion excitation can then be transferred to the Nd ions. For this to be an efficient transfer process the Cr ions need to be in the $^4T_2$ level rather than the ^2E level. In YAG the crystal field at the Cr site is large and the Cr population is therefore predominantly in the ^2E level. NdYAG co-doped with Cr does not achieve efficient transfer to Nd whereas the weaker-field in GSGG allows essentially 100% transfer efficiency. In addition to the improved efficiency compared to Nd:YAG, the crystal growth of GSGG has better characteristics allowing large, core-free boules to be grown. Scaling to very high power laser systems therefore looks encouraging.

Co-doping has also been used to good effect in the Ho:αβYLF laser, where the active Ho^{3+} ion, has co-dopants erbium and thulium in the LiYF$_4$ host. Emission is at 2.06μm and the dopants provide absorption bands extending from 0.2-1.8μm, thus allowing efficient pumping with a tungsten lamp. Results reported recently by Knights et al (31) are as follows. With tungsten lamp pumping, 100mJ 50nsec Q-switched pulses at 100Hz are obtained from a TEM$_{00}$ mode oscillator, with 'wall-plug' efficiencies between 2% and 3.5%. For this 100mJ, 100Hz system a projected lamplife allowing 10^{10} shots is predicted. Efficiencies approaching 5% are also anticipated.

Diode laser pumped solid state lasers

The good match between the emission wavelength of GaAs diode lasers and one of
the pump absorption bands of NdYAG was noted many years ago and a number of demonstration experiments were carried out in which Nd lasers were pumped by diode
lasers and LEDS (32-36). Improvements in the performance of diode lasers such
as long life operation, high power output, improved spatial coherence, and high
efficiency have combined to make the diode pumped NdYAG laser an attractive laser
source for a number of applications, including free space optical communications,
long distance fibre-optic communication, remote sensing, ring laser gyros. Recent
results indicate impressive efficiencies and output powers and intracavity
second harmonic generation in the NdYAG laser resonator has led to
11mW CW green output at a wall-plug efficiency of 1% (37). Other recent results
include the use of a diode pumped NdYAG laser to injection seed a high power
Q-switched NdYAG laser and thus achieve reliable single longitudinal mode operation
(38). Another interesting result is the demonstration of gain-switched operation of
a diode pumped NdYAG laser, leading to single longitudinal mode output from the
NdYAG laser, with a peak power greater than peak output of the diode laser (39).

The absorption band in NdYAG covers the range from ~790-820nm, with structure
as narrow as 1-2nm. The peak absorption, around 808nm has an absorption coefficient
of 0.8mm^{-1} in 1% doped NdYAG. For efficient pumping the diode wavelength must therefore be closely matched to the NdYAG absorption, and temperature tuning of the diode
emission wavelength (typically ~ 0.3nm/K) is used. The small absorption coefficient
also leads to best efficiency in the end-pumped configuration rather than side-pumped.
This end-pumped configuration also allows much lower threshold to be achieved, with
a few mw of diode output being sufficient to reach threshold (40). Using a 200mW
GaAlAs diode array to end pump a 10mm long x 5mm dia NdYAG rod, Sipes (41) has
achieved a 70mW NdYAG output. This corresponds to an overall efficiency of 7% as
the electrical input to the diode array was 1w. The diode efficiency was 20%
(electrical to light output) and the optical conversion efficiency 35% (810nm →
1.060nm). Overall efficiency can be expected to reach 10% and by pumping with
several arrays, NdYAG output powers in the range of 1w can be expected.

Baer and Keirstead (37) report intracavity frequency doubling of a diode pumped
NdYAG laser. The NdYAG was 5mm long, pumped by a 200mW GaAlAs array. The NdYAG
resonator contained a 5mm long KTP frequency doubling crystal and the second
harmonic radiation generated in both directions through the crystal was transmitted
through one resonator mirror only. 11mW of green radiation was generated for 1w
of electrical input to the diode. A dye laser, operating at ~ 810nm and 400mW, was
used to simulate the projected performance of future diode arrays. 40mW of green

output was achieved in this case. This promises to be an attractive source of coherent green light.

Single longitudinal mode (SLM) operation of high power NdYAG lasers is a necessary requirement for a number of applications. Techniques evolved so far have been based on prelase operation of the NdYAG laser until mode selection has been achieved and then followed by Q-switching (42, 43). An alternative technique is that of injection locking where the single frequency is provided by an ancillary laser (44). The injection power levels required are in the milliWatt range, well within the capability of a diode pumped YAG laser. Furthermore, single mode operation of the diode pumped YAG laser is easy to achieve since the short NdYAG rod implies a large mode spacing. Schmitt and Rahn report successful operation of such a scheme (38) with a few milliWatt power from the NdYAG laser. Stabilisation of the resonator length of the slave laser is required to ensure that the injected frequency is close enough to a mode frequency of the slave. A feedback control circuit has been devised to achieve this, based on the fact that the Q-switched pulse delay relative to the Pockels cell opening time is a function of the frequency mismatch between master and slave.

An interesting way to achieve higher power single mode operation from a diode pumped NdYAG laser has been demonstrated by Owyoung et al (39). Zhou et al (40) point out that the single mode output of the diode pumped NdYAG laser is limited by spatial hole burning effects (44) and is in the region of a few mW typically for a few mm crystal length. (Owyoung observed two mode operation at much lower power levels). Owyoung's technique for increasing the single mode power consists of pumping the diode laser at a low enough level to ensure SLM in the NdYAG laser and then pulsing the diode laser to a high power for a few microseconds. The single frequency in the NdYAG laser is thus amplified and a result of this 'gain-switching' technique is that SLM peak output powers have been obtained which are greater than the peak diode output power, for example 60mW of 1.06μm in a 180nsec pulse, for 45mW peak diode output (39). Clearly this ability to easily modulate the diode pump will have a number of interesting possibilities and the high peak powers achievable should provide scope for enhancement of nonlinear optical effects.

Literature

1. J L Emmett, W F Krupke and W R Sooy, 'The potential of High Average Power Solid State Lasers', Lawrence Livermore National Laboratory, Livermore, Calif. UCRL-53571 (1984)

2. L F Johnson, R E Dietz and H J Guggenheim, Phys. Rev. Lett 11, 318, (1963)

3. L F Johnson, H J Guggenheim and R A Thomas, Phys. Rev. $\underline{149}$ 179, (1966)

4. D E McCumber, Phys. Rev. 134, A299, (1964)

5. J C Walling, H P Jensen, R C Morris, E W O'Dell and O G Peterson, Opt. Lett. $\underline{4}$, 182 (1979)

6. P F Moulton, A Mooradian and T B Reed, Opt. Lett., $\underline{3}$ 164 (1978)

7. P F Moulton and A Mooradian, Appl. Phys. Lett $\underline{35}$, 838 (1979)

8. U Dürr, U Branch, W Knierim and C Schiller, p.20, in Tunable Solid State Lasers, Proc. 1st International Conf.   La Jolla, 1984, Springer Verlag 1985

9 . P F Moulton, IEEE J.Quantum Electronics QE-18, 1185 (1982)

10. P F Moulton, p.4, in Tunable Solid State Lasers, Proc. 1st International Conf. La Jolla, 1984, Springer Verlag, 1985

11. B C Johnson, P F Moulton and A Mooradian, Optics Letts, $\underline{10}$, 116 (1984)

12. Künzel, W Knierim and U Dürr, Optics Comm. $\underline{36}$, 383 (1981)

13. J C Walling, O G Peterson, H P Jensen, R C Morris, E W O'Dell, IEEE J. Quantum Electron. $\underline{16}$, 1302 (1980)

14. S T Lai and M L Shand, J Appl. Phys. 54, 5642 (1983)

15. M L Shand and J C Walling, IEEE. J of Quantum Electron. QE-18, 1829 (1982)

16. J Buchert, A Katz and R R Alfano, IEEE J of Quantum Electron. QE-19, 1477 (1983)

17. U Brauch and U Dürr, Optics Letts, 9, 441 (1984)

18. J Drube, B Stuve and G Huber, Optics. Comm. $\underline{50}$, 45 (1984)

19. G Huber and K Petermann, p.11 in Tunable Solid State Lasers, Proc. 1st International Conference, La Jolla 1984, Springer Verlag 1985

20. P T Kenyon, L Andrews, B McCollum and A Lempicki, IEEE J of Quantum Electron. QE-18, 1189 (1982)

21. H P Christensen and H P Jenssen, IEEE J of Quantum Electron. QE-18, 1197, (1982)

22. B Struve and G Huber, Appl. Phys. B, $\underline{36}$ 195, (1985)

23. J Druke, G Huber and D Mateika, Flashlamp pumped Cr^{3+}:GSAG laser. Post-deadline paper at CLEO '85

24. P Moulton, Optics News, (Nov/Dec (1982)) p.9

25. P Moulton, Laser Focus (May 1983) p.83

26. G F Albrecht, J M Eggleston, and J J Ewing, Optics Comm., 52, 401 (1985)

27. N P Barnes and D K Remelius, paper THE4, CLEO '85

28. P Lacovara, L Esterowitz and R Allen, paper THE3, CLEO '85

29. E V Zhorikov et al. Sov. J. Quantum Electron 12. 1652 (1982)

30. J A Caird, W F Krupke, M D Shinn, L K Smith and R E Wilder, paper THR3, CLEO '85

31. M G Knights, J Mosto and E P Chicklis, paper WJ1, CLEO '85

32. L J Rosenkrantz, J Appl. Phys. 43, 4603 (1972)

33. R B Chester and D A Dragvert, Appl. Phys. Lett., 23, 235 (1973)

34. J Stone and C A Burrus, Appl. Phys. Lett., 23 388 (1973)

35. K Washio, K Iwamoto, K Inoue, I Hino, S Matsumoto and F Saito, Appl. Phys. Lett., 29 720 (1976)

36. J Stone and C A Burrus, Fiber. Integ Opt. 2, 19, (1979)

37. T Baer and M S Keirstead, paper THZZ1, CLEO '85

38. R L Schmidt and L A Rahn, paper TUL4, CLEO '85

39. A Owyoung, G R Hadley, P Esherick, R L Schmidt and L A Rahn, paper THZZ2, CLEO '85

40. B Zhou, T J Kane, G J Dixon and R L Byer. Optics Letts. 10, 62 (1985)

41. D L Sipes, paper TUL2, CLEO '85

42. A J Berry, D C Hanna and C G Sawyers, Opt. Comm. 40, 54 (1981)

43. D C Hanna and Y-W J Koo, Optics Comm. 43, 414 (1982)

44. H G Danielmeyer, Progress in NdYAG lasers, in Lasers Vol. 4
     A K Levine and A J De Maria eds (Marcel Dekker, New York 1976)

Wirkungsgrad von Alexandritlasern

K. Mann
Fachbereich Physik
Universität Kaiserslautern
6750 Kaiserslautern

Einleitung

Alexandrit, ein mit Cr^{3+} dotierter Chrysoberyll stellt ein neues Lasermaterial dar,
dessen mechanische und thermische Eigenschaften mit denen von YAG vergleichbar sind,
und das zudem noch im Wellenlängenbereich zwischen 700 nm und 800 nm durchstimmbar
ist /1/. Diese Durchstimmbarkeit ist eine Folge der Kopplung von elektronischen Cr^{3+}-
Niveaus mit Phononen des Kristallgitters. Es bilden sich energetisch eng beieinander
liegende vibronische Zustände, die als untere Laserniveaus fungieren (s. Abb. 1). Das
obere Laserniveau, evenfalls in vibronische Zustände entartet, ist nur 0.1 eV vom
langlebigen-^{2}E-Niveau getrennt und wird von dort aus beim Pumpen thermisch besetzt.
Eine so entstandene Inversion wird im wesentlichen durch Übergänge zum unteren Laserniveau und durch Absorption in energetisch höher liegende Zustände ("Excited State
Absorption") /2/ abgebaut. Letzterer Vorgang und die Besetzung des oberen Laserniveaus sind temperaturabhängig und bestimmen über die Temperatur die Ausgangsleistung
des Lasers. Ebenfalls von der Temperatur abhängig ist die Brechkraft des aktiven Mediums, welche je nach Resonatorkonfiguration die Laserenergie entscheidend beeinflussen kann. Aus diesem Grunde wird im folgenden der Einfluß der Temperatur und der
Brechkraft auf den Wirkungsgrad von Alexandritlasern untersucht.

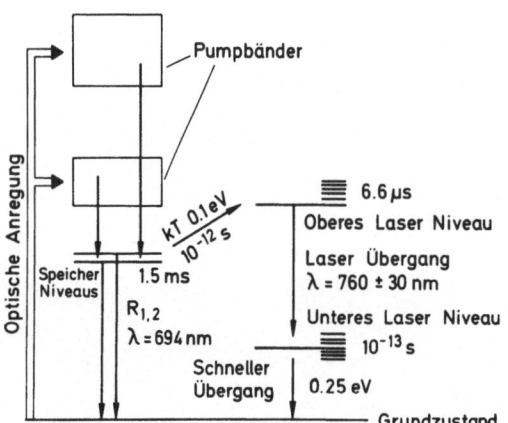

Abb. 1. Pump- und Laserniveaus beim
Alexandrit

Temperatur im Laserstab

Die Temperatur des Alexandrit Stabes kann mittels zweier Methoden variiert werden. Zum einen wird die Temperatur des Wassers, das den Stab umgibt, erwärmt und damit der Stab homogen aufgeheizt (s. Abb. 2), zum anderen wird bei jedem Pumplichtpuls das aktive Medium durch absorbiertes Pumplicht inhomogen erhitzt. Die im Stab aufgenommene Energie pro Puls beträgt 10% der vom Netzgerät abgegebenen Energie, wobei 8% in Wärme umgewandelt werden. Bei repetitiv gepumpten Systemen, bei denen die Zeit zwischen zwei Pulsen kleiner als die thermische Relaxationszeit (1.6 sec) ist, stellt sich eine stationäre Temperatur ein, die von der Stabmitte zum Rand parabolisch abfällt /3/. Die Temperatur zwischen Stabmitte und Rand kann mittels der temperaturabhängigen Absorption des Alexandrits bei 632 nm gemessen werden. Das Meßergebnis ist in Abb. 3 dargestellt. Man erkennt eine Proportionalität zwischen Pumpleistung und Temperatur. Die erreichte Temperaturerhöhung bei der zur Verfügung stehenden maximalen Pumpleistung von 1.5 kW beträgt 14 ^{O}C.

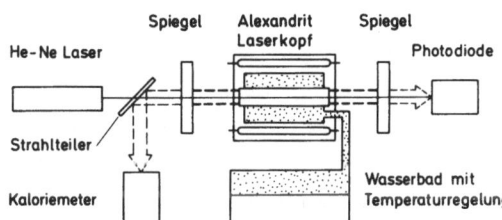

Abb.2. Experimenteller Aufbau zur Brechkraft-, Stabtemperatur- und Laserenergiemessung

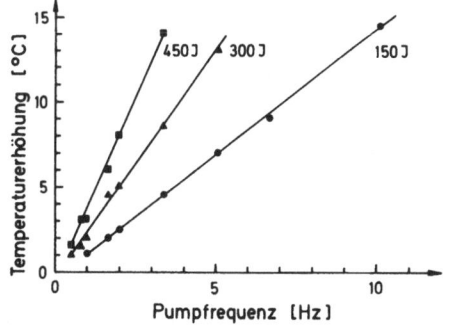

Abb.3. Temperaturdifferenz zwischen Stabmitte und Stabrand in Abhängigkeit von Pumpfrequenz und Pumpenergie

Brechkraft von Alexandrit

Die inhomogene Aufheizung des Laserstabes bewirkt ein parabolisches radiales Brech-
ungsindexprofil, das zu einer Linsenwirkung des Stabes führt. Die Stärke dieser Linsenwirkung wird durch die Brechkraft beschrieben. Die Brechkraft eines aktiven Fest-
körpermediums beeinflußt wiederum die Laseremission indem sie, je nach Resonatortyp
die Verluste der transversalen Moden bestimmt. Es werden nämlich nur die Moden anschwingen, deren Modenradien in den Stabradius passen. Möglichst hohe Effizienz erreicht man aber im stabilen Resonator durch hohe transversale Moden, da sie das Lasermedium am gleichmäßigsten ausfüllen und damit die Inversion am besten abbauen. Da
der Modenradius höherer transversaler Moden ein vielfaches des TEM_{00} Modenradius ist,
ist letzterer ein Maß für die Verluste höherer Moden. In Abb. 4 ist der Radius des
Grundmodes in Abhängigkeit von der Brechkraft aufgetragen. Der semikonfokale 1 m Resonator wird also trotz sich ändernder Brechkraft kaum eine Auswirkung auf die Effizienz des Lasers haben, während der Wirkungsgrad der plan-plan Resonatoren recht
stark von der Brechkraft beeinflußt wird. Bei geringer Brechkraft wird wegen zu hoher
Verluste kein höherer Mode anschwingen, die Effizienz des Lasers also gering sein.
Mit wachsender Brechkraft sinken die Verluste und der Radius des Grundmodes wird vergleichbar mit dem des semikonfokalen Resonators. Beide Resonatortypen werden ähnliche
Wirkungsgrade haben.

Die Brechkraft, die sich während des Laserbetriebes in der Stabmitte des Alexandrit
Kristalls einstellt, ist in Abb. 5 in Abhängigkeit von der Pumpenergie und der Pumpfrequenz dargestellt. Die Kurven wurden nach einer in /4/ beschriebenen Methode gemessen und erlauben eine Zuordnung von Eingangsenergie bzw. mittlerer Eingangsleistung zur Brechkraft. Mit Hilfe der Brechkraft lassen sich dann, wie oben beschrieben, aus Abb. 4 der Modenradius und damit die Verluste und der Wirkungsgrad abschätzen.

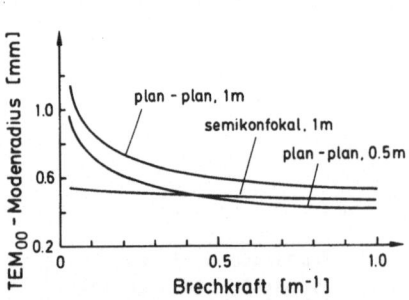

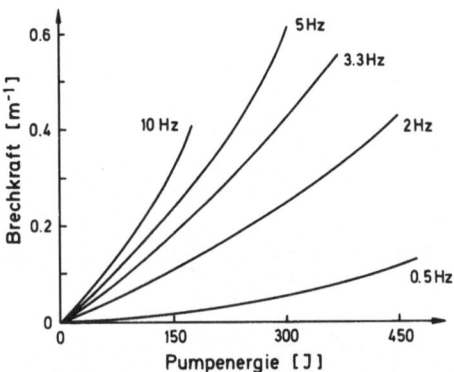

Abb. 4. Radius des TEM 00-Modes in Abhängigkeit von der Brechkraft
für verschiedene Resonatortypen (berechnete Kurven)

Abb. 5. Brechkraft in Abhängigkeit von
Pumpenergie und Pumpfrequenz
(gemessener Verlauf) bei einem
Alexandritstab von 4 x 1/4"

Wirkungsgrad von Alexandrit

Den folgenden Daten liegen Messungen an einem doppelelliptischen Laserkopf der Firma
Apollo zugrunde. Der Alexandritstab ist mit 3.3 x 10^{19} cm^{-3} Cr-Ionen dotiert und hat
die Größe 4 x 1/4".

Zunächst wird der Laser im Einzelschußbetrieb untersucht. Dabei stellt sich keine
mittlere Temperaturerhöhung durch das Pumpen ein. Eine homogene Erwärmung des Alexan-
dritkristalls erreicht man durch das ihn umgebende Wasserbad. In einem 0.5 m plan-
plan Resonator läßt sich dann der Wirkungsgrad von 0.44% bei 20 ^{O}C auf 0.67% bei
90 ^{O}C steigern. Dies entspricht bei 480 J Pumpenergie einer Laserenergie von 2.1 und
3.2 J. Die Schwellenergie sinkt während der Temperatursteigerung von 140 J auf 100 J.
Nach /5/ ist eine Steigerung der Laserenergie bis zu Temperaturen von 200 ^{O}C beobachtet worden.

Bei repetitiv gepumptem Betrieb, bei dem die Wassertemperatur konstant auf 30 ^{O}C gehalten wird, macht sich nun die inhomogene Temperaturerhöhung und je nach Resonatorkonfiguration auch die Brechkraft bemerkbar. Dies ist in Abb. 6 zu erkennen, wo die
Laserausgangsenergie eines 1 m plan-plan Resonators über der Pumpenergie dargestellt
ist. Die Laseremission dieses Aufbaus ist stark frequenzabhängig, da sich mit der
Pumpfrequenz die Brechkraft ändert und dadurch die Verluste höherer Moden beeinflußt
werden (s. Abb. 4 u. 5). Der totale Wirkungsgrad bei 480 J Pumpenergie beträgt bei
0.2 Hz lediglich 0.23%, während er bei 3 Hz auf 0.71% ansteigt.

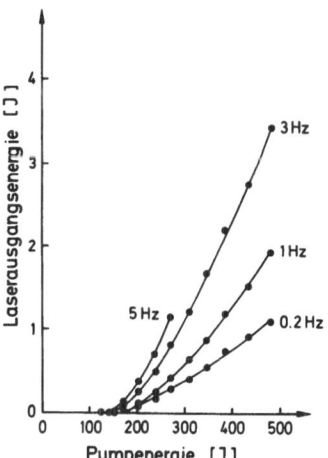

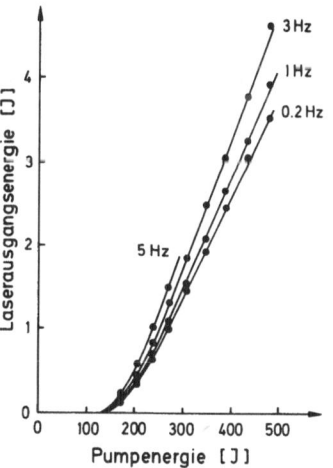

Abb. 6. Laserausgangsenergie in Abhängigkeit von Pumpfrequenz und
Pumpenergie für einen 1 m plan-
plan Resonator

Abb. 7. Laserausgangsenergie in Abhängigkeit von Pumpfrequenz und
Pumpenergie für einen 1 m semikonfokalen Resonator

Beim 1 m semikonfokalen Resonator variiert die Laserausgangsleistung dagegen nur wenig mit sich ändernder Temperatur (s. Abb. 7), was auf die Unempfindlichkeit dieses Resonatortyps gegen Brechkraftänderungen zurückzuführen ist. Da der TEM_{00}-Modenradius und damit auch die Verluste höherer Moden kleiner als im 1 m plan-plan Resonator sind, erreicht man hier einen Wirkungsgrad von 0.96% und eine Slope Efficiency von 1.6% bei 3 Hz , was einer maximalen Laserenergie von 4.6 J bei 480 J Pumpenergie entspricht.

Messungen am 0.5 m plan-plan Resonator zeigen die gleiche Frequenzabhängigkeit wie sie beim 1 m plan-plan Resonator auftritt. Die maximale Energie liegt jedoch mit 5.2 J bei 3 Hz etwas höher als beim semikonfokalen Resonator, wie dies nach Abb. 4 u. Abb. 5 auch zu erwarten ist.

Zusammenfassung

Die durchgeführten Messungen haben gezeigt, daß bei Alexandritlasern eine Steigerung des Wirkungsgrades durch Erwärmen des aktiven Mediums erreicht werden kann. Die Effizienz hängt außerdem von der Pumpfrequenz ab. Dadurch wird der Stab nämlich inhomogen aufgeheizt und durch die sich einstellende Brechkraft, je nach Resonatorkonfiguration, effektiver ausgenutzt.

Eine maximale Energieausbeute erreicht man danach beim Alexandritlaser durch Erwärmen des Stabes (bis 200 $^{\circ}$C), durch ein hohe Pumpfrequenz und durch einen Resonator mit geringen Verlusten für höhere Moden.

Literatur

/1/ J.C. WALLING, O.G. PETERSON, H.P. JENSEN, R.C. MORRIS, E.W. O'DELL: IEEE J. Quantum Electrn. QE-16, 1302 (1980)
/2/ M.L. SHAND, J.C. WALLING, R.C. MORRIS: J. Appl. Phys. 52, 953 (1981)
/3/ W. KOECHNER: J. Appl. Phys. 44, 3162 (1973)
/4/ H.P. KORTZ, R. IFFLÄNDER, H. WEBER: Appl. Optics 20, 4124 (1981)
/5/ S. GUCH, C.E. JONES: Optics Lett. 7, 608 (1982)

New Tunable Solid State Lasers

G. Litfin, P. Fuhrberg and W. Luhs
Spindler & Hoyer GmbH & Co.
Königsallee 23, D-3400 Göttingen

Introduction

The operation characteristics of laser materials based on Cr^{3+}-doped, low ligand field crystals were investigated. This new class of laser materials allows for tunable pulsed and cw operation at room temperature in the near IR. After the first demonstration of tunable operation in alexandrite (1), room temperature laser action in various Cr^{3+}-doped garnets, e.g. GSGG (2) and GSAG (3), and the perovskite $KZnF_3$ (4) has been reported. In this paper the laser properties as frequency behaviour, output power, slope efficiencies and thermal induced birefringence of GSGG and $KZnF_3$ are investigated using a krypton ion laser as the pump source.

Laser characteristics

As the emission lines of the crystals considered here are homogeneously broadened, a single frequency oscillating in the laser cavity is supported by the entire inversion. In a standing wave cavity, however, only parts of the spatial inversion profile coincide with the laser mode, so that spatial holeburning occurs. Therefore the output power is reduced compared with the power achievable in a travelling wave cavity. In order to investigate the main laser characteristics, both for standing wave and travelling wave operation, in our experiments we used linear and ring cavities. As an example the ring configuration is shown schematically in Fig. 1.

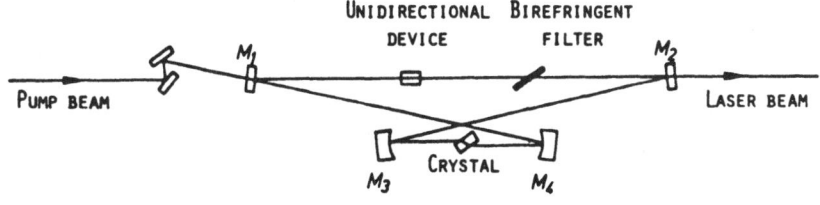

Fig. 1.   Schematics of the travelling wave cavity

The active material is pumped in a collinear geometry by focussing
the red lines of a krypton ion laser with a f= 50 mm lens or one of
the folding mirrors on the crystal. The crystal is mounted at
Brewsters angle on a convection cooled metal block. Due to the
relatively low gain of the new materials (effective emmission cross
section $\sigma_e \sim 10^{-20}$ cm^2) low loss components are required for the laser
cavity. The output coupler transmission is T= 1.1%.

Tuning is accomplished by a single plate birefringent filter. In the
linear cavity an additional etalon is introduced to suppress holeburning modes. Unidirectional travelling wave operation is achieved
by use of a Faraday rotator inserted into the ring cavity. A slight
tilt of the laser beams at mirror M2 with respect to the plane of the
other three mirrors produces a sufficient amount of back rotation for
one direction, so that the losses in the opposite travelling directions are different and the therefore the emission in the other
direction is suppressed. With KZnF$_3$. single mode operation is easily
achieved even without tuning element. The birefringent filter does
not influence the spectral width of the laser emission, however, it
is necessary as course tuning element. In Fig. 2 the frequency
spectrum of the laser is shown.

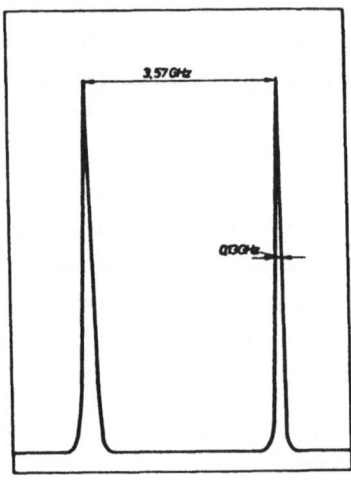

Fig. 2 . Frequency spectrum
of the laser output. The free
sprectral range of the spectrum analyzer is 3.57 GHz

The emission occurs as expected in a single mode. The width of 130
MHz is caused by the low finesse of the spectrum analyzer. Although
the laser linewidth has not yet been measured it can be expected that
the short term stability is better than 1 MHz.

The tuning curve in single mode operation is given in Fig. 3.

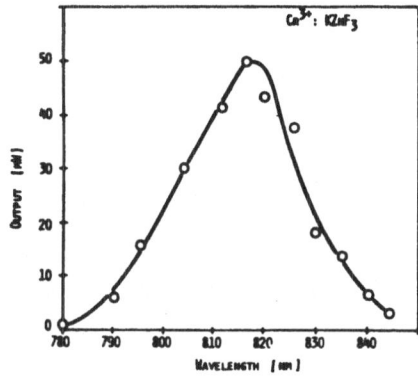

Fig. 3. Output power as a
function of emission wavelength for Cr^{3+}:KZnF$_3$
in single mode operation.

With Cr^{3+}-doped KZnF$_3$ the laser was tuned from 780 nm to 850 nm, at
3.5 W pump power. A power output of 50 mW has been achieved at the
peak of the gain profile. Fig. 4 shows the output characteristics of
this laser compared to Cr^{3+} GSGG in a three mirror cavity.

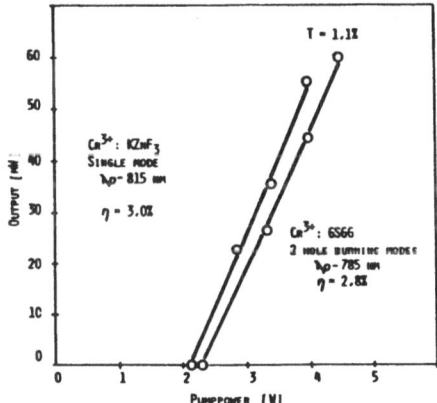

Fig. 4. Output characteristics
of KZnF$_3$ and GSGG

In both experiments 4 W pump power result in~60 mW output power.
Slope efficiencies of 3% with laser threshold values of ~2 W have
been measured with the lasers tuned to the peak wavelength of 815 nm
and 785 nm respectively. Total cavity losses are determined to be 2
to 4%.

Thermal induced birefringence

In contrast to the perovskite, GSGG and GSAG crystals exhibit a photoelastic anisotropy. Therefore the polarisation state of the laser mode is affected by induced birefringence in the active material. The degree of the influence on the laser mode strongly depends on pump and intracavity power and also on the crystal orientation. In order to obtain stable unidirectional operation with these crystals a higher loss ratio between forward and backward direction than for $KZnF_3$ is necessary. However, with GSGG and GSAG unidirectional operation is also possible although the power is reduced compared with the three mirror cavity.

Conclusions

For the first time tunable cw single mode laser oscillation has been demonstrated with Cr^{3+}-doped crystals. The wavelength ranges from 745 to 825 nm, from 760 to 830 nm and from 780 to 850 nm are covered with GSAG, GSGG and $KZnF_3$ respectively.

Output power levels on the order of several 100 mW are feasible with crystals of better optical quality. With our crystals power levels of 60 mW are obtained. These lasers show all the advantages of a stable solid state active material operating at room temperature.

Acknowledgements: The authors thankfully acknowledge helpful discussions and the growth of laser crystals from Prof., Huber, J. Drube and B. Struve, University of Hamburg and Prof. Dürr and U. Brauch, University of Stuttgart.

References

(1) S.C. Walling, H.P. Jenssen, R.C. Morris, E. W. O'Dell, O.G. Pentenon;
   Opt. Lett.4,182 (1979)

(2) B. Struve, G. Huber, W. V. Laptev, I.A. Shcherbakov, E. U. Zharikov;
   Appl. Physi. B 30, 117 (1985)

(3) J. Drube, B. Struve, G. Huber;
   Opt. Commun. 50,45 (1984)

(4) U. Brauch, U. Dürr;
   Opt. Commun. 49,61 (1984)

RT Stable F_2^- – Color Center Laser Tunable from 1.1 – 1.2 μm

D. Wandt, W. Gellermann and H. Welling

Institut für Quantenoptik, Universität Hannover

Welfengarten 1, D-3000 Hannover 1, Fed. Rep. Germany

F_2^--centers in alkali halides consist of two (110) neighboring anion vacancies binding three electrons. They are examples· for defects with weak electron – phonon coupling showing pronounced zero – and one – phonon line transitions in the absorption and emission bands up to relatively high temperatures. In the host LiF small concentrations of these centers have first been identified in heavily x-irradiated crystals (1).

Recently several Russian groups have reported pulsed and "quasi cw" laser activity of the F_2^--centers in LiF at room temperature, tunable in the 1.1 to 1.2 μm range (2-4). The used crystals were grown in air and the centers produced by ionizing radiation. However, the center concentrations were rather low, making it necessary to use laser crystals with long dimensions. This excluded the use of tightly focused resonators and resulted in laser operation with very high threshold pump powers.

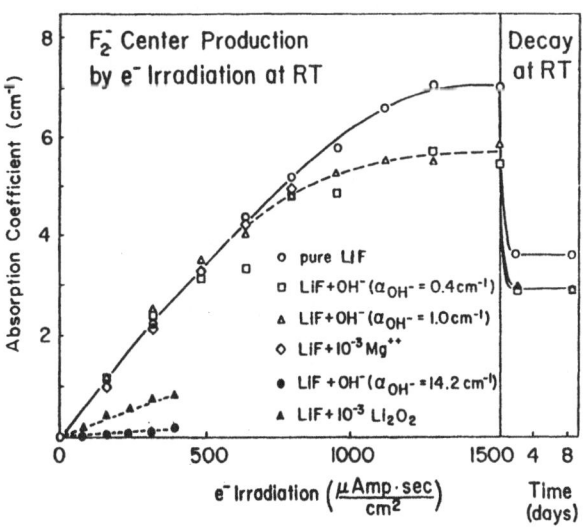

Fig. 1.

Stimulated by these laser properties of the F_2^--defects we started an investigation into the formation conditions for increased center densities which would result in a higher efficiency for the cw laser.

Comparing pure and air grown crystals (the latter containing OH⁻ impurities), we found that the concentration of F_2^--centers, formed under identical irradiation conditions, decreases drastically with

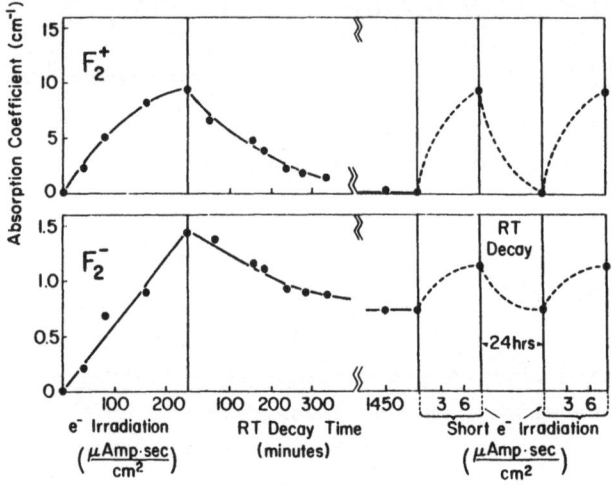

Fig. 2.

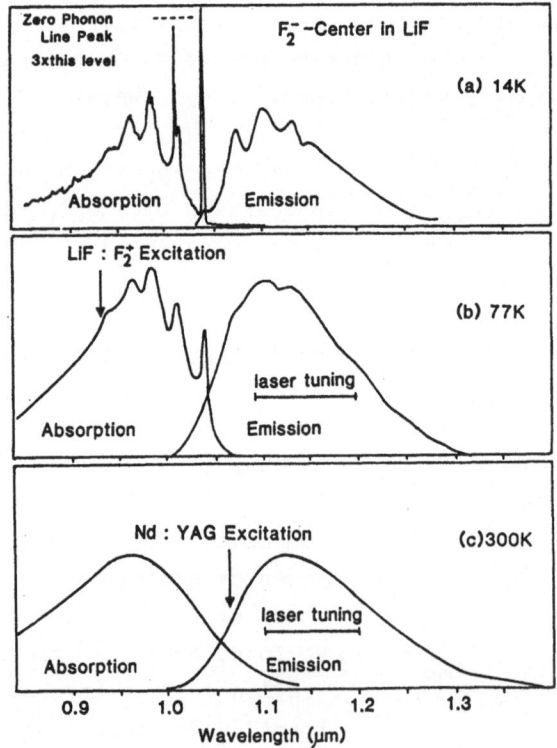

Fig. 3.

increasing OH^--impurity concentration. The same was observed, to a less drastic extent, also for O_2^-- and Mg^{++}-doped samples. In Fig. 1 the formation of F_2^--centers is plotted as a function of e^--irradiation dosis. The F_2^--center concentration builds up slowly with increasing dosis of irradiation and saturates at a fairly high irradiation level. It is clearly seen that the maximum F_2^--concentration is obtained in the pure sample (absorption coefficient 7 cm^{-1}). Storage of the irradiated crystals at room temperature leads in all samples to a decay of about half the initial F_2^--center concentration within twelve hours. In the same time interval the F_2^+-centers disappear totally, which are formed initially under the e^--irradiation. The remaining F_2^--centers are stable at RT for long periods (Fig. 2). We interpret this behaviour as follows: The F_2^--centers (present in about double the F_2^+-center concentration) are emitting electrons at RT and these electrons destroy the F_2^+-centers which leads to the above mentioned reduc-

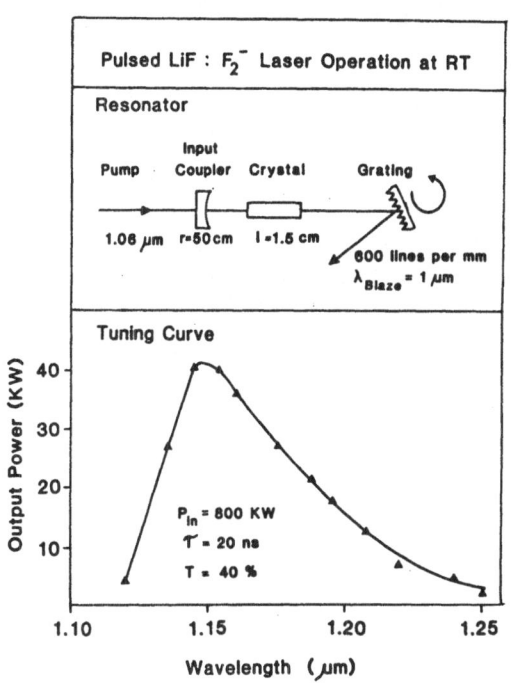

Pulsed LiF : F$_2^-$ Laser Operation at RT

Resonator

Input
Pump   Coupler   Crystal   Grating

1.06 μm   r=50cm   l =1.5 cm

600 lines per mm
λ$_{Blaze}$ = 1 μm

Tuning Curve

Output Power (KW)

P$_{in}$ = 800 KW
τ = 20 ns
T = 40 %

Wavelength (μm)

Fig. 4.

tion of the F$_2^-$-centers. The remaining half of the initial F$_2^-$-concentration then is stable at RT over long periods due to a complete lack of electron traps.

We started our first laser experiments at LNT by using pure crystals with optimized F$_2^-$-center concentrations. The optical cavity of the color center laser consisted of a conventional asticmatically compensated four mirror configuration. The laser was longitudinally excited by a LiF:F$_2^+$-center laser with a maximum output power of 1 W at 930 nm which is close to the absorption maximum of the F$_2^-$-center, as indicated in Fig. 3. The 4·4·2 mm-sized crystal absorbed about 70 % of the pump radiation. We obtained laser oscillation at a threshold pump power of 200 mW. The F$_2^-$-center laser was tunable from 1070 - 1200 nm at a pump power of 1 W. However, cw operation with F$_2^-$-centers is not possible. The output decreased within 30 - 60 ms. Constant laser operation is only possible if the pump beam is interrupted by a chopper, having a very small duty cycle (1:40). As the decrease of the output appears even at a very low pumping level, thermal effects can be excluded. The appearence of a decreasing power level can be repeated, which shows that the F$_2^-$-centers remained undestroyed. The explanation for this phenomena is the existence of metastable states which will be populated, preventing cw laser operation.

Although a realization of a cw F$_2^-$-center laser is not possible, this system is well suited for pulsed laser operation at RT. Under Q-switched excitation with a Nd^{3+}-YAG laser we obtained stable laser operation at RT. In the upper part of Fig. 4 the resonator is shown, consisting of a spherical input coupler and a grating in Littrow mount. The length of the crystal was 1.5 cm. It absorbed about 85 % of the pump radiation. The tuning curve of the F_2^--center laser is shown at the bottom of Fig. 4. The laser is tunable from 1.12 to 1.25 μm with about 40 kW output power in the peak at 1.15 μm. Taking this into consideration at a pump power of 800 kW the efficiency amounts to 5 %.

References
(1) J. NAHUM, Phys. Rev. 158, 814 (1967)
(2) Y.L. GUSEV, S.I. MARENIKOW, V.B. CHEBOTAEV, Appl. Phys. 14, 121 (1977
(3) Y. GUSEV and S.N. KONOPLIN, Sov. J. Quantum Electron. 14, 808 (1981)
(4) T.T. BASIEV, Sov. J. Quantum Electron. 12, 1125 (1982)

Laser Action in Stoichiometric Nd^{3+}, Pr^{3+} and Er^{3+} Compounds

W. Woliński, M. Malinowski and R. Wolski

Institute of Electron Technology

Warsaw Technical University, ul. Koszykowa 75,

00–662 Warsaw, Poland

Neodymium phosphate crystals, belonging to the group of so called stoichiometric compounds, have been extensively studied because of their excellent laser properties[1]. In these materials the neodymium enters as a constituent resulting in a very high concentration of active ions, of the order of $5 \times 10^{21} cm^{-3}$, and a very high optical gain allowing thus, effective miniaturization. In the neodymium phosphates laser emission has been obtained in laboratory experiments in the range of the lowest attenuation of the presently produced optical fibers[2].

However much less is known about the luminescence of other rare earth (RE) ions in the stoichiometric compounds. Because of our earlier studies of the $KNdP_4O_{12}$ system[3, 4, 5] which shows its value as a successful laser host for generation of the $\lambda = 1,05$ i $1,32 \mu m$, we have synthesized for the first time the $KPrP_4O_{12}$ crystals as well as crystals diluted with Y^{3+} ions [6]. Our interest in rare earth potassium tetraphosphates arises also from the fact that they have an accentric space group which may allow direct electrooptic modulation of the generated radiation. The crystals were obtained in a low temperature process from the orthophosphoric acid solution, their typical dimensions were 0,5 x 1 x 3 mm. X—ray crystallographic studies revealed that $KPrP_4O_{12}$ and $KErP_4O_{12}$ are isostructural to $KNdP_4O_{12}$ with slightly different cell parameters. The active ions concentration was determined to be 4.08, 4.05 and 4.54×10^{21} cm^{-3} in Nd, Pr and Er–potassium tetraphosphates respectively.

Absorption and fluorescence spectra were investigated at 10, 77 and 300 K in the UV, visible and near IR regions. This allowed us to determine the energy level diagram for Nd^{3+}, Pr^{3+} and Er^{3+} ions in examined crystals and to establish the most favorable excitation conditions. The crystals exhibit and intense absorption line structure Fig. 1. which makes the optical pumping by LEDs or laser diodes possible.

Energy levels of the three investigated compounds allows room temperature fluorescence at various wavelengths. The most intense transitions are marked on the simplified energy levels schemes in Fig. 2 Praseodymium and erbium tetraphosphates possess two metastable levels each. Howeyer, the excited $^3P_0$ state of Pr^{3+} in $KPrP_4O_{12}$ has a very short fluorescence litetime $\tau_f = 80$ ns allowing, despite relatively high effective emission cross section, only the pulsed laser action with the threshold energy which was estimated to be about 100 μJ.

We considered a 200 μm thick $KPrP_4O_{12}$ crystal placed in a semiconcentric resonator and excited by a 0,48 μm dye laser radiation.

In the diluted praseodymium crystals, for example $KPr_{0.03}Y_{0.97}P_4O_{12}$, most of the fluorescence is channeled into transitions from the $^1D_2$ excited level having much longer fluorescence lifetime $\tau_f = 205\ \mu s$. In this case continuous laser action can be expected at $\lambda = 1.02\ \mu m$ with the pumping power threshold of 100 mW (in the same excitation conditions as for $KPrP_4O_{12}$).

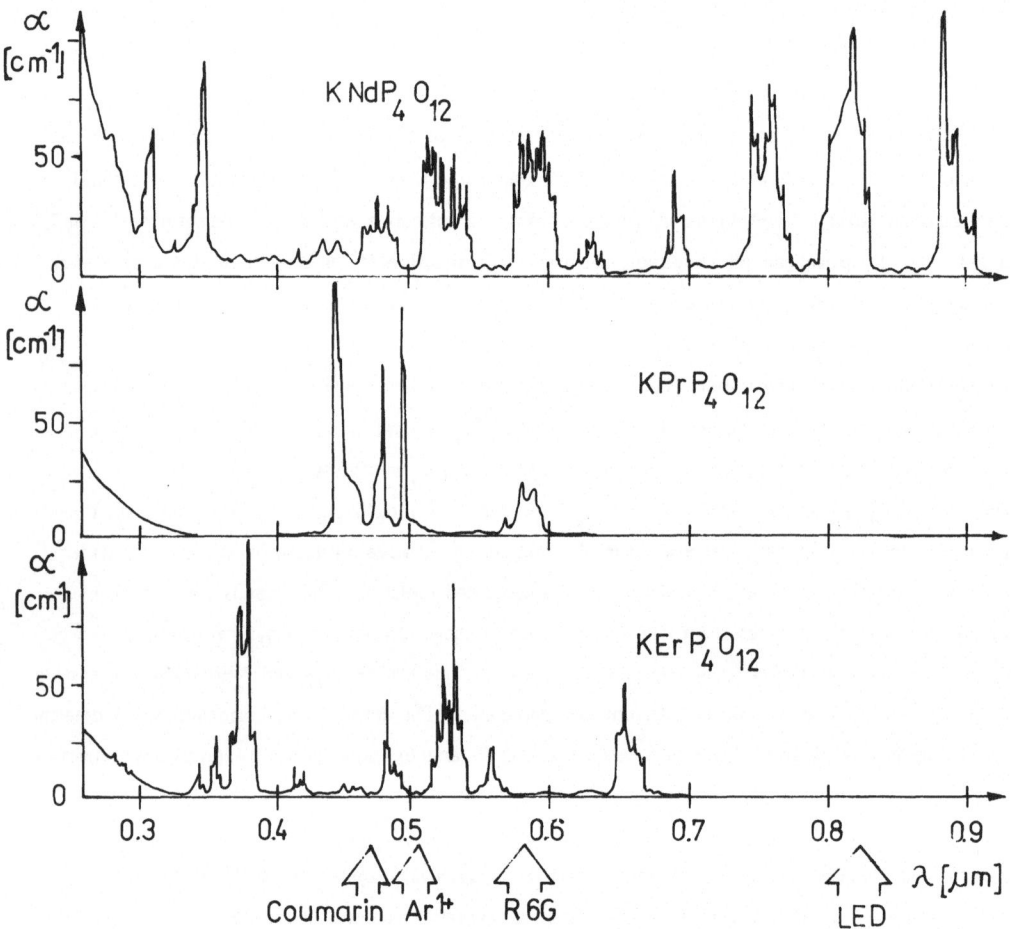

Fig. 1. Room temperature absorption spectra of $KNdP_4O_{12}$, $KPrP_4O_{12}$ and $KErP_4O_{12}$.

In the $KErP_4O_{12}$ the most interesting, from the point of view of the optoelectronic applications, is the $^4I_{13/2} \rightarrow {}^4I_{15/2}$ transition at $\lambda = 1.55\ \mu m$. C.w. lasing can be expected at slightly, lowered temperatures because of the three level laser configuration with relatively small splittiing of the $^4I_{15/2}$ ground state, $\triangle E = 277\ cm^{-1}$.

The most important spectroscopic parameters of the investigated materials are summarised in Table 1.

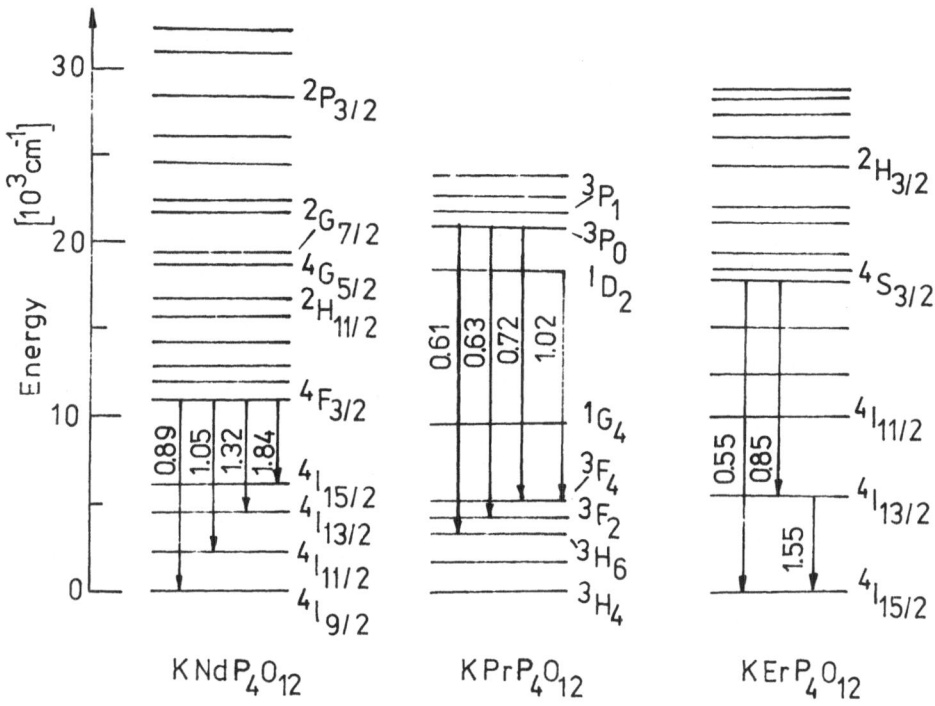

Fig. 2. Energy level schemes of the investigated crystals.

Table 1. Effective emission cross sections σ', fluorescence lifetimes τ and wavelengths λ of the most intense transitions in $KNdP_4O_{12}$, $KPrP_4O_{12}$ and $KErP_4O_{12}$ crystals

Material	Transition	λ [μm]	σ [$10^{-20}cm^2$]	τ [μs]
$KNdP_4O_{12}$	$^4F_{3/2} \to {}^4I_{11/2}$	1,05	9.0	100
	$^4F_{3/2} \to {}^4I_{13/2}$	1.32	2.7	100
	$^3P_0 \to {}^3F_4$	0.72	2.8	0.08
$KPrP_4O_{12}$	$^3P_0 \to {}^3F_2$	0.63	3.7	0.08
	$^3P_0 \to {}^3H_6$	0.61	3.7	0.08
$KPr_{0.02}Y_{0.97}P_4O_{12}$	$^1D_2 \to {}^3F_4$	1.02	6.5	205
	$^4S_{3/2} \to {}^4I_{15\,2}$	0.55	2.2	1.1
$KErP_4O_{12}$	$^4S_{3/2} \to {}^4I_{13\,2}$	0.85	0.1	1.1
	$^4I_{13/2} \to {}^4I_{15/2}$	1.55	~ 6.5	3400

After studying the crystall growth process and establishing the basic spectroscopic properties of the Pr^{3+} and Er^{3+} potassium tetraphosphates we work on determining the most favorable laser action conditions in this crystals utilising various excitation sources.

References

(1) G. Huber, Miniature Neodymium Lasers, in: Current Topics in Materials Sciences, Ed. E. Kaldis (North Holland, Amsterdam, 1980) p.1

(2) W.W. Krühler, R.D. Plättner and W. Stetter, Appl. Phys. 20, 329 (1979)

(3) M.Malinowski and W.Woliński, Acta Phys. Polon.A 65, 303 (1984)

(4) M. Malinowski and W. Woliński, J. Lumin. 29, 275 (1984)

(5) M. Malinowski, Ph. D. thesis, unpublished.

(6) M. Malinowski, R. Wolski and W. Woliński, submitted to J. Lumin.

Tunable Wavelength Selective Semiconductor Laser Systems

K.J. Ebeling*, W. Haessler, H. Fouckhardt, and M. Port
Drittes Physikalisches Institut, Universität Göttingen
Bürgerstr. 42-44, D - 3400 Göttingen
*present address: Institut für Hochfrequenztechnik, TU Braunschweig

1. Introduction

Recently several types of semiconductor lasers have been developed whose emission remains single-longitudinal mode even under gigabit-rate current modulation. Laser diodes with distributed Bragg reflector (DBR) [1] oscillate with extreme spectral purity at the wavelength determined by the Bragg condition. Distributed feedback (DFB) lasers [2] show similar behavior when a $\lambda/4$-phase shift is introduced for suppressing one of the two dominating modes of the grating resonator. As a result of the interaction of two Fabry-Perot type resonators cleaved or grooved coupled-cavity lasers [3,4] produce monomode oscillation and the emission wavelength is electronically tunable by controlling the relative pumping currents. Extremely narrow linewidths as low as 15 kHz can be obtained in external cavity systems [5] well suited for coherent optical communication. In the following, we present some properties of very simple hybrid coupled-cavity and external cavity semiconductor lasers.

2. Hybrid coupled-cavity laser

The coupled-cavity system schematically sketched in the inset of Fig. 1 consists of two commercial index-guided GaAs/GaAlAs laser diodes of slightly different lengths that are optically coupled via a narrow air gap. The housings of the diodes are removed and for alignment mechanical differential micropositioning devices are used. Feeding the light from one diode into the active channel of the second is faciliated by monitoring the induced photocurrent. As heat sinks protrude one diode is mounted upside down. For sufficient optical coupling the endfaces of the diodes should be aligned fairly parallel. Transverse alignment tolerances are of the order of one micrometer. The spectra shown are recorded for fixed pumping current $I_1 = 49.5$ mA of oscillator cavity 1 and various settings of the active etalon current I_2. The spectra are fairly clean monomode. The emission line is electronically tuned by changing current I_2 and thus the filter characteristic of the etalon cavity. The output remains single-mode under high-speed current modulation.

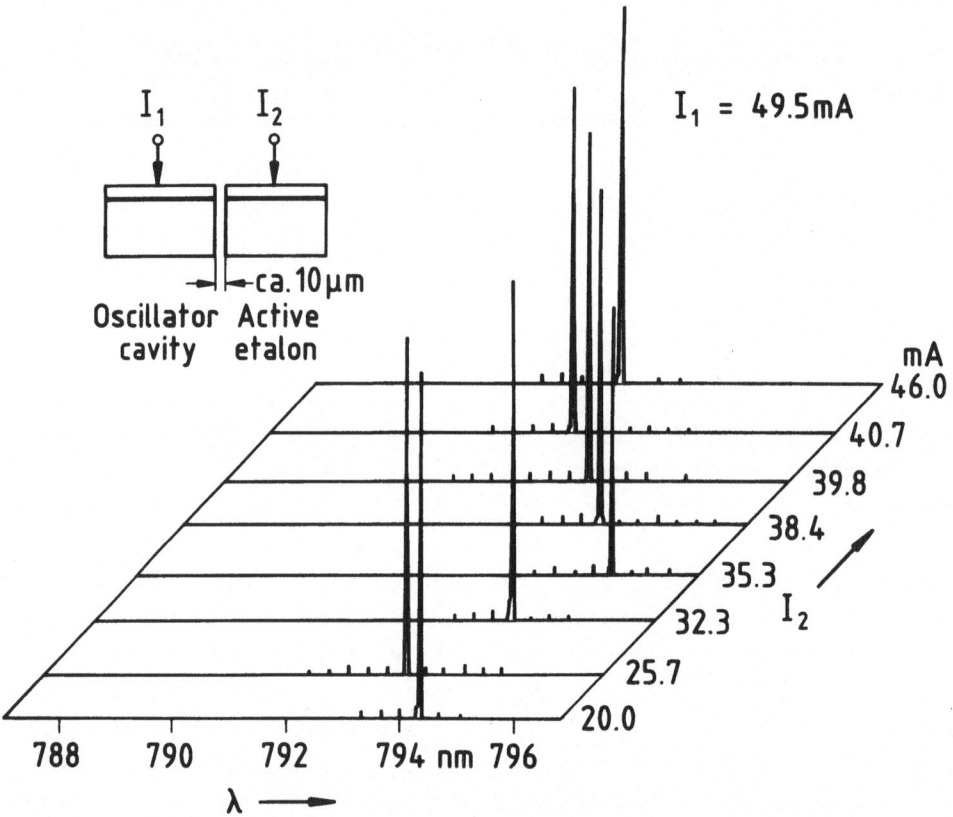

Fig. 1. Emission spectra of a coupled-cavity laser diode system.

Fig. 2 shows typical side mode suppression ratios S_R and emission wavelengths λ for fixed I_1 = 47.5 mA as a function of I_2. Mode suppression can be improved when the coupling strength is further increased by reducing the gap width. Optima are found for widths that are an integer multiple of half the wavelength. For strong coupling bistable operation in different modes or hysteresis of mode hop positions are readily observed depending on whether the etalon current is increased or decreased

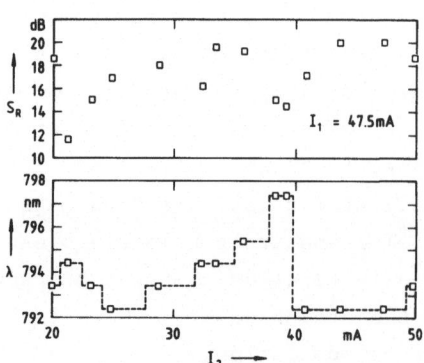

Fig. 2. Emission wavelength λ (bottom) and secondary mode suppression ratios S_R (top) as a function of the active etalon current I_2 when the oscillator cavity current I_1 is held constant at 47.5 mA above laser threshold.

3. External cavity laser

Fig. 3 shows a very simple arrangement of a laser diode with dispersive external resonator. The emitted light collimated by a microscope
objective passes a high-resolution holographic grating (1500 lines/mm,
ca. 70 % diffraction efficiency) and the first order diffracted light is
fed back into the laser with a mirror. Tilting the end reflector various
modes of the solitary laser diode can be excited as demonstrated by the
recorded spectra. The total tuning range is larger than 20 nm.

The performance critically depends on the mechanical stability of the
system. The transmission grating acts as a double-monochromator. The
resolution is large enough that individual modes of the external cavity
of 7 cm length are selected by fine tuning the angle of the end reflector. Fig. 4 shows a single (external) mode output spectrum of the system.

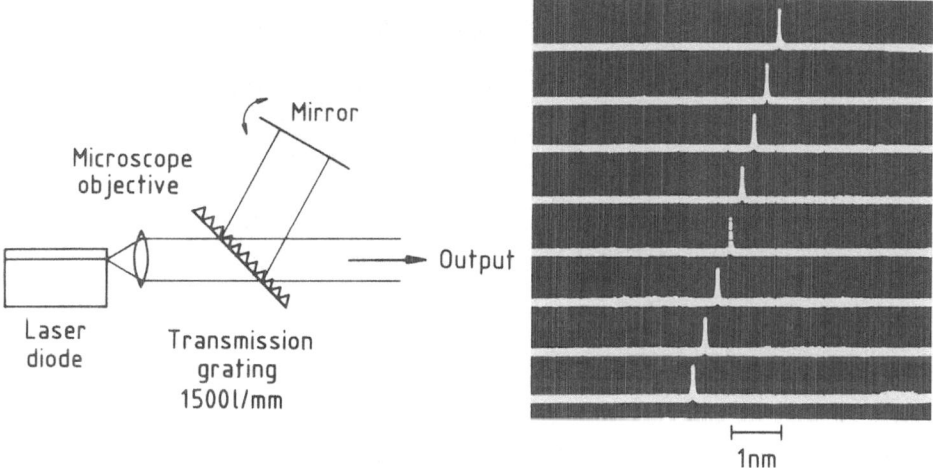

Fig. 3. External cavity system with transmission grating. The emission
wavelength is tuned by tilting the resonator end reflector.

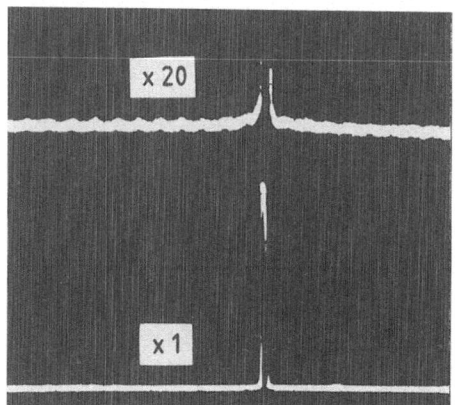

Fig. 4. Emission line of an external
cavity system. Side mode suppression
is larger than 27 dB. The frequency
halfwidth of the emission line is
less than 500 kHz as determined by
interferometric methods.

From the 20-fold enlarged display a side mode suppression ratio of 27 dB
is obtained in the particular case shown but more than 33 dB has already
been observed for optimum alignment. From interferometric and self-
heterodyne measurements the linewidth is determined to be less than 500
kHz as an upper limit. This is a considerable reduction compared to the
linewidth of typically 50 MHz of the solitary diode.

   Fig. 5 shows a slight modification of the system to produce two-wavelength emission. The oscillating modes are independently controlled by
the two end reflectors. Mode hopping between the two modes is not found
at frequencies up to 150 MHz. However, so far we cannot rule out that
hopping occurs at much higher rates corresponding to relaxation oscillation phenomena.

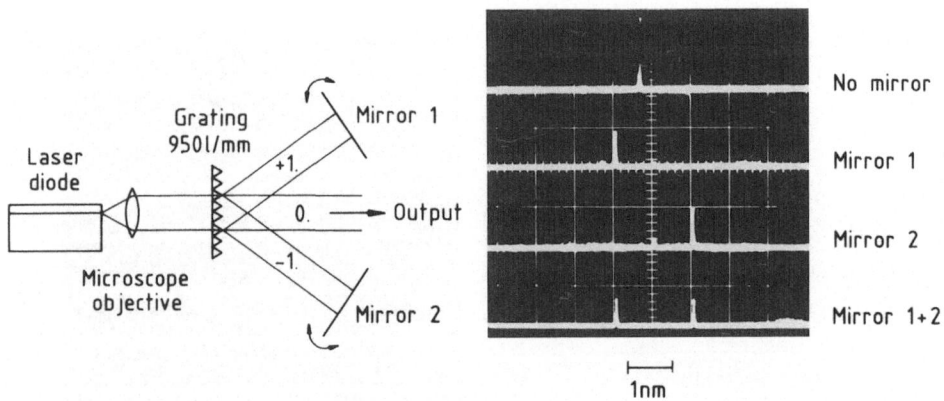

Fig. 5. Branched external cavity system for two-wavelength emission.

4. Conclusion

   The systems presented are well suited for fiber optic communication.
Wavelength division multiplexing and coherent optical communication are
techniques that might be realized. Tuning capabilities make the system
an interesting light source for spectroscopic applications. The excellent
space and time coherence also allows for interferometric, holographic,
or speckle measurement techniques replacing the HeNe laser in various
areas of coherent optical metrology.

Literature
[1] M. YAMAGUCHI et al.: Electron. Lett. 21 63-65 (1985)
[2] B. BROBERG et al.: Techn. Digest (post-deadline paper) of the Tenth
    European Conference on Optical Communication, Stuttgart 1984
[3] L.A. COLDREN et al.: Appl. Phys. Lett. 44, 169-171 (1984)
[4] L.A. COLDREN et al.: Appl. Phys. Lett. 44, 368-370 (1984)
[5] A. MOORADIAN: Physics Today, pp. 42-48, May 1985

A Compact Submillimeter Laser for Airborne Applications

R. Densing, P.B. van der Wal, H.P. Röser, R. Wattenbach
Max-Planck-Institut für Radioastronomie
Auf dem Hügel 69
D-5300 Bonn 1

Introduction

Research on coherent radiation sources in the submillimeter (submm)
wavelength region (100 µm - 1 mm, 3 THz - 300 GHz, 100 cm^{-1} - 10 cm^{-1}),
operating in a continuous wave (cw) mode led to the development of
klystrons, carcinotrons, glow discharge and optically pumped lasers.
Whereas klystrons and carcinotrons are tunable over a relatively wide
wavelength range, but only useful for wavelengths larger than 300 µm,
lasers however can only be operated on discrete transitions but on more
than 1000 laser lines in the whole submm region [1]. The advantage of
optically pumped submm lasers, compared with the other sources is high
output power, good signal to noise ratio and high spectral purity. This
paper reports the development of a compact submm laser system, consisting of a CO_2 pump laser and a submm laser, for astronomical airborne
observations [2]. The laser system has passed its first test flights
with a NASA science aircraft (Kuiper Airborne Observatory, type C-141A,
Lockheed) in April 1985. For this application, high mechanical stability and stable operation of the lasers, independent of orientation is
required over a flight duration of up to 8 hours. During a research
flight, the laser system will be mounted at a 91 cm aperture infrared
telescope.

1. The CO_2 pump laser

In the interest of structural compactness, the CO_2 pump laser is folded.
Two pyrex tubes of 10 mm bore diameter (Fig. 1) are sealed with ZnSe
brewster windows, which are attached to macor adapters with a flexible
silicon adhesive. The gas discharge is cooled by means of a water jacket and a 300 Watt cooling unit. Frequency tunability over more than 80
discrete laser lines in the 9 and 10 µm branch is provided by a grating
with 150 lines/mm. Varying the resonator length with a piezoelectric
transducer, each CO_2 line is tunable by about 60 MHz over its gain curve
[3]. Therefore a frequency offset up to 60 MHz between the pump- and
absorption frequency of the active molecules of the submm laser can be
accepted [4]. A partially transmitting, concave ZnSe mirror couples the

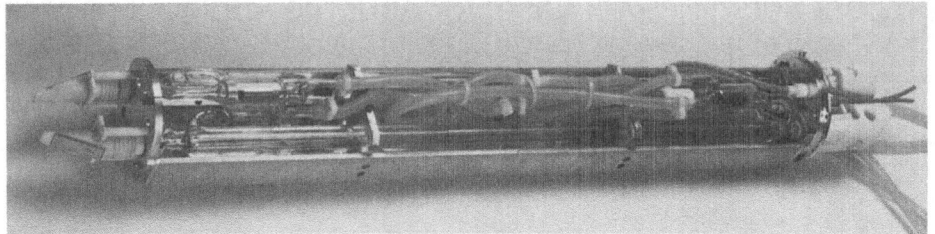

Fig. 1. Inner part of the folded CO_2 laser, showing the mounting
of the pyrex tubes.

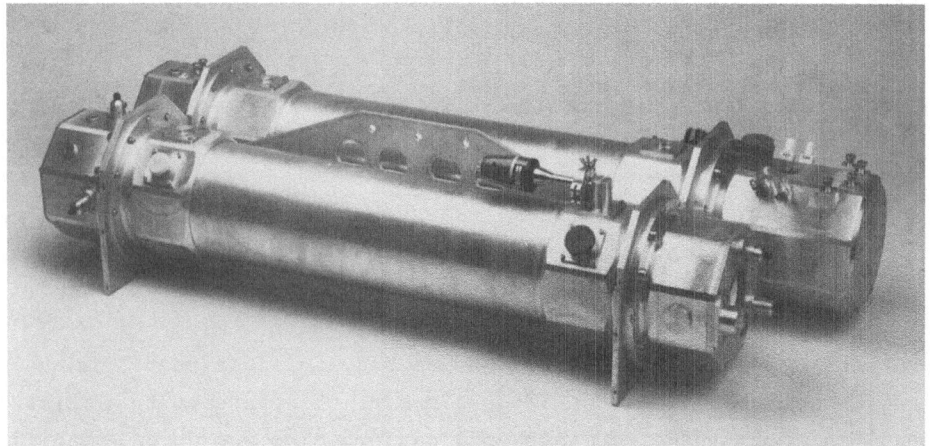

Fig. 2. Complete laser-head, enclosed in two aluminium tubes.

Fig. 3. Adjustable holder
for CO_2 laser op-
tics (Photos by
G. Hutschenreiter)

laser beam out. Output power of up to 40 W (P(20) 10 µm), with a He:
$N_2:CO_2$ gasmixture of 80:11:9, single mode operation and narrow bandwidth make this laser an effective pump source.

2. The submillimeter laser

After a beam compression of 10:1 by means of a dual lens system, the
CO_2 laser beam passes into the submm resonator, using a simple hole
coupled mirror. In order to get sufficient output power and to couple
out uniformly the submm resonator mode in a diffraction limited output
beam, a hybrid metallic-mesh output coupler [5] is used. This kind of
output coupler, which is highly reflective for the near infrared pump
laser signal improves the efficiency by reflecting the CO_2 laser beam
many times through the submm resonator, until it is fully absorbed by
the laser gas. At the same time, the meshcoupler has to be partially
transmitting (10-20 %) for the submm wavelengths. The plano-concave
submm resonator operates with metallic or dielectric waveguides of different diameters, depending on the required output power, mode behaviour and polarization [6]. Whereas a metallic waveguide provides high
output power, multi-mode operation in various polarizations, a dielectric waveguide delivers lower output power in a stable single mode operation with a defined polarization. A power level of more than 10 mW
cw has been achieved for a large number of laser lines. After a warm
up time of 30 minutes, the amplitude stability of the submm laser is
better than ± 30 % and the frequency stability is about ± 100 KHz in
an one hour time interval. There are more than 1000 discrete laser
lines known in the submm wavelength range, on which the laser can be
operated by choosing the laser gas and the pump frequency as required.

3. Mechanical aspects

The airborne environment is extremely noisy in mechanical terms: the
equipment is continuously exposed to high shock and vibration levels.
Therefore stringent control over the mechanical stability is required
of such a system. Moreover the use of any equipment attached to servo
positioned systems such as the 91 cm airborne telescope on board the
NASA Kuiper Airborne Observatory sets serious limits to its possible
weight and inertia.

Thin-walled cylinders show an optimal ratio of mechanical strength
(torsion and bending) to weight in comparison to other mechanical
structures. Making use of this principle the submillimeter laser and
the pump laser were designed to fit inside cylindrical structures of
aluminium (Fig. 2). When appropriately clamped together, the two cylinders create an indeal 'optical bench' with a mechanical strength
comparable to or even surpassing that of traditional laboratory ben-

ches. At the same time these cylinders serve as dust protection, thermal cover or as a vacuum compartment, as in the case of the submm laser. Special attention has been given to the design of adjustable holders for optical components so as to make them essentially insensitive to mechanical shocks and vibrations (grating, mirrors, lenses, etc.) (Fig. 3). Adjustable components are well balanced around their pivots to reduce their sensitivity to shock and vibrations. Moreover the thermal compensation technique is also applied to stabilize the resonator length of the laser cavities through a wide range of environmental temperatures by adequately combining invar spacers and aluminium parts. The compact laser system measures 1200 x 510 x 240 mm, with a weight of about 70 kg.

Conclusion

A submm laser system, consisting of a folded, dc excited CO_2 gas discharge laser and a low loss submm resonator has been developed. Compact size, low weight and excellent mechanical stability, which allows stable operation independently of direction in space, make this laser-system well-suited for airborne applications.

References

[1] D.J.E. KNIGHT: Nat. Phys. Lab., Teddington, Middlesex, U.K., NPL Rep. Qu. 45, 1st rev., Feb. 1981
[2] H.P. RÖSER, R. WATTENBACH, P. VAN DER WAL: NASA/A.S.P. Symposium on Airborne Astronomy, NASA conference publication No. 2353
[3] M.S. TOBIN: Proceedings of the IEEE, Vol. 73, No. 1, pp 61-85, 1985
[4] H.P. RÖSER, R. WATTENBACH: Laser und Optoelektronik, Vol. 16, No. 3 pp 165-174, 1984 (in German)
[5] E.J. DANIELEWICZ, P.D. COLEMAN: Appl. Opt., Vol. 15, No. 3, pp 761-767, 1975
[6] H.P. RÖSER, M. YAMANAKA, R. WATTENBACH, G.V. SCHULTZ: Int. J. IR and MM-Waves, Vol. 3, No. 6, pp 839-868, 1982

A New Compact 1.5235 µm HeNe Laser and its Characteristics

William W. Lee and Jeff W. Eerkens
Melles Griot Laser Product Division

A new compact HeNe laser having a 1.5235 µm single longitudinal mode output has been developed by Melles Griot as a significant source for coherent fiber optical communication (1,2). The laser is similar in structure to the standard .6328 µm HeNe laser produced by Melles Griot.

As shown in Figure 1, cylindrical symmetry is maintained in the design to provide mechanical and thermal stability. Other features in the design are intended to achieve performance optimization as well as volume production efficiency.

The laser is developed into various models in terms of polarization and packaging. The performance specification for each of these IReNe[tm] 1.5 µm lasers is summarized in Table 1.

The 1.5235 µm laser output is generated from the Ne $2S_2 \rightarrow 2P_1$ transition (3) as shown in the energy level diagram of Figure 2. Due to the fact that it is a gas laser transition, the linewidth of the 1.5235 µm HeNe laser is inherently much narrower than the linewidth of a laser diode emission. The instantaneous width of a laser line is given by Lax's relation (4), which in our notation reads:

Equation 1:
$$\delta \nu_L = \frac{\varepsilon_L}{P_L} \left(\frac{\pi cS}{2L}\right)^2 \nu \cdot \chi \cdot (1 + \alpha^2),$$

in which:
$$\nu = \Delta\nu_o / \left[\Delta\nu_o + cS/(4L)\right]$$

$$\chi = N_2/(N_2 - N_1) + \left[\exp\{h\nu_L/(kT)\} - 1\right]^{-1},$$

$$\alpha = (\nu_L - \nu_o)/\Delta\nu_o + cS/(4L),$$

and where:
$$\varepsilon_L = h\nu_L = \text{Photon Energy (Joules)}$$

$$\nu_L = \text{Laser Frequency (Hz)}$$

ν_o = Center Frequency of Resonance (Hz)

$\Delta\nu_o$ = Width of Transition Resonance (Hz)

P_L = Laser Power (Watts)

N_1, N_2 = Populations of lower (1) and upper (2) levels

S = Transmission (Spillage) of output coupling mirror

L = Laser Length (cm)

c = Velocity of light = 3×10^{10} cm/sec

For typical HeNe lasers, one finds from Equation 1 that $\delta\nu \sim 0.6$ Hz.

The laser frequency ν_{ℓ}, determined by the resonance constraint $\nu = \frac{nc}{2L}$, is affected by temperature on the cavity length ℓ. The cavity length, divided into segments of length ℓ_i and thermal expansion coefficients, α_i (Figure 3), will have net temperature effect on the laser frequency $\frac{d\nu}{dT}$ as follows:

$$\frac{d\nu}{dT} = -\nu\,\bar{\alpha} \text{ where } \bar{\alpha} = \sum_{i=1}^{5} \ell_i\alpha_i/\ell$$

The average expansion coefficient for the 1.5235 µm HeNe design is calculated to be $5.5 \times 10^{-6}/c^o$ which means:

$$\frac{d\nu}{dT} = -1.1 \text{ GHz}/c^o$$

Due to this temperature effect, the laser output as detected by a Fabry-Perot interferometer will trace the gain profile above threshold as the temperature changes (Figure 4). For a given temperature, if the resonance frequency is located near the transition line center, the laser output will be a single longitudinal mode (dashed line in Figure 4). On the other hand, if the frequency is moved away from the center, a second mode appears under the gain curve at a frequency that differs from the first mode by the free spectral range, i.e. $\frac{c}{2\ell}$ = 420 MHz.

The reason for two modes, or multiple longitudinal mode output, is that the transition gain is Doppler broadened and the threshold gain width is greater than the free spectral range.

Single mode or two mode output can also be identified from the optical signal measured by a Ge APD (Fujitsu) detector (Figure 5). The measurement set-up uses a Brewster window sealed 1.5 μm laser (Melles Griot IReNe[tm] 05-LIB-150). The external high reflector is adjusted for single mode or two mode output. The visible filter (Melles Griot 03 FCG 118) blocks the fluorescent light from the discharge, and the laser beam is focused onto the detector. For two mode output, the signal spectrum shows a beating frequency at 420 MHz and its second harmonic at 840 MHz (Figure 6a). This beating frequency disappears when the laser output is a single mode (Figure 6b).

To reduce dark current in the detector, an InGaAs photodiode (Epitaxx ETX 500) is used in the set-up for laser noise measurement. The spectrum of the 1.5 μm laser signal (Figure 7) shows that the wideband noise spreads at -60 dBm from low frequency to about 50 KHz (Figure 7a), drops to -70 dBm from 50 KHz to 250 KHz (Figure 7b) and then gradually drops to background levels around 300 KHz (Figure 7c). The DC laser power level is measured at -21 dBm. It is known that the wide band noise in a HeNe laser is primarily due to the fluctuation in the plasma density and current (5,6). Ways to eliminate or reduce such noise will be investigated in a future paper.

Lifetime is a primary concern for choosing the source in an optical communication system. Since HeNe lasers were first invented in 1961, the lifetime expectancy has been improved to more than 15,000 hours in continuous operation. Throughout the years, the failure mechanisms of HeNe lasers have been identified and minimized.

1. Seals - A hard (solder glass) seal between the metal mirror mount and the mirror has been developed to replace the early soft (epoxy) seal. Leakage and contamination of epoxy-sealed lasers was the primary cause of short laser lifetime.

2. Helium leak - Hard boresilicate glass used for the laser envelope sustains a helium leak rate of less than 0.01 torr per year, which essentially eliminates helium loss as a failure mechanism.

3. Contamination - A very clean vacuum system used in the manufacturing process reduces gas or volatile contaminates in the laser to extremely low levels. Vacuums of 10^{-7} torr are routine. In addition, an activated

"getter" in the laser further absorbs or traps unwanted gas.

4. Mirror Coating - Hard dielectric coatings enable the laser mirrors
to withstand the discharge and vacuum environment inside the laser tube.

5. Cathode Sputtering - The standard cold, aluminum cathode for the
HeNe laser has been optimized in many ways to extend the occurance of
sputtering in time. Typical HeNe laser lifetime is illustrated in
Figure 8. Note that sputtering is the main ultimate lifetime failure
mechanism.

The new compact 1.5 μm HeNe laser has been designed on the basis of
the standard Melles Griot HeNe laser, which has demonstrated long
lifetime and highly reliable performance. Various models developed
will provide the fiber optics industry with versatile laser sources
that provide 1.5235 μm single longitudinal mode TEM_{OO} output.

Acknowledgement

The authors wish to thank Dr. R. S. Vodhanel of Bell Communication
Research, Inc., D. Young and Prof. C. C. Lee of the E.E. Dept.,
University of California at Irvine, for their expert technical assist-
ance.

References

1. J. W. Eerkens, W. W. Lee, "Performance of a compact 1.523 μm
HeNe laser for IR fiber-optic applications", Proceedings of SPIE,
"Fiber Optics: Short Haul and Long Haul Measurements and Applications II"
August 21-22, 1984.

2. J. W. Eerkens, W. W. Lee, "New HeNe lasers with Green (543 nm) and
Fiber Optimum Infrared (1523 nm) outputs", Conference on Lasers and
Electro-Optics (CLEO '85). Poster Session, May 21-24, 1985.

3. R. A. McFarlane, C.K.N. Patel, W. R. Bennett Jr. and W. L. Faust,
"New Helium-Neon Optical Laser Transitions", Proc. IRE 50, p.2111,
Oct. 1962.

4. M. Lax, "Quantum Noise V: Phase Noise in a Homogeneously Broadened
Maser", Physics of Q.E., Conference Proceedings, Puerto Rico, June,
1965, McGraw-Hill Book Co.

5. J. A. Bellisio, C. Freed and H. A. Haus: Applied Physics, Letters 4 (1964) 5.

6. Takeo Suzuki: Japanese Journal of Applied Physics 9 (1970) 309.

7. G. S. Palecki, "Helium-Neon Lasers - Reliability and Lifetime", Lasers and Applications, p.73, Sept. 1982.

See Attachment for Figures.

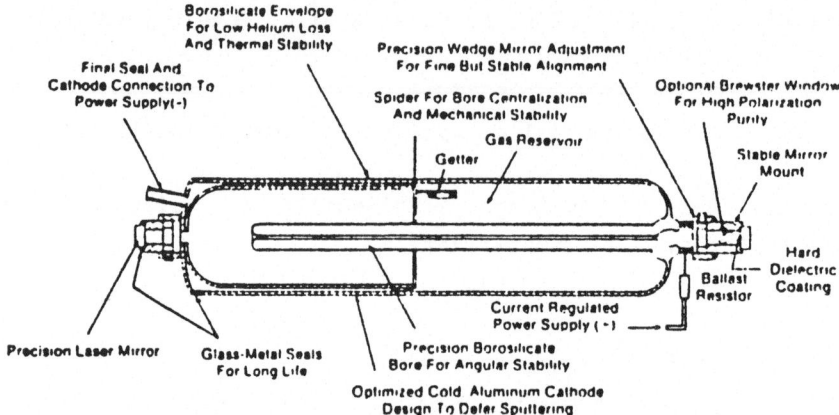

Figure 1. Structure design for 1.5235 μm HeNe laser (Melles Griot IReNe™ 05 LIR 150)

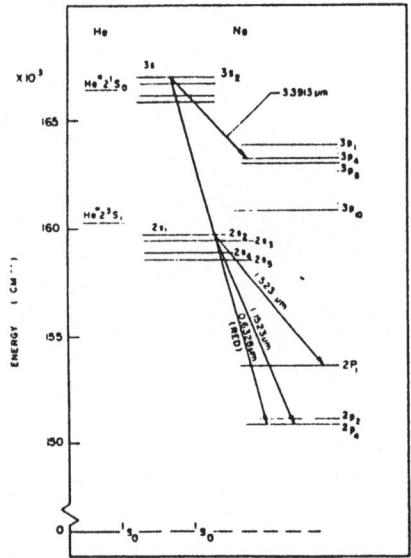

Figure 2. Energy levels of HeNe lasers for the 1.5235 μm Ne $2S_2 - 2P_1$ transition and other well known transitions.

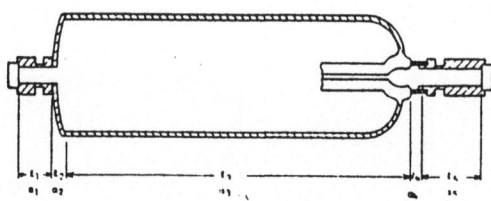

Figure 3. HeNe laser cavity divided into segments of length l_i's and thermal expansion coefficients α_i's.

**PRELIMINARY
SPECIFICATIONS**

IReNe™ 1.5 μm LASER

Specifications	Laser Tube		Cylindrical Laser Head	
	05 LIR 150	05 LIP 150 / 05 LIB 150	05 LIR 151	05 LIP 151
Minimum CW Power Output (mW) 1.5 μm (mW) TEM$_{00}$	>1.0	>0.8	>1.0	>0.8
Amplitude Stability %	±10	±10	±10	±10
Beam Diameter 1/e² (mm)	0.81	0.81	0.81	0.81
Beam Divergence (mrad)	2.10	2.10	2.10	2.10
Longitudinal Mode Spacing c/2L (MHz)	441	441	441	441
Polarization	Random	500:1	Random	500:1
Starting Voltage (Vdc)	>10K	>10K	>10K	>10K
Operating Voltage (Vdc) ±100	1900	1900	2350	2350
Operating Current (mA)	5.5	5.5	5.5	5.5
Recommended min Series Anode Ballast (K ohm)	75	75	Included	Included
Recommended Power Supply 115/230 VAC input	LPL 343	LPL 343	LPL 343	LPL 343

Table 1. Specifications for various 1.5 μm HeNe Laser Models

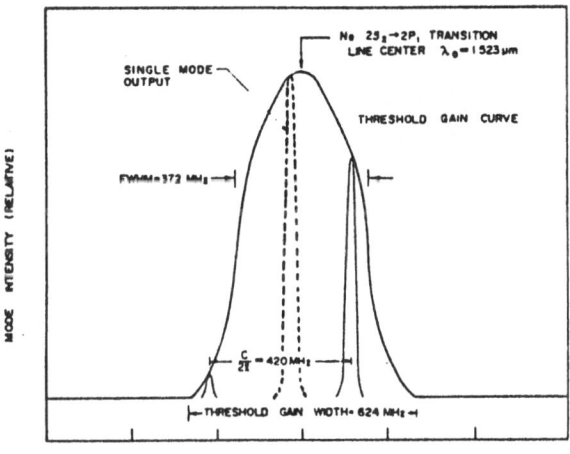

Figure 4. Threshold gain curve traced by 1.5 μm laser output detected with a Fabry-Perot interferometer; single longitudinal mode (dashed line) and two mode outputs under the gain curve.

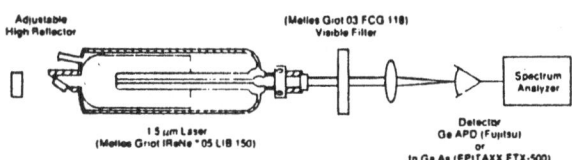

Figure 5. Mode Beating and Noise Spectrum Measurement Set-up.

108

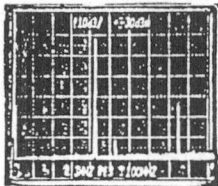 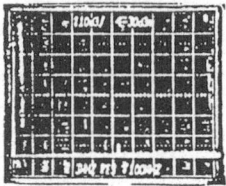

a) Two Mode Beating Frequency at 420 MHz and its second Harmonic, at 840 MHz   b) Single longitudinal mode 1.5235 μm output showing no beating frequency

Figure 6. Frequency Spectrum measured by a Ge APD detector (Fujitsu) 1.5235 μm laser
(Melles Griot IReNe™ 05 LIB 150) outputs = (a) Two mode and (b) single longitudinal mode.

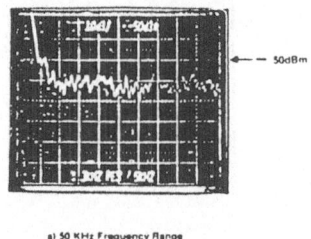

a) 50 KHz Frequency Range

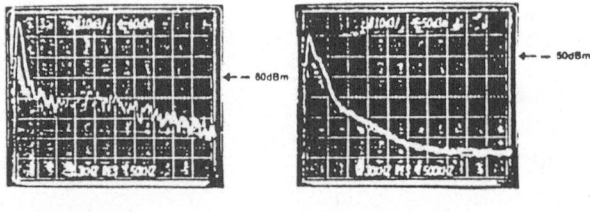

b) 500 KHz Frequency Range   c) 5 MHz Frequency Range

Figure 7. Noise spectrum of 1.5 μm laser measured by an InGaAs Detector (Epitaxx ETX-500) for
frequency ranges: (a) 50 KHz   (b) 500 KHz   (c) 5 MHz
• DC laser power: -21 dBm

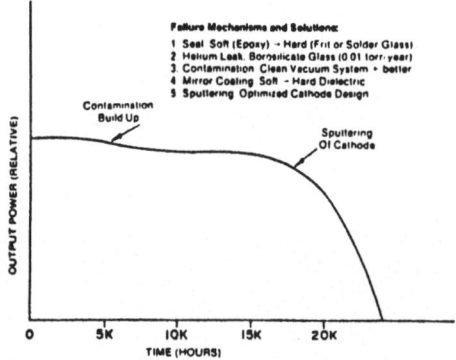

Figure 8. Typical HeNe Laser output power as function of time.

A New Picosecond Excimer Laser System

B. Burghardt, B. Nikolaus

Lambda Physik Forschungs- und Entwicklungs-GmbH

Hans-Böckler-Str. 12, D-3400 Göttingen

Short intense UV-pulses have proved to be usefully applicable in nonlinear optics, solid state physics, biophysical studies and photochemistry (1 - 3). Due to their unique abilities of large gain bandwidth and high output energy excimer lasers have been recognized very early as attractive candidates for generation and amplification of ultrashort light pulses with gigawatt peakpower.

Since straight forward mode locking of excimer lasers have not yet succeeded in generation of single background free picosecond pulses (4), we have decided for a more reliable approach via frequency conversion of a intermediate picosecond dye laser integrated in an uv-oscillator-amplifier system. Actually we take advantage of the Lambda Physik EMG 150 excimer laser which allows for an independent operation of two different laser gas mixtures.

The scheme of our system is shown in Figure 1. Pulse shortening takes place in a two stage cascade pumped dye laser system, applying quenching of resonator transients (5, 6). In the first stage the oscillator part of the EMG 150 operating at 308 nm pumps a broad band P-Terphenyl dye laser. The dye laser pulses having 1 mJ energy and 0.5 ns risetime are suitable to pump a short cavity dye laser (ca. 5 mm) containing Coumarin 307 respectively Rhodamin B depending, whether the final wavelength suppose to be at 248 nm or at 308 nm. The fast response of the high gain medium provides a pulse risetime of 20 ps. The long tail was quenched by an external low loss resonator, slightly tilted with respect to the dye cell, resulting in a pulse typically shorter than 40 ps, as measured by Hamamatsu C1370-01 streak camera. With a diffraction grating the wavelength was tuned to 497 nm respectively 616 nm, and amplified in a three stage amplifier, ending up with

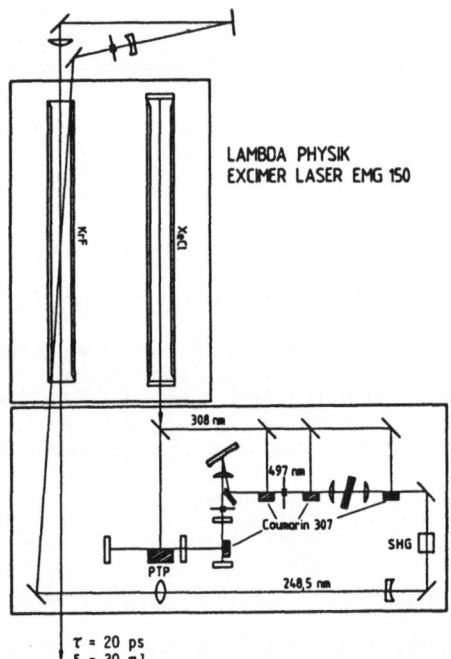

LAMBDA PHYSIK
EXCIMER LASER EMG 150

KrF

XeCl

308 nm

497 nm

Coumarin 307

PTP

248,5 nm

SHG

τ = 20 ps
E = 30 mJ

Fig. 1.

Scheme of the picosecond
excimer laser system

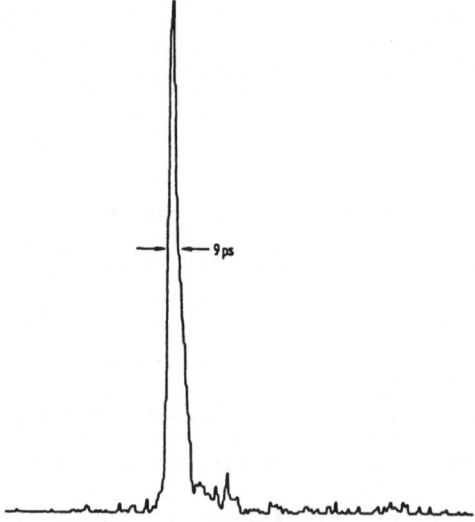

9 ps

Fig. 2.

Streak camera trace of a
picosecond excimer laser pulse

0.5 mJ output energy. Total ASE suppression was easily possible using saturable absorber between the second and third amplifier. These short pulses were frequency doubled (using KPB- or KDP-crystal) and injected through the excimer amplifier. In a double-pass configuration utilizing the whole aperture of the amplifier tube, up to 30 mJ single pulse energy was measured, in the case when KrF was chosen for amplification of the second harmonic at 248 nm. About 10 mJ final output was obtained at 308 nm in XeCl gas mixture. The integrated ASE background was less than 10 %.

On the timescale the pulses consisted in majority only of a single spike shorter than 10 ps as shown the streak camera recording in Figure 2, whereas a smaller part appeared with a double spike. Spiking phenomena turned out to be a consequence of spectral narrowing and mode beating in the picosecond dye laser part but, in average the pulses can be fitted by an 20 ps envelope.

In conclusion this tunable picosecond excimer laser system represents a reasonable compromise of simplicity and desired pulse shortening. It allows stable performance of picosecond uv-pulses with gigawatt peakpower at different excimer laser wavelength.

References

(1) T.SRINIVASAN, T.BOYER, D.F.MULLER, H.PUMMER, CH.RHODES:
    Picosecond Phenomena III, Springer Verlag 1982, p. 19
(2) J.BOKOR, P.H.BUCKSBAUM, L.EICHMER, R.H.STORZ:
    Techn. Dig. Topical Meet. Excimer Lasers, Incline Village, Nevada,
    Jan. 10 - 12, 1983, paper Tu A1
(3) B.NIKOLAUS, K.SCHMITT: Appl. Phys. B 36, 213 (1985)
(4) S.WATANABE, M.WATANABE, A.ENDOH: Appl. Phys. Lett. 43, 533 (1983)
(5) S.SZATMARI, F.P.SCHÄFER: Opt. Comm. 48, 279 (1983)
(6) Z.BOR, B.RACZ: App. Opt. (submitted)

Amplitudenmodulation und Pulsverstärkung des modengekoppelten Nd:YAG-Lasers durch Treiberfrequenzverstimmung

J. Krauser, Heinrich-Hertz-Institut für Nachrichtentechnik Berlin GmbH

H.J. Eichler, Optisches Institut, TU Berlin

Durch Modenkopplung eines kontinuierlichen Nd:YAG-Lasers werden Pulse hoher Leistung mit Halbwertsbreiten von 100-300 Pikosekunden erzeugt. Um eine weitere Pulsverstärkung zu erreichen, wird üblicherweise ein Güteschalter eingesetzt. Der dann entstehende Pulszug ist amplitudenmoduliert. Die Einhüllende des Pulszuges besteht aus einer Folge breiter Pulse /1/, die im folgenden Spikes genannt werden. Die Breite eines Spikes beträgt typischerweise 200 ns.

In dieser Arbeit wird eine neue Methode zur Pulsenergieverstärkung vorgestellt, bei der der Einsatz eines zusätzlichen Güteschalters nicht erforderlich ist /2/. Hierzu wird die Treiberfrequenz des akusto-optischen Modenkopplers ν_m = c/4L um einen kleinen Betrag $\Delta\nu$ verändert. Dies bewirkt eine zeitliche Änderung der Transmission des Modenkopplers. Nach n Umläufen des Pulses im Resonator ändert sich die Transmission entsprechend

$$T(t_n) = \cos^2 (\Theta_m \sin 2 \Delta\nu' t_n) \quad (1),$$

wobei Θ_m die Modulationstiefe des akusto-optischen Modulators bedeutet. $\Delta\nu'$ ist die Abweichung der Frequenz des Modenkopplers von der Resonatorgrundfrequenz: Diese ändert sich mit der Betriebsart des Lasers, da der Brechungsindex des Nd:YAG-Kristalls mit der Anregungsdichte variiert. Deshalb ist $\Delta\nu'$ verschieden von der Treiberfrequenzverstimmung $\Delta\nu$.

Die Cavity-Verluste nehmen nach Gl. 1 mit der Zeit zu, so daß schließlich die Laseraktion beendet wird. Um eine hohe Pulsenergie zu erhalten, sollte diese Verlustmodulation genügend langsam verlaufen, so daß sich der Puls zu einer hohen Amplitude aufbauen und die Energie dem oberen Laserniveau wirksam entziehen kann. Ist die Laseremission auf Null zurückgegangen, baut sich die Besetzung des oberen Laserniveaus wieder zu einer hohen Dichte auf und ein neuer Laserpuls kann sich entwickeln - ähnlich wie bei der konventionellen Güteschaltung.

Es gibt allerdings einen wesentlichen Unterschied der hier beschriebenen Methode zur üblichen Gütemodulation: Beim herkömmlichen Güteschalter wird die Transmission des Resonators plötzlich "eingeschaltet" und der Laserpuls aufgebaut. Der Abfall der Energie ist durch die Entleerung des oberen Laserniveaus gegeben. In der hier beschriebenen Methode baut sich die Pulsenergie auf, wenn die Besetzung des oberen

Niveaus oberhalb der Schwelle liegt. Die Pulsenergie nimmt mit zunehmenden Resona-
torverlusten ab - entsprechend der Verstimmung der Modenkopplertreiberfrequenz.
Mit anderen Worten: Die Pulsenergie kann also nicht stabilisiert werden, da der
Laser gleich nach dem Start seiner Tätigkeit wieder ausgeschaltet wird.

Oszillationen der Pulsenergie im modengekoppelten Nd:YAG-Laser sind auch von anderen
Autoren beschrieben worden /3/. Allerdings sind bisher nur gedämpfte Oszillationen
im Frequenzbereich zwischen 30 und 150 kHz beobachtet worden. Bei der hier vorgestellten Methode können durch geeignete Wahl der Verstimmung der Treiberfrequenz
ungedämpfte Oszillationen erzeugt werden, so daß der Laser periodisch Spikes emittiert, die als Einhüllende die Mode-Lock-Pulse einschließen.

Die Experimente wurden mit einem Nd:YAG-Laser vom Typ BLS 615ML/QS der Fa. Baasel /
München mit akustooptischem Modenkoppler mit einer Resonanzfrequenz von 50 MHz
durchgeführt. Der Resonator, dessen Länge ca. 1500 mm beträgt, ist mit einem Kera-
mikstab stabilisiert. Die Resonatorlänge kann mit einer Differentialmikrometerschraube um ca. 15 mm variiert werden. Das Mode-Lock-Band wird durch Längenabgleich
eingestellt. Die cw-Ausgangsleistung beträgt 3,5 W bei λ = 1,06 μm und 0,6 W bei
1,32 μm. Beim resonanten Mode-Lock-Betrieb ergeben sich Pulsbreiten von typischer-
weise t $\leqslant$ 100 ps bei 1,06 μm bzw. t $\leqslant$ 300 ps bei 1,32 μm (Abb. 1). Die kürzeste Pulslänge konnte bei 1,06 μm mit 63 ps ermittelt werden.

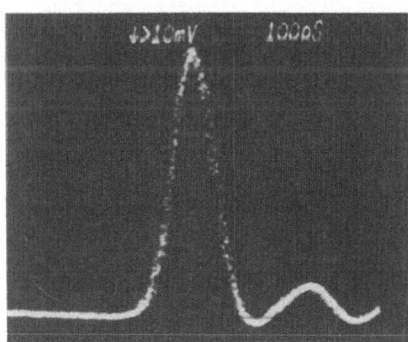 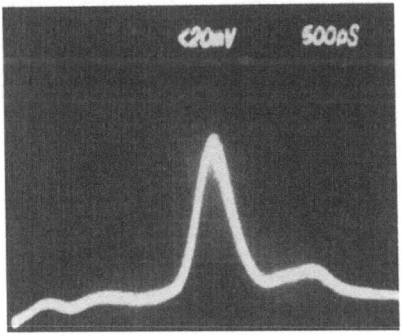

Abb. 1a. Mode-Lock-Puls
bei λ = 1,06 μm für
$\Delta\nu$ = 0 (τ_{FWHM} = 90 ps)

Abb. 1b. Mode-Lock-Puls
bei λ = 1,32 μm für
$\Delta\nu$ = 0 (τ_{FWHM} = 300 ps)

Durch kleine Änderung der Resonatorlänge bei fester Mode-Lock-Frequenz erhöht sich
zunächst die Pulsdauer. Bei weiterer Änderung werden unregelmäßige Spikes und gedämpfte Oszillationen beobachtet. Für λ = 1,06 μm treten bei einer Längenänderung
von L = 155 μm ungedämpfte Oszillationen im zeitlichen Abstand von t = 14 μs auf.

Dies entspricht einer Wiederholrate von ca. 70 kHz. Die Breite der "Quasi-Q-Switch"-Spikes beträgt t = 1800 ns mit 100 ps Einzelpulsdauer. Die Pulsverstärkung im Vergleich zum resonanten Mode-Lock-Betrieb ist typischerweise 20 bis 40-fach (Abb. 2).

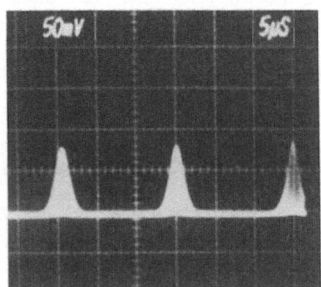

Abb. 2a. "Quasi-Q-Switch"-Spikes bei λ = 1,06 μm ($\Delta\nu \neq 0$)

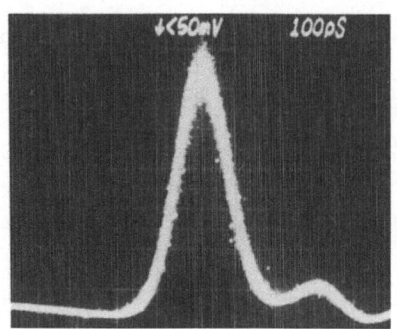

Abb. 2b. Einzelpuls aus einem Spike bei λ = 1,06 um ($\Delta\nu \neq 0$)

Für 1,32 μm erhält man bei geeigneter Resonatorlängenänderung eine Einzelpulsdauer von 700 ps, eine Einhüllende mit einer Spikedauer von 10 μs, eine Wiederholrate von etwa 20 kHz und einen bis zu 50-fachen Verstärkungsfaktor (Abb. 3).

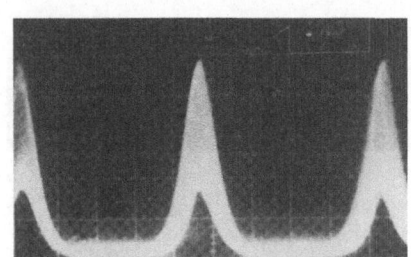

Abb. 3a. "Quasi-Q-Switch"-Spikes bei λ = 1,32 μm ($\Delta\nu \neq 0$)

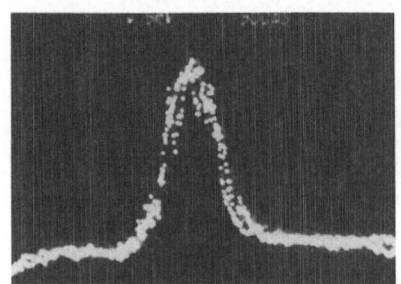

Abb. 3b. Einzelpuls aus einem Spike bei λ = 1,06 μm ($\Delta\nu \neq 0$)

In Tabelle (1) sind die Eigenschaften der durch Resonatorlängenänderung bzw. Treiberfrequenzverstimmung erzeugten Pulse zusammengestellt.

Durch den in der Länge veränderlichen Resonatoraufbau des Nd:YAG-Lasers kann die Einstellung der Treiberfrequenzverstimmung sehr genau vorgenommen werden. Eine Längenänderung von beispielsweise 1,5 μm entspricht einer Frequenzänderung von 50 Hz. Die gewählte Einstellung konnte über Stunden stabil gehalten werden.

Wellen- länge μm	Pulsdauer Pulsleistung ($\Delta\nu = 0$)	Pulsdauer Pulsleistung ($\Delta\nu \neq 0$)	Pulsbreite der Einhüllenden ($\Delta\nu \neq 0$)	Wiederhol- rate ($\Delta\nu \neq 0$)
1,06	100 ps 300 W	100 ps 12 kW	1,8 μs	70 kHz
1,32	250 ps 24 W	700 ps 1 kW	10 μs	20 kHz

Tab. 1: Übersicht über Pulsleistung und Pulsdauer bei $\lambda = 1,06\,\mu$m und $\lambda = 1,32\,\mu$m

Zusammenfassung

Es wurde eine neue Methode zur Pulsenergieverstärkung im cw - modengekoppelten Nd:
YAG-Laser vorgestellt: Durch Verstimmen der Resonatorlänge oder der Mode-Lock-Frequenz erhält man eine Pulsverstärkung gegenüber dem resonanten Mode-Lock-Betrieb
um den Faktor 20 - 40. Es ist sicherlich lohnend, diese Methode auch auf andere
Lasersysteme zu übertragen.

References

/1/ D.T. KUIZENGA, D.W. PHILLION, T. LUND, A.E. SIEGMANN, Optics Comm. 9, 221
     (1973)

/2/ H.J. EICHLER, J. KRAUSER, Optics Comm. 52, 129 (1984)

/3/ D.T. KUIZENGA, A.E. SIEGMANN, IEEE J. QE-6, 694 u. 709 (1970)

Erzeugung von VUV-Strahlung zur Plasmadiagnostik durch stimulierte Raman-Streuung an Wasserstoff

H.F.Döbele, M.Hörl, M. Röwekamp und B.Rückle[*], FB Physik Universität GH Essen

Zusammenfassung

Stimulierte Ramanstreuung an H_2 mit Pumpstrahlung bei 193 nm erlaubt die Erzeugung von schmalbandiger, abgestimmter VUV-Strahlung bis hin zu Spektralbe-
reichen unterhalb Lyman-alpha. Typische Leistungen bei 130 nm und größeren
Wellenlängen liegen im kW Bereich. Mit Farbstofflaserstrahlung im Sichtbaren
kann VUV-Strahlung bei kleineren Leistungen bis mindestens 138 nm kontinuierlich
abstimmbar erzeugt werden. Die so erzeugte Strahlung gestattet es, in künftigen
Fusionsexperimenten durch strahlungsinduzierte Fluoreszenz wichtige Plasmaverun-
reinigungen wie z.B. atomaren Sauerstoff und Kohlenstoff ortsaufgelöst in der
Randschicht des Plasmas nachzuweisen.

Problemstellung

In der Plasmadiagnostik sind Lasertechniken heute auf breiter Basis etabliert:
Elektronentemperaturen und -dichten werden routinemäßig orts-und zeitaufgelöst
über Thomsonstreuung mit Rubinlasern und neuerdings auch Nd:YAG-Lasern gemessen;
Infrarotlaser werden zur kontinuierlichen Bestimmung der Elektronendichte durch
Interferometrie und der poloidalen Magnetfeldstärke durch Faradaydrehung in
Tokamaks eingesetzt. Die Messung von Ionentemperaturen durch Thomsonstreuung ist
für dichte Plasmen bereits möglich; für die in Fusionsexperimenten vorliegenden
Plasmen werden entsprechende Entwicklungen verfolgt. Bei Experimenten zur Kern-
fusion kommt der Diagnostik von atomaren Verunreinigungen im wandnahen Bereich
wegen des kritischen Einflusses auf das Erreichen von Zündbedingungen eine
besondere Bedeutung zu. Für metallische Verunreinigungen ist das Problem durch
Einsatz von blitzlampengepumpten Farbstofflasern sehr weitgehend gelöst /1/. In
neuerer Zeit sind die sog. "Niedrig-Z-Verunreinigungen" wie z.B. Kohlenstoff,
Beryllium oder Bor deshalb besonders wichtig geworden, weil sich gezeigt hat,
daß durch absichtliche Verwendung derartiger Materialien im Wandbereich u.a.
wegen reduzierter Strahlungverluste deutliche Verbesserungen resultieren. Die
Resonanzlinien in diesem Zusammenhang besonders wichtiger Atome liegen unterhalb
2oo nm - z.B. C bei 156 nm und 166 nm und O bei 130 nm.

---

[*] Jetzt Fa. LAMBDA PHYSIK, Göttingen

Anforderungen an die VUV-Strahlung

Neben der richtigen spektralen Lage sind vor allem die Leistung und die Bandbreite der VUV-Strahlung von Wichtigkeit. Letztere soll wegen der Zeeman-
Aufspaltung der Resonanzlinien im Toroidalfeld typisch im Bereich um 0.01 nm liegen. Mit konservativen Daten für die Plasmas-Hintergrund-strahlung und die Verluste bei der Einstrahlung und Detektion ergibt sich die nebenstehende Abhängig-keit des Signal-Rauschverhältnis-ses von der eingestrahlten VUV-Leistung, die für große Leistungen deutlich den Einfluß der Sättigung zeigt. Die SNR-Analyse /2/ ergibt als Fazit, daß VUV-Leistungen im

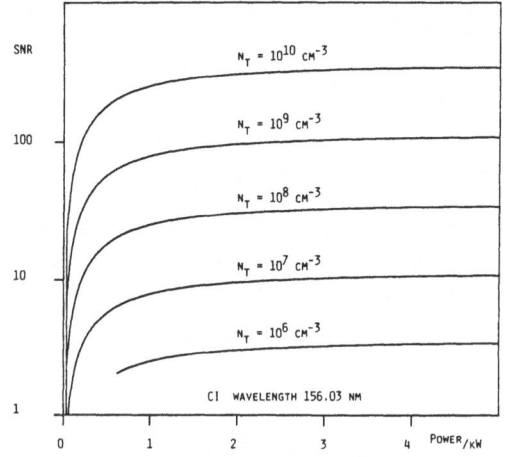

Bereich um 1 kW verfügbar sein sollten, um mit genügendem Spielraum die Untersuchung z.B. von Kohlenstoff zu erlauben.

VUV-Erzeugung durch SRS ausgehend von Farbstofflaserstrahlung

Wichtige Resonanzlinien liegen in der Nähe möglicher H_2-anti-Stokes Komponenten
des ArF*-Lasers. Dessen Abstimmbarkeit ist jedoch unbefriedigend /3/. Erheblich
größere Abstimmintervalle lassen sich bei Verwendung von ArF* als Verstärker-
medium erreichen /4/. Dies erfordert die unabhängige Erzeugung verstärkbarer Strahl-ung bei 193 nm. SRS von Farbstofflaser-strahlung in H_2 ist ein hierzu sehr ge-eigneter Weg. Fo-kussiert man die Strahlung eines im Bereich von 545 nm (Fluoreszein 27) etwa 100 mJ emit-tierenden, von einem frequenzverdoppelten

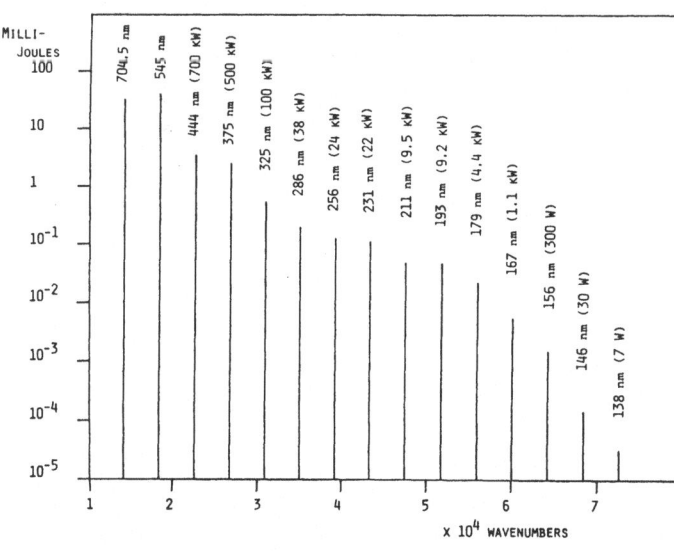

Neodymlaser longitudinal gepumpten Farbstofflaser in eine H_2-Ramanzelle /5/, so

118

lassen sich bis zu 13 AS-Komponenten mit den in der Abbildung aufgelisteten
Leistungen erzeugen. Übergang zu benachbart emittierenden Farbstoffen sichert
die lückenlose Abdeckung des Spektrum bis mindestens 138 nm. Etwa 10 kW stehen
zur Nachverstärkung in ArF* zu Verfügung. Mit einem XeCl-gepumpter Farbstofflaser (QUI) ergeben sich vergleichbare Resultate.

VUV-Erzeugung durch SRS von 193 nm - Strahlung

Durch Nachverstärkung der im Falle des XeCl-gepumpten Farbstofflasers etwa 30µJ
enthaltenden 193 nm-Ausgangsstrahlung in zwei ArF*-Stufen (EMG 100 und EMG 200)
lassen sich etwa 80 bis 100 mJ mit einer Bandbreite unter 0.5 cm^{-1} erzeugen.
Die Winkeldivergenz gestattet einen Luftdurchbruch mit einer MgF$_2$-Linse von 1 m

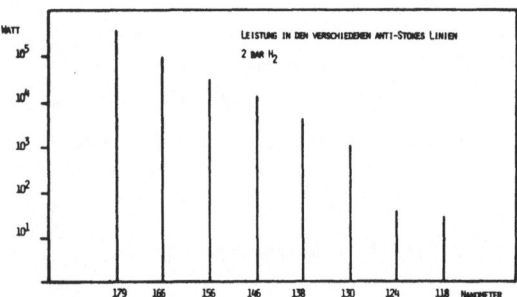

Brennweite. Fokussierung dieser
verstärkten Strahlung in eine
zweite Ramanzelle liefert weitere 8 anti-Stokes Komponenten,
deren Leistungen in der Abbildung dargestellt sind. Die optischen Wege zwischen den einzelnen Stufen sind vom Ausgang
der ersten Ramanzelle an in N$_2$
geführt, um Absorption durch
atmosphärischen Sauerstoff zu
vermeiden. Dies gestattet die
Abstimmung der 193 nm-Strahlung über maximal 870 cm^{-1}. Die erste AS-Komponente
ist maximal über etwa 8oo cm^{-1} abstimmbar; der entsprechende Wert liegt für die
höheren AS-Komponenten bei etwa 400 cm^{-1}. Die Pulsdauern betragen etwa 5 ns.
Zur Leistungsmessung wurde ein solar-blind Fotomultiplier in Verbindung mit
einem Doppelmonochromator eingesetzt. Die Bestimmung des Relativverlaufs der
Empfindlichkeit dieser Kombination wurde mit Hilfe eines stationären, für
Eichzwecke geeigneten /6/ Argon-Lichtbogens vorgenommen. Der Absolutanschluß
ergibt sich durch Messung der Energie bei 193 nm vor der Einstrahlung und
Vergleich mit dem Multipliersignal bei evakuierter Ramanzelle.

Erste Anwendungen und weiterführende Arbeiten

Die gemessenen Leistungen zeigen, daß auf diesem Wege VUV-Strahlung erzeugt
werden kann, deren Eigenschaften erfolgreiche Anwendungen zur Bestimmung von
Konzentrationen z.B. von C- und O- Verunreinigungen bei Tokamakexperimenten
versprechen.

In einer ersten Anwendung wurde die so erzeugte Strahlung bei 130 nm zur Fluoreszenzdiagnostik von atomarem Sauerstoff an einer stationären Quelle bekannter O-Konzentration eingesetzt /7/. Der Sauerstoff wird dabei durch eine Reaktion zwischen in einer Mikrowellenentladung hergestelltem atomarem Stickstoff und einer genau bemessenen Menge NO erzeugt. Den Aufbau für derartige Messungen zeigt die Abbildung.

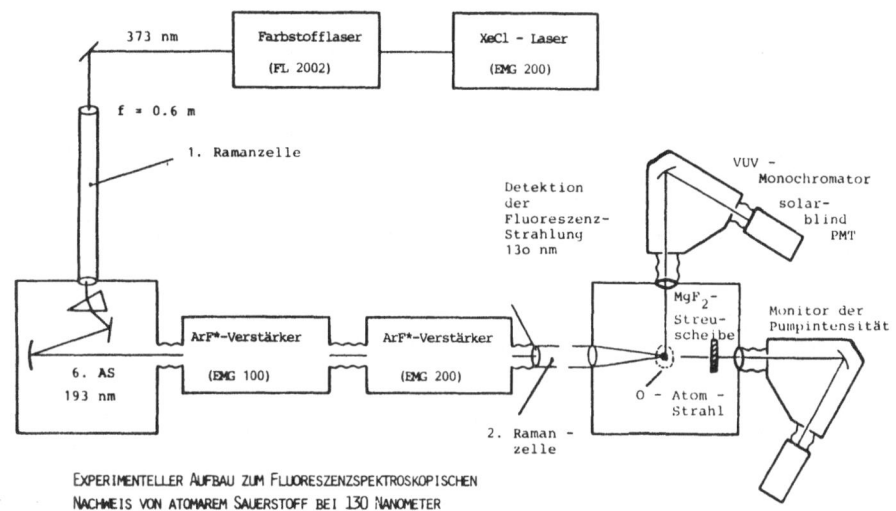

EXPERIMENTELLER AUFBAU ZUM FLUORESZENZSPEKTROSKOPISCHEN NACHWEIS VON ATOMAREM SAUERSTOFF BEI 130 NANOMETER

Es ist zu erwarten, daß sich bei Kühlung des Ramanmediums mit flüssigem Stickstoff - ähnlich wie in /8/ dargelegt- eine Erhöhung der Leistung ergeben sollte. Ebenso läßt die Verwendung von Para-H_2 eine weitere Verbesserung erwarten, da dann wegen der möglichen Besetzung des niedrigsten Rotationszustandes alle Moleküle am Konversionsprozess teilnehmen können.

Literatur

/1/ P.BOGEN und E.HINTZ, Comments Plasma Phys. Contr. Fusion 4,115 (1978)

/2/ H.F.DÖBELE, M.RÖWEKAMP und B.RÜCKLE, J.Nucl.Mater 128&129,986 (1984)

/3/ H.F.DÖBELE und B.RÜCKLE, Applied Optics 23,1040 (1984)

/4/ H.F.DÖBELE, M.RÖWEKAMP und B.RÜCKLE, IEEE J.Quant. El. QE-20, 1284 (1984)

/5/ H.SCHOMBURG, H.F.DÖBELE und B.RÜCKLE, Appl.Phys. B 30, 131 (1983)

/6/ J.M.BRIDGES und W.R.OTT, Applied Optics 16, 367 (1977)

/7/ H.F.DÖBELE, M.HÖRL, M.RÖWEKAMP und B.REIMANN Frühjahrst.DPG Bayreuth

/8/ D.J.BRINK und D.PROCH, Optics Letters 7, 494 (1982)

IR-Laser Spectroscopy of Free Radicals and Open-Shell Molecular Ions

W.Urban [*], A.Hinz, W.Bohle and D.Zeitz
Institut für Angewandte Physik, Universität Bonn, Wegelerstr. 8,
D 5300 BONN 1

Infrared laser spectroscopy of open shell molecules can take advantage
of the Zeeman effect for both modulating and tuning of the corresponding transitions into resonance with a fixed frequency laser line.
Thus it is very similar to EPR, however the tuning range is much smaller than the transition frequency, which can be either the rotational
or the vibrational energy of a molecule. This method, which is well
known as "Laser Magnetic Resonance" (LMR), has first been developed
and brought to perfection by K.M.EVENSON and coworkers /1/. They achieved sensitivies down to 10^6 diatomic molecules per cm^3 in the far
infrared region for pure rotational molecular transitions. An essential feature of EVENSON`s LMR system is the intracavity arrangement,
i.e. the absorption cell, where the shortlived species are generated
by a chemical reaction, is incorporated in the laser cavity. The high
sensitivity of this intracavity setup can be understood in terms of
the increased effective pathlength through the sample and the non-
linearity of the laser system, as has been pointed by H.E.RADFORD et
al. /2/.

In principle this enhancement in sensitivity by the intracavity arrangement should be the same in medium IR as well as in FIR, however our
experience shows, that the theoretically possible increase in sensitivity could not be achieved with a CO-Laser between 5-8 µm /3/. Finally
we found, that pick up from the modulation field into the resonator
was the limiting factor. This seems plausible since the laser cavity

----

[*] This paper is dedicated to Prof.Dr.H.G.Kahle, my previous super-
visor, for the occasion of his 60th birthday.

is certainly more sensitive to vibrations at 5μm compared to 100μm <u>and</u> moreover the modulation amplitude has to be much higher for transitions at 5μm due to the dominating Doppler-width (~30 MHz), compared to transions at 100μm, where the linewidth is mainly pressure broadered (~ 1 MHz) under the usual reaction conditions.

We have overcome this problem by a different approach, which has first been developed by HINZ et al /4/ in our laboratory and which has meanwhile been further improved to give a sensitivity of 10^7 NO-radicals per cm^3 /5/. We are using an extracavity arrangement and detect the paramagnetic species via their polarisation effects, e.g. the Faraday-effect, when tuned through resonance. With this polarisation type experiment, the sensitivity limiting laser noise is suppressed by the analyer in almost crossed position with respect to the incoming laser polarisation and the signal is detected upon a correspondingly small noise background. A block diagram of the apparatus is showen in Fig.1. Quantitative details of this method are described in /4/.

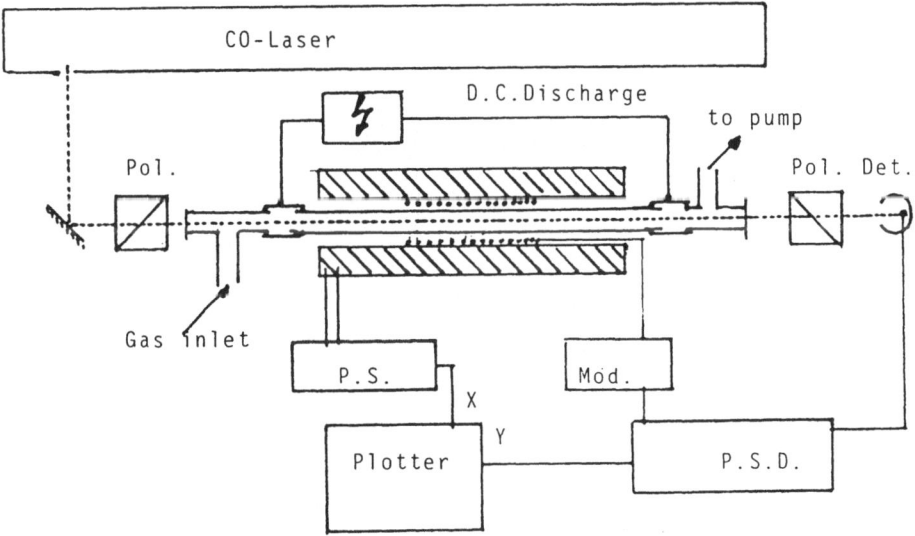

<u>Fig. 1</u>. Block diagram of Faraday-LMR system.

For spectroscopy of open shell molecular ions one should use an electric dischange, which again will form a means for strong crosstalk from the magnetic field modulation into the laser resonator both in medium <u>and</u> far infrared. The relatively poor signal to noise ratio of

122

FIR-LMR investigatons of e.g. HF$^+$ /6/ where the system by itself
should be more sensitive than our Faraday LMR /5/ seems to support
this extra source of noise for FIR-LMR with intracavity discharges.
With our system we have investigated vibration-rotation transitions of
open shell molecular ions achieving quite good signals for DCl$^+$ in the
first attempt /7/. Meanwhile, we found we could alter the generation
conditions and improve the ion concentration, thereby obtaining
signal to noise ratios of several thousands for the fundamental band
of DCl$^+$ X$^2\Pi$ v=1←0.

In this discharge we could also trace a series of hot band transitions
up to v=6←5, thus estiblishing that the extracavity Faraday-LMR is
most capable for this type of spectroscopy /8/. Fig.2 shows a signal
of the hot band v=5←4 in DCl$^+$ as an example. A series of other ionic
species having transitions in the spectral range of our CO-Laser
(5-8μm) is presently being investigated.

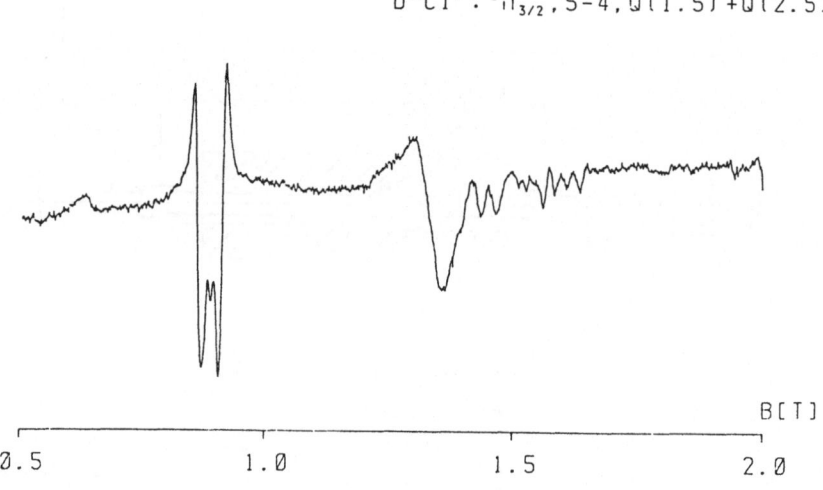

$^{12}C^{16}O$ P(15)18-17 1650.8108cm^{-1}
$D^{35}Cl^+$: $^2\Pi_{3/2}$,5-4,Q(1.5)+Q(2.5)

B[T]

0.5   1.0   1.5   2.0

Fig. 2. Faraday-LMR signal of hot-band v=5←4 in DCl$^+$

Acknowledgement:
This work is supported by the Deutsche Forschungsgemeinschaft through
Sonderforschungsbereich 42.

References

/1/ K.M.Evenson, H.P.Broida, J.S.Nells, R.J.Mahler, M.Mizishima
    Phys.Rev.Letters 21, 1038 (1968)

/2/ H.E.Radford, W.Rohrbeck, K.M.Evenson, R.W.McKellar, A.Hinz
    Proc. 15th Int. Symp. on Free Radicals, p.9 (1981)

/3/ W.Rohrbeck, A.Hinz, P.Nelle, M.A.Gondal, W.Urban
    Appl.Phys.B 31, 139 (1983)

/4/ A.Hinz, J.Pfeiffer, W.Bohle, W.Urban
    Molec. Phys. 45, 1131 (1982)

/5/ A.Hinz, D.Zeitz, W.Bohle, W.Urban
    Appl.Phys. B 36, 1 (1985)

/6/ D.C.Hovd, E.Schäfer, S.E.Strahan, C.A.Ferrari, D.Ray,
    K.G.Lubic, R.J.Saykally
    Molec. Phys. 52, 245 (1984)

/7/ A.Hinz, W.Bohle, D.Zeitz, J.Werner, W.Seebass, W.Urban
    Molec. Phys. 53, 1017 (1984)

/8/ W.Bohle, D.Zeitz, J.Werner, A.Hinz, W.Urban
    Molec. Phys. (to be published).

Laser Mass Spectroscopy – Diagnostics of Organometallic Compounds and Applications for Laser CVD

M. Stuke
Max-Planck-Institut für biophysikalische Chemie
Dept. Laserphysik
P.O. 2841
D-3400 Göttingen, Fed. Rep. Germany

Laser mass spectroscopy offers many advantages over conventional analytical techniques. Fast, sensitive, and selective fingerprint detection of organometallic precursors of III-V and II-VI semiconductors, with emphasis on CH_3TeCH_3, $CH_3TeTeCH_3$, $C_2H_5TeC_2H_5$, CH_3SeCH_3, $(CH_3)_3Ga$, $(CH_3)_3In$, AsH_3, and others is achieved using short and ultrashort pulse tunable dye laser controlled time-of-flight mass spectroscopy (1-4).

The experimental setup is schematically shown in Fig. 1. This differentially pumped time-of-flight mass spectrometer system (5) can analyze gaseous, pulsed jet or solid samples under complete computer control.

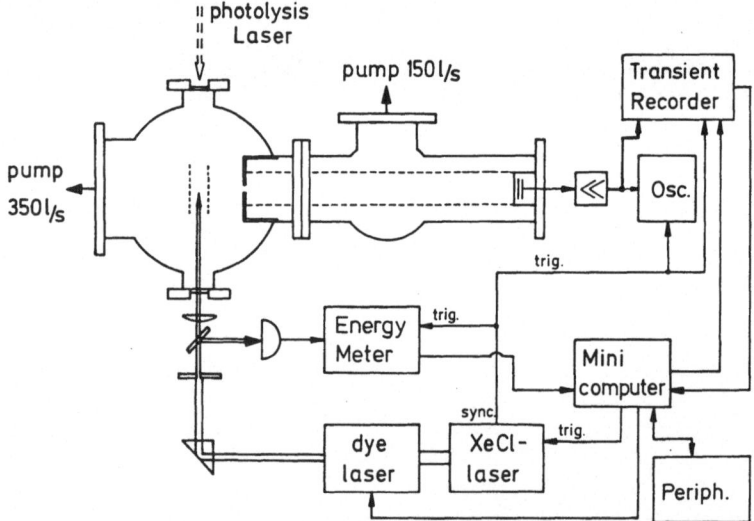

Fig. 1. Laser Mass Spectrometer, experimental setup (1,5).

Excellent spatial and time resolutions combined with high __single shot__ selectivity and sensitivity down to 10^{-7} Torr and below (corresponding to $3x\ 10^{19}$ cm^{-3} or about 3000 molecules in the observation region) __plus__ mass identification are obtained.

Therefore, this technique compares very favorably with other laser analytical techniques such as LIF and CARS, since laser mass spectroscopy gives both - rough information (i.e. the __mass__ of the species to be detected) - __and__ in addition spectroscopic information, which makes this method ideally suited for the characterization and understanding of MOCVD (Metal Organic Chemical Vapor Deposition) and LCVD (Laser Chemical Vapor Deposition) processes. Especially in cases, where spectroscopic information is lacking, for example in the case of adducts, the observation of the parent molecular ion is __essential__ for quick and reliable species identification. Often, however, especially in the case of organometallics, the parent ion has very low abundance. This problem can be overcome - if necessary - using picosecond laser excitation. The increase of the relative abundance of the parent molecular ion by more than one order of magnitude, when changing from nanosecond to picosecond excitation can be seen clearly in Figs 2-4. Even in cases, where ns laser excitation does not show __any__ parent ion abundance ($C_2H_5TeC_2H_5$ in Fig. 3 and $CH_3TeTeCH_3$ in Fig. 4), ps laser excitation - even for the same or higher intensities - reveals the parent molecular ion as the dominating ion in the mass spectrum.

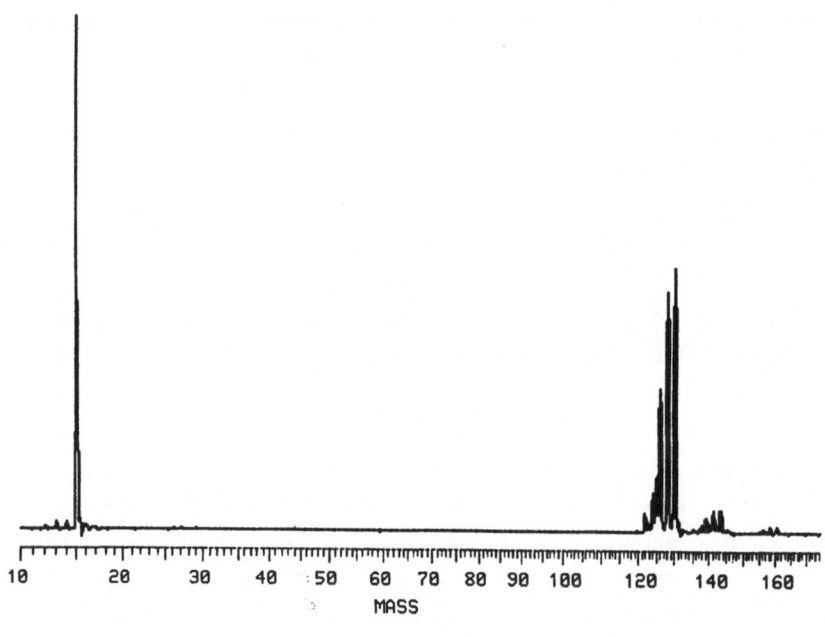

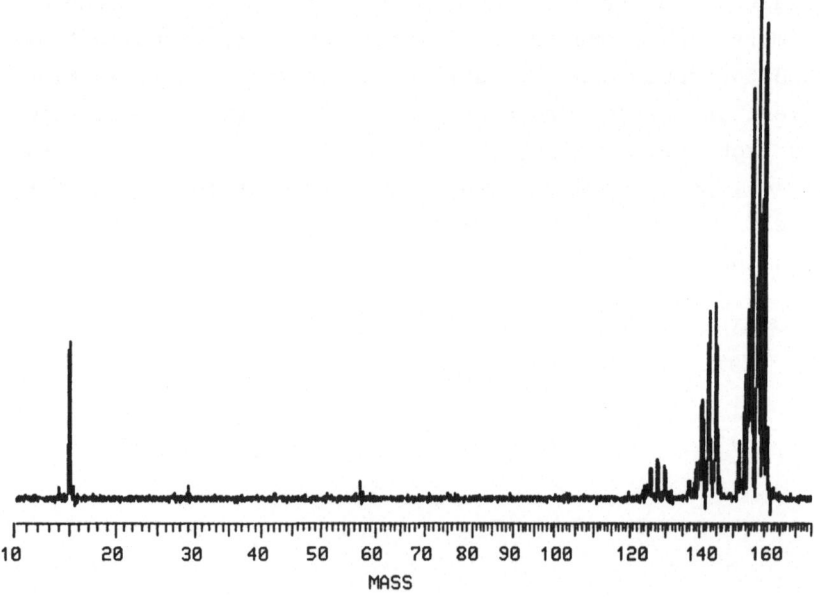

Fig. 2. Laser mass spectrum of CH_3TeCH_3 with nanosecond (top) and picosecond laser exciation (below). The parent ion abundance (m/e=160) is increased.

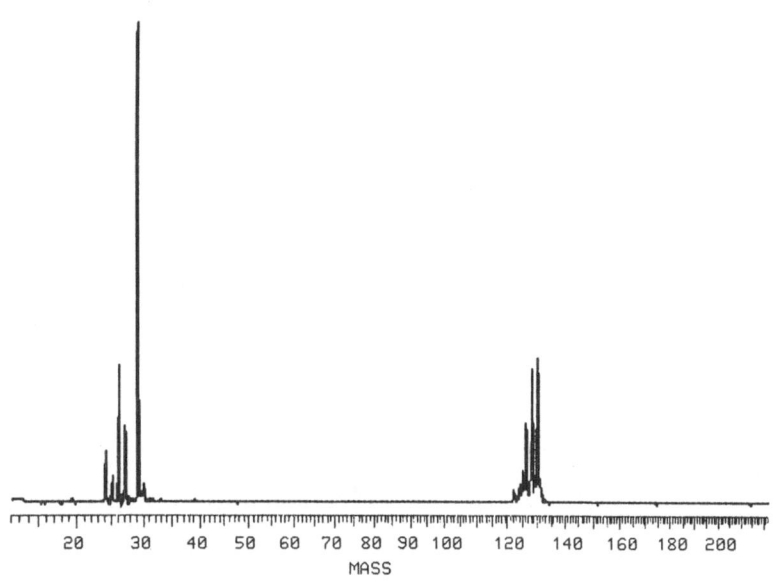

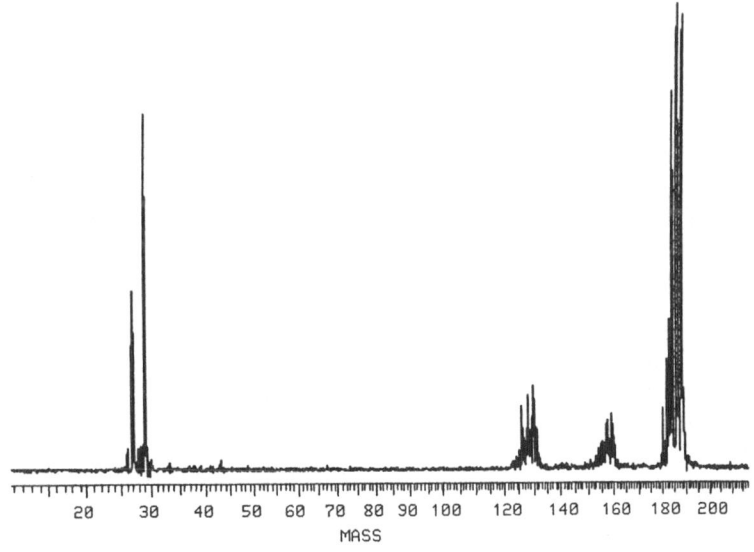

Fig. 3. Laser mass spectrum of $C_2H_5TeC_2H_5$ with nanosecond (top, <u>no parent ion</u>) and picosecond laser excitation, (below).

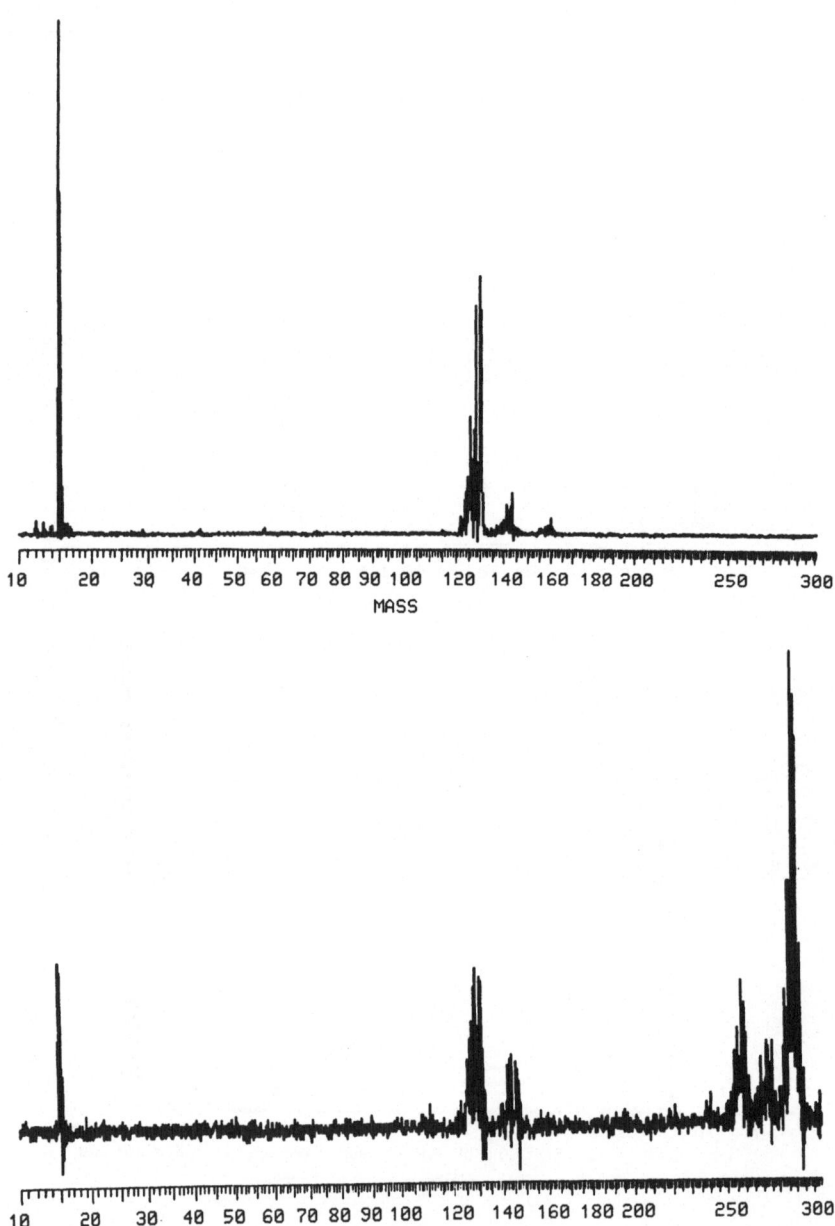

Fig. 4. Laser mass spectrum of $CH_3TeTeCH_3$ with nanosecond (top, <u>no parent ion</u> is observed at m/e=290) and picosecond laser excitation (below).

In summary, the technique of laser mass spectroscopy is ideally suited for sensitive and selective analysis of organometallic species. In general, tunable nanosecond laser excitation is sufficient for quick species identification and characterization. In cases, where the compounds to be detected are extremely labile, ps laser excitation will cause the parent molecular ion abundance to become the dominating ion in the mass spectrum.

References

(1) M. Stuke, Appl. Phys. Lett. 45, 1175 (1984)

(2) R. Fantoni, M. Stuke, Appl. Phys. B (1985) in press

(3) M. Stuke, R. Fantoni, in Springer Series in Optical Sciences "Laser Spectroscopy VII",Ed. T.W. Hänsch, Y.R. Shen p 414 (1985)

(4) M. Stuke, in "Oberflächentechnik", SURTEC 85, VDE-Verlag Berlin, p465 (1985)

(5) Technical details: SUMOTEK GmbH, P.O. 3311, D-3400 Göttingen

Laser Komponenten
Laser Components

The Research and Production of Laser Crystals in China

Zhang Yinxia

North China Research Institute of Electro-Optics

P.O.Box 8511, Beijing, People's Republic of China

Abstract: A general review about the research and production progress
of laser crystals in China. In the past quater centry, a considerable foundation and strength has been built. Through further effort
China's laser crystal will soon reach the world advenced level.

In the year 1960, China began to research laser crystals. The home
made laser crystals have ever played an improtant part in the birth
of China's first ruby lasers (September 1961) and their later development. Today in the whole country, there are about 30 institutes,
universities and factories with about 1,000 scientists, engineers
and workers developing about 30 kinds of laser crystals. Ruby and
YAG have already been put into production and widely used. Some new
and promising crystals are under intensive investigation, and their
applications will be found soon.

In this paper, we give a general review about the past and future
of the research and production progress of laser crystals in China.

I. RARE EARTH DOPED LASER CRYSTALS

1. Nd:YAG. In 1965, China began to grow YAG crystals by flux method.
The laser operation was obtained from crystals grown by Czochralski
method in 1967. Many laboratories and factories are using Cz. method
to grow Nd:YAG crystals by graphite resistance heating technique.

The flat interface crystals with no core and no dislocations can be
steadily grown in Southwest Technical Physics Institute and

Huaguan Instrument Factory of Sichuan province. The interference fringe of 0.25 per 25 mm rod long has been reached. After many year's hard work, our institute (NCRIEO) have been able to grow Nd:YAG crystals by induction heating Cz. method with iridium crucible being used. Crystals with convex interface and no scattering. The crystal diameter is 30-40 mm.

Jilin Laser Material Factory and Guanming Equipment Factory in Chengdu are making great effort on YAG crystals production.

Further research on Nd:Cr:YAG crystals in NCRIEO show that Cr^{3+} not only has the sensization effect in Nd:YAG. But also improves the deformation of crystal and increases the fluorescense lifetime of Nd^{3+} as well as the resistance against ultraviolet radiation.

2. $YAlO_3$. The crystal was grown in 1971, and lased in next year. Fujian Institute of Research on the Matter Structure persisted in the study on $YAlO_3$ and made a great deal of effective work on crystal growth, raw material purification and thermal effect as the specific design of lasers. The cw laser output of $Nd:Cr:YAlO_3$ is 162 W along b-axis at 1.079 um. Its cw output is 80 W at 1.34 um, which has reached the world advenced level. $Er:YAlO_3$ has also been grown.

3. $LiYF_4$(YLF). YLF has been grown in NCRIEO since 1977. An improved Cz. method with inert atmosphere and graphite growth equipment has been developed. It was lased under room temperature Nd:YLF at 1.047, 1.053 and 1.32 um, Er:YLF at 0.85 um and Ho:Er:Tm:YLF at 2.06 um. The pulse output of Nd:YLF is 838 mj, slope efficiency is 1.5%. In 1983, we obtained 1.047 and 1.32 um two color lasing in Nd:YLF, E_{in} = 50j, E_{out}=79.2mj (1.047 um) and 72.6 mj (1.32 um).

4. Stoichiometric laser crystals. Transparent NdP_5O_{14}, $(Nd,La)P_5O_{14}$ and $NdAl_3(BO_3)_4$ crystals has been grown in Shandong University by flux method with improved flux and seed rotation. The size of crystals is about 4 cm, which has reached the world's advenced level.

II. TRANSITIONAL ION DOPED LASER CRYSTALS

1. Ruby ($Cr:Al_2O_3$). After the high tide of developing Venuil's method at the beginning of 1960's, the research of ruby was suspended for a time. In the mid 1970's, Anhui Institute of Optics and Mechanics and Suzhou Crystal Device Factory grow some crystals with good homogeneity by Cz. method. With the assistance of Shanghai Institute of Ceramics and Jiaozhuo Research Institute of Laser, they have improved the quality of crystal by improving Vinuil's method. These two institutes are now making great efforts on Cz. method in order to produce high quality ruby used in holography.

2. Alexanderite ($Cr:BeAl_2O_4$). Because of the limitation of the toxic BeO, growth investigation is only developed in Shanghai and Anhui Institutes of Optics and Fine Mechanics. In 1981, they obtained laser performance and recently they also obtained the tunable laser output.

III. COLOUR CENTER LASER CRYSTALS

At the beginning of 1980's, there were about 10 laboratories in China developing color center laser crystals, such as doped and undoped LiF, KCl and NaF etc.. The Nd:YAG double-frequency laser pumped LiF: F_2^- and LiF:F_2^+ , pulse tunable color center laser operation was achieved at room temperature in 1981. These crystals are very stable though 42,000 pulse work. Li, Na doped KCl with $F_A(II)$ and $F_B(II)$ center will be used in Frequency-Standard-System and Photochemistry.

Joint efforts on the development of such crystals has been made by the University of Overseas Chinese and Shanghai Jiaotong University, etc..

IV. THE RESEARCH OF NEW LASER MATERIALS

In order to suit the demand of different kinds of solid state lasers. Some research institutes and universities in China have been trying to explore new laser materials since 1970's. The NCRIEO has grown some new laser crystals, such as $Nd:CaY_4(SiO_4)_3O$, $Nd:YVO_4$, $RE:Gd_2(MoO_4)_3$, $RE:Ca_5Y_{13}F_{49}$, $Nd:LaMgAl_{11}O_{19}$, $Nd:Cr:GSGG$ and a new type of molti-doped YAG, which has the effect of self Q-switching and self modelocking. A new crystal $Cr:YAl_3(BO_3)_4$ has been developed in Fujian Institute of Research on Matter Structure. A series of small size stoichiometric laser crystals, such as $KNdP_4O_{12}$, $Mn:CeP_5O_{14}$, $Tb_xDy_{1-x}P_5O_{14}$, etc. have been grown in Changchun Institute of Applied Chemistry. Congruent melting crystal $K_5Bi_{1-x}Er_x(MoO_4)_4$ is grown in Shanghai Institute of Ceramics. Beijing Institute of Physics has made the growth and property on Eu:GGG, Ho:GGG, Nd:GSGG and Cr:G(Ca)G(Mg,Zr)G, etc.. Some universities are making some foundamental researches in order to develop new materials.

In the past decades the development of laser crystal in China was quite fast, and considerable foundation and strength has been built. Comparing China's laser crystal with the world advanced level, there is still certain distance in the field of technique and quality. But we believe that through further effort and to strengthen international cooperation and intercourse. China's laser crystal will soon reach the world advenced level.

GaAlAs Power Laser-diodes by MO-VPE Technology

F.P.J. Kuijpers, G.A. Acket, H.G. Kock, Philips Research Laboratories, Postbus 218,
5600 Eindhoven / NL

Threshold current, output power and far field distributions were investigated on
MO-VPE grown gain guided laser-diodes, having a planar proton bombarded structure,
in their respective relation to the thickness of the active layer in the wavelength
regime 810-870 nm.
The lateral mode behaviour is non-Gaussian and consistent with a $\cosh^{-2}$ gain profile
even for thin active layers. Multi-mode fibre coupled devices with output powers as
high as 40 mW from the fibre (CW 30°C) were obtained.

High CW-Power Phase-locked Semiconductor Laser Arrays

F. Kappeler, H. Westermeier, R. Gessner, M. Druminski, C. Hanke
Siemens AG, Research Laboratories, Otto-Hahn-Ring 6, 8000 Munich 83

J. Luft
Siemens AG, Components Division, Balanstraße 73, 8000 Munich 80

Abstract

Very high coherent optical power can be achieved from recently developed GaAlAs laser arrays consisting of 10 to 40 closely spaced
laser stripes integrated on a common substrate. Phase-locked operation
of such a large number of lasers has been rendered possible by the
high uniformity of the expitaxial layers grown by metal-organic vapour
phase epitaxy (MO VPE). The achievable output power (up to 1.6 W per
facet from a 40-stripe array) and the high overall conversion efficiency (up to 36 %) considerably exceed the performance of single
stripe lasers. 12-stripe arrays delivering some hundred milliwatts at
an emission wavelength ranging from 805 to 880 nm have been processed
on a larger scale. They show very reproducible laser characteristics
with a standard deviation of less than 5 % both for threshold current
and differential efficiency across a 20 mm MO VPE wafer.

1) Introduction

Basically the output power available from a semiconductor injection
laser is limited by a critical optical power density where the facets
are irreversibly damaged by local heating. Reduction of the optical
power density by simply increasing the cross section of the laser resonator leads to instabilities due to the competition of higher order
optical modes. Thus a single semiconductor laser can deliver a maximum
output power of about 100 mW in CW operation. Much higher levels of
coherent radiation can be obtained from a laser array consisting of
several optically coupled semiconductor lasers fabricated on the same
substrate /1,2,3/. In this report design and performance of high-power
phase-locked GaAlAs laser arrays grown by metal-organic vapor phase
epitaxy (MO VPE) will be discussed.

2) Structural design

In an experimental study /3/ we have investigated laser arrays consisting of 3 to 40 laser channels with a center-to-center spacing
ranging between 8 and 40 µm. Both facet planes of the 400 µm long
devices are symmetrically coated with Al_2O_3 in order to improve

the level of catastrophic optical damage /4/. Fig. 1 shows an experimental 40-stripe array with a 10 μm center spacing. The chip dimensions are as small as 0.5 x 0.4 mm.

The vertical layer structure of the GaAlAs arrays has been grown by MO VPE /5/ and contains either a conventional double-hetero (DH) structure or a multiple quantum well (MQW) structure. In the latter case the 75 nm thick active layer consists of a number of extremely thin layers having different bandgaps. This structure helps to improve the optical coupling between neighboring laser channels.

Fig. 1. Experimental 40-stripe GaAlAs laser array

3) Experimental results

The light/current characteristics and the DC-to-light conversion efficiencies of two experimental MQW arrays operating CW are shown in Fig. 2. Best results achieved so far are maximum output powers of 0.86 W per mirror for a 10-stripe array and 1.65 W per mirror for a 40-stripe array, both emitting at 880 nm. The DC-to-light conversion efficiencies reach maximum values of 36 % and 33 % respectively. These high conversion efficiencies, which are the best values reported up to now for laser arrays /1,2/, result from the low threshold current (< 25mA per stripe), the good differential efficiency (up to 0.55 W/A per mirror) and above all from the low electrical series resistance (0.1 Ω for a 40-stripe array) realized in these devices.

Another prominent feature of optically coupled laser arrays is their phase-locked operation which is demonstrated in Fig. 3 for a 10-stripe MQW array. The spectrally resolved near field (that is, the spatial intensity distribution in the mirror plane vs. the spectral intensity distribution) shows that all ten laser channels operate with the same set of a few longitudinal modes. The output power of the individual lasers coherently combines into two sharp grating lobes of less than 2° FWHM. This double-lobed far field parallel to the active layer indicates that the emitters oscillate in antiphase with respect to each

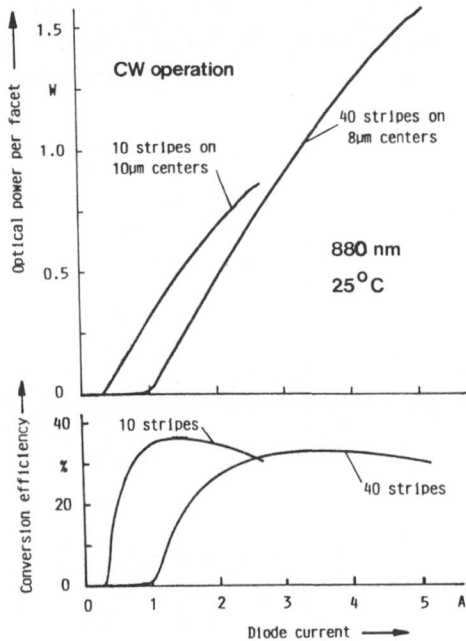

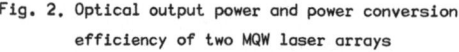

Fig. 2. Optical output power and power conversion
efficiency of two MQW laser arrays

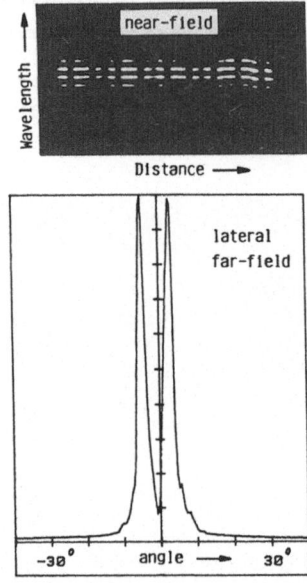

Fig. 3. Spectrally resolved near
field and far field of a
10-stripe MQW array

other, as it is expected for an uniformely spaced array /6/. The far
field in the vertical direction has the usual gaussian shape with a
FWHM of 35°.

For commercial applications long bars of 12-stripe DH arrays with an
emission wavelength ranging from 805 to 880 nm have been processed on
a larger scale. In order to investigate the uniformity of the array
parameters, 10 arrays were tested which have been cleaved from a 20mm
long MO VPE wafer designed for a wavelength of 808 μm (Fig. 4). The
arrays, which are regularly distributed in two groups over a total
distance of 14 mm as indicated by the position axis in Fig. 4, were
tested with 1 μs current pulses up to 2.5 A. All devices operated well
up to an optical output power of a least 0.75 W per mirror and showed
very similar light/current characteristics. This is documented by the
uniform distribution of the threshold current and the differential
efficiency having relative standard deviations as low as 4.8 % and
2.6 % respectively. Thus, an output power of a least 50 W at 1 μs
pulsewidth can be obtained from a 1 cm long bar containing about
70 arrays.

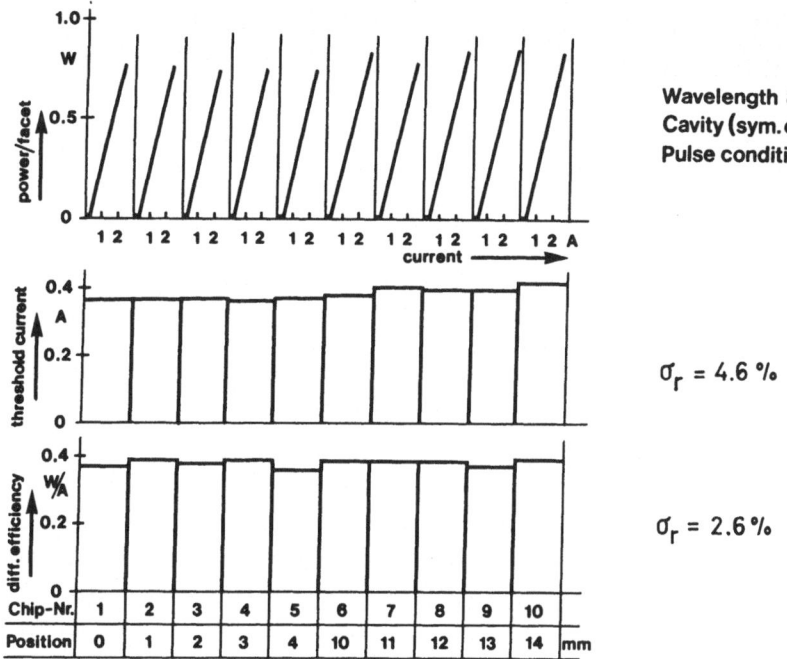

Wavelength 808 nm
Cavity (sym. coating) 300 µm
Pulse conditions 1 µs, 10 kHz

$\sigma_r = 4.6\%$

$\sigma_r = 2.6\%$

Fig. 4 Uniformity of 12-stripe DH laser arrays across a 20 mm MO VPE wafer

4) Conclusions

The presented results show that GaAlAs laser arrays grown by MO VPE
are capable of delivering high optical output power, that is more
than 1 W CW per mirror CW or some 50 W/cm in 1 µs pulses, with an
overall efficiency better than 30 %. Due to the perfect homogeneity
of the MO VPE layer structure, phase locked operation of all lasers
in an array as well as uniform operation of arrays distributed across
a distance of 14 mm has been achieved.

Thus, high power phase-locked laser arrays are highly attractive for
applications such as efficient pumping of Nd:YAG lasers, direct optical triggering of high-power thyristors or frequency doubling in
nonlinear crystals.

References

/1/ D.R. SCIFRES et. al., Appl.Phys.Lett. 42 (8), 645 (1982)
/2/ D.R. SCIFRES et.al., Electronics Lett. 19 (5), 169 (1983)
/3/ F. KAPPELER et. al., 9th IEEE Int. Semicond. Laser Conf.,
    Rio de Janeiro, Abstract of papers, 90 (1984)
/4/ F. KAPPELER et.al., IEE Proc. 129 (6), 256 (1982)
/5/ M. DRUMINSKI et.al., to be published in Jap.J.Appl.Phys.
/6/ S.R.CHINN et.al., IEEE J.Quantum Electr., QE-20 (4), 358 (1984).

Characterization of Diode Laser Mountings by only Electrical Measurements

Dr. Tamás SZIRÁNYI

VIDEOTON Development Institute

Budapest, Vörös Hadsereg utja 54.

H-1021, Hungary

Introduction

Light power and laser characteristics of diode lasers are strongly
depending on the laser chip temperature. The heat produced by current
flow through a diode laser results in a temperature rise in the active
region relative to the heat sink. The maximum temperature rise is
depending on the pulse length at short /shorter than 10 μsec / pulses.
The characteristics and quality of laser chip mounting and case can
be tested by measuring the voltage of diode laser at current pulses
with finite rising and falling time.

Operation characteristics

The threshold current of a diode laser is depending on the pn-junction
temperature in the next way:

$$I_{th}(T) = I_o \cdot \exp(T/T_o) \qquad (1)$$

where T is the absolute Kelvin temperature of the junction, I_o is a
constant, and T_o is a constant referred to as the characteristic
temperature.

The voltage of the laser (V) is depending on the temperature and
the current

$$V = V_{th} + (T-T_e) \cdot b + I \cdot R_s(T) + V_t(T) \qquad (2)$$

Here V_{th} = 1,45... 1,6 $[V]$ is the saturated threshold voltage at the
T_e enviromental temperature $[4]$, b is the temperature dependence of
V_{th} in the range of -1...-3 $[mV/K]$ $[4]$, $I > I_{th}$ is the current
through the pn-junction, R_s is the diode series resistance in the
range of 1 $[ohm]$. R_s is depending on the temperature in a similar
way as V_{th}.

V_t is the resultant thermoelectric voltage. Its temperature dependence is in the range of less than 0,1 $[mV/K]$ when the temperature gradient has risen to the steady-state distribution in the chip. The temperature dependence /the Seebeck coefficient/ of contact potential in the case of nondegenerated GaAs $[2,3]$:

$$S = \pm 0.0862 \cdot \left(C - \ln\left(n/n_i\right)\right) \left[\frac{mV}{K}\right] \tag{3}$$

where n_i is the intrinsic density of electrons, n is the density of the majority charge carrier. At T = 300 $[K]$ we get that C = 29.5 and the sign of S is the sign of the majority charge carrier. When $n_{electrons} > 10^{17} [cm^{-3}]$ or $n_{holes} > 10^{18} [cm^{-3}]$, the GaAs becomes degenerated $[Ref.1.\ p.\ A207]$. If $n = 10^{18}$ $[cm^{-3}]$, the Seebeck coefficients for p type $\left(n_i = 1.8 \times 10^6 [cm^{-3}] [Ref.1.\ p.\ A159]\right)$:

$$S_p = 0.21 \ [mV/K \ ] \tag{4}$$

For any metals or degenerated GaAs the Seebeck coefficients are not greater than some 10 $[\mu V/K]$. For a pn contact the resultant temperature dependence is $S_{pn} = S_p - S_n$.
The contact regions of the laser chip are at different temperatures. The thermoelectric voltage is important only at the boundaries of semiconductor parts. Since about 40 % of temperature drop between the center of the active region and the heat sink occurs within the active region $[Ref.\ 1.\ p.\ B234,\ Ref.\ 7.]$, the resultant thermoelectric voltage

$$V_t = |V_t| < S_{pn} \cdot \frac{T-T_e}{2} \tag{5}$$

So the temperature dependence of V_t is S_{pn} for short $\left(< 0.5 \ \mu sec\right)$ supply current pulses and less than $\frac{S_{pn}}{2}$ in the near steady state.

The pn junction temperature, T is depending on the electrical power V.I $\left(\text{reduced by the } P_{laser} \text{ light power}\right)$ and the R_t thermal resistance of the chip to the environment:

$$T = T_e + R_t \cdot \left(V.I - P_{laser}\right) \tag{6}$$

Here $\quad P_{laser} = V.\left(I - I_{th}\right) \cdot \eta_{\varphi} \tag{7}$

where η_{φ} is the characteristic quantum efficiency.

In the case of a given P_{laser} and given technical data, the numerical value of I can be computed approximately from the (1) (2) (6) (7) equations - if there is any solutions. However, the existance of a solution of the above equations can be determined in a much easier way. Simlifying the equations we get

$$T = T_e + R_t \cdot V_o \cdot I_{th} \tag{8}$$

where V_o is the voltage of the pulsed operation at the start of laser emission $\left(I = I_{th}\right)$.

In the case of average laser diode parameters the error from the simplification is less than 6 % at 100 mA.

(1), (8) equations can be solved for I_{th} in the only case when

$$I_{th}\left(T_e\right) \cdot R_t \cdot V_o \cdot e < T_o \tag{9}$$

where e = 2.718. If the inequality of (9) is not true, there is not a current to get a continuous wave laser light emission.

When a step drive current is applied, the junction temperature reaches the steady-state temperature in some microseconds (Figure 1. [5]).

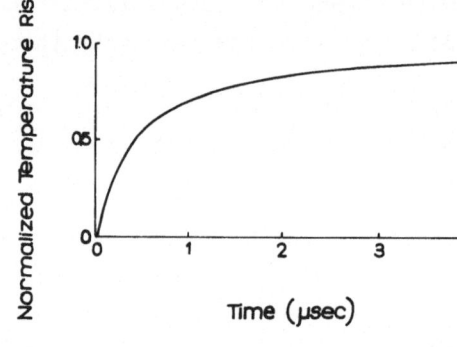

Figure 1.
Temperature variation at the active layer when a step drive current is applied.

This rising time characterizes the heat capacitance in the chip. If a laser diode must not be used in a continuous wave operation because of (9), the maximum usable pulsewidth τ_w can be determined by the help of the measured temperature variations. In this case

$$R_t = R_t\left(\tau_w\right) < R_t\left(\infty\right) \tag{10}$$

can be written into (6), where $R_t\left(\infty\right)$ is the steady-state thermal resistance.

Measuring a laser diode

The usual exact measurements of R_t need sophisticated equipment.

and/or much time (null techniques [7]).

Parameters of a laser diode can be measured quickly and simply by the following method.

At a given current the diode voltage depends on the temperature in the next way [4]:

$$\frac{dV}{dT}\bigg|_I = -1...-4 \quad [mV/K] \tag{11}$$

Tipically this value is -2 [mV/K]. It is nearly the same at a given type, so it should be measured only once for a series. This value can be obtained measuring the diode voltage at changing enviromental temperature.

The measured laser diodes are supplied by current pulses with $\tau_r <$ 0.2 µsec rising and $\tau_f \approx$ 2 µsec falling time. The pulse width $\tau_w >$ 20 µsec, the space factor $\eta <$ 50 %. During τ_r the diode hits its operating parameter values at a somewhat higher temperature than the environmental value. During τ_w the pn-junction temperature rises near to the steady-state value. Screening the diode current vs. voltage function on an x-y oscilloscope (Figure 2.), the curve r represents the rising slope, curve f represents the falling slope.

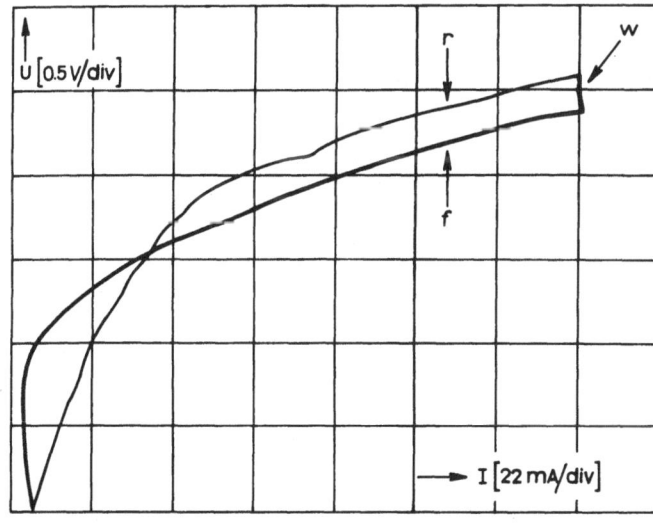

Figure 2.
The I-V function of a laser diode supplied by current pulse.
I_{max}=153 mA, U_{max}=2.5V

Explanation of Figure 2.

At the start of a current pulse through the diode the diode capacitance of about 20 [pF] slows down the voltage rising. After the threshold current the pn-junction voltage is saturated [Ref.1. p B225], so the diode capacitance becomes not too important. From this point the gradient is determined only by the series resistance in the range of

1 [ohm].

During the pulse lifetime the pn-junction is heated up to the steady-
-state temperature, while I is constant, but the voltage is decreasing
with the increasing temperature according to (11). So we get a vertical
line (curve w), with a stronging brilliance.

At the steady-state temperature the falling slope (curve f) is influenced only by the diode capacitance below the threshold current.

The Δ V voltage difference between curve r and curve f at about the
maximum current is directly proportional to the temperature rising in
the chip because of the electrical power.

The electrical power dissipated in the chip can be calculated from the
electrical data: $P_{el} = V.I - P_{laser}$. We get the R_t thermal impedance
from the next form:

$$R_t(\mathcal{T}_w) = \left| \Delta v(\mathcal{T}_w) \middle/ \left(P_{el} \cdot \frac{dV}{dT}\middle|_I \right) \right| \tag{12}$$

$\Delta v(\mathcal{T}_w)$ is increasing with $\mathcal{T}_w$ as it can be seen in Figure 1. for the
temperature. So $R_t(\mathcal{T}_w)$ is depending on the heating period when
$\mathcal{T}_w < 50$ µsec .

Since the junction temperature is changing very fast in the first
microseconds, the $\mathcal{T}_r$ rising time may give an error in the measurement.
Measuring the temperature variation in the first microseconds, the
$T(0$ µsec$)$ can be interpolated for about $\mathcal{T}_r/2$.

Errors of measurement

Seeing (4), (5), (11) the thermoelectric voltage gives 1-10 % error in
the measurement of the total thermal impedance because of the inexact
dV/dT value. The error from the finite $\mathcal{T}_r$ may be reduced by an interpolation written above.

Measuring the laser diode mountings quality

An 0 to 15 % overhang of the diode over the edge of the copper block
could easily result a substantial increase in thermal impedance [8].
Also, the defective facet coatings applied to the diode laser causes
increase in thermal impedance for certain areas that would not have
been wetted by indium solder. Measuring of the thermal impedance of
a diode can be used as a nondestructive technique for investigating
the quality of the bond.

In this measurement a longer $\mathcal{T}_r$ $(0.5$ µsec$)$ pulse rising time and
the thermoelectric voltage do not influence the result of qualifica-
tion because the temperature is changing only in the chip in about

the first half microsecond.

A series of the same type GaAsAl DH laser diodes were measured.
The thermal impedance was $R_t = 71 \pm 2$ K/W for the good quality
bond.
For the badly bounded diodes this value could be 5-10 K/W higher.
The heating-period dependence of R_t for different types of diodes
is in Table 1.

$\dfrac{R_t(\mathcal{T}_w) \cdot 100\%}{R_t(\infty)}$	50	90	95	99
$\mathcal{T}_w$ (µsec)	$2\ ^{+2}_{-0.5}$	8 ± 3	40 ± 10	about 80

Table 1.

$R_t(\mathcal{T}_w)$ can be got measuring the length of curve w in Figure 2. while
$\mathcal{T}_w$ is changing.
This measurement is not disturbed by the noises of the fast current
and voltage pulses.
The measurement outlined above is at least ten times faster than other
known thermal impedance measuring methods and gives a nondestructive
technique for investigating the quality of the bond.

References

[1] H.C. Casey, M.B. Panish: Heterostructure lasers, Academic Press
     /1978/
[2] Szirányi, T.: IC chip layout design in respect of thermal
     influences, M.Sc. thesis, Technical University of Budapest /1982/
[3] Szirányi, T.: Computer aided thermal mapping..., Proc. of the
     int. symp. on el. techn., Budapest /1983/, p. 371
[4] Szirányi, T.: Some questions of thermal stabilizing of injection
     laser diodes, Symposium Optika'84, Proc. SPIE 473, /1984/
     pp. 206-210
[5] Minoru Ito, Tatsuya Kimura: IEEE J. Q.E., VOL. QE-17, N.5. /1981/
     pp. 787-795
[6] Ryoichi Ito, Masua Suyama, Nagaatsu Ogasawara: Appl. Phys. Lett.
     40(3), /1982/, pp. 214-216
[7] J.S. Manning: J. Appl. Phys. 52(5), /1981/, pp. 3179-3184
[8] R.T. Lynch: Appl. Phys. Lett. 36, /1980/ p. 505

The Afterpulse Time Spectra of High-Speed Photon Detectors

Branko Leskovar

Lawrence Berkeley Laboratory

University of California

Berkeley, California 94720 U.S.A.

ABSTRACT

Recent progress of understanding of the afterpulse time spectra of high-speed photon detectors using photoemission and secondary emission processes is reviewed and summarized. Furthermore, the afterpulse time spectra of high-gain conventionally designed and microchannel plate photon detectors has been investigated. Specifically, the devices studied included RCA 8850, RCA 8854 and ITT F 4129f photomultipliers. Descriptions are given of the measuring techniques.

INTRODUCTION

The afterpulse time spectrum of high speed photon detectors has been the subject of intensive experimental and theoretical investigations. Over the years improvements in fabrication and activation techniques have significantly reduced afterpulses from most photomultipliers. However, for some applications afterpulses still may introduce serious error. For example, in photon counting systems for subnanosecond fluorescence lifetime measurements,[1-3] photomultiplier afterpulses can generate small amplitude late artifactual peaks in the sample fluorescence profile. Also, in particle physics measuring systems, which use large numbers of photomultipliers in a 3-dimensional array, a significant amount of afterpulsing could introduce serious error in measuring particle trajectories.[4-6] Similarly, in plasma diagnostic experiments,[7,8] afterpulses from fast photon detectors can seriously distort observed pulse waveshape.

In general, the afterpulses are mostly produced as a result of the ionization of residual gases, such as He^+, H_2^+, N_2^+ and CO^+, in the volume between the photocathode and the first dynode as well as between various dynodes of the electron multiplier. A simplified component arrangement of a high-gain photomultiplier is shown in Fig. 1. The positive ions formed are accelerated toward the photocathode by the focusing electric field. On impact these ions liberate secondary electrons which produce an afterpulse signal. The number of these electrons depends upon the momentum and species of the ions and the photocathode composition. These afterpulses generally occur from 20 ns to several microseconds after the main pulse. The time of occurrence of the afterpulses can be closely correlated with the mass-to-charge ratio of the residual gas inside the glass envelope.

The time interval between the main output pulse and afterpulse, which is approximately equal to the transit time of the positive ion to photocathode is nearly independent of its point of origin. Therefore, the photomultiplier is approx. acting as a time-of-flight mass spectrometer. The peaks in the time spectrum of the afterpulse time distribution, corresponding to a particular ion, generally show some fine structure. The positive ion transit time varies as $(M/Z)^{1/2}$ where M/Z is the mass-to-charge ratio. The mass-to-charge ratio for ions H^+, H_2^+, He^{2+}, He^+, N^+, N_2^+ and CO^+ is 1, 2, 2, 4, 14, 28 and 28, respectively. By neglecting the initial thermal velocity and the transit time of a photoelectron from the photocathode to the point of ionization, the following equation for the time-of-flight of the colliding ion can be obtained[10]

$$t_{ion} = 7.2 \times 10^{-5}(M/Z)^{1/2} \int_{s_0}^{0} [V(s_0) - V(s)]^{-1/2} ds \qquad (1)$$

where the integration of the electric potential function is over the path from s_0 (ion formation point) to the photocathode. $V(s)$ is the potential in volts and s is the distance in meters.

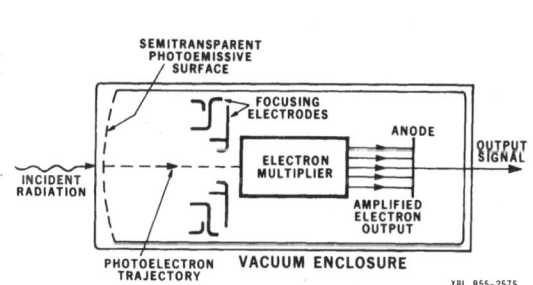

Fig. 1. Simplified component arrangement of a high-gain photomultiplier.

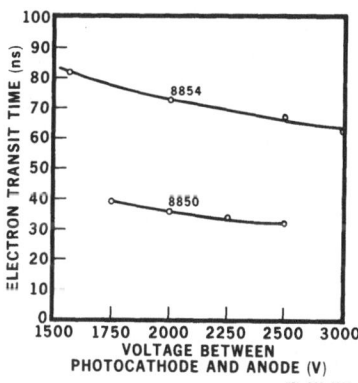

Fig. 2. Electron transit time as a function of the voltage between the photocathode and anode.

The afterpulse phenomena was studied systematically by several authors[9] who introduced trace amounts of various gasses into photomultipliers. Subsequently, in further works,[10-17] the physical origins of afterpulses were investigated, particularly with respect to the afterpulses which result from the diffusion of helium through the photomultiplier glass envelope.[18,19] Although the helium is present in small concentration in ambient air, it is sufficient so that its atoms, which can permeate readily through the glass envelope, can cause afterpulses. This effect may be enhanced often by the fact that photomultipliers are frequently used in laboratories where ambient concentration of helium is significantly increased by emissions from helium Dewar bottles or from gas Cherenkov counters. Other phenomena may also cause afterpulsing, such as dynode fluorescence, electrical fields over the exposed glass of the envelope, etc. In order to measure this characteristic, a signal induced time spectrum must be taken of the anode output pulses.

In this paper, an effort has been made to investigate and review the afterpulse time spectra of some new generation commercially available photomultipliers in order to further our understanding of afterpulsing. Other characteristics of the photomultipliers, generally not available from manufacturers, have been measured and previously presented.[20-22] Specifically, time spectra have been studied of the new generation RCA 8850 and 8854 photomultipliers, which use conventional multiplier structures. Furthermore, the spectra have been investigated on a new generation of ITT 4129f extended life microchannel plate photomultipliers.

The RCA 8850 2" 12-stage photomultiplier is a modified version of the RCA 8575. Both have identical electron multiplier structures, electron optics and semitransparent cesium-potassium-antimony photocathodes deposited on a pyrex entrance window. The only significant difference between 8850 and 8575 is that 8850 has a cesium-activated gallium-phosphide secondary emitting surface on the first dynode. This surface has a secondary emission ratio of 30-50 instead of the 5-8 ratio of the conventional dynode material. The high pulse-height resolution of the 8850 makes it possible to set the threshold of the signal discriminator in the valley between the single and double photoelectron peaks, eliminating most of the single photoelectron pulses which are mostly thermionic initiated from the photocathode. The anode output pulse (10-90%) rise time is 2.4 ns at an applied voltage between the anode and cathode of 2500 V.

The RCA 8854 photomultiplier is a variant of the RCA 4522. It has a high gain GaP (Cs) first dynode followed by 13 BeO dynodes. The new RCA designation for the photocathode is 35 ET (formerly 118), which has a peak response at 400 nm and a quantum efficiency of 27%. Its spectral response extends from 200 nm to 600 nm. The maximum useful photocathode diameter is 114 mm; the anode output pulse (10-90%) rise time is approx. 3.2 ns at the supply voltage between the anode and cathode of 2500 V. This photomultiplier is designed for experimental research instrumentation where good pulse-height resolution and large photocathode areas are important.

The electron transit time as a function of voltage between the photocathode and anode was measured for both photomultipliers. The reference pulse was the electrical signal from a pulsed mercury light source. The transit time was measured from 50% of the leading-edge amplitude of the reference pulse to 50% of the leading-edge of the photomultiplier output pulse. The results are shown in Fig. 2; the electron transit time was 32 ns and 67 ns at 2500 V for 8850 and 8854, respectively.

The new ITT F 4129f photomultiplier has an S-20 photocathode with a maximum usable diameter of 18 mm and 3 microchannel plates in cascade for the electron multiplication. The plates are in a Z-configuration to reduce the positive ion feed-back. The 3 plates are identical, having 12 μm diameter channels with length to diameter ratios of 40. Proximity focusing is used at the input and collector. The anode output pulse (10-90%) rise time is approx. 350 ps using microchannel plate voltage of 2500 V. In the ITT F 4129f device a protective film is provided between the photocathode and the microchannel plate which leads to a significant improve-ment in quantum efficiency stability and life expectancy.

DESCRIPTION OF THE MEASURING SYSTEM

The system described in Ref. 12 was used to measure afterpulse time spectra. A pulse generator was used to drive a light-emitting diode, type XP 21, which pro-duced light pulses for the photomultiplier. The trigger pulse from the pulse gen-erator was delayed and shaped and then used as a start pulse for the time-to-ampli-tude converter. The output pulse of the photomultiplier was used as the stop pulse for the converter after being processed by a constant fraction discriminator. In order to count the afterpulses which came immediately after the main photomultipli-er pulse, the main output pulse was delayed after the start pulse. The time at which the main photomultiplier pulse occurred was taken as time zero. However, in order to count the afterpulses which occurred significantly later in time and with a very low count rate, the trigger pulse from the pulse generator was purposely de-layed to come after the photomultiplier main pulse so that an output pulse from the converter would only occur when the photomultiplier generated an afterpulse. The output of the converter was then recorded and displayed on a pulse-height analyzer. In order to look for pulses many microseconds after the main pulse (the timing range of the converter being set accordingly) the operating frequency of the test system was quite low.

RESULTS AND DISCUSSIONS

The typical results of the measurements of afterpulse time spectra made on three RCA 8850 and 8854 photomultipliers are summarized in Table 1. Measurements were made using the voltage divider network suggested by RCA for fast pulse re-sponse. With 2500 V applied between the anode and cathode, the average gain was approx. 1.8×10^8, while the dark current was 4.6×10^{-8} A. With full photocathode illumination, under single photoelectron counting conditions at a rate of 100 kHz, afterpulses were detected in the typical time interval ranges of 18-22 ns and 342-415 ns, after the main output pulse for the RCA 8850.

When the light pulse intensity was increased so that pulses with three photoe-lectrons were produced by the 8850's, afterpulses were detected in the 19-22 ns and 342-413 ns time intervals, as before, in all three photomultipliers. Detailed results are given in Table 1. No afterpulse was observed beyond this range up to 68 µs under either the single or three photoelectron operating conditions. After-pulses in the 342–415 ns range had a peak amplitude of approx. 2–3 times that of the main pulse. The photomultipliers tested were 10 years old. During this time they had been used mostly for calibration purposes of various single photon count-ing systems and high-gain photon detectors under control conditions in a standard laboratory environment.

Figure 3 shows the time distribution of output pulses on a logarithmic scale in the time interval 0–150 ns after the main photomultiplier pulse under 100 kHz sin-gle photoelectron counting rate. The first distribution at the beginning of the spectrum is the main output pulse of the photomultiplier. The second distribution represents the afterpulses which occurred 18 ns after the main pulse.

Figure 4 shows the same afterpulse spectrum on a linear time scale over the range 60–680 ns under 100 kHz single photoelectron counting rate. Afterpulses were also present from 342–415 ns after the main photomultiplier output pulse.

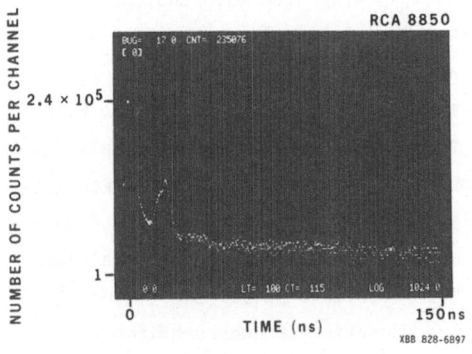
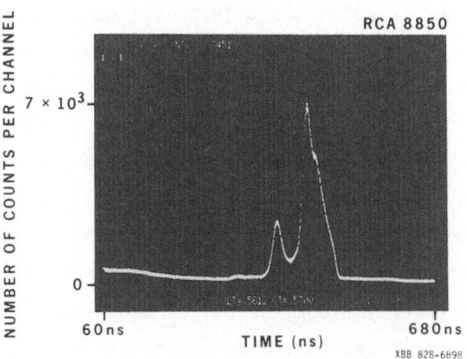

Fig.3. Time spectrum of anode output pulse between 0–150 ns for RCA 8850 with full photocathode illumination.

Fig. 4. Time spectrum of anode output pulse between 60–680 ns for RCA 8850 with full photocathode illumination.

The two groups of afterpulses occurring in the 342–415 ns range were attributed to He^+ ions which are created in the space between the photocathode and the first dynode and the second and third dynode. This conclusion is based on the time-of-flight of the helium ion from its point of origin to the photocathode and the ion mass-to-charge ratio. The two peak structure is clearly seen. The first peak of this distribution is caused by afterpulses generated by He^+ ions from the photo-cathode-first dynode region. The second peak of the distribution was studied by increasing the potential of each dynode by approx. 20 V. The effect of the voltage increase on afterpulse time distribution was observed. The time distribution was particularly sensitive to variations in the potential of the third dynode. This further substantiated the conclusion that the second peak is produced by afterpulses generated by ions which are formed between the second and third dynode. This conclusion is in agreement with the results of afterpulse measurements obtained by Coates[7] for 8852 photomultiplier. However, no report has been made in literature on afterpulses which occurred in the 18–22 ns range after the main signal pulse. These pulses appeared at a lower rate but were consistently present in all three 8850's. Furthermore, the amplitude of these afterpulses was approx. 1/2 of the single photoelectron pulse amplitude, unlike those at the 342–415 ns time range which were 3 to 4 times larger.

The afterpulses in the 18–22 ns range were the most probably generated by He^+ ions, produced between the first and second dynode, striking the first dynode and causing the secondary emission. This conclusion is based on the calculation of electron transit time between the photocathode and the first dynode and the photo-multiplier anode. Also the electric field distribution was taken into account. The 18–22 ns time range increased when the voltage between the photocathode and the first dynode/focusing electrode was decreased because of the change in the electric field intensity and distribution.

Afterpulse time spectra measurements were made on RCA 8854 photomultiplier us-ing the voltage divider network described in Ref. 12. With 2500 V applied between anode and cathode, the gain was 3.5×10^8, while the dark current was 1.3×10^{-7} A. These photomultipliers were new when tested.

Under single photoelectron counting at a rate of 100 kHz, afterpulses were de-tected in the typical time interval ranges of 52–58.5 ns and 190–198 ns, after the main output pulse in all three 8854's. In the 450 ns–68 μs time range, afterpulses were detected in two out of three 8854's. In one photomultiplier at a 10 kHz pulse rate, 0.8 afterpulses/s were observed at 933 ns and the other at 1 kHz pulse rate gave 3.4 afterpulses/s at 12.5 μs. When the light pulse intensity was increased so that pulses with three photoelectrons were produced by the 8854's, afterpulses were detected in four time intervals in all three photomultipliers: 54.5–59 ns, 197–198 ns, 956–987 ns and 10.6–12.8 μs. Typical results are given in Table 1.

Figure 5 shows the time distribution of output pulses in the time interval 0–150 ns after the main photomultiplier pulse under a 100 kHz single photoelectron counting rate. The first distribution at the beginning of the spectrum is the main output pulse of the photomultiplier. The second distribution represents the after-pulses which occurred in the 52–58.5 ns time interval after the main pulse.

Figure 6 shows the afterpulse spectrum in the time interval 4.5–68 μs under 10 kHz–three photoelectron counting rate (limited by the time-to-amplitude convert-er timing range setting). Afterpulses were present in all three photomultipliers at 8.5 μs, 9.2 μs 12.8 μs and 14 μs after the main photomultiplier output pulse.

154

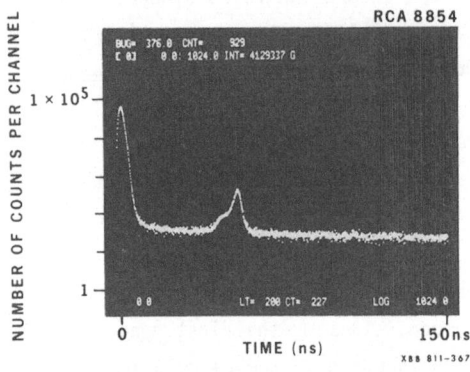

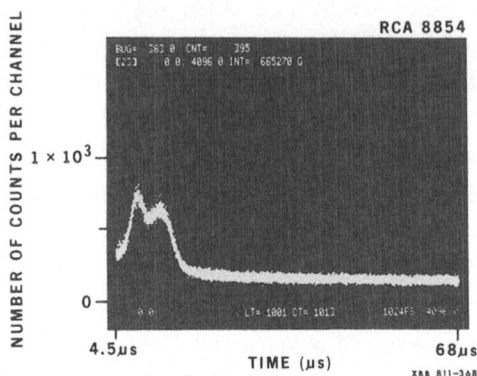

Fig. 5. Time spectrum of anode output pulse between 0–150 ns for RCA 8854 with full photocathode illumination.

Fig. 6. Time spectrum of anode output pulse between 4.5–68 μs for RCA 8854 with full photocathode illumination.

Based on considerations similar to those given above, afterpulses in the 956–987 ns range were attributed to He[+] ions which are created in the space between the photocathode and the first dynode. This conclusion is in agreement with the results of afterpulse measurement obtained by Bartlett et al[19] for a similarly designed photomultiplier 4522. However, no report has been made in literature on afterpulses which occurred in the 52–58.5 ns and 190–198 ns ranges after the main pulse. These pulses appeared at a higher rate. They were consistently present in all three 8854's. The afterpulses in the 52–58.5 ns range were generated by He[+] ions, in existence between the second and third dynode, striking the second dynode and causing the secondary emission. The afterpulses in the 190–198 ns range were caused by He[+] ions between the first and the second dynode, causing the secondary emission from the first dynode.

Afterpulse time spectra measurements were made on ITT 4129f using the microchannel plate voltage of 2500 V. The photomultiplier gain was 1.6×10^6. This new generation device has a 7 nm thick ion barrier film between the photocathode and the input face of the first microchannel plate to prevent the bombardment of the photocathode by positive ions. Introduction of the film has resulted in a significant improvement in the photocathode quantum efficiency stability and life expectancy. Furthermore, extensive studies of the performance characteristics of this device have shown that the introduction of the film has resulted in total elimination of the afterpulses. Similarly, no afterpulses were detected in the Hamamatsu R1564U microchannel plate photomultipliers with protective film. In this device the entrance part of the first microchannel plate is covered with a thin aluminum film. In contrast, all microchannel plate photomultipliers without protective film made by various manufacturers have demonstrated the strong afterpulsing phenomena and the photocathode quantum efficiency decrease during operating time.[7,8,11,18]

Afterpulses were detected in the 50 ns, 200 ns, 250 ns and 320 ns time regions. They were identified on the basis of the ion mass-to-charge ratio with the following species: H_2^+, H_2O^+, $N_2^+ + CO^+$ and CO_2^+. Furthermore, it was also shown that the afterpulse appearance probability increases proportionally with the number of photoelectrons in the pulse and the device gain. The appearance probability has shown a strong dependence on the residual gas pressure inside the vacuum envelope and the outgassing status of the microchannel plate.[11] The only disadvantage of the usage of the thin protective film between the photocathode and the input face microchannel plate is a reduction of photoelectron collection efficiency of the device input electron optics. According to preliminary calculations the collection efficiency can be reduced by approx. 20-30% depending on the particular photomultiplier and its operating conditions.

At present, physical processes responsible for the existance of the afterpulses in the 8.5-14 μs time range are not well understood. The very late appearance of afterpulses after the main output pulse eliminates a possibility of the optical feedback from the electroluminescence of the dynodes to the photocathode and the afterpulses generation at the photocathode or dynodes by positive ions. This conclusion is based on calculations of electron and positive ion transit times between the photocathode, the first, second and third dynode and the photomultiplier anode, taking into account the electric field distribution and the mass-to-charge ratio of various single ions and ionized molecules as well as the device geometry. The very late afterpulses appearing in the 5 μs and 40 μs time range after the main pulse were also reported in Ref. 23 in more recently manufactured 8850's. However, they were not observed in the same photomultipliers produced 10 years ago.[15] More experimental work is required before the nature of these sometimes observed, very late afterpulses can be determined.

In addition to the He^+ ions which are mainly responsible for the afterpulses, the H^+ and H_2^+ ions were also identified in some conventionally designed and microchannel plate photomultipliers[10,11,13] which were manufactured at a particular time. However, there is disagreement in literature concerning the afterpulses generated by oxygen positive ions. Results reported in Ref. 9 strongly suggest that no afterpulses are produced by oxygen. This conclusion is based on an experiment where small quantities of various gasses were introduced into the demountable photomultiplier to determine unambiguously the associated afterpulses. In this experiment afterpulses were produced and identified with various introduced gasses (Xe^+, A^+, N_2^+ and H_2^+) but not with oxygen. This result is in contrast with experimental data reported in Ref. 10 where presumably the O^+ ion afterpulses were observed in the 747-900 ns time range for the sealed 8852 photomultiplier. Later, results were reported in Ref. 13 which indicated difficulties to localize the time position of O^+ and O^{2+} afterpulses as well as to obtain agreement between theoretical considerations and experimental data. In our measurements,[12,15] we have not detected afterpulses which can be identified with oxygen ions. In general, the possibility of detecting the oxygen related afterpulses is extremely small because the oxygen molecules inside the photomultiplier envelope are immediately absorbed by the alkali metals of the photocathode or possible getter material. Consequently, afterpulses previously assigned to oxygen ions may be due to other gaseous species such as OH.

The experimental evidence mentioned above and in cited references clearly shows that photomultipliers have a long usable lifetime without serious increases in afterpulses in a typical laboratory environment. Necessary corrections should be made, however, in single photon counting systems to account for the existence of afterpulses when accurate results from the fluorescence lifetime measurements are required. This is particularly important in experimental situations where lifetime measurements are made over a dynamic range of several decades. Under some experimental conditions where ambient concentration of helium is artificially elevated by the emission from helium Dewar bottles or gas Cherenkov counters, the probability of generating afterpulses increases significantly.[19] In this case, the photocathode quantum efficiency decreases in a relatively short time period because the photocathode is continuously bombarded by highly concentrated helium positive ions. The effect of afterpulses can be reduced by introducing a dead time in the measuring system. In such cases, the deadtime should be adjusted to a value which will not compromise the system measuring accuracy.[24]

Table 1. Afterpulse Performance of 8850 and 8854 Photomultipliers

		8850 Photomultiplier			8854 Photomultiplier		
	Measurement Time Interval	0-680ns	450ns-7µs	4.5µs-68µs	0-680ns	450ns-7µs	4.5µs-68µs
Afterpulse Count Rate (cps)	Single Photoelectron Pulse Rate = 100kHz	171 at 18ns 155 at 351ns 310 at 415ns	b	b	106 at 58.5ns 5.5 at 190ns	b	b
	Single Photoelectron Pulse Rate = 10kHz	b	a	b	b	0.8 at 933ns	b
	Single Photoelectron Pulse Rate = 1kHz	b		a	b		a
	Three Photoelectron Pulse Rate = 10kHz	5.8 at 19ns 181 at 350ns 362 at 413ns	a	a	10.4 at 59ns 12 at 198ns	2 at 956ns	9 at 9.2µs 14 at 14µs

a - Afterpulses were not observed.

b - Measurement was not made due to a measuring system limitation.

ACKNOWLEDGMENTS

This work was performed as part of the program of the Electronics Research and Development Group of the Department of Instrument Science and Engineering of Lawrence Berkeley Laboratory, and was partially supported by the U.S. Department of Energy under Contract No. DE-AC03-76SF00098. The author would like to express his appreciation to Dr. George Morton for his clarifying comments.

Reference to a company or product name does not imply approval or recommendation of the product by the University of California or the U. S. Department of Energy to the exclusion of others that may be suitable.

REFERENCES

(1) B. LESKOVAR, C.C. LO, P.R. HARTING, K.H. SAUER: Rev. Sci. Instr. 47, 1113 (1976)
(2) P.R. HARTING, K.H. SAUER, C.C. LO, B. LESKOVAR: Rev. Sci. Instr. 47, 1122 (1976)
(3) B. LESKOVAR: IEEE Trans. Nucl. Sci. NS-32, 1232 (1985)
(4) B. LESKOVAR: Proc. of the 1980 Deep Underwater Muon and Neutrino Detection Signal Processing Workshop, A. Roberts, Ed., 21 (U. of Hawaii, Honolulu, 1980)
(5) A.G. WRIGHT: Ibid, 45
(6) C. CORY, D. SMITH, C. WUEST, J.G. LEARNED, H.A.W. TOTHILL, J. WARDLEY, A.G. WRIGHT: IEEE Trans. Nucl. Sci. NS-28, 445 (1981)
(7) L.P. HOCKER, P.A. ZAGARINO, J. MADRID, D. SIMONS, B. DARIS, P.B. LYONS: IEEE Trans. Nucl. Sci. NS-26, 356 (1979)
(8) P.R. LYONS, L.D. LOONEY, J. OGLE, R.D. SIMMONS, R. SELK, B. HOPKINS, L. HOCKER, M. NELSON, P. ZAGARINO: Los Alamos Scientific Lab. Report No. LA-UR-81-1028 (1981)
(9) G.A. MORTON, H.M. SMITH, R. WASSERMAN: IEEE Trans. Nucl. Sci. NS-14, 443 (1967)
(10) P.B. COATES: J. Phys. D: Appl. Phys. 6, 1159 (1973)
(11) K. OBA, P. REHAK: IEEE Trans. Nucl. Sci. NS-28, 683 (1981)
(12) C.C. LO, B. LESKOVAR: IEEE Trans. Nucl. Sci. NS-29, 184 (1982)
(13) S. TORRE, T. ANTONIOLI, P. BENETTI: Rev. Sci. Instr. 54, 1777 (1983)
(14) A.G. WRIGHT: J. Phys. E: Sci. Instr. 16, 300 (1983)
(15) C.C. LO, B. LESKOVAR: IEEE Trans. Nucl. Sci. NS-30, 445 (1983)
(16) C.C. LO, B. LESKOVAR: IEEE Trans. Nucl. Sci. NS-31, 413 (1984)
(17) C.C. LO, B. LESKOVAR: IEEE Trans. Nucl. Sci. NS-32, 360 (1985)
(18) W.C. PASKE: Rev. Sci. Instr. 45, 1001 (1974)
(19) D.F. BARLLETT, A.L. DUNCAN, J.R. ELLIOTT: Rev. Sci. Instr. 52 265 (1981)
(20) B. LESKOVAR, C.C. LO: IEEE Trans. Nucl. Sci. NS-19, 58 (1972)
(21) B. LESKOVAR: Proc. of the 5th International Congress, Laser ℓ²-Optoelectronics 381 (Munich, West Germany, 1982)
(22) B. LESKOVAR: Proc. of the 6th International Congress, Laser 83-Optoelectronics 68 (Munich, West Germany, 1984)
(23) M. YAMASHITA, S. TAKEUCHI: IEEE Trans. Nucl. Sci. NS-31, 438 (1984)
(24) B.H. CANDY: Rev. Sci. Instr. 56, 183 (1985)

Pyroelektrische Detektoren auf der Basis von PVDF als Sensoren für die Lasertechnik

W. Bohmeyer, H. Gündel, W. Kabel, H. Volkmann; ZIE, Berlin / DDR
R. Danz, B. Elling, W. Stark; IPOC, Teltow-Seehof / DDR

Seit einigen Jahren haben die pyroelektrischen Detektoren einen festen Platz unter den Sensoren für die Lasertechnik. Diese Detektoren gehören zu den thermischen Detektoren und besitzen einige Vorteile gegenüber den Quantendetektoren. Das Wirkprinzip dieser Sensoren besteht darin, daß die einfallende elektromagnetische Welle zu einer Erwärmung des aktiven Sensormaterials führt und daß diese Erwärmung über eine Änderung der Polarisation eine Oberflächenladung bewirkt. Zwei wesentliche Eigenschaften derartiger Sensoren sind zu nennen:
1. Aufgrund der Umwandlung der Strahlungsenergie in Wärmeenergie arbeiten diese Detektoren prinzipiell wellenlängenunabhängig.
2. Derartige Detektoren sind nur für den Nachweis von modulierter oder Impulsstrahlung geeignet (nur für Temperaturänderungen).
In Abhängigkeit vom Lastwiderstand sind zwei Betriebsweisen der Detektoren möglich, wobei die konstruktive Ausführung des Sensors der jeweiligen Betriebsweise angepaßt sein sollte (1). Bei einem hochohmigen Arbeitswiderstand können die erzeugten Ladungen während der Impulsdauer nicht abfließen, so daß die am Wandler zu messende Maximalspannung proportional der Temperatur des Wandlerelements, d. h. der Strahlungsenergie ist, wenn die Bedingung $t_r \ll \tau_{th}$ erfüllt ist (τ_{th} - thermische Zeitkonstante des Sensors, t_r - Impulsdauer). Bei geringem Arbeitswiderstand führt die erzeugte Ladung momentan zu einem Strom im Wandlerelement. Dieser Strom bzw. die Spannung am Lastwiderstand ist zur Leistung des Strahlungsimpulses proportional (siehe Abb. 1).

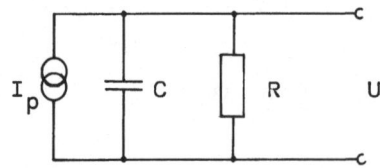

t_r - Impulsdauer
τ_{th} - thermische Zeitkonstante

Bedingung: $t_r < \tau_{th}$

Für $R \cdot C \ll t_r$ gilt $U \sim W_L$; W_L - Leistung des Strahlungsimpulses
für $R \cdot C \ll t_r$ gilt $U \sim E_L$. E_L - Impulsenergie

Abb. 1. Ersatzschaltung für einen pyroelektrischen Detektor

Neben kristallinen Materialien (z. B. TGS oder ferroelektrische Keramiken wie PZ) haben in den letzten Jahren die pyroelektrischen Po-
lymerfolien - hier insbesondere PVDF - große Bedeutung erlangt (2, 3).
Dieses Material liegt als dünne Folie vor und hat nach entsprechender
Formierung eine relativ hohe pyroelektrische Aktivität. Die etwa 5 bis
30 μm dicke Folie bietet den Vorteil der Großflächigkeit, und aufgrund
der Feuchteunempfindlichkeit ist die Herstellung fensterloser Detektoren möglich. Weiterhin kann die leicht formbare Folie der speziellen
Form des Sensors angepaßt werden.
Nachfolgend werden zwei Detektorkonzepte veranschaulicht:

<u>Zweischichtmodell</u>   <u>Dreischichtmodell</u>

| PVDF | Wärmesenke |

| Absorp-tions-schicht | PVDF | Wärmesenke |

Typische Anwendungsfälle für unterschiedliche Detektoren nach dem
Dreischichtmodell sind in den nachfolgenden Abbildungen dargestellt.
Abb. 2 zeigt die Energiemessung an einem N_2-Laser geringer Impulsenergie ($\sim$400 μJ) und hoher Wiederholfrequenz mit einem Empfänger mit
Schwarzschicht. In Abb. 3 wird ein typischer Fall der zeitaufgelösten
Energiemessung bei hoher Wiederholfrequenz und hohen Impulsenergien
(metallische Absorptionsschicht) gezeigt. Dargestellt ist die Impulsenergie eines CO_2-TEA-Lasers, wobei die Laserstrahlung unabgeschwächt
auf den Energiemesser trifft.

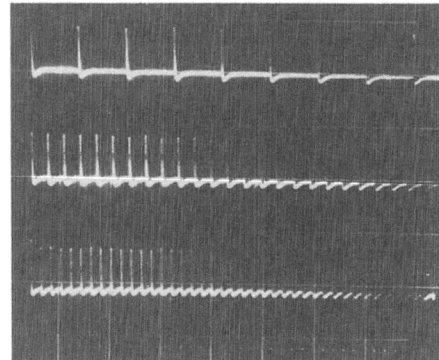
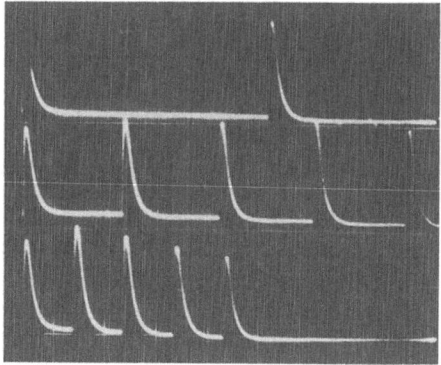

Abb. 2. Energiemessung an einem N_2-
Laser bei drei unterschiedlichen
Wiederholfrequenzen, Zeitablenkung:
50 ms/Einheit, vertikal 0,5 mJ/Ein-
heit; Dauer des Laserimpulses 3 ns

Abb. 3. Energiemessung an einem CO_2-
TEA-Laser bei drei unterschiedlichen Wiederholfrequenzen. Zeitablenkung: 10 ms/Einheit, vertikal
0,5 J/Einheit; Dauer des Laserimpulses 1,5 μs

160

Abb. 4 verdeutlicht die Parameter unterschiedlicher Energiemeßgeräte
im Zeit-Leistungs-Diagramm und deren Anwendungsbereich, wobei zusätzlich einige typische Daten für Impulslaser eingetragen sind. Es
sind die unteren Meßgrenzen (Rauschäquivalentenergie NEJ) für zwei
Empfängertypen und die oberen Meßgrenzen (Leistungsdichte, Energiedichte) sowie die thermischen Zeitkonstanten angegeben.
Dieses Bild liefert noch keine Aussagen über die zulässigen mittleren Leistungen, über mögliche Wiederholfrequenzen und über Anstiegs-
zeiten des Meßsignals.

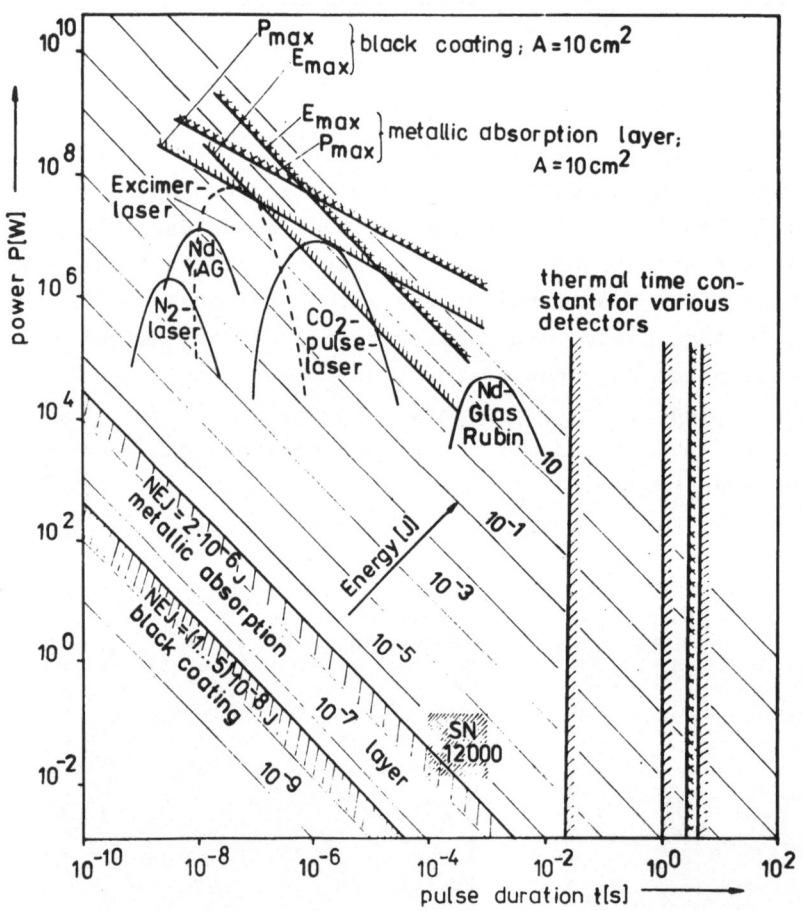

Abb. 4. Leistungs-Zeit-Diagramm zur Verdeutlichung des Einsatzberei-
ches und der Grenzen unterschiedlicher pyroelektrischer Energiemeßgeräte

Die Abb. 5 zeigt den relativen Anzeigewert für Energiemeßgeräte mit

Schwarzschicht und metallischer Absorptionsschicht in Abhängigkeit
von der Leistungsdichte für zwei Impulslängen. Die unterschiedlichen
maximalen Leistungsdichten sind deutlich zu erkennen.
Ein Sensor mit metallischer Absorptionsschicht kann auch zur Messung
der Energieeinkopplung in einen Festkörper in Abhängigkeit von der
Leistungsdichte verwendet werden.

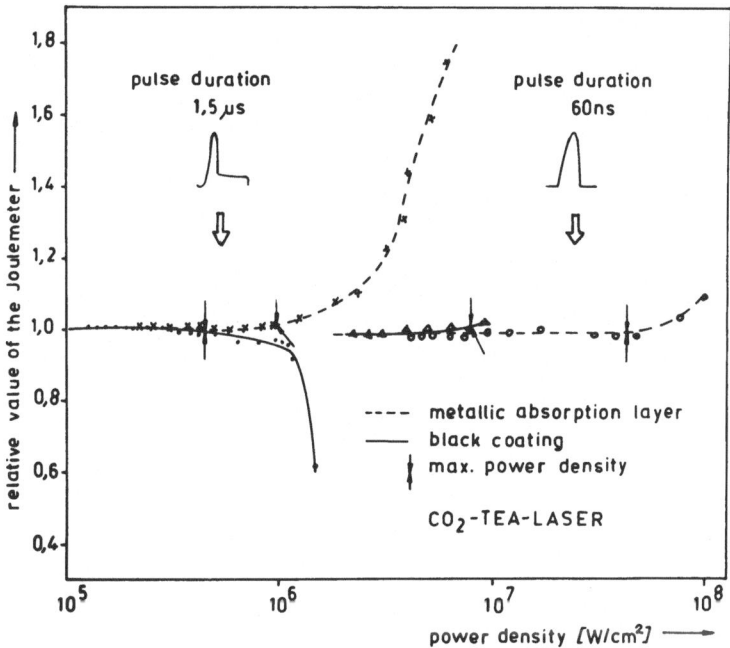

Abb. 5. Relativer Energiewert in Abhängigkeit von der Leistungsdichte
für zwei unterschiedliche Impulsdauern

Die hier vorgestellten Meßmittel besitzen einen Vorteil gegenüber
anderen Joulemetern, denn es ist möglich, auch bei relativ hohen
Wiederholfrequenzen die Energiemessung direkt mit einem Oszillo-
grafen oder einer elektronischen Anzeigeeinheit vorzunehmen. Durch
geeignete Wahl der thermischen und der elektrischen Zeitkonstante
können die Sensoren weitgehend an die jeweilige Meßaufgabe angepaßt
werden.

Literatur

(1) H. GÜNDEL u. a.: Laser 83 Optoelectronic, S. 90 (1984) Springer-
    verlag
(2) A. HADNI: J. Phys. E 14, 1233 (1981)
(3) H. LAI JULY u. a.: Solid State Technology, S. 165 (1984)

Laser- und optoelektronische Meßtechnik
Lasers and Optoelectronics in Measuring

Plant Evaluation of the Laser System for Paper Surface Measurement

P.Richter, E.Lőrincz, F.Engard, I.Péczeli
Technical University Budapest
1521 Budapest, Budafoki út 8, Hungary

Surface roughness is of great importance concerning printability of paper. A coherent optical method has been developed to measure this parameter on line. The basic ideas behind the measuring method and development of the instrument into its present, industrial version phase are shown.

Surface roughness of paper is one of the most important parameters that determine its applicability. Its measurement and control is therefore of decisive importance.

Traditional ways to measure surface roughness of paper; the Bendtsen, Beck, Parker, Printsurf, FOGRA KAM /1-4/ methods all require sample taking therefore are not suitable for continuous monitoring that is however highly desirable for process controll and feedback. For this purpose recently new optical measuring systems have been developed /5,6/. A coherent optical method to measure surface characteristics of rolling material, e.g. paper had been developed and reported /7/ by us.

After laboratory experiments an experimental system for paper surface measurement was tested at the pilot plant of the Finnich Pulp and Paper Research Institute.

In cooperation with the Research Institute of the Hungarian Paper Industry a prototype industrial system was built and is now under test on the supercalander of the Füzfő Paper Mill.

The theoretical principle of the measurement has been described in detail earlier /7,8/. The moving paper surface is illuminated by a focused laser beam and the backscattered light is detected by optical heterodyne technique. Variations in the backscattered light intensity are measured observing different parameters of the fluctuating signal. It was shown both experimentally and theoretically /7,8/ that a "correlation length" deduced from the signal fluctuations gives good indication of the roughness of the surface.

Setup of the instrument is shown diagrammatically in Fig 1, and the experimental istrument mounted on the pop-roll of the Finnish pilot calander in Fig.2.

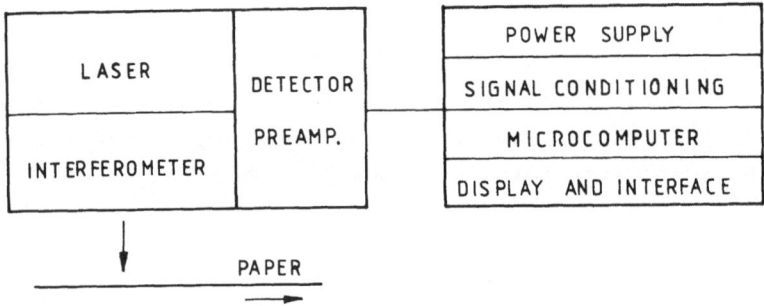

Fig.1. Block diagram of the instrument

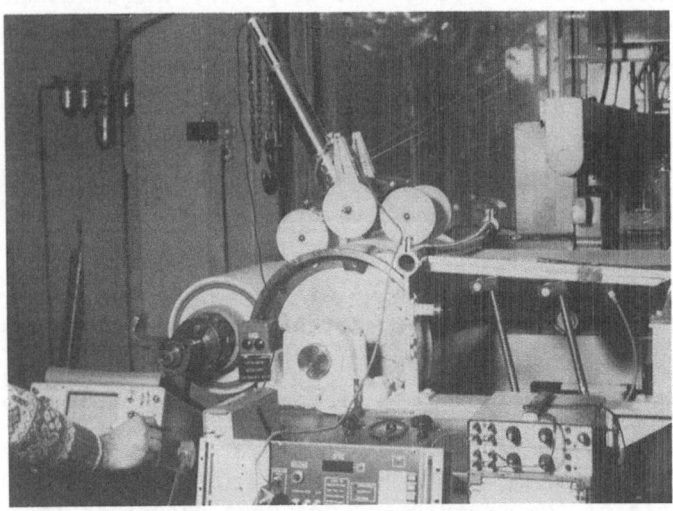

Fig.2. The experimental instrument

The experimental instrument followed very well variations in surface roughness induced by changing nip pressure.
Reproducibility was good and subsequent measurements on samples taken from different parts of the rolls showed good correlation between our method and the traditional /1-4/ measurements.

During design of the prototype instrument for the paper mill problems of heavy vibrations large electric disturbances had to be faced. Positioning the measuring head was also not without problems. The solution is shown in Fig.3.

Fig.3. Industrial paper surfaoc roughnese measuring instrument

Literature
/1/ DIN 53lo8
/2/ DIN 53lo7
/3/ J.R.PARKER: Tappi 64, 56 /1981/
/4/ M.BRUNE, K.HALLER: Papierfabrikat 96, 731 /1968/
/5/ S.SMIDT: Paper 17, 24 /1982/
/6/ J.LUCAS, S.GRACOVETSKY, J.MARCUS: Pulp and Paper Canada, 79,T164
                                                                1978
/7/ P.Richter, E.Lőrincz, I.Péczeli, F.Engard: Proc.Spie 398,258/1983/
/8/ E.Lőrincz, P.Richter, I.Péczeli, F.Engard: Proc.SPIE 473,142/1984/

Ein neues Konzept für ein hochgenaues und vielseitiges Laserdistanzmeter und Optisches Radar

A New Concept for a Precise and Versatile Laser Range Finder and Optical Radar

R. Schwarte, V. Baumgarten, B. Bundschuh, R. Dänel, W. Graf, K. Hartmann, F. Heuten, O. Loffeld.

Institut für Nachrichtenverarbeitung, Universität-GH Siegen

Abstract

At present strong efforts and high investments can be observed, trying to find better ways from the blind to the seeing robot. Fast non tactile distance measuring capabilities are required e.g. in industrial production scenes (robot vision and control, recognition and handling of 3D-objects etc.) and in space applications (automatic rendezvous and docking maneuver).
Today available sensors using ultrasonics, microwaves, triangulation and mono or stereo CCD-cameras are appropriate to distinct problems but lack of versatility.
Basing on the time-of-flight evaluation of reflected laser pulses we have investigated a new distance measuring concept, finding promising features: High resolution (mm), high angular resolution (some mrad), a measuring range from some cm to about 100 m to passive objects or up to 10 km with retroreflectors, a special time gate and amplitude gate for target selection, self-test and self-calibration facilities due to an internal reference fiber, a precise range and velocity evaluation using Kalman filtering and last not least remote sensing capabilities through 10 to 100 m of glass fiber. A small optical sensor is connected to the measuring system only by two fibers and can be mounted - save from electromagnetic interference - e.g. in a robots hand, enabling the robot to see its surroundings.
The most critical problem to be solved is to attain an accuracy in the millimeter range corresponding to a time accuracy in the 10 picosecond range.
Important aspects of the new concept and first experimental results are given.

1. Das optische Konzept des neuen Laserdistanzmeters

Bild 1 veranschaulicht die Struktur und Komponenten der optischen Seite eines Distanzmeters, das geeignet ist, Abstände mit mm-Genauigkeit bzw. Pulslaufzeiten im 10 Piko-

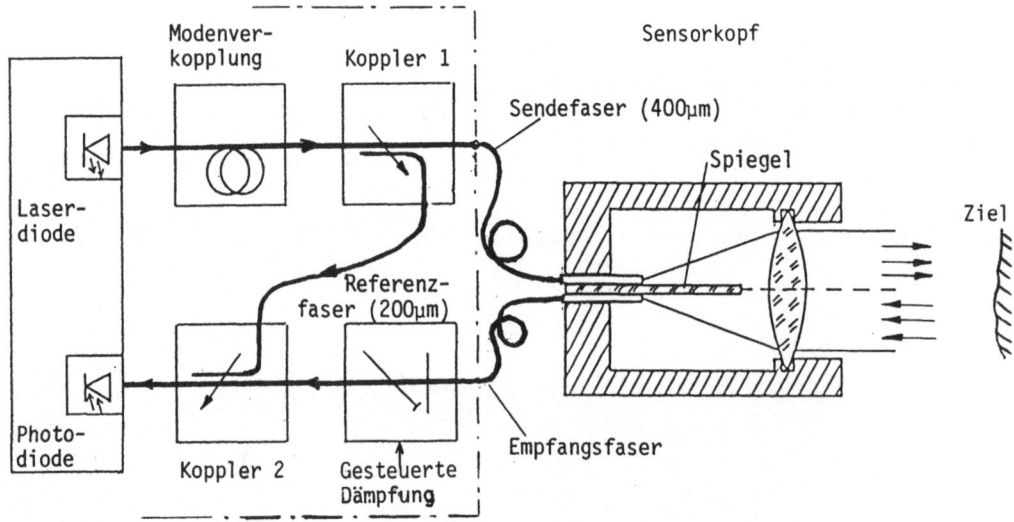

Bild 1. Das faseroptische Konzept und der Sensorkopf mit Spiegeloptik

sekundenbereich zu messen. Eine Infrarot-Laserdiode wird im 10 kHz-Takt mittels La-
winentransistoren betrieben und erzeugt bei einer Wellenlänge von 900 nm Lichtimpulse
der Dauer 10 ns mit einer Anstiegszeit von 500 ps und einer Leistung von ca. 30 Watt.
Über eine 10 bis 20 m lange Glasfaser (400 µm - PCS-Faser) werden diese Impulse zum
Sensorkopf geführt und gerichtet abgestrahlt.

Das von einem reflektierenden Ziel empfangene Streulicht wird in eine Empfangsfaser
etwa gleicher Länge eingekoppelt und auf eine PIN- oder Lawinen-Photodiode gegeben.
Zur Erzielung höchster Genauigkeiten sind weitere Komponenten erforderlich:
Ein Modenkoppler hinter der Laserdiode dient der Bildung einer einheitlichen, von den
räumlichen und zeitlichen Einkoppelverhältnissen der Laserdiode weitgehend unabhängigen Phasen- bzw. Impulsfront. Mit dem darauffolgenden Koppler 1 wird ein Bruchteil
des Sendeimpulses aus der Faser ausgekoppelt und über den Koppler 2 in die Empfangsfaser eingekoppelt. Der so entstehende Referenzimpuls simuliert ein Referenzziel mit
hoher Konstanz, auf das die Zielreflektionen bezogen werden.
Große Amplitudenschwankungen der reflektierten Zielimpulse werden mit einer
elektronisch gesteuerten Dämpfungseinheit in 10dB-Schritten ausgeglichen.
Der Sensorkopf besteht aus nur einer Linse und einem Spiegel. Die Endflächen der Sen-
de-und Empfangsfaser befinden sich in der Brennebene so nah wie möglich am Brennpunkt,
lediglich durch einen dünnen Spiegel (100 µm) voneinander getrennt. Auf diese Weise
liegt die eine Apertur jeweils im Spiegelbild der anderen und Sende- und Empfangsfaser
sind automatisch richtig fokussiert.
Ein weiterer Vorteil der Spiegeloptik ist der bis an die Linse heran erweiterte Meßbereich und die z.B. im Verhältnis zu einer Koaxialoptik geringe Amplitudenänderung
der Empfangsleistung im Nahbereich bis etwa 10 m, wie in Bild 2 dargestellt.

2. Das Gesamtkonzept und die elektronischen Komponenten

Bild 3 zeigt ein Blockschaltbild des gesamten Entfernungsmeßsystems im Zusammenhang
mit einem möglichen Robotereinsatz. Der Sensorkopf befindet sich z.B. auf der Roboter-
hand. Durch Schwenken erhält man ein Schnittbild. Das Distanzmeter liefert relative Abstände, Geschwindigkeiten und ggf. Grauwerte. Der Roboter kennt die absoluten Positionen und die Meß-richtungen.

Die reflektierten Impulse werden nach der Lawinenphotodiode mit einem Trans-impedanzverstärker hoher Bandbreite (500 MHz) mit Bootstrapping der Pho-todiode und einer Transimpedanz von 18 kOhm verstärkt. Eine Laufzeitdetek-tion ist bis zu minimalen Leistungen von 10 nW möglich.

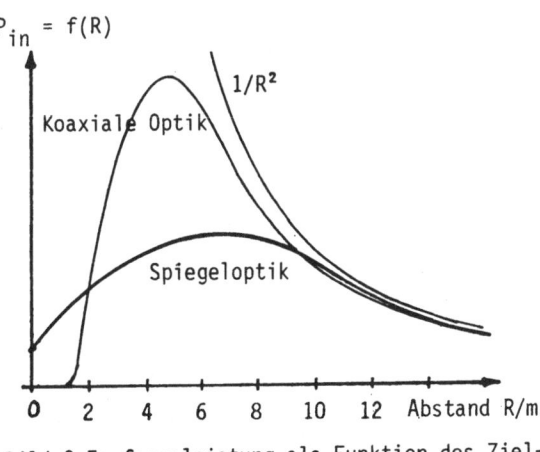

Bild 2. Empfangsleistung als Funktion des Ziel-
abstandes

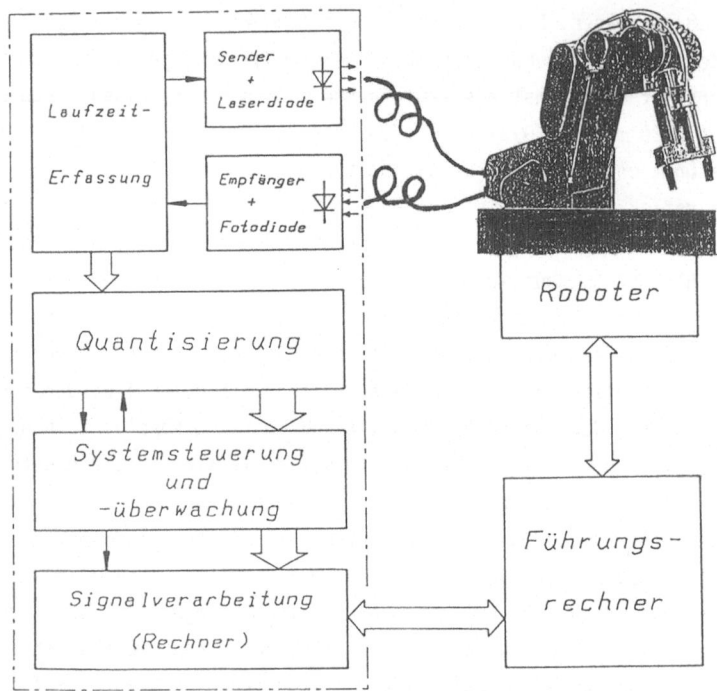

<u>Bild 3</u>. Blockschaltbild des Gesamtsystems eines Distanzmeters
mit einem Anwendungsbeispiel "Robotereinsatz"

Die Laufzeiterfassung hat die Aufgabe, die Start- und Ankunftszeiten der Referenz- bzw.
Zielimpulse in Form eines Rechteckimpulses mit einer Genauigkeit von wenigen Pikosekunden darzustellen. Die Trennung von Referenz- und Zielsignal geschieht durch ein Zeit-
fenstersignal, das den Komparator K1 in Bild 4 über den Latch/Enable-Eingang aktiviert.
Bei der Zeitbestimmung der Zielimpulse kommen die Amplitudenschwankungen erschwerend
hinzu. Konventionelle Constant-Fraction-Triggerschaltungen wiesen trotz Optimierung
bei Amplitudenschwankungen von 1:10 bereits Laufzeitfehler von 400 ps auf. Bei der in
Bild 4 dargestellten Schaltung wird lediglich der sog. Offset-Komparator hinzugefügt,
der im geeigneten Zeitpunkt ein Offsetsignal zuschaltet und ein "schnelleres" Schalten
bewirkt. Ergebnis: Amplitudenschwankungen von 1:40 führten auf Laufzeitfehler von nur
$\pm$25 ps.
Darüberhinaus trifft die Zeiterfassungsstufe eine Amplitudenauswahl. Neben dem Offset-
komparator, der das Amplitudenminimum bestimmt, existiert ein sog. Maximum-Komparator
für die maximal zulässige Signalamplitude. Mit Hilfe weiterer Amplitudenschwellen
oder eines geeigneten Analog-Digital-Wandlers kann den Zielen eine Grauwertskala zugeordnet werden.
Die Zeitquantisierung erfolgt in zwei Stufen, einer Grobquantisierung in 20ns-Schritten
und einer Feinquantisierung des Restwertes nach einer Zeitdehnung mit einem modifizier-
ten Dual-Slope-Verfahren um den Faktor 1024.

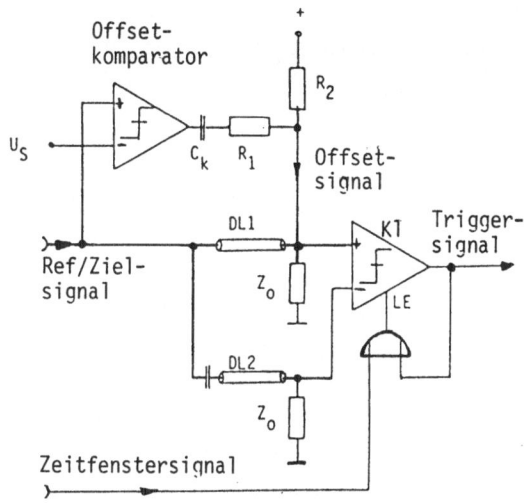

Bild 4. Constant-Fraction-Trigger mit gesteuerter Offset-Spannung

Die von der Zeitquantisierungsstufe gelieferten Rohdaten werden in der Signal- bzw. Meßwertverarbeitung hinsichtlich einer Eliminierung systematischer Fehler und Reduzierung statistischer Fehler aufbereitet.

Die in diesem Zusammenhang entwickelten Filteralgorithmen realisieren eine adaptive Kalmanfilterung zur Ermittlung der Abstands- und Geschwindigkeitswerte, wobei ein spezielles, in [1] beschriebenes "geschaltetes Kalmanfilter" eine Verarbeitung im 10 kHz-Takt mit einem 16 Bit-Mikro-/Signal-Prozessor trotz des großen Meßwertebereichs gestattet.

3. Übersicht der Eigenschaften

Das hier nur in einigen Teilaspekten skizzierte Konzept eines Distanzmeters und Optischen Radars weist eine Reihe interessanter Spezifikationen auf, die hier zusammengefaßt und z.T. kommentiert werden:

1) Entfernungsauflösung 1 mm entsprechend einer Zeitauflösung von 6,6 Pikosekunden. Angabe nur zusammen mit der Meßzeit sinnvoll. Die Standardabweichung für Einzelschüsse liegt entsprechend dem elektronischen Zeitjitter bei mehreren Millimetern.

2) Winkelauflösung einige Milliradiant. Vor allem eine Frage der Optik. Durch Beugungsbegrenzung gilt $\sin \alpha = 1,2 \cdot \lambda/D$; λ = Wellenlänge, D = Linsendurchmesser.

3) Meßbereich von einigen cm bis ca. 100 m für passive Ziele, ca. 10 km mit Retroreflektoren. Reichweite wesentlich vom Linsendurchmesser (Empfangsfläche) bestimmt.

4) Ca. 10 000 Messungen pro Sekunde. Begrenzung durch Wärmeentwicklung der Laserdiode. Diese Repetierrate reicht für viele Scan-Aufgaben aus.

5) Kleiner, flexibler Sensorkopf. Durchmesser je nach Anwendung 1 bis 10 cm.

6) Remote Sensing über einige 10 m Stufenindex-Faser bzw. über einige km über Gradientenindex-Faser.

7) Selbstbeleuchtend im Infrarotbereich. Ein Interferenzfilter schützt vor Störlicht.

8) Elektronisches Zeitfenster ermöglicht Wahl eines Zielmeßbereich. Ein Messen durch optische Hindernisse (Staub, Nebel) hindurch wird möglich. Ferner Unterdrückung von Mehrfachreflexionen. Multiplexbetrieb mehrerer Sensorköpfe möglich.

9) Erfassung der Grauwerte (Reflexionsfaktoren) über den Abstand möglich.

10) Die interne Referenzfaser erlaubt eine Selbstüberwachung und -kalibrierung.

4. Anwendungsbereiche

Die wichtigsten Anwendungen seien kurz aufgezählt: Fertigungsautomatisierung; Robotervision; Füllstandmessung; Fernüberwachung; Navigation im Nahbereich; Rendezvous und Docking in der Raumfahrt[2]; Blindenhilfe.

Literatur

[1] O. LOFFELD: "A Switched Kalman Filter ...", IASTED Proceedings, Int. Symposium on Applied Signal Processing, June 1985, Paris

[2] R. SCHWARTE: "Performance Capabilities of Laser Ranging Sensors", Proc. ESA Workshop on Space Laser Applications and Technology (ESA SP-202, May 1984)

Längenmessungen mit einem Laserdiodeninterferometer

A. Abou-Zeid

Physikalisch-Technische Bundesanstalt

Bundesallee 100, D-3300 Braunschweig

Es wird über ein einfaches Verfahren zur Stabilisierung der Temperatur der aktiven Zone eines Diodenlasers unabhängig von seinem Injektionstrom berichtet. Damit wird die Frequenz eines Longitudinalmode über den gesamten zulässigen Strombereich der Diode stabilisiert. Mit solch einer stabilisierten Laserdiode wurden Längenmessungen an zwei Maßstäben mit Hilfe eines aufgebauten Verschiebungskomparators durchgeführt und erläutert.

1. Einleitung

Der Diodenlaser stellt im Verhältnis zu dem in der Längenmeßtechnik meist verwendeten Gaslaser eine kleine, leistungsstarke und energiesparende Lichtquelle dar. Jedoch liegt die relative Frequenzstabilität ($\pm 5 \times 10^{-7}$ innerhalb einer Stunde) eines strom- und temperaturstabilisierten Diodenlasers (1) fast zwei Größenordnungen unterhalb der eines stabilisierten He-Ne-Zwei-Frequenz-Lasers (2). Dabei kann der Diodenstrom i um Werte weniger als ± 60 µA variieren, die Temperatur der Wärmesenke T_{ws} um ± 6 mK. Um höhere Stabilitäten zu erreichen, muß der Diodenlaser an ein externes Fabry-Perot-Etalon (3) oder an eine bekannte Absorptionslinie (4) angeschlossen werden.

2. Eignung eines Diodenlasers zum Einsatz in der Längenmeßtechnik

Trotz der oben genannten Vorteile eines Diodenlasers gegenüber einem Gaslaser muß ein Diodenlaser zunächst einige Kriterien erfüllen, um in der Längenmeßtechnik eingesetzt zu werden. Als 1. Kriterium muß der einmodige (sowohl transversal als auch longitudinal) Diodenlaser eine kinkfreie Strahlungsleistungs/Diodenstrom-Kennlinie, eine stabile Abstrahlungscharakteristik und keine Alterungserscheinungen aufweisen. Das 2. Kriterium besteht darin, den Diodenlaser in einem sprung- und hysteresefreien Bereich in Betrieb zu nehmen. Weiterhin muß der Diodenlaser möglichst bei relativ hohen Strömen ($\geqslant 1,2\ i_{th}$, i_{th} = Schwellstrom) betrieben werden, bei denen der Rauschanteil und die Linienbreite relativ niedrig sind. Ferner muß das von äußeren optischen Elementen reflektierende Rücklicht in den Diodenlaser beseitigt werden, z.B. durch Verwendung optischer Isolatoren.

3. Frequenzstabilisierung eines Diodenlasers unabhängig von seinem Injektionsniveau

Messungen der emittierten Wellenlänge eines Diodenlasers in Abhängigkeit seiner Parameter (i und T_{ws}) zeigen im Durchschnitt (5), daß der Stromkoeffizient der Modenwellenlänge (λ_m) etwa 0,006 nm/mA beträgt, der Temperaturkoeffizient 0,06 nm/K. Aus diesen beiden Werten der Parameterkoeffizienten läßt sich ableiten, daß die Temperatur der Wärmesenke bei einer Stromzunahme von 1 mA ($i > i_{th}$) etwa um 0,1 K abgekühlt werden muß, um die Temperatur der aktiven Zone und somit die emittierte Frequenz des Diodenlasers zu stabilisieren. Nach diesem einfachen Verfahren wurde eine Laserdioden-Versorgungseinheit mit einem je nach Diodentyp einstellbaren Strom/Temperatur-Rückkopplungsgrad aufgebaut. Dabei hielt eine Stromquelle den Strom auf Werte besser als $\pm$ 1 µA konstant. Ein durch den eingestellten Strom rückgekoppelter PI-Temperaturregler sorgt für die Temperaturstabilisierung der Diodenwärmesenke ($\pm$ 0,005 K). Messungen mit dieser Einheit an einem einmodigen Diodenlaser zeigen (s. Abb. 1), daß die Wellenlänge eines Fabry-Perot-Mode m unabhängig vom Diodenstrom ($i > i_{th}$) stabilisiert werden kann. Die relative Frequenzstabilität des Diodenlasers beträgt etwa $\pm$ 1 x 10^{-6} zwischen 1,05 i_{th} und 1,65 i_{th} (6).

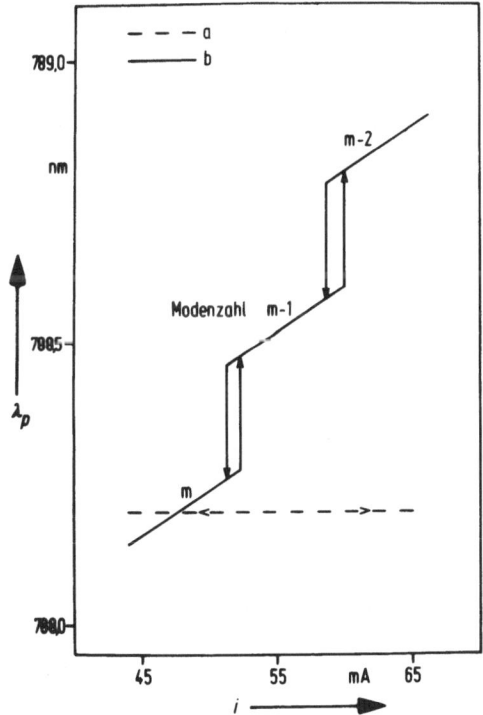

Abb. 1. Änderung der Schwergewicht-wellenlänge λ_p mit dem Injektionsstrom i.
a: konstante Temperatur der aktiven Zone (25,34 °C) durch eine Strom/Temperatur-Rückkopplung.
b: konstante Temperatur der Wärmesenke (20,39 °C).

Außer der Stabilisierung der Modenfrequenz unabhängig vom Strom ermöglicht das Verfahren der Strom/Temperatur-Rückkopplung die Bestimmung des mittleren Wertes des thermischen Widerstands zwischen der aktiven Zone und der Wärmesenke (50 $\pm$ 1 K/W) durch Messung der Spannung/Strom-Kennlinie (6).

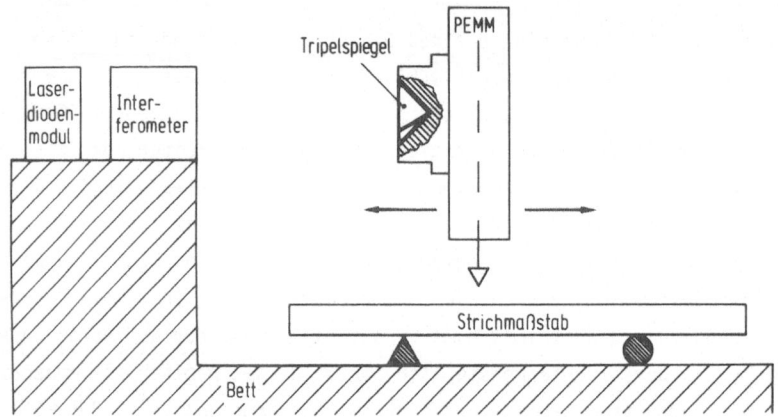

Abb. 2. Verschiebungskomparator

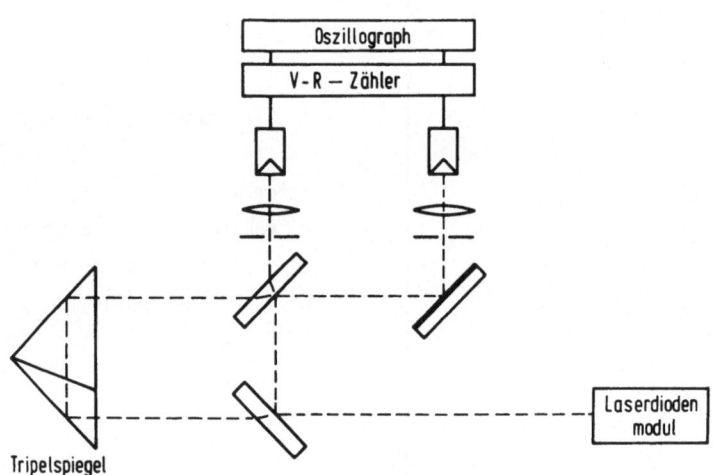

Abb. 3. Interferometerstrahlengang

4. Interferometrische Längenmessungen mit einem stabilisierten Diodenlaser

Der aufgebaute Verschiebungskomparator (s. Abb. 2) besteht aus dem Laserdiodenmodul, dem Interferometer, dem zu prüfenden Maßstab, dem Tripelspiegel und dem photoelektrischen Meßmikroskop (PEMM). Die ersten drei Teile sind fest miteinander verbunden. Auf einem Verschiebungstisch ist der Tripelspiegel zusammen mit dem PEMM befestigt. Der Laserdiodenmodul (1) besteht aus einer einmodigen Laserdiode (Hitachi 7801, i_{th} = 41,5 mA bei T_{ws} = 20 ^{O}C), Peltier-Kühleinrichtung und Mikroskopobjektiv zum Parallelisieren des emittierten Laserlichtes. Der Strahlengang des vom Twyman-Green- -Typ verwendeten Interferometer wird in Abb. 3 dargestellt. Der Interferometerzähler zeigt die Tischverschiebung in $\lambda/2$-Schritten an. Das PEMM sorgt für die Meldung des Stricheinfanges an den VR-Zähler.

Zur Längenmessung von Maßstäben wurde zunächst ein in der PTB bereits geprüfter Glasmaßstab als Normal (Gesamtlänge 500 mm) verwendet, um die emittierte Wellenlänge (788,168 nm) des stabilisierten Diodenlasers zu bestimmen. Ausgehend von dieser Wellenlänge wurden Messungen der gesamten Länge (500 mm) von 2 Maßstäben (einem aus Glas und einem aus Stahl) durchgeführt. Dabei wurde die Temperatur, der Druck und die Feuchtigkeit der umgebenden Luft gemessen und mit Hilfe der Edlen-Formel (7) der Brechungsindex approximiert sowie die Länge bei 20 $^{\circ}$C berechnet. Zahlreiche Messungen in Rück- und Vorwärtsrichtung und auch bei umgedrehtem Maßstab sowie nach Aus- und Wiedereinschalten der Laserdiode zeigen, daß die Länge um etwa 4 μm ($\pm$ 4 x 10^{-6}) schwankt. Aus den Unterschieden der Längenmessung vor und nach dem Wiedereinschalten der Laserdiode wird die Reproduzierbarkeit der Laserwellenlänge zu etwa $\pm$ 2 x 10^{-6} abgeschätzt. Der restliche Anteil der Schwankungen ist im wesentlichen auf die mangelnde mechanische Stabilität des aufgebauten Versuchskomparators in Verbindung mit der Verletzung des Abbeschen Prinzips zurückzuführen. Über den systematischen Fehler, der im Mittel 2 μm bis 3 μm beträgt, sind noch keine Untersuchungen ausgeführt worden.

5. Literatur

(1) A. Abou-Zeid, G. Leppelt: PTB-Bericht Me-67, Braunschweig, 1985

(2) R. Balhorn, H. Kunzmann, F. Lebowsky: Appl. Opt. 11, 742 (1972)

(3) T. Okoshi, K. Kikuchi: Electron. Lett. 16, 179 (1980)

(4) H. Tauschida, M. Ohtsu, T. Tako: Jap. J. Appl. Phys. 21, L1 (1982)

(5) A. Abou-Zeid: PTB Mitt. 94, 163 (1984)

(6) A. Abou-Zeid: in Vorbereitung

(7) B. Edlén: Metrologia 2, 71 (1966)

Fast Interferometric Measurement of Velocity and Velocity Deviations

W. Balzer,T. Tschudi

Institut für angewandte Physik,TH Darmstadt

Laser interferometers provide measurement of object positions with a resolution of, depending on system design, some fractions of light wavelength.The principle of measurement is to count intensity changes due to the change of optical path length in an interferometer arm. Fig. 1 shows the well-known interferometer setup we used. The direction of object displacement is determined by exploiting the quadrature property of the linear components of circularly polarized light.

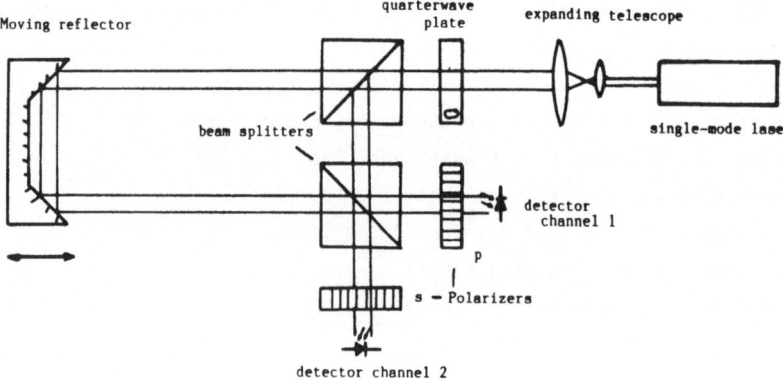

Fig.1. Interferometer optical setup

With the same interferometer, measurement of object velocity can also be performed by measuring the frequency of intensity changes in moving fringes. In this paper, some methods and questions of temporal resolution of velocity measurement are discussed. We will present a new method of processing interferometric information to solve practical measuring problems.

Velocity v follows from measured frequency by $\qquad v = \lambda \cdot f / C$
with λ laser source wavelength and f measured frequency. C is an interferometer-specific constant which represents the number of counts per wavelength of object displacement.

In general, object motion will be disturbed by drive imperfections and mechanical coupling from other moving parts of a device. This leads to a velocity
distribution instead of a sharply determined velocity. Fig. 2 shows an example
of a velocity distribution in the frequency domain.

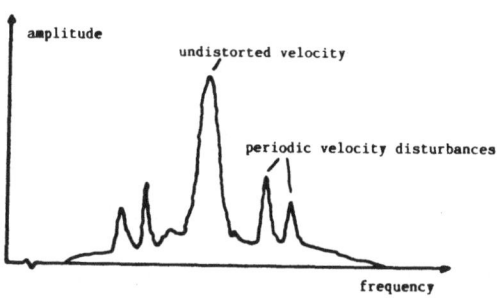

Fig.2. Example of velocity distribution due to motional disturbances

Usually, frequency will be determined by counting fringe rates. The
problem is that for a given resolution, a certain gate time is required which limits temporal resolution
of measurement. Thus, fast velocity
disturbances are not detected.

In a practical example, given an
average velocity of 10 mm/s, a
source wavelength of 633 nm, a full
scale measuring range of +/- 10 % of
average velocity and a desired
resolution of 1/256 of full scale, a center frequency of approx. 31 kHz occurs
and a gate width of at least 80 ms is required.

Another possibility is measurement of intensity signal period time. In this
case, as, although not to this extent, in the case of frequency counting a large
amount of data is produced. This data has to be stored and processed to obtain
interpretations of object motion.

We developed a new type of signal processing which doesn't posess the disadvantages mentioned above. Object motion disturbance can be understood as temporal
modulation of object position and velocity described by

$$x = x_0 + \sum a_i \sin(\omega_i t + \alpha_i) \qquad \text{and}$$

$$v = v_0 + \sum b_i \cos(\omega_i t + \alpha_i) \qquad \text{,respectively.}$$

ω_i, α_i are the frequency and phase components of disturbance. So motion disturbance analysis can be done by performing a frequency demodulation, i.e. trans-
forming the average velocity to the origin of a new coordinate system.

Use of a phase locked loop (PLL) as frequency demodulator is a good solution to
this problem. A PLL is a circuit where the frequency of a reference oscillator
is shifted until the phase difference between input and reference frequency is
minimum, i. e. both frequencies are the same. Output signal of a PLL is a

178

voltage which is linearly dependent on input frequency.

For the PLL output voltage with disturbed velocity signal as input we get

$$U_{PLL} \propto v_o + \sum b_i \cos(\omega_i t + \alpha_i)$$

which is the desired demodulated signal. If the PLL is tuned to a center frequency corresponding to desired center velocity, deviations from this velocity appear as analog voltages and periodic velocity disturbances are represented by their frequency spectrum.

The parameters for dimensioning a PLL are frequency-to-voltage conversion ratio, which determines the full scale measuring range, dynamic behaviour and demodulator bandwidth.

For our application, we chose the response time to be 1 ms, an adjustable center frequency range equivalent to a center velocity range from 0.5 mm/s to 160 mm/s, and a full scale measuring range of +/- 15 % of center speed. Output voltage is proportional to relative deviation from center velocity. Full scale measuring range can easily be chosen to below 1 % of center velocity.
Absolute accuracy is limited by thermal stability and linearity of our VCO to approx. 0.5 % of measuring range. Resolution is better than 0.1 % of measuring range.

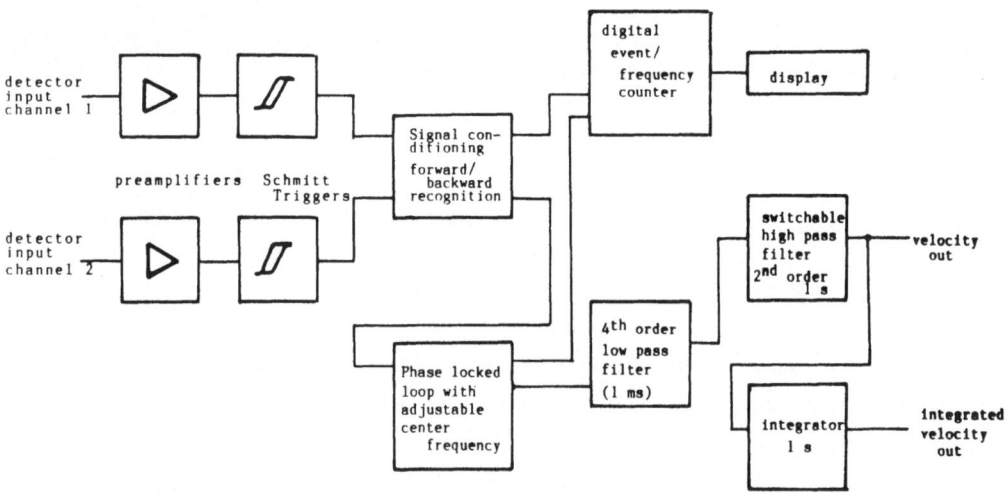

Fig.3. Interferometer electronics block diagram

Fig.3 shows the electronics block diagram. Also included are the standard digital measuring circuitry and some analog circuitry for additional PLL signal processing.

Specific advantage of the PLL system is the possibility of simple registration and interpretation of measurements. In particular, the PLL signal can be frequency-analyzed to identify sources of periodic velocity disturbances by their frequencies, the need of which is often encountered in mechanical systems.

We demonstrated the performance of our system in an industrial application.

The PLL signal was obtained from interferometric measurement of a machine part which should have moved with constant velocity.

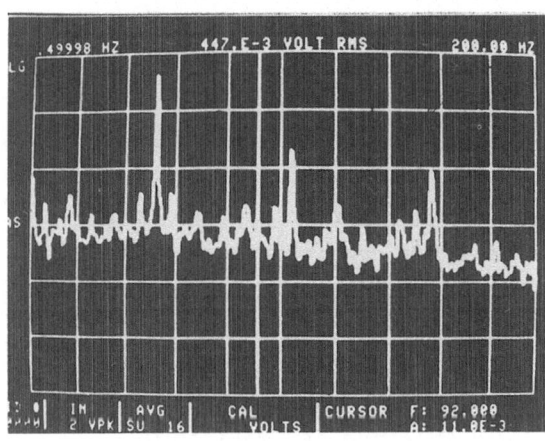

Fig. 4 shows a sample screen output of the spectrum analyzer. The spectral peaks are clearly visible. The sources of motion disturbances could be identified by their characteristic frequencies.

Interferometric velocity analysis can also be used for a quality monitoring during drive system development since overall modulation amplitude is dependent on degree of motional disturbances.

Fig.4. Sample velocity analysis

In conclusion, we have shown that an interferometer, extended by PLL circuitry, is a tool to determine qualitatively the characteristics of linearly moving devices as well as to perform quantitative analysis of motion disturbance. Little additional circuitry for PLL is needed. The system provides efficient and user-oriented measuring.

Since the source wavelength has not to be known exactly in interferometric motion-disturbance analysis, use of commercially available single-frequency semiconductor lasers can be made to replace the cost-effective conventional single-frequency lasers needed due to coherence length requirements. This would also improve compactness and ruggedness of field-measuring units.

Ein neues Zweikomponenten Laser-Doppler-Anemometer mit elektro-optischem-Modulator und erste Anwendungsbeispiele

R. Bahnen
K. Köller
Lehrstuhl für Technische Thermodynamik
RWTH Aachen, Schinkelstr. 8

Zur Messung der Strömungsgeschwindigkeit in unzugänglichen, gegenüber äußeren Störungen sensiblen oder extreme Bedingungen für den
Meßsensor aufweisenden Objekten wird in letzter Zeit zunehmend die
Laser-Doppler-Anemometrie eingesetzt. In vielen Fällen ist dabei
die gleichzeitige Messung mehrerer Geschwindigkeitskomponenten zur
Minimierung der Meßzeit und zur Erfassung der turbulenten Spannungen des Strömungsfeldes erforderlich /1/.

Prinzipiell wird mit der Laser-Doppler-Anemometrie nur eine Komponente der Geschwindigkeit gemessen, die bei dem in letzter Zeit
üblichen Zweistrahlverfahren in der von den beiden Strahlen aufge-
spannten Ebene und senkrecht zur optischen Achse liegt.

Zur Messung einer weiteren Geschwindigkeitskomponente orthogonal
zur optischen Achse wird dann ein dazu um 90° gedrehtes zusätz-
liches Meßvolumen aufgebaut, wobei die Trennung der Meßinformation aus beiden Meßvolumina üblicherweise über die Laserfrequenz
(Farbe) oder die Polarisationsrichtung geschieht. Nach dieser optischen Trennung ist dann für jede Geschwindigkeitskomponente ein
Detektions- und Auswertekanal notwendig, bestehend aus Photomultiplier und Frequenzanalysator.

Im folgenden wird ein Verfahren vorgestellt, welches zur Verringerung des Aufwandes bei der Signalverarbeitung kürzlich entwickelt
wurde. Die Signaltrennung geschieht hierbei durch eine sehr
schnelle zeitliche Umschaltung der Meßvolumenorientierung. Dies
gestattet dann eine nur noch einkanalige Auswertung mit ähnlichem
Aufwand wie bei Einkomponentenmessungen. Ist diese Umschaltung wesentlich schneller als die Durchflugzeit eines Streuteilchens
durch das Meßvolumen, so kann die Messung beider Komponenten am
selben Streuteilchen vorgenommen werden und ist quasisimultan.

Meßprinzip

Um zeitlich versetzt zwei zueinander gedrehte Meßvolumina zu erzeugen, wie in Bild 1 dargestellt, muß man den Laserstrahl innerhalb sehr kurzer Zeit umschalten. Als Umschalter wird für den linear polarisierten Laserstrahl aus Ar-Ion oder He-Ne Laser eine Kombination aus elektrooptischem Modulator (Pockels-Zelle) und polarisierendem Strahlteiler verwendet. Die Polarisationsrichtung des Laserstrahls wird dabei bei geeigneter Gleichspannung an der Pockels-Zelle um 90° gedreht und der Strahl dann im polarisierenden Strahlteiler aus der optischen Achse herausgelenkt. Mit den beiden Strahlen I (bei ausgeschalteter Pockels-Zelle) und II (bei eingeschalteter Pockels-Zelle) wird dann je ein Meßvolumen gebildet, wobei beide um 90° zueinander gedreht sind /2/. Der optische Aufbau ist in Bild 2 dargestellt.

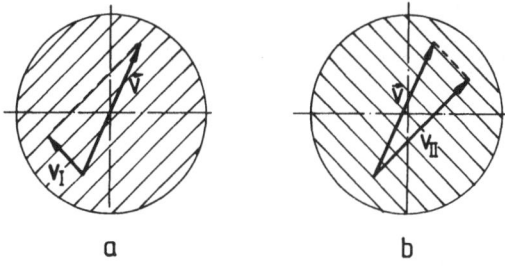

a   b

Bild 1.   Meßvolumen vor (a) und nach (b) dem Umschalten

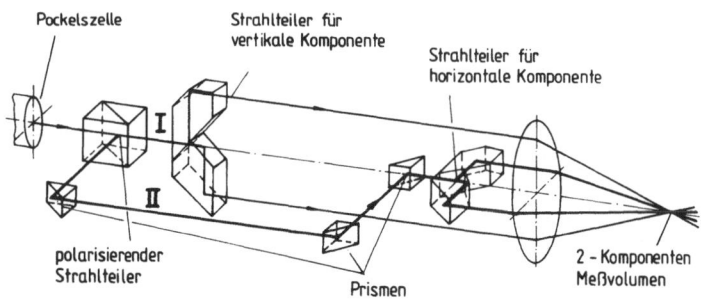

Bild 2.   Optischer Aufbau für Zweikomponentenmessungen

Betreibt man die Optik zusammen mit einem Counter als Signalanalysator, so kann dieser die Geschwindigkeitskomponente bei ausgeschalteter Pockels-Zelle messen und mit einem Signal, daß das Ende der Messung an-

zeigt, die Pockels-Zelle schalten, um sodann die zweite Komponente zu
erfassen. Eine ausgeführte Optik in Modulbauweise ist in Bild 3 dargestellt.

In Verbindung mit dieser Optik kann der Counter die eingeschaltete Geschwindigkeitskomponente messen, die Pockels-Zelle nach beendeter Messung einschalten und dann am selben Partikel die zweite Geschwindigkeitskomponente bestimmen.

<u>Bild 3.</u> Zweikomponentenoptik an einer Methan-Diffusionsflamme

Der wesentliche Vorteil dieser Art der Zweikomponentenmessung liegt
in der drastischen Reduktion der Auswerteelektronik, die jetzt im wesentlichen der der Einkomponentenmessung entspricht und ohne den
sonst üblichen zweiten Signaldetektor (z. B. Counter) auskommt.

<u>Meßsignalauswertung</u>

Der "Counter" als Frequenzdetektor liefert im sogenannten "fixed mode"
(Messung über eine konstante Streifenzahl) ein Signal bei beendeter
Messung. Dieses kann dann zur Umschaltung der Pockels-Zelle herangezogen werden und gleichzeitig zusammen mit einer Festfrequenz zur Messung der Zeit zwischen den Einzelsignalen dienen.

Aus der Frequenz, der Zeit zwischen den Messungen und dem Schaltzu-
stand der Pockels-Zelle kann dann eine vollständige Zweikomponenten-
messung abgeleitet werden, die, wenn folgende Kriterien erfüllt sind,
als simultan (innerhalb desselben Dopplerbursts gemessen) angesehen
werden kann.

- Pockelszellenstatus für zwei aufeinanderfolgende Messungen ungleich
- Zeit zwischen diesen Messungen etwa gleich der Zeit für 8+1 Zyklen
  der Meßfrequenz für die zweite Komponente

Alle benötigten Daten werden dann einem Rechner zugeführt und nach den oben erwähnten Kriterien sortiert. Das Blockschaltbild ist in Bild 4 dargestellt.

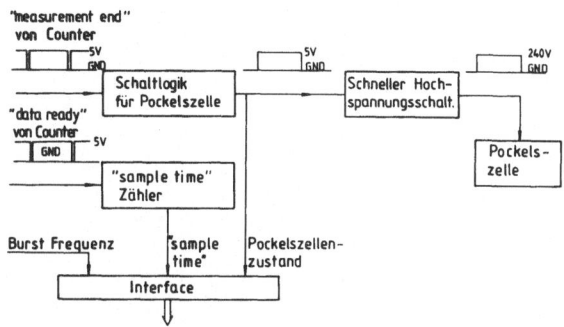

Bild 4.   Blockschaltbild zur Signalverarbeitung

Als Ergebnis wird aus der Summe aller Zweikomponentenmessungen der Mittelwert $\bar{U}$ und $\bar{V}$ für die axiale bzw. radiale Komponente der Geschwindigkeit, die Schwankungsgrößen $\overline{u'^2}$ und $\overline{v'^2}$ und die Reynoldsche Schubspannung $\overline{u'v'}$ errechnet.

Meßbereiche

Betrachtet man die benötigten Zeiten für Messung und elektrooptische Umschaltung, so muß die Summe aus

- Messung für die erste Komponente (8+1 Streifen)
- Umschaltung für die Pockels-Zelle (geschieht innerhalb einer counterspezifischen internen Verarbeitungszeit in der keine Messung durchgeführt werden kann) in 1,2...1,4 µs.
- Messung für die zweite Komponente (8+1 Streifen)

kleiner als die Partikeldurchflugzeit sein.
Unter Berücksichtigung der Streifenzahl und einer eventuell notwendigen Braggverschiebung dieses Streifensystems ergibt sich für den möglichen Erfassungsbereich eines Geschwindigkeitsvektors der in Bild 5 dargestellte Winkelbereich β, bei um den Winkel $\alpha = 45^{\circ}$ zur Hauptströmungsrichtung gedrehter Optik.

Beispiel

f_B = 10 MHz

N = 80 Streifen

$|v|$ = 96 m/s

Δx_{Fr} = 3,2µm

also:

v/d = 30 MHz

→ ß = ± 47°

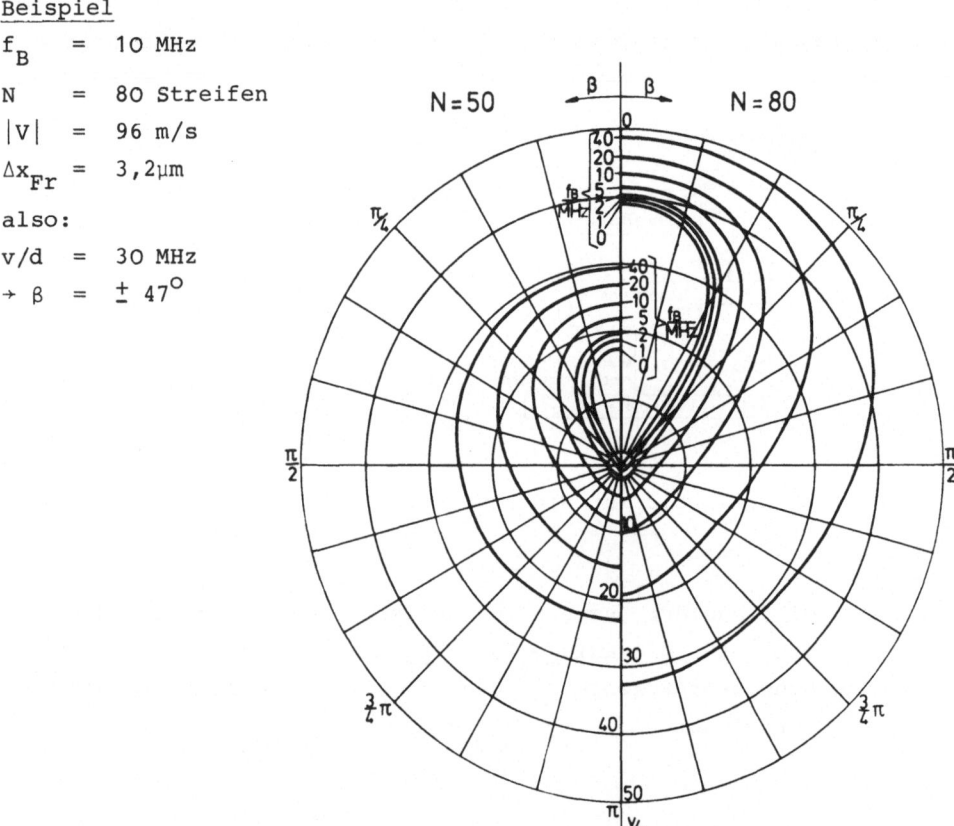

Bild 5. Erfaßbarer Winkelbereich ß eines mit dem Streifenabstand normierten Geschwindigkeitsvektors in Abhängigkeit von Streifenzahl N und Braggfrequenz f_B bei um 45° zur Hauptströmungsrichtung gedrehter Optik.

Man erkennt, daß es mit Braggverschiebung des Meßvolumens ohne weiteres möglich ist, den schon bei üblichen Streifenzahlen N großen Winkelbereich ß bis auf 360° auszudehnen.

Messungen

Erste Messungen an offenen Flammen zeigten eine eindeutige Kanaltrennung für beide Meßinformationen bei Verwendung des sogenannten "measurement end" (im Counter verfügbar), da dann daß Meßvolumen vollständig umgeschaltet ist, ehe der im "fixed mode" betriebene Counter eine neue Messung beginnen kann (nach 1,2...1,4 μs).

Besonderer Wert ist dabei auf die sorgfältige Ausrichtung der Polarisationsrichtung relativ zu Strahlteiler und Pockels-Zelle und auf eine Pockelszellenausrichtung exakt auf der optischen Achse zu legen. Die dann erreichbaren Intensitätsverhältnisse zwischen dem jeweils aktiven und dem ausgeschalteten Meßvolumen sind mit etwa 1:100 (40 dB) weitaus größer als die zur sicheren Kanaltrennung benötigten Intensitätsverhältnisse von 1:30 (30 dB).

Bild 6 zeigt einen Zweikomponentenburst aus einer CH_4-Diffusionsflamme und den zu diesen Messungen gehörenden analogen Counterausgang, der die beiden Komponenten mit je 0,8 und 1,4 MHz deutlich erkennen läßt.

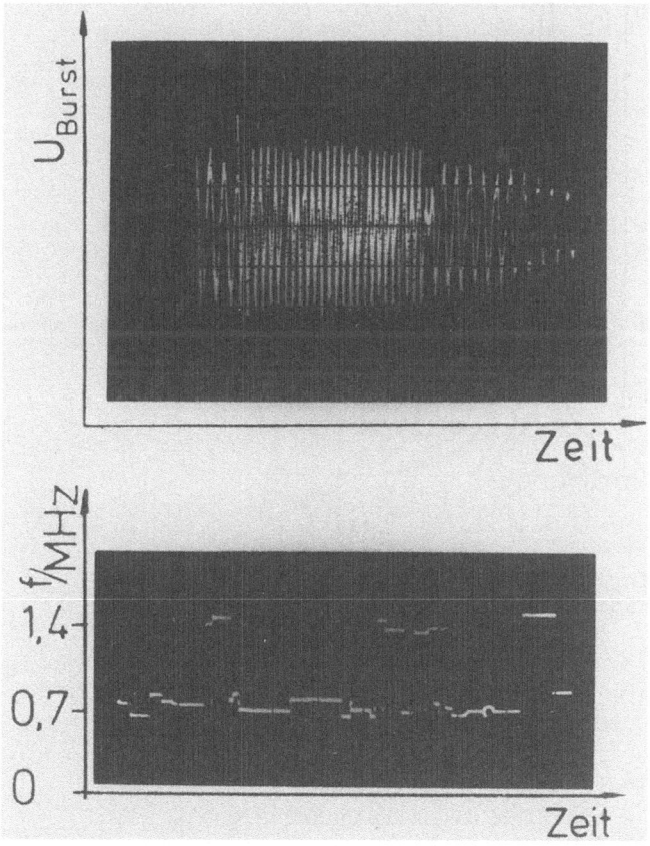

Bild 6. Zweikomponenten-Dopplerburst und analoger Counterausgang für denselben Meßort

Ein Beispiel für ein vollständiges Geschwindigkeitsprofil an einer
CH$_4$-Diffusionsflamme ist in Bild 7 dargestellt. Die Optik wurde bei
den Messungen um 45 ^{O} zur Hauptströmungsrichtung gedreht (vgl. Winkel α in Bild 5) um zu gewährleisten, daß beide Komponenten die
gleiche Größenordnung haben, also für die zweite Messung noch genügend Zeit zur Verfügung steht. Aus jeder Zweikomponenteninforma-
tion werden dann die Komponenten in axialer und radialer Richtung
berechnet und daraus nach abgeschlossener Messung Mittelwert und
Schwankungsgrößen ermittelt.

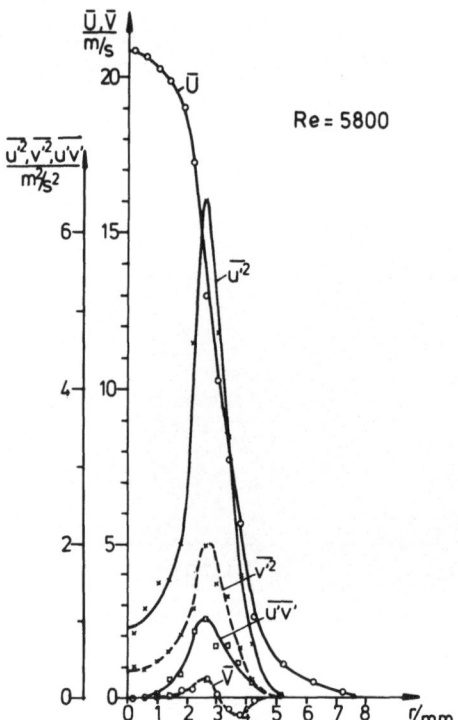

<u>Bild 7.</u>   Mittlere Geschwindigkeit und Schwankungsgrößen für die
         axiale (U) und radiale (V) Richtung in einer CH$_4$-Diffu-
         sionsflamme

Messungen in Rückstreuung an Luftfreistrahlen, als vorbereitende
Messungen zur Geschwindigkeitsbestimmung innerhalb eines Diesel-
Motors, liefern ebenfalls eindeutige Zweikomponentensignale wie
in Bild 8 zu sehen ist. Das Signal entstammt einem Luftfreistrahl
und wurde mit 40 MHz Braggverschiebung und anschließender elektronischer Mischung mit einem 43 MHz Sinus erhalten.

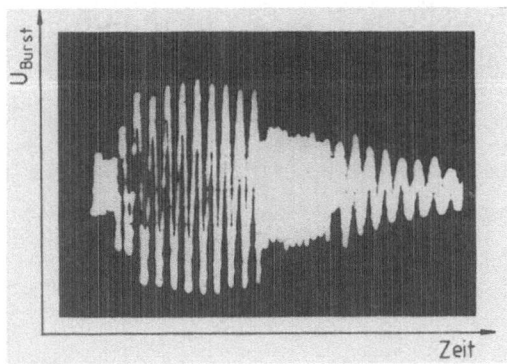

Bild 8. Zweikomponentenburst aus Rückstreumessungen

Erste Messungen am geschleppten Diesel-Motor zeigten auch bei diesen extremen Bedingungen die Einsetzbarkeit der Optik zur Zweikomponenten-messung.

Zusammenfassung

Das vorgestellte Laser-Doppler-System mit elektrooptischer Umschaltung des Meßvolumens ist in der Lage, eine simultane Zweikomponenten Laser-Dopplermessung durchzuführen. Der elektronische Aufwand bei der Signal-detektion und Verarbeitung entspricht dem der Einkomponentenmessung. Die Nachteile, die sich bezüglich des Meßbereiches aus der Umschaltung ergeben, sind geringfügig größer im Vergleich mit den bisher üblichen Optiken zur Zweikomponenten Laser-Doppler Messung. Mehrere Messungen an diversen Meßobjekten in Vor- und Rückstreuung zeigten eine sichere Kanaltrennung und damit eine eindeutige Eignung des Verfahrens für die "simultane" Zweikomponentenmessungen.

Literatur

/1/ T. S. Durrani, G. A. Greated
    Laser Systems in Flow Measurement
    Plenum Press, New York 1977

/2/ R. H. Bahnen, K. H. Köller
    Rev. Sci. Instr.; Vol. 55, No. 7, (1984), 1090-1093

A Laser System for Automated Positioning of Constructions

A. M. Woźniak and W. Knapczyk
Ship Research Institute
Technical University of Gdańsk
Majakowskiego 11/12,   PL - 80 952 Gdańsk

1. Introduction

In the final assembly process of the ship's hull, hydraulic or electro-mechanical supporting-shifting units for positioning of large sections and blocks are implemented. This raises the necessity of simple and accurate determination of dynamically changing spatial orientation of structural element being positioned. The traditional measurement methods are in this case not applicable. A proposition of a laser measuring system which makes it possible to follow changes in the position of the construction in course of positioning operation is presented in the paper. Together with developed electronic control system, the pro - posed measuring system enables to automatize the positioning operation carried out by hydraulic supporting-shifting units.

The work is a part of the research programme on the development of ship-building metrology.

2. Conception of the system

It is assumed that a spatial orientation of the positioned element, re-garded as a rigid body, can be explicitly defined by giving the co-ordinates of three check-points P1, P2, and P3 situated on the element. It is acceptable provided that the following initial requirements are satisfied:

- all technological operations which can cause some deformations of the element are finished,
- rigidity of the element is sufficiently large enough to avoid any deformations during transportation and positioning,
- real dimensions and shape of the positioned element are measured and requirements for its spatial orientation are verified with re-gard to determined manufacture distortions,
- the above mentioned check-points on the element are fixed according to an arrangement of the supporting-shifting units and requirements concerning spatial positions of these points are stated.

A lay-out of the proposed system is shown in Fig.1.

The instantaneous positions of the check-points, and consequently - the
structural element, are determined in relation to the local reference
system OXYZ, which is represented by a laser reference lines system.
A laser beam, appropriately split and refracted,as illustrated in Fig.2
is used for setting up the system of laser reference lines. They are
spatially oriented on the stand by use of reference detectors.

Position errors of the check-points are measured by photodetectors placed as shown in Fig.1. Error signals are of three-state type /-, 0, +/.
Regardless of the error signals, each detector provides an additional
two-state signal indicating whether the detector is illuminated by the
laser reference line. Error signals from the detectors are utilized as
a feed-back for automatic control of the supporting-shifting units according to the block-diagram in Fig.3. A flow-chart of a positioning algorithm is presented in Fig.4. It has been found out that the proposed
algorithm is the easiest one to be realized if an additional detector
is placed at check-point P4 and the error signals are arranged as shown
in Fig.1.

The positioning operation proceeds in two stages:
- the element of construction is placed upon the supporting-shifting
  units by crane into a position possibly close to the desired one,
  and, if necessary, is initially positioned by manual control till
  all indicating signals D1, D2, D3, and D4 appear.
- the automatic controlled positioning starts and is realized in
  three steps: levelling, symmetrization, translation.
  If the levelling procedure has been completed by all the hydraulic
  units, the symmetrization is initiated to be followed by translation operation till all error signals from the detectors reach "0"
  state.

3. Laboratory model and experiment
A general view of the experimental stand and measuring devices is shown
in Fig.5. There was performed a laboratory model of the positioning equipment which is used in the Gdynia Shipyard. The positioning equipment
comprises of three hydraulic supporting-shifting units. Each of them
consists of three hydraulic cylinders, corresponding to X,Y,Z axes. The
operational movement of the hydraulic cylinders is controlled by use of
three-state electro-magnetic velves in the following way in terms of:
- Z direction error signals - independent movement of each of the
  three cylinders of that direction,
- Y direction error signals - concurrent movement of that direction
  cylinders of units 1 and 2, and independent movement of cylinders

of unit 3,
- X direction error signals - concurrent movement of all the three
  cylinders of that direction.

In order to control the model as well as the future full scale positioning equipment there was designed and constructed a special electronic
measuring-control instrument. It enables to perform manual or automatic
control of positioning operation based on the above presented conception
As it is shown in Fig.5, all the elements necessary to control the positioning operation are situated on the front panel of the instrument.

To optimize the control parameters a number of experiments with regard
to variable speed of operational movements, insensitivity zone of detectors, etc. was carried out. The achived accuracy of the model positioning under laboratory conditions is within $\pm$ 1 mm.
Fig.6 presents, as an example, error signals supplied by detectors and
corresponding to them operational movements of hydraulic cylinders.

4. Conclusion

The performed experiments have proved the conception and the applica -
bility of the elaborated measuring equipment under laboratory conditions
The implementation of the system in shipyard should be, however, preceed
by full scale experiment. Complementarily to the existing three-state
control, the application of proportional control /within the range of
positioning corresponding to the diameter of the laser beam/ can be
considered.

5. Acknowledgement

This work was performed at the Ship Research Institute of the Gdańsk
Technical University within Project MR I-27 and was in part included in
the work plan of research co-operation agreement with Rheinisch-Westfä-
lische Technische Hochschule Aachen. Appreciations should be express to
the Authorities of the TU Gdańsk and the RWTH Aachen for their support
and creating a possibility to attend the Conference.

6. Literature

/1/ A. WOZNIAK, W. KNAPCZYK: Possibility of the Ship Assembling Auto-
    mation Using Laser Technique, Proceed.of 2nd National Conference
    on Geodesy in Marine Industry, Gdańsk 16-17.09.1979, /1979/
/2/ A. WOZNIAK: Dimensional Control System for Ship Constructions
    Based on Laser Technique, Proceed. of Electro-Optics/Laser 82 UK
    Intern.Conference, Brighton 23-25.03.1982, /1982/
/3/ A. WOZNIAK, W. KNAPCZYK, K. ŁUTOWICZ: Microprocessor Controlled
    Laser System for Measurements of Constructions, Proceedings of
    Electro-Optics/Laser 84 UK Intern.Conf.,Brighton 20-22.03./1984/
/4/ B. WITTE - Personal Contacts /1984/

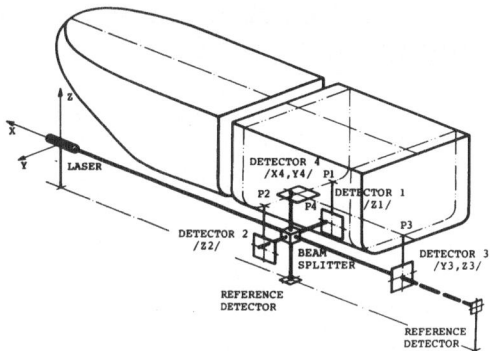

Fig.1 A lay-out of the proposed
system.

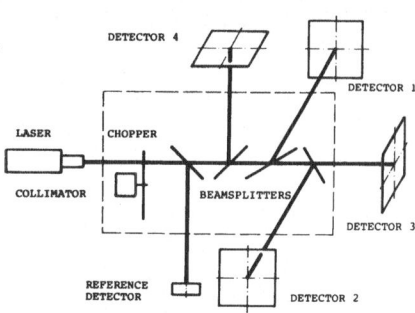

Fig.2. An idea of a beam-
splitter.

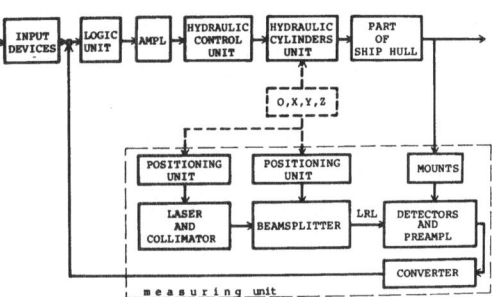

Fig.3. A block-diagram
of the system.

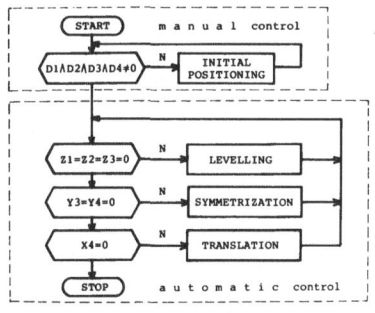

Fig.4. A flow-chart
of a positioning
algorithm.

Fig.5. A general view
of the experimental stand
and measuring devices.

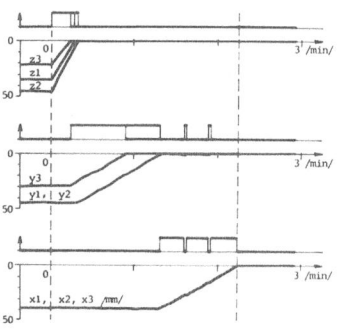

Fig.6. Operational movements
of hydraulic cylinders
versus error signals.

Ein neuartiger Laserscanner mit gekrümmter Abtastbahn zur Strichcode-Erfassung in der Materialflußtechnik

H.-H. Spratte

Vorentwicklung der Leuze electronic GmbH + Co

Postfach 1111, 7311 Owen-Teck, BR Deutschland

1. Einleitung

Fortschreitende Automatisierungsbestrebungen, vor allem im Bereich der Materialfluß-
technik, erfordern in zunehmendem Maße den Einsatz der Strichcodierung auf z.B. Lager-
behältern, KFZ-Teilen, Fluggepäck usw. Dabei gewinnen die Strichcode-Lesegeräte,
deren Funktion auf Laserscannern [1] beruhen, immer mehr an Bedeutung, da derartige
Geräte eine relativ große Leseentfernungsvariation (Tiefenschärfe) ermöglichen und
gleichzeitig durch eine relativ hohe Scanrate Mehrfachlesungen erlauben und damit
eine hohe Lesesicherheit garantieren.

Zur Lichtablenkung eignen sich ganz allgemein alle physikalischen Effekte, die direkt
oder indirekt irgendwie eine Lichtquelle beeinflussen. Nach [1] werden gewöhnlich aber
nur diejenigen "makroskopischen" Effekte zur Lichtablenkung angeführt, die den Wellen-
vektor direkt durch Beeinflussung der räumlichen Phase ändern, und zwar durch Reflexion, Beugung und Brechung (siehe auch Abb. 1). Die vier weiteren zur Scanner-Reali-
sierung geeigneten Parameter einer ebenen Lichtwelle, nämlich Wellenlänge, Amplitude,
Polarisation und absolute zeitliche Phasenlage, bleiben häufig unberücksichtigt.

Abb. 1 zeigt eine Übersicht mechanisch bewegter Scanner nach [2] sowie Erweiterungen
durch den Autor. Wie oben schon angedeutet, wird auch hier nur der erste und einzige
Parameter einer ebenen Lichtwelle, nämlich die direkte Beeinflussung des Wellenvek-
tors durch Reflexion, Beugung und Brechung ausgenützt.

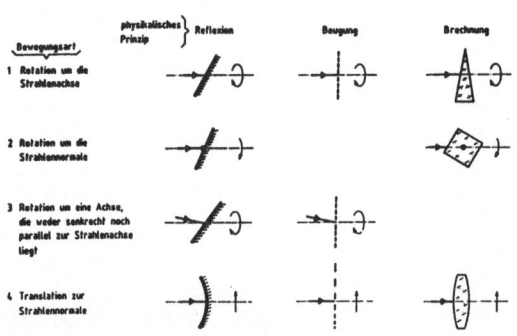

Abb. 1.

Mechanische Scan-
ner-Konfiguratio-
nen nach [2] sowie
Erweiterungen
durch den Autor

---

1) Laserscanner bestehen prinzipiell aus einer Laserquelle und einem Lichtablenker

Die in Spalte 1 dargestellten Reflexionsablenker sind Untersuchungsgegenstand der vorliegenden Abhandlung und sollen in den Abschnitten 2 und 3 diskutiert werden. Unter den in Spalte 2 skizzierten Beugungsablenkern versteht man holografische Lichtablenker, die die Beugung von räumlich-kohärentem Licht an einem bewegten Hologramm ausnutzen. Da beugende Elemente spektrale Dispersion aufweisen, verwendet man im Normalfall auch monochromatisches, also zeitlich-kohärentes Licht. Die in Spalte 3 gezeigten Brechungsablenker bestehen in der Regel aus bewegten Glaskörpern, deren Ablenkprinzip auf dem Brechzahlsprung bzw. Brechzahlgradient beruht.

2. Theorie und Experiment

Die Reflexionsablenker nach Spalte 1 in Abb. 1 sind Scanner relativ hoher mechanischer Trägheit und haben bei leider nur seriellem Zugriff eine hohe Scan-Punkt-Anzahl (aufgrund großer Aperturen und großer Scanwinkel), eine relativ gute Ablenklinearität und eine hohe Scangeschwindigkeit (definiert als Pixel pro Zeit).

2.1 Scanpfad und Laserspotgeschwindigkeit für den Reflexionsablenker nach Zeile 2 in Abb. 1

Aus dem Berechnungsschema in Abb. 2 folgt für die Scanpfadlänge in x-Richtung

$$x_{SL} = z_0 \tan\delta - r \left(\frac{1}{\cos\psi} - 1 \right) \frac{\sin(\delta/2 + \psi)}{\sin(\pi/2 - \gamma - \delta/2)} \tag{1}$$

wobei δ der Scanwinkel, ψ der Einfallswinkel für $\vartheta = 0$, z_0 die Scanpfadentfernung, ϑ der Drehwinkel und r der Radius (halbe Schlüsselweite) des Reflexionsablenkers ist. Da die Einfallsebene für beliebige ϑ senkrecht auf der Drehachse steht, hat der Scanpfad geradlinige Form. Den für $\psi = \pi/4$ gültigen Scanwinkel $\delta = 2\vartheta$ in Gl. (1) eingesetzt und dann Gl. (1) unter Berücksichtigung von $\vartheta = \dot{\vartheta} t$ nach der Zeit differenziert liefert die Laserspotgeschwindigkeit mit

$$v_x = \frac{dx_{SL}}{dt} = \dot{\vartheta} \left[\frac{2 z_0}{\cos^2 2\vartheta} - r \, \frac{1 - \cos\vartheta + \sin\vartheta \tan(\pi/4 - \vartheta)}{\sin(\pi/4 - \vartheta) \tan(\pi/4 - \vartheta)} \right] \tag{2}$$

unter der Voraussetzung $\dot{\vartheta} = \text{const.} \neq f(t)$.

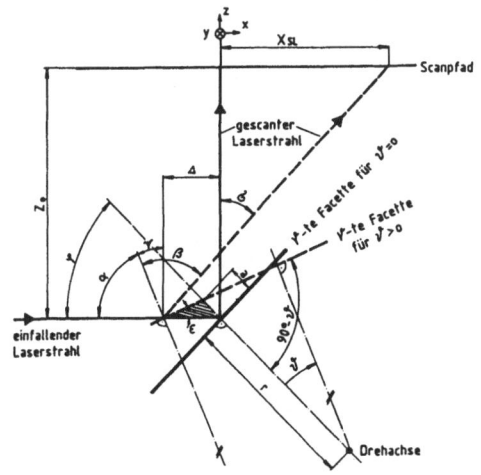

Abb. 2.

Berechnungsschema zur Bestimmung der Laserspotgeschwindigkeit in x-Richtung für den Reflexionsablenker nach Zeile 2 in Abb. 1

In Abb. 3 ist für die Scanpfadentfernung z_o = 200 mm die Laserspotgeschwindigkeit v_x
als Funktion des Reflexionsablenker-Drehwinkels mit Parameter r aufgetragen. Die dargestellten Abhängigkeiten haben Gültigkeit für die konstante Drehwinkelgeschwindig-
keit $\dot\vartheta$ = 314 rad/s und zeigen eine größere Geschwindigkeitszunahme für r = o
(Reflexionspunkt fällt mit Drehpunkt zusammen) als für r > o.

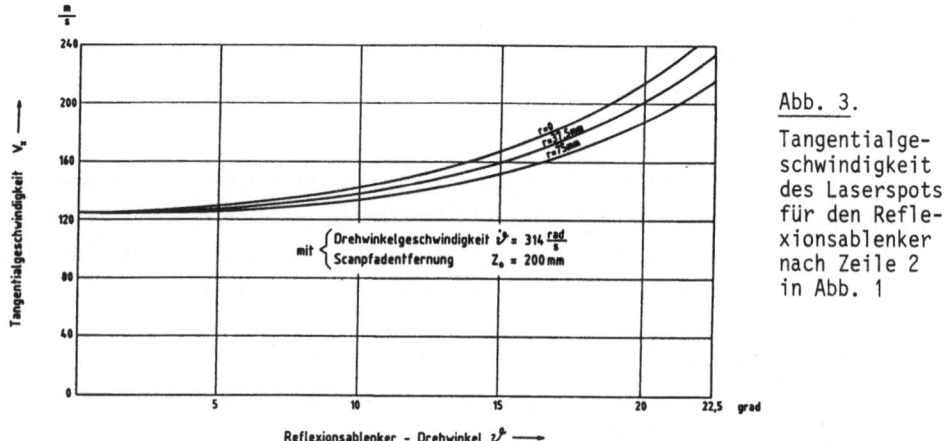

Abb. 3.
Tangentialge-
schwindigkeit
des Laserspots
für den Refle-
xionsablenker
nach Zeile 2
in Abb. 1

Da aber für die hier vorliegende Applikation die Erkennungsinformation eng mit der
Laserspotgeschwindigkeit verknüpft ist, muß sichergestellt sein, daß die Geschwindig-
keitsänderung entlang des Scanpfades bestimmte Grenzwerte nicht überschreitet. Daraus
folgt sofort, daß der Reflexionsablenker einen bestimmten Mindestradius haben sollte.

2.2 Scanpfad und Laserspotgeschwindigkeit für den Reflexionsablenker nach Zeile 3 in Abb. 1

Das in Abb. 4 skizzierte Berechnungsschema liefert für den in die y,z-Ebene projizierten Pyramidalwinkel

$$\mathcal{J} = \arc\tan\left[\cos\vartheta\,\tan\xi\right] \tag{3}$$

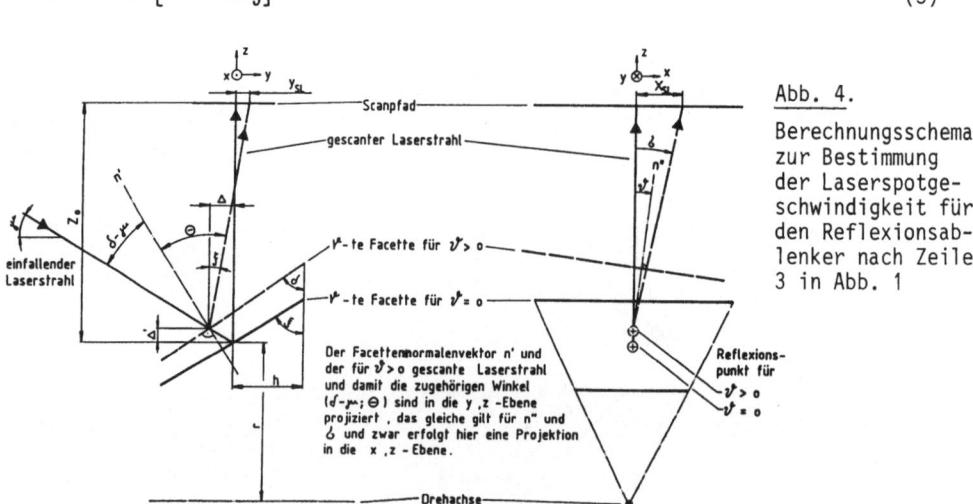

Abb. 4.
Berechnungsschema
zur Bestimmung
der Laserspotge-
schwindigkeit für
den Reflexionsab-
lenker nach Zeile
3 in Abb. 1

Der in der Einfallsebene liegende Einfallswinkel α_e lautet

$$\alpha_e = \text{arc} \cos\left[\cos\vartheta \cos(\delta-\mu)\right] \tag{4}$$

mit μ gleich Winkel zwischen einfallendem Laserstrahl und Drehachse. Damit folgt für den in die y,z-Ebene projizierten Reflexionswinkel

$$\Theta = \text{arc} \cos\left\{\frac{\cos\vartheta}{\cos\delta} \cdot \frac{\sin(\phi-\alpha_e)}{\sin(\pi-\phi)}\right\} \tag{5},$$

wobei gilt

$$\phi = \text{arc} \sin\left\{\frac{\sin\alpha_e}{\cos(\delta-\mu)}\left[\tan^2\vartheta + \tan^2(\delta-\mu)\right]^{-\frac{1}{2}}\right\}$$

Aus Gl. (4) und (5) läßt sich nun der Scanwinkel b zu

$$b = \text{arc} \cos\left\{1-\left[\frac{\sin\vartheta}{\sin(\pi-\phi)}\left\{\sin(\phi-\alpha_e)+\frac{\sin\alpha_e}{\cos\vartheta}\left[\tan^2\vartheta+\tan^2(\delta-\mu)\right]^{-\frac{1}{2}}\right\}\right]^2\right\}^{\frac{1}{2}} \tag{6}$$

herleiten. Der Scanwinkel b ist dabei in die x,z-Ebene projiziert.

Abb. 5 zeigt den Pyramidal- δ , Einfalls- α_e, Scan- b und Reflexionswinkel Θ als Funktion des Drehwinkels ϑ . Die Interpretation der Abb. 5 liefert ein Ergebnis in der Form, daß der Scanpfad eine Krümmung aufweist, und zwar in Richtung der + y-Achse.

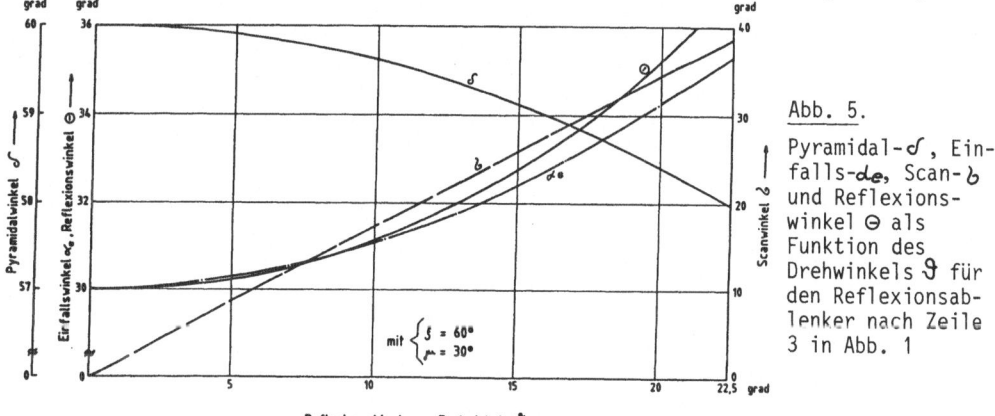

Abb. 5.

Pyramidal- δ , Einfalls- α_e, Scan- b und Reflexionswinkel Θ als Funktion des Drehwinkels ϑ für den Reflexionsablenker nach Zeile 3 in Abb. 1

Der entscheidende Vorteil des gekrümmten Scanpfades ist die geringere Geschwindigkeitsvariation des Laserspots entlang der x-Achse gegenüber den Reflexionsablenkern

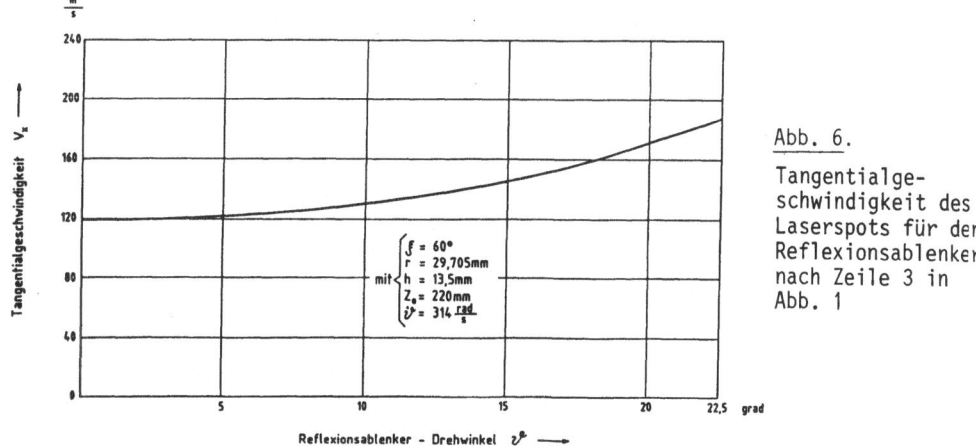

Abb. 6.

Tangentialgeschwindigkeit des Laserspots für den Reflexionsablenker nach Zeile 3 in Abb. 1

nach Abschnitt 2.1. Dieser Sachverhalt soll durch Abb. 6 dokumentiert werden.

2.3 Sendeseitiger Fleckradius als Funktion der Leseentfernung

Nach [4] arbeitet man bei Fokussierung eines Laserstrahls in großer Entfernung (Fern-bereich)[1] vorteilhaft mit einem teleskopischen Linsensystem, bei dem die benötigte kleine Primär-Strahltaille w_0' vom Okular erzeugt wird. Allgemein gilt bei der Trans-formation eines Gaußstrahles durch eine Linse [2] der Brennweite f

$$W_o'^{-2} = W_{ol}^{-2} (1 - \frac{d}{f})^2 + f^{-2} (\frac{\pi \, W_{ol}}{\lambda})^2 \tag{7},$$

$$d'-f = \frac{f^2 (d-f)}{(d-f)^2 + (\frac{\pi \, W_{ol}^2}{\lambda})^2} \tag{8}$$

mit w_{ol} gleich Laserstrahltaille, d Abstand zwischen Linse und w_{ol}, d' Abstand zwi-schen Linse und w_0' und λ Laserwellenlänge. Aus Gl. (7) und (8) folgt unter Berück-sichtigung des teleskopischen Linsensystems die Sekundär-Strahltaille w_0 nach Abb.7. Für den Zusammenhang im Lesegebiet läßt sich dann schreiben

$$W_{(z)} = W_o \left[1 + (\frac{\lambda \, Z}{\pi \, W_o^2})^2 \right]^{\frac{1}{2}} \tag{9}$$

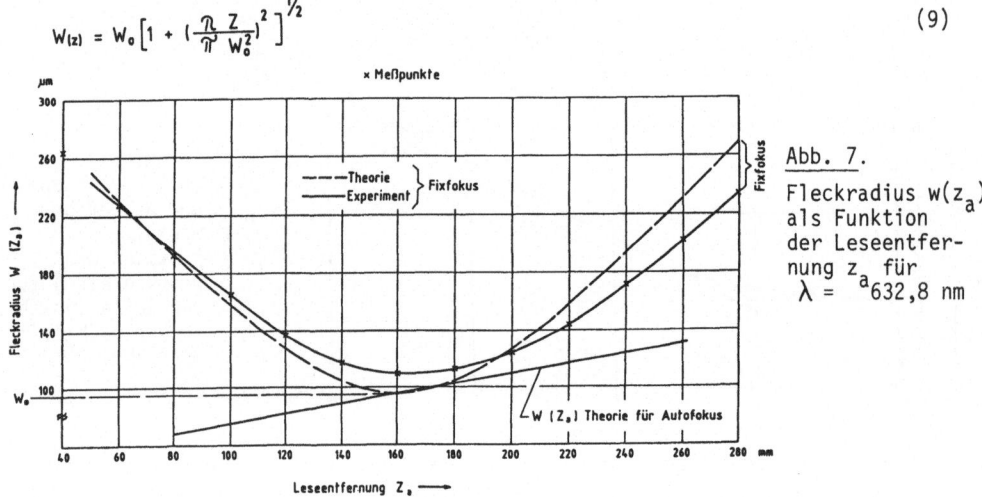

Abb. 7.
Fleckradius $w(z_a)$ als Funktion der Leseentfer-nung z_a für $\lambda = 632,8$ nm

Die in Abb. 7 angegebenen Meßpunkte für die Fixfokusoptik wurden mit dem Beam Scan-Meßgerät der Fa. Photon Technology ermittelt. Der theoretische Zusammenhang für die Autofokus-Optik zeigt einen erheblich kleineren Fleckradius bei Entfernungsvariation, was bedeutet, daß deutlich bessere Leseeigenschaften als bei einer Fixfokusoptik zu erwarten sind.

2.4 Empfangsleistung als Funktion der Leseentfernung

Ausgangspunkt der folgenden Abschätzung ist die Annahme, daß das eingestrahlte Laser-bündel auf der Bar-Code-Ebene eine reflektierte Streulichtverteilung erzeugt, die in erster Näherung bezüglich der Richtcharakteristik einen Lambert-Strahler darstellt

---

1) Die Grenze zwischen Nah- und Fernbereich liegt bei Annahme realistischer Werte unter Berücksichtigung sphärisch korrigierter Linsen bei einigen 10 mm.
2) Da am Linsenrand keine Beugung auftreten darf, sollte der Linsendurchmesser $D_L \gtrsim 4$ w sein.

und zum anderen als Punktlichtquelle approximiert werden kann. Bezogen auf ein Kugel-koordinatensystem strahlt diese Punktlichtquelle nach [5] in den Halbraum $o \leq \ell \leq \pi/2$ in das Raumwinkelelement $d\omega = \sin\ell \, d\ell \, d\varphi$ (φ = Azimutwinkel, ℓ = Winkel zwischen z-Ach-se und Radiusvektor) die Leistung

$$dP = J_o \cos\ell \sin\ell \, d\ell \, d\varphi \qquad (10)$$

ab, wobei I_o die Strahlstärke in z-Richtung charakterisiert. Damit folgt für die max. detektierbare Leistung

$$P_e = \int_{\varphi=o}^{2\pi} \int_{\ell=o}^{\ell_o} J_o \cos\ell \sin\ell \, d\ell \, d\varphi \qquad (11)$$

mit ℓ_o gleich halber Öffnungswinkel der Empfangsoptik [1]. Die Integration der Gl. (11) liefert

$$P_e = \pi \, J_o \sin^2 \ell_o \qquad (12)$$

In Abb. 8 ist dieser Zusammenhang in Abhängigkeit von der Leseentfernung z_a für den Reflexionsablenker-Drehwinkel ϑ = o aufgetragen. Die Meßwerte stimmen recht gut mit den theoretischen Werten überein.

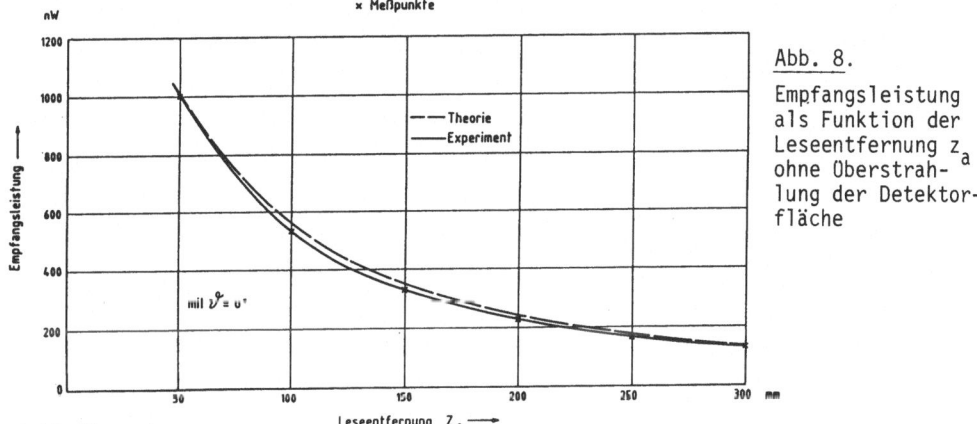

Abb. 8.

Empfangsleistung als Funktion der Leseentfernung z_a ohne Oberstrahlung der Detektor-fläche

3. Schlußbemerkungen

Die Untersuchungen haben ergeben, daß mit der Konzeption des vorgestellten Laser-Scanners wesentliche Vorteile gegenüber bekannten Systemen erreicht wurden. Die wichtigsten Vorteile sind:

- kleinere Geschwindigkeitsvarianten entlang des Scanpfades,
- Autofokus und damit größere Entfernungsvariation bezüglich der Leseebene,
- relativ hohe Empfangsleistung bei niedriger Laserleistung, da das von der ent-sprechenden Facette des Reflexionsablenkers erfaßte Empfangslicht unmittelbar auf die Empfangsoptik gespiegelt und von dort auf den Detektor fokussiert wird.

4. Literatur

[1] V.GERBIG: Dissertation, Erlangen (1981)
[2] O.BRYNGDAHL, W.-H. LEE: Applied Optics 15, 183 (1976)
[3] U. IKEDA, M. ANDO, T. INAGAKI: Applied Optics 18, 2166 (1979)
[4] D. ROSENBERGER: Technische Anwendungen des Lasers, Springer-Verlag (1975)
[5] M. UHLE: Dissertation, Bochum (1974)

1) Bei dieser Abschätzung wird die Empfangsleistung nicht durch die Detektorfläche, sondern durch den Durchmesser der Empfangsoptik begrenzt.

Untersuchungen zur Schwingungsmessung an großflächigen Bauteilen mit Lichtleitfaser-Abstandssensoren

W. Jüptner, H. V. Fuchs*, H. Kreitlow
BIAS - Bremer Institut für angewandte Strahltechnik, D-Bremen
*Fraunhofer-Institut für Bauphysik, D-Stuttgart

Einleitung

Es wird über die Entwicklung und Anwendung eines Lichtleitfaser-Sensors berichtet, der für die berührungslose Messung kleinster Abstandsänderungen im Nanometer-Bereich geeignet ist. Der Sensor besteht aus einer Multimode-Gradientenfaser zur Beleuchtung der Objektoberfläche und 6 weiteren konzentrisch darum herum angeordneten Fasern zur Reflexionsmessung an der Objektoberfläche. Aus der Kennlinie "Sensorabstand zur Oberfläche - Ausgangsspannung an der Empfängerdiode" können neben Meßgrößen der Fertigungsmeßtechnik insbesondere Schwingungsamplituden hochgenau gemessen werden. Der Meßbereich beträgt im linearen Teil der Kennlinie 10 - 500 μm und erreicht im nichtlinearen Teil bis zu mehreren Millimetern. Die Meßauflösung liegt bei bis herunter zu einem Nanometer.

Aufbau des Lichtleitfaser-Sensors (LFS)

Der LFS besteht in seinem Kern aus einer Sende- oder Beleuchtungsfaser, in die Licht einer inkohärenten oder kohärenten Lichtquelle eingekoppelt wird unter Beachtung der faserabhängigen numerischen Apertur für den Lichteintritt. Das Licht tritt am anderen Ende der Faser, dem Meßkopfende des Sensors, kegelförmig wieder aus und fällt auf die streuende Meßoberfläche. Das reflektierte und/oder zurückgestreute Licht wird von einer Empfangsfaser (oder mehreren konzentrisch zur Sendefaser angeordneten Fasern) aufgenommen und zu einem optoelektronischen Wandler, z.B. einer Fotodiode, geleitet. Die Empfangsfasern können in einem oder mehreren Ringen um die Beleuchtungsfaser angeordnet sein, s. Bild 1.

Der LFS für die Untersuchung des Abstrahlgrades von leichten Bauteilen (s.u.) wurde aus Kunststoff-Stufenindexfasern (Polymethylmethacrylate) mit einem Durchmesser von 1 mm in einer ähnlichen Weise wie in (1) beschrieben gebaut. Die 6 äußeren Fasern sind mit einem photoelektrischen Wandler, Typ OSI 5k, verbunden. Die innere Faser ist mit einer Lichtquelle (3,2 Volt/100 mA Gleichstrom-Lampe) verbunden. Für die ersten Versuche mit diesem LFS wurden der photoelektrische Wandler und die Lampe in LWL-Verbindungsbauelemente eingebaut.

Ermittlung der Kennlinie

Variiert der Abstand zwischen dem Auskoppelende des Sensors und der Meßoberfläche, so wird die an der Fotodiode gemessene, von der Intensität des dort ankommenden Lichtes abhängige Spannung, verändert. Die Abhängigkeit der gemessenen Spannungsänderung von der Abstandsänderung wird durch die typische Kennlinie in Bild 2 wiedergegeben:

Für die Anwendung wichtig ist der in den Bereichen 2 und 4 annähernd lineare Zusammenhang zwischen dem Ausgangssignal an der Empfängerdiode und dem zu messenden Abstand zwischen Sensor und Oberfläche des Objektes. Im ansteigenden Bereich 2 der Kennlinie ist die Meßempfindlichkeit jeweils höher als im abfallenden Bereich 4.

Der im Bild 2 gezeigte grundsätzliche Verlauf der Kennlinie ist mit Hilfe der in den Abstandsbereichen 1 bis 5 unterschiedlichen Einkopplungsverhältnisse des von der Oberfläche gestreuten Lichtes in die Empfangsfasern zu erklären, s. Bild 3. Zwei Effekte bestimmen den Kennlinienverlauf: Der sich verbreiternde Sendelichtkegel und die sich ändernden Reflexionsbedingungen zu den Empfangsfasern. Danach hängt das Ausgangssignal an der Empfängerdiode neben dem zu messenden Abstand im wesentlichen noch von folgenden Größen ab:

- Durchmesser des Sendefaserkerns
- Durchmesser des Empfangsfaserkerns
- Abstand zwischen beiden.

Darüberhinaus hängt die auf das Maximum des Empfängersignals normierte Steigung in den linearen Bereichen der Kennlinie hauptsächlich von der Oberflächenbeschaffenheit, insbesondere von den durch die Mikrogeometrie beeinflußten Reflexionseigenschaften des Meßobjektes ab.

Die Reflexionseigenschaft der Meßoberfläche ist i.a. abhängig vom Werkstoff, s. Bild 4, und kann darüberhinaus bei Verwendung desselben Werkstoffs, abhängig von der Mikrogeometrie der Oberfläche, von Meßpunkt zu Meßpunkt, variieren und hat somit einen großen Einfluß auf die Empfindlichkeit.

Dieser Einfluß wurde untersucht, indem die Empfindlichkeit an über 100 verschiedenen Punkten an einer Meßoberfläche ermittelt wurde. Hierzu wurde die Positioniervorrichtung programmiert, um den Bereich "2" der Kennlinie schrittweise aufzunehmen und daraus die resultierende Steigung E zu berechnen. Die Standardabweichung von E wurde ermittelt für Messungen auf einer gereinigten Aluminiumoberfläche. Die mittlere Empfindlichkeit lag bei 20,8 mV/µm mit einer Standardabweichung von 2,6 mV/µm. Die Wiederholung desselben Vorgangs auf nur einem Meßpunkt ergab eine mittlere Empfindlichkeit von 23,9 mV/µm mit einer Standardabweichung von 0,47 mV/µm. Für genaue Messungen ist es also notwendig, den Kalibriervorgang bei jedem Meßpunkt erneut durchzuführen.

Anwendungen

1. Messung von Abstandsänderungen

Aufgrund der in Bild 2 gezeigten Kennlinie können Abstandsänderungen hochgenau gemessen werden, wobei im linearen Teil der Kennlinie über einen Meßbereich von mehreren 100 µm eine Auflösung bis herunter zu 1 nm erreicht wurde und die zeitliche Auflösung von der Empfänger- und Auswerteelektronik begrenzt wird.

Bild 5 zeigt das Meßergebnis für eine periodische Abstandsänderung zwischen Sensorkopf und einem rotierenden Werkstück in einer Dreh-maschine.

2. Oberflächeninspektion

Durch eine translatorische Bewegung des Sensorkopfes parallel zu einer zu prüfenden Oberfläche können neben Rauheitseigenschaften Oberflä-chenfehler wie Riefen, Kratzer u.ä. erkannt werden, s. Bild 6. Die Auf-lösung hängt dabei vom Aufbau des Sensors und dessen Kennlinie ab (s.o.)

3. Untersuchung des Abstrahlgrades

Für die experimentelle Untersuchung des Abstrahlgrades von leichten Bauteilen muß die Schwingungsamplitude und die abgestrahlte Schall-leistung gemessen werden. Der Leichtleitfasersensor wird dazu in eine rechnergesteuerte 3-Koordinaten-Positioniereinrichtung eingebaut, so daß eine große Anzahl von Meßpunkten auf der Bauteiloberfläche in einem programmierten Meßraster in kurzer Zeit angefahren werden kann. In jedem Meßrasterpunkt wird die Sensorkennlinie automatisch neu auf-genommen, um Einflüsse der Oberfläche zu vermeiden. Nach dem Anfahren des Arbeitspunktes im linearen Teil der Kennlinie wird die Messung der Schwingungsamplitude freigegeben (2).

Durch breitbandige Schwingungsanregung des Bauteils und mit Hilfe der Fourieranalyse des gemessenen Schwingungsamplitudenfeldes werden die Schwingungsformen des Bauteils ermittelt, s. Bild 7.

Schrifttum

(1) S. Shaw: IBP Bericht BS 99/84 (1984)

(2) S. Shaw, H.V. Fuchs, H. Kreitlow, W. Jüptner:
    DAGA'84, Tagungsband (1984)

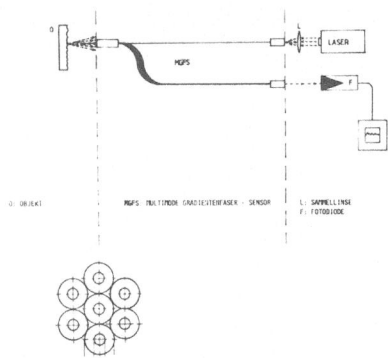

Bild 1.
Prinzip und Aufbau des Lichtleitfaser-Sensors (LFS)

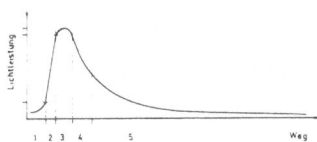

Bild 2.
Kennlinie des LFS

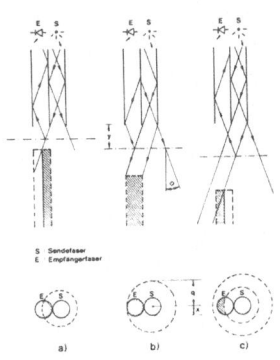

Bild 3.
Einfluß des Sensorab-
standes von der Meß-
oberfläche auf die
Lichtleistung in der
Empfängerfaser (1)

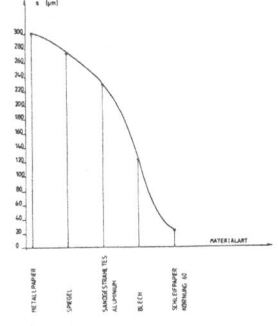

Bild 4.
Einfluß der Werk-
stoffoberfläche
auf die Länge des
linearen Kennlinien-
bereichs 2

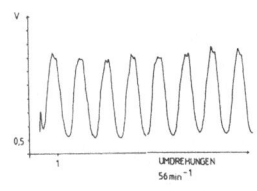

Bild 5.
Periodische Ab-
standsänderung
zwischen LFS und
rotierendem Werk-
stück

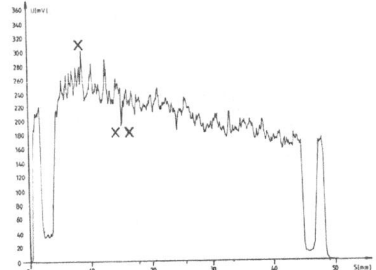

Bild 6.
Messung der Mikrogeometrie,
Nachweis von Kratzern und
Riefen (z.B. x bzw. xx)

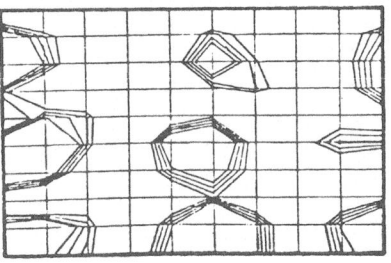

Bild 7.
Höhenlinien der (3,3) Schwin-
gungsform eines 1m x 2m Alu-
Blechs, Meßraster 100 Punkte,
Frequenz 67 Hz, Amplitude 0,9
- 1,2 µm

An All Fiber Current Sensor Operating without Linkage on Three-Phase Lines

V. Annovazzi Lodi, S. Donati
Dipartimento di Elettronica
Università di Pavia
27100 - PAVIA, ITALY

Summary

We demonstrate a current sensor made of monomode fiber allowing operation without linkage to the HV-line. This is obtained by combination of non-reciprocal Faraday birefringence in a fiber coil of properly-choosen geometry, which, in addition, allows a very compact design and a good immunity to environmental magnetic fields.
The theory of operation is discussed in detail and experimental results are reported. A current meter has been built employing a multimode laser diode as the source and a polarization analyzer readout of the output. The instrument sensitivity performance is about 1Arms (S/N = 1, f = 50 Hz, τ = 1s), the dynamic range is three decades, while the maximum available bandwidth is about 1KHz.

1 Introduction

Power transmission lines are expected to operate at ever increasing voltages in the coming years; in addition, higher short-circuit current values will be managed. Consequently, the design and installation of conventional current transformers for metering, control and protection purposes will be difficult and costly mainly for insulation requirements.
To circumvent these problems, a number of unconventional methods for current measuring have been proposed and most of them are based on electrooptical approaches [1,2]. Many authors have developped measuring schemes in which a fiber optic link is used to connect an active sensing unit at line potential and a metering apparatus at ground potential. This approach allows a high insulation level and a very good immunity to electromagnetic interference, though requiring a supplementary power supply for the circuitry at the line voltage.
Another approach exploits the interaction between the magnetic field generated by the electric current to be measured and a proper material to cause a change in its optical properties, such as the refractive index; this change is detected by interferometry or other suitable method.
This approach allows to build completely passive sensors. By exploiting the Faraday effect, an optical fibre coil can be used as a sensing element, thus giving a compact, low cost current sensor. The sensitivity can be high even with a standard fiber since the interaction length can be inherently long.

2 The current sensor

The Faraday rotation in a fiber path 1 due to a magnetic field H can be expressed as:

$$(1) \qquad \Phi = \int_l V\overline{H} \cdot \overline{dl}$$

where V is the Verdet constant of the material ($V = 3.4 \ 10^{-4}$ deg/A for silica). Obviously, if H is the magnetic field of a current I flowing in a straight wire, the integral (1) vanishes on any closed path external to the wire.

From Fig. 1a it can be seen, for a particular geometry, how the contribution on opposite straight sides cancel each other in an ideal fiber. However, if the bending birefringence phase delay of the semicircular edges between the straight sides is not negligible and amounts to a semiinteger value of half-waves, the Faraday contributions can be summed up (Fig. 1b), allowing the current measurement without linkage between the conductor and the fiber coil [3,4] .

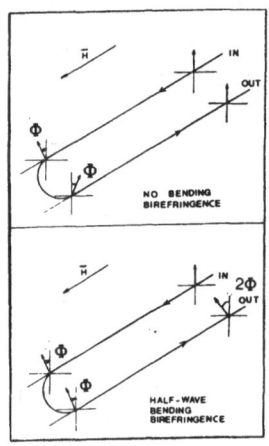

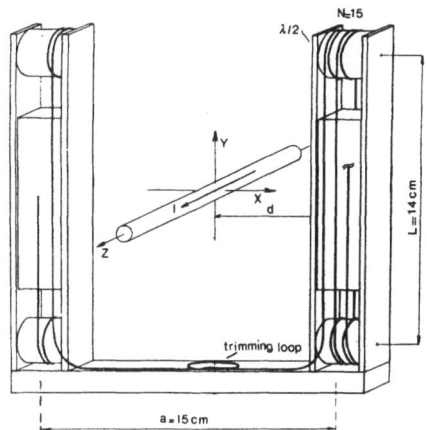

Fig. 1. Effect of bending birefringence   Fig. 2. Fiber sensor geometry

A sensor might be implemented simply with one-half of the configuration reported in Fig. 2 (one coiled-half) [3,4] . The required bending radius of the retarders is calculated in Refs. [5,6] ; in most fibers, the bending attenuation is too high to implement the retarders with a half turn only and we shall employ a semiinteger number of turns, which can be of larger radius and weaker attenuation. In fact, the dependence of the phase delay on the bending radius is of the form [5,6] :

$$(2) \qquad \Delta \psi \propto \frac{1}{R} \qquad\qquad\qquad \text{(for a whole loop)}$$

while for the attenuation constant α :

$$(3) \qquad \alpha = \frac{A}{\sqrt{R}} \ \exp \ (-BR)$$

An accurate trimming of the bending radius is advisable to minimize spurious effects, though independent trimming of each loop is not mandatory.

The effect of the magnetic field on the edge loops can be found in Ref. 7 where the superposition of reciprocal bending birefringence and non reciprocal Faraday birefrin

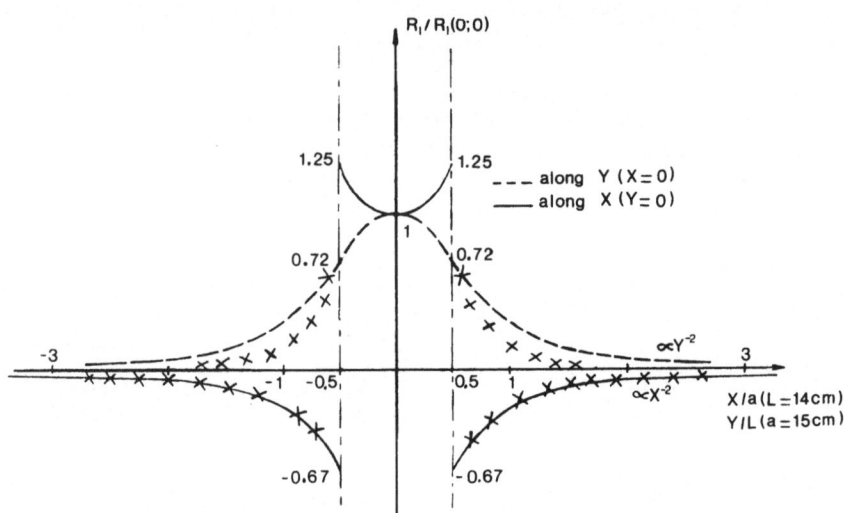

Fig. 3. Responsivity vs.position; theoretical curves and experimental points (X).

gence on an arc of fiber has been studied in detail. For example, when the $\lambda/2$ retarder is made by a one-turn-and-half loop (as we used in our sensor),its contribu
tion to the Faraday effect is zero. Incidentally, this theory allows to build a non
-linked fiber current sensor made of a circular coil of proper radius, though, this
way, the responsivity is not optimized for a given length of fiber [7,8] . The current responsivity of the one-arm sensor can be derived as [4] :

(4) $R_I = \Phi/I = [NVL/(\pi d)]$ $(2d/L)$ atan $(L/2d)$

where N is the number of turns while the geometrical parameters d and L can be found
in Fig. 2.
Such a basic sensor, however, is not well-suited for current detection on a three-
-phase system, because of superposition of the three current contributions, though
weighted inversely with the relative distance.
A better immunity to the currents in the other two conductors can be achieved with
the two-arm geometry of Fig. 2.
In this sensor the Faraday rotation is summed up on the two arms for current contri
butions between the arms while it is canceled out for current contributions external
to both arms.
The responsivity of this sensor as a function of position for a wire parallel to the
z-axis can be derived as:

(5) $R_I = \dfrac{NV}{\pi}$ atan $\left[\dfrac{L + 2Y}{a + 2X} + \text{atan} \dfrac{L - 2Y}{a + 2X} + \text{atan} \dfrac{L - 2Y}{a - 2X} + \text{atan} \dfrac{L + 2Y}{a - 2X} \right]$

and is reported in Fig. 3 for L = 14 cm, a = 15 cm, i. e., the values we choosed
for our prototype. As can be seen, the immunity to external contributions is good
(R_I $(XY)^{-2}$ for large X and Y) and allows to make measurements on three-phase lines
with small error (e.g., less than 1% for a system where the distance between   the
conductors is l = 1m).

The immunity to current contributions along the axis X and Y is also high as it can
be deduced by the geometry of fig. 2.

Also the saddle point in X = Y = 0 makes it easy to position the sensor.

Starting from this current sensor, we have developed a current measuring instrument
whose block scheme is given in Fig. 4. The fiber coil is fed by the linear polarized
radiation emitted at 830nm by a multimode laser diode and the Faraday rotation   is
detected by a photodiode through a polarizer oriented at $\lambda/4$ with respect to the in
put polarization direction giving a signal proportional to sin 2 Φ.

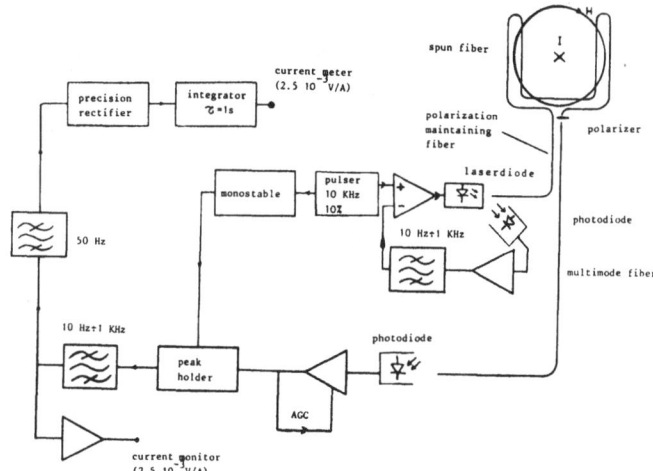

Fig.4. Block scheme of
the instrument

The responsivity in X = Y = 0 is R_I = 85 µrad/A for N = 15, L = 14 cm, a = 15 cm; the
sensitivity is I = 1A (S/N = 1; $\tau \doteq$ 1s) and the dynamic range extends to I = 2400A
(Φ= .20 rad) for an integral linearity error less than 2%. Though some improvements
could be achieved by doubling the detection channel ore by on interferometric measurement scheme , the present perfomances are adequate for the actual application.

A good immunity to environmental disturbances (vibrations, temperature drift) has
been obtained using low-birefringence (spun) fiber for the coil, polarization maintaining fiber for the adduction trunk and multimode fiber for the abduction trunk.

The source is pulsed to improve the long term reliability without introducing an ac
tive temperature control. An electrooptic feedback loop is implemented, which uses a
photo-diod placed near the laser-to-fiber joint to reduce the source noise. An AGC
block stabilizes against slow fluctuations due to launch, and mechanical and thermal
transients affecting the fiber; these control loops can improve sensitivity (S/N = 1)
of about 40 dB.

Two outputs were made available: one giving the rms value of the fundamental compo
nent (50 Hz) of the measured current, the other being a monitor for the current wave
form (10 ÷ 1000 Hz).

In Fig. 5 the minimum detectable signal is reported for both outputs while   Fig. 6
shows a switching transient as transduced by the current sensor.

Fig.7 shows the linearity error and the baseline drift of the instrument.

206

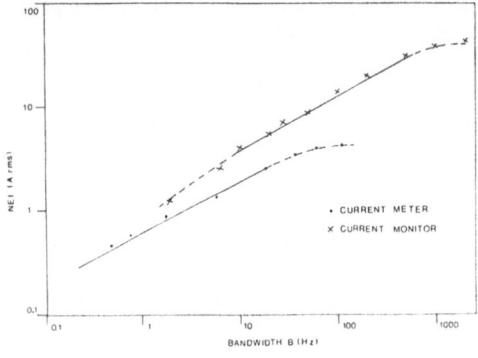
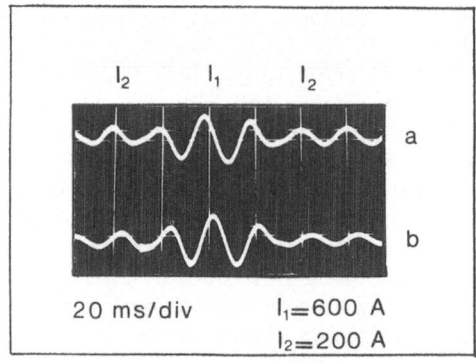

Fig. 5. Noise equivalent current (NEI)   Fig. 6. Current transient monitoring:
         vs. measure bandwidth B   a) by the fiber sensor, giving I
                                                    b) by a solenoid, giving $j\omega MI$

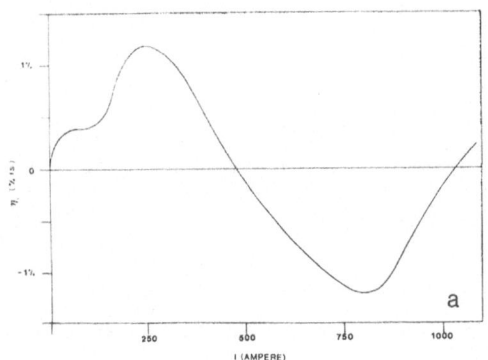
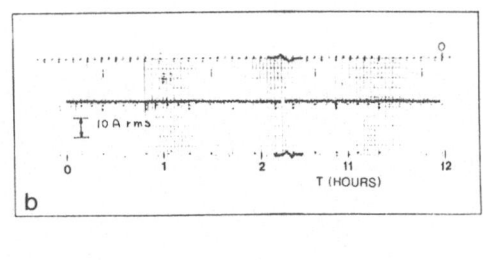

Fig.7. Integral linearity error vs the measured current (a); baseline drift of the
instrument (b).

References

1 M.N. Rzewuski, M.Z. Tamawecky, IEEE Trans on Instr. and Meas., IM-24 N° 1 (1975),
   43-51.
2 R.E Hebner, R.A. Malewski, E.C. Cassidy, Proc of the IEEE vol 65 N° 11 (1977),
   1524-1548.
3 R.A. Berg, H.C. Lefevre, H.J. Shaw in "Fiber Optic Rotation Sensors", S. Ezekiel,
   H.J. Arditty (eds.), Springer Verlag 1982, p 400-403.
4 S. Donati, V. Annovazzi Lodi, Alta Freq. 6-LIII (1984), 310-314.
5 V. Annovazzi Lodi, S. Donati, Alta Frequenza, 3-LI (1982), 159-163.
6 R. Ulrich, S.C. Rashleigh, W. Eichoff, Opt. Lett. 5-6 (1980), 273
7 V. Annovazzi Lodi, S. Donati, Opt. and Quant. El. 15 (1983), 381-388.
8 G.W. Day, D.N. Payne, A. J. Barlow, J.J. Ramskir-Hansen, Opt. Lett. 7 (1982),
   238-240.

Aufbau einer breitbandigen IR-Pulslichtquelle und deren Anwendung zur Charakterisierung von Bauelementen der integrierten Optik

J. Krauser, Heinrich-Hertz-Institut für Nachrichtentechnik Berlin GmbH,
Einsteinufer 37, D-1000 Berlin 10

1. Einleitung

Der Wellenlängenbereich zwischen 0,8 μm und 1,6 μm gewinnt aufgrund der stürmischen Entwicklung der optischen Nachrichtentechnik immer stärkeres Interesse. Bei der Suche nach einer geeigneten Pulslichtquelle für diesen Spektralbereich bietet ein Faser-Raman-Emitter /1/ die gewünschten Eigenschaften, die zur Charakterisierung von Komponenten der optischen Nachrichtentechnik benötigt werden.

2. Ramaneffekt in Glasfasern

Bei der Raman-Streuung werden Elektronen des Streumediums (hier Glasfaser) durch einen Pumplaser vom Grundzustand E_0 auf ein virtuelles Niveau angeregt. Die emittierten Photonen besitzen entweder die gleiche Energie wie die Pumpphotonen (Rayleigh-Streuung), niedrigere (Stokes) oder höhere (Anti-Stokes) Energie, wobei ein optisches Phonon erzeugt bzw. vernichtet wird. Bei genügend hoher Pumpintensität kommt es zu einer stimulierten Emission.

Die Ramanverstärkung in Glasfasern ist proportional

$$\exp\left(\frac{g\,PL}{A}\right) \quad (1),$$

mit g = Gainkoeffizient, P = Pumpleistung, A = effektive Fläche des Glasfaserkerns, L = Wechselwirkungslänge. Die Wechselwirkungslänge ist eine Funktion der Dämpfung

$$L = \frac{1-e^{-\alpha\ell}}{\alpha} \quad (2).$$

Für die Schwellenleistung zur Erzeugung der stimulierten Ramanemission gilt /2/

$$P_S \approx 17 \cdot \frac{A}{g\,L} \quad (3)$$

Das Maximum des Gainkoeffizienten liegt bei ca. $2 \cdot 10^{-11}$ cm/W bei einer Frequenzverschiebung von etwa 450 cm^{-1} /3/.

Aus (3) errechnet sich die Schwelle für die 1. Stokeslinie mit

α = 0,4 dB/km Faserdämpfung /4/ zu $P_S \approx 25$ W.

3. Experimenteller Aufbau

Als Pumplaser wird ein kontinuierlicher Nd:YAG-Laser (Baasel 615 ML/QS) mit der Möglichkeit der Modenkopplung und Güteschaltung verwendet. Die Resonatorlänge beträgt ca. 1500 mm. Die Stabilisierung des Resonators erfolgt mit einer Cerudurstange. Der Feinabgleich der Resonatorlänge wird mit einer Differentialmikrometerschraube vorgenommen, mit der auch das Mode-Lock-Band eingestellt wird. Mit einem 10-fach Mikroskopobjektiv wird das Licht in die Faser eingekoppelt (Abb. 1).

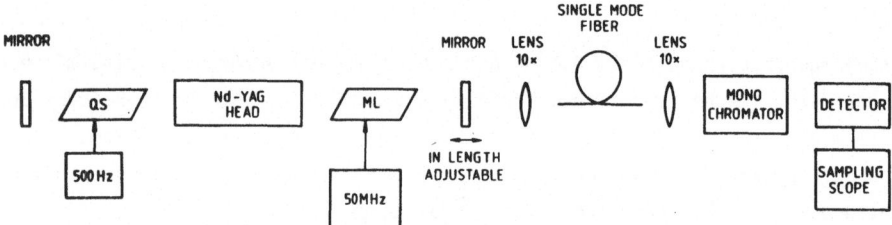

Abb. 1. Experimenteller Aufbau

Der Koppelwirkungsgrad liegt zwischen 10% - 20%. Der Faserausgang wird auf den Eintrittsspalt eines 30-cm-Gitter-Monochromators abgebildet. Am Ausgangsspalt erhält man dann das in seine Stokes bzw. Antistokes-Komponenten zerlegte Spektrum, das mit einer großflächigen Si- oder Ge-Diode oder einer schnellen Ge-PIN-Diode (τ_{FWHM}= 75 ps) detektiert wird.

Folgende Möglichkeiten zur Erzeugung der erforderlichen Pulsleistung stehen zur Verfügung:

1. Modenkopplung (Nd:YAG-Pulsleistung 300 W / 100 ps)

2. Güteschaltung (Nd:YAG-Pulsleistung 20 kW / 200 ns)

3. Verstimmen der Mode-Lock-Frequenz
   (Nd:YAG-Pulsleistung 12 kW / 150 ps)

4. Modenkopplung und Güteschaltung
   (Nd:YAG-Pulsleistung 75 kW / 150 ps)

4. Ergebnisse

In den Abb. 2-5 sind die in der Faser durch verschiedene Pumpleistungen erzeugten Spektren dargestellt. Bei einer Eingangsleistung von 30 W wird die 1. Stokeslinie mit etwa 100 ps Pulsdauer erzeugt (Abb. 2).

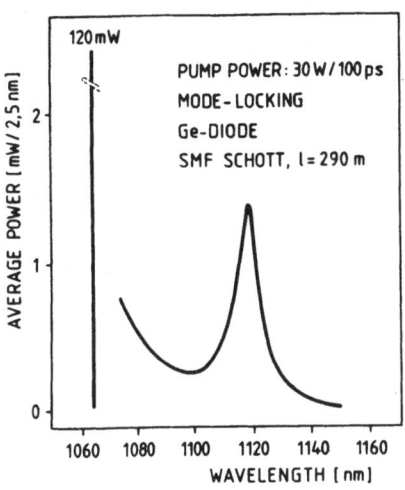

Abb. 2. Erzeugung der
ersten Stokes Linie
(nur Modenkopplung)

Im Falle der Güteschaltung mit 2 kW/200 ns Puls-Eingangsleistung erscheinen im Stokes-Spektrum mehrere Linien. In Abb. 3b ist der zeitliche Verlauf des Eingangspulses und Ausgangspulses bei λ= 1,06 μm dargestellt.

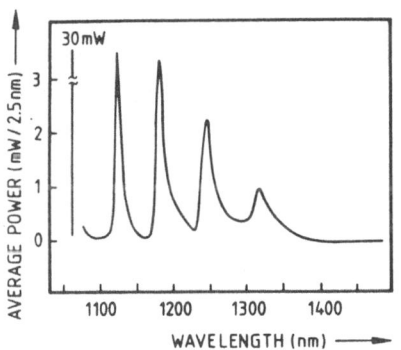

Abb. 3a. Raman-Spektrum,
mit einer Pumpleistung von
2 kW/200 ns und 800 Hz
Wiederholrate erzeugt (nur
Güteschaltung)

Abb. 3b. Fasereingangs- (oben)
und Faserausgangspuls (unten)
bei λ= 1,06 um (nur Güteschaltung)

Die im Ausgangspuls fehlende Leistung wird für den Aufbau der Stokeslinien höherer Ordnung verwendet. In dem in Abb. 3a dargestellten Spektrum wurde die Durchschnittsleistung mit einer großflächigen Ge-Diode gemessen.

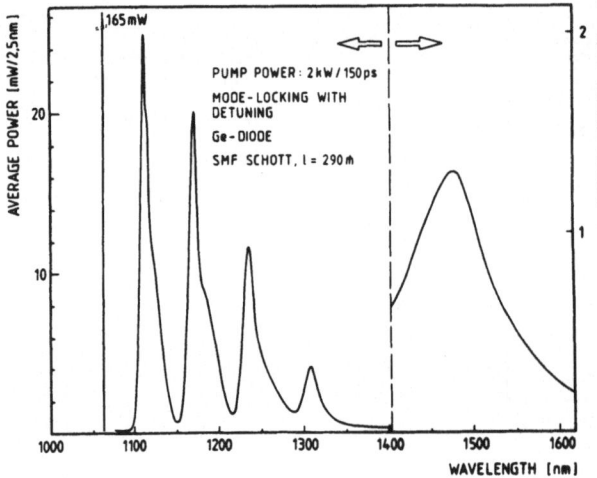

Abb. 4. Ramanspektrum nur
mit Modenkoppler erzeugt
(Verstimmung der Mode-Lock-
Treiberfrequenz)

Das in Bild 4 dargestellte Ramanspektrum ist ohne Güteschalter, nur mit dem Modenkoppler erzeugt worden. Hierzu wird die Mode-Lock-Frequenz bzw. die Resonatorlänge
geringfügig geändert /5/. Dadurch kann eine bis zu 40-fache Pulsüberhöhung und hohe
Wiederholrate erreicht werden. Die Pulsbreite beträgt weniger als 200 ps.

In Abb. 5a ist das mit Modenkopplung und simultaner Güteschaltung erzeugte Ramanspektrum dargestellt (Fasereingangsleistung = 12 kW/150 ps). Neben den Stokes- treten
auch Antistokeslinien und die Summenfrequenz von Pump- und erster Stokeslinie auf
(Abb. 5b), wobei die Phasenanpassung durch intermodale Wechselwirkung erreicht wird.

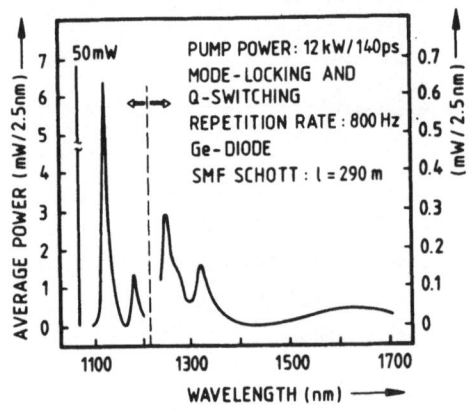

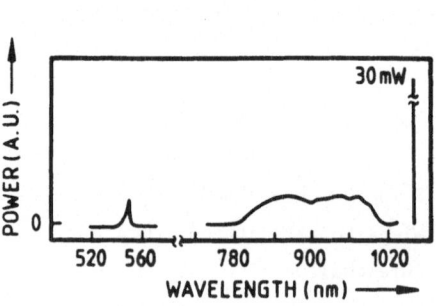

Abb. 5a. Stokesspektrum

Abb. 5b. Antistokesspektrum

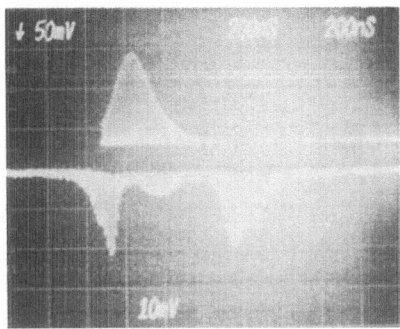

Abb. 5c. Fasereingangs- (oben) und Faserausgangspuls (unten, invertiert) bei
λ = 1,06 μm (Güteschaltung und Modenkopplung)

In Abb. 5c ist der Fasereingangs- und Ausgangspuls bei λ = 1,06 μm dargestellt. Man
erkennt wiederum deutlich die Verarmungszone im mittleren Teil des Pulses.

5. Anwendungen

Die Eigenschaften des Faser Raman Lasers - Abstimmbarkeit im nahen IR-Bereich und
Erzeugung kurzer Pulse - werden zur Charakterisierung der Komponenten der optischen
Nachrichtentechnik benötigt. Ein wichtiges Anwendungsgebiet ist die Untersuchung der
Dispersion von Glasfasern. Von den zahlreichen Möglichkeiten für den Einsatz der
Durchschnittsleistung zur Charakterisierung von Bauelementen der Integrierten Optik
sei eine näher erläutert.

Die Kopplung eines LiNbO$_3$-Richtkopplers wurde wellenlängenabhängig untersucht.
Abb. 6 zeigt die normierte Ausgangsintensität in Abhängigkeit von der Wellenlänge
für TE- und TM-Polarisation. Bei λ = 1,190 μm ist der Koppler unabhängig von TE- und
TM-Polarisation auf I$_\otimes$ durchgeschaltet.

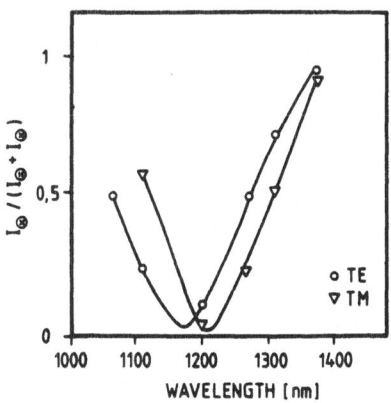

Abb. 6. Normierte Ausgangsintensität in Abhängigkeit von der Wellenlänge für einen LiNbO$_3$-Richtkoppler

6. Zusammenfassung

Es wurde der Aufbau einer Faser-Raman-Puls-Lichtquelle für den Wellenlängenbereich 0,75 μm bis 1,6 μm beschrieben. Die Pulslichtquelle ist ein Nd:YAG-Laser mit Güteschalter und Modenkoppler. Es wurden verschiedene Möglichkeiten der Erzeugung der erforderlichen Pulsleistung aufgezeigt. Als Ramanmedium wurde eine 290 m lange Monomode-Faser verwendet. Der Faser-Raman-Emitter stellt eine flexible, leistungsfähige, insbesondere zur Charakterisierung von Komponenten der optischen Nachrichtentechnik geeignete Pulslichtquelle dar.

References

/1/   L.G. COHEN, C. LIN, IEEE J. Quantum Electron. QE-14, 855 (1978)

/2/   R.H. STOLEN, Fiber and Integrated Optics, 3,1,21 (1980)

/3/   R.H. STOLEN, E.P. IPPEN, A.R. TYNES, Appl. Phys. Lett, 20,2,62 (1972)

/4/   Die Glasfaser wurde freundlicherweise von Herrn Kunstmann, Fa. Schott Mainz, zur Verfügung gestellt

/5/   J. KRAUSER, H.J. EICHLER "Amplitudenmodulation und Pulsverstärkung des modengekoppelten Nd:YAG-Lasers durch Treiberfrequenzverstimmung" in diesem Band

Limiting Influence of the Atmosphere upon the Accuracy and the Laser Alignment Method

M. Kašpar and J. Pospíšil

Department of Special Geodesy of the Faculty of Civil Engineering
Technical University of.Prague
Prague 6, Thákurova 7, ČSSR

Special kinds of work in the sphere of engineering surveys, e.g. in the construction of atomic power plants, in the observation of deformations of dams, etc., require an ever increasing accuracy and effectiveness. Modern methods making use of the laser technique fulfil in general the respective criteria CH~RZANOWSKI, JARZYMOVSKI, KAŠPAR (1976), KAŠPAR (1977). A limiting factor, however, is represented by the effect of the atmospheric elements on the accuracy and the range of the method KAŠPAR, POSPÍŠIL (1981, 1982), PELIKÁN (1982), POSPÍŠIL (1980).

Experimental measuring took place on the geodetic comparison base at the park enclosure Hvězda in Prague. The length of the base has been determined by repeated measuring and it equals 960.874545 m $\pm$ 0.17 mm. The mentioned length of the base has been reduced to the level of the mean stabilization (underground) mark, with the elevation of 368.45 m above the Mediterranean sea level. The eastern part (pillars No.0-20) is of a length of 480.460184 m while the western part (pillars No. 20-40) is 480.414361 m long BÖHM, HORA, KOLENATÝ (1982). The main characteristic of the examined area is a continuous level territory throughout the entire length of the comparison base.

The experiment was carried out at the western part of the geodetic comparison base, delimited by the pillars No. 20-40.

The lines of sight on the base were formed by a source of laser radiation, namely the Czechoslovak laser Tesla TKG 205 and a laser theodolite consisting of a laser and a laser eyepiece (produced by the firm Kern), of a special transition unit according to the Czechoslovak standard ZN 11/82, and of an onesecond theodolite of the type Zeiss Theo 010 A. For comparison's sake the third line of sight was realized with the help of a levelling instrument of the type Zeiss Ni 002. The displacements of the laser spots and of the end points of the line of sight of the levelling instrument were measured every 15 minutes, with a quadruple reading at the distance of 480.4 m. The arrangement of the measuring instruments at the pillar No. 20 of the geodetic comparison base is shown in Fig. 1.

Fig. 1

The temperature was measured by means of 9 temperature sensors NTC with the help of a digital temperature measuring device THERM 3440, in connection with anautomatic switsch of the measured places THERM 3435, the results being recorded at the THERM 3465 printer. The relative air humidity was measured with a digital psychrometer of the type THERM 2246. (All the mentioned instruments were produced by the firm Ahlborn, Holzkirchen, FRG). The barometric pressure was measured by two aneroids, type Metra. The wind velocity was measured by a summarizing Metra anemometre. The degree of cloudiness was determined according to a ten-grade meteorological scale. The temperature was measured at three vertical temperature profiles the distance of which, from the pillar No. 20 in the direction toward the end point of the sight line, was 1 m for the first temperature profile, 73 m for the second and 255 m for the third temperature profile. The vertical distance of the outer tempertature sensors from the central one was 0.5 m. The average height of the line of sight above the territory was 1.45 m. The remaining atmospheric elements were measured at the level of the sight line at the pillar No. 20 and they were determined every hour.

The temperature was always measured twice at an interval of 15 minutes. The arithmetic means of these values, rounded off to 0.1 $^\circ$C, was entered into the processing of the measuring results. The relative humidity was determined with an 1 per cent accuracy. In the measuring of the atmospheric pressure with help of the aneroids the arithmetic mean was determined with an accuracy of 0.1 Torr. Wind velocity was measured with an accuracy of 0.1 m/sec.

Fig. 2 shows an example of the time interval be tween 21.30 - 13 (summer time) in respect to the relative vertical displacements.

In the results processing we stared from the theoretical fundamentals

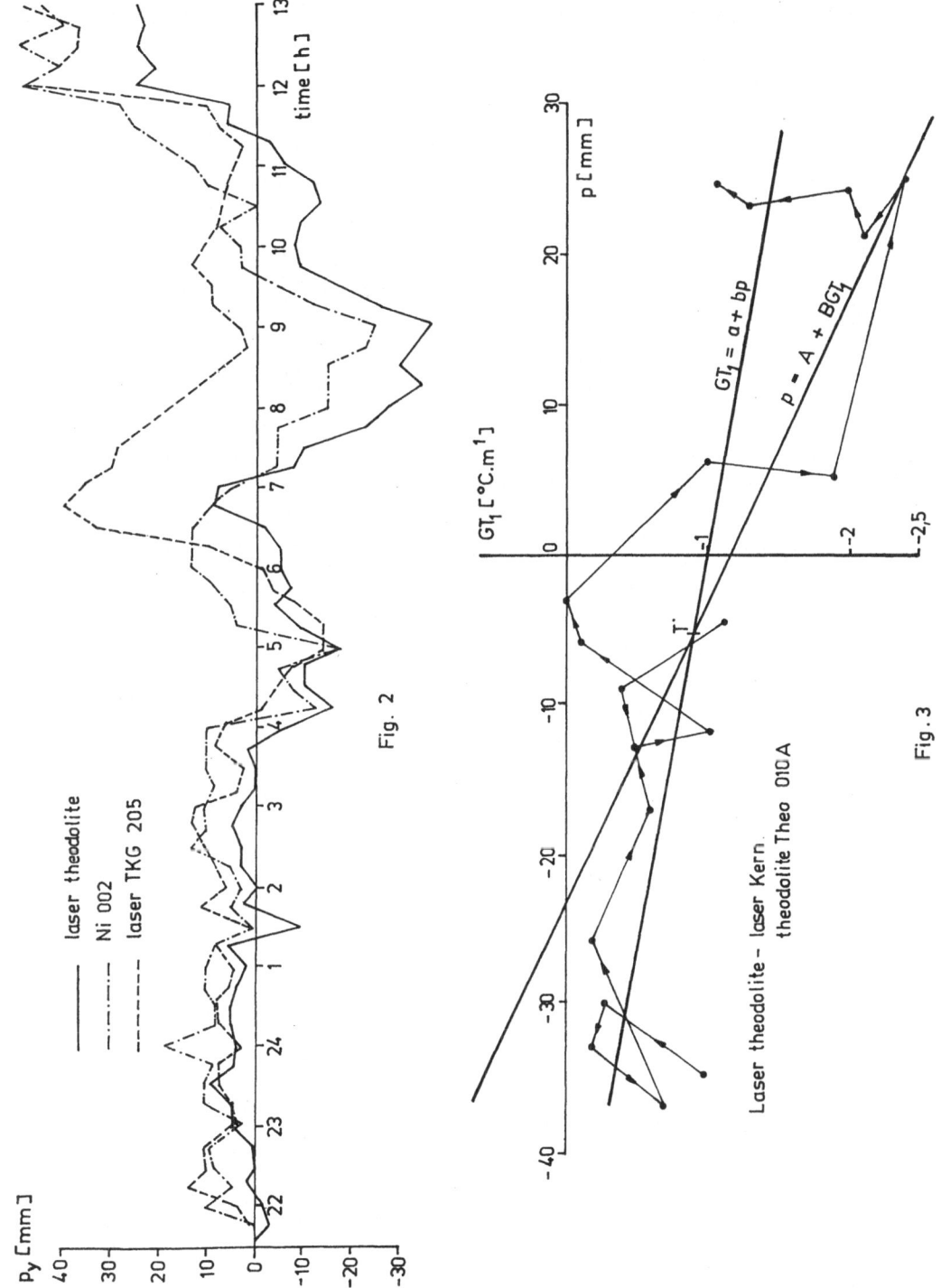

Fig. 2

Fig. 3

of the regression and correlation analysis.The points of the correlation field followed an approximately linear course in the majority of sets, and for this reason we used the simplest model of linear regression. The regression relationships were investigated for the measured displacements in dependence on the measuring instruments producing the lines of sight, namely for the values of the temperature gradients and the temperatures in the 1st, 2nd and 3rd temperature profile. Fig. 3 shows, by was of an example, the dependence of the measured vertical displacements on the temperature gradient GT_1 for the laser theodolite. The same figure illustrates two regression lines with the equations $GT_1 = a + b \cdot p$ and $p = A + B GT_1$.

In the first case we consider the measured displacements as free from errors and the temperature gradients as subjected to errors. In respect of the accuracy of the determination of temperatures, temperature gradients and vertical displacements the very opposite case may be said to be justified, when the measured temperature values and the ensuing computed values of the temperature gradients may be considered as free from errors while the measured values of the vertical displacements may be regarded as subjected to errors.

In the experimental measuring of the influence of the environment on the laser instrument as well as the levelling device we have achieved the following accuracies: laser theodolite $m_{oh} = 1.7$ mm, $m_{ov} = 1.6$ mm laser TKG 205 $m_{oh} = 2.4$ mm, $m_{ov} = 2.2$ mm, Ni 002 $m_{oh} = 4.0$ mm, $m_{ov} = 4.8$ mm, where m_{oh} (m_{ov}) are the main unit errors in the adjustment of the target to the centre of the laser beam spot in the horizontal (vertical) direction in the case of the laser instrument, and in the reading of the horizontal (vertical) displacements of the line of sight in the case of the levelling instrument, respectively. On the basis of a comparison of the model of linear regression with the measured values of the displacements it was found that the laser theodolite and the levelling instrument reacted in a similar way to the influence of atmospherc elements while the laser TKG 205 behaved differently. The laser TKG 205 is of a lower stability than the laser theodolite and the influence of the environment has a direct effect on the construction of the instrument. Thus the laser TKG 205 has a different temperature range as compared both with the laser theodolite and the levelling device. This system is therefore in a different interaction with its surroundings.

As regards the special microclimate of the geodetic comparison base Hvězda in Prague, the time between 21.30 - 4.00 (sumer time) appears as the best suited time interval for the measuring, with regards to the effects of atmospheric elements. In case that a higher accuracy of measuring were required the length of the line of sight ought to reduced to 200 m.

Literature
A. CHRZANOWSKI, A. JARZYMOVSKI, M. KAŠPAR: The Canadian Surveyor.30, 81 (1976)
M. KAŠPAR: Procc. of XV. FIG Congress. 610.4, 335 (1977)
M.KAŠPAR , J. POSPÍŠIL: XVI. FIG Congress (1981)
M. KAŠPAR, J. POSPÍŠIL: Procc. of III. Int. Symp. FIG ü. Deformationsmess. mit geod. Meth. 161 (1982)
M. PELIKÁN: Acta Polytechnica 3 (I, 3), 79 (1982)
J. POSPÍŠIL: GaKO, 26/68, 9 (1980)
J. BÖHM, L. HORA, E. KOLENATÝ: Vyšší geodézie (Geodetic Surveying). Part I. Technical University Prague (1982)

An Interactive System for Digital Optical Image Processing

P.J.S. Hutzler, S. Berber, W. Waidelich
Gesellschaft f. Strahlen- & Umweltforschung, D-8042 Neuherberg

Introduction

The coherent optical processor (COP) was introduced an ultra fast image and and general data processor working with the 'speed of light', or with a realistic estimate, performing about 10^{14} multiplications per second /1/. On the other side the COP shows two major handicaps. It has a rather poor signal to noise ratio of about 20 to 4, depending on the quality of the lenses and surfaces /1/. Another problem arises at the data input stage where the data should be presented on transparencies coded as amplitude transmittance in a range between 0.0 and 1.0. To avoid the photographic process various incoherent/coherent and electro/optical transducers have been developed /2/.

Another way to overcome the problems of the coherent optical processor was indicated by Harry C. Andrews as early as 1970 /3/. He proposed "digital optical processing" (DOP), using spatial and spectral clipped images, properly sampled and quantized, and a complex two-dimensional Fast Fourier Transform (FFT) algorithm implemented at a digital computer.

At that time big main frames were necessary to perform image transform operations. In the meantime powerful micro processors have been developed. In addition micro computer systems can be expanded by special processors like vector processors. A low cost vector processor e.g. speeds up a 16 bit micro to about 1 MegaFLOPS (Floating Point Operations Per Second). A digital optical processor (DOP) is presented, which is based on 16 bit micro computer including a plug in floating point vector processor.

Implementation

An image f(x,y) which is space-band limited as well as frequency-band limited /4/ is represented by the quantized signal values at N by N discrete points. Its two-dimensional discrete Fourier transform F(u,v) is definded as

$$F(u,v) = 1/N \sum_{y=\emptyset}^{N-1} \sum_{x=\emptyset}^{N-1} f(x,y) \exp\{(-2\ i/N)(xu+yv)\} \qquad (1)$$

$$= 1/N \sum_{y=\emptyset}^{N-1} F(u,y) \exp\{(-2\ i/N)yv\}$$

Due to the separability of formula (1) the 2-dim FT can be broken into a series of 1-dim FTs. First e.g. a 1-dim FT is performed over each of the N lines in x direction, then a 1-dim FT is performed over each of the N lines in y direction.

In practice the complex values of a 2 dimensional image are arranged 1-dimensional in sequential order at the computers random access memory (RAM). Therefore transformation in x-direction means running the 1-dim FFT over a vector of N elements with a certain start address, and an increment of one beween two vector elements. Transformation in y-direction runs over a vector with N elements too, but there is an increment of N between two vector elements, the elements of the vector are distributed over the whole range of the data field. That means if the RAM of the computer cannot hold the complete complex image at once, then additional time consuming sorting and loading algorithms must be run.

For filtering the coherent transfer function (CTF) H(u,v) is multiplied point by point with the transformed image. In gerneral the CTF might be stored with 8 bit quantization at a peripheral video frame store, and data are transferred line by line when they are needed. So no additional RAM for data storage is necessary.

Thus the representation of the output of a coherent optical processor with filter function h(x,y) is given by

$$g(x,y) = 1/N \sum_{u=\emptyset}^{N-1} \sum_{v=\emptyset}^{N-1} F(u,v)\ H(u,v) \exp\{(-2\ i/N)(ux+vy)\} \qquad (2)$$

System Hardware

The digital optical processing system is configured around a 16 bit DEC LSI11/23 micro computer with 22 bit addressing capabilty. It is equipped with 768 kbytes of memory. Therefore it can hold a 256 x 256 complex*8 image, needing 512 kbytes, plus operating system, program, etc. All vector processing is done by a plug in vector processor SKYMNK, which reaches a speed of about 1 megaflops. Its MOVE instruction is able to address the full 22 bit address space of the Q-bus.

But unfortunately processing instructions like FFT are not (under RT11 operating system)! Another important component for the user's interactive work is a peripheral video frame store holding 4 images with 512 x 512 pixels each. It is coupled to the Q-bus via a 16 bit parallel DMA interface.

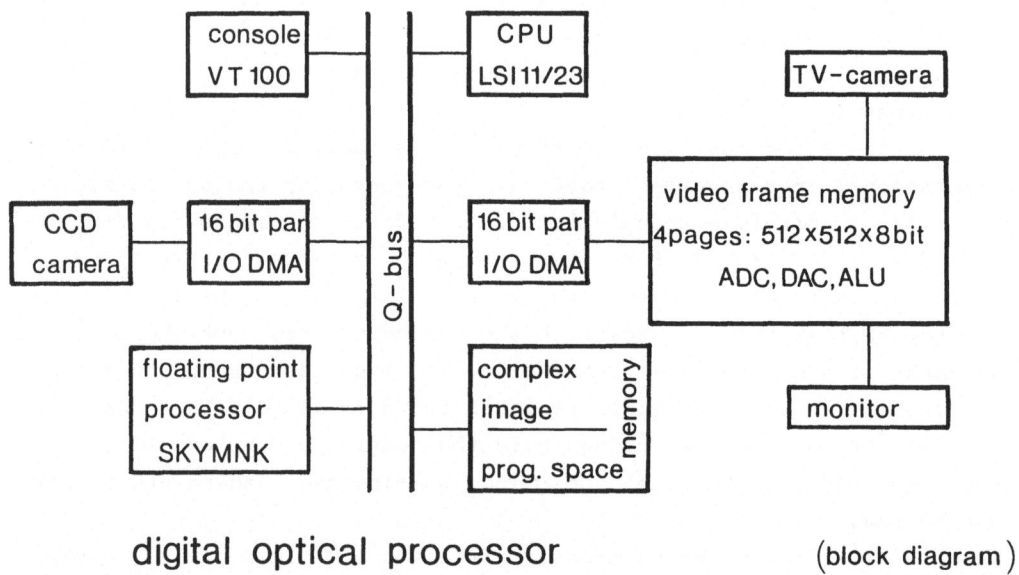

digital optical processor (block diagram)

System Software

The software for digital optical processing has a menue structure in several levels. All sub-menues and processing programs are coupled by a simple chaining technique. There are three main categories of routines: I/O routines, mask generation routines, and FFT routines.

- Input routines build up a 256 x 256 complex image in the RAM from real data resident on video frame store or disk in 8 or 16 bit quantization. Time needed is about 4 s.

- 2-dim complex in place FFT including bit reversal is performed in 17 seconds, under the restrictions given above.

- mask generation programs produce CTFs on the video frame store. The CTFs are mostly expressed as analytic functions, and contain one or more degrees of freedom. E.g. a Gaussian high-pass filter has two independant continuously variable parameters, i.e. the transmittance at zero frequency, and the half value width of the Gaussian. The Laplacian operator with Gaussian apodization has one continuous variable, i.e. the half value width of the Gaussian /5/.

- Output routines produce a real extract with 8 bit grey scale resolution from the complex image. Real and imaginary part can be shown
with arbitrary offset and scaling, as well as magnitude and magnitu-
de-squared values. Especially for images at FT space the logarithm
onto the operations mentioned above is provided. For FT images an automatic matrix transpositioning is supplied to obtain the well known
power spectra with zero frequency at the center. Outputting images
onto the video frame store takes from 4 to 16 seconds, depending on
the operations used.

Thus, within a dialogue session it is possible to produce on the video
display the input image and its power spectrum, a proposed CTF of a
filter and the result of its application, and all at one glance and
within a few minutes. After a first pass through the filter might be
corrected for a second pass through etc.

Conclusion

Application of a low-cost vector processor within a 16 bit micro processor system reduces calculation time in simulating the coherent
optical processor to the range of seconds. A comfortable menue system
allows optical filtering in dialogue sessions on a micro computer work
station with direct visual control of the results. Further gain in
speed of a factor 10 is possible if much faster 32 bit floating point
processors are used within 32 bit micro computers.

References

/1/ K. PRESTON JR.: Proc. IEEE, 60-10, 1216-31, (1972).

/2/ M. P. PETROV: Acta Politechnica Scandinavica, Ph 149, Vol. 1, 43
    (1985).

/3/ H. C. ANDREWS: Computer Techniques in Image Processing, Academic
    Press, New York, 31-54 (1970).

/4/ B. R. FRIEDEN: Image Enhancement and Restauration. In: Picture
    Processing and Digital Filtering (Ed. T. S. Huang) Berlin:
    Springer, 181 (1975).

/5/ P.J.S. HUTZLER: Optical Filtering Techniques in Digital Image En-
    hancement. IEEE Computer Society Press: Proc. 7ICPR 2, 899 - 901
    (1984)

Optoelectronic Measurement of Surface Flaws

L R Baker

Sira Ltd

South Hill, Chislehurst, Kent BR7 5EH, UK

INTRODUCTION

Although the presence of flaws such as digs, scratches and poor polish on consumer products such as camera lenses and spectacles is usually disliked for cosmetic reasons, similar defects on laser and opto-electronic components such as mirrors, windows, modulators, filters and focusing optics used with laser radiation often have a direct influence on the function of the component. The level of radiation they diffract and scatter into the image plane can be comparable with that from the object itself. Whilst the function of the principal image forming system is to transfer as much information as possible from the object to the image plane, flaws on or under the surfaces of the optical components contribute misinformation, the nature of which is unknown. If the source is incoherent, then the light scattered by flaws may just prevent the detection of weak object signals, but if the source is coherent, scattered light may interfere actively with the object signal and degrade the quality of the received image to a quite unacceptable degree.

Methods for quantifying and specifying these flaws are therefore urgently required to ensure fitness for purpose and avoid costly overspecification of their severity.

METHODS FOR MEASUREMENT OF SURFACE FLAWS

Scratches and digs less than a micron in size can be seen and will influence adversely the performance of some components. They can moreover exist anywhere over a surface of perhaps up to 100 mm diameter or more so that the task of locating and then measuring them in a few seconds of time is obviously formidable. The most widely adopted approach is to illuminate the component and view the light scattered by the defect, seen against a dark background, in comparison with the intensity of light scattered from a standard defect viewed under the same conditions. This method has the disadvantage of being subjective and since the visibility of the scratch depends on several factors such as the direction of illumination and viewing - factors not usually considered important enough to specify - the results obtained by different inspectors are rarely consistent. In spite of the above problems which are now becoming generally recognised, this method is embodied in a number of national specifications such as BS-4301 (1982), DIN-3140 (1980), ANSI-PH3-617 (1980) and MIL-013830A.

Attempts [1] to make the measurement more objective by recording the polar scatter patterns obtained when the defect is illuminated with a beam of laser radiation have also run into difficulties due to the fact that the scatter from a fine scratch can be less than that due to the polish or contamination of the surface on which it occurs.

As an alternative, in an attempt to simplify a complex problem, just the width of the defect has been measured on the assumption that the depth is less important in influencing scatter. This is clearly wrong, as indicated by the British Standard defects which cover a significant range of visibilities but are all of nominally the same width but vary in their depth. Moreover this measurement is difficult in practice due to the high numerical aperture and small depth-of-focus of a lens needed to resolve sub-micron defects.

A flaw image comparator [2] has been designed recently to overcome some of these problems. It combines the good features of imaging and scattering methods and permits the quantification of both amplitude and phase defects by measuring the amount of light they scatter out of an illuminating beam.

FLAW IMAGE COMPARATOR

A schematic representation of the principle of operation is shown in Figure 1. Light from a tungsten lamp at the focus of lens L_1 is imaged through the polariser Z_1 by lens L_2 into the plane of the lens L_3 after reflection at the polarising beamsplitter B. The field iris diaphragm at P_1 is at the focus of L_2 so that it is imaged by L_3 through the quarter-wave plate Q_1 on to the plane of the specimen under examination T at the focus of L_3. P_2 is an aperture diaphragm used to control the amount of light available. The light reflected by T passes again through Q_1 and L_3 and after transmission by B is brought to a focus by L_5 on to the TV camera, having passed through the analyser Z_1. The remaining portion of the light illuminates L_4 and after passing through a quarter-wave Q_2 it then illuminates an internal scratch reference graticule R placed at its focal point. As in the test channel, the light reflected from R is brought after reflection at B to focus on the TV camera. Z_1 is rotated to equalise the light transmission of the two channels and the analyser Z_2 is used to disturb this balance by a measurable amount so as to equalise the visibility of the test and internal reference scratches. The plates Q_1 and Q_2 are rotated to maximise the intensity of the images of T and R superimposed on the TV.

It can be shown that the ratio of the visibilities of the test and reference defects is proportional to the tangent squared of the angular setting of Z_2 needed to equalise their visibilities on the TV monitor. As the visibility of a particular defect depends on many different parameters such as veiling glare and lens aberration, the method is not absolute. It can however be used to compare in a consistent, reproducible way the severity of a defect on a test piece with a reference defect.

TYPICAL RESULTS

The sensitivity of the instrument can be appreciated from the photographs of the TV monitor display shown in Figure 2. A white line across the centre of the photograph indicates the track of a video line profile analyser, the output of which is shown in the dark space at the top of each photograph. Figure 2A shows two parallel scratches, invisible to the naked eye, of rectangular cross-section of width 16 µm and depth 23 nm etched in a glass substrate and Figure 2B is a similar structure but the depth has been increased to 115 nm. The visibility of all such defects can be readily quantified in comparison with the reference defect as described above. An indication of the sensitivity of the method when examining reflecting surfaces can be judged from Figures 2C and 2D which show a grating of depth profile 9 nm and a diamond turned surface of R_a = 50 nm.

CONCLUSIONS

The ability to quantify objectively surface flaws on either transmitting or reflecting surfaces in comparison with reference defects has been demonstrated. When the objective relationship between such flaws and system performance has been demonstrated, it should be possible to draft, for the first time, standards which are usable and credible to both manufacturer and customer.

REFERENCES

(1)   Scattering in optical materials, SPIE, Vol 362, 1982. Ed: Solomon Musikant.

(2)   Baker, L R., 1984, Optica Acta, <u>31</u>, 611-614.

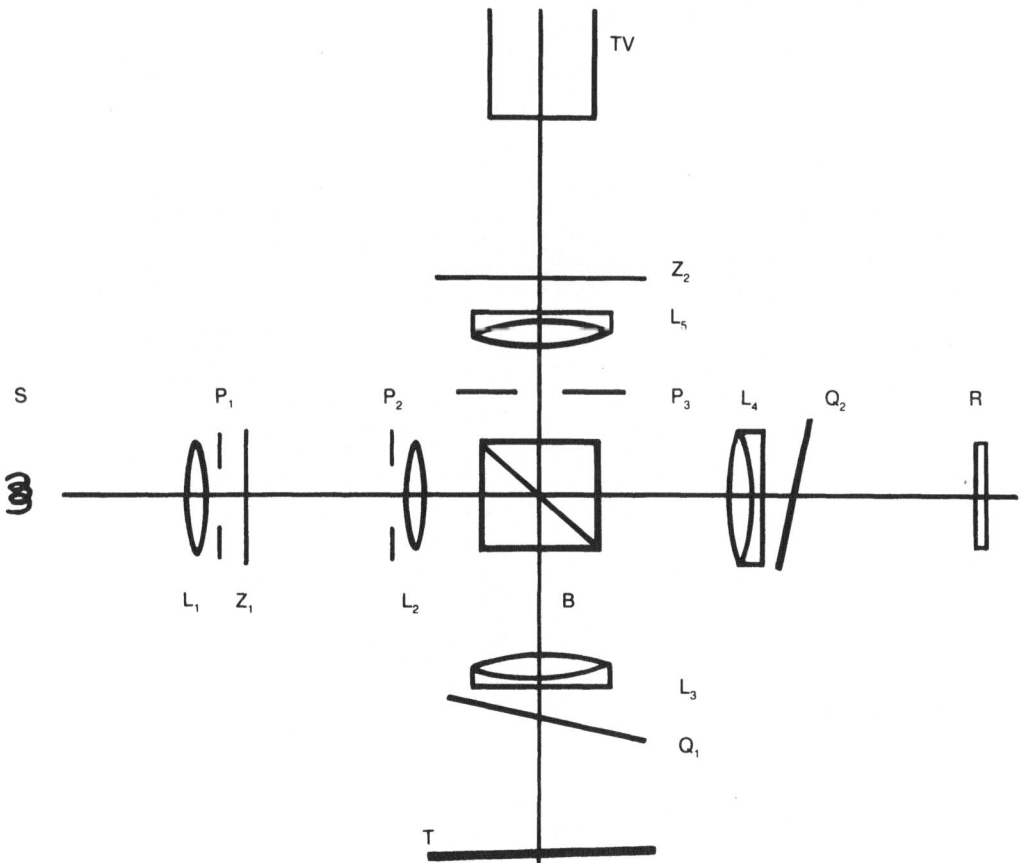

FIGURE 1. OPTICAL SYSTEM FOR QUANTIFYING SURFACE FLAWS

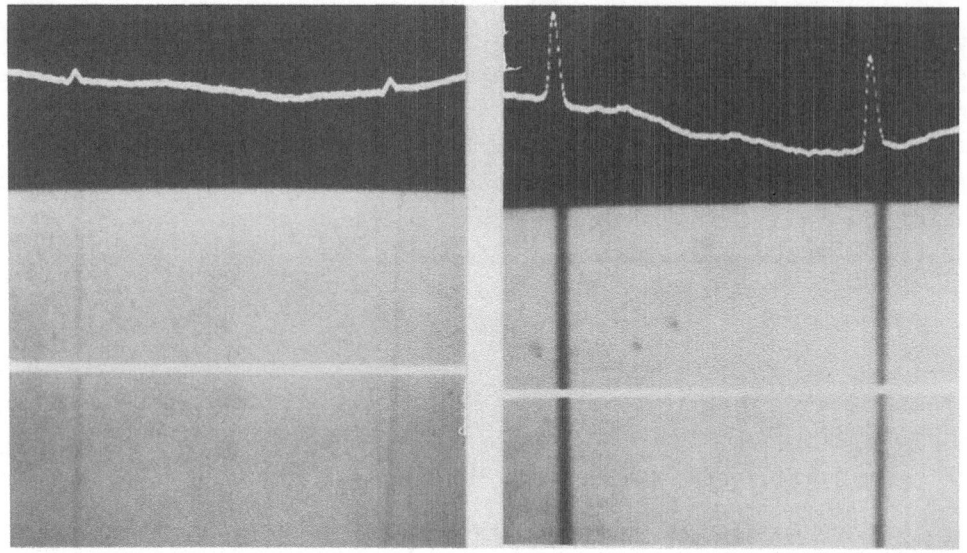

A   B

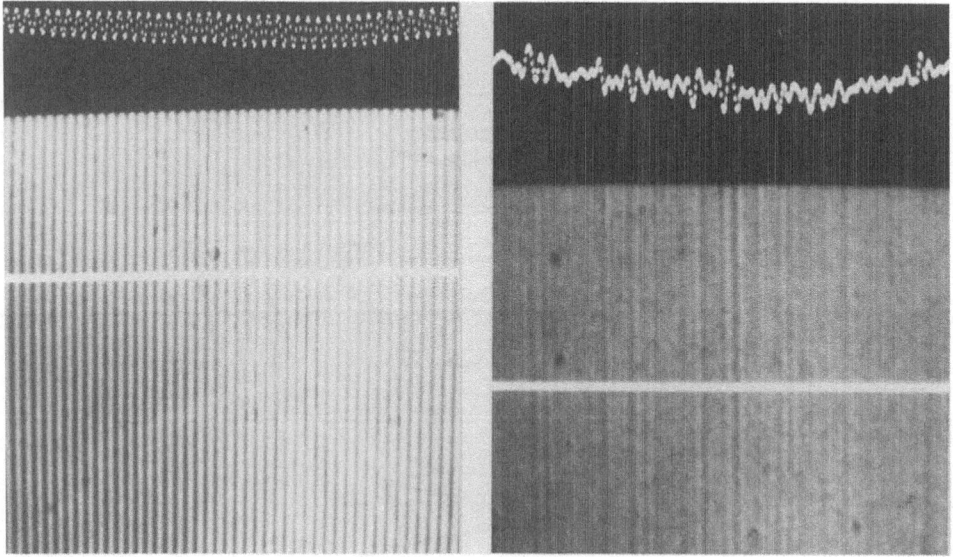

C   D

Figure 2.

TV DISPLAYS OF VARIOUS SURFACES

High Resolution Fourier Transform Spectroscopy in Real Time

B. Reuter, P. Hutzler

Gesellschaft für Strahlen- und Umweltforschung mbH München, Abteilung Angewandte
Optik, Ingolstädter Landstr.1 , D-8042 Neuherberg

Holographic Fourier transform spectroscopy

Fourier transform spectroscopy (FTS) in the visible region of spectrum has been
demonstrated first by STROKE and FUNKHOUSER /1/ in 1965, by use fo an interferometer with no movable parts, photographic recording and coherent optical reconstruction. Since this method is quite similar to holography it has been named holographic spectroscopy /2/.

The use of a stationary interferometer solves the stability problems and allows
application of FTS even in the UV-region of the spectrum. For this in principle
any kind of source doubling interferometer is suitable. If the two equivalent
source images are separated by an angle of 2θ as seen from the recording plane the
phase difference between the waves from any two equivalent source points can be
written as

$$\Delta\Phi = 4\pi\sigma x \sin\theta$$

where the wave number σ has the reciprocal value of the wavelength and x is the
spatial coordinate in the recording plane.

Since the phase difference is a linear function of x, a sinusoidal interference
pattern is produced for every wave number contained in the spectral intensity distribution $g(\sigma)$ of the source and the resulting intensity distribution $I(x)$ in the
recording plane corresponds to the cosine Fourier transform of $g(\sigma)$.

$$I(x) = \int_0^\infty g(\sigma) \left[1 + \cos(4\pi\sigma x \sin\theta) \right] d\sigma \tag{1}$$

the spectrum can be reconstructed by coherent optical or digital Fourier transformation.

The main advantage of this method is a high optical throughput since an extended
source can be used and the resolving power R is only given by the number of recorded fringes N and not dependent on the size of the entrance slit. Drawbacks on the
other hand are the limited dynamic range of linear recording and the time consuming wet processing in photographic recording as well as the noise problem in coherent optical reconstruction and its limited flexibility and accuracy. These drawbacks have prevented practical applications so far. Due to improvements in opto-

electronic detectors with spatial resolution and the availability of powerful microcomputer systems these drawbacks can be overcome. Recently a Fourier transform spectrometer with a self-scanning photodiode array and a microcomputer for reconstruction of the spectrum was proposed /3/. This concept seems to be very promissing since it has the capability of near real time processing.

Optical filtering technique for improvement of spectral resolution

In order to increase the resolution and enhance the sensitivity in photographic recording of spectral holograms a Moiré method was proposed /4/. The same optical filter technique can be used to allow for high spectral resolution even when a photodetector with a low number of elements is being used for recording the interferogram. This is due to the fact that by application of the Moiré technique the resolving power is no longer limited by the number of recorded interference fringes. For this purpose the high spatial frequencies of the original interferogram are mixed with a similar frequency of a grating. From the resulting Moiré pattern the difference frequency component can be recorded with a photoelectronic detector with low spatial resolution. The similarity to the heterodyning technique well known in electronics shall be emphasized (Figure 1). In both cases the principle permits conversion of high input frequencies to predetermined lower frequency values which can be detected, amplified and processed more efficiently.

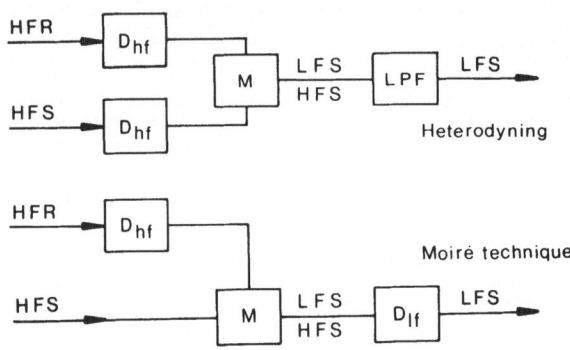

Figure 1.

Schematic diagram of heterodyning technique and Moiré technique. HFS: high frequency signal, HFR: high frequency reference signal, LFS: low frequency signal, M: multiplication, D_{hf}: detector with high cutoff frequency, D_{LF}: detector with low cutoff frequency, LPF: low pass filter.

If a grating is being used as a filter the filter function will be

$$F(x) = 1 + \cos (4 \pi \sigma_F x \sin \theta_F) \qquad (2)$$

where $\sigma_F \sin \theta_F$ is the spatial frequency of the grating. The intensity distribution detected behind the filter plane is given by multiplication of Eq.1 and Eq.2. Fourier transformation leads to a convolution of the Fourier transform of the filter function $f(\sigma)$ and the spectral intensity distribution of the source $g(\sigma)$.

$$g^F (\sigma') = F^{-1} \{F(x) \; I(x)\} = f(\sigma') * g(\sigma') \tag{3}$$

From the low frequency component the spectrum will be reconstructed at a wavenumber range that is shifted by an amount of σ_F towards shorter wavenumbers. The resolving power R is still given by the number of fringes in the original interferogram N but this can be realized by a smaller number N^F of recorded fringes

$$R = N = N^F \; \sigma_{max}/(\sigma_{max} - \sigma_{min}) \tag{4}$$

The resolving power thus depends on the width of the spectrum. The smaller the spectral width the larger the gain in resolving power will be. In the case of the Na D-lines with a spectral width of about 20 cm^{-1} the gain in resolution will be 848 and a detector with 512 resolution elements will allow for a spectral resolution of 0.078 cm^{-1}.

Experimental results

For experimental demonstration we have used a Michelson interferometer with tilted mirrors. It was realized by a special beam splitter cube shown in Figure 2 which leads to a compact, stable and self aligning set-up. The fringes of equal thickness were imaged onto the filter plane with a magnification of 2. As a filter either a Ronchi ruling (grating period 8 μm) or a bleached grating, recorded photographically with the same set-up was used.

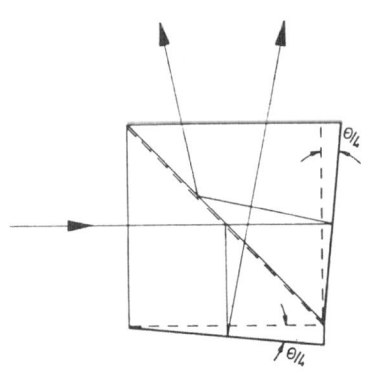

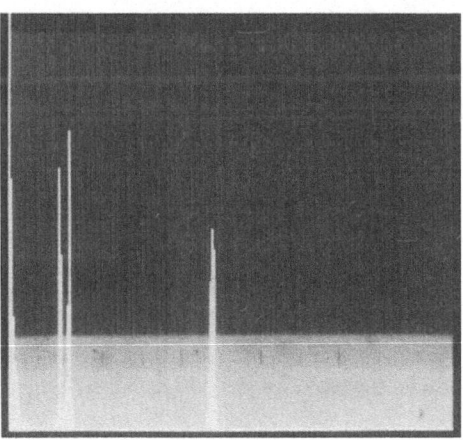

Figure 2.
Path of rays in a compact
Michelson interferometer
based upon a beam splitter
cube with tilted mirrors.

Figure 3.
Interferogram and spectrum of the light from a
cold mercury arc lamp. The interferogram contai
ning five beats was filtered with a Ronchi ruling and recorded with a TV-camera. In the
overlay the calculated spectrum shows the yellow lines at 546 nm and 579 nm and the green
line at 436 nm.

Interferograms of a cold mercury arc lamp have been recorded with a high resolution CCD line sensor camera /5/ (512 elements at 26 µm distance) and alternatively with a TV-camera equipped with a 1-inch Newicon tube. The data were digitized with 8bit resolution and read into a video frame memory of size 512 x 512. This was done in real time. For signal correction for linear detection, apodisation and fast Fourier transform (FFT) a pipelined floating point processor (SKYMNK) which reachs a speed of about 1 megaflops is used. Complex in place FFT is done line by line within a preselectable range of lines. From this the spectrum is obtained either by adding complex amplitudes from different lines and taking the magnitude (coherent case) or by adding the magnitudes of the spectra calculated from different lines (incoherent case). The complete signal processing system is controlled by a 16 bit DEC LSI 11/23 microcomputer /6/.

Filtered spectral holograms and reconstructed spectra are shown in Figures 3 and 4·

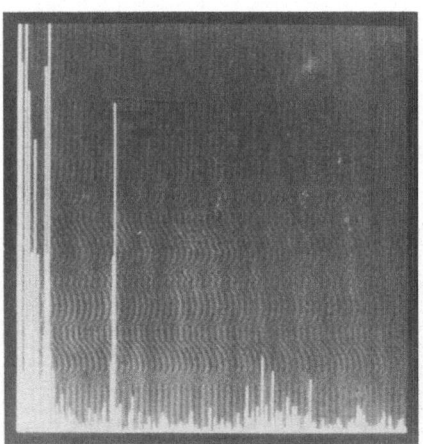 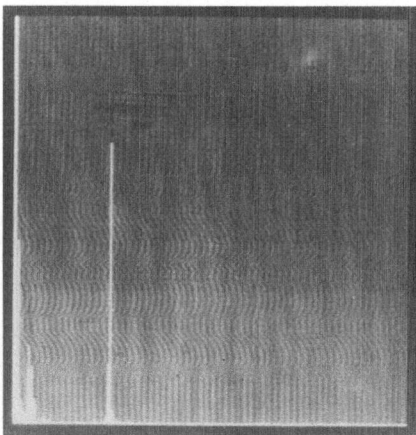

Figure 4:
Interferogram and spectrum of the light from a HeNe laser used for calibration. The interferogram was recorded by use of the CCD-camera. The spectra were calculated from the distorted part of the interferogram between lines 90 to 110 by coherent (left) and incoherent (right) superposition.

These results demonstrate the improvement in spectral resolution as a result of the filtering technique. In addition the property of real time storage of time varying interferograms should be mentioned which enables time resolved spectroscopy.

References

/1/ G.W. STROKE, A.R. FUNKHOUSER: Phys. Lett. 16, 272 (1965)
/2/ H.J. CAULFIELD: Spectroscopy, in Handbook of Optical Holography (J.H. Caulfied, Ed.) Academic Press, New York, 587 (1979)
/3/ T. OKAMOTO, S. KAWATA, S. MINAMI: Appl. Opt. 23, 269 (1984)
/4/ F. LANZL, B. REUTER, W. WAIDELICH: Opt. Commun. 5, 354 (1972)
/5/ H. SEIDLITZ, S. BERBER, P. HUTZLER: These proceedings (1985)
/6/ P. HUTZLER, S. BERBER, W. WAIDELICH: Acta Polytechn. Scand., Appl. Phys. Ser. 150, 144 (1985)

Optical Time Domain Reflectometry Using the Heterodyne Principle

R. Knoechel*, S. Heckmann+, J. Rybach+, E. Brinkmeyerx

* Philips GmbH Forschungslaboratorium Hamburg, Vogt-Koelln-Str. 30, 2000 Hamburg 54
+ F&G Nachrichtenkabel und -anlagen, Geschäftsbereich N, Abteilung NMV,
  Postfach 80 50 04, 5000 Köln 80
x Technische Universität Hamburg-Harburg, Harburger Schloßstr. 20, 2100 Hamburg 90

Introduction

Optical time domain reflectometry (OTDR) is well established to measure the transmission characteristics (attenuation, splice loss, fault location) of optical-fibre communication channels. A short light-pulse (typical duration 1 µs) is launched into one end of the fibre and the resulting backscatter response is evaluated. Present OTDR-systems are mostly using direct-detection receivers.

However, future single mode communication systems operating at 1300 nm and 1550 nm will call for OTDR-measurement equipment offering the utmost in sensitivity and dynamic range, because with increasing wavelength
- the level of Rayleigh-backscatter decreases,
- detector sensitivity becomes worse, and
- optical sources are less powerful,
whereas the distance between repeaters increases.

One possibility to overcome the noise limitation of direct detection systems is the application of heterodyne detection, first introduced in OTDR by /1/ and then refined and modified by others /2-4/. The present paper reviews the advantages and shortcomings of heterodyne OTDR, and proposes a system configuration which optimizes launch power into the fibre and improves receiver sensitivity.

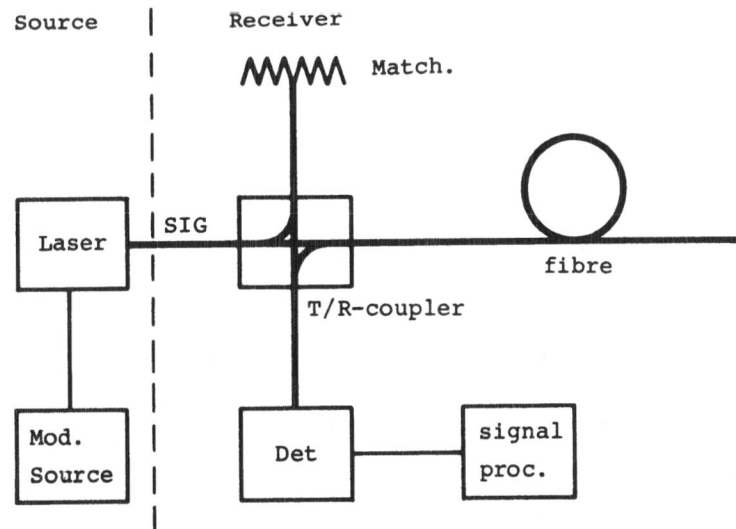

Fig. 1. Schematic of direct detection OTDR

Heterodyne OTDR-System

For comparison, a schematic of a direct detection OTDR is sketched in fig. 1. Such a system shows the following features:
- relatively simple optical arrangement,
- spectral characteristics of the laser source unimportant,
- laser can be directly pulsed by the modulation source, and
- detector is insensitive to polarization and phase of the backscatter response.

The direct detection OTDR (see fig. 1) is optimized by
- maximizing the output SIG of the laser source,
- implementing a transmit/receive (T/R)-coupler with low insertion loss, and
- choosing the most sensitive detector, where the smallest detectable power is
  given by the noise equivalent power (NEP).

The best direct detection OTDRs use acousto-optic modulators (AOM) as T/R-couplers and PINFET-receivers. The received signal is proportional to the backscattered power. After detection it is processed by post-detection averaging. Using the maximum one-way fibre attenuation range, $L\,[dB]$, as introduced by /1/ as a figure of merit, the best direct detection OTDRs arrive at $L \approx 20{-}25$ dB.

Using heterodyne reception, a beat-signal at an intermediate frequency (IF) is produced between the incoming backscatter wave and a local oscillator (LO). As is well known, heterodyne detection leads to an improved receiver sensitivity, which is only limited by the shot noise of the detector /5/. The signal-to-noise-ratio S/N is given by

$$\frac{S}{N} = \frac{\eta\,P_D}{h\,\nu\,B}\ ,\tag{1}$$

where P_D means the average received power, η means the quantum efficiency of the detector, $h\nu$ is the photon energy and B is the receiver IF-bandwidth. Assuming $B=1$ MHz, $\eta \approx 0.7$, and a wavelength of $\lambda=1.5$ μm, a comparison with a very good direct detection system (NEP=$5 \cdot 10^{-14}$ $W/\sqrt{Hz}$) gives an advantage of ~ 24 dB for the heterodyne detection, corresponding to a 12 dB improvement in L.

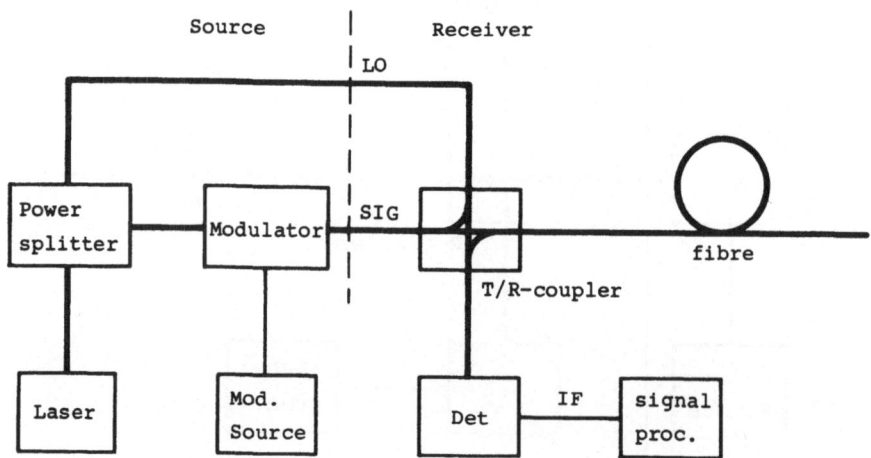

Fig. 2. General presentation of a heterodyne OTDR

A general presentation of a heterodyne OTDR is shown in fig. 2. In contrast to other heterodyne systems, e.g. in optical communication, the LO and the excitation signal SIG are derived from the same laser source. Features of the heterodyne system are summarized as follows:
- comparatively complicated optical system configuration,
- an external modulator has to be provided in order to pulse the excitation signal and to produce the IF-frequency shift,
- extremely narrow spectrum for the laser source required to meet the narrow IF-bandwith,
- detection depends on polarization and phase of the backscattered wave, but
- high reception sensitivity due to amplification of the incoming signal with the LO, and
- high dynamic range due to the proportionality of the IF-signal and the amplitude of the backscatter electric field (and not the power).

The crucial points in the manufacture of heterodyne OTDRs are
- laser stabilization; a linewidth of less than 1 MHz is required,
- system implementation with minimum insertion loss.

Considering the system configuration of existing heterodyne OTDRs /1,2,4/, a beam splitter or a fused coupler are used as LO-power splitter and T/R-coupler and an AOM acts to pulse and frequency shift the launch signal (see fig. 2). Hence a considerable system insertion loss in the transmit- and receive-path is produced. Comparing with the system of fig. 1, such a system suffers from an additional loss of 10-12 dB, thus deteriorating the advantage of heterodyne detection.

A system implementation, which offers a significantly lower insertion loss is proposed in fig. 3. Two AOMs are used as LO-power splitter and T/R-coupler. The first AOM takes the function of a modulator likewise. The second AOM can be used to gate the backscatter wave in order to blank out fresnel reflections. Both acoustic frequencies ω_{A1} and ω_{A2} can be chosen to produce an arbitrary IF with the frequency

$$\omega_{IF} = \omega_{A1} \pm \omega_{A2} , \qquad (2)$$

allowing for more freedom in the realization of the signal processing electronics. The additional circuit loss (compared to fig. 1) can be as low as 3-5 dB. The disadvantage of the system of fig. 3 is seen in the difficulty to adjust four output ports to an AOM.

A problem, which presently seems not to be completely solved, is the linewidth-stabilization of the required semiconductor laser. An attractive solution was recently proposed by /6/: A standard semiconductor laser was coupled to a length of single-mode-fibre, the Rayleigh backscatter than acting to line-narrow the laser to a linewidth of some kHz. The center frequency of such a source is not constant but jumps some 100 MHz on a time scale of some milliseconds (this is not a disadvantage in the application as an OTDR-source). With regard to the instability of the backscatter signal, the stabilization mechanism seems to be highly sensitive to acoustically induced frequency jumps and linewidth deteriorations.

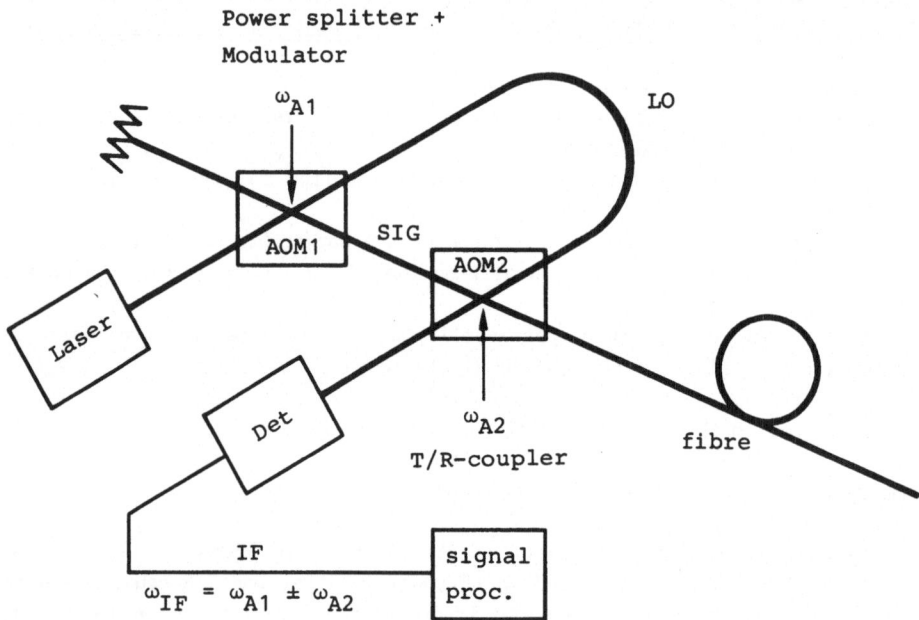

Fig. 3. A heterodyne OTDR optimized with respect to minimum system insertion loss

Conclusions

Heterodyne detection in OTDR is an attractive means to improve sensitivity and dynamic range. However, the heterodyne advantage of about 24 dB in detector sensitivity will only lead to an increase in measurable one-way fibre attenuation range, if a relyable and field portable solution for laser stabilization is found and if the system insertion loss is minimized.

Acknowledgement

The advice and information given by P. Healey from BTRL, Ipswitch, UK, is gratefully acknowledged.

Literature
/1/ P. HEALEY, D.J. MALYON: Electron. lett., 18 (1982), 862-863.
/2/ S. WRIGHT, K. RICHARDS, S.K. SALT, E. WALLBANK: Proc. 9th ECOC (1983), 177-182.
/3/ P. HEALEY, R.C. BOOTH, B.E. DAYMOND-JOHN: Electron. lett., 20 (1984), 360-362.
/4/ S. WRIGHT, R.E. EPWORTH, D.F. SMITH, J.P. KING: Proc. OFS (1984), 347-350.
/5/ M.C. TEICH: Applied Phys. lett., 14 (1969), 201-203.
/6/ R.E. EPWORTH, D.F. SMITH, S. WRIGHT: Proc. 10th ECOC (1984), 132-133.

Neues Verfahren zur Bestimmung der Wellenlänge und Leistung monochromatischen Lichtes

J. Seidenberg, V. Blazek, K.-J. Krath, H. J. Schmitt
Institut für Hochfrequenztechnik, RWTH Aachen
Melatenerstr. 25, D - 5100 Aachen

Es wird ein Lichtmeßverfahren vorgestellt, das ohne Prismen, Gitter etc. die Wellenlänge und die Leistung monochromatischen oder näherungsweise monochromatischen Lichtes bestimmt.

Einleitung

In der Laser- und Lichtwellenleitertechnik wird man häufig mit dem Problem konfrontiert, die Wellenlänge und die Leistung monochromatischen Lichtes zu messen. Die bisher verwandten Verfahren zur Messung der Wellenlänge sind mit einem recht großen apparativen Aufwand verbunden. So benötigt man in der Regel Monochromatoren mit Prismen oder Gittern, die sich jedoch nicht für kompakte Meßgeräte eignen. Das gleiche gilt für Filtersätze oder Meßverfahren mit rotierenden Prismen. Das hier vorgestellte neue Meßverfahren arbeitet daher gänzlich ohne Prismen, Gitter etc. und ist damit insbesondere für den Einsatz in handlichen Meßgeräten geeignet.

Die Leistungsmessung monochromatischen Lichtes wird i.a. mit Photodetektoren vorgenommen, denen ein Korrekturfilter, ein sog. radiometrisches Filter, vorgeschaltet ist, das die Abhängigkeit der Empfindlichkeit des Photodetektors von der Wellenlänge des zu messenden Lichtes kompensieren soll. Der Nachteil dieser Meßmethode ist, daß die Genauigkeit in einem größeren Wellenlängenbereich dennoch nicht besser als etwa +/- 7% ist. Andere Meßgeräte sind bei einigen diskreten Wellenlängen kalibriert, die vor der Messung eingestellt werden müssen. Das setzt allerdings die Kenntnis der Wellenlänge und eine dazu passende Einstellmöglichkeit voraus. Bei den heutzutage üblichen Kompakt-Meßgeräten sind in der Regel nur zwei bis drei Wellenlängen vorwählbar. Somit kann auch nur bei diesen Wellenlängen exakt gemessen werden, während Messungen bei davon abweichenden Wellenlängen zu Meßfehlern bis etwa 30% führen können.

236

Aus diesen Gründen wurde ein Meßverfahren entwickelt, mit dem zugleich die Wellenlänge als auch die Leistung monochromatischen oder näherungsweise monochromatischen Lichtes gemessen werden kann.

Messprinzip

Über eine Aufteilvorrichtung wird ein Teil des zu messenden Lichtes zwei Photodetektoren aufgeprägt. Die Photodetektoren haben zwei leicht unterschiedliche spektrale Empfindlichkeitsverläufe (Bild 1). Die Ausgangssignale der beiden Photodetektoren werden über einen A/D-Wandler einem Mikrorechner zugeführt, der das Verhältnis der beiden Signale ermittelt. Dieses Verhältnis ist charakteristisch für die Wellenlänge des gemessenen Lichtes. Es ergibt sich ein eineindeutiger Zusammenhang zwischen dem Verhältnis und der Wellenlänge, der in Bild 2 dargestellt ist.

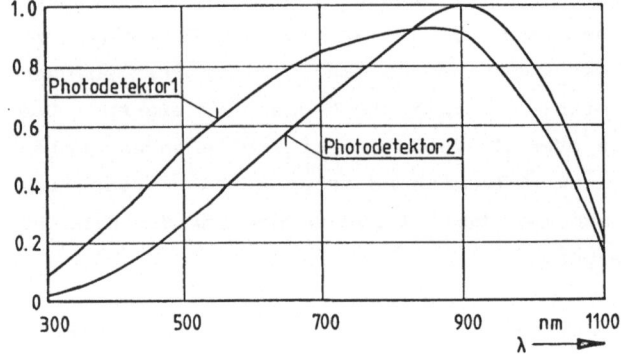

Bild 1.
Spektrale Empfindlich-
keit der Photodioden

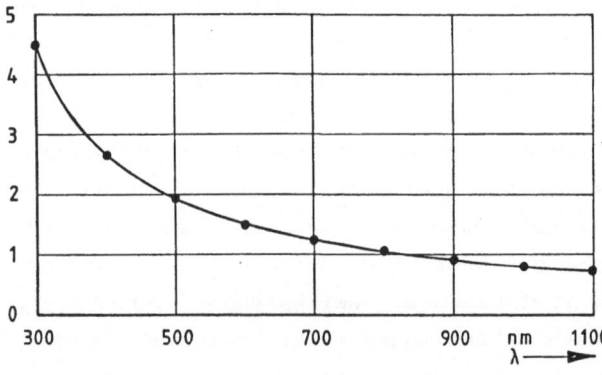

Bild 2.
Verhältnis der bei-
den Photoströme

Wenn diese Verhältniskurve im Meßgerät abgespeichert ist, kann somit die Wellenlänge ermittelt werden. Die Abspeicherung erfolgt in einem einmaligen Kalibriervorgang: Ein Monochromator wird rechnergesteuert über den gesamten Wellenlängenmeßbereich in Schritten von z.B. 10 nm durchgestimmt. Dabei ermittelt der Mikrorechner des Meßgeräts zu jeder Wellenlänge das betreffende Verhältnis und speichert dieses in einer nichtflüchtigen Speichereinheit ab. Durch diese einmalige Kalibrierung werden Streuungen in der spektralen Empfindlichkeitskurve der Photodetektoren aufgefangen. Nach der Kalibrierung kann der Mikrorechner durch Vergleich des Verhältnisses des zu messenden Lichtes mit den abgespeicherten Verhältniswerten und ggf. Interpolation die aktuelle Wellenlänge ermitteln.

Mit bekannter Wellenlänge sind dann sehr genaue Leistungsmessungen möglich. Auch hierbei wird zuvor ein einmaliger Kalibrierprozeß durchgeführt, wobei dem Meßgerät die aktuelle Leistung und Wellenlänge der Meßstrahlung über die Rechnerschnittstelle mitgeteilt wird. Der Mikrorechner ermittelt dann einen für die jeweilige Wellenlänge notwendigen Korrekturfaktor, der die wellenlängenabhängige Empfindlichkeit der Photodetektoren und die Transmissionscharakteristik der Aufteilvorrichtung erfaßt. Diese Korrekturfaktoren werden ebenfalls für hinreichend viele Wellenlängen des Wellenlängenmeßbereiches in dem Mikrorechner abgespeichert. Bild 3 zeigt eine typische Leistungskorrekturfunktion.

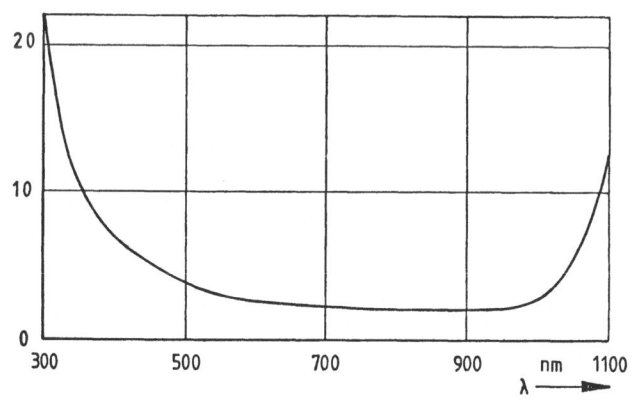

Bild 3.
Leistungskorrektur-
funktion für einen
Photodetektor

In einem Meßvorgang wird dann nach dem Ermitteln der aktuellen Wellenlänge der hierfür spezifische Leistungskorrekturfaktor ausgewählt und somit die Leistung wellenlängenkorrigiert zur Anzeige gebracht.

Realisierung

Ein nach diesem Meßverfahren arbeitender Prototyp eines kompakten Wellenlängen- und Leistungsmeßgerätes für Lichtwellenleiteranwendungen ist zunächst für den Wellenlängenmeßbereich 500 nm – 1100 nm konstruiert worden. Dabei werden zwei Silizum-Photodioden verwendet, die sich gemäß Datenblattangaben in ihren spektralen Empfindlichkeitskurven geeignet unterscheiden. Das aus dem Glasfaserstecker austretende Licht gelangt über eine Kugellinse und einen halbdurchlässigen Spiegel zu den zwei Photodioden. Zur Meßwerterfassung und -verarbeitung wird im Gerät ein Mikrorechner mit IEC-Bus Anschluß eingesetzt. Das Gerät ist batteriebetrieben und mit einer alphanumerischen Flüssigkristall- anzeige ausgestattet. Die Außenmaße des Meßgerätes betragen 18 cm x 10 cm x 4.5 cm. Die Auflösung der Wellenlängenmessung beträgt 1 nm, die der Leistungsmessung 100 pW.

Schlussfolgerungen

Mit dem Prototyp ist gezeigt worden, daß es ohne Prismen, Gitter etc. möglich ist, auf elektronischem Wege die Wellenlänge monochromatischen Lichtes zu bestimmen. Zur Zeit können noch keine abschließenden Anga- ben über die Genauigkeit des Meßverfahrens gemacht werden. Es werden jedoch Genauigkeiten im nm-Bereich erwartet. Um für das Meßgerät ein größeres Anwendungsgebiet im Glasfasersektor zu erschließen, ist es notwendig, den Meßbereich der Wellenlängenmessung in den Infarotbe- reich auszudehnen. Hierzu müssen andere, z.B. Ge- oder InGaAsP-Photo- dioden, verwendet werden. Damit wird vor allem die Temperaturdrift der spektralen Empfindlichkeit der Photodetektoren eine zunehmende Rolle spielen.
Diese und andere mit der Aufteilung des Lichtes zusammenhängende Probleme sind derzeit Gegenstand weiterer Untersuchungen.

Literatur

J. SEIDENBERG, V. BLAZEK: DPA P3429541

Industrielle Qualitätskontrolle mit hochauflösenden, farberkennenden Spektralfotometern im Auflicht- und Durchlichtverfahren

P. Glatz, W. Eberle, F. Freier und U. Rohner

Datacolor AG

Brandbachstrasse 10, CH-8305 Dietlikon (Schweiz)

1. Einleitung

Die industrielle Farbmetrik wird in vielen Industriezweigen eingesetzt, so auch mit den Schwerpunkten Farbqualitätskontrolle und Farbrezeptierung. Als trägheitsloses und praktisch berührungsfreies optisches Prüfverfahren wird sie an festen oder flüssigen industriellen Erzeugnissen angewendet, z.B. an Textilien, Papieren, Kunststoffen, Pigmenten und Lacken. Die folgenden Ausführungen beschreiben zusammenfassend die apparativen Voraussetzungen für die Farbmetrik , die messtechnischen Ergebnisse sowie die Auswertung in ausgewählten Anwendungsfällen.

2. Gerätetechnik

Ein Farbmessgerät umfasst minimal die in Fig. 1 dargestellten Baugruppen. Die Probenbeleuchtung (genormt nach DIN, ISO und anderen, industriespezifischen Normen) schliesst eine Beleuchtungsgeometrie ein, die die Probenoberfläche oder das Probenvolumen entweder diffus oder gerichtet beleuchtet. Die Beleuchtung ist polychromatisch, unpolarisiert und gleichmässig, um allfällige Unregelmässigkeiten der Probe auszugleichen. Die Verwendung gepulster Lichtquellen verlangt eine sorgfältige Auslegung der Kanäle des remittierten Lichts (S) und der Lichtquelle (R). Die Lichtmessung bezieht den Spektralbereich des sichtbaren Lichts ein sowie die angrenzenden Bereiche des Ultravioletts und des Infrarots. Die Wandlung der analogen elektrischen Signale in digitale leistet ein hochauflösender Analog-Digitalwandler mit kurzer Konversionszeit. Die erwähnten Baugruppen bilden zusammen ein simultan arbeitendes Zweistrahlspektralfotometer für die Lichtmessung von 300 bis 1100 Nanometer. Im Detektionsteil des Monochromators (Fig. 2) finden wir zwei Reihen parallel messender Fotodioden, deren Fotoströme nach Blitzauslösung von den n-fach vorhandenen ersten Verstärkerstufen integriert werden. Die Integrationszeit entspricht der Blitzdauer. Die Fotodioden in Si-Planartechnik zeichnen sich durch kleine Dunkelströme (typisch 5 pA/mm2 bei U=1 V) aus. Die nach Wellenlängen zerlegten, simultan erfassten Lichtmengen werden mit Multiplexern dem A-D-Wandler (Auflösung 16 bit) zugeführt und seriell verarbeitet. Die erreichbare Dynamik des Verhältnisses zwischen voll ausgesteuertem Nutzsignal ('weiss') und minimalem Nutzsignal ('schwarz') überdeckt in der farbmetrischen Anwendung 3 bis 4 Dekaden. Das Oeffnen der n-fach vorhandenen Analogschalter erzeugt messbare Spannungshübe der

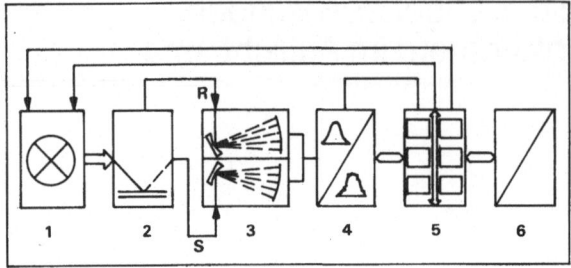

1: Gepulste Lichtquelle
2: Probenbeleuchtung
3: Zweistrahl-Monochromator
4: Analog-Digitalwandler
5: Mikroprozessoreinheit
6: Serielle Schnittstelle
S: Signalkanal
R: Referenzkanal

Fig. 1. Blockschema Farbmessgerät

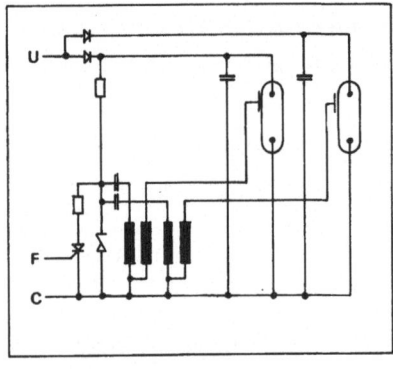

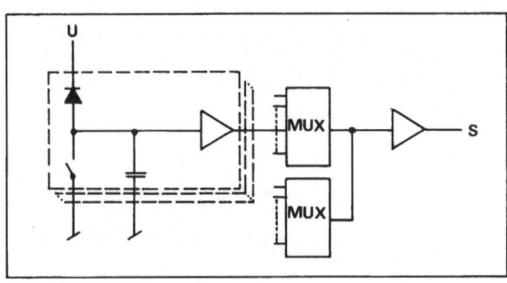

Fig. 2. Blockschema Zweistrahl-Monochromator

Fig. 3. Schaltschema Lichtquelle (vereinfacht)

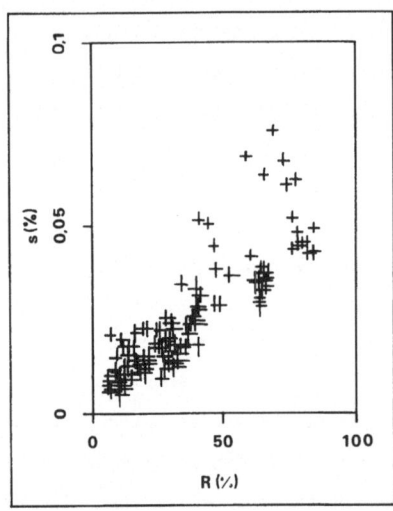

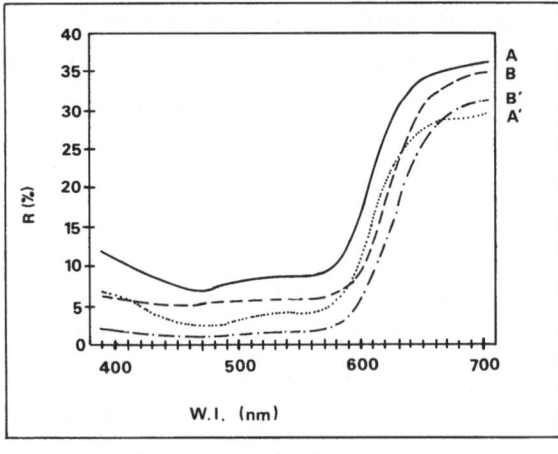

Fig. 5. Mittlerer Messfehler in
Abhängigkeit des Messwerts; R:
Messwert, s: mittlerer Messfehler

Fig. 4. Remissionsspektren metallisierter
Lacke mit und ohne Glanzmessung (Geometrie d/8)
Das Spektralfotometer unterscheidet objektiv
zwischen farblichem Eindruck und Glanzeffekt

Probe	Glanzmessung	L*,a*,b*-Koordinaten		
A	mit	L*=33,61	a*=22,9	b*=10,9
A'	ohne	31,21	29,8	17,6
B	mit	L*=40,23	a*=23,7	b*=11,8
B'	ohne	23,30	33,4	21,6

Integrationskondensatoren.   Vor dem Auslösen des Lichtblitzes wird das Signal jedes
Kondensators   mehrmals gemessen und digital gespeichert. Der Beleuchtungseffekt
(Blitz) führt zu einem zur Lichtmenge proportionalen Spannungshub. Durch wiederholte
Abtastung vor,während und nach der Blitzauslösung erfasst man die Spannungshübe und
die Drifte dynamisch, was mit Mikroprozessor (8085A) zu den Nutzsignalen und letztlich zur spektralen Verteilung des von der Probe remittierten Lichts führt. Für die
farbmetrische Anwendung wird das Remissionsspektrum (bzw. das Transmissionsspektrum)
mit normierten Gewichtsfunktionen multipliziert und integriert. Die Transformation
der Spektren in einen der dreidimensionalen Farbräume (Farbkoordinatensysteme) er-
leichtert die Interpretation der Ergebnisse und die Festlegung der Toleranzen für
das zu prüfende Farbgut. Die vereinfacht wiedergegebene Schaltung der Blitzeinheit
zeigt zwei Entladekreise mit Parallelzündung. Pro Blitzpaar werden mit Wiederholfrequenz 0,5 Hz ca. 20 J elektrischer Energie umgesetzt. Die Xenon-gefüllten Lampen
reichen für einige 10'000 Blitze aus.

3. Geräteleistungen

Der mittlere Fehler der Einzelmessung ist vom Messwert linear abhängig (Fig. 5). Auf
Grund der oben erläuterten Abtastung des Signals der einzelnen Kondensatoren werden die
Signaldrifte erfasst und korrigiert. Die praktisch erreichbare Dynamik der Messung
liegt zwischen 60 und 80 dB und wird durch geringe, doch messbare Variationen der
Lichterzeugung begrenzt. Ein praktisch erreichbares Linearitätsmass an festen Proben
(Keramiken oder Farbgläsern) ist 0,5 R% und an Flüssigkeiten (dispergierten Farbstof-
fen) 0,02 AE.

4. Anwendungen in der Industrie

- Helligkeit und Farbe metallisierter Lacke im sichtbaren Spektralbereich (400-700 nm)
  Durch Transformation der Remissionsspektren (Fig. 4) erhalten wir die Reflexionsfak-
  toren Rx, Ry, Rz, die Normfarbwerte X, Y, Z, die Normfarbwertanteile x,y,z oder die
  Koordinaten des Farbraums L*, a*, b*. Das Spektralfotometer mit ungerichteter Proben-
  beleuchtung und gerichteter Probenbeobachtung (d/8-Geometrie) erzeugt Farbkoordina-
  ten, deren Werte unabhängig sind von der eingesetzten Messöffnung. Durch Wahl der
  Messbedingung mit und ohne Glanzeinschluss gelingt es, Farbeindruck und unbunten
  Glanzeffekt zu trennen, objektive Farbdifferenzen zu bestimmen.
- Konzentrationsbestimmung dispergierter Farbstoffe in Lösung (400-700 nm)
  Durch Konzentrationsbestimmung der einzelnen Komponenten eines Gemischs von mehreren
  Farbstoffen in wässriger Lösung (Färbeflotten) kann der Prozess des Aufziehens auf
  Mischfasern im Minutentakt zeitaufgelöst mit hoher Genauigkeit verfolgt werden, um
  optimale Färbebedingungen zu erkennen. Das Farbmessgerät liefert hiezu die spektral
  aufgelösten Transmissionswerte. Die Färbeflotte wird mit einer rechnergesteuerten

Misch- und Dosiereinheit verdünnt und durch einen Kapillarschlauch der Transmissionsmessstelle zugeführt.

- Lumineszenzanregung optischer Aufheller (300-400 nm)

  Der Ultraviolettanteil der gepulsten Probenbeleuchtung regt die Lumineszenz optischer Aufheller an. Industriespezifische Kenngrössen (Weissgrade) wurden definiert, um die Qualität der optisch aufgehellten Erzeugnisse (Papiere, Textilien, Kunststoffe) zu erfassen und einzustufen.

- Infrarottransmission von lichtdurchlässigen Kunststoffen (560-1100 nm)

  Als Fenstermaterialien in infrarotoptischen Sendern/Empfängern verwendet, kommt es bei den Kunststoffen auf die eng tolerierte IR-Transmission an. Das Messgerät ermittelt die Transmissionsspektren.

- Infrarotremission von Militärtextilien (600-1200 nm)

  Diese besonderen Textilien bzw. Beschichtungen zeichnen sich durch eine an die Hintergrundumgebung angepasste Infrarotremission aus, die vom Spektralfotometer quantitativ beurteilt wird.

5. Zusammenfassung

Die untenstehende Zusammenfassung gibt Aufschluss über den Stand der optoelektronischen Messdatenerfassung mit nachfolgender farbmetrischer Auswertung in der Industrie.

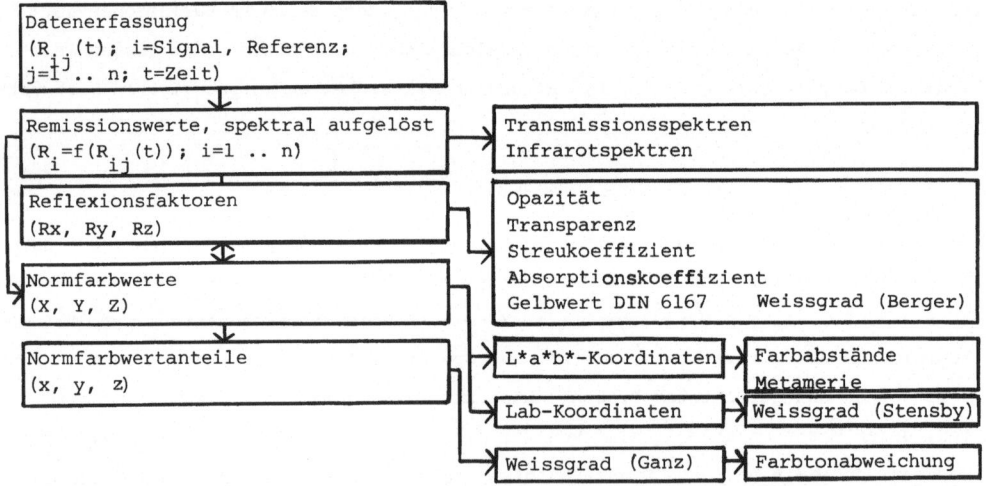

Die industrielle Farbmetrik bildet überdies eine entscheidende Grundlage für die automatisierte Sichtkontrolle und für die Bildverarbeitung im Fabrikationsprozess.

Literatur
(1) G. Buchsbaum: Proc. of the IEEE 69, 772-786 (1981)
(2) D.J. Granrath: Proc. of the IEEE 69, 552-561 (1981)
(3) C.J. Bartleson: Colorimetry, in: Optical Radiation Measurements, Vol. 2
    'Color Measurement', Academic Press, 33-148 (1980, New York)
(4) B.G. Batchelor, D.A. Hill, D.C. Hodgson (editors): Automated Visual Inspection,
    North-Holland (Elsevier) and IFS Ltd. (to be published 1985)

Messung der Wärmeleitfähigkeit unter Benutzung eines Lasers

H.U. Fritsch, U. Luft
Institut für Werkstoffkunde und Werkstofftechnik, TU Clausthal
Agricolastr. 2, D-3392 Clausthal-Zellerfeld

Einleitung

Die thermischen Eigenschaften metallischer Werkstoffe, wie Wärmeleitfähigkeit und spezifische Wärme, haben einen großen Einfluß auf
die Ergebnisse beim Oberflächenumschmelzen, z.B. auf die Einschmelztiefe und die Form der Wärmeeinflußzone, und werden auch für theoretische Beschreibungen solcher Prozesse benötigt /1/.
Es wird über ein Verfahren berichtet, das die Bestimmung der Wärmeleitfähigkeit über ein instationäres Meßverfahren ermöglicht /2, 3/.
Die möglichen Auswirkungen auf z. B. das Laseroberflächenumschmelzen
werden am Beispiel eines Gußwerkstoffes dargestellt.

Experimentelles

Ein Gußwerkstoff GGG 40 (3.6% C, 2.6% Si, 0.35% Mn) wurde zur Erzielung möglichst unterschiedlicher Gefüge wärmebehandelt, siehe Tabelle 1. Durch die unterschiedliche Verteilung des Kohlenstoffs konnten
so Veränderungen in der Wärmeleitfähigkeit erwartet werden. Aus diesen Materialien wurden dann Proben mit einem Durchmesser von 5 mm
und einer Länge von 10 bis 30 mm hergestellt. Durch Belichten der
einen Zylinderstirnseite mit einem Laserimpuls von 10 - 50 W Leistung
bei einer Pulslänge von 100 ms Dauer wurde ein Wärmepuls durch die
Probe geschickt und die Temperaturerhöhung an der entgegengesetzten
Stirnseite mit Hilfe eines angeschweißten Thermoelementes gemessen.
Durch Vorwärmen der Probe in einem Ofen konnte die Wärmediffusivität a in einem Temperaturbereich bis zu 800°C in Schritten zu 50°C
ermittelt werden, vgl. auch die prinzipielle Darstellung in Bild 1.

244

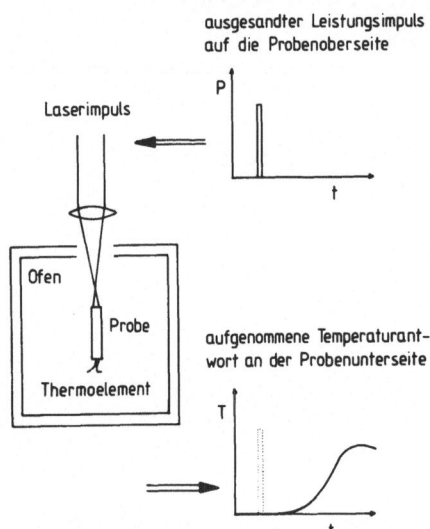

Bild 1. Schematische Darstellung des Meßprinzips

Theoretische Arbeiten

Durch Simulation des oben beschriebenen Aufheiz- und Abkühlprozesses des Zylinders mit Hilfe eines mathematischen Modells unter Berücksichtigung der 3-dimensionalen Wärmeleitung kann auf theoretischem Wege der Temperaturverlauf berechnet und mit der experimentell ermittelten Kurve verglichen werden. Durch die Ermittlung desjenigen Wertes für die Wärmeleitfähigkeit, bei dem die Abweichungen zwischen experimenteller und theoretischer Kurve minimal sind, läßt sich die Wärmeleitfähigkeit bestimmen, sofern das benutzte Modell den tatsächlichen Vorgang ausreichend gut beschreibt.

Aufgrund der Achsensymmetrie der Proben kann durch Übergang vom kartesischen Koordinatensystem x, y, z zu Zylinderkoordinaten r, ϕ, z der Vorgang über die 2-dimensionale partielle Differentialgleichung

$$\frac{\partial T}{\partial t} = a \left(\frac{\partial^2 T}{\partial r^2} + \frac{1}{r} \frac{\partial T}{\partial r} + \frac{\partial^2 T}{\partial z^2} \right)$$

beschrieben werden, die mit wesentlich weniger Rechenzeit numerisch gelöst werden kann, als die entsprechende 3-dimensionale Variante.

Die benutzten Randbedingungen sind:

$$\lambda \frac{\partial T}{\partial z} = - Q + \alpha \ (T_{Oberfl} - T_{Umg})$$

für z = 0 und $t \le t_b$, Q = 0 für $t > t_b$.

$$\lambda \frac{\partial T}{\partial z} = - \alpha \ (T_{Oberfl} - T_{Umg}) \qquad \text{für z = 1}$$

$$\lambda \frac{\partial T}{\partial r} = 0 \qquad \text{für r = 0}$$

$$\lambda \frac{\partial T}{\partial r} = - \alpha \ (T_{Oberfl} - T_{Umg}) \qquad \text{für r = 1}$$

Die Anfangsbedingung lautet:

$$T \ (r,z) = T_{Umg}$$

Hierbei sind:

Q = Leistungsdichte des Lasers
λ = Wärmeleitfähigkeit
α = Wärmeübergangszahl
T_{Oberfl} = Oberflächentemperatur
T_{Umg} = Ofentemperatur
a = Wärmediffusivität

Die Wärmediffusivität a setzt sich über die Gleichung

$$a = \frac{\lambda}{c_p \ \rho}$$

aus der Wärmeleitfähigkeit λ, der spezifischen Wärme c_p und der Dichte ρ zusammen.
Für die Lösung der oben angegebenen Differentialgleichung der Wärmeleitung wurde das numerische Differenzen-Verfahren "Alternating direction implicit methods" nach /4/ angewandt, das sich als sehr vorteilhaft für die Lösung dieses zweidimensionalen Problems herausgestellt hat.

Ergebnisse

Tabelle 1. Wärmebehandlungen des Gußeisens

Probe	Behandlungszustand
1	900°C, 1/2 h, Ölabschreckung, 3 x 270°C, 2 h
2	900°C, 1/2 h, Ölabschreckung auf 400°C, 2 h
3	900°C, 1/2 h, Luftabkühlung
4	1000°C, 1/2 h, Ölabschreckung
5	900°C, 1/2 h, Ofenabkühlung

Um die Brauchbarkeit des benutzten Modells zu zeigen, wurde zu Vergleichszwecken /3/ die Wärmediffusivität von Armco-Eisen untersucht, siehe Bild 2.

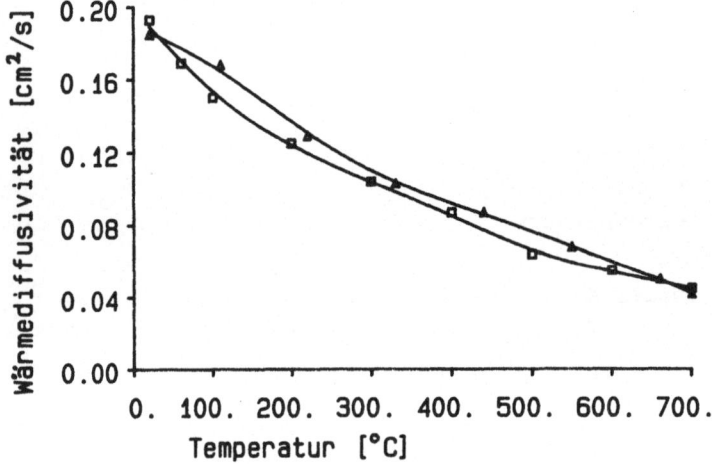

Bild 2. Wärmediffusivität von Armco-Eisen (□) als Funktion der Temperatur im Vergleich zu /3/ (▲)

Die festgestellten Abweichungen sind klein und durch eine Verfeinerung der Meßmethoden sicher noch verringerbar. Das Ziel dieser Arbeiten war jedoch ein möglichst niedriger Aufwand, d.h. z.B. Verzicht auf eine Vakuumkammer, die Abkühleffekte aufgrund eines Wärmeübergangs zwischen Probe und Luft vermeiden könnte. Diese Einflüsse sollen vielmehr durch eine möglichst reale Beschreibung mit Hilfe des Modells mit einbezogen werden, weshalb auch nicht eine analytische Lösung der oben angegebenen Wärmeleitungsgleichung benutzt werden konnte.

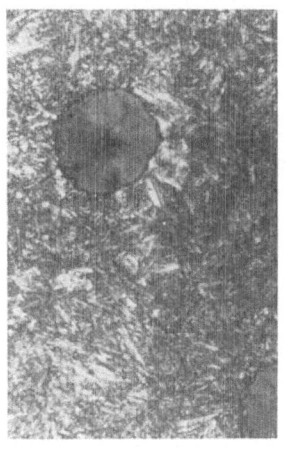 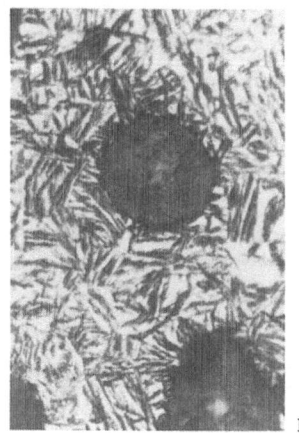 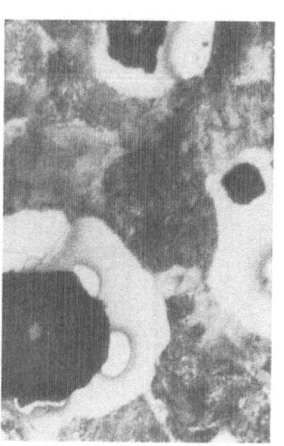

 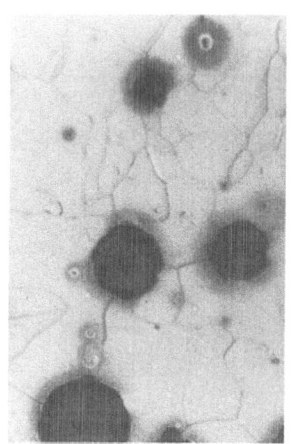

Bild 3. Gefüge des Gußeisens in Abhängigkeit von der Wärmebehandlung
    a) martensitisch   b) bainitisch
    c) perlitisch   d) martensitisch (hoher
    e) ferritisch   Restaustenitgehalt)

Die Wärmediffusivität a als Funktion der Temperatur ist als Ergebnis
der experimentellen Messungen und der zugehörigen theoretischen Berechnungen für die unterschiedlichen Wärmebehandlungszustände in
Bild 4 dargestellt. Man erkennt, daß sich die Wärmediffusivitäten bis
zu 100% unterscheiden, wobei das ferritische Grundgefüge den größten,
das perlitische Gefüge den niedrigsten Wert zeigt. Die Werte des bai-
nitischen bzw. des martensitischen Gefüges liegen zwischen diesen
Extremen. Um die Wärmeleitfähigkeit der Materialien bestimmen zu
können, wurden separat mit Hilfe eines Kalorimeters die spezifischen
Wärmen bestimmt. Diese Ergebnisse zeigt Bild 5.

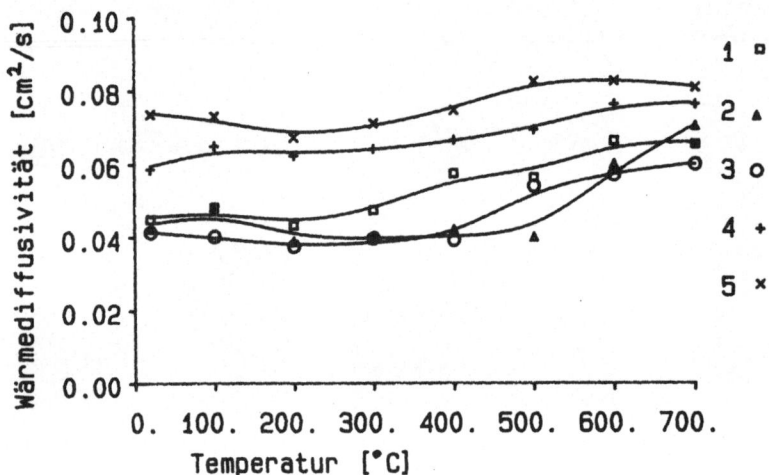

Bild 4. Gemessene und berechnete Wärmediffusivität des wärmebehandelten Gußeisens als Funktion der Temperatur

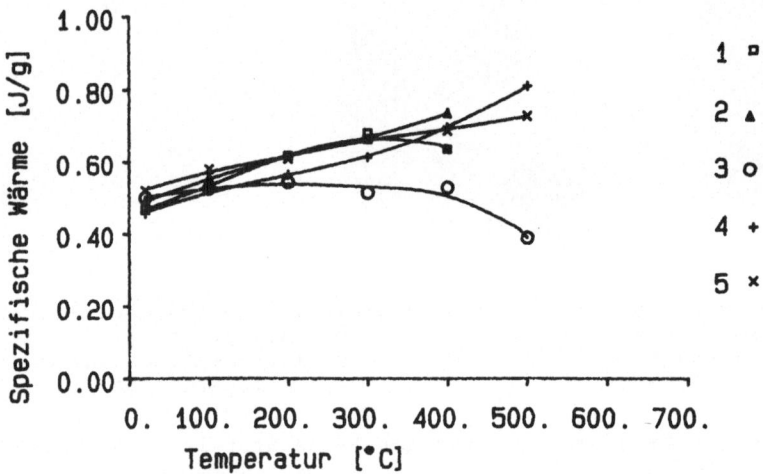

Bild 5. Spezifische Wärme c_p des Gußeisens als Funktion der Temperatur

Hieraus konnte dann die Wärmeleitfähigkeit als Funktion der Temperatur berechnet werden, siehe Bild 6.

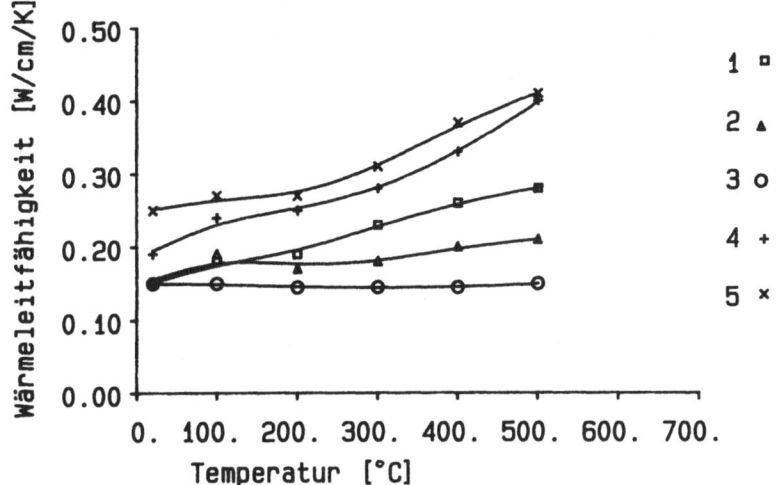

Bild 6. Berechnete Wärmeleitfähigleit des Gußeisens als Funktion der
Temperatur

Die Ergebnisse dieser Messungen sollen für verfeinerte theoretische
Modellrechnungen zur Beschreibung des Oberflächenumschmelzens mit
Hilfe eines Lasers benutzt werden /5/, da praktische Versuche hierzu
deutlich unterschiedliche Einschmelztiefen und Wärmeeinflußzonen
zeigen, was hauptsächlich durch die unterschiedlichen Wärmeleitfä-
higkeiten verursacht wird. Für gleiche Bearbeitungsparameter während des Laserumschmelzens ergaben sich für die 5 verschiedenen
Wärmebehandlungszustände des Gußeisens z. B. Einschmelztiefen
zwischen 250 und 350 μm bei einer Vorschubgeschwindigkeit von
0,5m/min.

Literatur

/1/ H.W. BERGMANN, G.HUNGER, H.U. FRITSCH, J. Mat. Sci. 16
    (1981) 1935
/2/ W.J. PARKER, R.J. JENKINS, C.P. BUTLER and G.L. ABBOTT
    J. Appl. Physics 32 (1961) 1679
/3/ H.-J. SÖLTER, Thermochimica Acta 83 (1985) 125
/4/ D.U. von ROSENBERG, Methods for the Numerical Solution of
    Partial Differential Equations, American Elsevier Publ. Comp.,
    New York, 1969
/5/ H.U. Fritsch et al, in Vorbereitung

Photovoltaisches Referenzelement für bewertete Empfindlichkeit und hohe Bestrahlungsstärke

J. Metzdorf, H. Kaase

Physikalisch-Technische Bundesanstalt

Bundesallee 100, D-3300 Braunschweig

1. Einleitung

Das einfachste und schnellste Verfahren für die routinemäßige Kalibrierung von Solarelementen basiert auf dem Einsatz eines Sonnensimulators, dessen Bestrahlungsstärke E_S in der Meßebene - mit Hilfe eines sogenannten Referenzelementes - auf den Standardwert von $E_o = 1000$ Wm^{-2} eingestellt wird, indem E_S so variiert wird, daß der Photostrom des Referenzelementes $I^R_{sc}(E_S)$ mit dem für das gewünschte AMn-Sonnenspektrum gültigen Kalibrierwert $I^R_{o,AMn}$ übereinstimmt /1/.

$$I^R_{sc}(E_S) = s^R_S(E_S) \cdot E_S = I^R_{o,AMn} \tag{1}$$

Das Referenzelement sollte vorher nach einer absoluten, unabhängigen Methode kalibriert werden, wobei entweder die Empfindlichkeit $s^R_{AMn}(E_{AMn} = E_o)$ oder der Kurzschlußstrom

$$I^R_{o,AMn} = s^R_{AMn}(E_{AMn} = E_o) \cdot E_o \tag{2}$$

anzugeben sind. Für die AMn-Spektren (Air Mass n; n = 0 oder 1.5) gibt es verschiedene Empfehlungen /1/.

Ein derartiges Verfahren bietet sich allgemein immer dann an, wenn die bewertete Empfindlichkeit eines photovoltaischen Elementes (oder beliebigen Strahlungsempfängers) für eine bestimmte integrale Bestrahlungsstärke gemessen werden soll. Dabei sind jedoch einige Bedingungen einzuhalten, die dann nicht selbstverständlich sind, wenn die gesamte Bestrahlungsstärke groß ist.

2. Anforderungen an das Referenzelement

Der mit der "Referenzelement-Methode" bestimmte Photostrom des Testelementes $I_{sc}(E_S)$ ist nur mit dem Wert $I_{o,AMn}$ identisch, wenn die relativen spektralen Bestrahlungsstärken von Simulator und Bezugsspektrum oder die relativen spektralen Empfindlichkeiten der Elemente gleich sind.

$$\left(\frac{\partial E_{AMn}(\lambda)}{\partial \lambda}\right)_{rel} = \left(\frac{\partial E_S(\lambda)}{\partial \lambda}\right)_{rel} ; \quad s(\lambda)_{rel} = s^R(\lambda)_{rel} \tag{3}$$

Andernfalls muß die zweifache spektrale Fehlanpassung mit einem Korrekturfaktor c^R gemäß /1/ korrigiert werden.

Zwischen Simulator- und Bezugsspektrum besteht - insbesondere für hohe Bestrahlungsstärken - im allgemeinen eine deutliche spektrale Fehlanpassung, deshalb bleibt als lösbare Aufgabe, die spektrale Empfindlichkeit des Referenzelementes an die der Testelemente anzupassen. Dabei ist es nicht trivial, die Linearität des Referenzelements zu verlangen oder zu kontrollieren, daß die Empfindlichkeit unabhängig von der Bestrahlungsstärke ist (s. Abb. 1); außerdem sollte die Empfindlichkeit zeitlich stabil und über einen genügend großen Bereich frequenzunabhängig sein (für Element R1b hier bis > 10 kHz).

3. Realisierung eines Referenzelementtyps

Abb. 2 zeigt den Prototyp eines Referenzelements mit einer aktiven Fläche von 2 x 2 cm² und mit einem großen Öffnungswinkel ($\gtrsim 160^O$). Um die bestmögliche Linearität einzuhalten, wurde (i) ein Element aus einkristallinem Si mit hochreiner, p-dotierter Basis (FZ-Material[1]) eingesetzt, (ii) durch Wahl des Präzisionsmeßwiderstands (100 mΩ bzw. 50 mΩ) gewährleistet, daß der Spannungsabfall bei der Kurzschlußstrommessung unterhalb von 10 mV bleibt, (iii) durch guten Wärmekontakt zum gesandstrahlten, schwarz eloxierten Al-Gehäuse die Temperaturdifferenz zwischen Element und thermostatisiertem Gehäuseboden minimiert (< 0,5 K). Diese drei Bedingungen sind bei dem in Abb. 1 zum Vergleich gezeigten kommerziellen Solarex-Referenzelement sämtlich nicht erfüllt. Die Kalibrierungen wurden nach der Methode der differentiellen, spektralen Empfindlichkeit durchgeführt.
Die in der Praxis einzusetzenden Solarelemente aus weniger reinem kristallinen oder aus amorphem Si sind im allgemeinen nichtlinear und eignen sich deshalb nicht für den Einsatz als Referenzelement, insbesondere sind a-Si-Elemente außerdem nicht zeitlich stabil genug. Da für die Minimierung der spektralen Fehlanpassung gemäß Gl. (3) nur die relativen spektralen Empfindlichkeiten übereinstimmen müssen, reicht es aus, das übliche Quarzglasfenster (Dicke 1 mm) durch ein geeignetes Farbfilter (hier Fa. Schott) zu ersetzen (s. Abb. 3). Die relative spektrale Empfindlichkeit des Elementes R1b mit Filter BG39 (Dicke 1 mm) sollte z. B. ausreichend an die beiden eingezeichneten a-Si-Elemente (a-Si 2 und a-Si 3) angepaßt sein, um mit üblichen Sonnensimulatoren eine spektrale Fehlanpassung $|c^R - 1| \leq 0,03$ zu erhalten.

Literatur

D. Hahn, H. Kaase und J. Metzdorf (Herausgeber) PTB-Bericht Opt-17 (1984)

---

[1] Hersteller: Telefunken Electronic in Heilbronn

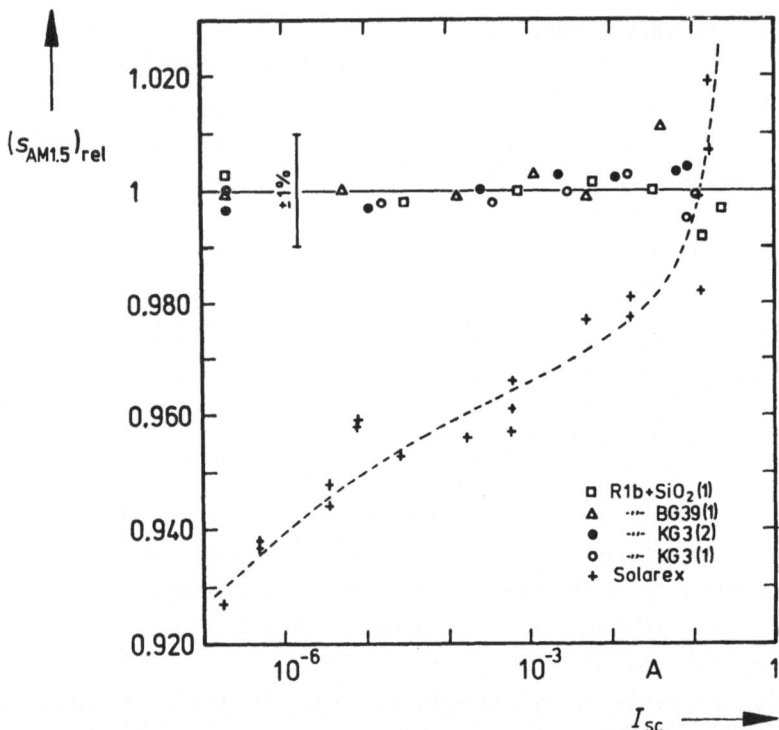

Abb. 1. Relative Empfindlichkeit bezüglich AM1.5-Bestrahlungsstärke $(s_{AM1.5})_{rel}$ des Referenzelementes R1b mit unterschiedlichen Filtern ($\square$, $\triangle$, $\circ$, $\bullet$) und die eines kommerziellen Referenzsolarelementes (Solarex) in Abhängigkeit vom Kurzschlußstrom I_{sc}.

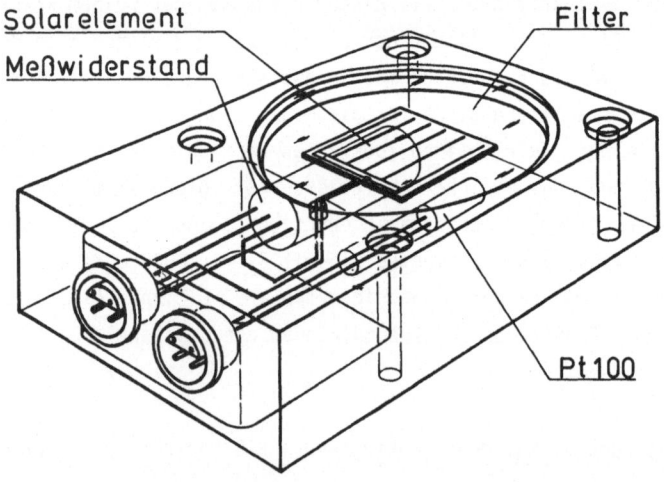

Abb. 2. Darstellung des Referenzelementes.

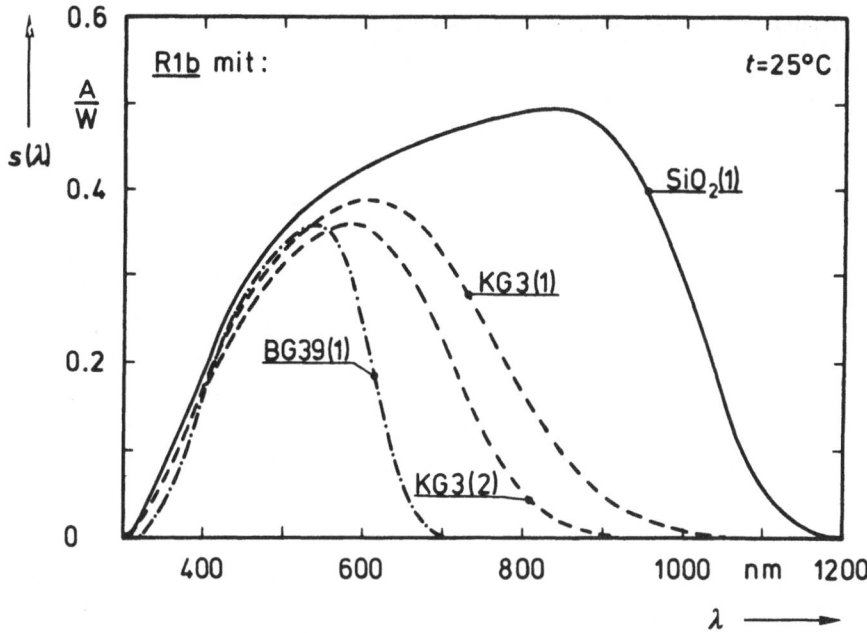

Abb. 3a. Spektrale Empfindlichkeit s(λ) des Referenzelementes R1b mit unterschiedlichen Filtern in Abhängigkeit von der Wellenlänge λ bei der Temperatur
t = 25 °C.

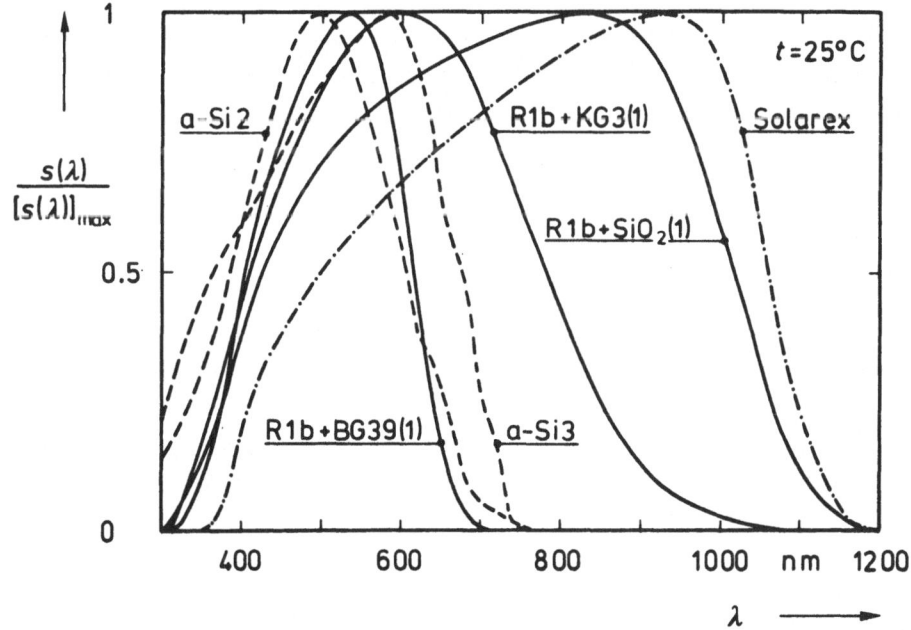

Abb. 3b. Vergleich der relativen spektralen Empfindlichkeit s(λ)/[s(λ)]max des Referenzelementes R1b mit unterschiedlichen Filtern und der des kommerziellen
Referenzelementes (Solarex) bzw. der von amorphen Si-Solarelementen (a-Si 2
und a-Si 3).

Ein Präzisionsradiometer für Laserstrahlung

K. Möstl, Physikalisch-Technische Bundesanstalt, Braunschweig/D

1. Einleitung

Seit längerer Zeit werden als Normale zur Messung optischer Strahlungs-
leistungen thermische Strahlungsempfänger benutzt (1). Bei diesem Meß-
prinzip wird die Strahlungsenergie in einem Absorbermedium, oft einem
Schwarzlack, in Wärme umgewandelt. Ist der Absorber nur lose an eine Wär
mesenke angekoppelt, so erhöht sich bei konstanter Einstrahlung seine
Temperatur so lange, bis die Wärmeverluste durch Wärmeleitung, Konvektic
und Wärmeabstrahlung ebenso groß werden wie die absorbierte Strahlungsleistung. Nachdem die Gleichgewichts-Temperaturerhöhung mit einem geeigneten Sensor, meist einer Thermosäule, gemessen worden ist, wird die
Strahlungsheizung durch eine elektrische Widerstandsheizung ersetzt. Zu
diesem Zweck ist der Absorber mit einem Heizwiderstand ausgerüstet. Re-
gelt man den Heizstrom so ein, daß für beide Heizungsarten das Sensorsignal gleich groß ist, so sind auch in erster Näherung die Leistungen
gleich. Die Strahlungsleistung ist damit auf eine leicht mit guter Genau
igkeit meßbare elektrische Leistung zurückgeführt. Für Präzisionsmessun-
gen muß man allerdings noch die Prozesse untersuchen, die das Sensorsignal beeinflussen. Dies führt vor allem zu folgenden Korrekturen:

* Die Strahlung wird nicht vollständig absorbiert. Der Absorptionsgrad α
  ist als Korrektur anzubringen.
* Strahlungs- und elektrische Heizung werden durch den Temperatursensor
  nicht vollständig gleich bewertet. Ursache ist eine unterschiedliche
  Verzweigung der Wärmeströme für beide Heizungsarten. Diese sogenannte
  Nichtäquivalenz wird durch einen Korrekturfaktor f_n berücksichtigt.
* Die Inhomogenität beschreibt, daß nicht alle Teile der Empfängerfläche
  die gleiche Strahlungsempfindlichkeit haben. Die Hauptursachen sind be
  thermischen Empfängern ungleichmäßiger Kontakt zum Temperatursensor
  oder ein ungleichmäßiger Absorptionsgrad (Korrekturfaktor: f_h).
* Ein Korrekturfaktor f_l für die Zuleitungsheizung berücksichtigt, daß
  auch in den Stromzuführungen des elektrischen Heizers Joulesche Wärme
  erzeugt wird, die mehr oder minder stark zum Sensorsignal beiträgt.
* Vom Temperatursensor eines elektrisch kalibrierbaren Empfängers muß nu
  gefordert werden, daß seine Empfindlichkeit während der Messung konstant bleibt. Kann dies nicht erreicht werden, so muß man die Einfluß-

größen untersuchen, um Korrekturen anbringen zu können (A und B in der Tab.).

Nennt man das Verhältnis aus Thermospannung U_t und der zugeführten elektrischen Leistung P die elektrische Empfindlichkeit s_e, so ergibt sich daraus die Empfindlichkeit s_s für Strahlung zu

(1) $s_s = \alpha \cdot f_h \cdot f_n \cdot f_l \cdot s_e$.

Von der Genauigkeit, mit der die Korrekturen bekannt sind, hängt wesentlich die Meßunsicherheit des Radiometers ab.

2. Aufbau des Empfängers

Der Empfänger wurde so konstruiert, daß die notwendigen Korrekturen möglichst klein und gut meßbar sind. Außerdem war zu beachten, daß die Bestrahlungsstärke im Laserstrahlungsbündel sehr hoch ist. Um starke Oberflächenaufheizungen zu vermeiden, die einerseits zu Beschädigungen des Absorbers, andererseits zu schwer quantifizierbaren Nichtäquivalenzen führen können, war ein Absorber mit guter Wärmeleitfähigkeit erforderlich. Hier konnte auf den für Dauerstrich-Laser bewährten, innen glänzend schwarzvernickelten Hohlkegel aus Kupfer (s. Abb. 1) zurückgegriffen werden (2). Der Kegel hat einen Öffnungswinkel von 45°, so daß Strahlung,

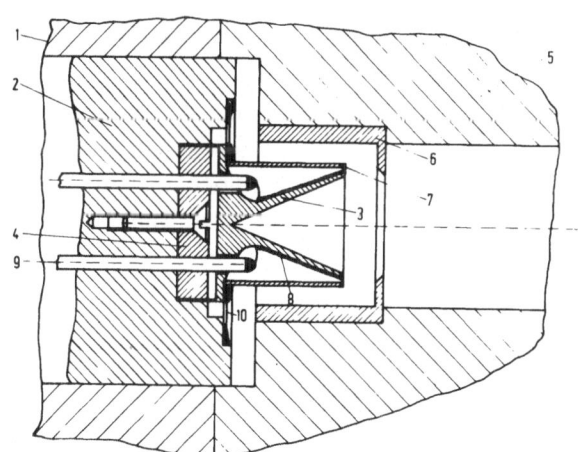

Abb. 1. Detailzeichnung des Laserradiometers 1,2,4 Teile der Wärmesenke (vgl. Abb.2); 3 Kegel; 6 Schutzblende, 7 vergoldeter Abschirmzylinder; alle aus Kupfer. 8 Heizer; 9 massive Cu-Stäbe für Heizstrom und -spannung; 10 Thermoelement.

die parallel zur Kegelachse einfällt, vier spiegelnde Reflexionen erleidet, die zur praktisch vollständigen Absorption der Laserstrahlung ausreichen sollten. Wegen eines unvermeidlichen Anteils an diffuser Reflexion, ergibt sich jedoch ein Absorptionsgrad unter 100%, nämlich etwa 99,8%. Dieser Wert ist aber so hoch, daß es ausreicht, den Reflexionsgrad auf 10% genau zu messen, um den Absorptionsgrad auf 0,02% genau zu kennen. Da von der Kegelspitze zum Kegelrand hin der Öffnungswinkel für diffus reflektierte Strahlung zunimmt, ist auch der Absorptionsgrad

leicht abhängig vom Ort des Auftreffens des Laserstrahls bzw. vom Bün-
deldurchmesser. Dem kann Rechnung getragen werden, indem man einerseits
das Laserbündel mit Justierhilfen gut zentriert und andererseits den effektiven Reflexionsgrad des Kegels mit den gleichen Bündeldurchmessern
bestimmt, die auch bei späteren Strahlungsmessungen vorkommen. Andernfalls ist eine kleine Korrektur anzubringen (s.Tab.).

Die im Kegel erzeugte Wärme fließt über die Kegelgrundplatte und einen
ringförmigen Steg (Länge: 1 mm, Wandstärke: 0,1 mm) zur Wärmesenke, die
aufgrund ihrer großen Wärmekapazität während der Messung ihre Temperatur
nicht meßbar ändert. Der Steg stellt wegen seiner geringen Wandstärke
einen erhebliche Wärmewiderstand dar. Der hieran von dem Wärmestrom erzeugte Temperaturabfall wird mit Hilfe von radial um die Kegelgrundplatt
angebrachten Thermoelementen (Ni/Cr-Ni-Draht von 0,1 mm Durchmesser) gemessen. Um den Themoelementen einen guten und alterungsarmen Wärmekontak
zu verschaffen, sind sie mit gut wärmeleitendem Epoxidharz in Nuten ein-
geklebt, die nur 0,2 mm breiter als die Lötstellen sind. Da weit mehr al
90% des Wärmestroms den genannten Weg nimmt und nur ein sehr kleiner
Bruchteil durch Wärmeleitung der Luft und Konvektion abgeführt wird, ist
die Empfindlichkeit s (Verhältnis aus Thermospannung und eingestrahlter
Laserstrahlungsleistung) nicht meßbar von Luftfeuchte und Luftdruck abhängig. Um die Konvektion zu kontrollieren, ist um den Kegel ein vergoldeter Abschirmzylinder angebracht und gut wärmeleitend mit der Kegelgrundplatte verbunden. Auf diese Weise nimmt der Abschirmzylinder stets
die Temperatur der Grundplatte an und schirmt den Kegel, dessen Temperaturverteilung von den Einstrahlbedingungen abhängt, gegen die konvek-
tierende Luft ab. Diese Maßnahme sorgt dafür, daß außer der oben beschriebenen, auf unterschiedlicher Reflexion beruhenden Inhomogenität
keine thermisch bedingte auftritt.

3. Die Nichtäquivalenz

Der Abschirmzylinder hat aber noch eine zweite Funktion: Er reduziert di
Nichtäquivalenz zwischen Strahlungsheizung und elektrischer Heizung des
Absorberkegels. Der für die Kalibrierung erforderliche Heizwiderstand au
Manganindraht ist außen um den Kegel gewickelt. Ein gut wärmeleitendes
Epoxidharz sorgt für guten thermischen Kontakt zum Kegel. Dennoch gibt
die Heizwicklung sicher mehr Wärme an die umgebende Luft ab als dies bei
einer Strahlungsheizung an der Innenfläche des Kegels der Fall ist. Soweit die Überschußwärme in Form von (langwelliger) Wärmestrahlung emittiert wird, wird sie fast vollständig vom vergoldeten Abschirmzylinder
zum Kegel hin reflektiert. Wärme, die durch Wärmeleitung der Luft ab-
fließt, wird dagegen großenteils auf den Abschirmzylinder übertragen und
von diesem wieder großenteils dem Meßwärmewiderstand zugeführt. Die ver

bleibende Nichtäquivalenz beträgt etwa 0,3 % (s.Tab.). Damit ist sie noch immer die größte notwendige Korrektur. Einer genauen Bestimmung ihres Wertes kommt deshalb große Bedeutung zu. Da die Nichtäquivalenz auf der unterschiedlichen Wärmeableitung durch die Luft beruht, kann man gemäß Gl.(1) aus dem Unterschied der elektrischen bzw. Strahlungsempfindlichkeiten in Luft (L) und Vakuum (V) den Nichtäquivalenz-Korrekturfaktor bestimmen:

$$(2) \quad f_n = (s_s^L/s_s^V) \cdot (s_e^V/s_e^L)$$

Zur Messung des ersten Verhältnisses aus Gl.(2) muß das Radiometer in einem Rezipienten mit einem amplitudenstabilisierten Laser (3) bestrahlt werden. Nach Ablesung der Thermospannung wird der Rezipient evakuiert und nach Einstellung des Gleichgewichtes wieder die Thermospannung abgelesen. Eine leichte Drift der Laserleistung wird über eine Referenzmessung korrigiert. Trotzdem ergab sich bei diesem Korrekturexperiment eine im Vergleich zu den anderen Meßunsicherheiten vergleichsweise hohe relative Unsicherheit von 0,05% (95%-Vertrauensniveau), die ihre Ursache in mechanischen Spannungen infolge der Evakuierung haben dürften.

4. Weitere Korrekturen

Zur Messung des Absorptionsgrades wird der Reflexionsgrad des Kegels an einer Ulbrichtkugel mit dem bekannten Reflexionsgrad eines Bariumsulfatpreßlings (4) verglichen.

Die Korrektur für die Zuleitungsheizung läßt sich aus drei Thermospannungsmessungen bestimmen, wenn bei gleicher Stromstärke der elektrische Heizer erstens über die Stromzuleitungen (Thermospannung U_i), zweitens über die Leitungen zur Messung des Spannungsabfalls am Heizwiderstand (U_u) beheizt werden. Bei der dritten Messung werden nur die Zuleitungen für sich, also ohne den Heizer, mit gleicher Stromstärke beheizt (U_1). Der Korrekturfaktor ist dann:

$$(3) \quad f_1 = (U_i + U_u - U_1)/2U_i .$$

Die elektrische Empfindlichkeit s_e ist in geringem Maße von der Umgebungstemperatur t des Empfängers abhängig, da die Wärmeleitfähigkeit von Kupfer und die Thermokraft der verwendeten Thermoelemente temperaturabhängig sind. Dies hat auch eine leichte Abhängigkeit der elektrischen Empfindlichkeit von der Heizleistung P zur Folge. Gemäß

$$(4) \quad s_e = s_0 (1 + A \cdot t + B \cdot P)$$

kann dies gut korrigiert werden (s.Tab.). s_0 ist die auf 0°C reduzierte elektrische Empfindlichkeit. Die hierfür angegebene Meßunsicherheit beruht auf der Unsicherheit der Strom- und Spannungsmessung sowie der Wiederholbarkeit des Kalibrierexperimentes. Insgesamt ergibt sich mit den in der Tabelle genannten Korrekturen und deren Meßunsicherheiten eine Meßunsicherheit der Strahlungsempfindlichkeit von $\pm 0,08\%$ bei einem Vertrau-

Tabelle. Korrekturgrößen des Radiometers und Meßunsicherheiten

Größe	Meßwert	Bemerkungen
α	$0,9984 \pm 0,0002$	$d = 5$ mm; 465 nm $\leq \lambda \leq$ 633 nm
	$0,9966 \pm 0,0012$	$d = 5$ mm; $\lambda = 1064$ nm
f_h	$0,99907 \pm 0,00019$	$d = 1$ mm
	$1,00000 \pm 0$	$d = 5$ mm
f_n	$1,00318 \pm 0,00053$	
f_1	$0,99829 \pm 0,00016$	
s_o	$1,05416 \pm 0,00032$	mV/W
A	$(7,6 \pm 0,1)\ 10^{-4}$	$1/^{\circ}C$
B	$(2,75 \pm 0,1)\ 10^{-4}$	W^{-1}

ensniveau von 95%. Geeignet ist das Radiometer für Leistungsmessungen im
Bereich von 1 mW bis 10W.
Ein nach dem gleichen Prinzip aufgebautes Radiometer zeigt Abb. 2. Es ist
zusätzlich mit einem kühlbaren Wärmeschutzschild und einer Meßblende ausgerüstet, und eignet sich deshalb für die Messung hoher (über die Meßblende homogener) Bestrahlungsstärken im Bereich von 0,01 bis 50 Solar-
konstanten.

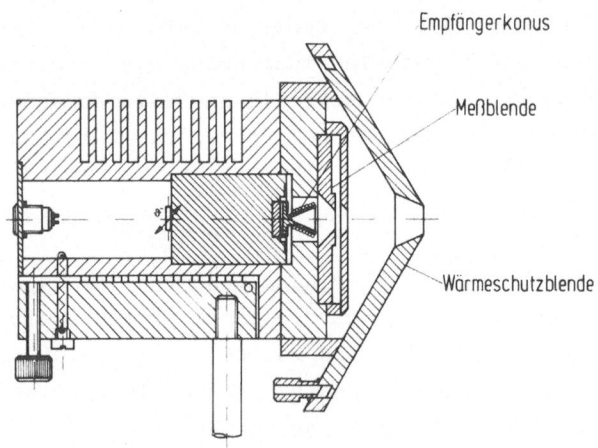

Empfängerkonus

Meßblende

Wärmeschutzblende

Abb. 2. Radiometer für
hohe Bestrahlungsstär-
ken; Details des Kegels
siehe Abb.1.

Literatur

(1) K. BISCHOFF: Optik 28, 183 (1968)
(2) K. MÖSTL: Feinwerktechn. 86, 72 (1978)
(3) K. MÖSTL: Optik 52, 167 (1978/79)
(4) W. ERB: PTB-Bericht PTB-Opt-3 (1975)

Holographische Interferometrie
Holographic Interferometry

Flat Holographic Combiner for Avionic Display

Giancarlo C. Righini and Stefano Gasperini*
Istituto di Ricerca sulle Onde Elettromagnetiche, CNR
50127 Firenze, Italy
* AGUSTA Spa, Divisione Sistemi, 00187 Roma, Italy

Holographic optical elements are attracting an ever increasing interest. A recent field of application, where the great potential value of their unique properties is confirmed, is constituted by the diffractive optics systems for avionic displays ([1]).

Electronic displays are being increasingly used in aircrafts for the presentation of information to the pilot and other aircrew such as navigator and flight engineer. A very effective way of presenting the pilot the most significant flight data while allowing him to pay attention to the outside world is represented by the Head-Up Display (HUD) architecture. In its most simple form, sketched in Fig.1, a conventional HUD includes a symbol generator (monitor), collimating optics (L), and a combiner. The monitor is the source of display information, and is located at the focal plane of the lens L, so that the CRT's screen is imaged at infinity. The combiner is a beam splitter that folds the optical beam from the monitor so that it is superimposed on the outside world's image seen by the pilot.

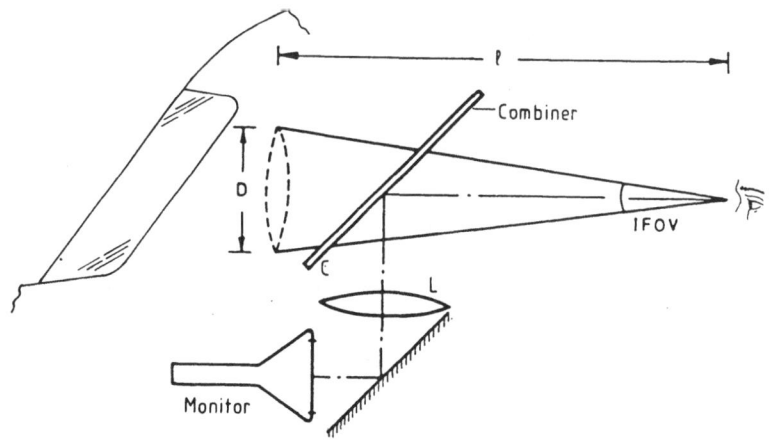

Fig. 1. Artist's sketch of a conventional head-up display

A critical parameter of the combiner is its optical efficiency. It must not severely attenuate the light passing through it from outside. On the other hand, the display information should be read under almost all ambient lighting conditions. Therefore the combiner must be designed to exhibit a high reflectivity only over a narrow spectral range, correspon ding to the CRT's peak spectral output. The reflectivity must also be high over a relatively large range of angles of incidence. All this make quite difficult to fabricate very efficient combiners using conventional thin-film reflective coatings. Typical reflectivity (quite broadband) of conventional combiners is about 25%, and average transmission is of the order of 70%. Holographic combiners, i.e. Bragg reflectors, can be expected to exhibit narrow-band reflectivity up to 90% and transmission over 80%. Moreover, holographic combiners can be fabricated which have optical power as well, so allowing larger fields of view than convention al HUDs do ([1]).

In this paper we report some preliminary results obtained in the fabri- cation of volume phase holograms to be used as flat combiners in HUDs. Dichromated gelatin is the only material that can allow us to get the required values of transparency and diffraction (reflection) efficiency. The present results have been obtained working on gelatin plates derived from Kodak 649F plates with preparation and development processes very similar to those described by Chang and Leonard ([2]). However, we are planning to deposit gelatin layers from solution, in order to get the proper thickness for optimum efficiency.

Figure 2 shows the sketch of the recording configuration of the volume (Bragg) hologram, with a desired 45° reflection angle. The reference

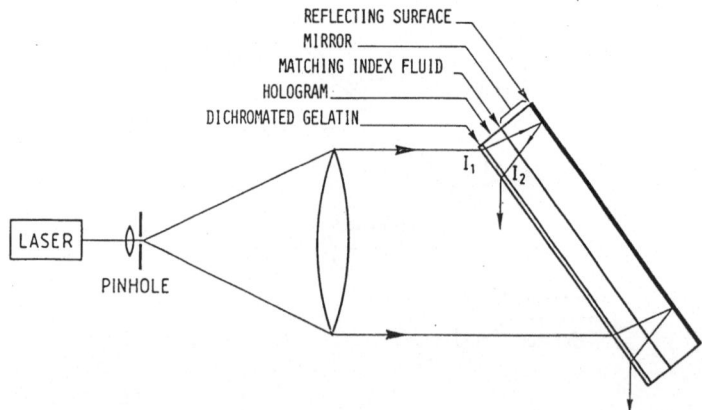

Fig. 2. Recording configuration of a reflection volume hologram to be used as a flat combiner in a HUD

beam is obtained by reflection of the collimated incident beam at the mirror placed behind (in contact with) the hologram plate.

A photo of the experimental system to record holograms up to 9 x 12 cm^2 in size is shown in Fig.3. The laser is a Spectra Physics 165 argon laser with etalon inserted in the cavity. Exposures have ranged from 200 to 500 mJ/cm^2 at the wavelength of 514,5 nm; exposure times were from 150 to 500 seconds.

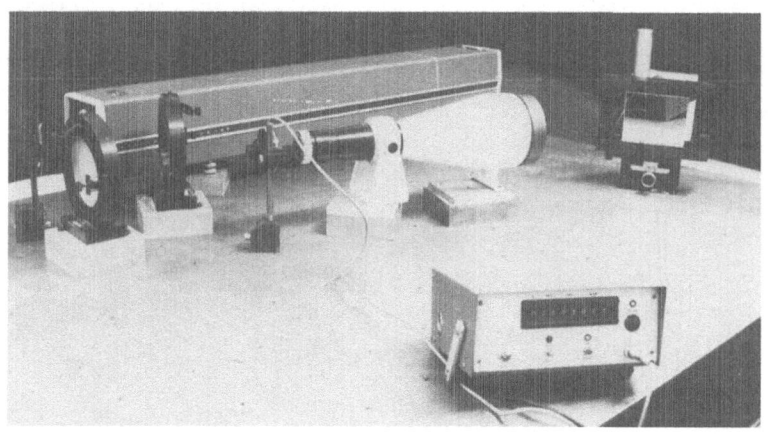

Fig. 3. Experimental recording system. The Spectra Physics argon laser, the Tropel collimator, and the electronic shutter are clearly visible

The efficiency and spectral bandwidth of the holographic combiners have been measured using a Perkin Elmer spectrophotometer. The combiner is placed in one of the sample beams and oriented such that the beam is incident at an angle of 45 degrees. The transmission of the combiner is thus measured and plotted as a function of the wavelength, as shown in Fig. 4 for a typical sample (continuous line).

The use of an holographic combiner in a HUD aboard the airplane requires sealing it against atmospheric moisture. However, we wanted to test the combiner reliability after a long storage even without having encapsulated it: therefore we have repeated the spectrophotometer measurements after a ten month storage in laboratory, with peak-to-peak changes in temperature and humidity equal to 16 °C and 45% respectively. The broken line in Fig. 4 indicates the result of the latter measurement on the sample considered: one can notice only a very slight shift (about 3 nm) of the wavelength corresponding to the peak reflectivity.   Similar results have been obtained for most of the samples tested, which thus exhibit a good resistance, at least in usual environment.

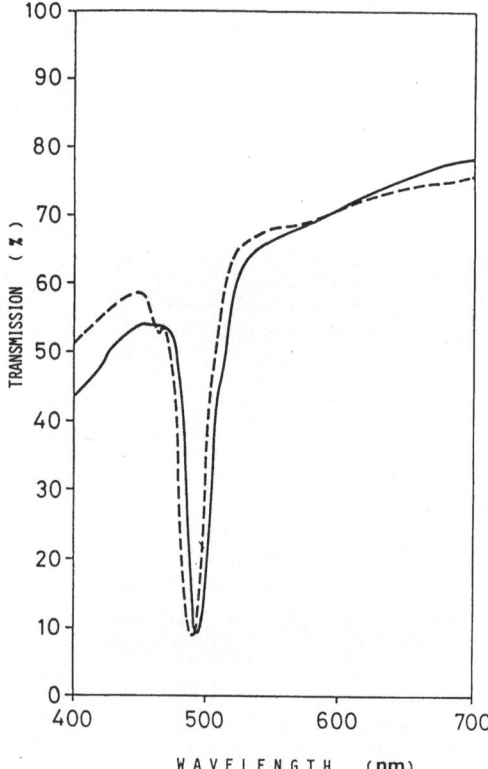

Fig. 4. Typical spectrophotometer transmission curve of a holographic combiner, for a beam incident at an angle of 45°. The solid line refers to the measurement carried out just after the fabrication of the holo- gram, while the broken line refers to a measurement performed ten months later

In conclusion, volume phase holograms to be used as flat combiners in HUD systems have been succesfully fabricated in dichromated gelatin. Typical results obtained so far are: white-light transmission up to 80% and narrow-band reflectivity up to 60%. The 3-dB spectral bandwidth is about 25 nm, centered around 500 nm for samples recorded at 514.5 nm. These values, however, are not yet repeatedly achieved in different batches, due to the lack of steady controlled environment condition both in the recording and processing rooms: work is in progress to get more reliable results.

References

(1) D.W. SWIFT: AGARD Conference Proc. N. 329, paper 31 (1982)
(2) B.J. CHANG and C.D. LEONARD: Appl. Opt. 18, 2407 (1979)

This work was promoted by Ottico Meccanica Italiana OMI Spa.

Possibilities of Difference Hologram Interferometry in Deformation/Shape Measurement

Z. Füzessy, F. Gyimesi
Institute of Physics
Technical University Budapest
H-1521 Budapest, Hungary

1. Introduction

Hologram interferometry has provide a versatile tool for measuring
displacement and shape of objects and analyzing the temperature
and mass density distribution of transparent objects. The hologram
interferometric techniques compare different states of the same
object and display them by interference pattern.

Nevertheless, in industrial applications there is often a need to
measure the difference in deformation of two objects in response to
the same load, to determine the shape deviation of an object from
the master one or, finally, to calculate the difference e.g. in
temperature distribution of two phase objects.

Traditionally the comparision can be made by elvaluating double
exposure interferograms belonging to the given object and the numerical data can only be compared.

Difference hologram interferometry gives a new possibility for comparison of two different objects. On the other hand, in case of
large displacement field, or complicated surface the interference
pattern can be too dense to be observed. The measuring range can
be extended by difference hologram interferometry /DHI/.

The idea of DHI /with the name of comparative holography/, as it
came to the author's attention lately, was formulated by Neumann
/1/ in 1980. Nevertheless, prior to this, the principle was realized by Denby at al /2/ in their electronic speckle pattern interferometer in 1976 in two-wavelength contouring with optically
polished surface of master object.

The first experimental results of DHI with no surface treatment at
all were presented by authors on deformation measurements in 1983
/3,4/. Extension of DHI towards conturing was recently realized
in two-refractive-index method /5/.

In the present paper further applications of DHI will be presented
with special respect to phase object measurements.

2. Principle and realization of the technique

Because of its comparative character DHI works with master and test objects. Master wavefronts with orbitrary phase difference, corresponding to the given displacement, refractive index variation or shape, can be produced and stored recording a double exposure interferogram of a master object. Single or two reference beams can be used. The test object is illuminated holographically: illuminating wavefronts can be produced by wavefront reversal, i.e. reconstructing the master interferogram by conjugate reference beam/s/ used during the recording process for master object.

The second step in applying DHI is recording the difference interference pattern. A new double exposure interferogram is made. Using master wavefronts for illumination of the test object in the original sequence the new interference pattern displays only the difference in deformation, refractive index variation or shape.

There are different ways for recording and following reconstruction of the master object. It is possible to record the two wavefronts on different plates with a single reference beam. However, in practice the thickness and the wedge angle of holographic plates are different and therefore the mutual reference wave can not be reversed in the same way for the two plates.

If the two wavefronts of the master object are recorded on a single plate and with two different reference waves the interferometric coincidence in backprojection can easily be achived.

As an alternative to the preceding arrangement the two wavefronts of the master object can be recorded on the same plate with one mutual reference wave as well, if they arrive at the hologram plate from different directions.

Dropping the requirement for seperate reconstructions the usual simple holographic arrangement with one plate and one reference wave can also be used.

3. Experimental results

In the deformation measurement published in 1983 /1/ the object was a pressure chamber. The pressure was increased in it between the two exposures. The corresponding master interference pattern contained concentric interference fringes displaying the bulge of the bottom. For the sake of simplicity the same chamber was used as the test object. During recording the difference interference pattern the

master hologram was placed in its original position, the test object
was illuminated holographically as discussed above and a new hologram plate was placed in the direction where master object had been
illuminated from.

The recorded new double exposure interferogram displays the difference in bulge.

Another interesting application of DHI was the shape measurement of
two aluminium membranes /master and test/ of 60 mm diameter which
were contoured with 10 µm sensitivity.

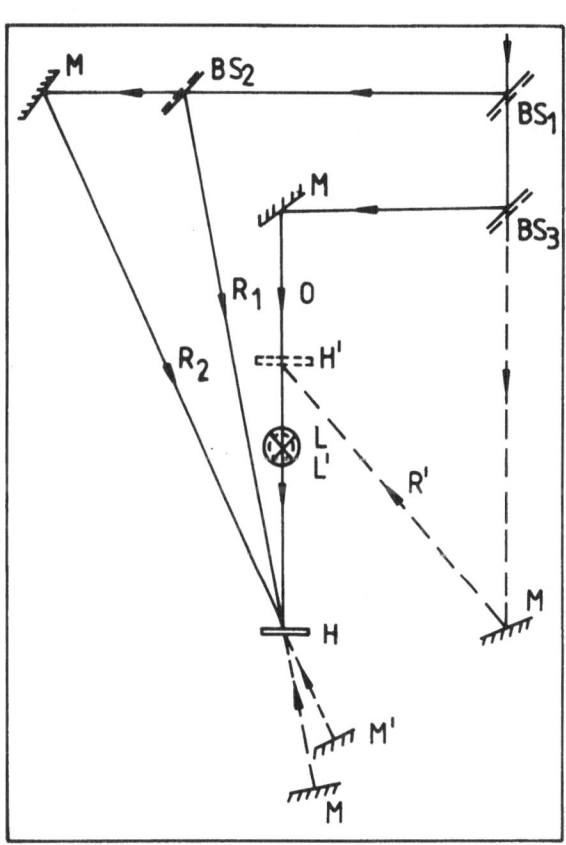

Fig. Simplified experimental
set-up

Lately the DHI was applied to investigation of phase objects. A simplified version of the experimental set-up without beam expanding elements is shown on the figure. Symbols BS are beamsplitters, M mirrors, R_1 and R_2 denote reference beams for recording the master holograms. The object beam O for the lamp L is the reflected light on the beamsplitter BS_3. Parts of the set-up denoted by dotted lines do not work when the master hologram is recorded. The objects were two filament lamps. As first step the master interferogram H was recorded. The lamp L was illuminated through a diffusor. The first master hologram was made without any voltage by reference beam R_1, the second one, when the lamp was loaded by e.g. 100 V, by reference

beam R_2. The master interferogram displays the change of temperature
and gas density distribution in response the load. After developing
the master interferogram was placed back in its previous position,

the master lamp L was taken away, a new lamp L' was placed where the master one was before. A new hologram plate H' was placed in the direction where master lamp was illuminated from.

The test lamp L' was illuminated by reconstructed real images of the master lamp L and a double exposure interferogram was made. R_1 and R_2 were plane waves, so the complex conjugates $R_1^{\times}$ and $R_2^{\times}$ were simply produced by mirrors M'. The first hologram was recorded without any voltage while the first real image /illuminating beam/ was reconstructed by reference beam $R_1^{\times}$. The second exposure was made with the lamp subjected to a given voltage while the another real image of master object /illuminating beam/ was reconstructed by reference beam $R_2^{\times}$.
This new interference pattern corresponds to the difference in changes of temperature and gas density distribution of the two lamps. In case of no differences interference fringes will not be formed.

4. Conclusions

Differences hologram interferometry can be applied for comparative measurement of displacement, shape of diffusely reflecting surfaces and as well as for comparative investigations of phase objects.

Literature

/1/ D.B. NEUMANN, Tec. Digest, Topical Meeting on Hologram inter-
    ferometry and speckle metrology, Opt. Soc. Am., MB 2-1/1980/
/2/ D. DENBY, G.E.QUINTANILLA, J.B. BUTTERS, Proc. Strathclyde
    Conf., 323/1975/.
/3/ Z. FÜZESSY, F. GYIMESI, Proc. SPIE 398, 240/1983/
/4/ Z. FÜZESSY, F. GYIMESI, Opt. Eng., 23, 780/1984/
/5/ F. GYIMESI, Z. FÜZESSY, Opt. Commun. 53, 17/1985/

Holographische Untersuchung von Industriellen Ölbrennern

J.J. Timkó

Forschungsinstitut für Technische Chemie der
Ungarischen Akademie der Wissenschaften
Pf. 98. Budapest 1502. Hungary

Die holographische Messmethode eignet sich vorzüglich um schnell
ablaufende Prozesse – wie z.B. die Tropfenbildung – störungsfrei,
in der Phase der Bildung, festzuhalten und in einer beliebigen
Zeit die Tropfen wieder herzustellen zum Zweck der Auswertung.
Unsere Methode wurde seit Jahren bei verschiedenen Sprühnebeln
verwendet, um Grösse, Grössenverteilung räumliche Lage von den
erzeugten Tropfen zu bestimmen.

Von dem Zwang, auch die industrielle Anwendung anzustreben,
wurde eine patentierte Aufnahme-Anordnung gebildet; die nicht nur
im Laboratorium zu verwenden ist.

Als Belichtungsquelle dient ein Impulsrubinlaser, mit Blitzdauer
von 20 ns; die Hologramme werden auf AGFA-Gevaert "Holotest" Filme
aufgenommen; gleichzeitig mit der Geradeaus-Methode und mit ge-
trenntem Referenzbündel. Dem abbildenden Prozess gemäss werden die
Aufnahmen entweder mit einer Streulinse aufgeweitetem Laserstrahl,
oder mit parallelem Laserstrahl durchgeführt.

Die wiedergabe erfolgt durch einem He-Ne-Laser /50 mW/. Das räumliche Bild erscheint am Monitor einem geschlossenen TV-System,
Ebene auf Ebene scharfgestellt. Dabei können überlagerte Teilchen
getrennt werden. Die Auswertung der nacheinanderfolgenden Bild-
reihe kann auch elektrooptisch, automatisch vorgenommen werden.
Von dem optischen Aufbau betrachtend sind auch hier mehrere Möglichkeiten.

Schon früher haben wir Aufnahmen von Heizöltropfen gemacht; im
kalten Zustand, aber auch brennend.

Die Düsen haben radiale oder tangentiale Innenkanäle; wie es am
ganzen austretenden Tropfenschar gut zu beobachten ist. In industriegrössen Brennofen haben sich die Düsen mit tangentialen
Kanälen bewährt.

Nachdem, dass wir die Heizöltropfen in einem Ofen von Grossla-
boratoriummassstab abgebildet hatten, kam die zweite – und

schwerere - Versuchsreihe. Wir nahmen dieselben Düsen, mit denselben mechanischen Parametern, aber brennend. Das Problem der
selbstleuchtenden Objekts kam noch zu den bisherigen.

Die Raumabschnitte im Ofen sind so gewählt, dass zusammengleich-
bare Resultate entstehen können. Es zeigte sich, dass die Tropfenbildung während der Heizung die gleichen Anomalien zeigt; die
Grössenverteilung der erzeugten Tropfen ändert sich auch wenig.
Natürlich ist die mittlere Partikelgrösse im brennenden Zustand
kleiner.

Die Verteilungskurven aufeinander gezeichnet zeigen deutlich, wie
im Heizungsprozess der Mittelwert der Tropfen minder wird.

Die Ergebnisse haben grosse Bedeutung bei dem Entwurf von neuen
Ölbrennern, da mit ihnen die Anomalien im Sprühbereich qualitative
und auch quantitative zu zeigen sind.

Nach erfolgsreicher Versuchsreihe haben wir an einem ölgefeurten
Ofen in Industriegrösse diese Messmethode erprobt. Viele Schwierigkeiten entstanden nicht nur von der Grösse des abzubildenden Objekts, sondern auch von den Bedingungen /wie Staub, Hitze,Vibra-
tion/, die die Arbeit entschwerten.

Eine neue Aufnahmevorrichtung mit Holokamera wurde gebaut, um bei
dem Ofen arbeiten zu können.

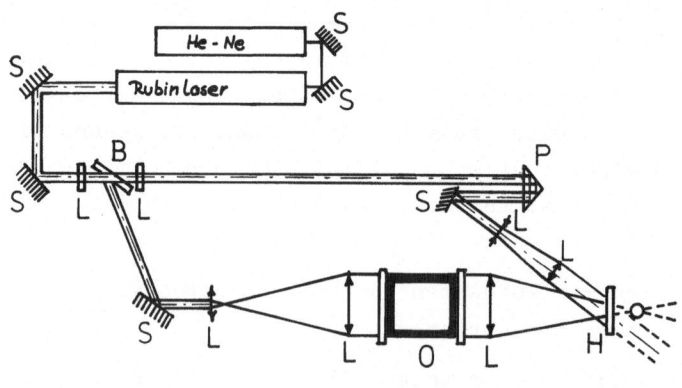

S Spiegel   P Prizma
L Linse   H Hologram
B Beam-Splitter   O Ofen

Aufnahmenscheme

Die Versuche hatten den Ziel, die Tropfenbildung von einzelnen Düs-
en und die Einwirkung von mehreren Düsen aufeinander auszuwerten.

Um bei dem Brennofen arbeiten zu können, musste ein paralleler
Laserstrahl durch ein Guckloch geführt werden, der Referenzstrahl
wurde um den Ofen geleitet. Der optische Aufbau musste einen Spielraum möglich machen, weil mehrere, in der Energiewirtschaft ge-
brauchte Düsen miteinander zu vergleichen waren. Die Düsen zer-
sprühten ein Dampf- Öl - Mischung.

Dampfund Öl treffen aufeinander in einem Mischraum des Zerstäubers.
Der Druck wurde von 2 bar bis 8 bar in vier Stufen geregelt. Also
ungefähr 300 Hologtamme wurden aufgenommen mit verschiedenen Druckverhältnissen um die Wirkung auf die Tropfenbildung beurteilen zu
können. /2 bar Dampf - dazu 2-4 - 6-8 bar Öl; 4 bar Dampf -2-4-6-8
bar Öl u.s.w. Danach wurde der Öldruck festgehalten und dazu der
Dampfdruck stufenweise zugeordnet./

Von dieser Versuchsreihe haben wir festgestellt, dass der Tropfen-
diameter von dem Dampfdruck mehr abhängig ist, als von dem Öldruck.
Apparative gibt die Länge des Mischraumes den grössten Unterschied
in der Tropfengrösse.

Die Bilder zeigen einen Nebel von 4 bar und 8 bar paarweise. Der
Tropfengrössenunterschied ist schon qualitative zu sehen .

Öltropfen

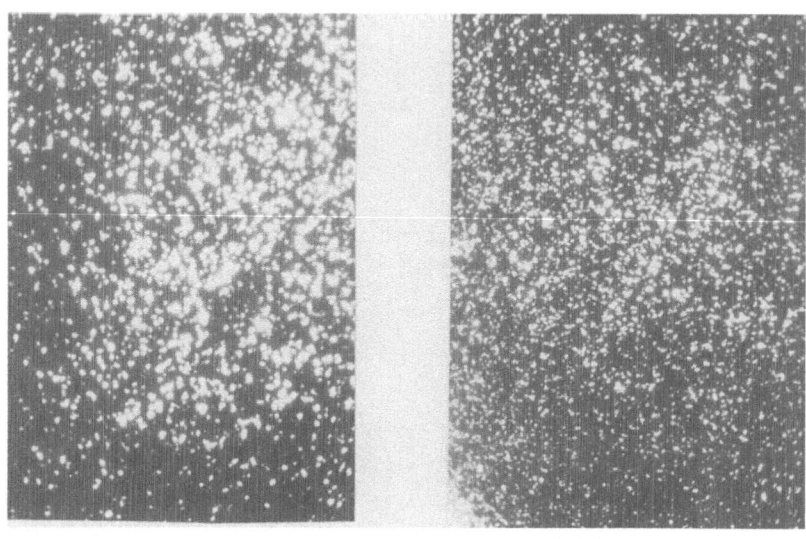

4-4 bar   8-8 bar

Angewandte Holographie als eine Messmethode in Industriegrössen ist noch immer ein interessantes Gebiet der Holographie.

Die Versuche in Industriegrössen wurden bisher in einem isothermen System gemacht. In kurzer Zeit werden die Tropfen auch in dem industriegrössen Ofen brennend abgebildet, um unsere Versuchsreihe vollständigen.

Literatur
/1/ TIMKÓ, J.J.: Diss. Th. 1974.
/2/ TIMKÓ, J.J.: Holography in chemical engineering.
        CHISA'75 Praha, 1975. G.2.4.
/3/ TIMKÓ, J.J. BLICKLE, T. NÉMETH, J.: Angewandte Holographie bei
        der Auswertung von zerstäubten Flüssigkeitstropfen.
        ACHEMA'76; Frankfurt am Main, 1976. jun.
/4/ TIMKÓ, J.J.: Holographic Study of the Dispersoids in Technical
        Chemistry. Proceedings 2-nd Europaen Congress on Optics
        Applied to Metrology "METROP". /M.GROSSMANN - P.MEYRNEIS/
        SPIE, Vol. 210. 1979. p.p. 140-143.
/5/ TIMKÓ, J.J.: The Investigation of Transport Phenomena by
        Applied Holography. FLOW VIZUALIZATION II.
        /Ed.: W. MERZKIRCH/ Heimspere Publishing Corporation,
        Washington, New-York, London, 1982. /p.p. 535-541/.
/6/ Patents: Hung. Patent 175.498., USA Patent 4,278.319.
        UK Patent GB 2,042.754B.

Holografische Interferometrie spiegelnder Oberflächen (Referenzebenen-Methode)

G. Schönebeck

Kraftwerk Union AG

Wiesenstraße 35, D-4330 Mülheim/Ruhr

Einleitung

Die holografische Interferometrie erlaubt die Ausmessung von Deformationen und Verschiebungen von diffus reflektierenden Oberflächen. Die Deformationen spiegelnder Oberflächen können so, nach bisherigen Erkenntnissen, holografisch nicht vermessen werden. In diesem Beitrag soll beschrieben werden, wie unter Ausnutzung der physikalischen Möglichkeiten statische und dynamische Deformationen vorwiegend ebener Spiegel doch holografisch vermessen werden können. Das hierfür angewendete Prinzip soll "Referenzebenen-Verfahren" genannt werden. Durch dieses neue Verfahren wird gleichzeitig die Empfindlichkeit der Deformationsmessung verdoppelt.

Prinzip des "Referenzebenen-Verfahrens"

In Bild 1 ist das Prinzip des neuen Verfahrens aufgezeichnet. Der Spiegel Sp sei das zu untersuchende Objekt. Der Spiegel Sp und die Referenzebene RE ist so anzuordnen, daß die feste Referenzebene RE durch die Hologrammplatte H vom Beobachterstandpunkt B zu sehen ist. Deformiert sich die Spiegeloberfläche (Bild 2), so wird die Lichtwegdifferenz zwischen den beiden Positionen 1 und 2 beim Doppelbelichtungsverfahren gegenüber den Verfahren mit diffus reflektierenden Oberflächen verdoppelt.
Entsprechendes gilt auch für andere Verfahren, wie z.B. Time-average und Real-time bei schwingenden Oberflächen. Durch die Lageänderung des zu untersuchenden Spiegels wird die Lichtweglänge zwischen den beiden betrachteten Positionen 1 und 2 geändert. Die Lichtwegänderung (Differenz) führt zur Bildung von Interferenzlinien, die ein Maß für die vorhandenen Deformationen sind.

Auswertung der Interferenzlinien

Die grundsätzliche Auswertung erfolgt nach dem Ellipsoidenverfahren von N. ABRAMSON [1]. Der Spiegel Sp (Bild 2) verschiebe sich von Position 1 nach 2 um

$$\Delta |\bar{n}| = \frac{\lambda}{2} \tag{1}$$

Bei der Holografie diffus reflektierender Objekte bildet sich hierbei 1 Interferenzlinie. Bei Objekten mit spiegelnder (ebenen) Oberfläche bilden sich dagegen 2 Interferenzlinien, wie nachfolgend gezeigt wird, weil die Wegdifferenz verdoppelt wird.

$$|\overline{B\ P_{Sp1}\ P_o}| - |\overline{B\ P_{Sp2}\ P_o}| = 2 \cdot \Delta|\overline{n}|$$

$$+|\overline{P_o\ P_{Sp1}\ L}| - |\overline{P_o\ P_{Sp2}\ L}| = 2 \cdot \Delta|\overline{n}| \tag{2}$$

Gesamtverlängerung: $\Sigma\Delta|\overline{n}| = 4\ \Delta|\overline{n}|$

G. (1) und (2) ergeben

$$\Sigma\Delta|\overline{n}| = 4\ \frac{\lambda}{2}$$

$$\tag{3}$$

$$\Sigma\Delta|\overline{n}| = 2\ \lambda$$

D.h., es bilden sich hier 2 Interferenzlinien. Damit ist die Empfindlichkeit gegenüber den konventionellen Verfahren um den Faktor 2 vergrößert worden.

Man kann das Referenzebenen-Verfahren in folgender Weise auf die üblichen Verfahren
der Holografie diffus reflektierender Objekte zurückführen. In Bild 3 ist das Ersatzbild zu Bild 2 aufgezeichnet. Die Referenzebene befindet sich an ihrem virtuellen
Ort, d.h. hinter dem hier weggedachten Spiegel. Der Lichtweg gehe von L zu P_{o1} und B.
Der Winkel dazu sei α. Die scheinbare Lage der Referenzebene RE verschiebe sich von
P_{o1} nach P_{o2}. $\Delta|\overline{n}| = \frac{\lambda}{2}$ sei die gedachte Bewegung des Spiegels Sp. Es werden also die
beiden Strahlen $L\ P_o$ und $P_o\ B$ um je $2\ \Delta|\overline{n}|$ verlängert. Es ist daher aus der Zeichnung
direkt ersichtlich, daß

$$P_{o1} - P_{o2} = 2\ \Delta|\overline{n}| \tag{4}$$

ist. Die Lichtwegdifferenz ist dann

$$|\overline{L\ P_{o1}\ B}| - |\overline{L\ P_{o2}\ B}| = 4 \cdot \Delta|\overline{n}| \tag{5}$$

d.h., es bilden sich 2 Interferenzstreifen.

Nach der allgemeinen Auswertegleichung der "Spiegelmethode" [2] und [3] ist N die
Zahl der Interferenzlinien, die zur Deformationsrechnung eingesetzt werden. Es ist:

$$\overline{n_\eta^o} \cdot \overline{D} = |\overline{n_\eta}| \qquad \eta = 1, 2, 3 \tag{6}$$

mit

$$|\overline{n_\eta}| = \frac{\lambda}{2} \cdot \frac{1}{\cos\frac{\alpha\eta}{2}} \cdot N_\eta$$

Es bedeuten:

$\overline{\eta^o}$ = Richtungscosinus

$\overline{D}$ = räumlicher Verschiebungsvektor

α = Winkel zwischen Beleuchtungs- und Beobachtungsrichtung

N = Anzahl der zur Deformationsrechnung einzusetzenden Interferenzlinienzahl.

Hier benötigen wir, weil die Deformationsrichtung eindeutig ist, nur eine Gleichung mit $\eta = 1$. Wenn $\cos \frac{\alpha}{2} \to 1$ geht, d.h. $\alpha < 15^o$ ist, wird der Fehler kleiner als 1 %. Es ist praktisch meist möglich, α kleiner als ca. 15° zu machen. Oft kann der Winkel noch kleiner ausgeführt werden, so daß α nicht berücksichtigt zu werden braucht. Es macht aber auch keine Schwierigkeiten, α rechnerisch mit einzubeziehen, wenn man es für nötig hält.

Transformationsgleichungen

In [4] wird gezeigt, daß die in die Rechnung einzuführende Interferenzlinienzahl N identisch ist mit der Interferenzlinienzahl i, die auf dem Interferenzlinienbild der Deformation ausgezählt wird. Z.B. bei Doppelbelichtungsaufnahmen. i kann aber bei den verschiedenen holografischen Verfahren auf N transformiert werden [4], so daß die Grundgleichung (6) für alle bekannten holografischen Verfahren unverändert angewendet werden kann. Durch das Referenzebenen-Verfahren ist ersichtlich, daß sich bei statischer Belastung (Doppelbelichtung) die Interferenzlinienzahl i (auf dem Bild gezählt) bei gleicher Deformationsgröße gegenüber der konventionellen Doppelbelichtungsholografie verdoppelt. Die Transformationsgleichung muß also lauten:

$$2 i = N \qquad \text{Doppelbelichtung} \qquad (7)$$
$$i = \frac{N}{2} \qquad \text{(statische Belastung)}$$

Bei diffus reflektierenden Oberflächen wäre

$$i = N$$

Gleichung (7) ist auch gültig für stroboskopische Beleuchtung. Es ist bei Time-average:

$$i = \frac{N}{2} = \frac{1}{2} \frac{\text{Arg } (J_o^2)}{\pi} \qquad \text{Time-average} \qquad (8)$$
$$\text{(dynamische Belastung)}$$

In Real-time bei schwingender Belastung ist zu setzen:

$$i = \frac{N}{2} = \frac{1}{2} \frac{\text{Arg } (1+J_o)}{2\pi} \qquad \text{Real-time} \qquad (9)$$
$$\text{(dynamische Belastung)}$$

276

Die Werte der Besselschen Funktionen $\frac{\text{Arg } (J_0{}^2)}{\pi}$ und $\frac{\text{Arg } (1+J_0)}{2\pi}$ sind in [2] und [4] ta-
belliert. Man kann sie sich aber auch selbst errechnen.

Beispiele

Das Foto Bild 4 zeigt einen unten eingespannten Spiegel, der oben durch eine statische Einzellast ausgelenkt wird. Der Rand des Spiegels ist hier für die Demonstration mit einem diffus reflektierenden Band beklebt. Man sieht, daß sich im Spiegel
die Zahl der Interferenzlinien verdoppelt. Bild 5 zeigt die Grundtorsionsschwingung
des Spiegels. Bild 6 gibt eine Oberschwingung mit einer Knotenlinie wieder. Auch
hier ist die Verdopplung der Interferenzlinien deutlich zu erkennen. Aus Demonstra-
tionsgründen ist die Referenzebene RE kleiner als der Spiegel Sp gemacht worden.

Anwendungen

Es können z.B. Deformationen von Spiegeln kontrolliert werden, wenn sie z.B. durch
Befestigungselemente oder durch Temperaturfelder verspannt werden. Bei sehr kleinen
Objektdeformationen kann durch Aufbringen von kleinen ebenen Spiegeln auf diffus reflektierenden Oberflächen die Empfindlichkeit verdoppelt werden. Das gilt besonders
im Bereich von Schwingungsknoten. Die Schwingungssteifigkeit von Spiegeln kann kontrolliert werden und evtl. Maßnahmen zur Verbesserung der Formsteifigkeit von Spiegeln z.B. in optischen Geräten getroffen werden. Bei Anwendung der allgemeinen Auswertegleichung (Spiegelmethode) nach [2], [3] kann es zur Kontrolle der Spiegel erforderlich sein. Weiterhin kann man sehr genau den Verlauf von Deformationsänderungen
verfolgen, die infolge z.B. örtlicher nichtstationären Temperaturfelder auftreten.
Die Referenzebene soll selbst steif und örtlich fest aufgebaut sein. Das kann in
einem konventionellen Aufbau holografisch kontrolliert werden.

Zusammenfassung

1. Spiegelnde ebene Oberflächen sind holografisch meßbar geworden.
2. Die Meßgenauigkeit ist verdoppelt worden.
3. Die bisher bekannten Auswertegleichungen bleiben gültig.
4. Einfache Transformationsgleichungen für die verschiedenen holografischen Verfahren.

Literatur

[1] N. ABRAMSON: Appl. Optics 8 (1969), 1235; 9 (1970), 97, 2311; 10 (1971), 2155;
    11 (1972), 1143.
[2] G. SCHÖNEBECK: Diss. TU München (1979).
[3] G. SCHÖNEBECK: VDI-Bericht 313, 155 (1978).
[4] G. SCHÖNEBECK: Laser 79, Proceedings 576 (1979).

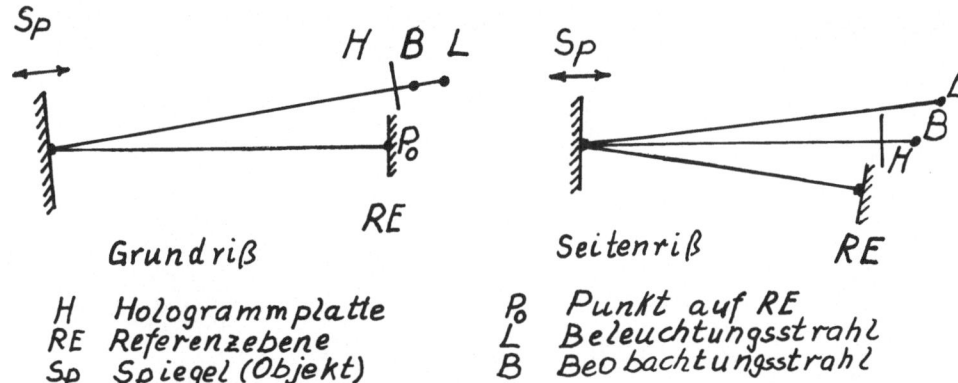

H   Hologrammplatte
RE Referenzebene
Sp Spiegel (Objekt)

P_0 Punkt auf RE
L   Beleuchtungsstrahl
B   Beobachtungsstrahl

Bild 1. Prinzip des Referenzebenen-Verfahrens
für Objekte mit spiegelnden Oberflächen

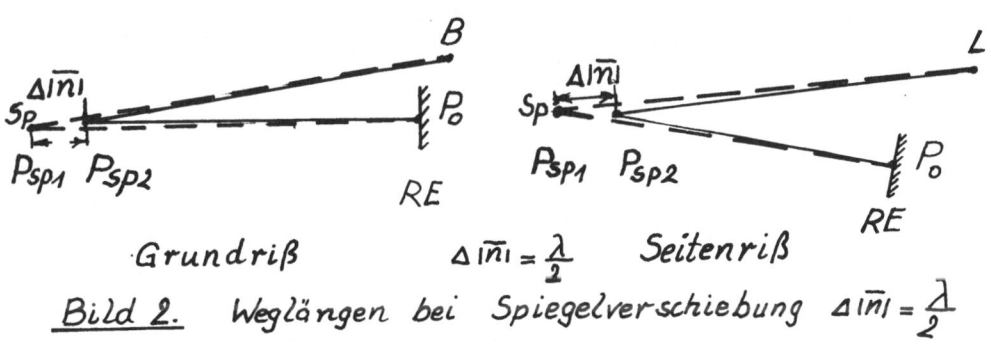

Grundriß $\Delta |\overline{m}| = \frac{\lambda}{2}$ Seitenriß

Bild 2. Weglängen bei Spiegelverschiebung $\Delta |\overline{m}| = \frac{\lambda}{2}$

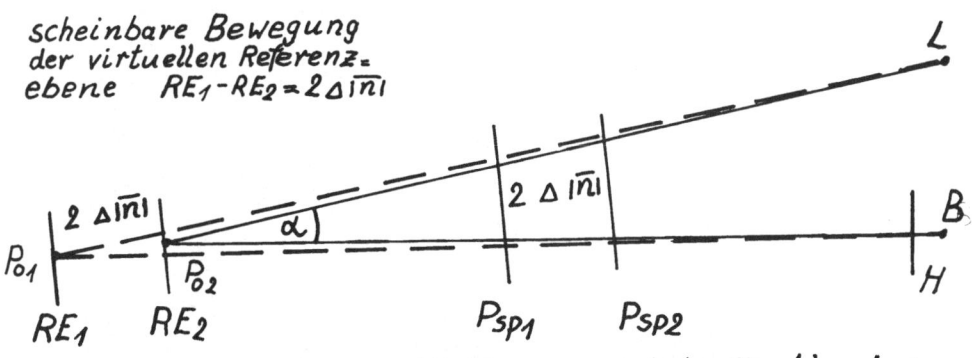

scheinbare Bewegung
der virtuellen Referenz-
ebene $RE_1 - RE_2 = 2\Delta|\overline{m}|$

Verlängerung jeden Strahles bei
Spiegelbewegung um $\Delta|\overline{m}| = \frac{\lambda}{2}$
$P_{Sp_1} - P_{Sp_2} = 2\Delta|\overline{m}|$

Bild 3. Ersatzbild (Rückführung auf Objekte mit diffus
reflektierenden Oberflächen)

278

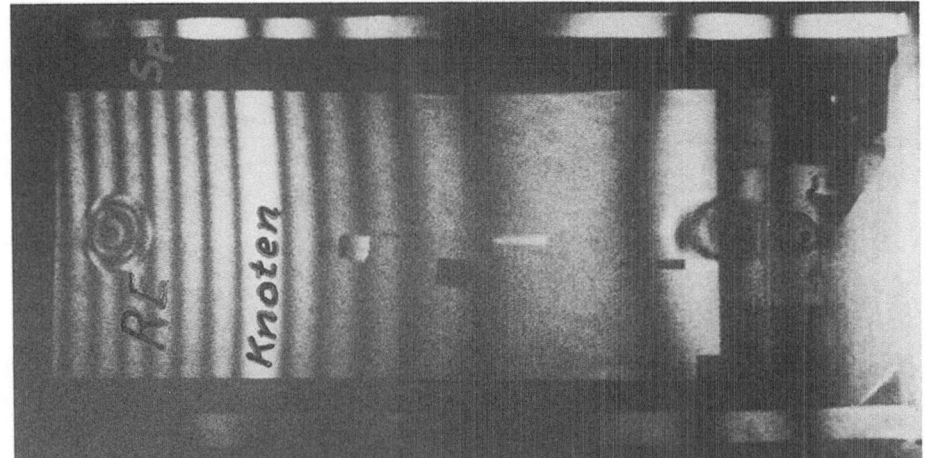

Bild 6. Biege-
schwingung

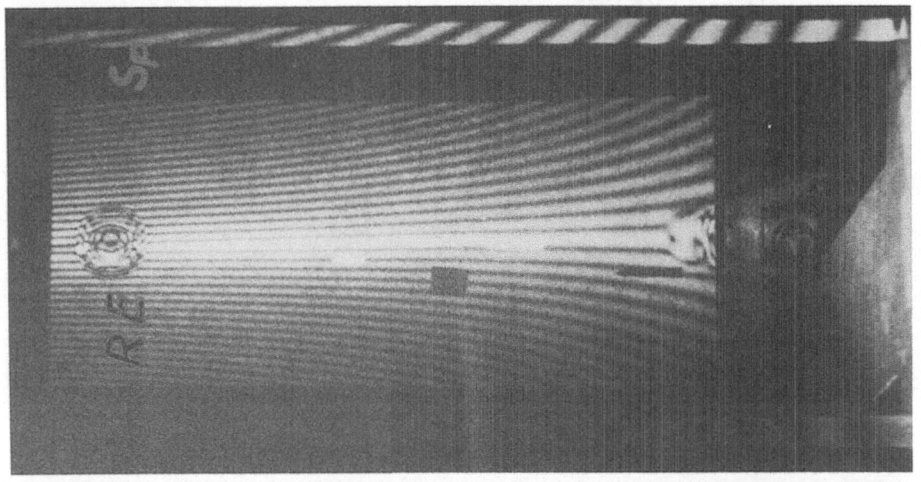

Bild 5. Torsions-
schwingung

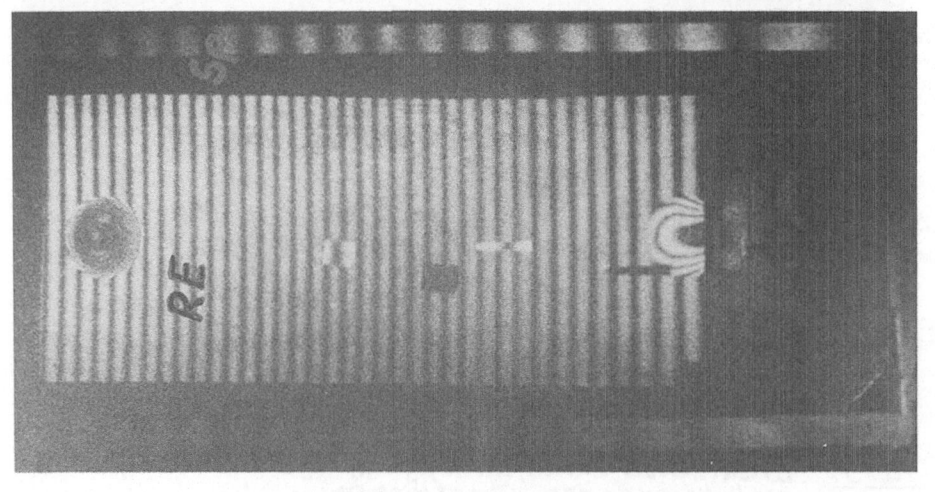

Bild 4. statische
Einzellast

Ein rechnergestütztes Holografie-System

B. Breuckmann, W. Thieme
M.A.N. - Neue Technologie, Maschinenfabrik Augsburg-Nürnberg AG
Dachauer-Str. 667
D-8000 München 50

1. Einleitung

Die Holografische Interferometrie hat in den letzten Jahren in vielen Gebieten der industriellen Anwendung stark an Bedeutung gewonnen. Der wesentliche Vorteil dieses Meßverfahrens liegt in der Tatsache, daß der Anwender ein 2-dimensionales Bild der Meßgröße erhält. Die darin enthaltene hohe Informationsdichte ist jedoch quantitativ nur mit einer rechnergestützten Auswertung auszunutzen.

In den vergangenen Jahren wurden in einer Reihe von Arbeiten die theoretischen Grundlagen für eine rechnergestützte Hologrammauswertung erarbeitet /1,2,3,4,5,6/. Verschiedentlich wurde auch bereits der prinzipielle Funktionsnachweis der unterschiedlichen Auswerteverfahren erbracht. Ein den Erfordernissen der industriellen Praxis genügendes rechnergestütztes Holografiesystem, welches eine Auswertung kompletter Interferenzmuster gewährleistet und damit den besonderen Vorteil dieses bildhaften Meßverfahrens erhält, war bisher nicht verfügbar.

In diesem Vortrag wird ein solches rechnergestütztes Holografie-System vorgestellt, welches eine praxisgerechte, hochgenaue Auswertung des gesamten Interferogrammes gewährleistet. Die Möglichkeiten des Verfahrens werden an Beispielen aus der industriellen Praxis demonstriert.

2. Das Phasenshift-Verfahren

Die wesentlichen Schwierigkeiten bei der rechnergestützten Auswertung von Interferenzstreifensystemen beruhen auf der Tatsache, daß die lokale Intensität $J(x,y)$ eine Funktion von drei im allgemeinen unbekannten Größen ist, nämlich

$$J(x,y) = H(x,y) * \left[1+K(x,y) * \cos \varphi(x,y)\right] \qquad (1)$$

mit $H(x,y)$: Hintergrundintensität

$\quad K(x,y)$: Streifenkontrast

$\quad \varphi(x,y)$: zu messende Interferenzphase.

Die direkte Berechnung der eigentlichen Meßgröße mit Mustererkennungsalgorithmen erweist sich bei starken Helligkeitsschwankungen des Bildhintergrundes - insbesondere mit hoher Raumfrequenz - oder schlechtem Streifenkontrast häufig als unpraktikabel.

In der klassischen Interferometrie wird daher das seit langem bekannte Phasenshiftverfahren eingesetzt /7,8,9/: Durch eine gezielte Variation des Phasenwinkels der miteinander zu vergleichenden Objektwellen kann Gleichung (1) zu einem Gleichungssystem erweitert werden:

$$J_i(x,y) = H(x,y) * \left[1+K(x,y) * \cos(\varphi(x,y) + \alpha_i) \right] \qquad (2)$$

Durch Messung der Intensitäten J_i für mindestens 3 verschiedene Phasen-verschiebungen α_i lassen sich jetzt die 3 Unbekannten H, K und φ für jeden Bildpunkt getrennt mit hoher Genauigkeit berechnen. Gleichzeitig führt die unabhängige Bestimmung von H, K und φ zu einer großen Un-empfindlichkeit gegenüber Störungen wie Objektkanten, Helligkeitsdif-ferenzen oder Fremdmustern.

Eine Möglichkeit, dieses Verfahren für die holografische Interferome-trie praxisgerecht nutzbar zu machen, wurde von Dändliker et.al. auf-gezeigt /6/. In der dabei verwendeten 2-Referenzstrahl-Holografie wird jede der beiden zu vergleichenden Objektwellen mit einem eigenen Refe-renzstrahl aufgenommen. Bei der Rekonstruktion wird die Phasenverschie-bung zwischen den Objektwellen dadurch erreicht, daß der optische Weg eines der beiden Referenzstrahlen gezielt verändert wird.

3. 2-Referenzstrahl-Holografie

Eine schematische Darstellung eines für die Aufnahme von 2-Referenz-strahl-Hologrammen geeigneten optischen Aufbaus ist in Abb. 1 darge-stellt. Gegenüber dem üblichen holo-grafischen Aufbau ist ein zusätz-liches leicht dejustiertes Interfero-meter in den Referenzstrahlweg einge-baut, durch welches die beiden Refe-renzstrahlen erzeugt werden, die typischerweise einen Winkel von ca. 0,05° einschließen. Daraus resultiert eine große Unempfindlichkeit des 2-Referenzstrahl-Hologrammes gegenüber fehlerhafter Repositionierung /6,10/. Gleichzeitig ergibt sich aus dem klei-nen Winkel zwischen den Referenzstrah-len die Möglichkeit, das Hologramm mit einer anderen als der Aufnahme-Wellenlänge zu rekonstruieren, wodurch das Verfahren auch für die Verwendung von Pulslasern geeignet wird.

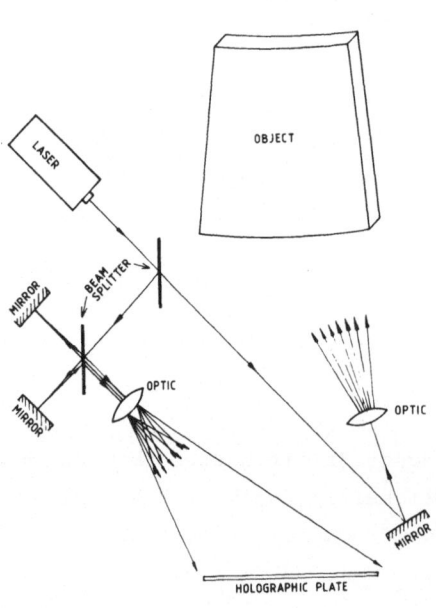

Abb. 1. Optischer Aufbau für 2-Referenzstrahl-Holografie

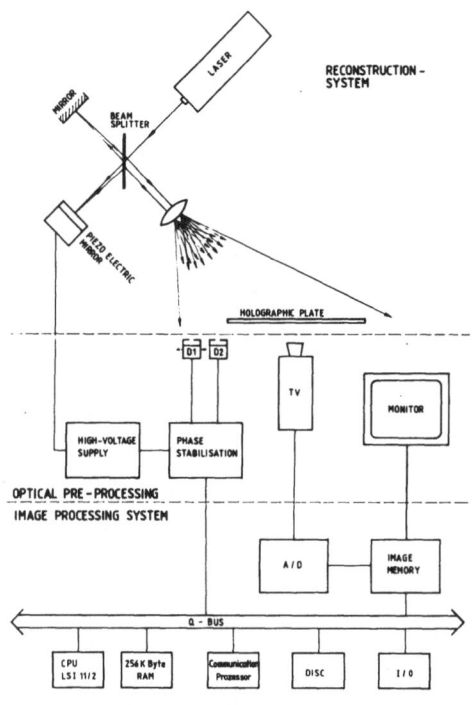

Abb. 2 zeigt eine Skizze eines Rekon-
struktionsaufbaus mit nachgeschalteter
rechnergestützter Auswertung. Die beiden Referenzstrahlen werden in einem
kompakten Interferometer erzeugt. Durch
einfache Justierung dieses 2-Referenz-
strahl-Modules erfolgt die Anpassung des
Winkels zwischen den beiden Strahlen.
Die Phasenverschiebung wird mittels des
Piezospiegels durchgeführt. Die phasen-
verschobenen Interferogramme werden mit
einer TV-Kamera aufgenommen und an ein
Bildverarbeitungssystem übergeben, wobei
die Bilder maximal in 512x512 Bildpunkte
à 256 Grauwerte (8 bit) digitalisiert
und weiterverarbeitet werden.

Abb. 2. Rechnergestützte Hologrammauswertung

4. Ergebnisse

Der Einsatz dieses rechnergestützten Holografiesystems soll im folgenden anhand verschiedener Beispiele aus der industriellen Praxis demonstriert werden:

a) Verformungsanalyse

Ein ca. 1.5 m^2 großes Segment einer Radioteleskopantenne wurde holografisch bei Druckbelastung vermessen. Dadurch sollte ermittelt werden,
welche mittlere Abweichung von der Ausgangsform dieses Bauteil aus CFK-
Werkstoff bei simulierter Windlast erleidet. Abb. 3 zeigt links die Rekonstruktion eines bei dieser Messung aufgenommenen 2-Referenzstrahl-
Interferogramms sowie das daraus rechnerisch ermittelte Verformungsbild
des CFK-Segments. Letzteres wird sowohl als 3D-Plot (rechts) als auch
als Grauwertdarstellung (Mitte) wiedergegeben. (Farbdarstellung möglich)

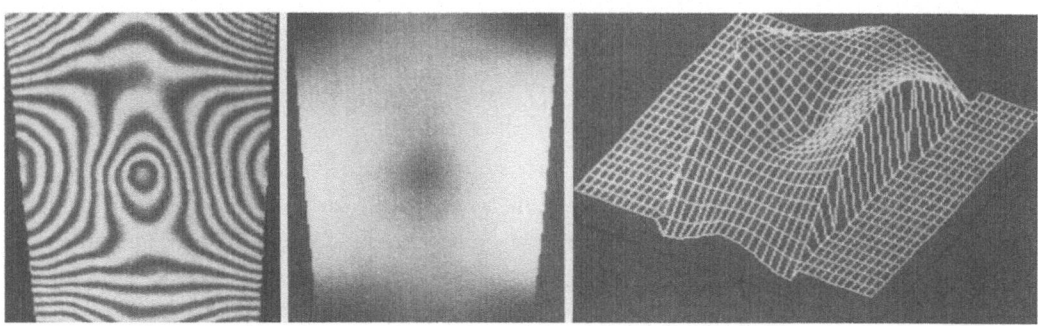

Abb. 3. Rechnergestützte holografische Verformungsanalyse

b) Schwingungsanalyse mit Pulslaser

Meßtechnische Gründe erfordern in einer Vielzahl von Anwendungszwecken
- insbesondere um Maschinen im Betriebszustand zu untersuchen - den
Einsatz von Pulslasern. Dazu wird in der 2-Referenzstrahltechnik ein
schnelles Schalten zwischen den beiden Referenzstrahlwegen notwendig.
Hierzu wurde im Labor Dr. Steinbichler ein gepulster Rubinlaser mit
entsprechend schnell arbeitenden elektrooptischen Schaltern ausgerüstet.
Abb. 4 zeigt ein mit diesem Laser aufgenommenes 2-Referenzstrahl-Ho-
logramm einer bei 1244 Hz schwingenden Ölwanne, das mit einem He-Ne-
Laser rekonstruiert wurde. Es sind wiederum die Grauwertdarstellung
und der 3D-Plot der Schwingung dargestellt. Da mit dem Phasenshiftverfahren zwischen steigenden und fallenden Interferenzordnungen eindeutig unterschieden werden kann, wird die Schwingung über das gesamte
Bauteil phasenrichtig wiedergegeben.

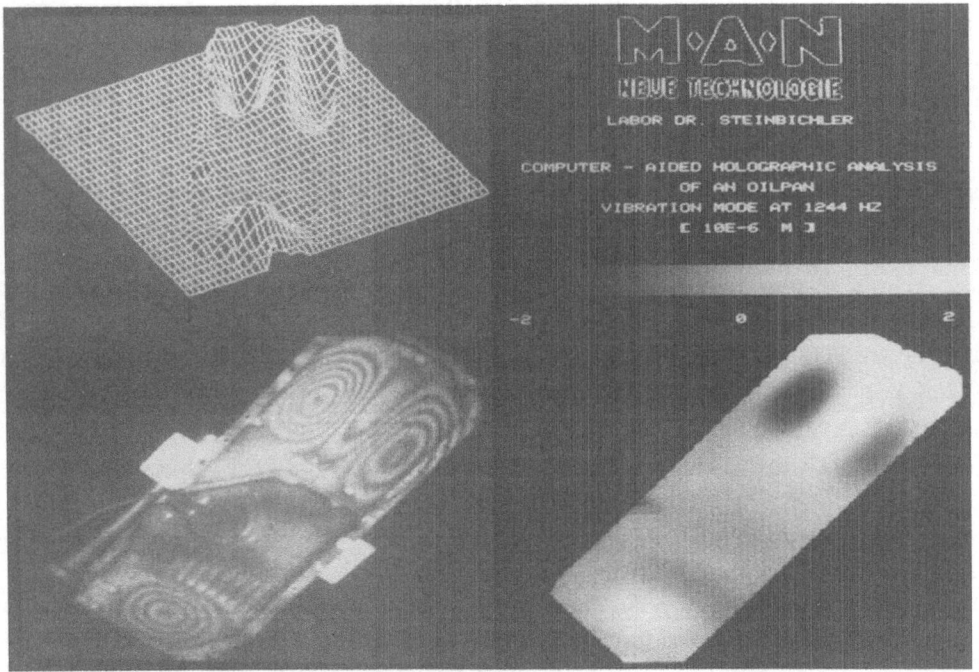

Abb. 4. rechnergestützte holografische Schwingungsanalyse

c) Vermessung von Phasenobjekten

Für die hier gezeigte Messung zur Ermittlung der Druckverteilung in
einem künstlich im Labor erzeugten Tornado wurde ein Pulslaser im Ein-
zelpulsbetrieb verwendet. Nach der ersten Belichtung mit Referenzstrahl 1 bei ausgeschaltetem Tornado erfolgte die zweite Belichtung
ca. 1 Minute später mit Referenzstrahl 2 bei eingeschaltetem Tornado.
Das Interferogramm wurde mit einem He-Ne-Laser rekonstruiert.

In Abb. 5 ist ein rekonstruiertes 2-Referenzstrahl-Interferogramm des
Tornados dargestellt (links). Aufgrund von relativ kontrastreichen Störungen, die vom optischen Aufbau erzeugt wurden, sind die Interferenzlinien, welche durch den geänderten Brechungsindex der zu vermessenden
Luftsäule erzeugt werden, weitgehend überdeckt und können vom Auge kaum
erkannt werden. Die rechnergestützte Auswertung läßt nach Elimination
von Hintergrund und Streifenkontrast das eigentliche Interferogramm erkennen (Mitte), aus dem Lage und Druckverteilung des Tornados berechnet
werden können (rechts). Dabei werden aufgrund der hohen Auflösung des
Phasenshiftverfahrens Druckunterschiede sichtbar, die einer Phasenverschiebung von ca. 15 Grad entsprechen.

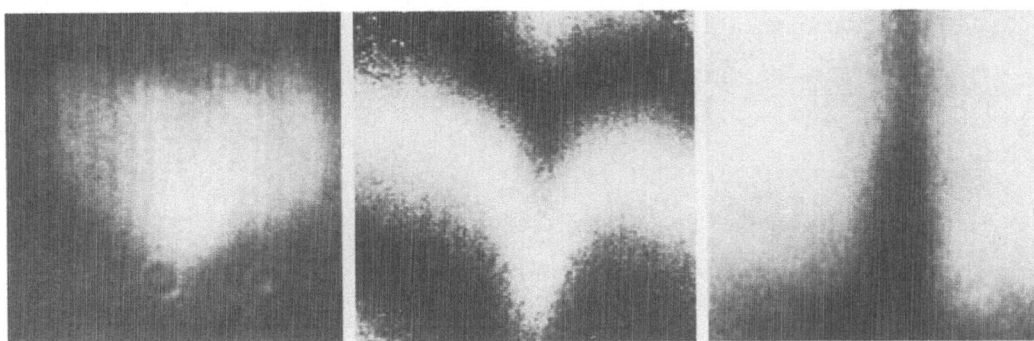

Abb. 5. Rechnergestützte holografische Vermessung eines Phasenobjektes

Zusammenfassung

Es wurde ein rechnergestütztes Holografiesystem vorgestellt, welches
holografische Doppelbelichtungsinterferogramme mit einem Phasenshiftverfahren auswertet. Damit wurde erstmals eine praxisnahe Auswertung
mit hoher Zuverlässigkeit und Genauigkeit demonstriert. Mit dem angewandten Verfahren bleibt - im Gegensatz zu punktförmiger oder auf Linien beschränkter Auswertung - die bildhafte Darstellung des Meßergeb-
nisses erhalten.
Das vorgestellte Verfahren hat mittlerweile seine Tauglichkeit in der
industriellen Praxis unter Beweis gestellt. Die guten Erfahrungen, die
bisher mit dem rechnergestützten Holografiesystem gesammelt wurden,
lassen eine Absenkung der Hemmschwelle für den Einsatz der Holografie
erwarten.

Die Verfasser danken Prof. Dändliker für die Hilfe bei der Verfahrensentwicklung sowie dem Bundesministerium für Forschung und Technologie
für die Unterstützung des Forschungsvorhabens.

Literatur

1. BELLANI, V. F. and Sona, A.
   Appl. Opt. 13 (1974) 1337

2. BIEDERMAN, K. and Ek, L.
   Appl. Opt. 16 (1977) 2535

3. LANZL, F. and Schluter, M.
   SPIE 136 (1977) 166

4. FISCHER, B. GELDMACHER, J. and JÜPTNER W.
   Proc. Laser 79, Optoelectronic Conf. Munich (1979) 412

5. KREITLOW, H. and Kreis, T.M.
   SPIE 210 (1979) 196

6. DÄNDLIKER, R., THALMANN, T. and WILLEMIN, J.F.
   Optics Communications Vol. 42 (1982) 301

7. BRUNING, J. H., HERRIOTT, D.R. GALLAGHER, J.E.
   ROSENFEHLD, D.P. WHITE A.D., BRANGACCIO, D.J.
   Appl. Opt. 13 (1974) 2693

8. DORBAND, B., Optic 60 (1982) 161

9. WYANT, J.C. Laser Focus May (1982) 65

10. DÄNDLIKER, R., Progress in Optics. Vol 17
    ed. E. WOLF (North-Holland, Amsterdam, 1980)

Holographische Untersuchungen an Lötverbindungen

W. Neumann und F. Krause
Fachhochschule Hagen, Abt. Iserlohn
Frauenstuhlweg 31, D - 5860 Iserlohn

Funktion und Lebensdauer von Schaltschützen mit aufgelöteten Kontaktplättchen, die elektrischen, mechanischen und thermischen Beanspruchungen ausgesetzt sind, hängen entscheidend von der Qualität der Lötverbindung ab. Bei den Herstellern von Schaltelementen besteht deshalb ein
großes Interesse an geeigneten zerstörungsfreien Prüfverfahren und es
stellte sich die Frage, ob die holographische Interferometrie auch zur
Ermittlung der Qualität von Lötverbindungen eingesetzt werden kann. Über
entsprechende Untersuchungen wird nachfolgend berichtet.

Bild 1 zeigt das Prüfobjekt, ein Anschlußschaltstück aus einem Schalt-
schütz, in zwei Ansichten. Es besteht aus einem abgewinkelten Bügel aus
Elektrolytkupfer mit einem aufgelöteten rechteckigen Kontaktplättchen aus einer Ag-CdO-Legierung. Die Maße des Bügels und des Plättchens betragen 110×25×6 mm bzw. 23×16×1 mm. Die Lötung erfolgte mit dem silberhaltigen Hartlot L - Ag45Cd. Bei den Untersuchungen war der Kupfer-bügel in einer stabilen Halterung ein-gespannt, während auf das freiliegende Kontaktplättchen mit einem hydraulisch bewegten Stempel Scherkräfte F von 1 kN bis 8 kN übertragen wurden. Diese Scher-kräfte erzeugten in der Lötverbindung eine Vorspannung, bei der die erste Be-lichtung des holographischen Films zur Erzeugung von Doppelbelichtungs-Inter-ferogrammen erfolgte. Eine Erhöhung von F um $\Delta F = 0,2$ kN ergab den erforderlichen neuen Spannungszustand für die zweite

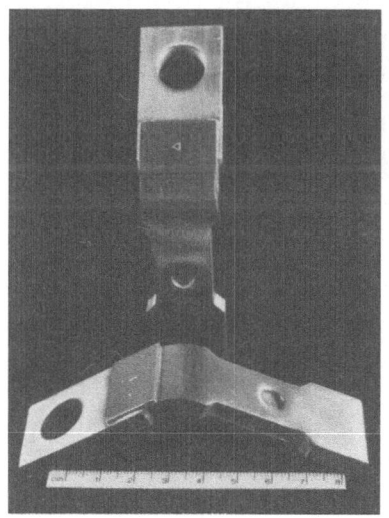

Bild 1. Das Prüfobjekt, ein
Schaltstück, in 2 Ansichten.

holographische Aufnahme. Die hierdurch erzeugten holographischen Interferogramme stellen die Änderungen des Prüfobjektes unter dem Einfluß von
ΔF dar. Sie können in drei Kategorien eingeteilt werden:

Interferogramme des Typs "A" entsprechend Bild 2, mit nahezu vertikalen und äquidistanten Interferenzlinien, die im Unendlichen lokalisiert sind über das ganze Prüfobjekt verlaufen und auf dem Schaltbügel sowie dem Kontaktplättchen jeweils die gleiche Dichte aufweisen.

Interferogramme des Typs "B", die ebenfalls nahezu vertikale und äquidistante Interferenzlinien zeigen, aber mit höheren Liniendichten auf dem Kontaktplättchen als auf dem Kupferbügel. Ein Beispiel für ein derartiges Interferogramm ist auf Bild 3 dargestellt.

Interferogramme des Typs "C" entsprechend Bild 4, mit einem komplizierten Verlauf der Interferenzlinien im Bereich des Kontaktplättchens im deutlichen Gegensatz zu den nahezu geraden Linien auf dem Bügel.

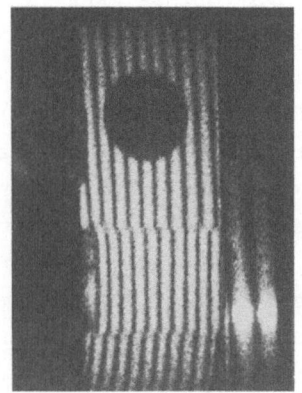 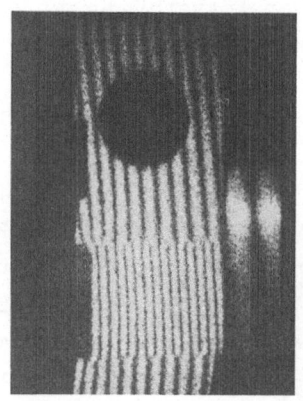 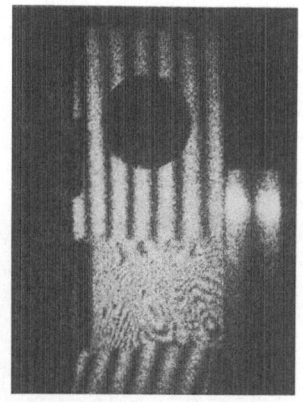

Bild 2. Interfero-
gramm des Typs "A".

Bild 3. Interfero-
gramm des Typs "B".

Bild 4. Interfero-
gramm des Typs "C".

Wie sind diese Ergebnisse zu erklären? Bekanntlich stellen parallele, vertikale, äquidistante und im Unendlichen lokalisierte Interferenzlinien eine Objektverschiebung in horizontaler Richtung dar, wobei die Liniendichte proportional zur Verschiebung ist. Die Interferogramme "A" bedeuten deshalb eine Verschiebung des ganzen Prüfobjektes als eine starre Einheit, während auf den Interferogrammen "B" eine Eigenbewegung des Kontaktplättchens auf dem Kupferbügel festzustellen ist. Bei einer weiteren Steigerung der Vorbeanspruchung F erhält man bei den kleineren Lötflächen A Interferogramme des Typs "C". Diese stellen im Bereich des Kontaktplättchens keine einfachen Verschiebungen mehr dar, sondern Verformungen, die aufgrund der starken Vorbeanspruchung der Lötverbindung durch die Scherkraft F unter dem Einfluß von ΔF auftreten.

Die Interferogramme des Typs "B" und "C" stellen somit die Ergebnisse dar, die Aussagen über die Güte der Lötverbindung liefern können, wobei die Interferogramme "B" eher als "C" für eine quantitative Auswertung in Frage kommen. Hierfür wurden an den Interferogrammen "B" zunächst am oberen und am unteren Rand des Kontaktplättchens die mittleren Interferenzliniendichten d_{1p}, d_{1b}, d_{2p} und d_{2b} auf dem Plättchen und dem Bügel bestimmt. Hieraus ließen sich die relativen Verschiebungen des Plättchens bezüglich des Bügels $n_1 = d_{1p}/d_{1b}$ und $n_2 = d_{2p}/d_{2b}$ berechnen, deren Mittelwert n die auf Bild 5 dargestellte Abhängigkeit von der Vorbeanspruchung F bei konstantem ΔF für verschiedene Werte der Lötfläche A zeigt. Bei der größten untersuchten Lötfläche $A_1 = 368$ mm^2 (Kurve 1) erhält man eine Abhängigkeit n(F), die nur in sehr geringem Maße vom Wert 1 abweicht. Das Prüfobjekt stellt somit unter den genannten Bedingungen eine starre Einheit dar. Wird die Fläche A auf 276 mm^2 verringert, so ergibt sich Kurve 2 von Bild 5, die im Vergleich zu Kurve 1 leicht höhere Werte

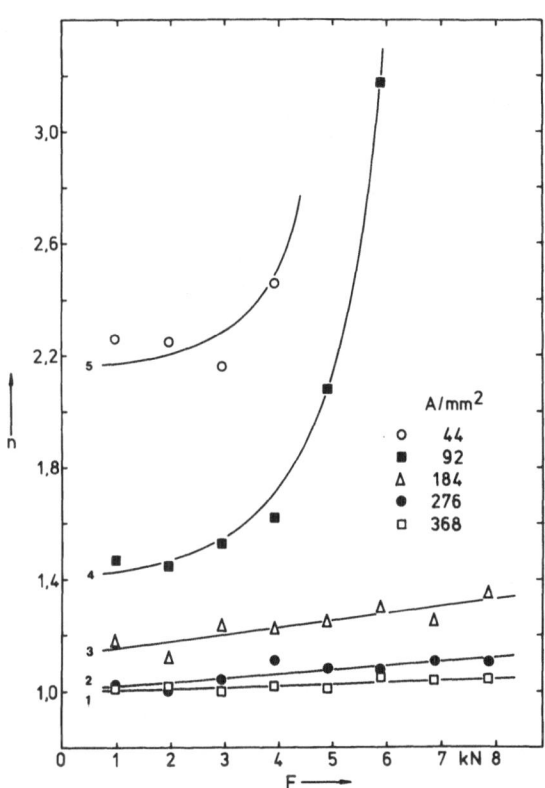

Bild 5. Die relative Verschiebung n in Abhängigkeit von der Vorbeanspruchung F für ΔF = const und verschiedene Werte der Lötfläche A.

von n aufweist. Dies läßt bereits auf eine gewisse Eigenbewegung des Plättchens auf dem Bügel schließen. Diese Tendenz setzt sich bei Kurve 3 mit $A_3 = 184$ mm^2 fort und man erkennt, daß die Werte von n nun deutlich > 1 sind. Sie liegen im Mittel bei 1,2 und steigen mit wachsendem Wert von F leicht an. Für Kurve 4 mit $A_4 = 92$ mm^2 liegen die relativen Verschiebungen n für F < 4 kN bereits bei 1,4 - 1,6. Eine Erhöhung von F auf 5 - 6 kN führt zu einem starken Anstieg von n auf Werte bis zu 3,2. Für F > 7 kN treten Interferenzmuster des Typs "C" auf dem Plättchen auf, die in der vorherigen Weise nicht mehr auszuwerten sind. Bei Kurve 5 für $A_5 = 44$ mm^2 erhält man bereits für F = 1 kN hohe Werte für n, die eine beträchtliche Eigenbewegung des Kontaktplättchens erkennen lassen.

288

Aus Bild 5 läßt sich die Abhängigkeit der relativen Verschiebung n von
der Größe der Lötfläche A bei konstanten Werten von F und ΔF ermitteln.
Diese Abhängigkeit zeigt Bild 6 für F = 0,98 kN und ΔF = 0,2 kN. Da F
und ΔF nun konstant sind, ergibt sich mit abnehmender Größe der Lötfläc̣
A eine Erhöhung der Scherspannung in der Lötverbindung und damit ein
Anstieg der relativen Verschiebung n. Die Darstellung n(A) gemäß Bild 6
kann für gleichartige Prüfobjekte als Eichkurve zur Ermittlung der relativen Qualität der Lötverbindung verwendet werden. Dies geschieht

folgendermaßen:

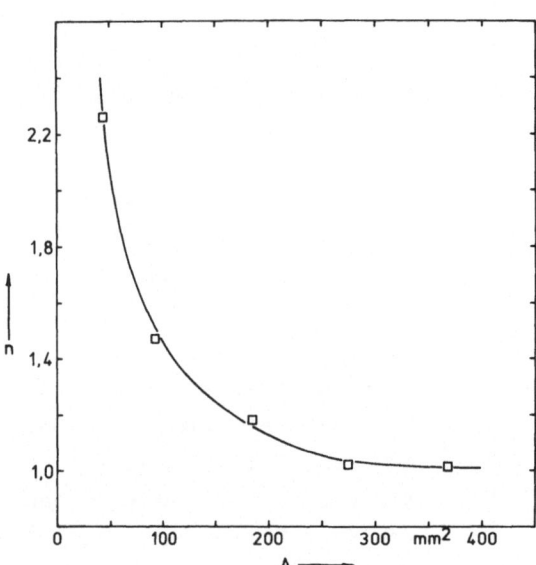

Bild 6. Die relative Verschiebung n
in Abhängigkeit von der Lötfläche A
für F = 0,98 kN und ΔF = 0,2 kN.

Bei bekannter Größe der Lötfläche A_s ergibt sich aus dem
Diagramm ein Sollwert für die
relative Verschiebung n_s. Unterscheidet sich die Qualität
der Lötverbindung des Prüfobjektes von derjenigen der
Referenzprobe, so wird nun als
relative Verschiebung statt n_s
der Istwert n_i experimentell
ermittelt, der mit n(A) gemäß
Bild 6 zu dem Istwert A_i für
die Größe der Lötfläche führt.
Das Verhältnis A_i/A_s = q liefert nun eine Aussage über die
Güte der Lötverbindung bzw.
den Wert des Bindungsanteils.
q = 1 wird für eine Lötung von
der gleichen Qualität wie die

der Referenzprobe erhalten. q < 1 stellt eine schlechtere Lötung dar und
liefert einen Richtwert für den relativen Bindungsanteil des Kontaktplättchens auf dem Träger. q > 1 würde man bei einer Lötung erhalten,
deren Qualität besser als die der Referenzlötung ist.

Damit ist an einem konkreten Beispiel aus der industriellen Praxis gezeigt, daß die holographische Interferometrie prinzipiell zur zerstörung
freien Prüfung der Güte von Lötverbindungen geeignet ist. Bei der praktischen Anwendung dieses Prüfverfahrens ist selbstverständlich für jede
Art des Prüfobjektes eine eigene Eichkurve mit optimal angepaßten Werten
von F und ΔF aufzunehmen.

Ortsfrequenzanalyse holographischer Interferenzstreifensysteme zur automatischen Fehlerdetektion

H.-A. Crostack*, W. Jüptner**, Th. Kreis**, A. Krüger*, K.-J. Pohl*
* Fachgebiet Qualitätskontrolle, Universität Dortmund
** Bremer Institut f. angewandte Strahltechnik (BIAS), Bremen

1. Einleitung und Problemstellung

Ein häufiges Problem der zerstörungsfreien Prüfung ist nicht nur die Erkennung und Lokalisation, sondern vor allem die Bewertung von Bauteilfehlern. Hierzu erweist sich insbesondere die holographische Interferometrie als geeignetes Verfahren zur berührungslosen und flächigen Erfassung der Verformungsfelder belasteter Bauteile.

Unter Belastung ergibt sich im Bereich eines Fehlers eine inhomogene Verformung, die im holographischen Interferogramm als Unregelmäßigkeit im Streifenverlauf erkennbar ist. Während eine rein qualitative Auswertung zur Fehlererkennung hochqualifiziertes Personal voraussetzt und zudem subjektiv ist, erfordert eine vollständige quantitative Analyse einen hohen technischen Aufwand und ist i.a. zusätzlich sehr zeitaufwendig.

Deshalb wurden Untersuchungen zur Erarbeitung eines schnelleren und einfacheren Verfahrens auf der Basis einer Ortsfrequenzanalyse des Interferenzmusters durchgeführt, deren Ergebnisse vorgestellt werden.

2. Optimierung des holographischen Aufbaus

Erfahrungsgemäß weisen holographische Interferenzstreifensysteme von sich aus wechselnde Streifendichten auf, so daß der Einsatz der Ortsfrequenzanalyse nicht unmittelbar zum gewünschten Ziel der Fehlererkennung führt. Um die Auffindung von fehlerbedingten Inhomogenitäten im Streifensystem zu ermöglichen, gilt es, das Grundstreifenmuster, wie es z.B. auf einer fehlerfreien Probe auftritt, so homogen wie möglich zu gestalten. Dies setzt für jeden Anwendungsfall eine vorherige Optimierung der Versuchsbedingungen, insbesondere der Geometrie des holographischen Meßaufbaus voraus. Deshalb müssen zunächst optimale holographische Aufbaudaten theoretisch hergeleitet werden, mit

denen dann in einem zweiten Schritt interferometrische Untersuchungen
zur Fehlererfassung erfolgen.

Bild 1 gibt im wesentlichen die Größen zu einer Optimierung der Geometrie des Versuchsaufbaus im Hinblick auf die Prüfung einer Flachzugprobe wieder. Dargestellt ist die Änderung des Empfindlichkeitsvektors
$\vec{G}(P)$ beim Übergang des betrachteten Probenpunktes P nach P*; dabei ist
$\vec{G}(P)$ als Differenzvektor aus den Einheitsvektoren der Beobachtungs-
und Beleuchtungsrichtung definiert.

Im rechten Teil von Bild 1 sind die Abmessungen der untersuchten Alu-
Flachzugproben angegeben. Für eine derartige Versuchskonfiguration
liegen aus der Elastizitätstheorie umfangreiche Erkenntnisse vor, so
daß Ergebnisse verglichen und sicher beurteilt werden können. Die
nach Belastung auftretenden Interferenzstreifen werden an der Stirnseite der Probe zwischen den Punkten $-X_O$ und $+X_O$ abgenommen. Sämtliche
Auswertungen und Berechnungen werden in der Ebene y = o durchgeführt.
Alle Winkelangaben beziehen sich auf die positive Z-Achse; dabei
werden Winkeländerungen im Uhrzeigersinn als negativ definiert.

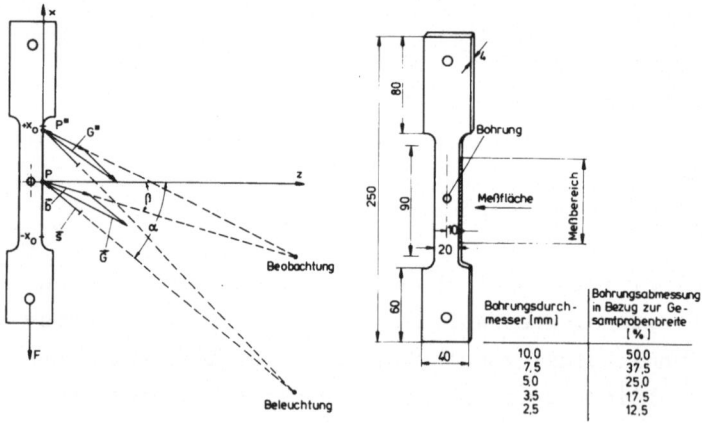

Bild 1. Geometrie des Versuchsaufbaus und Abmessungen der Aluminium-
Flachzugproben

Die Grundgleichung (1) der holographischen Interferometrie /1/ verknüpft die im Punkt P auf der Probe beobachtete Interferenzstreifen-

ordnung N(P) über den Empfindlichkeitsvektor $\vec{G}(P)$ mit dem Verschiebungsvektor $\vec{d}(P)$:

$$N(P) = \frac{1}{\lambda}\vec{d}(P)\vec{G}(P) \qquad (1)$$

Speziell bei einachsiger Zugbelastung in x-Richtung setzt sich $\vec{d}(P)$
aus Translationen und überlagerten Dehnungen zusammen (2):

$$\vec{d}(P) = (d_x(x_0) + (x-x_0)\,\varepsilon(x),\ d_y,\ d_z(x_0) - \frac{1}{2}bv\varepsilon(x)) \qquad (2)$$

$\varepsilon(x)$ = Dehnung

v = Poissonzahl

b = Probenbreite

$+x_0, -x_0$ = Anfangs- und Endpunkt der betrachteten Meßstrecken

Durch Differentiation der Grundgleichung (1) nach dem Weg x erhält man
die Streifendichte und somit die Ortsfrequenz der Interferenzstreifen
(3) entlang der Belastungsrichtung:

$$N'(P) = \frac{1}{\lambda}\left\{\left[d_1 + (x-x_0)\,\varepsilon(x)\right]G'_x + \left[d_2 - \frac{1}{2}\,bv\varepsilon(x)\right]G'_z\right.$$
$$\left. + \left[\varepsilon(x) + (x-x_0)\,\varepsilon'(x)\right]G_x - \frac{1}{2}bv\varepsilon'(x)G_z\right\} \qquad (3)$$

Um Richtlinien zur Wahl eines optimierten holographischen Aufbaus auf-
stellen zu können, wird der Einfluß der einzelnen, die Streifendichte
bestimmenden und durch die Geometrie vorgegebenen Terme in Gl. (3)
zunächst hinsichtlich der Größenordnung abgeschätzt. Dabei gilt für
realistische holographische Aufbauten:

$d_1 G'_x/\lambda$	$\cong 0,5$	$1/mm$	für $x = -x_0$
$(x-x_0)\,\varepsilon G'_x/\lambda$	$\cong 0,05$	$1/mm$	
$d_2 G'_z/\lambda$	$\cong 0,1$	$1/mm$	für $x = 0$ (Fehler mittig) folgt
$(1/2)bv\varepsilon G'_z/\lambda$	$\cong 0,0003$	$1/mm$	$\varepsilon' \sim 0,1 \cdot \varepsilon\left[\frac{1}{mm}\right]$ im Bereich
$\varepsilon G_x/\lambda$	$\cong 0,01$	$1/mm$	größter Dehnungsänderung
$(x-x_0)\,\varepsilon'G_x/\lambda$	$\cong 0,5$	$1/mm$	
$(1/2)bv\varepsilon'G_z/\lambda$	$\cong 0,1$	$1/mm$	

Dabei haben nur die Terme, die betragsmäßig größer als $0,1 \cdot 1/mm$ sind,
einen wesentlichen Einfluß auf die Streifendichte. Offensichtlich
gehen die Probentranslationen d_x und d_z mit der Empfindlichkeit von

G'_x und G'_z, Probendehnungen ε und Dehnungsinhomogenitäten ε' mit der Empfindlichkeit von G_x und G_z ein.

Daraus ergeben sich erste praktische Anforderungen an den zu wählenden holographischen Aufbau: Terme, die hauptsächlich durch Dehnungsinhomogenitäten zur Streifendichte beitragen, sollten a) betragsmäßig sehr groß und b) über die Meßstrecke betrachtet, konstant sein, damit bei homogener Dehnung, d.h. fehlerfreier Probe, die Streifendichte über die Meßstrecke konstant bleibt.

Daraus folgt, daß die Komponenten G_x und G_z des Empfindlichkeitsvektors möglichst groß und über den Meßbereich konstant zu wählen sind.

Insbesondere die zweite Anforderung wird, wie man sich leicht überlegen kann, durch Beleuchtung mit parallelem Licht sowie Beobachtung aus großer Entfernung weitestgehend erfüllt. Eine Beleuchtung der Probe mit parallelem Licht ist technisch einfach realisierbar. Da hinsichtlich der Beobachtung aus großer Entfernung einerseits und einer geforderten hohen lokalen Auflösung des Bildes andererseits aufgrund einschränkender Rahmenbedingungen des experimentellen Aufbaus (Kameraobjektiv, Probenzugänglichkeit) ein Kompromiß geschlossen werden muß, ist der Beobachtungspunkt aufzufinden, bei dem der Empfindlichkeitsvektor möglichst wenig variiert, und der zudem innerhalb des durch das Kameraobjektiv vorgegebenen Bereiches liegt. Um diesen optimalen Beobachtungspunkt aufzufinden, wurden sowohl die Komponenten des Empfindlichkeitsvektors, als auch deren erste Ableitungen nach dem Weg X unter Variation der Beobachtungs- und Beleuchtungsrichtung, rechnerisch ausgewertet.

2.1 Optimierung der Beobachtungsrichtung

Bild 2 stellt die Ableitungen der Komponenten des Empfindlichkeitsvektors, die ein Maß für die Translationsmeßempfindlichkeit sind, über der Meßstrecke bei unterschiedlichen Beobachtungsgeometrien dar. Aus dem oberen Bild geht hervor, daß Translationen in Belastungsrichtung bei senkrechter Beobachtungsrichtung, also 0°, zwar sehr empfindlich erfaßt werden, jedoch zur geringsten Variation der Streifendichte entlang der Meßstrecke führen.

Mit betragsmäßig größer werdendem Beobachtungswinkel sinkt zwar die Meßempfindlichkeit für Translationen, jedoch nimmt die Streifendichte-

variation dann entlang der Meßstrecke stark zu, so daß hinsichtlich
einer zu erzielenden homogenen Streifendichte die senkrechte Sicht zu
bevorzugen ist.

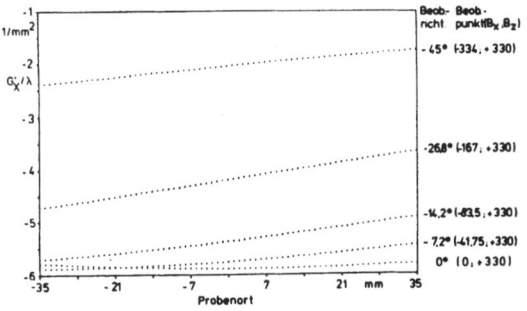

Bild 2.
Ableitungen G'_x und G'_z
der Komponenten des
Empfindlichkeitsvektors
über dem betrachteten Probenort bei unterschiedlichen
Beobachtungsgeometrien

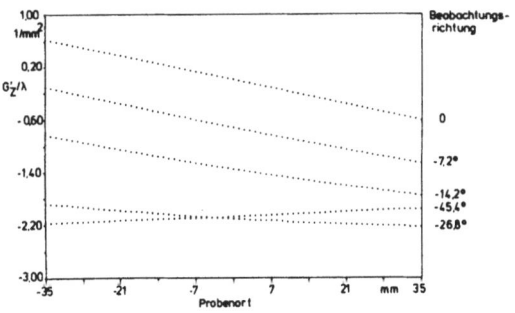

Im Vergleich zur X'-Komponente weist die Z'-Komponente gegenläufige
Tendenz auf. Hier nimmt die Variation entlang der Meßstrecke mit zunehmender Schrägsicht ab, gleichzeitig wird die Empfindlichkeit für
Translationen größer. Wenn man berücksichtigt, daß reine Translationen
in Belastungsrichtung aufgrund der Nachgiebigkeit des Restprobenab-
schnittes größer als mögliche, aber hier ungewollte Translationen der
Probe in Z-Richtung sind, so sollte ein holographischer Aufbau mit
senkrechter Beobachtungsrichtung optimal sein, da dann kleine Translationen in Z-Richtung noch relativ unempfindlich erfaßt werden.
Werden diese kleinen Translationen in Z-Richtung jedoch größer, so ist
wie noch im folgenden ausgeführt wird, nach einem geeigneteren Beobachtungswinkel zu suchen.

2.2 Optimierung der Beleuchtungsrichtung

Über die Komponenten des Empfindlichkeitsvektors G_x und G_z fließen die durch Werkstoffehler verursachten Dehnungsinhomogenitäten in die lokale Streifendichte ein. In beide Komponenten geht neben der Beobachtungs- auch die Beleuchtungsrichtung ein. Um Fehler möglichst empfindlich zu erfassen, müssen deshalb beide Komponenten des Empfindlichkeitsvektors betragsmäßig möglichst groß gewählt werden. Dazu wurde der funktionale Zusammenhang zwischen dem Vektor und der Beleuchtungsrichtung bei vorgegebener Beobachtungsrichtung untersucht.

Wie in Bild 3 dargestellt, weisen beide Funktionen gegenläufige Tendenz auf: Mit betragsmäßig kleiner werdendem Beleuchtungswinkel nimmt die Empfindlichkeit für Dehnungsinhomogenitäten in X-Richtung ab, die Empfindlichkeit für Inhomogenitäten in Z-Richtung nimmt jedoch zu. Als sinnvoll erweist sich hier, einen möglichst großen Beleuchtungswinkel zu wählen, da dann zwei Terme in Gl. (3), d.h. $\left(\varepsilon G_x/\lambda\right)$ und $\left\{(x - x_0)\right.$ $\left.\varepsilon'(x)G_x/\lambda\right\}$ empfindlich erfaßt werden und zusätzlich die Empfindlichkeit von G_z noch relativ wenig herabgesetzt ist. Als Beobachtungsrichtung wurde hier die senkrechte Sicht gewählt.

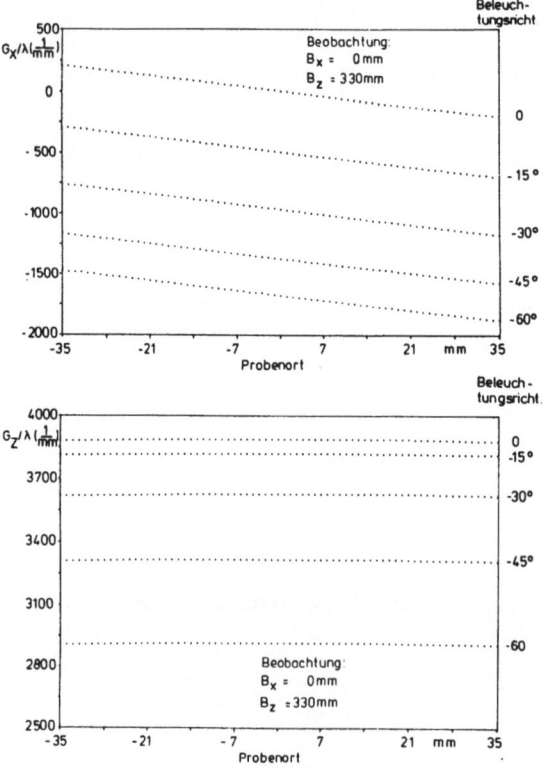

Bild 3.
Komponenten G_x und G_z des Empfindlichkeitsvektors über dem betrachteten Probenort, bei unterschiedlichen Beleuchtungsrichtungen

2.3 Feinabstimmung des holographischen Meßaufbaus für eine fehler-
freie Probe unter Zugbelastung

Nach der unter 2.1 und 2.2 dargestellten Voroptimierung wird eine
Feinabstimmung des holographischen Meßaufbaus vorgenommen. Wie bereits
erwähnt, ist eine senkrechte Beobachtung nicht immer als optimal anzusehen, da letztlich das Zusammenspiel der Translationen und Dehnungen die Streifendichte entlang der Meßstrecke bestimmt.
Aus diesem Grund wurden für eine Anzahl unterschiedlicher Transla-
tionswertesätze bei konstanter Dehnung jeweils optimale, holographische Aufbaugeometrien berechnet. Als Beispiel sind in Bild 4 zwei
konkrete Fälle aus der Vielzahl der Kombinationen dargestellt.

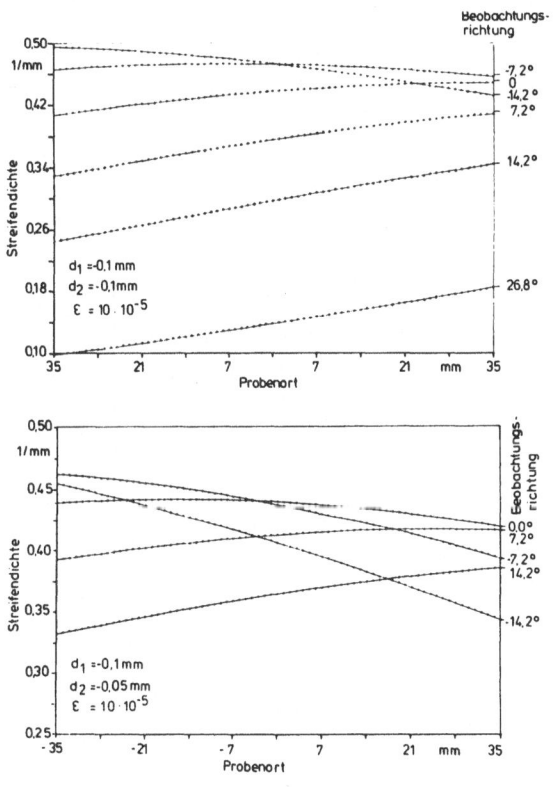

Bild 4.

Feinabstimmung des holo-
graphischen Aufbaus für eine
gedehnte Probe mit überlagerten X-Z-Translationen
a) Verhältnis der Trans-
   lationen d_1/d_2 = 1 : 1
b) Verhältnis der Trans-
   lationen d_1/d_2 = 2 : 1

Im oberen Teil des Bildes ist der Translation in X-Richtung eine betragsmäßig gleichgroße in Z-Richtung überlagert. Abweichend vom Ergebnis der Voroptimierung ergibt sich hierfür die optimale Beobachtungsrichtung zu -7,2°.

Werden die Translationen in X-Richtung größer als jene in Z-Richtung,
unteres Bild, so wird die senkrechte Sicht wieder zur optimalen Be-

obachtungsrichtung.

Grundsätzlich läßt sich immer ein Aufbau finden, bei dem die Streifendichte nur um wenige Prozent von der maximalen Streifendichte abweicht.
Für jede Wertekombination ist somit der Aufbau, d.h. die Beobachtungsrichtung neu auszulegen. Dabei sind folgende Tendenzen zu beobachten:
- mit zunehmender Dehnung in Belastungsrichtung wandert die optimale
  Beobachtungsrichtung zu größeren positiven Winkeln
- bei betragsmäßig etwa gleichen Translationen in X- und Z-Richtung
  sollten Aufbauten mit kleinerem bzw. negativem Beobachtungswinkel gewählt werden.

Einen anfänglichen Kompromiß stellt jedoch ein Aufbau mit senkrechter
Beobachtungsrichtung dar.

3. Experimentelle Untersuchungen

3.1 Fehlerfreie Flachzugproben

Zur Überprüfung der theoretisch hergeleiteten optimalen Aufbaugeometrie
bezüglich konstanter Interferenzstreifendichte wurden Experimente an
Aluminium-Flachzugproben durchgeführt. Die Simulation von Fehlstellen
wurde durch in die Mitte eingebrachte Bohrungen mit Durchmessern von
2,5 - 10 mm erreicht. Die Proben wurden mit Zugspannungen von 2,5 -
12,5 N/mm² bzw. 0,01 - 0,08 $R_{p0,2}$ beaufschlagt. Mit einem den theoretischen Ergebnissen der letzten Kapitel entsprechend optimierten holographischen Aufbau wurde zunächst an einer fehlerfreien Flachzugprobe
das Streifensystem analysiert. Bild 5 zeigt das Interferogramm einer
fehlerfreien Probe (nach einer aufgebrachten Zugspannung von 7,5 N/mm²),
das Signal einer Bildzeile aus der Bildmitte, sowie das Frequenzspektrum dieser Zeile. Die Blickrichtung wurde hier für eine möglichst geringe Variation der Streifendichte zu -13,4° gewählt.

Mit dieser Optimierung des holographischen Aufbaus wurde somit eine
weitgehend konstante Streifendichte entlang der dargestellten Bildzeile
erreicht. Das relativ schmale Frequenzspektrum bestätigt dies. Weiter-
gehende Spektralanalysen /2/ belegen, daß die Linienbreiten fehlerfreier Proben in Relation zur Grenzfrequenz (im folgenden relative
Linienbreite genannt) bei optimierten Aufbauten immer eindeutig kleiner
als 10% dieser Grenzfrequenz sind. Im günstigsten Fall beträgt sie
ca. 7%.

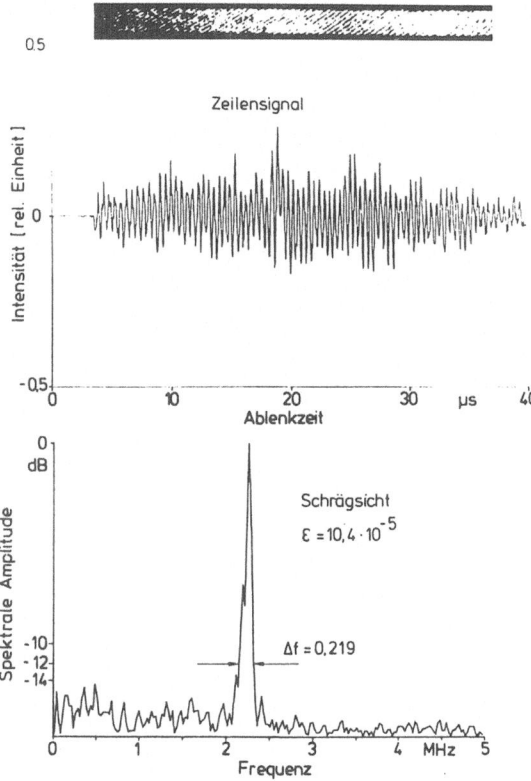

Bild 5.
Interferogramm, Zeilesignal
und Frequenzspektrum bei
einer fehlerfreien Probe.
Aufgebrachte Zugspannung:
7,5 N/mm²

Linienbreite und obere Grenzfrequenz werden dabei 12 dB unterhalb des
spektralen Maximums bestimmt, da, wie die durchgeführten Untersuchungen /3/ belegen, an dieser Stelle die beste Übereinstimmung zwischen
Frequenzgehalt des Interferenzstreifensystems und Frequenzbereich der
Anzeige beobachtet wird. Diese Beobachtung ist rein empirisch und läßt
sich derzeit noch nicht theoretisch vollständig begründen.

3.2 Untersuchung von Proben mit Modellfehlern unter Zugbelastung

Bild 6 zeigt Interferogramm, Zeilensignal und Spektrum für eine Zugprobe mit einer Bohrung von 7,5 mm Durchmesser, wobei die aufgebrachte
Zugspannung 10 N/mm² beträgt. Die Störung des Interferenzmusters über
dem Fehlerbereich ist sowohl im Interferogramm als auch im Zeilensignal bei diesem relativ großen Fehler deutlich auszumachen. Auch im
Spektrum sind markante Veränderungen im Vergleich zum Spektrum der
fehlerfreien Probe, Bild 5, zu beobachten.

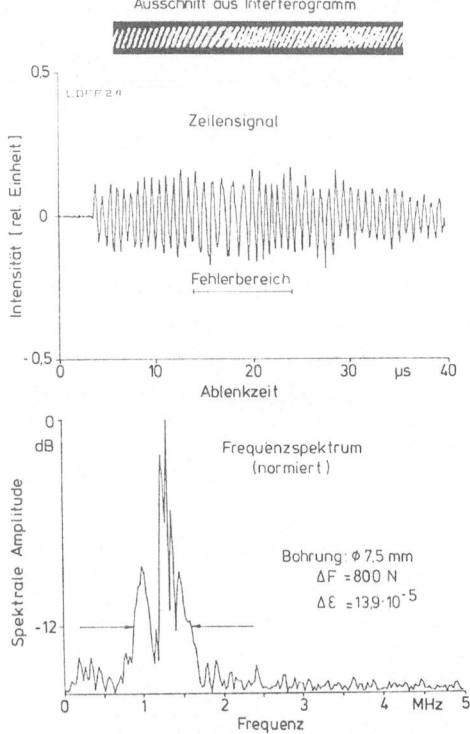

Bild 6.

Interferogramm, Zeilensignal und Spektrum einer zug-
belasteten (10 N/mm²)
fehlerhaften Probe
Fehler-∅: 7,5 mm

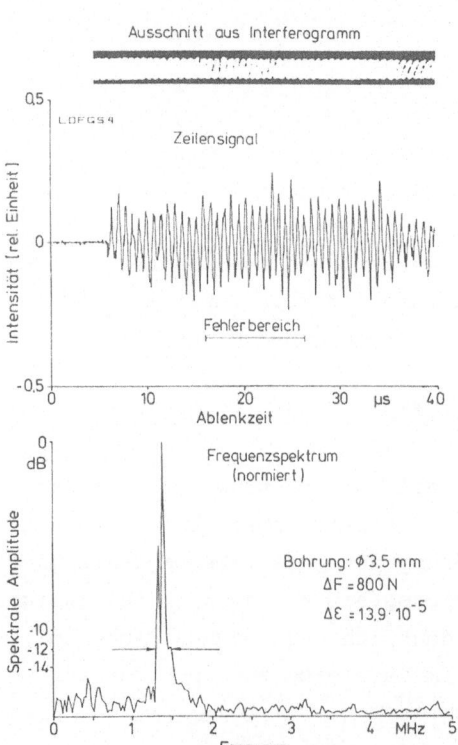

Bild 7.

Interferogramm, Zeilensignal und Spektrum einer zug-
belasteten (10 N/mm²)
fehlerhaften Probe
Fehler-∅: 3,5 mm

Die relative Linienbreite beträgt hier 44,6% und liegt damit deutlich über jener der fehlerfreien Probe.

Bild 7 zeigt Interferogramm, Zeilensignal und Spektrum für eine mit 10 N/mm² beaufschlagte Flachzugprobe. Der Bohrungsdurchmesser beträgt 3,5 mm.

Sowohl im Interferogramm als auch in der Bildzeile ist der Fehler mit dem Auge nicht mehr auszumachen. Demgegenüber beträgt die relative Linienbreite ca. 11% bei -12 dB und liegt damit über den Werten, die für fehlerfreie Proben ermittelt werden.

In Bild 8 sind alle Untersuchungergebnisse an den gelochten Flachzugproben zusammengefaßt dargestellt. Angegeben sind die bei den jeweiligen Fehlergrößen beobachteten relativen Linienbreiten, wobei die Balken die auftretende Schwankungsbreite der Meßwerte innerhalb einer Meßreihe bei unterschiedlichen Belastungszuständen darstellen. Extrapoliert man von den gemessenen Werten zu kleineren Fehlern hin, so kann die Fehlernachweisgrenze abgeschätzt werden: bei optimalem Meßaufbau und einer Grundstreifendichte im Bereich von 2,5 MHz beträgt die relative Linienbreite einer fehlerfreien Probe max. 7%. Damit liegen Fehler im Bereich von ca. 12% der Wandstärke an der Grenze der Nachweisempfindlichkeit.

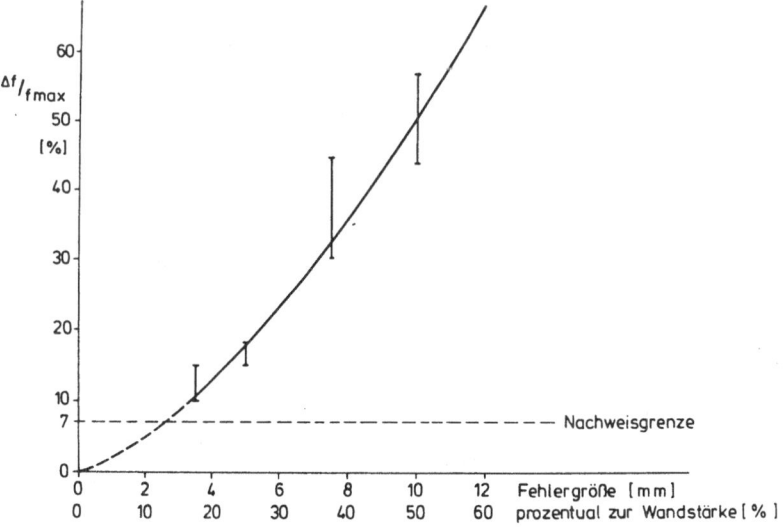

Bild 8. Fehlernachweis in Abhängigkeit von der Fehlergröße

4. Zusammenfassung

Zur schnellen on-line-Auswertung holographischer Interferenzstreifensysteme bietet sich die Frequenzanalyse an. Die Untersuchungen belegen, daß die Ortsfrequenzanalyse erfolgreich zur Fehlerdetektion eingesetzt werden kann, wenn gewisse Bedingungen bezüglich des holographischen Meßaufbaus und der Signalauswertung erfüllt werden:

1) Parallele Beleuchtung des Objekts unter möglichst großem Winkel
2) Optimierung der Beobachtungsrichtung anhand der zuvor an einer
   Stelle punktuell vermessenen Probentranslation und Dehnung oder
   des zu erwartenden Bereichs dieser Werte.

Durch diese Maßnahmen lassen sich bei fehlerfreien Proben Interferogramme erzielen, deren Streifendichte über dem Bild nur um wenige Prozent schwankt. Bei Proben mit Fehlern werden deutlich höhere relative
Linienbreiten festgestellt, so daß die Linienbreite als trennscharfes
Kriterium zum Fehlernachweis herangezogen werden kann.

Damit wird es möglich, Fehler, deren Einfluß auf das holographische
Interferenzstreifensystem mit dem Auge weder erkannt noch beurteilt
werden kann, sicher zu detektieren. Dies gilt für Fehler mit einer
Größe bis hinab zu 12% der Wandstärke. Darüber hinaus wurde innerhalb
der durchgeführten Arbeit festgestellt, daß bei Einsatz hochwertiger
Videosysteme mit besserem Auflösungsvermögen und geringerer Bildver-
zerrung noch kleinere Fehler auffindbar sein sollten.

Die Untersuchungen wurden im Rahmen eines durch die Deutsche Forschungsgemeinschaft (DFG) geförderten Forschungsvorhabens durchgeführt, der an dieser Stelle gedankt sei.

Literatur

/1/ C.M. VEST: Holographic Interferometry; John Wiley & Sons, New York,
           Toronto, 1979
/2/ H.-A. CROSTACK, A. KRÜGER: Zwischenbericht zum DFG-Vorhaben Cr 4/6
           Mai 1982
/3/ H.-A. CROSTACK, A. KRÜGER: Abschlußbericht zum DFG-Vorhaben Cr 4/6
           Januar 1985

Die holografische Interferometrie als Dienstleistung für eine wirtschaftliche experimentelle Spannungsanalyse

Th. Kreis, H. Kreitlow, W. Jüptner
BIAS - Bremer Institut für angewandte Strahltechnik
Ermlandstraße 59, D-2820 Bremen 71

Einleitung

Eine experimentelle Spannungsanalyse läßt sich in vielen Anwendungsfällen in ihrer Leistungsfähigkeit wesentlich erhöhen oder wird überhaupt erst einsetzbar durch eine Verformungsmessung mit Hilfe der holografischen Interferometrie. Die besonderen Vorteile der holografisch interferometrischen Meßmethode liegen darin, daß hiermit

- berührungslos und rückwirkungsfrei im Mikrometerbereich mit interferometrischer Genauigkeit gemessen werden kann,
- am Originalobjekt mit diffus reflektierender rauher Oberfläche gemessen werden kann
- an nahezu beliebig komplex geformten Objektoberflächen mit Abmessungen von wenigen Millimetern bis zu mehreren Metern gemessen werden kann,
- flächige und nicht nur punktweise Informationen geliefert werden, so können Stellen mit hoher lokaler Verformung nicht übersehen werden,
- auch an bewegten oder rotierenden Objekten gemessen werden kann,
- durch transparente Fenster hindurch, z. B. in Druck- oder Vakuumkammern gemessen werden kann,
- zeitlich wesentlich voneinander getrennte Zustände verglichen werden können
- zeitlich und örtlich getrennte Prozesse der Aufnahme und der Auswertung der Meßinformation möglich sind,
- Schwachstellen, Fehler oder Spannungskonzentrationen im Originalbauteil bei Belastungsintensitäten weit unterhalb der im Betrieb auftretenden erkannt und lokalisiert werden können.

Die Belastung der zu untersuchenden Struktur kann sowohl mechanisch wie auch thermisch erfolgen, wobei besonders die Betriebsbelastung geeignet ist, die für den späteren Einsatz des Bauteils kritischen Schwachstellen aufzuzeigen.

Die Auswertung der holografischen Interferenzmuster kann qualitativ oder quantitativ vorgenommen werden. Bei einer qualitativen Auswertung werden charakteristische Teilmuster gesucht, die auf das Auftreten von Werkstoffehlern oder lokalen Verformungs- oder Spannungsspitzen hinweisen. In einer quantitativen Auswertung wird das Veränderungsvektorfeld der betrachteten Objektoberfläche numerisch bestimmt. Aus den so gewonnenen Daten lassen sich die Dehnungen und Spannungen ableiten. Diese Aufgabe erfordert in der Regel den Einsatz der elektronischen Datenverarbeitung /1,2/.

Im folgenden werden Möglichkeiten der holografischen Interferometrie in der experimentellen Spannungsanalyse aufgezeigt und dabei auftretende Anforderungen an das Verfahren diskutiert. Anhand von praktischen Anwendungen wird der erfolgreiche Einsatz der Methode dokumentiert.

Die holografische Interferometrie zur Unterstützung von DMS-Messungen

Das am weitesten verbreitete Verfahren zur experimentellen Spannungsanalyse ist neben, der Spannungsoptik die Dehnmeßstreifenmethode. Zur Spannungsanalyse an großflächigen Bauteilen wird diese Methode aufgrund der großen Zahl der notwendigen DMS sowie dem Umfang der Auswerteelektronik jedoch unwirtschaftlich, etwa ab 50 DMS, und insbesondere bei komplex geformten Oberflächen auch unsicher, da Spannungsspitzen zwischen den Meßpunkten übersehen werden. Darüberhinaus ist die Methode weder vollständig berührungslos noch rückwirkungsfrei.

Hier bietet sich die holografische Interferometrie mit qualitativer Auswertung an, um die Orte mit den höchsten Verformungsamplituden zu bestimmen. An diesen Stellen können sodann gezielt mit Hilfe von w e n i g e n DMS die Dehnungen und Spannungen ermittelt werden. Hiermit wird sowohl eine Erhöhung der Aussagesicherheit über das Bauteilverhalten als auch eine Verbesserung der Wirtschaftlichkeit erreicht.

Holografische Verformungsmessung für eine Spannungsanalyse

Eine genaue Spannungsanalyse ist möglich über eine holografisch interferometrische Verformungsmessung mit anschließender numerischer Ableitung der Spannungen. Der dabei durchgeführte numerische Differentiationsprozeß setzt genaue Verformungsmeßwerte in einem dichten Netz von Auswertepunkten voraus. Herkömmliche quantitative Auswerteverfahren der holografischen Interferometrie bestimmen nur die Orte der Interferenzstreifenextrema /1/, zwischen diesen Punkten wird mit Polynomen niedriger Ordnung interpoliert. Hiermit wird jedoch in der Regel nicht die für eine aussagekräftige Spannungsanalyse erforderliche Genauigkeit erreicht.

Um die gewünschte Genauigkeit zu erhalten, ist die Interferenzphase direkt an den Netzpunkten zu bestimmen. Dies erreicht man mit Heterodynetechnik /3/, die jedoch nur punktweise mißt und einen hohen technischen Aufwand erfordert oder mit dem Verfahren mit phasengeschobener Referenzwelle /4,5,6/. Dieses Verfahren zeichnet sich durch hohe Meßgenauigkeit und hohe Auflösung aus und ist auch in der Lage, nichtmonotone Verformungsänderungen festzustellen. Störungen durch variierenden Hintergrund und Specklerauschen werden in dem Verfahren inhärent berücksichtigt und korrigiert, während sie bei Anwendung anderer Auswerteverfahren aufwendig und oft unvollständig mit Hilfe einer Bildvorverarbeitung eliminiert werden müssen. Damit ist das Verfahren mit phasengeschobener Referenzwelle das für eine experimentelle Spannungsanalyse mit Hilfe der Holografie am besten geeignete Verfahren und Voraussetzung für eine wirtschaftliche Spannungsanalyse.

Herleitung der Spannungsverteilung aus den Oberflächenverformungen

Über die holografische Interferometrie werden die Veränderungsvektoren $\vec{u}(\vec{x})=(u_1(\vec{x}),$ $u_2(\vec{x}),u_3(\vec{x}))$ für die Punkte $\vec{x}=(x_1,x_2,x_3)$ der betrachteten Oberfläche bestimmt, wobei x_1 und x_2 Koordinaten in tangentialer, x_3 die Koordinate in normaler Richtung sind. Numerisch bildet man die Ableitungen $\mu_{ik}=\text{grad }\vec{u}$, aus denen sich die Dehnungen $\varepsilon_{ik}=(\mu_{ik}+\mu_{ki})/2$ und die Spannungen $\sigma_{ik}=E(\varepsilon_{ik}+\delta_{ik}(\varepsilon_{11}+\varepsilon_{22}+\varepsilon_{33})/(1-2\nu))/(1+\nu)$ berechnen lassen.

Aus dem holografisch bestimmten Verformungsfeld lassen sich die μ_{ik} mit k=1,2 direkt berechnen. Für k=3 wird vorausgesetzt, daß die betrachtete Oberfläche frei von äußeren Kräften ist und somit ein ebener Spannungszustand vorliegt. Dann sind $\mu_{13}=-\mu_{31}$, $\mu_{23}=-\mu_{32}$ und $\mu_{33}=-\nu(\mu_{11}+\mu_{22})/(1-\nu)$. Somit sind die Ableitungen in Normalenrichtung durch die berechenbaren Ableitungen in tangentialer Richtung ersetzt.

Hierauf aufbauend existieren numerische Verfahren, durch Iteration aus den für die Oberfläche bestimmten Spannungen auch auf den Spannungszustand im Innern eines homogenen isotropen Werkstoffs zu schließen /7,8/.

Holografische Dehnungsmessung an einem Rohrbehälter

Zu bestimmen war das Verformungs- und Dehnungsfeld eines 4 m langen Rohres mit einem Durchmesser von 60 cm und einer Wanddicke von 8,2 mm unter Innendruckbelastung, Bild 1, /9/. Aufgrund der Abmessungen und des Gewichts bei Wasserfüllung konnte der Behälter nicht auf einem üblichen schwingungsisolierten holografischen Aufbau plaziert werden, sondern mußte mit einem nicht-schwingungsisolierten Aufbau auf dem Boden liegend untersucht werden. Hierzu wurde die Referenzwelle über einen auf der Rohroberfläche befestigten und mitbewegten Spiegel aus der Objektbeleuchtungswelle herausreflektiert. Mit einem derartigen Aufbau, Bild 2, werden störende Starrkörpertranslationen kompensiert. Das holografische Interferenzmuster wurde nach Doppelbelichtungstechnik

Bild 1. Rohrbehälter, für holografische   Bild 2. Nichtschwingungsisolierter holo-
       Messung auf dem Boden liegend   grafischer Aufbau am Rohr

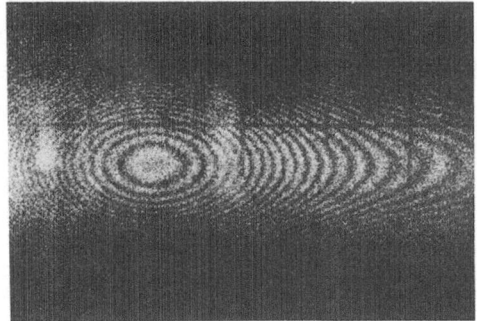

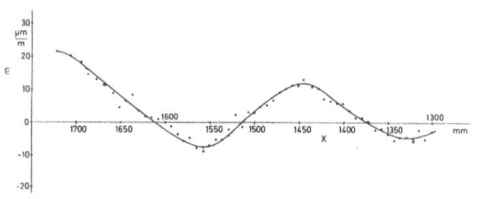

Bild 3. Holografisches Interferenzmuster
   im Beulenbereich des Rohres

Bild 4. Dehnungsverteilung im Beulenbereich

bei einem Differenzdruck von 0,2 bar zwischen den Belichtungen aufgenommen, ausgehend von einem Grunddruck von 10 bar. Innerhalb des Beulenbereichs ergab sich das in Bild 3 gezeigte Muster, die daraus bestimmte Dehnungsverteilung zeigt Bild 4.

Holografische Verformungsmessung an einem Druckkessel

Ebenfalls mit einem nicht-schwingungsisolierten holografischen Aufbau wurde die Verformung des Deckels eines zylindrischen Druckkessels gemessen, Bild 5, /10/. Der Deckeldurchmesser betrug dabei 1,1 m. Die quantitative Auswertung ergab eine Auswölbung von maximal 18 μm bei einer Innendruckdifferenz von 0,5 bar, Bild 6. Ein Vergleich der hieraus berechneten Dehnungen mit nach der DMS-Technik bestimmten Dehnungswerten ergab Abweichungen von generell unter 10%.

Aus den Messungen am Druckkessel wie am Rohrbehälter läßt sich folgern, daß auch an Großobjekten mit hoher Aussagesicherheit Verformungen holografisch interferometrisch bestimmt werden können, wobei bei niedrigen Belastungen im Bereich linear-elastischen Werkstoffverhaltens gemessen wird.

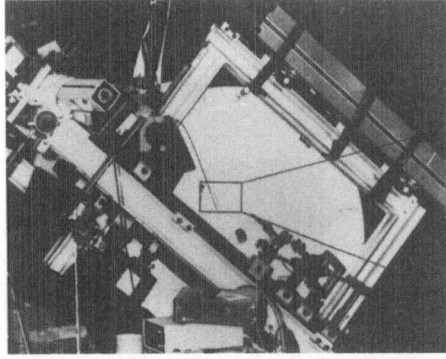

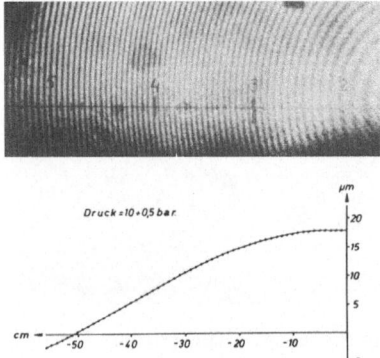

Bild 5. Mobiler Holografieaufbau für
   Messungen an Großobjekten

Bild 6. Interferenzstreifen und Auswölbung entlang dem Deckelradius

Holografische Spannungsbestimmung an Zugproben

Untersuchungen an Zugproben mit mittigem Riß hatten zum Ziel, nachzuweisen, daß die Oberflächenverformungen hinreichend genau und engmaschig bestimmt werden können, um hieraus auf die Spannungsintensitäten im Innern und daraus auf Werkstoffehler zu schließen /9,11,12/. Hierzu wurden die Verformungen und Spannungsverteilungen für rißbehaftete Strukturen mit variierenden Rißparametern nach der Methode der Finiten Elemente berechnet und mit den Messungen verglichen. Über eine iterative Anpassung der in die Rechnung eingegebenen Fehlerparameter konnte der real vorliegende Fehler approximiert werden.

In Bild 7 ist die Zugprobe schematisch gezeigt. Aus den FEM-Berechnungen ergaben sich bei variierender Rißlänge die in Bild 8 angegebenen Verformungskurven. In Bild 9 ist der Versuchsaufbau, das holografische Interferenzmuster sowie die ausgewertete Verformung zu sehen. Ein Vergleich der theoretischen mit den experimentellen Werten ist in Bild 10 gezeigt. Die Ursache für die Abweichungen liegt darin, daß in der Praxis im Gegensatz zu den Berechnungen kein idealer Riß vorliegt, sondern vielmehr vor der Rißspitze eine plastische Zone mit endlichem Radius auftritt /11/. Hierdurch wird ein größerer Riß vorgetäuscht.

Für eine Fehlerfindung mit der Methode der Finiten Elemente hat sich besonders folgendes Vorgehen als trennscharf erwiesen: Die gemessenen Verformungen werden als Zwangsverschiebungen in die FEM-Berechnung eingegeben und die damit errechnete Spannungsverteilung mit der bei gleicher Struktur und Belastung ohne Zwangsverschiebung berechneten verglichen. Bei mit der realen Zugprobe übereinstimmender Struktur ergibt sich ein scharfes Minimum der über alle Elemente integrierten Spannungsdifferenzen.

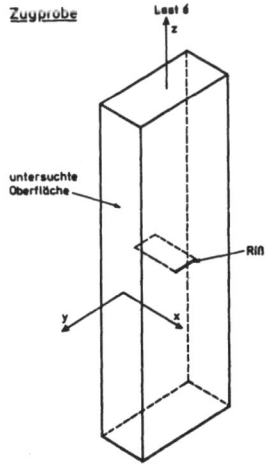

Bild 7. Untersuchte Zugprobe

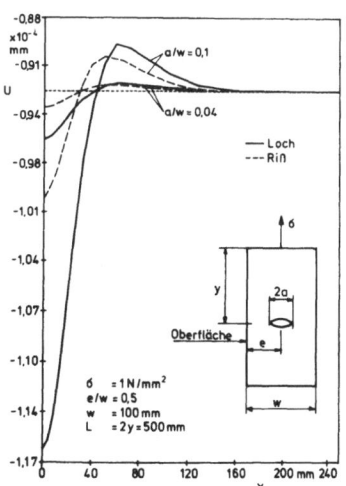

Bild 8. Oberflächenverformung, FEM-Berechnung

306

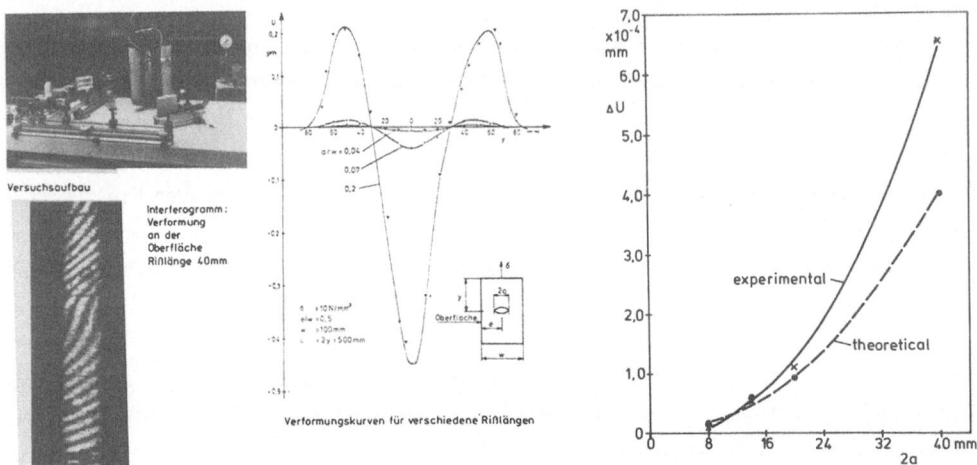

Bild 9. Experimentelle Spannungsanalyse einer
Zugprobe

Bild 10. Vergleich der gemessenen
und berechneten Verformung

Zusammenfassung

Es wurden Möglichkeiten aufgezeigt, die holografisch interferometrische Verformungsmessung für eine experimentelle Spannungsanalyse zu nutzen. Die Beispiele zeigten,
daß diese Methode auch an großflächigen schweren Objekten mit hoher Genauigkeit durchgeführt werden kann. Verglichen zur herkömmlichen Spannungsanalyse hat die Methode neben anderen deutliche wirtschaftliche Vorteile aufzuweisen.

Literatur

(1) H. KREITLOW, TH. KREIS: Proc. of Laser 79, 426 (1979)

(2) TH. KREIS, H. KREITLOW: Proc. of SPIE 210, 196 (1979)

(3) R. DÄNDLIKER, B. INEICHEN, F. M. MOTTIER: Opt. Comm. 9, 412 (1973)

(4) TH. KREIS, B. FISCHER, W. JÜPTNER, G. SEPOLD: Proc. of Laser 81, 105 (1981)

(5) W. JÜPTNER: DPG/DGaO-Frühjahrsschule 78, Hannover (1978)

(6) W. JÜPTNER, TH. KREIS, H. KREITLOW: Proc. of SPIE 398, (1983)

(7) K. A. JACOB: VDI-Berichte 313, 335 (1978)

(8) R. DÄNDLIKER: VDI-Berichte 313, 163 (1978)

(9) W. JÜPTNER, H. KREITLOW, TH. KREIS, P. STEINLEIN: Ber. zu BMFT-RS 15006628 (1984)

(10) B. FISCHER, K. GRÜNEWALD, H. KREITLOW, A. NOELKER: DVS-Berichte 58, 51 (1978)

(11) H. WACHUTKA, H. KORDISCH, B. FISCHER: VDI-Berichte 366, 71 (1980)

(12) G. SEPOLD, B. FISCHER, TH. KREIS, H. KREITLOW: Proc. of Laser 81, 100 (1981)

Laser in der Materialbearbeitung
Lasers in Material Processing

High Power Lasers and Their Industrial Applications

Alberto SONA

Center for Information, Studies and Experiments , (CISE)
Via Reggio Emilia , 39 - 20C9C SEGRATE (Milan) Italy

INTRODUCTION

Lasers applications in materials processing can be subdivided into two
major groups:

A- Applications requiring small but carefully controlled amount of
energy such as micromachining ,semiconductor annealing,phototherapy,
 microsurgery etc.

B- Applications requiring substantial amounts of energy to induce the
required phase transformation in the workpiece for processes such as
cutting ,welding, heat treating and cladding.

   Laser Efficiency and Power are not so important for processes A)
whereas they are specially relevant for processes B).In both cases
beam quality is a much desired feature with a different rôle in
various processes.
   Due to the less stringent requirements the lasers suitable for
class A)processes are much numerous and provide a broad choice of
wavelengths which are often an added constraint for the specific
application.

A non exhaustive list of lasers suitable for the first group of
applications is the following:
Excimers;Ion (Argon - Krypton) ; Metallic Vapours (Cadmium - Selenium
- Copper -Gold) Neodimium in YAG or in Glass ; Semiconductors ;
Erbium in YAG ; Carbon Oxide ; Carbon Dioxide.

Efficiency and Power scalability requirements have restricted up
to now the lasers usable for the second group of processes to Carbon
Dioxide and Neodimium in YAG lasers .However not all the
potentialities of Neodimium in Glass (especially in the slab
configuration) and of the Carbon Oxide lasers (which have high
efficiency and can be transmitted through low loss chalcogenide glass
fibers) have been exploited and many researches are in progress on
this line(1,2).

In addition Excimers lasers seems to be very promising for the
nexth future due to their high efficiency ,the relatively high average
power levels (in the range of several hundreds watts) and the higher
absorption by most metallic materials at these wavelengths.The present
state of the art and perspectives were presented in the previous
paper.

On the short term however Carbon Dioxide and Neodinium in YAG
lasers will be the workhorses for the large majority of class B)
processess.

No industrial applications are expected on the other hand for
chemical lasers such as HF or DF or for gasdynamic CO or CO2 lasers in
spite of the extremely large amount of power they can deliver.Actually
both of them are not closed cycle and in addition toxic chemicals are
released by HF and DF lasers thus rendering industrial applications in
a factory absolutely unpractical.

CARBON DIOXIDE LASERS - STATE OF THE ART AND EXPECTED EVOLUTION

The evolution of industrial high power CO2 laser tecnology is
subjected to contradictory requirements.On one side the growing
expertise of the users requires an increase in performances such as

beam quality ,power stability,reliability which are the basic
requirements for successfull industial applications.On the other side
one of the major limitation to the diffusion of laser technology is
set by the cost of the investment especially for products with low
added value.The manufacturers have tried to compromise between the two
requirements.The efforts have been directed towards a cost limitation
resulting by simpler technical solutions and the use of new less
expensive materials and components.In addition they have tried to cut
operational costs by a careful microprocessor controlled management of
the laser source.This results in a better exploitation of the
consumable and in a reduced downtime due to more appropriate
maintenance schedules made possible by the new self diagnosis systems.

Additional requirements emerged recently from the need of
integrating lasers with robots or flexible manufacturing
systems.Lightweigth compact movable units would be desirable at least
up to the one kW power level;for higher power levels the requirements
are shifted onto the beam delivery system which has to be flexible
and accurate.A second generation of lasers was consequently developed
with substantial technical innovations in the following interacting
areas:

A) - GAS FLOW CONFIGURATION

B) - OPTICAL POWER EXTRACTION

C) - MICROPROCESSOR CONTROL AND SELF DIAGNOSIS

D) - ACTIVE VOLUME EXCITATION

It is important to notice that the evolution towards more advanced
technological solutions occurs slowly due to the always present
constraint of the cost limitation.

A) - GAS FLOW CONFIGURATION

One of the major problem of gas transport lasers is to provide a
fast flow in the discharge channel with typical velocities of 50 m/s
for the transverse flow systems and of 300 m/s for the fast axial flow
lasers.

A relatively new solution for the transverse flow lasers is now
being considered by the manufacturers namely the use of tangential
blades cylindrical blowers.It allows a more compact and effective
geometry for the laser head and possible modular arrangements for
optically cascading multiple heads.This configuration was used since
many years in closed cycle excimer lasers but only recently adopted
for $CO2$ .Turbulence control by positioning a suitable size mesh grid
upstream is also important to avoid instabilities which can start
arcing in transverse flow lasers.

Fast axial flow lasers have been using from the beginning the Roots
pump blowers which are rather cumbersome and noisy.Turbine lightweigth
blowers are now being increasingly used although they require usually
high frequency motors.The most significant example of this trend is
probably the 20 kW laser prototype built by Hitachi.With appropriate
fluidodynamical components a turbulent velocity field can be generated
which allows discharge stabilization in larger volumes and higher
loading without arcing.To achieve this discharge regime again very
efficient blowers are required and the high speed turbine arrangement
seems to be an appropriate solution (3).

B) - OPTICAL POWER EXTRACTION

Fast axial flow lasers usually have active volumes with relatively
small Fresnel numbers (around 10).This allows efficient power
extraction in low order modes with conventional stable cavities.The
output beams can be easily focused.Multimode emission for uniform
irradiation can also been obtained with stable optical cavities.When
maximum power per unit length is required larger cross sections are

excited with an higher Fresnel number which requires extraction by unstable cavities providing annular beams.If the gain is large high magnification factors can be used thus allowing an excellent focusability quite similar to the gaussian beam of the stable cavity.In all cases the fast axial flow discharge has the very remarkable property of cylindrical symmetry both of the medium and of the mode volume which allows an easy geometrical matching and an effective extraction of the available power.

The transverse flow discharges usually have an higher Fresnel number (around 50) and a rectangular cross section of the active medium.To optimize the extraction a multipass stable optical cavity can be used, designed in order to have a good superposition of the mode volume and of the active material ,however the achievable filling factor does not allow a full exploitation of the available power.In addition the regions where the optical paths are superimposed can give rise to mode selection due to the different level of gain saturation.Further problems can arise from the more critical alignement of the multiple mirrors arrangement.The other possible approach for optimal power extraction make use of an unstable resonator.To match at best the active medium and the mode volume however a set of low loss modes with a rectangular cross section has to be generated;this can be done using a scraper mirror with a rectangular aperture inside an unstable cavity or a special set of mirrors (toroidal - spherical or two crossed cylindrical as provided in lasers made by HERAEUS).

An unstable cavity with circular mirrors can of course be used reducing the power extraction.The gain distribution in the cross section is in general non uniform, the cathode fall region being dominated by faster electrons.A symmetry plane is present only in AC excited discharges ,in DC excited channels where the electric field is coincident with the flow direction (biaxial lasers made by Toshiba) or in specially designed DC discharges (4).

The gain distribution in the active medium is non uniform and this affects the intensity distribution in the laser beam even when optical cavities with circular symmetry are used.A partial compensation can be

obtained cascading optically an even number of channels with reversed
gain distribution.The problem of generating a transverse flow
discharge with uniform and symmetrical gain distribution in a volume
well matched with a low order mode has not yet been solved in a
practical way and new geometries are to be considered.

C) - MICROPROCESSOR CONTROL AND SELF DIAGNOSIS

Microprocessors are now currently used to provide assistance for
automatic start up procedures with full control of the single step and
for power programming.Automatic process control can be implemented wth
a feedback loop acting on the laser emission according to the signal
provided by a suitable sensor.In addition microprocessors can provide
full autodiagnosis of the laser unit to allow the optimization of the
operating parameters including gas consumption an other running costs
and of the maintenance schedule to reduce downtime.Microprocessor
assisted operation is most effective in reducing operational costs.One
of the first lasers including this facility was the UTRC modular
transverse flow laser with unitary power of 3kW (up to four moduli can
be cascaded providing power in excess of 12 kW).

D) - ACTIVE VOLUME EXCITATION

1) - Self Sustained Discharge

The self sustained discharge has the advantage of simplicity and
has been adopted up to now in the two most popular configurations
namely the fast axial flow and the transverse flow lasers.However it
is well known that it does not allow optimum efficiency because of the
lack of independent control of the ionization degree and of the
electrons temperature which cannot be tuned to the value providing the
maximum excitation efficiency of the upper level (5).In addition the
negative resistance of the glow discharge do require ballast resistors
where a substantial amount of power is lost.Even if the efficiency
problem is disregarded the self sustained discharge has rather severe

limits as regards uniformity in the active volume and maximum allowed
excitation before arcing starts.The uniformity problem is definitely
less severe in the fast axial flow scheme where axial symmetry is
present and axial non uniformities can be averaged out.

2) - Radio Frequency Excitation

The nexth step for a uniform discharge stable under heavy loading
conditions is the use of radiofrequency excitation.The E/N ratio in
this case stays above the value for the avalanche ionization regime
only for a limited part of the cycle ;for most of the time the
discharge is operated in the recombination regime with electrons
temperature closer to the optimum value for upper level pumping .This
excitation can be provided by using conductive or dielectric
electrodes and has the advantage to exploit ballast capacitors without
additional power losses.Frequencies in the ten megahertz range would
be preferable to reduce ripple in the output power due to
recombination processes but ,due to electromagnetic interference
problems and to the possibility of using solid state components for
the AC supply, frequencies in the ten kilohertz range are preferred in
practice (6).This appears to be the most convenient solution for
compact lightweigth units with an output power at the one kW level
such as those from Laser Corp.of America (7) and Laser Innovation
(8);however units up to the 10 kW level with this technology are
commercially available from Metalworking Lasers International (9).

3) - Double Control Discharges

A further step providing an independent control of the ionization
degree and of the electrons temperature (the " two knobs controlled
discharge") would result in an improved efficiency ,in a more uniform
excitation ,in the suppression of the ballast resistors and in added
control capabilities of the laser power of special importance for the
modern closed loop controlled machining processes.

The most effective ways to provide the two independent controls
are the following:

- Electron Beam preionization:it was first introduced by AVCO for CW lasers and it has been exploited up to the 20 kW level in commercially available units.It has been one of the first example of the double control discharge (10).

- Photoinitiated Impulse Enhanced discharge : it was first studied by UTRC with a long train of pulses and subsequently developed for cw operation in Canada where now a 10 kw commercial unit is manufactured by Majestics Lasers Systems (11,12).

- Radiofrequency assisted DC excitation:many configurations have been tested with different roles and relative importance of RF vs DC power.One of the most significant was developed by Mitsubishi (the "SAGE" discharge).Laboratory units with CW emission in excess of 20 kW and commercial units up to 10 kW are available (13).

Other possible approaches exploit UV or X-rays preionization but up to now they did not generate lasers suitable for industrial applications.

In summary many different approaches with different degree of complexity and of added performances are available for stable and reliable operation at power levels up to 10 kW (20 kW at the prototype level).Added complexity results in general in higher laser cost which can be accepted only if the added performances are really relevant to the process economy. This is usually the case for high added value products already manufactured by sophisticated fully automated systems. A better space and time control of the laser emission ,the improved efficiency are really needed for the future flexible manufacturing systems or robots assisted laser machining.

Many additional technical improvement can be accomplished.Single mode TEMoo emission can be obtained with Master Oscillator - Power Amplifier (MOFA) configurations.Beam intensity distribution and focusability can be substantially improved in high power oscillators with unstable cavities by using adaptive optics systems similar to those developed for military applications.Sophisticated beam monitoring and real time alignement systems can be borrowed from lasers systems developed for nuclear fusion experiments.This is the

confirm that the present industrial lasers for material processing
are not exploiting the best available solutions .The tecnology of high
power lasers for the forementioned applications is actually much more
advanced.In practice the complexity and related performances of lasers
are limited only by the acceptable level of cost which is at present
high enough to limit the use of lasers only to the products with high
added value.Further efforts and ingenuity are needed to reach better
comprorises.

NEODYMIUM IN YAG CR GLASS LASERS

The technical level of the industrial solid state laser is far
from providing the best state of the art performance.High average
power units are available up to the 600 watt level usually with
multimode emission resulting in a focused spot typically one order of
magnitude larger than the diffraction limit.The advantages of the
better focusability dependent on the shorter wavelength cannot in
practice be exploited.The main motivation for this discrepancy is the
power dependent thermal lensing effect induced in the material by the
pumping process.Again with added complexity this problem can be solved
but not in a cost effective way.With different geometrics such as the
slab laser configuration this effect can be compensated and beams
divergence close to the diffraction limit can be achieved.

More efficient solid state flashlamp pumped laser materials will
be available in a relatively short time. Of special interest are the
Gadolinium Scandium Gallium Garnets (GSGG) doped with Neodimium and
Chromium with a reported slope efficiency up to seven percent (14).The
recent developments in solid state lasers were presented in the
previous paper.

BEAM DELIVERY SYSTEMS

The requirement of integrating a laser in a Flexible Manufacturing

System or in Robot Assisted Laser Machining poses the problem of beam delivery to the workpiece which has in general five degrees of freedom with respect to the beam focusing head.In addition a sixth one has to be considered for the auto focus feature.Depending on the mass and size of the workpiece the five degrees of freedom can be shared in different ways between the laser focusing head and the worpiece itself. The simpler beam delivery system is ,of course,the one required when the workpiece is given all the degrees of freedom.The opposite case is the one ,typical of robotic systems, where the workpiece is fixed and all the five degrees of freedom are given to the focusing head thus implementing a five axis machine.The robotic system can be polar or of the gantry type ,in any case it has to provide relative positioning with an accuracy of the order of the focal spot size (typically of 100 microns) which is not usual for conventional robotic tooling systems.

In general a simple solution can be provided by positioning with a robot the beam focusing head which has to be connected to the laser head by a suitable flexible delivery system namely an articulated arm or an optical fiber.This solution allows the use of already existing robots ,provided they have the necessary accuracy.However it requires a rather sophisticated software able to avoid mechanical interferences which can damage the flexible optical link.In addition more or less severe limitations to the kinematics of the robot are introduced with both arrangements.

Neodymium in YAG laser beams with an average power up to 600 watts have been conveyed in a fused silica optical fiber with a 1.2 mm core diameter and 15 meters length.The losses ,in the system implemented by Toshiba,were smaller than 10 percent and were mainly due to the end faces.Smaller core diameters could be used at lower power levels the limit being set by the beam divergence (15).Under this respect a substantial improvement is expected by lasers with slab geometries providing low divergence beams due to optical index gradients compensation.Researches along this line are in progress at General Electric in USA (2,16).

On the long term the best solution will be the use of specially

designed hollow robots with a inner free optical path providing
flexible optical guiding without added kinematic constraints and
suitable positioning accuracy.

Flexible Manufacturing System will require also appropriate beam
delivery systems for feeding the different workstations.Two different
approaches can be considered depending on the specific manufacturing
processes (17).

One possible approach makes use of a single very high power laser
unit delivering the beam to different workstations on a time sharing
basis.All the power can be conveyed on a single workstation at a time
or more workstations can be fed simultaneously with a fraction of the
total power.

A second possible approach consists in using a few smaller power
laser beams which can be combined by suitable optical systems
providing a single beam with a total power equal to the sum of the
individual ones or two or more beams having a power resulting from the
superposition of the selected beams.The generated beams can then be
forwarded to the different workstations with the appropriate power
level selected according to the process requirement.The
combining-switching matrix can be implemented by mobile computer
controlled optical elements (18).

The main advantages of the second approach is the use of smaller
power units ,in general more reliable,by the possibility of adding
further lasers in a modular fashion ,by the substantial reduction of
the risk of a complete stop because of the multiple laser heads in
comparison with the case of the single high power laser going out of
service.Finally a back-up unit is not as expensive.The disadvantages
are the major investment cost for the same total power installed.

From the application point of view the first approach is more
appropriate when most manufacturing processes do require the full
power whereas the second is more useful when most processes require a
fraction of the total power and only a few need the total.

CONCLUSIONS

A continuous growth of high power lasers applications in industry
is expected.On one side the diffusion of the laser manufacturing
technology for low added value products will require low cost
dedicated laser systems.For mass production and /or for high added
value components high performances such as good beam quality
,reliability , computer controlled emission are required to allow
fully automated processing.The added requirements set by the use of
lasers in connection with robots and flexible manufacturing systems
need further steps in industrial lasers technology.Sophisticated
technologies for generating high power high quality laser beams with
accurate delivery systems are already available for other
applications.Their use in industry is limited by economical problems
which will become less severe in time as the use of laser will be
further expanded.A key point to speed up this process is the
development of standard modular subsystems for high power laser beams
generation and delivery.This would result in a cost reduction and in
reliable and controlled performances as it occurred already in the
machine tooling industry.Possible future applications are expected in
nuclear and aerospace industries in the short term and on a longer
time scale in shipbuilding and iron industries when higher power laser
units will be available for practical uses.

REFERENCES

1) - H.Saito.t al.   Proc. CLEC '85,paper WM30,pg.120.

2) - W.B.Jones Laser Focus / Electro Optics Sept. '83.

3) - G.H.Sugawara et al. Proc. CLEC '84 , paper TUC3,pg.54.

4) - M.Kamasatsu et al.   Proc. CLEC '81 , paper WE7,pg.4

5) - W.L.Nighan   Phys.Rev.A2,1989,1970

6) - S.Yagi,N.Tanaba   Proc. CLEO ' 81 ,paper We5 ,pg.132.

7) - J.S.Eckersley   Proc. ICALEC '84 , paper No.4.6.

8) - P.Hoffman Proc. Laser in Manufacturing '85 ,pg.201.

9) - P.Agmon,E.Hoch et al.Proc. CLEC '85,paper FP3 ,pg.302.

10) - E.D.Hoag,H.Pease et al. Appl.Opt. Vol.13,1959 (1974)

11) - Alan E.Hill Appl. Phys. Lett. Vol.22 ,670,(1973)

12) - H.J.J.Seguin et al. Appl.Opt. Vol.10 ,2233 (1981)

13) - M.Hishi et al.   Proc. CLEC '83 ,paper WC5,pg. 132 .

14) - J.A.Caird et al.   Proc. CLEC '85 ,paper ThF3 ,pg.232.

15) - A.V.La Rocca et al. Proc. LASEROECTICS I 1985 ,paper No.4.

16) - M.G. Jones et al. Proc. ICALEO '83 ,pg.148.

17) - D.Plankenhorn et al. Proc. LASEROECTICS I 1985 ,paper No.15.

2,5 KW CO$_2$ CW Modular Laser for Industrial Applications

V. Fantini, L. Garifo, G. Incerti, W. Cerri
CISE S.p.A., P.O. Box 12081 - 20134 Milano, Italy

In the frame of Italian National Research Council Special Project on High Power Lasers, a 2.5 kW CW CO$_2$ laser source of self sustained discharge type has been built.

The mechanical structure of the laser allows to double the output power, using two 2.5 kW modules connected together.

With reference to Fig. 1, the 2.5 kW source consists of laser head (a), control and power supply unit (b), laser beam diagnostic unit (c), control consolle (d) and cooling water unit (not shown in Fig. 1).

A 3000 rpm axial blower of $\sim$ 60 cm diameter makes to flow the gas through the discharge region at a velocity of 50 m/s. The laser mixture is CO$_2$ - O$_2$ - N$_2$ - He: 5% - 2.5% - 32.5% - 60% at a nominal pressure of 40 mbar. The gas consumption is about 100 Nl/h.

Fig. 1.   2.5 kW laser source: (a) laser head; (b) control and power supply unit; (c) laser beam diagnostic unit; (d) control console

The cathode is a deionized water cooled copper pipe, while anode is a set of resistively ballasted uncooled copper pads. The interelectrode gap is 5 cm and the discharge length along the optical axis is 85 cm. Approximatively 10% of the supplied electrical power is dissipated by the ballast resistors.

The laser is equipped with a confocal unstable resonator, suitable for welding and cutting, and a multimode stable resonator, suitable for heat treatment processes. The beam from the unstable resonator has the diameter of 45 mm. The magnification ratio M is 2 and the divergence in about 2 mrad. The multimode Laguerre-Gauss beam from the stable cavity has the diameter of 40 mm and a flat top intensity profile. The beam divergence is about 6 mrad.

The long term power stability of the laser source is less than $\pm$ 5%. At the output power of 2.5 kW the electrooptical efficiency, defined as the ratio between the output power and the electrical power supplied to the discharge, is 13.5% and the overall efficiency of the source is about 8%.

The control and power supply unit consists of the gas control, the alarm panel and the high voltage power supply.

The laser beam diagnostic unit provides for control diagnostic and delivery to the workstation of the beam. The propagation height of the output beam from the unit is variable between 2 and 2.2 m, depending on the height of the workstation optics and the CO_2 beam alignment to the workstation is accomplished by two beam steering mirrors. The diagnostic unit incorporates a beam shutter, having the closing or opening time of 0.1 s, and an on-line beam power measuring system, which utilizes a chopper delivering 2% of the beam power to the power meter. A He-Ne laser provides for the prealignment of the optical cavity of the laser head and the alignment of the laser beam handling optics up to the workpiece. Moreover the diagnostic unit is equipped with an on-line mode viewing and optical cavity alignment systems. During laser working, the operator can watch, through a window of the diagnostic unit, the intensity distribution of the laser beam, using an IR sensitive plate and control the displayed beam power. The optical cavity alignment of the laser head resonator is achieved by one motor driven mirror. Making use of the optical cavity alignment system the operator can easily accomplish the final alignment of the resonator. He can control, at the same time, the laser beam intensity distribution and power, using the mode viewing and power measuring systems.

Some welding tests have been carried out. Fig. 2a shows the bead-on-plate penetration t on AISI 304 vs. the workpiece velocity v. The laser

beam, at a power of 2 and 2.5 kW, has been focused with a 5" focal
length ZnSe lens and a coaxial He shielding gas flow at a rate of 15
1/min has been used. Fig. 2b shows the cross-section of the bead-on-
plate penetration for a thickness of 4 mm and v = 1 m/min. As shown,
a narrow weld without any crack or porosity has been obtained.

Some results of surface heat treatment on AISI 4137 H are reported in
Fig. 3a and 3b. Treatments have been carried out at an incident laser
intensity of 1300 W/cm^2 and a workpiece velocity of 0.4 ÷ 0.8 m/min.
The workpieces have been coated with graphite and He shielding gas flow
has been used. Fig. 3a (5x) shows the heat treatment results when the
workpiece velocity is 0.6 m/min and Fig. 3b reports the hardness profi
le of the treated region, which is uniform in a thickness of 1.2 mm.
The average value of the hardness in the treated region is 630 HV and
corresponds to a complete hardening transformation of the material.

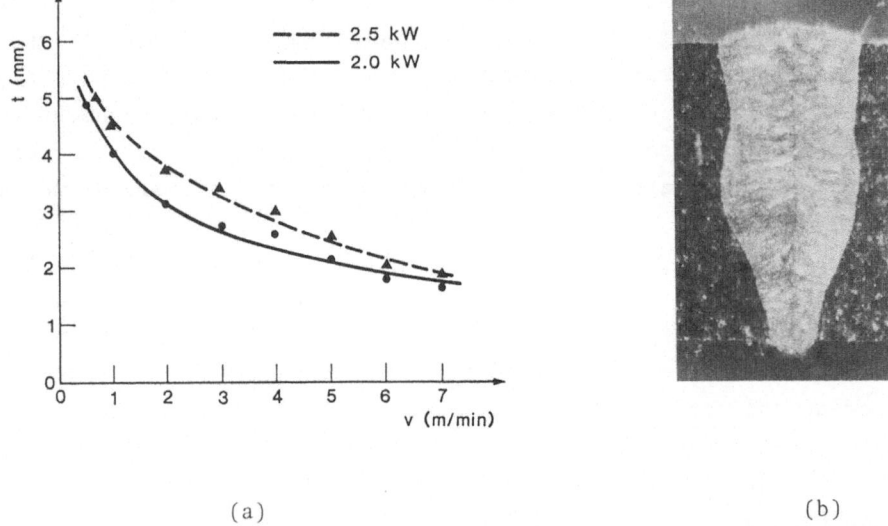

(a)   (b)

Fig. 2.   (a) bead-on-plate penetration on AISI 304 vs. the workpiece
              velocity v. Focal length 5", He flow rate 15 1/min.

          (b) cross-section of bead-on-plate penetration on AISI 304.
              Laser power 2.0 kW, workpiece velocity 1 m/min, thickness
              4 mm, focal length 5", He flow rate 15 1/min.

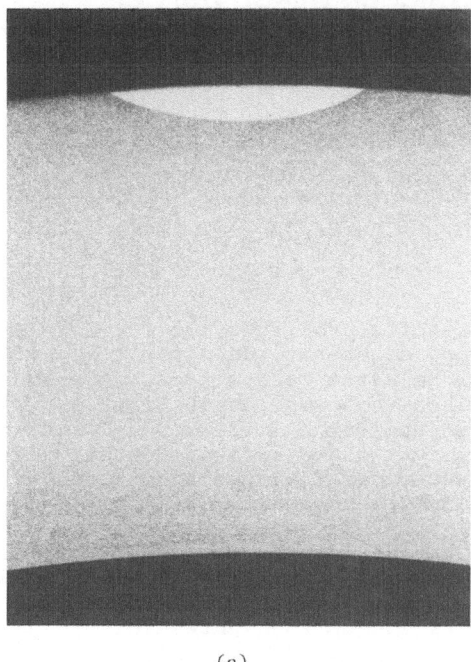

(a)

(b)

Fig. 3.   Laser heat treatment on AISI 4137 H:
          (a) treated zone profile;
          (b) hardness values HV vs. the depth z from the workpiece sur-
              face, taken at the center of the treated region.

Erfahrungen mit Spiegeloptiken bei der Werkstoffbearbeitung mit Hochleistungslasern

Rüdiger Rothe, Werner Jüptner, Gerd Sepold

BIAS - Bremer Institut für angewandte Strahltechnik

Ermlandstr. 59, D-2820 Bremen 71

1. Einleitung

Optiken für die Werkstoffbearbeitung mit CO_2-Hochleistungslasern sind hohen Beanspruchungen ausgesetzt, da sie mit hohen Strahlleistungen und Leistungsflußdichten belastet werden. Besonders hoch ist die Belastung der Bearbeitungsoptiken, da diese durch Spritzer, Gase und Dämpfe vom Werkstück verschmutzt und durch Reflexion und Streuung von Laserstrahlung zusätzlich belastet werden können. Bearbeitungsoptiken aus Halbleitermaterial (Linsensysteme) werden unter solchen Bedingungen schnell zerstört. Daher sollten transparente Optiken - wenn immer es geht - vermieden und Spiegeloptiken aus Metall gewählt werden. Der Stand der Technik ist so weit, daß alle erforderlichen Spiegeloptiken kostengünstig mit hoher Güte gefertigt werden können. Es wird über ein industriell einsetzbares optisches System zur Laser-Materialbearbeitung berichtet, das mit folgender Zielsetzung entwickelt wurde:

1. Geringe Verluste von Laserleistung und damit hoher Wirkungsgrad der Laseranlage.
2. Gewährleistung der Beständigkeit der Spiegel unter den Bedingungen des Einsatzes im Materialbearbeitungsprozeß.
3. Gleichbleibende optische Qualität der Spiegel-Eigenschaften (Reflexionsvermögen) bei möglichst hoher Lebensdauer.
4. Einsetzbarkeit für verschiedenartige Bearbeitungsaufgaben.

Die gewünschten Eigenschaften werden durch eine geeignete konstruktive Gestaltung der abbildenden Oberfläche sowie teilweise durch zusätzliches Beschichten (Vergüten) erzielt. Der Entwicklungszustand und die Erfahrungen dieses im Bias entwickelten Laser-Bearbeitungssystems wird beschrieben.

2. Spiegelherstellung

Metallspiegel für die Werkstoffbearbeitung mit CO_2-Hochleistungslasern werden heute überwiegend durch Bearbeitung mit Diamantwerkzeugen hergestellt und bestehen zur Sicherstellung guter Wärmeableitung in der Regel aus Kupfer oder Aluminium /1/. Die Bearbeitung erfolgt auf schwingungsfrei angetriebenen Maschinen, die z. B. Luftlagerspindeln enthalten. Die Fertigung von Planspiegeln, Zylinderspiegeln, konischen oder prismatischen Flächen sowie von Kugelspiegeln ist relativ umkompliziert und sicher durchzuführen, da hierbei die Zerspanungsbedingungen über der gesamten Fläche konstant gehalten werden können und da diese Geometrien ohne Rechnersteuerung erzeugt werden können. Rotationsparaboloidspiegel erfordern jedoch besonderen Aufwand. Üblich sind numerisch gesteuerte Drehmaschinen. Mit derartigen Maschinen werden hochgenaue Rotationsparaboloidspiegel gefertigt, Bild 1. Abweichungen von der Sollgeometrie liegen unter 0,4 µm. Messungen des Reflexionsvermögens dieser Oberfläche ergaben teilweise mehr als 99 %. (für Cu, λ = 10,6 µm) /2/. Derartig gefertigte Spiegelflächen zeigen eine hohe Konturgenauigkeit, jedoch nur geringe Beständigkeit gegen thermisches Anlaufen und gegen mechanische Beschädigung, wie z. B. durch Reinigen von Staub oder Metallspritzern beim Einsatz in Bearbeitungsköpfen. Hierdurch wird die

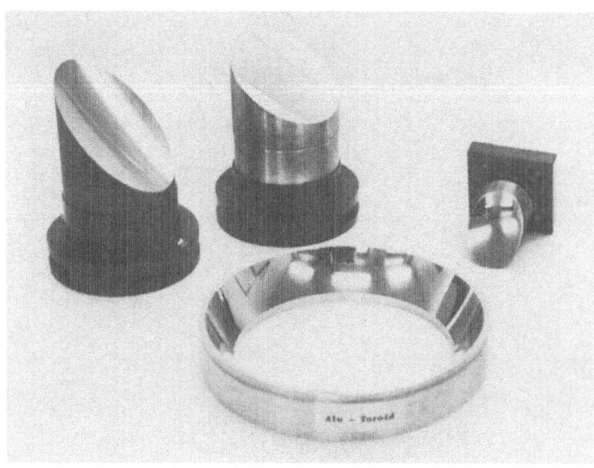

<u>Bild 1.</u> Rotationsparaboloidspiegel aus Reinkupfer und Aluminium, diamantgedreht

Oberfläche zerkratzt. Für den industriellen Einsatz von Werkstoffbearbeitungsoptiken werden jedoch verschleißbeständigere Spiegelflächen gefordert /3/. Ein geeigneter Werkstoff ist Molybdän. Molybdän hat bei relativ hoher Härte ein hohes Reflexionsvermögen (97 % bis 98 %) für CO_2-Laserlicht. Entsprechend hoch ist auch die Wärmeleitfähigkeit, wodurch die Spiegel mit hohen Strahlleistungen belastbar sind. Allerdings läßt sich dieser Werkstoff nicht mit Diamantwerkzeugen fertig drehen oder fräsen. Molybdänspiegel werden daher durch Schleifen und Polieren hergestellt. Mit dieser Methode lassen sich kaum Rotationsparaboloidflächen mit der erforderlichen Präzision fertigen. Die Spiegelgeometrie wird daher besser in Reinkupfer gefertigt und anschließend mit einer ausreichend dicken Molybdänschicht bedampft /4/.
Schichtdicken von 2 µm haben sich im Einsatz bewährt. Als Beschichtungsverfahren kommt die PVD-Methode "Sputtern" infrage /5/. Hiermit lassen sich festhaftende Schichten mit hohem Reflexionsvermögen reproduzierbar auftragen.

Laserspiegel für Bearbeitungsstationen mit bewegten Optiken, die wegen der hohen Beschleunigung besonders leicht sein sollen, werden aus Reinaluminium hergestellt. Auch Aluminiumspiegel in Bearbeitungsköpfen lassen sich durch Molybdänschichten gegen Verschleiß schützen. Zur Verbesserung der Reflexionsfähigkeit (99 %) werden Laser-Optiken häufig mit Gold beschichtet. Sehr erfolgversprechend sind auch dielektrische Mehrfachbeschichtungen auf Metallspiegeln, die sich heute noch in Erprobung befinden (6/.

3. Bearbeitungssysteme
Mit der reproduzierbaren Herstellung von Spiegeloptiken hoher Güte ist heute die Voraussetzung gegeben, komplette Bearbeitungssysteme zu fertigen, die der jeweiligen Bearbeitungsaufgabe optimal angepaßt werden können. Bild 2 zeigt ein System, das für das Laserschweißen, Laserschneiden und Oberflächenveredeln mit Laserstrahlen entwickelt wurde und in langjährigem Einsatz erprobt ist. /7/. Es wurde so konstruiert, daß es zur Durchführung von Wartungsaufgaben oder beim Wechsel einzelner Komponenten nicht nachjustiert werden muß. Die Systemkomponenten lassen sich schnell und reproduzierbar umrüsten, und die Optikkomponenten sind ebenfalls schnell und reproduzierbar austauschbar. In diesem

Bild 2. Universal-Bearbeitungskopf mit Umlenk- und Fokussier-
spiegel und Härteadapter

System sind die Spiegel schnell wechselbar montiert, wobei die Präzision in der Fertigung so hoch ist, daß z. B. der Brennfleck des Fokussier-
spiegels in immer gleicher definierter Position liegt. Durch Anbau von
Vorsätzen kann dieser Universal-Bearbeitungskopf zum Schweißen, Schneiden und Oberflächenveredeln ausgerüstet werden. Der Anschluß einer
Schweißrauchabsaugung an einen Absaugstutzen schützt die Spiegeloptiken
auch gegen Spritzer und Dämpfe, wie sie z. B. beim Schweißen von unbe-
ruhigtem Stahl auftreten können.

Durch einen Schweißvorsatz, Bild 3b, entsteht ein Schweißkopf. Hier wird
ein Schutzgasschleier durch eine Ringdüse auf die Werkstückoberfläche
geleitet, so daß die Schweißnaht ausreichend gegen Atmosphäreneinflüsse
geschützt ist. Ein zusätzlicher Gasstrahl reduziert das Schweißplasma
oberhalb des Laser-Auftreffpunktes. Zum Schneiden lassen sich Vorsätze

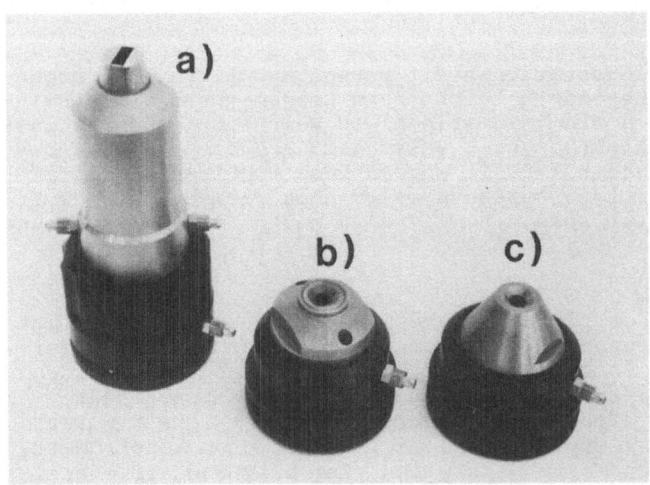

Bild 3. Bearbeitungs-
adapter zum Laser-
a)Härten, b)Schweißen,
c)Schneiden

Vorsätze montieren, Bild 3c, die den Schneidstrahl koaxial zum Laserstrahl mit Überschallgeschwindigkeit zuführen. Die koaxiale Anordnung
eines Überschallstrahles wird durch mehrere schräg verlaufende Gas-
strahlen aus Lavaldüsen erreicht, die in einem Punkt auf der Strahlachse konvergieren und den Schneidstrahl erzeugen. Der Vorsatz zum
Oberflächenveredeln, Bild 3a, formt aus einem Laserstrahl mit beliebi-
gem Querschnitt einen Strahl mit rechteckigem Querschnitt und homogener Verteilung der Leistungsflußdichte. Dieser Vorsatz eignet sich besonders zum Umwandlungshärten. Die Tiefenschärfe reicht je nach Auslegung bis zu 20 mm. Aus der Mündung strömt Schutzgas, um die Werkstückoberfläche vor Atmosphäreneinflüssen zu schützen und eine Verschmutzung
der Optik zu verhindern. Durch Austausch von Strahlformungsspiegeln im
Vorsatz lassen sich die verschiedenen Strahlquerschnitte einstellen. Das
hier vorgestellte Bearbeitungssystem wird für unterschiedlich große
Laserstrahlquerschnitte gefertigt, so daß es für alle üblichen CO_2-
Lasersysteme einsetzbar ist. Durch Austausch der Komponenten läßt es
sich jeder Aufgabe optimal anpassen.

4. Schlußbemerkung

In dem vorliegenden Artikel wurde gezeigt, daß Spiegeloptiken aus metallischen Werkstoffen für die Laser-Materialbearbeitung geeignet sind
und heute mit hoher Güte kostengünstig angeboten werden. Als Vorteil
der heute üblichen Fertigungsmethoden, Diamantdrehen und Diamantfräsen,
ist die hohe Reproduzierbarkeit der Spiegelgeometrien genannt. Dies war
eine Grundlage zur Entwicklung und Fertigung von kompletten Bearbei-
tungssystemen mit unempfindlichen Komponenten, die außerdem reproduzierbar austauschbar sind und dabei keiner aufwendigen Justierung bedürfen. Durch Einführung dieser Systeme zum Schweißen, Schneiden und
Oberflächenveredeln mit Laserstrahlen ist es möglich, einen Laserstrahl
jeder Bearbeitungsaufgabe optimal anzupassen. Damit ist ein Schritt in
Richtung erhöhter Zuverlässigkeit der Anlagen getan, um den Lasereinsatz für industrielle Aufgaben attraktiv zu machen.

5. Dank

Die Autoren danken dem BMFT und dem Projektträger VDI-Physikalische
Technologien für die Förderung dieser Arbeit.

Schrifttum

/1/ Watt, G.J.: Diamonds, Air Bearings, and Optics, SPIE Bd. 159,
    S. 18/24, 1978
/2/ Decker, D.L. et.al.: Surface and Optical Studies of Diamond-
    Turned and Other Materials, Optical Engineering, Vol 17, No. 2,
    1978, p. 160-166
/3/ Newman, B.E.: Optical Materials for High Power Lasers, Laser
    Focus 2 (1982) S. 55
/4/ Carver, G.E. u. Seraphin, B.O.: Chemical-Vapor Deposited Molybdenum
    Films of High Infrared Reflectance Appl. Phys. Lett 34(4),
    15. Febr. 1979, p. 279 - 281
/5/ Vossen, J.L. u. Kern W.: Thin Film Processes, Acad. Press,
    New York 1978
/6/ Nichols, D.B. u. Hall, R.B.: Large-Spot DF Laser Damage of
    Dielectric-Enhanced Mirrors, Proc. Symp. "Laser Induced Damage
    in Optical Materials", 1977, Boulder, Colorado, p. 325-341
/7/ Mitteilung: Universelles Bearbeitungssystem für Hochleistungs-
    laser Materialbearbeitungsaufgaben

On-Line Diagnostic System for CO_2-Lasers to Control Production Quality

Gerd Sepold, Rüdiger Rothe, Jan Telepski

BIAS - Bremer Institut für angewandte Strahltechnik

Ermlandstr. 59, D-2820 Bremen 71

1. Introduction

Laserwork in industry demands reproducible results over long time periods. One requirement to match this needs is to guarantee always the same beam qualities as laser power, beam direction, mode structure and beam caustic. Especially high power lasers may change these beam qualities by aging of the output coupler or other optical components, e.g. by temperature changes in the laser. To avoid an influence to material processing by such changes a real time laser diagnostic system has been developed and will be described which uses a simple technique to convert the infrared laser beam into electrical signals while material processing. These signals can be calibrated and further processed for control and inspection.

2. Principle of the Diagnostic System

The diagnostic system consists of a sensor head, an image analysis system, and displays, fig. 1.

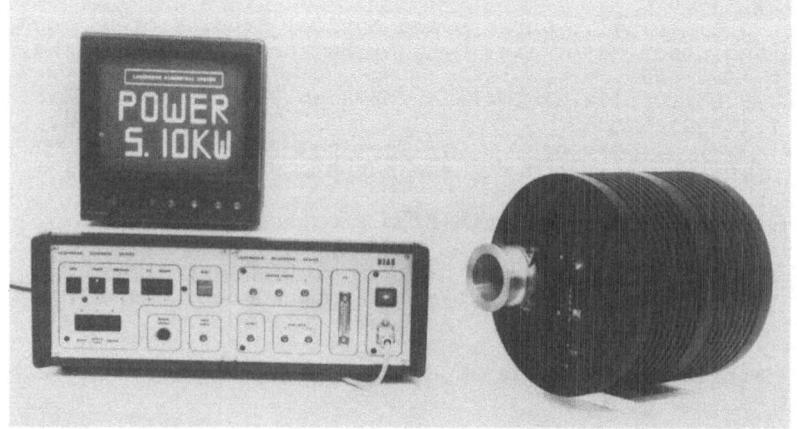

Fig. 1. Diagnostic system with a) sensor head, b) image analysis, c) TV display

The sensor head detects directly the high power laser beam. It comprises reflector segments cabable to evaluate via pyroelectric detectors the beam cross section. This measurement device consists of a so called spoke-wheel with eight mirror stripes arranged spokelike at an angle of 45° to the beam axis, fig. 2.

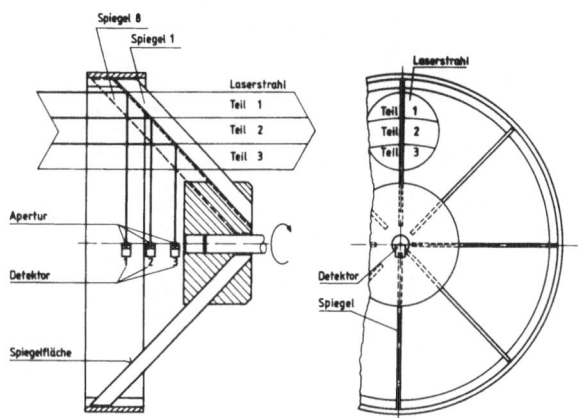

Fig. 2.
Principle of sensor head

Neighbouring stripes are displaced 2 mm in axis direction. The thin sides of the stripes are provided as plan reflectors. At each time only one spoke is in the area of the beam. This spoke reflects one stripe of the laser beam at an angle of 90° onto the revolution axis.

On this axis three detectors are mounted in a distance relative to each other of 1/3 of the beam diameter. On the detectors there are fixed slot apertures which allow only a small portion of the laser beam to pass onto the detectors. The slots are arranged in a vertical position relative to the revolution axis. Through these slots passes that part of the laser beam coming from the mirror-point perpendicular to the axis.

With the fixation of the detector components on the revolution axis it is not required to have additional optical devices. The three signals are amplified in parallel and recorded at once. Each following spoke reflects those parts (lines) of the laser beam into the three detectors, which are displaced by 2 mm in radial direction. One revolution of the wheel scans the picture consisting of 3 times 8 lines, (total 24 lines). The processing of the signals and image analysis is done by microprocessors after a digitalisation process. One type of processing is to transform them into a grey scale picture on a monitor. This is done respecting TV standards. Another type of processing is performed by integrating the electrical signals, so that the total beam power is evaluated.

Display can be done by TV monitor, by oscilloscope, by printer or by external computer. The TV display enables several visualisation modes: a grey scale mode, fig. 3, a 3-D-mode, fig. 4 and a power indication mode, fig. 1.

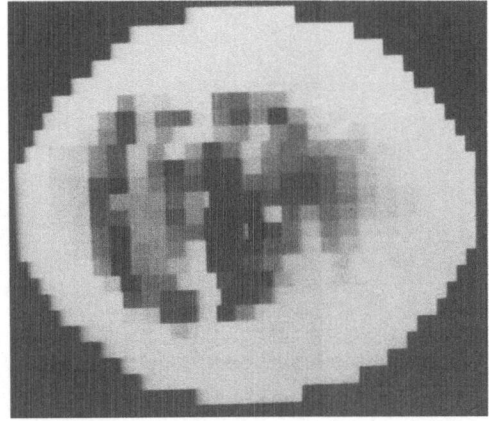

Fig. 3. Grey scale display of a CO_2 laser beam in real-time, total power P = 4,5 kW

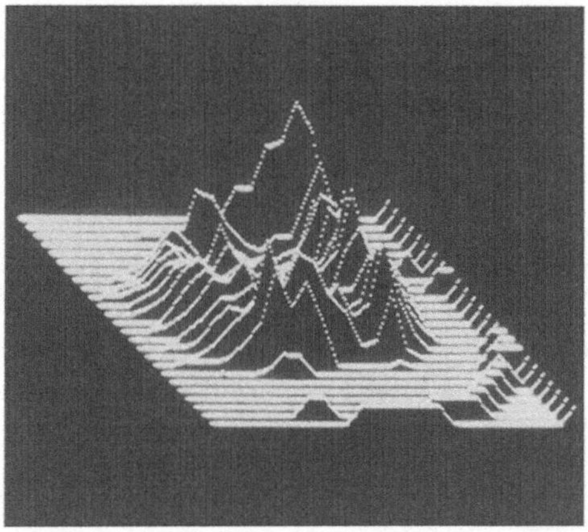

Fig. 4. 3-D-display of the intensity distribution, total power 5 kW

Monitoring by an oscilloscope, fig. 5, uses directly the analog signals of the three detectors. In this case the highest temporal and spatial resolution is possible (10 μs = 0.25 mm).

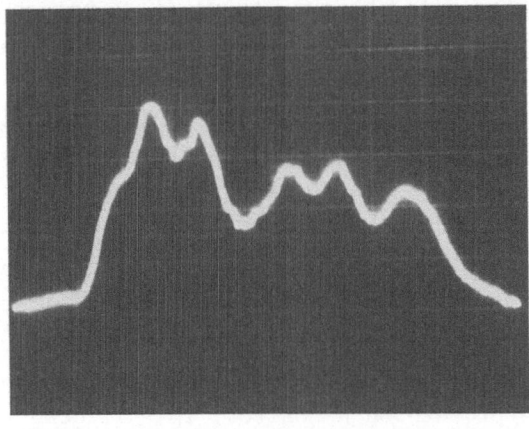

Fig. 5.
Monitoring by oscilloscope central line of a high powe laser beam, total power P = 5.3 kW; horizontal axis diameter; vertical axis: intensity (arbitrary scale)

The printer mode of monito-ring allows documentations of the intensity distributi over the beam cross section The intensity at each point is transformed linearly int numbers in the range 0 (no intensity) to 85 (highest i tensity), analog to the grey scale, fig. 3.

The computer mode allows to transfer all information into an external processor to be stored or to be processed for other applications.

3. Possibilities of Application

By using the laser diagnostic system it can be chosen several modes depending on the application: The grey scale mode is advantageously used to align a laser resonator, because intensity distribution and the total power is indicated in real-time simultaneously. This avoids the prints into acrylics and enables a faster and more precise procedure. Furthermore, by processing the signals it should be possible to do the alignment automatically. The same possibility is given, to control and change the laser beam direction which is indicated by deviations of the evaluated beam center from the center of the monitor.

The 3-D-mode of monitoring can be used advantageously to interprete and evaluate the shape of the beam cross section and - if it is drawn out - for documentation of the power distribution. This is to guarantee and document reproducibility of beam datas while material processing. The printing mode leads to the same support of reproducibility.

The power indication mode can be used to show the total power as digital read-outs, so that power may be controlled visually. This visual control can be supported by acoustical or optical signals if the deviation of the power exceeds a given tolerance (e.g. 5 % or 10 %). If the analogous power signal is red out directly it can be used to regulate the beam power. In this case the laser beam diagnostic system with high frequency response.

All these possibilities should be used today during industrial application of CO_2 lasers in the range of material processing, starting from service into process control.

4. Acknowledgement

The investigations in laser beam diagnostic were sponsored by the Deutsche Forschungsgemeinschaft (DFG). We want to thank the DFG for their financial support.

Ein Gerät zur automatischen Überwachung des Lasermodes bei Leistungslasern während der Materialbearbeitung

Dr. R. Aratari, A.L.L. GmbH, Hans Graessel Weg 1, 8000 München 70 / D

Das Gerät "Laser Beam Monitor" ermöglicht in Zusammenhang mit einem "Laser Beam Analyser" (vgl. "Ein Gerät zur Analyse der Intensitätsverteilung im Laserstrahl während der Materialbearbeitung" von Dr. P. Arnold, LASER 83 OPTOELEKTRONIK) sowohl die Digitalisierung und Speicherung des Modenprofils und den Vergleich des augenblicklichen Profils während einer Materialbearbeitung mit einem früher gespeicherten Profil. Wichtige Werte wie Leistung, Halbwertbreite, Position, usw. des Modes werden fast "realtime" berechnet. Nach Angabe beliebiger Toleranzwerte wird ein Alarm ausgelöst, wenn der aktuelle Mode nicht mit dem gespeicherten Mode innerhalb der Toleranzgrenze übereinstimmt.

Das Gerät ist mit einem Z 80 Rechner ausgerüstet, es kann nach entsprechender Programmierung im ständigen Vergleich der Profile von "Ist" und "Soll" die Laserspiegel für optimalen Mode steuern.

Nontactile Clearance and Seam Tracking Sensors of Power Laser Cutting and Welding Purposes

K. H. Schmall, precitec GmbH & Co KG, Entwicklungs KG, Waldstr. 20,
7570 Baden-Baden / D

Advanced developments of existing capacitive clearance sensors used for years in industrial applications of CO_2 power lasers have been carried out.

Now an appropriate range of integrated nozzle/sensor systems for several focal length is available, offering versatile combinations for machines and robots especially used for thin and heavy sheet metal cutting.

Inductive 3-axis-sensors for seam tracking and handling purposes have also been developed. In laser welding some types may be placed concentrially around the nozzle.

They are qualified for guiding the nozzle along butt joints as well as overlap joints and edges, also for sensor guided positioning of nozzles with high accuracy.

In Process Laser Beam Characterisation

V.M. Weerasinghe and W.M. Steen
Metallurgy Department,
Imperial College, London, SW7 2BP

Introduction

The need for in process monitoring of high powered laser beams in welding, cutting and surface treatment is becoming strongly felt by those working in scientific research or those using lasers in production.

This need is not because lasers are unstable, they are not. Infact they would rank amongst the most stable high energy sources available to industry today. It is because they are being used with a precision which requires a high level of reproducibility. If this level is not achieved then the operator is faced with a mind stretching array of variables which could be the cause of the problem.

A new technique is described in this paper whereby in-process signals can be gained with no additional beam interference. The technique is based on analysing the acoustic signals from the mirrors reflecting high powered laser beams. It is seen that these mirrors are ringing under the photon stress imposed on them.

Acoustic Method:

A new method of beam analysis which has recently been invented (1) measures the acoustic signal which can be detected from the mirrors and other optical elements in the laser cavity or beam guidance train of certain high powered laser systems. It is perhaps surprising to find that mirrors reflecting high intensity radiation are effectively 'ringing' with a detectable signal. The preliminary investigation of these signals, reported here, shows the basic characteristics of the event and so indicates the possible cause of the signal.

The general experimental arrangement is shown in fig. 1. Fig. 2 shows the sort of signal achieved from an instrumented mirror mounted in the beam guidance train. Fig.3 shows the signal from an instrumented mirror in the laser cavity. Figs. 4,8 show the variation of this signal intensity with beam diameter. Fig. 5 shows its variation with laser power. Fig. 6 shows the variation of the signal with the position of the reflected beam on the mirror.

From this data a number of observations can be made:

1. The signal depends on the power density rather than the power.
2. The response time is faster than the chart recorder (.1 s).
3. There is a long frequency oscillation of the signal strength outside the cavity (approx: 36s in Fig. 2) but not inside the cavity.
4. There is some directionality of the signal as it passes through a mirror block made of copper.

5. The signal has a frequency spectrum in the high frequency range of the order of several MHz.
6. The signal strength falls as the system warms up as shown in fig. 2 and the low frequency oscillation of the signal becomes slower. The second break in the record of fig. 2 shows that the signal strength is not altered if the detector only is allowed to cool while the rest of the beam guidance system and the laser remain warm. This indicates that this change in signal is independent of the detector system.

Discussion

The signal poses one or two problems and it is probably simplest to discuss them one at a time.

1. What is the property of the beam which is being measured?

Since the response time is so fast and the signal shows the final value almost instantly the signal is not due to a capacitive or heating effect which might cause, for example, a strain on the piezo electric crystal of the detector due to thermal distortion.

Since the signal is constant over long periods of time, apart from the slow frequency oscillation, it is not due to a single pressure from, say, the photons. A single pressure on the piezo electric crystal would produce a signal which would decay with time through the amplifying circuit.

It must therefore be due to some oscillating signal within the laser beam generating an oscillating photon pressure or photon/phonon oscillation at the mirror surface. The frequency spectrum of the signal shows that it has peaks at very high frequencies of the order of a few MHz. The data shown here is primarily derived from a Control Laser Fast Axial Flow 2kW CO_2 Laser. This machine has a cavity length of around 6m and therefore longitudinal cavity mode oscillations would be separated by a frequency difference, $\Delta\nu$, of:

$$\Delta\nu = c/2l = 25 \text{ MHz}$$

where c = velocity of light and l = cavity length.

This is the expected beat frequency from such a laser. The observed peak frequencies are considerably lower than this. It is hard to conceive of any lower beat frequency being capable of lasing. However it should be noted that the doppler shift due to the fast axial flow gases would, if they lased, be capable of a beat frequency within the observed range. A velocity difference of around 50 m/s would have a possible beat frequency of 5 MHz, for example. But such a beat mechanism is against the requirement that the lasing wavelength fits the cavity an integral number of times for lasing to occur.

A possible alternative could be that there is a fast mode switching occurring in this high powered beam. Preliminary studies on a slow flow TEM00 laser (Ferranti MF400) suggested that there was either no signal or a greatly reduced signal, for the tranverse flow CL5 5kw laser it appeared that a stronger signal was obtained by slightly detuning the cavity. These experiments need to be confirmed but they do indicate that our observed signal may be due to mode switching.

2. What is the cause of the low frequency oscillation?

The low frequency oscillation, seen in fig. 2, is not a quirk of the electronics of the detector. If it were it would show even when the detector is mounted on the laser cavity, which is not so, (fig.3). It is almost certainly due to the laser cavity heating up and the lasing action passing through a changing series of long- itudinal modes quickly to start with and slower as it warms up. The frequency is also quicker when the laser is operating at higher powers.

3. Why should the signal fall with time for constant beam power as shown in fig. 2?

One explanation for this observation is that mirrors 'up-stream' from the detector heat up thus causing the mirrors to expand very slightly and so become convex, this will enlarge the beam further down stream. This enlargement of the beam would cause a reduction in the signal but without there being any reduction in the beam power, as indicated by the signal from the flowing cone calorimeter and the weld penetration experiments. Fig. 4, showing the variation of signal with beam diameter, suggests that over a period of 1.3 hrs the beam diameter might vary by as much as 2 to 3 mm for a 22 mm diameter beam. Such variations have been noted before with the laser beam analyser (LBA).

4. For what applications could this instrument be considered?

The signal depends not only on the incident laser radiation but also on any high powered radiation shining on the optical element. In particular the back reflected radiation as shown in fig. 9. From this figure it is seen that there is the potential here for an in-process monitor of the 'keyhole' stability. If the 'keyhole' fails for any reason then the back reflected signal would rise sharply and the detector would signal such an event. This, in turn, could trigger some message to quality control that some special inspection of that part is required.

In most production applications of the laser a combined signal which indicates that one of the main beam parameters has changed could be extremely useful as a front line warning that some attention to the optics may soon be required - such as the mirrors are becoming dirty.

In scientific research this instrument could be used, in association with other devices, to measure the beam power, or diameter, or level of back reflection or simply some variation in the beam while an experiment was proceeding.

In system design any feed back signal which is consistent and does not interrupt the beam in any way must have a use in developing 'intelligent' laser processing.

Conclusion

1. A new device is illustrated which measures the acoustic signals from optical
    elements.

2. The signal varies with beam power, diameter, position and level of back reflection.
    It also has a characteristic frequency spectrum.

3. The signal is gained without any additional beam interference.

4. The applications of this device are seen to be in the field of in-process beam

sensing for quality control purposes in production; beam data collection in
scientific research and in in-process feed back control in laser system design.

Acknowledgements:

The authors would like to express their thanks to Quantum Laser Corp., 99, Second Ave.,
Long Island, New York, USA for support for this research and their continued interest
in manufacturing this instrument.

References

1. Weerasinghe V.M., Steen W.M. UK patent app 84 12832 Feb. 1984.

VARIATION WITH POWER DENSITY   Fig. 4.

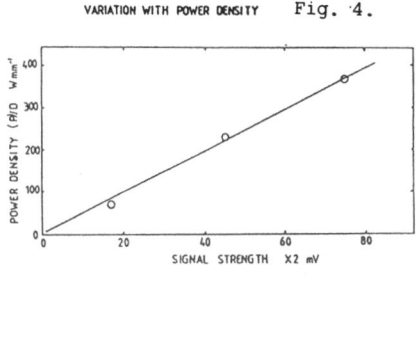

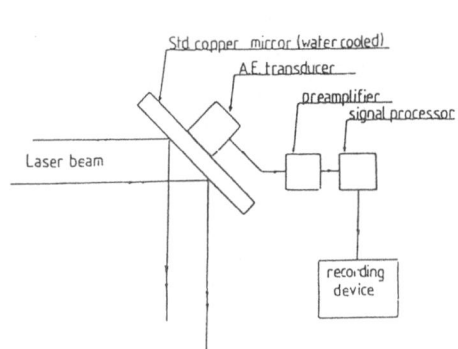

A.E. Monitoring of high power laser beam characteristics   Fig. 1.

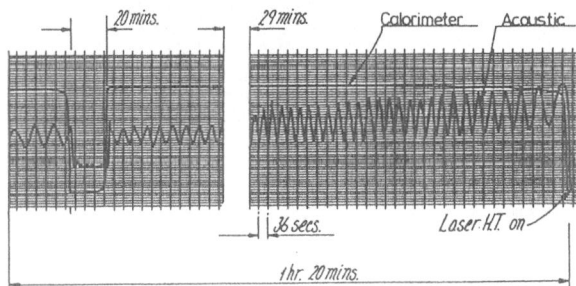

Fig. 2.

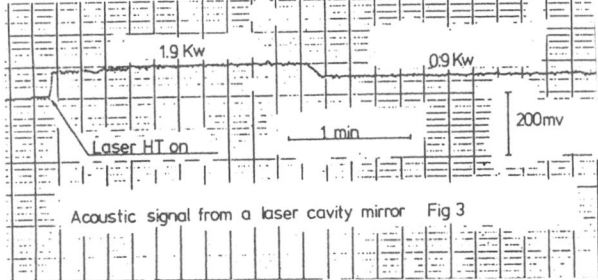

Fig. 3.

340

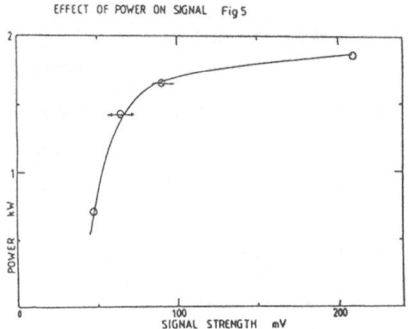

Fig. 5.

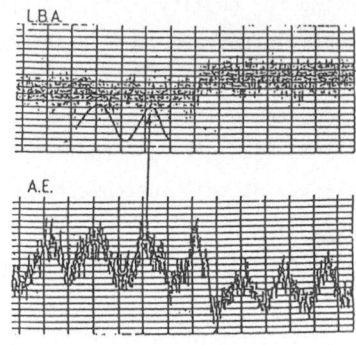

Effect of moving the beam by 2mm.   Fig. 6.

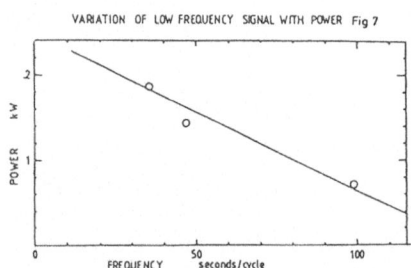

Fig. 7.

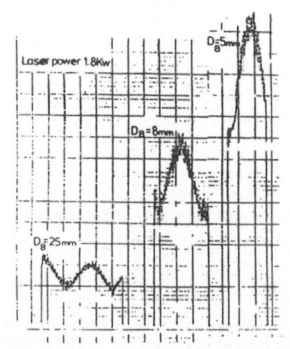

Effect of varying beam diameter   Fig. 8.

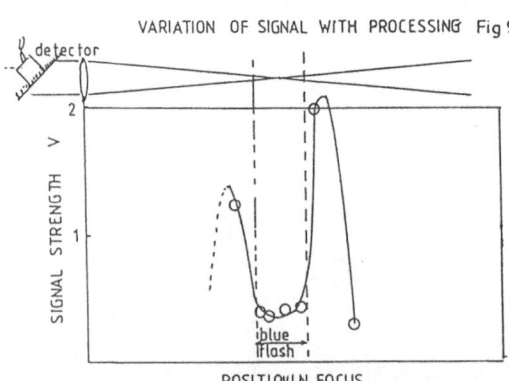

The Method of the Control of Laser Absorbing Films Processing

L.P.Boruc

Institute of Precision and Electronic Instruments Engineering,
Warsaw Technical University
Ul. Chodkiewicza 8, PL 02-525 Warszawa

I. Introduction

The technological optimization of laser machining of the multi-layer structures require the complete understanding of the phenomena accompanying laser irradiation on matter [1]. This study concern effects appearing during the process of removing of absorbing films deposited on a substrate. The laser intensity is sufficient for evaporation of material, but it does not exceed the threshold intensity, which causes a laserinduced gas breakdown [2].

Presented idea of coupling the theoretical model with the corresponding experiment is suitable for investigation of the real process, because experimental insight does not provide full knowledge from a short time and a small area of the occurence of the process [4].

II. Model of the process of the absorbing films removing

In order to describe the process of thermal conductivity the one-dimension equation of the heat flow was applied:

$$\frac{\partial T}{\partial t} = \frac{\partial}{\partial z}\left(\frac{K(T)}{\rho(T)\ c(T)}\ \frac{\partial T}{\partial z}\right) + \frac{1}{\rho(T)\ c(T)}\ S_T$$

The source term S_T is found to be:

$$S_T = \frac{4E}{\pi\ t_p\ r_F^2}\ \tau(t)\ \exp\left[-2(r/r_F)^2\right]\ A\ \alpha\ \exp(-\alpha z)$$

where r_F, t_p, E, $\tau(t)$, A and α denote the Gaussian beam radius, the pulse duration, the incident laser pulse energy, the temporal profile of the pulse, the absorption coefficient and the attenuation coefficient, respectively. Nonlinear dependence of the physical properties of the material, namely the specific heat $c(T)$, mass density $\rho(T)$ and thermal conductivity $K(T)$ on temperature, was used in this formulation. In further considerations melting and vapourization equations [8,9] were also applied. Due to the associated pressure $p(t)$ the liquid target material is pressed out of the processing zone [3] at velocity W (Fig. 1), therefore the effective depth of removal is deepened by h_W (Fig. 2).

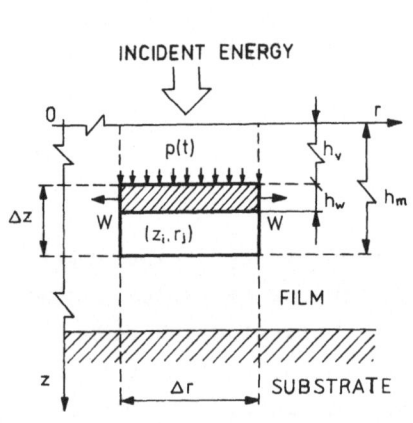

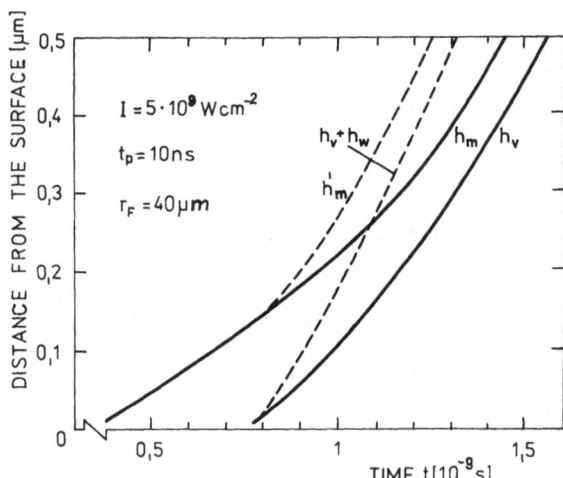

Fig.1. The geometry of the element (z_i, r_j) under conditions of an explosive removing of the molten material

Fig.2. Removing depth h_v and the melting depth h_m as a function of time for vapourization exclusively and for additionally accompanying explosive removing of the molten material $h_v + h_w$ and h'_m respectively

To determine W the motion equation of a flat boundary layer [10] has been solved [4] and has been found to be:

$$ W = \frac{\varepsilon}{2\rho_m \xi} \left[\delta - \left(\delta^2 - 2\xi^2 \rho_m \, p(t)/\varphi \right)^{\frac{1}{2}} \right] $$

where:

$$ \varepsilon = 1 + \beta(h_m - h_v)/3\mu_d; \quad \psi = \beta(h_m - h_v)/2\mu_d $$
$$ \varphi = 1 + \psi + 2\psi^2/5; \quad \delta = \rho_m B \nu \beta; \quad \xi = 2\varphi \mu_d (h_m - h_v) $$

Here ρ_m is the averaged density of melting film, β is the coefficient of friction, proportional to adhesive forces of liquid layer to solid phase, μ_d is the absolute viscosity, ν is the kinematic viscosity and B is the averaged flow length.

The above model has been taken as the basis for numerical computations of the complete removal of the film by means of the finite differences method with special improvements [5].

III. An example of the process control

The aluminium film of thickness H = 0,2÷1,0 μm deposited on a glass substrate has been analyzed, and the experiments were carried out using Nd: YAG laser $\left(\lambda = 1,06 \text{ μm, } t_p = 5\div25 \text{ ns, laser intensity}\right.$ $\left. I = 0,25\div100\cdot10^9 \text{ W/cm}^2\right)[6,7]$.

The absorbtivity coefficient A in the examined range of I, cannot be accepted as a constant value independently of the laser pulse parameters. The good approximation of A arising from macroscopic considerations, can be described in terms of the formula:

$$A = a + b I t_p^2$$

where in the case of Al it was found out that a = 0,095 and b = 8 10^{-5} $\left(\text{Fig. 3}\right)$. For this describing reflectivity formula, the computation results are in close correspondence with the experimental findings.

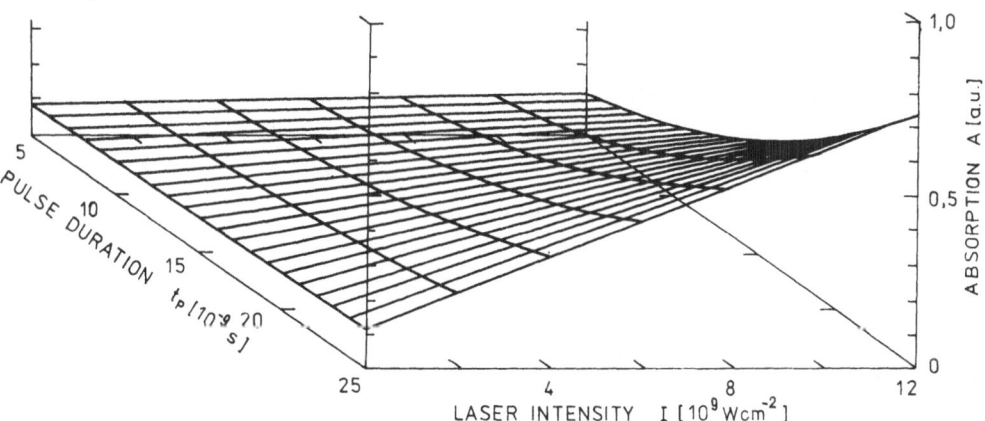

Fig.3. Absorption coefficient A of Al film as a function of the laser intensity I and the pulse duration t_p

The simulation becomes a reliable method for the study of the film removal process and allow for its control.

As an example of the achieved results is shown the effect of the dynamic removal of the molten film and its significant contribution to the process $\left(\text{Fig. 4}\right)$. For a typical laser pulse $\left(\text{i.e. } I=5\cdot10^9 \text{ W/cm}^2,\right.$ t_p = 10 ns $)$ about 70% of H removed in the z axis is due to dynamic removal. For $r/r_F > 0,7$, for the reason of absence of the resultant pressure $p\left(t\right)$, the removal results are achieved only from vapourization.

At time t_{bm} $\left(\text{Fig. 5}\right)$ melting of the incident Al surface begins and at t_{bv} vapourization occurs additionally, ending at time t_{ev}. For $r/r_F > 0,90$ evaporation has not taken place. Finally, the dynamic removal takes place between time t_{bw} and t_{ew}.

344

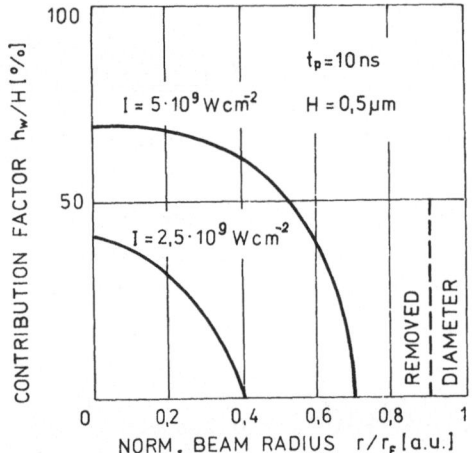

Fig.4. Calculated estimation of
the contribution factor of the
depth h_w gained exlusively by
liquid expulsion, to thickness
H of removed film, as a function
of normalized beam radius

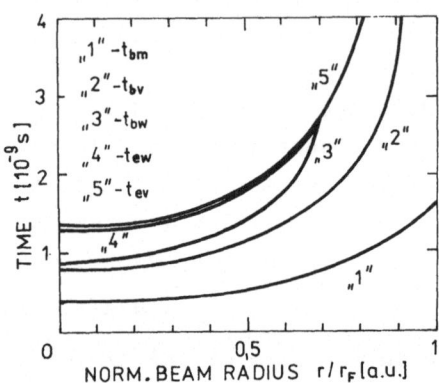

Fig.5. The time dependence at the
thermal processes vs normalized
beam radius

The effect of dynamic removal has been found to have no influence
on the extension of the removed diameter, however, it is contributive
to an earlier disclosure of the substrate and as a result, a direct
exposure to the incident laser energy.

In the region of removal by evaporation only $(0,7 < r/r_F < 0,9)$,
there remains a film thickness of about 20 nm on the substrate which
is not able to absorb sufficient energy to evaporate.

IV. Conclusion

The obtained knowledge concerning the quantitative aspects of the
process, allows to carry out research into the control and optimization
of the films removal by means of the laser beam. The procedure describe
above found applications in studies of technological processes such as
hybrid resistors trimming, capacitors trimming and others.

References
1 Drozd Z. and Boruc L., Proc. 6th Congr. "Laser 83", p.410.
   Springer-Verlag, Berlin (1984).
2 Poprawe R. et al., ibid., p. 361.
3 Treusch H.-G. et al., ibid., p. 383
4 Boruc L. Dissertation, Warsaw Technical University (1984).
5 Boruc L. and Wawrzyniak Z., Proc. AMSE Conf. on Modelling and
   Simulation, Athens, Ga. Vol. 3.4, p. 297 (1984).
6 Boruc L. and Wawrzyniak Z., Infrared Phys. 25, 145 (1985).
7 Boruc L. and Wawrzyniak Z., Proc. CIRP3 Conf. on Infrared Phys.,
   p. 509 Zurich, Switzerland (1984).
8 Baeri P. et al., J.appl. Phys. 50, 788 (1979).
9 Jain A.K. et al., Appl. Phys. 25, 127 (1981).
10 Veiko V.P. et al., Kvantovaya Elektron. 7, 34 (1980).

Physikalisches Modell des Laserschneidvorgangs

D.Becker, W.Schulz, G.Simon, H.M.Urbassek, M.Vicanek
Institut für Theoretische Physik
I.Decker, Institut für Schweißtechnik
Technische Universität, D-3300 Braunschweig

Beim Laserschneidvorgang sind verschiedene physikalische Vorgänge miteinander gekoppelt /1,2/. Hier sollen beispielhaft die Teilprozesse Strahlungsabsorption und Schmelzaustrieb näher untersucht werden. Um handhabbare mathematische Formulierungen zu finden, werden jeweils vereinfachte geometrische Verhältnisse betrachtet.

Zur Berechnung der Absorption beschränken wir uns auf ein zweidimensionales Modell (s. x,z-Ebene in Fig.1), bei dem die Schneidfront durch eine Funktion $z=z(x)$ angegeben werden kann. Es wird ferner angenommen, daß die Schmelze völlig ausgeblasen wird und dadurch die Oberflächentemperaturen für eine Plasmabildung nicht ausreichen. Aus der Energiestrombilanz kann man für den stationären Schneidprozeß die folgende Bestimmungsgleichung für $z(x)$ formulieren:

(1) $z'(x) \cdot (\underline{w}_x - S/A) - \underline{w}_z = 0$,

wobei $\underline{w} \cdot = \underline{w}(x,z)$ die vorgegebene Intensitätsverteilung des Laser-

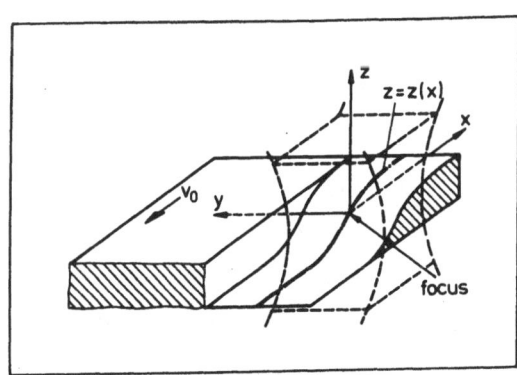

Figur 1. Die vereinfachte Schneidgeometrie

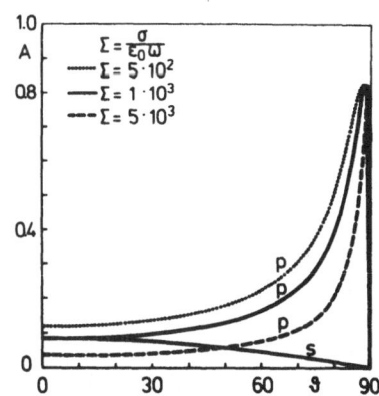

Figur 2. Der Fresnelsche Absorptionskoeffizient A als

Funktion des Einfallwinkels ϑ und der Leit-

fähigkeit σ für s- sowie p-Polarsation

strahles, A der Absorptionskoeffizient für p- bzw. s-Polarisation bezüglich der Metalloberfläche und S die zum Materialabtrag not-

346

wendige Schneidenergiestromdichte sind. Die Energiestromdichte S
setzt sich hier aus dem zum Aufschmelzen notwendigen Energie-
strom $\rho \cdot H \cdot v_0$ und der x-Komponente des Wärmestroms q zusammen.

Im Fall der Fresnelschen Absorption ist A vom Einfallwinkel des
Lichtes auf die Schneidfront abhängig, d.h. von der Steigung z'(x)
und der Richtung von $\underline{W}$. Gleichung (1) ist damit eine nichtlineare,
implizite Dgl. und kann nur in Sonderfällen analytisch gelöst werden. Für eine parallel zur Schneidrichtung polarisierte Welle ist
die Absorption stark winkelabhängig mit einem ausgeprägten Maximum
bei fast streifendem Einfall (Fig.2). Daher ist das Modell eine gute
Approximation für das Schneiden mit einem p-polarisierten Strahl.
Der Wirkungsgrad für die Energieeinkopplung η_A ist definiert als
das Verhältnis von absorbierter zu eingestrahlter Laserleistung. Die
wichtigsten, auf numerischem Wege gefundenen Ergebnisse der Lösung
von (1) sind: Infolge der stark winkelabhängigen Absorption zeigt
η_A ein Schwellenverhalten als Funktion des Parameters W_0/S
(W_0: Fokusintensität). Bei höheren Laserleistungen erzielt man mit
einem höheren Lasermode einen größeren Wirkungsgrad als mit dem
Grundmode (s.Fig.3). Die erreichbare Schneidtiefe variiert mit der
Fokuslage und besitzt ein ausgeprägtes Maximum. Der Wirkungsgrad
nimmt mit wachsender Strahldivergenz ab.

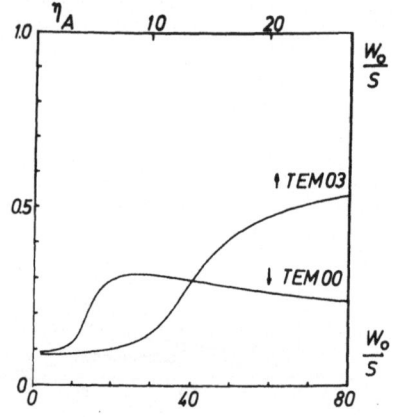

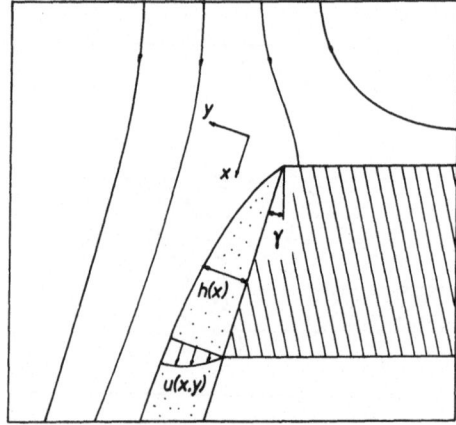

Figur 3. Der Wirkungsgrad für die Energieeinkopplung
als Funktion der Fokusintensität für den
TEM00 und TEM03 Mode.

Figur 4. Vereinfachte Schneidfrontgeometrie
zur Berechnung des Schmelzaustriebs

Die Schmelze wird durch einen Gasstrahl aus der Schnittfuge ausgetrieben. Die antreibenden Kräfte sind der Druckgradient $p_x = \partial p/\partial x$
des Gasstrahls längs der Schneidfront und die durch die Grenzschicht
des Gasstrahls erzeugte Scherkraft τ (s.Fig.4). In dem zweidimensionalen Modell konnte der Gasstrahl mit der Methode der freien
Stromlinien über mehrere konforme Abbildungen und eine numerische
Integration im Komplexen auf eine Parallelströmung abgebildet und
damit auf bekannte Strömungsverhältnisse zurückgeführt werden. In
Fig.5 sind die auf diese Weise berechneten Antriebskräfte $-p_x$ und τ
dargestellt. Die Schmelzströmung wurde mit Hilfe der Navier-Stokes-
schen Gleichungen im Grenzfall schleichender Strömung behandelt. Das
Ergebnis für das tangentiale Strömungsprofil und die Schmelzfilmdicke ist mit den in Fig.4 eingeführten Koordinaten:

$$(2) \quad u(x,y) = v_o \sin\gamma + \frac{1}{\eta}\tau y - \frac{1}{\eta}p_x\left(hy - \frac{1}{2}y^2\right) \quad ,$$

wobei $h = h(x)$ aus $(h/h_p)^3 + (h/h_\tau)^2 = 1$ zu bestimmen ist mit

$$h_\tau^2 = 2\eta v_o \cos(\gamma)x/\tau \quad , \quad h_p^3 = -3\eta v_o \cos(\gamma)x/p_x \quad .$$

Dabei sind v_o die Vorschubgeschwindigkeit und η die Viskosität der
Schmelze. Setzt man für das Laserschneiden typische Zahlenwerte ein,
so ergeben sich Schichtdicken in der Größenordnung von 10^{-5}m und
Ausströmungsgeschwindigkeiten im Bereich von $1\ ms^{-1}$. Angenommen
wurde dabei , daß die Schmelze frei ausströmen kann und nicht unter
dem Einfluß der Oberflächenspannung am unteren Rand haftet.

Beim Schneiden von Metallen entsteht auf den Schnittflächen eine
unerwünschte Riefenstruktur, deren Ursache noch nicht geklärt ist.
Dazu wurde im Rahmen der Orr-Sommerfeld-Theorie die differentielle
Stabilität der Schmelzströmung untersucht, indem die zeitabhängigen
Bewegungsgleichungen linearisiert und Lösungen der Form $\exp(ikx+\lambda t)$
mit einer reellen Wellenzahl k untersucht wurden. Für die Stromfunk-
tion $\Psi(y)$ ergibt sich in unserem Fall eine etwas modifizierte
Orr-Sommerfeld-Gleichung (in dimensionslosen Größen):

$$(3) \quad \Psi''' - \lambda\Psi' - ik\left(u(y)\Psi'-u'(y)\Psi\right) = \frac{k^4}{\lambda+ik}\,\Gamma\cdot\Psi(y=1)$$

mit den Randbedingungen $y=0: \quad \Psi = \Psi' = 0$

$$y=1: \quad (\lambda+ik)\Psi'' - iku''\Psi = 0 \quad .$$

Die Lösungen hängen neben dem Grundströmungsprofil $u(y)$ von dem
Parameter

$$(4) \quad \Gamma = \frac{\sigma_o \cdot \eta^2}{\rho\,h^3\,u_o^4}$$

348

(σ_o = Oberflächenspannung, u_o = Grundströmungsgeschwindigkeit an der
Oberfläche) ab. Die Integration von (4) führt auf eine komplexe
Frequenz λ . Das System ist instabil, wenn es Werte von k gibt, für
die λ einen positiven Realteil hat. Die numerische Lösung ergibt,
daß die rein druckgradientenkontrollierte Strömung (τ =0) immer
instabil, die rein scherkraftkontrollierte Strömung (p_x=0) dagegen
stabil ist. In Fig.6 ist Re(λ) in Abhängigkeit von k dargestellt.

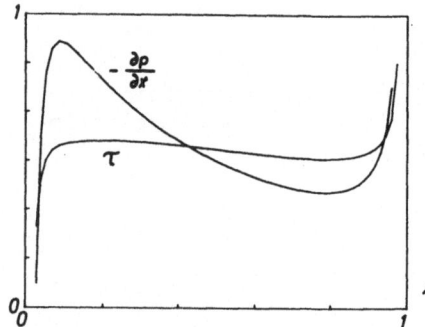

Figur 5. Verlauf der Scherkraft und des Druck-
gradienten längs der Schneidfront

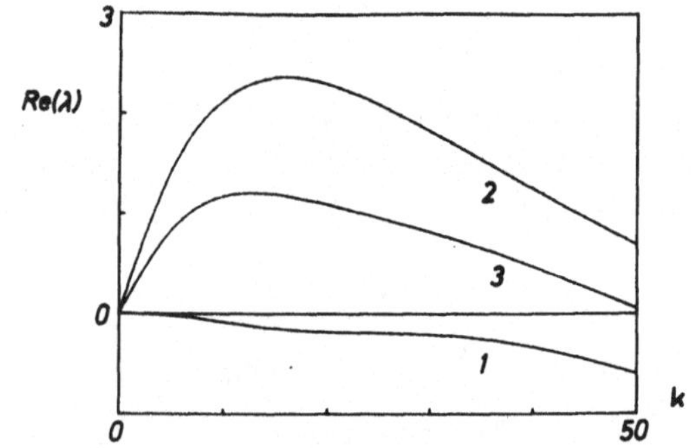

Figur 6. Instabilität der Schmelze. 1: scherkraftkontrolliert, 2: druckgradientenkontrolliert. Kurve 3
ergibt sich unter Berücksichtigung beider Kräfte.

Die mittlere Kurve entspricht repräsentativen Werten von p_x und τ.
Zu dem maximalen Wert von Re(λ) korrespondiert eine Frequenz, die zu
einem Riefenabstand der Größenordnung von 30 µm führt. Um die
Riefenbildung gering zu halten, sollte nach diesem Modell die
Schmelzströmung möglichst scherkraftkontrolliert sein. Das ist der
Fall, wenn folgende Bedingung erfüllt ist:

(5) $v_o^2 d/U \ll \eta_g^3/\rho_g n^2$.

Dabei sind U die Anblasgeschwindigkeit, η_g und ρ_g die Zähigkeit
und Dichte des Gases. Die Berechnungen erklären die Riefenbildung im
oberen Teil der Schnittfläche. Im unteren Teil dagegen werden
größere Riefen beobachtet, die noch durch andere Effekte beeinflußt
sind.

<u>Literatur</u>

/1/ I.DECKER,J.RUGE,U.ATZERT, Physical Models and Technological
    Aspects of Laser Gas Cutting. SPIE Vol.455, 1984, 81-87
/2/ D.SCHUÖCKER,B.WALTER, Inst.Phys.Conf.Ser.No.72,5th GCL
    Symp.,Oxford 20. 24.August 1984,111-116

Untersuchungen zur Wechselwirkung Laser – Werkstück

W.L. Bohn, A. Giesen, R. Nowack, M. Schellhorn

DFVLR - Institut für Technische Physik

Pfaffenwaldring 38-40, 7000 Stuttgart 80

1. Einleitung

Die Bedeutung der Wechselwirkungsphänomene für die Qualität und Effizienz von Laser-
fertigungsverfahren ist allgemein anerkannt. Trotz zahlreicher Arbeiten auf diesem
Gebiet ist es bislang nicht gelungen, diese Phänomene hinreichend zu verstehen und
quantitativ zu beschreiben, so daß eine verläßliche Vorhersage für das jeweilige Bearbeitungsverfahren nicht möglich ist. Von besonderem Interesse ist dabei die Interpretation der Leuchterscheinung an der Oberfläche des Werkstücks, die sehr häufig
und kontrovers diskutiert wird. Eine Korrelation mit den Laserbetriebsparametern erscheint hier erforderlich, weil Rückwirkungen vom Werkstück bzw. von der Wechselwirkungszone in den Laser nicht ausgeschlossen werden können und experimentell bereits
hinreichend belegt sind. Die vorliegende Arbeit versucht über eine Reihe optischer
diagnostischer Methoden einen Beitrag zum Verständnis der Laserwechselwirkung zu
liefern.

Alle Messungen sind mit einem kommerziellen 1000 Watt CO_2-Laser durchgeführt worden,
der auch über eine Pulsoption verfügt. Als Werkstücke sind Baustahl (St 37) und Aluminium von 2 mm Dicke benutzt worden. Die Bestrahlung der Werkstückoberflächen ist
sowohl senkrecht als auch unter einem Einfallswinkel von 45° durchgeführt worden.
Diese unübliche Anordnung hat den Vorteil, den einfallenden Laserstrahl vom senkrecht zur Metalloberfläche abströmenden Metalldampf praktisch zu entkoppeln.

2. Spektroskopie

Mit einem optischen Vielkanalanalysator sind zeitintegrierte Emissionsspektren der
Wechselwirkungszone mit Luft als Umgebungsgas aufgenommen worden (Plangitterspektro-
graph f = 2078 nm und 7 Å/mm Dispersion). Ein für Baustahl St 37 typisches Spektrum
ist in Bild 1 aufgezeichnet. In allen untersuchten Fällen konnten neben intensitäts-
starken FeI-Linien nur sieben schwache FeII-Linien identifiziert werden. Eine Auswertung dieser Spektren mit Hilfe des üblichen Boltzmann-Diagramms führt zu einer
Temperaturabschätzung zwischen 5000 und 10000 K. Unter der Annahme von Gleichgewichts-
verhältnissen und nach Auswertung der Saha-Gleichung ergibt sich der in Bild 2 dargestellte Zusammenhang zwischen Elektronendichte und Temperatur für verschiedene Drücke
des Gesamtgases. Ausgehend von den oben angeführten Temperaturen liegt der Grenzbe-

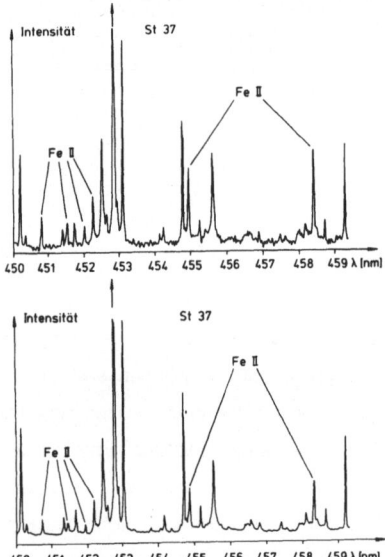

Bild 1. Zeitlich integrierte
Spektralmessungen

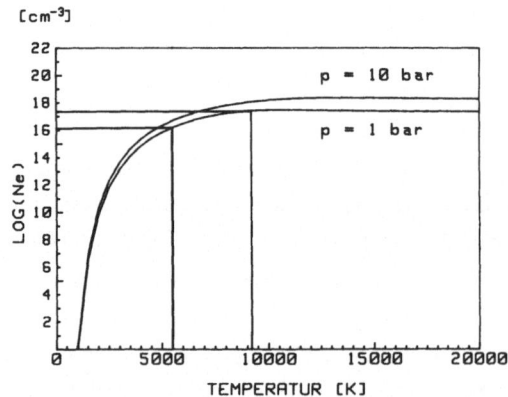

Bild 2. Berechnung der Elektronendichte
im Gleichgewicht (Saha)

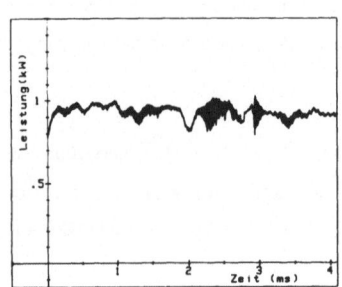

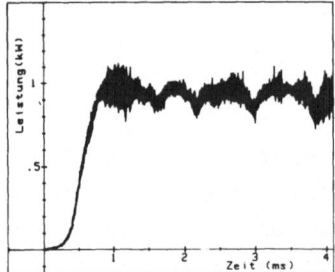

Bild 3. Fluktuationen des
Laserlichts im Frei-
lauf (oben) und mit
Werkstück (unten)

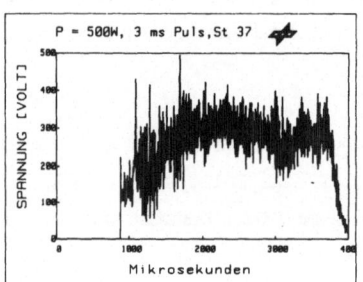

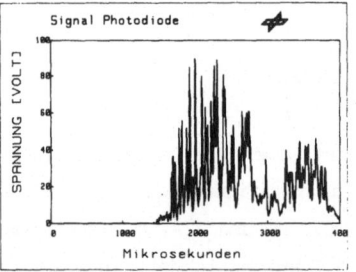

Bild 4. Fluktuationen des Laserlichts (oben) und des
Gesamtlichts (unten)

reich für die Elektronendichte zwischen $10^{16} cm^{-3}$ und $10^{17} cm^{-3}$. Aus der Linienhalb-
wertsbreite von AlI folgt ebenfalls eine obere Grenze von $10^{17} cm^{-3}$. Somit liegt in
den Fällen höchster Laserstrahleinkopplung (ca. 10^6 W/cm²) ein Metalldampf mit einem
Ionisationsgrad < 1 % vor (blaue Leuchterscheinung wie unter Punkt 4 beschrieben).
Diese Aussage bezieht sich allerdings nur auf das räumlich und zeitlich integrierte
Verhalten der Wechselwirkungszone, da Messungen mit entsprechender Auflösung noch
ausstehen.

3. On-line Messung der Laserstrahlung

Die am Endspiegel des Lasers transmittierte Strahlung (0,5 %) wurde mit einem HgCdTe-
Detektor (10 MHz Bandbreite) aufgenommen. Bild 3 zeigt das Meßergebnis bei freilau-
fendem Laser (oben) sowie bei Bestrahlung von Stahl St 37. Im Leerlaufbetrieb weist
der Laser eine Intensitätsschwankung im kHz-Bereich auf, die von der Energieversor-
gung herrührt und eine um 200 kHz liegende Intensitätsfluktuation, deren Ursprung
noch nicht geklärt ist. Bei Werkstückbestrahlung wird durch optische Rückwirkungen
der hochfrequente Anteil deutlich erhöht. Bei Aluminium wird dieses Verhalten noch
verstärkt. In Bild 4 ist zusätzlich zur Laserintensität die Gesamtlichtstrahlung aus
der Wechselwirkungszone mit Hilfe einer Photodiode gleichzeitig aufgenommen worden.
In diesem Fall handelt es sich um einen 3 msec-Puls. Ein Verzug zwischen Beginn des
Laserpulses und Einsetzen der Gesamtlichtabstrahlung wird festgestellt. Die Anlauf-
bzw. Abklingphase des Diodensignals werden dem Leuchten der Schmelze, die starken
Fluktuationen dem wechselnden Aufleuchten des schwach ionisierten Metalldampfes zugeordnet. Eine direkte Korrelation zwischen Laser- und Gesamtlichtfluktuationen konnte
nicht hergestellt werden.

4. Hochgeschwindigkeitsphotographie

Die Wechselwirkungszone ist mit 10000 Bildern/sec gefilmt worden: Eine intensive blaue
Leuchterscheinung (siehe Spektroskopie) tritt in unregelmäßigen Abständen auf und
zeigt in der Reihenfolge der Schutzgase Argon, Stickstoff, Helium eine immer geringere
Ausdehnung von der Oberfläche. Wir führen das einerseits auf die entsprechenden Unterschiede in der Wärmeleitfähigkeit und andererseits auf die verschiedenen Expansions-
geschwindigkeiten des heißen Metalldampfes in ein Umgebungsgas mit unterschiedlichem
Molekulargewicht (40, 28, 4) zurück. Aus der Analyse der Bildsequenzen mit senkrechtem und 45°-Lasereinfallswinkel ergibt sich, unter Berücksichtigung der Ergebnisse
aus den Abschnitten 2 und 3, folgende Interpretation:

- Bei geringen Leistungsdichten erzeugt die Laserstrahlung einen schwach (rot/gelb)
  leuchtenden Metalldampf, der immer senkrecht zur Werkstückoberfläche abströmt.

- Erhöht sich die Laserintensität auf ca. 10^6 W/cm², so findet eine verstärkte Absorption dicht über der Werkstoffoberfläche im Metalldampf statt (ca. 10 % pro mm), der blau leuchtet und zumindest teilweise ionisiert ist. Dieser stark angeregte Dampf strömt ebenfalls senkrecht zur Oberfläche ab.

- Bei senkrechtem Einfall und hoher Laserintensität kommt es hin und wieder zur Ablösung des Metalldampfes von der Oberfläche.

- Bei schrägem Einfall strömt der angeregte Metalldampf vorwiegend senkrecht von der Oberfläche ab. Dabei wird eine Ausbuchtung der blauen Leuchterscheinung in Richtung des einfallenden Laserstrahls (dicht oberhalb der Oberfläche) beobachtet. In sehr seltenen Fällen löst sich die Leuchterscheinung ab und bewegt sich in Richtung des einfallenden Laserlichtes. Möglicherweise handelt es sich hierbei um das Einsetzen einer LSC-Welle.

- Die Abströmungsgeschwindigkeit des Metalldampfes liegt oberhalb 100 m/sec.

Die Dynamik der Schmelze im Puls- und Dauerstrich-Betrieb sowie die Bildung von Metalltröpfchen, die mit steigender Laserintensität zunimmt, werden anhand des 16 mm-Films aufgezeigt.

Einfluß der optischen Rückwirkung auf die Lage der Polarisationsebene beim Laserschweißen und -schneiden

P. Loosen 1), E. Beyer 1), R. Küchler 2), R. Kramer 1)

1) Fraunhofer-Institut für Lasertechnik
   Drosselweg 87, D-5100 Aachen
2) VDI-Technologiezentrum
   Graf-Recke-Str. 84, D-4000 Düsseldorf

1. Einleitung

Im Vergleich zu herkömmlichen Werkzeugen weisen Laser in der Werkstoffbearbeitung eine wichtige Besonderheit auf: Die Strahleigenschaften wie Leistung, longitudinale Modenstruktur (Frequenz), transversale Modenstruktur und Polarisationszustand sind während der Bearbeitung stochastischen Änderungen unterworfen. Dies wird in Bild 1 erläutert. Das System aus Laserstrahlquelle, Bearbeitungsoptik und Target läßt sich zurückführen auf ein System zweier gekoppelter optischer Resonatoren. Durch die vom Target in den Laserresonator zurückreflektierte Laserleistung ergibt sich eine Verkopplung der in Bild 2 dargestellten Prozesse im Laser und in der Bearbeitungszone. Die Strahleigenschaften sind nicht mehr allein durch den Laserresonator definiert, sondern starken Änderungen aufgrund der dynamischen Vorgänge in der Bearbeitungszone unterworfen. Die ersten Arbeiten zur optischen Rückwirkung, die prinzipiell bei allen Lasersystemen auftritt, finden sich für Nd-YAG und He-Ne-Laser in /1-4/. Stochastische Änderungen der Laserleistung bei Hochleistungs-CO_2-Lasern während der Werkstoffbearbeitung werden in /5-7/ beschrieben.

2. Beeinflussung des Bearbeitungsergebnisses durch die Lage der Polarisationsebene

Neben der Laserleistung ist der Polarisationszustand ein wichtiger Strahlparameter bei der Werkstoffbearbeitung mit Laserstrahlung. Je nach Winkel zwischen Polarisationsebene und Vorschubrichtung (φ) ergeben sich z.T. stark unterschiedliche Bearbeitungsergebnisse.

Dies wird zunächst am Beispiel des Laserstrahlschweißens erläutert. Laserstrahlschweißen erfolgt vorwiegend mit Multi-KW-CO_2-Lasern, die statistisch polarisiert (d.h. die Lage der Polarisationsebene wird nicht durch polarisierende Elemente im Resonator festgelegt) oder die durch Resonator-Faltungsspiegel linear polarisiert sind. Bild 3 zeigt für einen solchen linear polarisierten Schweißlaser die Einschweißtiefe als Funktion der Vorschubgeschwindigkeit bei S-Polarisation ($\varphi = 90^O$) und P-Polarisation ($\varphi = 0^O$) /8/. Während bei kleinen Schweißgeschwindigkeiten der Unterschied zwischen den beiden Polarisationsrichtungen gering ist, zeigt sich bei höheren Geschwindigkeiten ein charakteristischer Unterschied. Da die Laserstrahlung mit S-Polarisation eine geringere Absorption in der Werkstückgeometrie aufweist als die Laserstrahlung mit P-Polarisation, wird die Schwelle für die Ausbildung des laserinduzierten Plasmas /9/ früher unterschritten; damit nimmt der Bearbeitungswirkungsgrad und die Einschweißtiefe um mehr als Faktor 3 ab.

Der Einfluß der Polarisation auf den Laserschneidprozeß ist seit langem bekannt /10/. Bestimmender Faktor für das Schneidergebnis ist auch hier der Winkel γ zwischen Polarisationsebene und Vorschubrichtung: bei $\gamma = 0^o$ ist die Vorschubgeschwindigkeit beim Schneiden metallischer Werkstoffe um bis zu 50% größer als bei $\gamma = 90^o$; bei $0^o < \gamma < 90^o$ ist die Schnittfuge nicht normal zur Metalloberfläche. Die maximale Abweichung wird bei $\gamma = 45^o$ beobachtet. Richtungsunabhängige Ergebnisse liefert die Bearbeitung mit zirkular polarisierter Laserstrahlung. Neben diesen Lasern werden ebenfalls statistisch polarisierte Laser zum Schneiden eingesetzt.

3. On-Line Diagnostik des Polarisationszustandes

Aus dem bisher Gesagten wird deutlich, daß zur Kontrolle und Optimierung des Bearbeitungsergebnisses die Kontrolle des Laser- Polarisationszustandes einen wichtigen Beitrag liefert. Da der Polarisationszustand wie auch alle anderen Strahleigenschaften während der Bearbeitung stochastischen Änderungen unterworfen ist, wird die Messung der Polarisation On-Line, d.h. während der Bearbeitung vorgenommen.

Bild 4 zeigt schematisch den Meßaufbau. Als Beispiel ist die On-Line Diagnostik an einem Laser dargestellt, der durch einen 90°-Faltungsspiegel linear polarisiert ist. Am Endspiegel, der eine Transmission von 1% aufweist, wird ein kleiner Teil der Laserstrahlung ausgekoppelt und mit zwei schnellen pyroelektrischen Detektoren (Bandbreite 2 MHz) sowohl die Gesamtleistung als auch die Gesamtleistung hinter einem Polarisationsanalysator (Brewster-Platten Analysator) zeitaufgelöst gemessen. In Durchlaßrichtung des Polarisationsanalysators sind beide Detektoren auf gleichen Signalpegel kalibriert. Bedingt durch den Aufbau weist der Analysator in Sperrichtung eine Resttransmission von ca. 10% der Transmission in Durchlaßrichtung auf. In diesem Aufbau ist es möglich, die Laserleistung und den Polarisationszustand während des Bearbeitungsvorgangs (hier Schneiden und Schweißen) zu erfassen.

Im freilaufenden Laserstrahl, d.h. ohne Bearbeitung, ergibt sich für den linear polarisierten Laser der in Bild 5 dargestellte Fall. Das Bild zeigt den zeitlichen Verlauf der Leistung (obere Spur, Detektor 1) und der Leistung hinter dem in Sperrichtung justierten Polarisationsanalysator (untere Spur, Detektor 2). Die Leistung weist eine nur geringe Modulation von etwa 10% auf, die durch

Instabilitäten von Entladung, Gasströmung und Restwelligkeit des Entladungsstroms verursacht wird. Durch die Resttransmission des in Sperrichtung justierten Polarisationsanalysators zeigt auch Detektor 2 ein allerdings nur kleines Signal.

Während der Bearbeitung ergeben sich zum Teil erhebliche Abweichungen von dem Fall des freilaufenden Laserstrahls. Dazu zeigt zunächst für eine Laserschweißanwendung mit einem linear polarisierten Laser Bild 6 den zeitlichen Verlauf der Leistung, gemessen mit den Detektoren D1 und D2. Wie in vorangegangenen Veröffentlichungen schon dargestellt /5,6/, tritt bei zeitaufgelöster Messung der Gesamtleistung während der Bearbeitung eine im allgemeinen starke Fluktuation der Laserleistung auf. Neben der Laserleistung ist jedoch auch der Polarisationszustand starken Schwankungen unterworfen. Teilweise steigt das Signal des Detektors 2 bis auf das Signal des Detektors 1 an, was bedeutet, daß der Laser durch die optische Rückwirkung für kurze Zeiten in einen Polarisationszustand übergeht, dessen Ebene um 90° in die Durchlaßrichtung des Analysators gedreht ist. Lediglich im zeitlichen Mittel zeigt sich noch eine Bevorzugung der Polarisationsrichtung, die durch die Faltung des Resonators ausgezeichnet ist. In diesem Sinne der zeitlich gemittelten Aussage über den Polarisationszustand sind die in Bild 3 eingezeichneten Kurven über S- und P-Polarisation zu verstehen. Der Fall der zeitlich stabilen S- oder P-Polarisation der Laserstrahlung ist im allgemeinen aufgrund von Störungen durch die optische Rückwirkung nicht realisierbar.

Beim Laserschneiden muß zwischen dem Rückwirkungsverhalten bei statistisch polarisierter und zirkular polarisierter Lasersystrahlung unterschieden werden. Bild 7 zeigt ein Meßergebnis für das Schneiden von Stahl mit Laserstrahlung bei statistisch polarisiertem Laser, in dem die Beeinflussung der Polarisationsebene durch die optische Rückwirkung besonders deutlich zum Vorschein kommt. Während das Signal des Detektors D1 keine Rückwirkungen zeigt, hat die hier gezeigte Polarisationsrichtung unregelmäßige Modulationen mit Spitzen bis zu einem mehrfachen des Grundsignals. Bild 8 zeigt eine analoge Messung mit linear polarisiertem Laser. Die Laserstrahlung wird durch einen λ/4-Phasenschieberspiegel zirkular polarisiert. Die Signale der beiden Detektoren zeigen keine Rückwirkungserscheinungen. Die Kombination von λ/4-Phasenschieber und linear polarisiertem Laser wirkt bei einer Werkstückoberfläche, die sich wie ein idealer Spiegel verhält, wie eine optische Einwegleitung. Bei realen Oberflächen erfolgt jedoch keine vollständige Unterdrückung der reflektierten Laserstrahlung, sondern lediglich eine je nach Laserkonfiguration und Werkstückoberfläche mehr oder weniger starke Dämpfung der reflektierten Strahlung. Bild 9 zeigt eine Messung für das Schneiden von Aluminium, wo die Anordnung aus linear polarisiertem Laser und λ/4-Phasenschieber im Gegensatz zu dem in Bild 8 gezeigten Beispiel nicht mehr ausreicht, Rückwirkungserscheinungen zu unterdrücken; Rückwirkungssignale können hier sowohl in der Leistung als auch hinter dem in Sperrichtung justierten Polarisator gemessen werden.

4. Zusammenfassung

Das Ergebnis der Werkstoffbearbeitung mit Laserstrahlung ist neben anderen Faktoren von dem Polarisationszustand des Lasers abhängig. Dies wird für das Schweißen und Schneiden mit Laserstrahlung diskutiert. Zur Optimierung und Kontrolle des Bearbeitungsergebnisses ist die Diagnostik des Polarisationszustandes ohne Bearbeitungsvorgang nicht ausreichend, da der Polarisationszustand während der Bearbeitung zum Teil starken Änderungen unterworfen ist. Dazu werden einige Meßergebnisse beim Schweißen und Schneiden mit Laserstrahlung gezeigt und diskutiert.

Literatur

/1/ H.EICHLER, G.HERZIGER: Z. Angew. Phys. 23 (1967) 297

/2/ H.EICHLER, W.WIESEMANN: Z. Angew. Phys. 28 (1969) 125

/3/ H.EICHLER, G.HERZIGER: Z. Angew. Phys. 12 (1964) 193

/4/ R.DÄNDLIKER, T.TSCHUDI: Appl. Optics 8 (1969) 1119

/5/ P.LOOSEN, L.BAKOWSKY, G.HERZIGER: Feinwerktechnik und Meßtechnik 92 (1984) 11

/6/ E.BEYER, A.DONGES, P.LOOSEN, G.HERZIGER: Proc. Laser 83, P. 259, München 1983

/7/ A.DONGES: Optische Rückwirung am Beispiel der Materialbearbeitung mit Laserstrahlung, Dissertation, TH Darmstadt 1985

/8/ E.BEYER, W.SOKOLOWSKY: zur Veröffentlichung vorgesehen

/9/ E.BEYER: Einfluß des laserinduzierten Plasmas beim Schneiden mit CO2-Lasern, Dissertation TH Darmstadt, 1985

/10/ F.O.OLSEN: Cutting with polarized laser beams, DVS-Berichte 63, 197, Deutscher Verlag f. Schweißtechnik, 1980

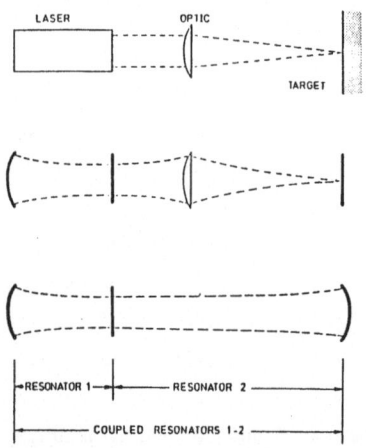

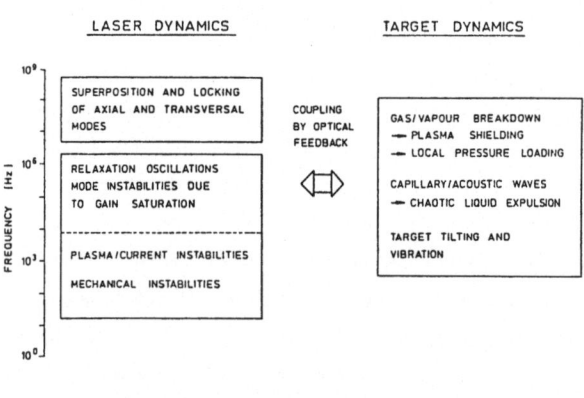

Bild 1.
Schematische Darstellung der Ausbildung eines Systems gekoppelter optischer Resonatoren während der Materialbearbeitung mit Laserstrahlung

Bild 2.
Dynamische Vorgänge im Laser und in der Bearbeitungszone, die in dem System gekoppelter Resonatoren wechselwirken

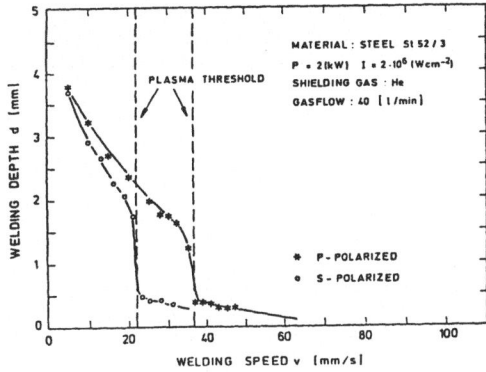

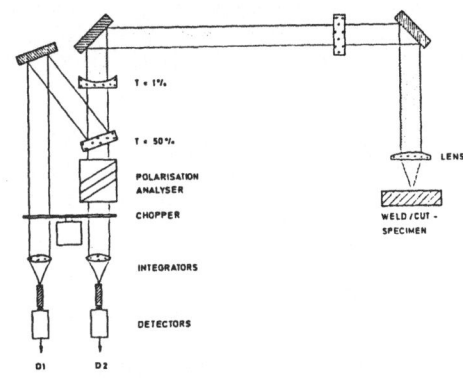

Bild 3.
Beispiel zum Einfluß der Lage der Polarisationsebene auf das Laserschweißen: Einschweißtiefe als Funktion der Schweißgeschwindigkeit bei P-Polarisation (Winkel zwischen Polarisationsebene und Vorschubrichtung $\gamma = 0^{\circ}$) und bei S-Polarisation ($\gamma = 90^{\circ}$)

Bild 4.
Schematische Darstellung des Meßaufbaus zur On-Line Diagnostik der Laserleistung und des Polarisationszustandes, mit Detektor 1 wird die Laserleistung, mit Detektor 2 die Leistung hinter dem Polarisationsanalysator gemessen

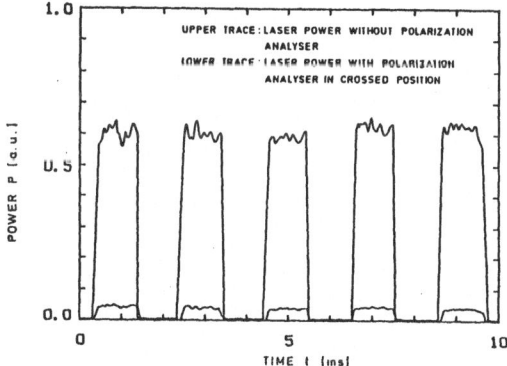

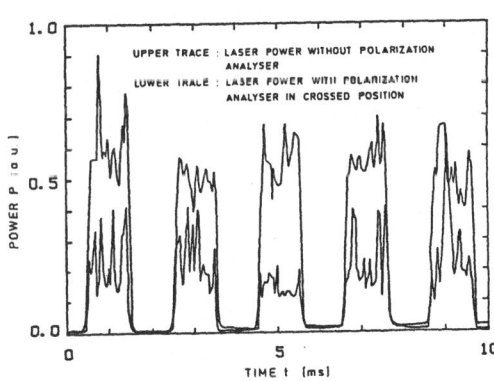

Bild 5.
Diagnostik der Laserleistung (D1:obere Spur) und des Polarisationszustandes (D2:untere Spur) für den freilaufenden Laserstrahl, d.h. ohne Bearbeitungsvorgang. Der Laser ist durch eine 90°-Faltung linear polarisiert. (Die Signale werden mit einer Frequenz von etwa 500 Hz gechoppt)

Bild 6.
On-Line Diagnostik während des Schweißens mit einem linear polarisierten Laser. Die obere Spur (D1) zeigt die Laserleistung, die untere Spur (D2) die Leistung hinter dem in Sperrichtung justierten Analysator (Die Signale werden mit einer Frequenz von etwa 500 Hz gechoppt)

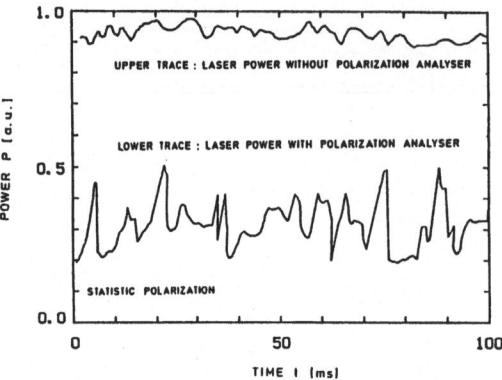

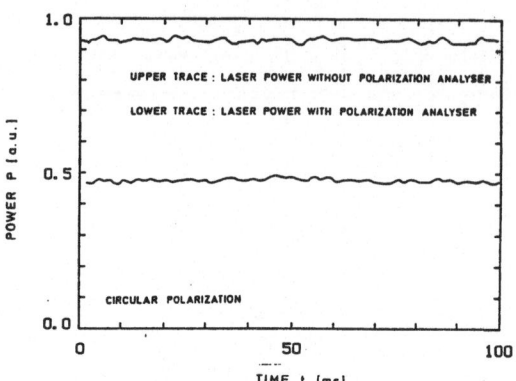

Bild 7.
On-Line Diagnostik während des Schneidens von Stahl mit einem statistisch polarisierten Laser. Die obere Spur zeigt die Laserleistung, die unter Spur die Laserleistung hinter dem Analysator

Bild 8.
On-Line Diagnostik während des Schneidens von Stahl mit zirkular polarisierter Laserstrahlung, die zirkulare Polarisation wurde mit einem linear polarisierten Laser und einem $\lambda/4$ Phasenschieber erzeugt; die obere Spur zeigt die Laserleistung, die untere Spur die Leistung hinter dem Analysator

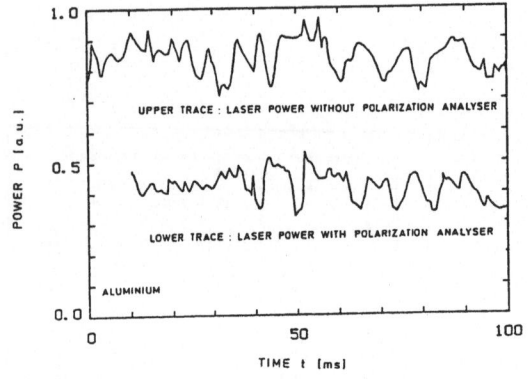

Bild 9.
On-Line Diagnostik während des Schneidens von Aluminium mit zirkular polarisierter Laserstrahlung, Angaben sonst wie in Bild 8

Theoretisches Modell der Riefenbildung beim Laserschneiden

D. Schuöcker und B. Walter
Technische Universität Wien
1040 Wien - Mostgasse 3

1. Einleitung

Die Qualität des Laserschnittes wird durch das charakteristische Riefenmuster, das eine gewisse Rauhigkeit der Schnittflächen bedingt, beeinträchtigt. Um Maßnahmen zur Verringerung der Rauhigkeit ergreifen zu können, muß man den Mechanismus der Riefenbildung verstehen.

Nun existieren zwar bereits seit längerer Zeit Modellvorstellungen für die Ausbildung solcher Störungen der Schnittflächen, die allerdings nur für besondere Fälle Gültigkeit haben.

Ein neueres, in allgemeinerer Weise gültiges Modell /2/ geht nun davon aus, daß die geschmolzene Masse, die sich unter der Einwirkung der fokussierten Laserstrahlung bildet, Eigenschwingungen des Volumens und der Temperatur ausführen kann, die durch die stets vorhandenen zeitlichen Fluktuationen der Laserleistung angefacht werden können, womit durch das Fortschreiten der geschmolzenen Zone in Schneidrichtung eine periodische Veränderung der Schnittflächen erfolgen muß. In der Folge soll nun der Einfluß der Parameter des Schneidvorganges, wie etwa der Schneidgeschwindigkeit, auf diesen Mechanismus der Riefenbildung untersucht werden, um Wege zur Optimierung der Schnittqualtität zu finden.

2. Stationärer Zustand - Schneidgeschwindigkeit

Der auf die Oberfläche des Werkstückes gebündelte Laserstrahl tritt am jeweiligen Ende des Schnittspaltes in diesen ein und breitet sich dann in die Tiefe des Werkstückes aus, wobei bei geeigneter Polarisation /1/ das momentane Ende der Schnittspalte, eine nahezu senkrechte Fläche, die eingekoppelte Laserstrahlung kräftig absorbiert. Dadurch bildet sich an der Grenzfläche zwischen dem bereits hergestellten Schnitt und dem noch ungeschnittenen Material eine dünne geschmolzene Zone aus. Zusätzlich zum Laserstrahl dringt auch ein gebündelter

Sauerstoffstrahl in den Schnittspalt ein, der zum Teil von der
geschmolzenen Zone absorbiert wird und dort durch exotherme
Reaktion mit dem flüssigen Werkstoff zu einer zusätzlichen
Erwärmung der geschmolzenen Zone führt. Diese Zone erreicht damit
Temperaturen bis zur Nähe des Siedepunktes. Dadurch findet an
ihrer Oberfläche eine kräftige Verdampfung statt. Außerdem wird
durch die Reibung zwischen dem Sauerstoffstrahl und der Oberfläche
der geschmolzenen Zone flüssiges Material an der Unterseite des
Werkstückes ausgetrieben. Die geschmolzene Zone kann einerseits
durch die Energiebilanz, die Erwärmung durch absorbierte
Laserstrahlung und durch exotherme Reaktion sowie Kühlung durch
Wärmeleitung und Verdampfung umfaßt, und andererseits durch die
Massenbilanz, die den Massengewinn durch Aufschmelzen und
Massenverluste durch Verdampfung und Reibung mit dem Schneidgas
berücksichtigt, beschrieben werden. Die analytische Formulierung
dieser Bilanzgleichungen ist in /1/ angegeben. In diese
Gleichungen geht unter der Annahme optimaler Schneidgaszufuhr
die Laserleistung Pl, die Schneidgeschwindigkeit v, die
Werkstückkdicke d, die mittlere Dicke der geschmolzenen Zone s
sowie deren Temperatur T ein. Abb. l zeigt nun im Temperatur-
Geschwindigkeitsfeld die Kurven für Gleichgewicht der Energie- und
Massenbilanz für eine Laserleistung von 1200 W und für eine Werk-
stückdecke von d=5 mm. Die beiden Gleichgewichtskurven begrenzen
jeweils den Bereich überfüllter Energie- und Massenbilanz nach
oben, also zu höheren Temperaturen hin. Die Energiebilanzkurve
hängt dabei nicht von der Dicke der Schmelzzone ab, während die
Massenbilanzkurve sich nicht unwesentlich bei Veränderung der
Dicke der Schmelzzone verschiebt. So liegt die Kurve für s=o bei
höheren Temperaturen, weil dann keine Materialabtragung durch
Austreiben flüssigen Materials erfolgt und die gesamte
Materialabtragung, die durch die Schneidgeschwindigkeit bestimmt
wird, durch Verdampfung erfolgen muß ("Sublimationsschneiden").

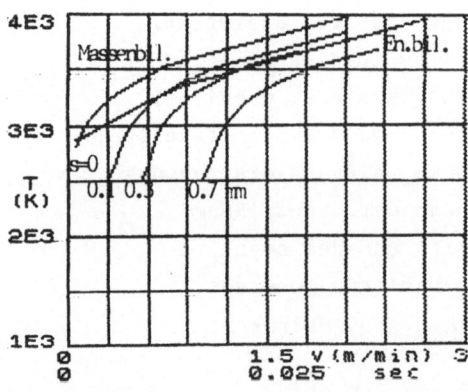

Abb.1. Energiebilanz und Massen-
bilanzen für s=0,0.1,0.3 und
0.7 mm.
(Pl=1200 W,Stahl,d=5mm).

Andererseits sinkt die Massenbilanzkurve mit steigendem s
zu niedrigeren Temperaturwerten ab, weil dann der Anteil der
Materialabtragung durch Reibung mit dem Schneidgas größer wird,
womit der Anteil der Verdampfung abnimmt und so die Temperatur
sinken kann. Allerdings kann s nicht beliebig groß werden, weil
oberhalb einer maximalen Dicke der Schmelzzone s max (hier etwa
0,5 mm) die Massenbilanzkurve die Energiebilanzkurve nicht mehr
schneidet. Die Schnittpunkte der Massenbilanzkurven für s=o und
für s max bestimmen nun die minimale und die maximale Schneidgeschwindigkeit.

3. Dynamisches Verhalten - Riefenbildung

Um das dynamische Verhalten der Schmelzzone zu untersuchen, muß
man davon ausgehen, daß Energie und Massenbilanz nicht völlig im
Gleichgewicht sind. Wenn dabei der Energiegewinn größer ist als
die Summe der Verluste, so muß ein zeitlicher Anstieg der
Temperatur erfolgen und ebenso muß ein Überwiegen des Massenge-
winnes gegenüber der Summe der Massenverluste zu einem zeitlichen
Anstieg der Masse und damit des Volumens der Schmelzzone führen.
Man erhält damit aus den Bilanzgleichungen für den Gleichge-
wichtsfall /1/ zwei gekoppelte Differentialgleichungen für die Tem
peratur und das Volumen der Schmelzzone. Diese Gleichungen sind in
/2/ angegeben und beschreiben Oszillationen der Temperatur und des
Volumens der geschmolzenen Zone. Wie in /2/ gezeigt wurde, beträgt
die Frequenz fo dieser Schwingungen:

$$f_0 = \frac{T_o \, V_o}{E_o \, M_o} \frac{\partial p_1}{\partial T} \frac{\partial m_1}{\partial V} \qquad (1)$$

(pl Energieverluste, ml Massenverluste der Schmelzzone pro Zeiteinheit, To Temperatur, Mo Masse, Vo Volumen, Eo Energieinhalt
der geschmolzenen Zone in stationärem Zustand). Die Dämpfungs-
konstante beträgt:

$$\alpha = \left(\frac{1}{E_o} \frac{\partial p_1}{\partial T} - \frac{1}{M_o} \frac{\partial m}{\partial T} \right) \qquad (2)$$

Diese Oszillationen können nun durch Fluktuation etwa der Laserleistung angestoßen werden, wenn deren Frequenzspektrum die oben
angegebene Eigenfrequenz der geschmolzenen Zone enthält, was
stets der Fall ist, da die untere Grenze des Frequenzspektrums

der Strahlleistungsschwankungen durch die Netzfrequenz gegeben
ist und fo nach /2/ typisch in der Größenordnung 1000 Hz liegt.
Es läßt sich nun leicht zeigen, daß die Amplitude dieser
Oszillationen dem Quotienten der Eigenfrequenz und der Dämpfungs-
konstante proportional ist. Diese Schwingungen der geschmolzenen
Zone führen nun in Folge der Fortbewegung der geschmolzenen
Zone in Schneidrichtung zu einer periodischen Störung der
Schnittflächen, deren "Wellenlänge"   durch den folgenden Aus-
druck gegeben ist:

$$\lambda = \frac{v}{fo} \tag{3}$$

Die maximale Tiefe der damit gebildeten Riefen muß der oben erwähnten Amplitude der Oszillationen proportional sein:

$$h\ max \sim \frac{fo}{\alpha} \tag{4}$$

Pl	d	b	s	v	T	$\frac{\partial pl}{\partial T}$	$\frac{\partial ml}{\partial T}$	$\frac{\partial ml}{\partial V}$
W	10^{-3}	10^{-3}	10^{-3}	10^{-2}	K	$\frac{W}{K}$	$\frac{kg}{s \cdot K}$	$\frac{kg}{s \cdot m^3}$
	m	m	m	m/S				
			0	0,15	2850	0,723	$0,844 \cdot 10^{-7}$	$9,975 \cdot 10^5$
1200	5	0,5	0,1	1,2	3300	2,98	$4,138 \cdot 10^{-7}$	$1,543 \cdot 10^5$
			0,3	2,25	3500	5,04	$7,199 \cdot 10^{-7}$	$0,905 \cdot 10^5$

Tab. 1.   Zahlenwerte zum stationären Zustand der Schmelzzone

Mit Hilfe der in Tab. 1 angegebenen Zahlenwerte wurden nun die in
Tab. 2 angeführten Werte der Eigenfrequenz und der Dämpfung sowie
der resultierenden Wellenlänge und des der maximalen Tiefe der
Riefen proportionalen Faktors f_o/α für drei verschiedene Werte
der Schneidgeschwindigkeit innerhalb des erlaubten Bereiches berechnet. Aus Tab. 2 ist ersichtlich, daß sowohl die Wellenlänge
der Riefen wie auch der der Tiefe der Riefen proportionale
Quotient f_o/α mit sinkender Schneidgeschwindigkeit sinken, sodaß
mit sinkender Schneidgeschwindigkeit die Riefen sowohl feiner als
auch weniger tief werden, was insgesamt eine verbesserte Schnittqualität bedeutet.

v	s	f_0	α	λ	$\frac{f_0}{\alpha}$
m/s	m	1/s	1/s	m	-
0,0015	0	$\sqrt{\infty}/s$	$\sim 1/s$	0	0
0,012	10^{-4}	5994	$1{,}72.10^{5}$	2.10^{-6}	0,035
0,0225	3.10^{-4}	2620	$2{,}24.10^{4}$	$8{,}59.10^{-6}$	0,17

Tab. 2   Zahlenwerte zum dynamischen Verhalten der Schmelzzone

Der Quotient f_0/α , der die Amplitude der Oszillationen der geschmolzenen Zone angibt, geht bei Annäherung an die minimale
Schneidgeschwindigkeit und damit an s=o gegen Null, sodaß die
hier beschriebenen Oszillationen beim reinen "Sublimationsschneiden" nicht vorhanden sind und dabei jedenfalls die bestmögliche Schnittqualität zu Stande kommt, allerdings erkauft mit
der kleinstmöglichen Schneidgeschwindigkeit.

Schlußbemerkungen

Es konnte gezeigt werden, daß die Wellenlänge und die Tiefe der
Riefen, die generell beim Laserschneiden auftreten, stark von der
Schneidgeschwindigkeit beeinflußt werden. Dabei wird beim
maximalen Wert der Schneidgeschwindigkeit für gegebene Laserleistung und Werkstückdicke eine besonders grobe Riefung erzielt,
während beim Minimalwert der Schneidgeschwindigkeit die Riefenbildung durch den hier beschriebenen Mechanismus verschwinden
müßte. Damit zeigt sich hier vielleicht ein Weg, die Schnittqualität zu verbessern, allerdings auf Kosten der Schneidgeschwindigkeit.

Literatur

/1/   D. Schuöcker u. W. Abel
      "Material removal mechanism of laser cutting",
      "Industriel Applications of High Power Lasers"
      Dieter Schuöcker, Editor, Proc. SPIE 455, p. 88 - 95, 1984

/2/   D. Schuöcker und B. Walter
      "Theoretical model of oxygen assisted laser cutting"
      Inst. Phys. Conf. Ser. No. 72 (5th GCL Symp. Oxford 1984)
      p. 111 - 116, 1984 (Adam Hilger)

Surface Finish and Laser Materials Processing

E.W. Kreutz[1], M. Krösche[1] and H.G. Treusch[2]

[1]Institut für Angewandte Physik, Technische Hochschule Darmstadt
Schloßgartenstr. 7, D-6100 Darmstadt, Fed. Rep. Germany

[2]Fraunhofer-Institut für Lasertechnik
Drosselweg 87, D-5100 Aachen, Fed. Rep. Germany

1. Introduction

Use of lasers and laser systems continues to grow steadily with important application
areas as materials processing, information handling, audio and video equipment, optical fiber communication in addition to measurement and laboratory science. The
expansion into industrial processes and production plants requires overall improvement of the trade-offs between performance characteristics, such as dynamic range,
temporal stability, and power control.

A particular combination of power density and interaction time defines a specific
operational regime within the various interaction processes between a laser beam
and a substrate surface resulting either in directed energy processing /1/ or in materials machining /2/. The laser-solid interaction, the hydrodynamics of the molten
pool, the optical feedback, the absorption and the recoil of the laser-induced plasma for example govern the efficiency and the surface finish of laser materials processing. The present status of the achievable precision is described for several
examples of actual application. Some critical improvement needs are included on the
base of the illustrations used.

2. Periodic Surfaces Structures

Laser-induced periodic ripple structures (LIPRS) have been observed in laser-genera-
ted damage at surfaces of liquids, dielectrics, semiconductors, and metals for vari-
ous laser wavelength λ_L. For intensities $I_M \leqslant I \leqslant I_V$ above the intensity I_M of melting
LIPRS are generated by pulsed illumination showing a complex dependence on polariza-
tion and angle of incidence α of the laser radiation /3/. For p-polarized light the
laser-induced surface corrugations consist of two superimposed sets of periodic
structures with the spacings $\lambda_L/(1 \pm \sin \alpha)$ µm /3,4/. For s-polarized light the laser-induced surface corrugations consist of a single set of periodic structures
with a spacing $\lambda_L/\cos \alpha$ µm /3,4/. The periodic structures are in any case perpendicular to the polarization.

Fig. 1 shows the range $I_M<I<I_V$ of the single shot threshold intensity I_T as a function of interaction time t_L for generation of LIPRS on Si (111) surfaces. The polished surfaces of mirror-like finish exhibit a higher threshold intensity I_T than the grounded surfaces /3/. Qualitatively similar results are obtained for steel surfaces with the threshold depending on the thermophysical and metallurgical properties /3/. For $I<I_M$ the phonon density seems only sufficient to produce locally an increase of temperature resulting in internal stresses, which lead to fractures (Fig. 1) along cleavage planes /5/. For $I>I_V$ the vaporized material becomes partially ionized by the laser radiation resulting in the generation /6,7/ of a laser-induced plasma (LIP) with abrasive processing (Fig. 1).

LIPRS result from an interference /4,8/ between the incident laser beam and a surface scattered wave, which originates from scattering of the incident polarized coherend radiation at the microroughness of the surface /9,10/. The interference effects account for all the structure spacings, orientations, polarization, material, and microroughness dependences /3,4,8,10/.

Following these considerations LIPRS are related to a fundamental property of laser radiation, namely, coherence. In order to avoid the generation of LIPRS the processing has to be performed by a laser source of less coherence with the intensities and interaction times choosen according to Fig. 1.

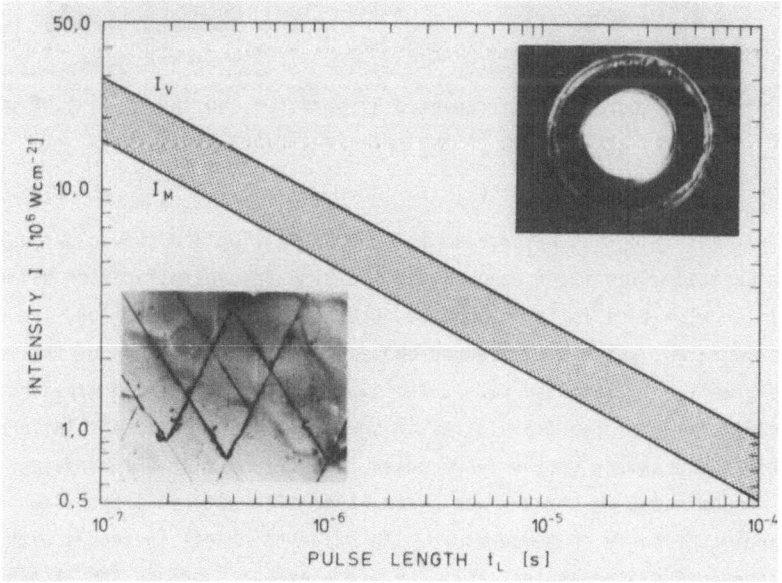

Fig. 1. Single-shot threshold intensity versus pulse length (Nd:YAG, λ_L = 1.06 μm) for the generation of LIPRS on Si (111) surfaces and optical micrographs of the irradiated surfaces for $I<I_M$ and $I>I_V$

3. Concentric Surface Structures

For $I > I_V$ Fig. 2 shows for Al concentric surface structures after single-pulse laser illumination as a function of pressure and composition of the ambient. The structures broaden spatially with increasing fluence /3/. The structures are obliterated with increasing pressure (Fig. 2) with a material-dependent speed of reaction. At very high fluence levels the structures are destroyed with the onset of abrasive processing.

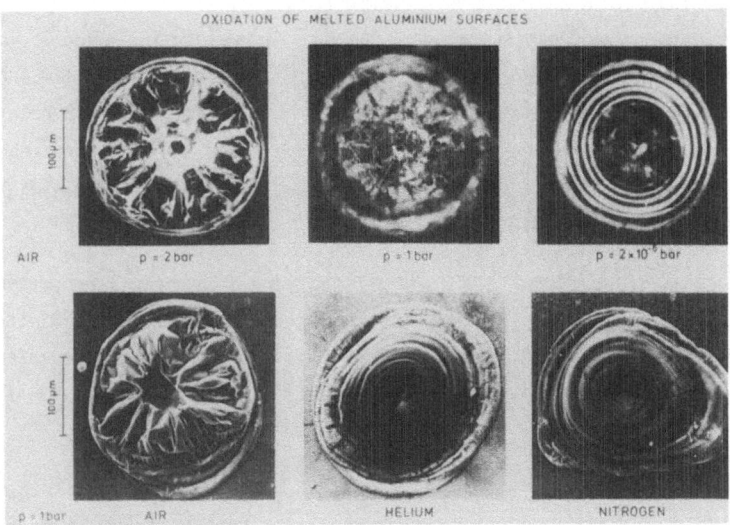

Fig. 2. Optical micrographs of irradiated Al surfaces (Nd:YAG, λ_L = 1.06 µm, t_L = 100 µs) as function of ambient and ambient pressure

Material removal by pulsed laser beam is done by melting and subsequent vaporization for $I > I_V$. Excited waves are frozen-in, if the time for solidification after switching off the laser radiation is far below the duration of the wave period. The concentric surface structures, which are observed above a critical thickness of the melt, might be due to gravity or capillary waves. The experimentally observed wavelengths ($1 < \lambda |\mu m| < 15$) claim for capillary waves in agreement with model calculations /9/ developed for the coupling of the laser power into the melt at the surface. The energy coupling is dominated by the surface oxide layer with optical properties and thickness depending strongly on temperature. The oxide thickness increases with temperature and pressure of the ambient air. The oxide exhibits cracks and lateral displacements within the interaction zone preventing the generation of capillary waves (Fig. 2). Inert gases reduce the oxide growth rendering possible the formation of capillary waves (Fig. 2).

The thickness of the molten layer decreases with increasing intensity I. High surface finish might be achieved either by intensities $I \gg I_V$ with very short t_L or by intensities $I > I_M$ with prolonged t_L depending on the operative materials-processing effect. In the former case the melting depth is well below the structural spacing, in the latter case the duration of the structural transformations is well above the time constant of freezing. Hence, after switching off the laser irradiation attenuation of surface structures will occur.

4. Surface Hillocks

Pulsed laser irradiation can lead for $I > I_V$ in Si to hillock-like deformations /9,11/ in the surface as to be seen from investigations of surface morphology (Fig. 3).

Fig. 3. REM micrograph of a irradiated Si (111) surface (Nd:YAG, λ_L = 1.06 μm, $I = 6 \times 10^5$ W/cm², t_L = 100 μs)

Following the outline of other authors /11/ the deformations are due to bubble formation by the impressed inhomogeneous temperature distribution.

However the hydrodynamics of the molten material and the vapor/plasma-melt interaction also may originate in hillock-like deformations. The recoil of vaporized material and generated plasma creates a depression of the liquid surface beneath the incident beam with a subsequent ridging in the surrounding liquid surface. Surface tension of the molten layer and shear stresses on the molten layer caused by gradients in surface tension of the liquid layer also push the liquid layer down in a nonplanar concave meniscus shape. As the height difference between the liquid under the beam and away from the beam increases, the developing pressure is most pronounced. At the end of the laser pulse this enhanced pressure raises the liquid layer up in a nonplanar convex meniscus shape. Because of the large temperature gradients and rapid solidificaiton rates the distortion of the liquid layer is frozen-in in hillock-like deformations (Fig. 3).

In order to avoid the hillock formation, the surface temperature has to be reduced slowly compared to heat flow induced gradients of surface tension by suitable shaping of the laser pulse /11/ with reducing the laser power continuously to zero.

5. Surface Cratering

Fig. 4 shows REM micrographs of holes drilled in the plasma-assisted mode of operation for $I>I_v$. Hole geometry /12,13/, hole depth /14/, and drilling efficiency /14/
for I = const. depend on the interaction time governing the energy density in the interaction zone.

During abrasive processing the LIP absorbs partially the incoming radiation limiting
the depth of penetration due to the limited amount of laser energy absorbed, thus,
yielding a thicker layer of solidified material within the hole /12,13/. A second
limitation occurs in the recondensation of material around the entrance of the hole,
which provides a crater-like lip for the hole (Fig. 4). This arises because material
vaporized and ejected from the hole can easily recondense on the first cool surface
that it strikes. The development of a hole within the matter changes the boundary
conditions of plasma expansion. The emission of radiation, as main loss of a plasma
plume in front of the processed material, contributes to the enhancement of the processing efficiency by coupling of the plasma radiation into the perimeter of the
hole in combination with plasma losses by heat flow and condensation. The melt expulsion becomes accelerated by the plasma within the hole, i.e. hole depth and drilling efficiency increase /14/ yielding simultaneously in a cylindrical hole geometry
/12,13/. At higher hole depth the melt partially resolidifies in the upper half of
the hole /14/ because of the insufficient energy coupling from the plasma, which
might be overcome by suitable tailoring of the laser pulse shape /14/.

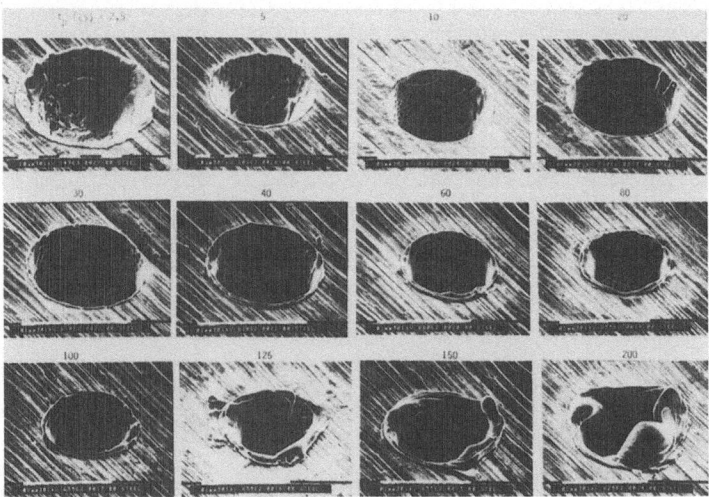

Fig. 4. REM micrographs of holes drilled in steel as function of pulse length
(Nd:YAG, λ_L = 1.06 µm, $I_v<I=10^8$ W/cm²) in the plasma-assisted mode of
operation

For intensities $I>I_D$ (threshold intensity of detonation) and resulting smaller absorption length the limiting case of an laser supported detonation wave will occur.

The laser radiation is absorbed substantially in a thin plasma sheat travelling with supersonic speed towards the incident laser beam. According to the strong plasma absorption the target is completely shielded and the machining is interrupted. Due to the recoil of the expanding plasma sheet the pressure at the surface is strongly enhanced, which may result in a rigorously uncontrolled expulsion of molten material, which is splattered around the crater edges of the processed area (Fig. 5). The laser-induced plasma changes dramatically the processing conditions. In the limit of weak absorption the plasma plume lenses the original beam geometry. In the limit of strong absorption the plasma completely absorbs the incoming radiation. In any case, the processing is no longer controlled by the laser parameters, rather by the generated plasma plume influencing the processing results and - by optical feedback - intensity and mode structure of the laser itself. The working process may become totally irregular. For controlled and reproducible machining the laser parameters have to be matched to the dynamic and transport properties of the induced plasma in processing diagrams /15/. By proper shaping of melt depth, intensity,and pressure

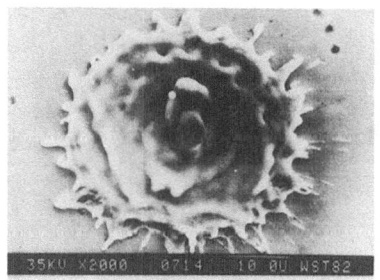 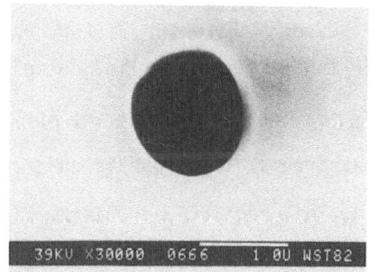

a   b

Fig. 5. REM micrographs of irradiated Cu surfaces (KrF excimer laser, λ_L = 248 nm) for uncontrolled plasma heating with chaotic melt expulsion (a) in the regime of laser supported detonation waves ($I>I_D$) and for controlled streaming plasma jet with regular hole formation (b) in the regime of laser supported combustion waves ($I<I_D$)

these unwanted effects mainly can be suppressed (Fig. 5).

6. Conclusion

The spatial and temporal intensity distribution at the surface of the processing material has to be matched carefully to the special application by pulse shaping within processing diagrams in order to obtain optimum surface finish during laser processing. Coupled with the availability of tremendous computer power lasers of greater stability and more precise power control have to be developed in order to be optimized with respect to the characteristic time constants of the physical processes involved following the outline of this paper. The precision and repeatability of automated laser control and monitoring in addition with automated instrumentation and positioning promise to refine present-day laser applications and accelerate technology into new areas through faster data collection and reduction.

Literatur

(1) J.M. Poate, J.W. Mayer, Eds.: Laser Annealing of Semiconductors (Academic Press, New York) 1982

(2) M. Bass, Ed.: Laser Materials Processing (North Holland, Amsterdam, New York, Oxford) 1983

(3) E.W. Kreutz, M. Krösche, H.G. Treusch, G. Herziger: Proc.Int.Conf. on Laser Processing and Diagnostics (Linz) 107 (1984) and references therein

(4) J.E. Sipe, J.F. Young, J.S. Preston, H.M. van Driel: Phys. Rev. $\underline{B27}$, 1141 (1983)

(5) M. Birnbaum: J.Appl.Phys. $\underline{36}$, 3688 (1965)

(6) D.M. Roessler, V.G. Gregson: Appl.Opt, $\underline{17}$, 992 (1978)

(7) E. Beyer, L. Bakowsky, R. Poprawe, G. Herziger: Optoelectronic in Engineering (Springer Verlag, Berlin, Heidelberg, New York, Tokyo) 367 (1984)

(8) P.M. Fauchet, A.E. Siegman: Appl. Phys. $\underline{A32}$, 135 (1983)

(9) E.W. Kreutz: to be published

(10) M.J. Soileau: IEEE J. Quantum Electr. $\underline{QE-20}$, 464 (1984)

(11) W. Lüthy, K. Affolter, M. Fuhrer: Phys. Lett. $\underline{72A}$, 60 (1979)

(12) H.G. Treusch: Schweizer Maschinenmarkt $\underline{25}$, 28 (1984)

(13) A. Gillner, G. Herziger, E.W. Kreutz, H.G. Treusch, K. Wissenbach: Proc. Materials Research Society (Europe) 1985, to be published

(14) H.G. Treusch: Schweizer Maschinenmarkt $\underline{26}$, 29 (1984)

(15) E. Beyer, L. Bakowsky, P. Loosen, R. Poprawe, G. Herziger: Proc. Industrial Appl. of High Power Lasers (Linz), SPIE $\underline{455}$, 75 (1984)

Micro Machining by Pulsed Laser Radiation

R. Poprawe and H.G.Treusch,

Fraunhofer Institute of Laser Technology, D-5100 Aachen, FRG

1.Abstract

Nowadays the most widely used lasers in micro machining are CO_2- and Nd-YAG-lasers. High technical reliability combined with good beam quality made possible many industrial applications. However, especially high precission processing of metals in the µm-range still is a point of laboratory investigation.

Therefore, in this paper characteristic elements of micro processing by laser radiation such as plasma absorption and processing velocity are discussed. In addition to CO_2- and Nd-YAG lasers the properties of excimer lasers are included. Because of the short wavelength of UV-excimer laser radiation and the resulting high focussability these lasers might play a dominant role in processing structures in the sub-µm-region.

2. Target absorption of CO_2-, Nd-YAG- and KrF-laser radiation

The absorption of radiation strongly depends on resonances of the photon energies with band-transition-energies in the solid. Fig.1 shows characteristic absorption curves /1/ of a metal (silver) and an insulator (sapphire) for a wavelength range from $0.1<1(µm)<20$. The typical increase of absorption for wavelengths below $\lambda \approx 0.3$ µm is due to band-band-transitions /2/. Fig.1 therefore shows high

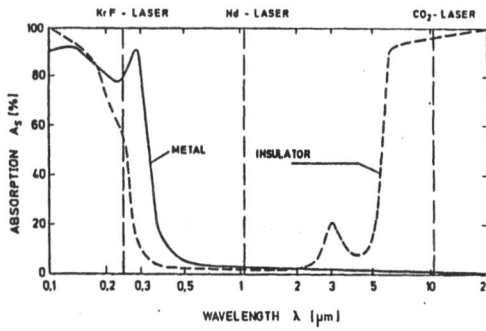

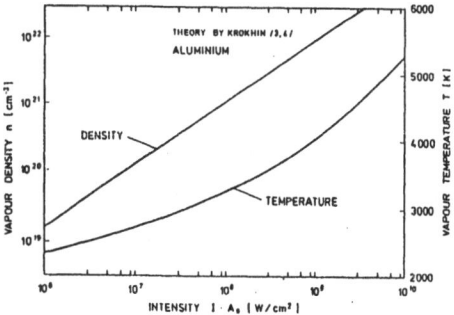

Fig.1. Absorption in % of electromagnetic radiation; shown is the typical behaviour of a metal and an insulator; included are the wavelengths of typical lasers used for micro machining: CO_2-, Nd-YAG-and KrF-laser

Fig.2. Density n (cm^{-3}) and temperature T(K) of aluminium vapour at vaporization front calculated according to Krokhin's model /3,4/

absorption for KrF-laser radiation in metals as well as in insulators. At wavelengths around $\lambda = 1\mu m$ (Nd-YAG-laser) the natural absorption is only a few percent
for metals. The behavior of insulators is not as uniform as for metals in this wavelength range. Glass and jewels are mostly transparent, whereas - for example - ceramics show a strong absorption at around $\lambda=1\mu m$. For the insulator example shown in
Fig.1 the phonon interaction increases at wavelengths above $\lambda\approx5$ μm, so there is a
strong absorption of CO_2-laser radiation. On the other hand, metals show high reflectivity at $\lambda=10.6$ μm (CO_2-laser).

3. Plasma absorption of CO_2-, Nd-YAG- and KrF-laser radiation

When the absorbed energy density in the target is high enough to heat the surface to
temperatures above the evaporation threshold, a vapour jet emerges from the target.
Due to its temperature the vapour is partly ionized, and therefore has plasma properties. The laser radiation can be absorbed within the plasma and increase the
coupling of laser radiation to the target surface. The full description of the process has to encounter the hydrodynamics of the plasma jet as well as absorption and
heat conduction characteristics simultaneously.

To give a first order approximation of critical quantities like electron density,
vapour temperature and absorption length in the plasma, a one-dimensional approach
by Krokhin /3,4/ will be used as a basis. In this equilibrium formulation of the
phase transition at the boundary solid-gas saturation conditions (Clausius Clapey-
ron), ideal-gas behaviour and expansion with the speed of sound (Chapman-Jou-
guet-condition /5/ are assumed. Including the three hydrodynamic conservation equations of mass, momentum and energy gives six equations for the variables density n
(cm^{-3}), vapour pressure P (bar), saturation pressure P_S(bar), energy density
ϵ(J/kg), vapour velocity v(m/s), and the velocity of the vapourization front v_{DR}
(cm/s), the drilling velocity. Fig.2 shows some results for aluminium. Plotted are
vapour density n and temperature T versus the absorbed laser intensity IA_s, where A_s
is the solid absorption . The data represent minimum values, since absorption of
radiation in the vapour is not included, which increases the coupling and hence the
values of density and temperature.

On the basis of Krokhin's model the electron density n_e (cm^{-3}) and, consequently,
the absorption coefficient of laser radiation $\alpha(cm^{-1})$ can be estimated. Keeping the
frame of equilibrium formulation the saha-equation /6/ gives the electron density as
a function of temperature T, density n and ionization energy E_i (eV):

$$\frac{n_e^2}{n} = 2 \cdot \frac{U_+}{U_0} \frac{(2\pi m kT)^{3/2}}{h^3} \exp\left\{-\frac{E_i}{kT}\right\} \qquad \frac{(2\pi m k)^{3/2}}{h^3} = 2.4\cdot10^{15} \frac{1}{cm^3 K^{3/2}} \qquad (3.1)$$

where U_t and U_o are the partitition functions for ion and atom. Fig.3 shows a plot of (3.1) for aluminium vapour with $U_t \simeq 1$, $U_0 \simeq 5.8$ /7/. Also included is the estimated absorption coefficient $\alpha(cm^{-1})$ for the wavelengths of the CO_2, Nd-YAG and KrF-laser following an expression for inverse bremsstrahlung absorption by Mulser /8/.

$$\alpha = \frac{2\omega}{c_o} \left[\frac{1}{2} \left\{ \sqrt{\left[1 - \left(\frac{\omega_p}{\omega}\right)^2 \frac{1}{1+(\nu_c/\omega)^2}\right]^2 + \left[\frac{\nu_c}{\omega}\left(\frac{\omega_p}{\omega}\right)^2 \frac{1}{1+(\nu_c/\omega)^2}\right]^2} \right. \right.$$
$$\left. \left. - \left[1 - \left(\frac{\omega_p}{\omega}\right)^2 \frac{1}{1+(\nu_c/\omega)^2}\right] \right\} \right]^{1/2} \tag{3.2}$$

Here ω_p (s^{-1}) is the plasma frequency and ν_c (s^{-1}) is the electron-atom/ion frequency estimated following Pirri /9/. At laser intensities below $I=10^9$ W/cm^2 a strong dependence of the absorption coefficient α on the laser wavelength can be seen: At $I=10^8$ W/cm^2 the absorption of infrared CO_2-laser radiation ($\lambda=10.6$ μm) is about three orders of magnitude larger than the value for UV-KrF-laser radiation ($\lambda=0.25$ μm). Hence, one expects a strong coupling of infrared radiation at lower laser intensities compared to UV-lasers. This effect has been observed experimentally (Fig.4): Due to the formation of laser supported detonation waves /10/ the

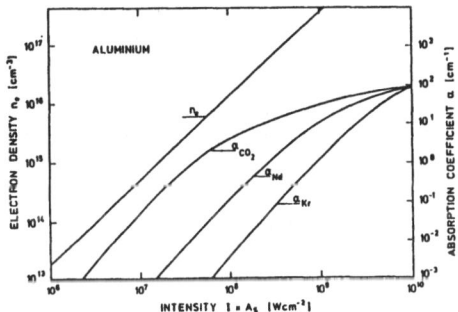

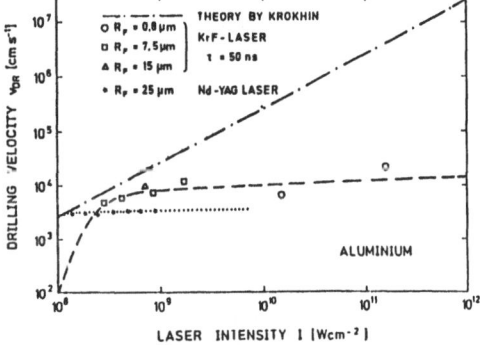

Fig.3. Electron density n_e calculated with equilibrium theory for aluminium vapour at vaporization front; included are estimations of the absorption coefficient α for CO_2-, Nd-YAG and KrF-laser; the curves represent minimum values, since plasma absorption has been omitted in the calculation

Fig.4. Drilling velocity v_{DR} of aluminium for KrF and Nd-YAG-laser radiation; the higher value of v_{DR} for the limit of high intensities for the KrF-laser indicates laser absorption in the shielding plasma at the target surface

drilling velocity v_{DR} shows saturation behaviour in the limit of high intensities /11/ . However, the saturation value for Nd-YAG-laser radiation ($\lambda=1.06$ μm) is well below the value for KrF-laser radiation ($\lambda= 0.25$ μm) indicating higher plasma transparency with smaller wavelength. As a consequence, the optimum processing intensities increase with increasing laser frequency.

4. Hole formation in laser drilling - influence of plasma

Once a laserinduced plasma is established a strong coupling of laser radiation to the workpiece is observed /12/. However, when the laser intensity is chosen too high a laser supported detonation wave (LSD-wave /5/) will begin to formate and shield off the workpiece, so the laser intensity has to be matched carefully to plasma absorption. Once a hole of depth $z_H \approx 2 r_H$ (r_H(cm):radius of hole) is drilled, the plasma energy density decreases due to coupling to the hole wall. Hence, all typical threshold phenomena, such as plasma formation threshold intensity I_C or LSD-wave formation threshold I_D /10/, shift to higher values. Fig.5 illustrates the process by means of a streak image of the expanding plasma and corresponding micrographs of the hole-structure. A temporally rectangular laser pulse has been used in the experiment. After about 20 µs drilling time the plasma changes from a shielding element into an energy coupler. The streak image shows expulsed melt for the latter case.

Quantitative results are shown in fig.6: Here, again for temporally rectangular laser pulses, the drilling efficiency has been measured for different intensities with increasing time. At low intensity $I=6 \cdot 10^7$ W/cm^2 a relatively high efficiency is observed in the beginning of the drilling process. With increasing intensity the efficiency decreases at that time due to decoupling via formation of a LSD-wave.

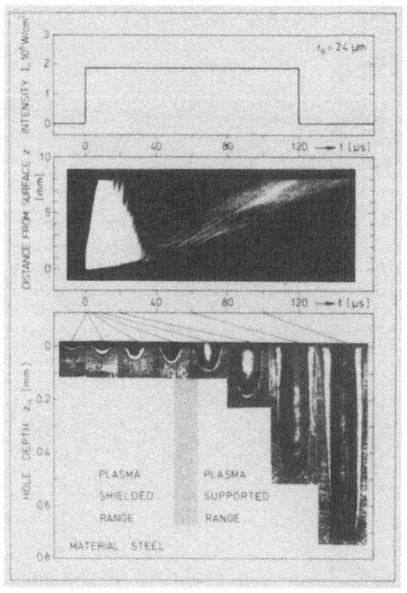

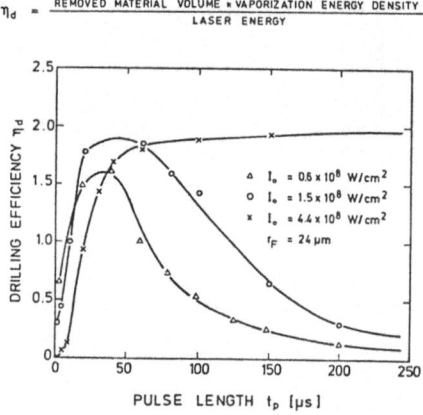

Fig.6. Drilling efficiency η_d versus laser pulse length using Nd-YAG laser radiation; at high laser intensity the energy density in the plasma is sufficient to yield efficient melt expulsion ($\eta_d \approx 2$)

Fig.5. Streak image, corresponding micrographs, and temporal shape of Nd-YAG-laser pulse to illustrate the drilling process; after about 20 µs the laserinduced plasma changes from a shielding to a coupling element

However, once the hole is established, the efficiency is drastically higher at high intensities (t>20μs). This is due to the increase of melt expulsion.

As a consequence, an optimum shaped laser pulse for drilling with maximum efficiency has to begin at the plasma threshold intensity I_c. After the plasma has vanished into the hole the intensity must be increased to yield maximum melt expulsion through strong coupling of the laserinduced plasma to the hole wall. Fig.7 shows an example to illustrate this concept: shown is the top view of a bore, a micrograph and corresponding pulse shape for two extreme pulse forms. Part A shows a high intensity in the beginning associated with LSD-wave formation and chaotic melt expulsion at the workpiece surface. In part B the temporal behaviour is matched to the dynamics of ablated target material to yield a cylindrical hole geometry. High efficiency and quality of laser drilled bores can be achieved by applying the concept of matching the laser parameters to workpiece material dynamics.

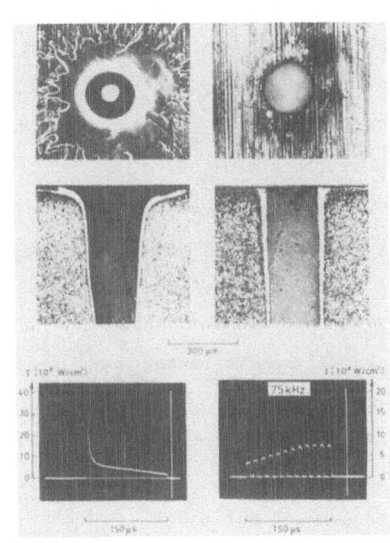

Fig.7. Top view, micrograph, and pulse shape for a Nd-YAG-laser drilling; A: Example for processing in the laser supported detonation regime; B: The laser pulse shape has been matched to the target material dynamics yielding high quality, high efficiency drillings

Literature
/1/ M.v.Ardenne: Tabellen zur Angewandten Physik, VEB
     Deutscher Verlag der Wissenschaften, Berlin (1973)
/2/ P. Drude, Ann.Phys. (Leipzig) 1,566 (1900)
/3/ O.N.Krokhin, in: Laser Handbook 2, Eds.: F.T. Arecci and E.O. Schulz-Deboise,
     North Holland publ. Comp., Amsterdam (1972)
/4/ Y.V.Afanas'eV und O.N.Krokhin, Sov.Phys. JETP 24(4),639 (1967)
/5/ Yu.P.Raizer, Sov.Phys. JETP 21(5),1009 (1965)
/6/ W.Lochte Holtgreven: Plasma Diagnostics, North Holl. Publ. Comp., Amst. (1968)
/7/ H.W.Drawin und P.Felenbok: Data for Plasmas in local thermodynamic equlibrium
     Gauthier-Villars, Paris (1965)
/8/ P.Mulser, IPP 3/95 (1969)
/9/ G.Weyl, A.Pirri, R.Root, AIAA 19 (19),461 (1981)
/10/ R.Poprawe et al., Proceedings Laser '83, München 1983
/11/ R.Poprawe: Materialabtragung und Plasmaformation im Strahlungsfeld von UV-
     UV-Lasern, Dissertation, Darmstadt 1984
/12/ H.G.Treusch, Dissertation (to be published)

Kennzeichnung von Kunststoffen mittels Laserstrahlen

J. Junghans

GRETAG AKTIENGESELLSCHAFT

Althardstrasse 70, CH-8105 Regensdorf

Obwohl Leserbeschriftungsmaschinen im Vergleich zu herkömmlichen Beschriftungsarten ziemlich teuer sind - eine Lasermaschine kostet je nach Optionen zwischen 150 kDM und 200 kDM und kann sogar bei aufwendigem Handling noch teuerer sein - bieten sie bedeutende Vorteile, da

- berührungsfreie Bearbeitung,

- Beschriftung auf unterschiedlichen Ebenen,

- kein Werkzeugverschleiss,

- keine mechanischen Kräfte,

- Beschriftung auch an unzugänglichen Stellen,

- freie Programmierung und Schriftwahl,

- Graviertiefe und -art lassen sich über die Intensität der Strahlung und Frequenz
  des Q-switches einstellen.

Aus der freien Programmierbarkeit ergibt sich die Möglichkeit auch komplizierte Firmenzeichen sowie auch laufenden Nummern ohne Probleme zu schreiben. Die einstellbare Intensität gestattet bei Metallen

- die Eindringtiefe vom tiefen Gravieren bis

- oberflächlichem Verfärben durch Anlauffarben oder

- legieren

sowie bei Kunststoffen

- gravieren,

- verbrennen und zerstören,

- verdampfen und

- thermisches Verfärben des Grundmaterials

- erzielen eines Farbumschlages des eingelagerten Füllstoffes oder Farbpigmentes,
  sowie

- thermisches Aufschäumen des Grundmaterials, hervorgerufen durch Strahlungsabsorption des Füllmittels, welches dann das Grundmaterial zum gasen bringt,
  durchzuführen.

Anforderungen an den beschriftenden Laser

Liegt der reine Kunststoff vor, so ist er meist im sichtbaren Licht transparent oder opak. Diese Kunststoffe lassen sich dann nur mit einem CO2-Laser oberflächlich bedingt gravieren (schwer, da die entstandene Furche wieder zu läuft) oder verbrennen. Eine einwandfreie Beschriftung mit dem Nd:YAG-Laser ist nicht möglich. Anders sieht es jedoch bei Kunststoffen mit beigemischten Füll- und/oder Farbstoffen aus.

Da die Wärmeleitfähigkeit und die Verdampfungs-/Zesetzungstemperatur von Kunststoffen verglichen mit den von Metallen niedrig ist, muss für eine sauberen Schrift ein Laser mit einem möglichst steilen Instensitätsanstieg, aber kleiner Pulsenergie verwendet werden. Wir haben deshalb einen periodisch gepulsten Nd:YAG-Laser verwendet. Der steile Pulsantieg gestattet uns möglichst schnell in das Gebiet maximaler Absorption zu kommen. Folgende Laserdaten bestimmen die Qualität einer Beschriftung:

- Wellenlänge des Lasers

- Modulationsfrequenz und Pulsdauer der verwendeten Laserpulse

- Ausgangsleistung des Lasers

- Schreibgeschwindigkeit

Beschriftungseigenschaften von Kunststoffen

Bei einer Abklärung der Beschriftbarkeit unterschiedlicher Kunststoffer wird in erster Linie immer von der Art des Grundmaterials ausgegangen. Um allerdings definierte Aussagen machen zu können, sollte man nicht nur wissen um welchen Kunststoff, sondern um welchen Füllstoff es sich handelt.

Für die Laserbeschriftung sind folgende Werkstoffdaten von Bedeutung:

- Wärmekapazität

- Wärmeleitfähigkeit

- die Absorption der einfallenden Strahlung durch den Kunststoff, bzw. den beigemischten Füllstoff

- die Temperatur, bei der der Kunststoff erweicht, bzw. zerstört wird oder gast

- Brandverhalten des Kunststoffes.

Duroplaste bestehen aus einer dreidimensionalen, vernetzten Molekülstruktur, die verhindert, dass sie bei Temperatureinwirkung erweichen oder schmelzen; sie werden bei höheren Temperaturen (zwischen 300 und 400 Grad C) zersetzt. Aufgrund ihrer Molekühlketten erweichen oder schmelzen Thermoplaste und beginnen dann zu fliessen.

Je nach Konstitution des Polymeren ergeben sich unterschiedliche Zersetzungsprodukte:

- Es bilden sich fast ausschliesslich gasförmige Produkte (z.B. Polymethylmethacrylat)

- gasförmige Produkte und kohleartige Rückstände (PVC)

- feste ausschliesslich kohleartige Rückstände (Polyacrylnitril, hochtemperaturbeständige Kunstoffe, wie Polyimide)

Für die Beschriftungsversuche wurden einige der gängisten Kunststoffe ausgewählt:

Polymerisate (Thermoplast)

 Polyäthylen (Lupolen)

 Polypropylen (Novolen)

 Polystyrol (Terluran, Luran)

 Polyacetal

 Fluorkunststoffe (Delrin, Ultraform)

Polykondensate (Thermoplast, (Duroplast))

 Polyamid (Ultramid, Zytel)

Polyaddukte (Duroplast)

 Epoxydharz, Polyester

abgewandelte Naturprodukte ((Thermoplast, Duroplast) wurden noch nicht untersucht.

Die Beschriftungseigenschaften lassen sich gut über das Brandverhalten beschreiben. Obwohl einige Kunststoffe bei Zündinitialisierung selbständig brennen, kann meistens bei der Laserbeschriftung auf einen scharfen löschenden Luftstrahl verzichtet werden. Ein Brennen tritt in der Regel nur bei breiter Beschriftung (Schriftbreite grösser 0,5 mm) und tiefer Gravur ein.

Bei Thermoplasten, wenn man vom Verbrennen absieht, eignen sich ungefüllte (reine) Kunststoffe nicht zur Beschriftung; bei gefüllten lassen sich sehr kontrastreihe Beschriftungen erzielen, wenn das Füllmittel Russ und Quarzsand ist. Der Russ verbrennt und der Quarzsand bleibt übrig oder schäumt mit dem Grundmaterial auf; es erscheint auf schwarzem Grund, bei richtiger Lasereinstellung, ein weisser Schriftzug. Bei farbigen Pigmenten tritt meistens ein Umschlag gegen dunkel ein.

Bei der Beschriftung von Duroplasten wird der Kunststoff geringfügig oberflächlich zerstört und abgetragen. Die Beschriftung ist eine, oftmals noch mit einem Farbumschlag unterlegte Gravur.

Oberflächengeschaffenheit nach dem Beschriften

Bei den oben geschildertem Beschriftungsverfahren wird die Oberfläche des Kunststoffes immer beeinträchtigt, sei es dass sie wie bei Duroplasten geringfügig abgetragen wird oder bei Thermoplasten aufschäumt oder aufquillt. Bei gut dosierter Laserleistung beträgt die Eindringtiefe einige Mikrometer. Da diese Beeinträchtigung sehr klein ist, ergeben sich auch für den eingebetteten Körper (z.B. Chip) keine negativen Einflüsse wie durch erhöhte Wasseraufnahme, welche bei diesen Kunststoffen im unbearbeiteten Zustand schon bei einigen Gewichtsprozenten liegt, noch eine erhöhte Gasdurchlässigkeit.

Die periodischen Laserpulse beschriften mit einigen kHz. Diese Frequenz wird in der Regel durch den umgebenden, verbleibenden Kunststoff gut gedämpft.

Neben der Kennzeichnung von Folien und Karten zur Datenspeicherung, auf die hier nicht eingegangen wurde, werden hauptsächlich folgende Kunststoffbeschriftungen durchgeführt:

- Beschriftung von Kabeln
- Beschriftung von Tasten
- Beschriftung von elektronischen Halbleitern wie

      . Halbleitern

      . vergossenen Spulen

      . Kondensatoren

- Beschriftung von Frontplatten und diversen Abdeckungen

Zusammenfassung

Bei einer Abklärung der Beschriftbarkeit unterschiedlicher Kunststoffe wird in erster Linie immer von der Art des Grundmaterials ausgegangen.

Um definierte Aussagen machen zu können, sollte man nicht nur wissen um welchen Kunststoff, sondern zuerst um welchen Füllstoff es sich handelt.

Für die Anfertigung der Fotografien möchte ich mich bei Herrn M. Huber und A. Faust, GRETAG bestens bedanken.

Literatur

H. DOMININGHAUS, Die Kunststoffe und ihre Eigenschaften, VDI-Verlag 1976

J. TROITZSCH, Brandverhalten von Kunststoffen, Carl Hanser Verlag München 1981

Kunststoff Lexikon, Carl Hanser Verlag München 1975

Einschreiben von Information in das Volumen transparenter Plastmaterialien mittels Laserstrahlung

E. Heumann, J. Kleinschmidt, K. Vogler, G. Wiederhold

Friedrich-Schiller-Universität Jena, Sektion Physik

DDR-6900 Jena, Max-Wien-Platz 1

1. Einleitung

Bei den Verfahren zur automatischen Beschriftung, Markierung oder Strukturierung von Materialien mittels Laserstrahlung werden thermische Effekte genutzt, die infolge der Absorption der Laserenergie im Oberflächenbereich des Materials auftreten und meist einen Materialabtrag bewirken /1/. Eine interessante Weiterentwicklung der Lasermarkierung ergibt sich durch die Möglichkeit, entsprechende Informationen (Zeichen, Symbole, Strukturen usw.) in das Volumen bestimmter Materialien einzubringen, ohne dabei die Materialoberflächen zu beeinflussen. Auf diese Weise kann die Information gegen nachträgliche Veränderungen gesichert werden und bleibt auch bei eventueller Beschädigung der Oberfläche erhalten.

Aus der Literatur ist ein Verfahren bekannt, bei dem aus mehreren Schichten zusammengesetzte Plastwerkstoffe zum Einschreiben von Informationen unter die äußere Oberfläche verwendet werden /2/. Dabei ist die äußere Schicht für die Laserstrahlung völlig transparent, während die darunter liegende Schicht an ihrer Oberfläche die Strahlung vollständig absorbiert. Die absorbierende Schicht erfährt eine sichtbare und bleibende Veränderung (Verfärbung oder Verbrennung). Der große Nachteil dieser Methode liegt in der aufwendigen Technologie bei der Herstellung entsprechender Mehrschicht-Plastwerkstoffe.

Im vorliegenden Beitrag wird ein Verfahren angegeben, das geeignet ist, Zeichen, Symbole, Skalen und andere Strukturen in das Volumen homogener transparenter Plastmaterialien einzuschreiben, wobei die Materialoberflächen völlig unbeeinflußt bleiben /3/.

2. Charakterisierung des Verfahrens

Als Strahlungsquelle wird ein cw-Nd-YAG-Laser mit der Ausgangsleistung von 50 W im Multimode-Betrieb verwendet. Die Güteschaltung erfolgt mit einem akustooptischen Modulator und liefert Impulse von 150 - 200 ns Dauer mit Impulsspitzenleistungen von 40 - 50 kW bei Folgefrequenzen zwischen 1 und 3 kHz. Über das von einem Mikrorechner gesteuerte Ab-

lenksystem (Ablenkspiegel oder verfahrbare Optik in x-y-Richtung) gelangt der Laserstrahl durch eine Fokussieroptik in die Bearbeitungs-
ebene. Der Abstand zwischen Fokussiersystem und dem zu beeinflussenden
Werkstoff muß so gewählt werden, daß der Fokusbereich im Volumen des
transparenten homogenen Plastmaterials liegt (s. Abb. 1). Als Plastwerkstoffe werden vorzugsweise Materialien aus Celluloseester, PVC
oder Polyolefinen verwendet (Dicke des Materials d = 3...10 mm), die
für die Wellenlänge des Nd-YAG-Lasers (1,06 µm) nahezu vollständig
transparent sind oder nur sehr schwach absorbieren. Die Kleinsignal-
transmission der Proben liegt zwischen 95 % und 80 %. Oberhalb einer
materialspezifischen Schwellintensität der verwendeten Laserstrahlung
zeigen sich scharf begrenzte irreversible Veränderungen, die sich kon-
trastreich von der unbeeinflußten Umgebung im Innern des homogenen Materials abheben. Die Wahl der Parameter Laserleistung und Brennweite
der Fokussieroptik muß so erfolgen, daß die Schwellintensität im Material nur im Fokusvolumen überschritten wird, d. h., bei kurzbrennweitiger Fokussierung muß die Laserleistung entsprechend kleiner gewählt
werden als bei langbrennweitiger Fokussierung. Die Breite der erzeugten
Strukturen wird durch den Fokusdurchmesser entsprechend der Näherungs-
formel

$$w_0 = \Theta \cdot f \tag{1}$$

bestimmt. Die Ausdehnung der Strukturen in Strahlrichtung bzw. in die
Tiefe des Materials ist durch die Fokustiefe

$$\Delta z \simeq 4 \Theta^2 f^2 / \lambda \tag{2}$$

gegeben (Θ - Divergenz der Laserstrahlung, f - Brennweite der
Optik, λ - Laserwellenlänge). Durch die Wahl der Brennweite der Optik
ist es möglich, Breite und Tiefe der Strukturen in einem bestimmten Bereich zu variieren (vgl. Abb. 1a, b).
Bei Verwendung eines Lasers im transversalen Grundmodus (beugungsbe-
grenzte Strahldivergenz) können mit geeigneten Optiken Strukturen erzeugt werden, die hinsichtlich ihrer Ausdehnung in der Größenordnung
der Wellenlänge liegen.

Ein weiterer Vorteil des Verfahrens besteht darin, daß bei entsprechend
kurzbrennweitiger Fokussierung (Fokustiefe Δ z viel kleiner als die
Dicke d des Materials) die Information wahlweise in unterschiedlichen
Tiefen des homogenen Plastmaterials eingeschrieben werden kann, ohne
dabei die Oberflächen des Materials zu beeinflussen. Auf diese Weise
können auch echt dreidimensionale Strukturen im Material erzeugt werden (Abb. 2).

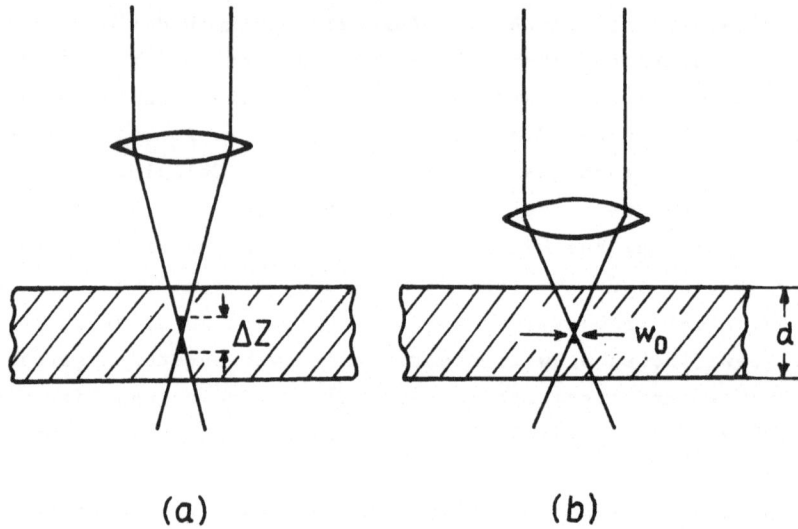

Abb. 1. Schematische Darstellung zum Verfahren mit Lage der beeinflußten Zone im Materialvolumen bei langbrennweitiger Fokussierung (a) und bei kurzbrennweitiger Fokussierung (b)

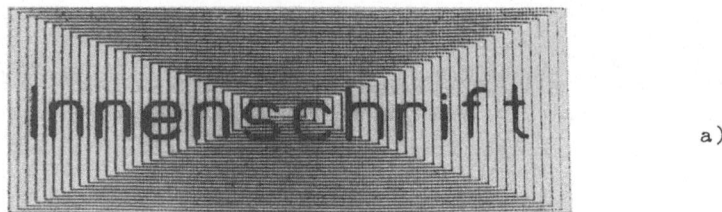

a)

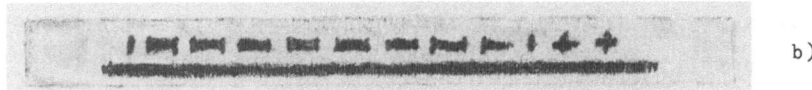

b)

Abb. 2. Probe mit zwei verschiedenen Strukturen, die übereinander
in unterschiedlicher Tiefe des Materials eingeschrieben
wurden. a) Draufsicht, b) Seitenansicht

3. Diskussion

Die physikalischen Ursachen des beschriebenen Prozesses sind bisher
nicht eindeutig geklärt. Intensitätsabhängige Messungen der Beeinflussung im Volumen der verwendeten Materialien haben gezeigt, daß eine
sichtbare Strukturierung erst oberhalb einer bestimmten Schwellintensität erfolgt und daß im untersuchten Intensitätsbereich kein linearer
Zusammenhang zwischen dem Grad der Beeinflussung und der Intensität
der Laserstrahlung besteht. Die Schwellintensität für die hier verwendeten Plastmaterialien konnte zu 4 bis 5 MW/cm² bestimmt werden. Beim
Erreichen bzw. bei Überschreiten der Intensitätsschwelle wird im Material im Fokusbereich des Laserstrahles eine Lichtemission im sichtbaren Spektralgebiet beobachtet. Diese kann z. B. als Fluoreszenz aus
durch Mehrquantenprozesse besetzten Zuständen der Moleküle des Materials
gedeutet werden. Die bisherigen experimentellen Befunde zur irreversiblen Beeinflussung des Volumens homogener transparenter Plastwerkstoffe
legen die Vermutung nahe, daß die beobachteten Effekte durch Mehrquan-
ten-Anregungsprozesse,verbunden mit einer nachfolgenden chemischen Umsetzung des Materials im Fokusvolumen des Laserstrahles hervorgerufen
werden. An einer genaueren wissenschaftlichen Klärung der Prozesse
wird zur Zeit noch gearbeitet.

4. Zusammenfassung

Der Beitrag behandelt ein Verfahren zum irreversiblen Einschreiben von
Information in das Volumen homogener transparenter Plastmaterialien.
Dabei wird die Strahlung eines gütegeschalteten cw-Nd-YAG-Lasers über
ein rechnergesteuertes Ablenksystem und eine Fokussieroptik in das Materialvolumen fokussiert. Die Laserleistung und die Brennweite der Optik werden so gewählt, daß die zur Beeinflussung notwendige material-
spezifische Schwellintensität nur im Fokusvolumen des Laserstrahles
überschritten wird. Die Kennzeichnung kann in beliebiger Tiefe des Ma-
terialvolumens erfolgen. Bei kontinuierlicher Bewegung des Fokus in
3 Koordinaten können dreidimensionale Strukturen erzeugt werden.
Das Verfahren kann insbesondere zum Einschreiben von Zeichen, Symbolen,
Skalen und anderen Strukturen in das Volumen des Materials verwendet
werden, wobei die Oberflächen völlig unbeeinflußt bleiben. Es kann
überall dort Verwendung finden, wo Informationen fälschungssicher und
vor Beschädigung geschützt sein sollen. Weiterhin umfaßt das Anwendungsgebiet die analoge oder digitale Speicherung von Informationen,
die eine optische Abtastung ermöglichen (z. B. Ton- oder Bildträger)
und ebenfalls weitgehend gegen mechanische Beschädigung der Oberfläche
unempfindlich sind. Das Verfahren ist ebenfalls anwendbar bei der Her-

384

stellung von modischen, mit unterschiedlichen Designs versehenen Artikeln.

Literatur

/1/ J. BUCHHOLZ: "Laserbeschriftungs- und Markierungsarbeiten",
    Informationsmaterial von Laser-Optronic GmbH, München
/2/ T. MAURER, H. J. HOLBEIN: Offenlegungsschrift DE-OS 30 48 735 A1
/3/ E. HEUMANN, J. KLEINSCHMIDT, W. KRAMER, E. SAUER, K. VOGLER,
    G. WIEDERHOLD, W. ZSCHOCKE: Offenlegungsschrift DE-OS 34-25 263 A1

Verfahren zur Lasermarkierung von Metallen

U. Sowada

Lambda Physik

Hans-Böckler-Str. 12, D-3400 Göttingen

Für Metallmarkierung benutzt man gepulste Laser, da nur so die in das zu markierende Teil eingebrachte Wärmemenge klein zu halten ist. Gerade darin liegt einer der Vorteile von Lasermarkierung; denn anderenfalls muß ein Präzisionsteil eventuell noch nachgearbeitet werden. Sehr wichtig ist auch bei kohlenstoffreichen Stählen, daß die erwärmte Oberflächenschicht so dünn wie möglich gehalten wird, um Veränderungen der metallurgischen Mikrostruktur zu vermeiden. Weil die Wärmeeindringtiefe bei Laserbestrahlung direkt mit Wärmeleitvermögen und Laserpulsdauer korreliert ist, erzielt man nur mit gepulsten Lasern die gewünschten Ergebnisse.

Es gibt drei verschiedene Verfahren, um mit einem Laser eine Oberfläche zu markieren. Das einfachste davon ist die sogenannte Maskentechnik, bei der der Laserstrahl eine Maske durchstrahlt, welche wie eine Schablone das zu markierende Symbol enthält. Meist wird die Maske dann verkleinert abgebildet, um auf der Oberfläche die für eine Markierung erforderliche Leistungsdichte zu erzeugen. Das Symbol wird also in einem Vorgang ohne bewegte Teile geschrieben. In den anderen beiden Verfahren, der Matrixtechnik und dem Laserschreiben, wird das Symbol rasterartig mit punktförmig gebündeltem Laserstrahl zusammengesetzt oder gar mit überlappenden Brennflecken geschrieben. Das ist zwar gleichbedeutend mit großer Flexibilität, aber die Laserstrahl-Steuerung ist aufwendig, und es sind viele Pulse nötig. Am schnellsten und am billigsten geht es mit der Maskentechnik.

Diese wird aber bisher bei Markierung von Metallen nicht eingesetzt. Der wesentliche Grund dafür ist aus Tabelle 1 ersichtlich.

Tabelle 1. Absorption reiner Metalloberflächen (in %)

bei senkrechtem Einfall für unterschiedliche Wellenlängen

Metall	A (10,6 μm)	A (1,06 μm)	A (0,25 μm)
Aluminium	2	10	18
Eisen	4	35	60
Kupfer	1	8	70
Molybdän	4	42	60
Nickel	5	25	58
Silber	1	3	77

Es ist deutlich, daß bei den infraroten Wellenlägen von CO_2-Laser (10,6 μm) und Nd:YAG-Laser Metalle kaum die auffallende Energie absorbieren. Daher muß für Markierung mit diesen Lasertypen der Strahl gebündelt werden, um die erforderliche Oberflächenerwärmung zu erreichen.

Für Excimerlaser hingegen gilt mit wenigen Ausnahmen (z.B. Aluminium), daß mehr Energie absorbiert als reflektiert wird. Interessanterweise ist Excimerlasermarkierung besonders vorteilhaft bei Kupfer und Silber, wo die Maskentechnik bisher überhaupt nicht einsetzbar war.

Metallurgische Untersuchungen an einem kohlenstoffreichen Stahl haben gezeigt, daß die Dicke der martensitischen Schicht bei Excimerlaser-Markierung nur 2,5 μm beträgt. Sie ist damit bei besserer Sichtbarkeit der Markierung um den Faktor 10 kleiner als bei Nd:YAG-Lasermarkierung. Diese Tatsache eröffnet der Lasermarkierung neue Möglichkeiten insbesondere dort, wo es auf die Erhaltung der metallurgischen Eigenschaften ankommt.

A Programmable Solid State Laser Marking System

D.I. Greenwood, J.J. Harris, N.T. Stock, L. Bademian
Isomet Laser Systems Limited
18 Llantarnam Industrial Park,
Cwmbran, Gwent. NP44 3AX. U.K.

INTRODUCTION

Pulsed lasers are now in common use for date stamping and batch coding of products. The most common technique is mask marking, wherein a specially prepared stencil mask is illuminated by the laser and imaged onto the target material. To change the inscription, the stencil mask must be changed either manually or electro-mechanically, and thus becomes unwieldy in applications requiring individual inscriptions such as serialisation or the recording of quality control data.

In this paper we describe systems based on a novel acousto-optic technique wherein a row of dot matrix alpha-numerics can be generated from a single laser pulse (reference 1). By this means any alpha-numeric mark can be produced on a shot by shot basis under micro-computer control. The technique is applicable at most laser wavelengths and in the following discussions we outline both a Nd:YAG and a Carbon Dioxide marking system.

NEODYMIUM:YAG LASER SYSTEM

Principle of Operation

The system is based around a multi-channel acousto-optic cell, where each acoustic column represents a column or row of a dot matrix alpha-numeric. A five channel cell, capable of producing a single character, is illustrated in figure 1.

There are two methods by which a character can be created. The first is analogous to a stencil mask. Here a single frequency is used to drive all the columns, but for each column the drive is modulated to generate the pixels necessary to produce the required alpha-numeric. The deflected light is then collected by simple optics and the cell is imaged onto the target. While very simple in principle, this technique suffers from two drawbacks; the laser pulse must be short enough to 'freeze' the acoustic columns, and the jitter on the laser timing must be held extremely low in order to ensure that the pixels are in the correct position.

In the second approach, the cell operates as a deflector. Into each column a sequence of 7 frequencies is injected, the duration at each frequency corresponding to one seventh of the aperture of the cell. In this case the character is generated by inhibiting or enabling each frequency in each column.

Unlike the 'active mask' technique above, anamorphic optics are required to realise the character at the target plane: the alphanumeric comprises a transform plane in the deflection direction, together with an orthogonal image plane. This is shown in figure 2.

By repeating the 7 frequency sequence continuously all 7 frequencies will be contained in the cell at any time, and since the position of each frequency is immaterial, accurate timing of the laser pulse is unnecessary (reference 2).

This principle can be extended by increasing the number of acoustic columns or frequencies to obtain multiple characters.

The Laser

The Nd:YAG source is a fixed Q flashlamp pumped oscillator with an output energy of 3J in a pulse length of approximately 200µs. The laser is multimode in order to obtain maximum energy extraction and to present a suitable energy distribution to the cell.

The selection of a fixed Q source followed a series of comparative tests with the oscillator, both Q-switched and with fixed Q. It was found that for a given pump energy, although a greater range of target materials could be marked with a Q-switch, the higher output energy of the fixed Q version gave a better quality of mark on the majority of surfaces including inked and painted materials and plastics.

There are additional benefits from using a long pulse length which will become apparent in the following section.

The Character Generation Cell

A six character string from a single laser pulse is realised by using a cell with seven electrodes each generating 30 frequencies (figure 3). This configuration offers lower cell and electronic complexity than the inverse.

With overlap of adjacent pixels, the required resolution of the cell is 35, corresponding to the 30 frequencies and 5 inter-character spaces. This resolution is realised by a bandwidth of 34MHz and a run length at each frequency of 1µs.

With a short laser pulse the acoustic column length of the cell would need to be equivalent to 30µs in order to contain all the frequencies in the cell. The 200µs laser pulse used eases this requirement, since during the course of the laser pulse, all the frequencies can sequence through the cell several times. A cell aperture equivalent to 5µs is used, which is sufficient to minimise pixel broadening through the filling and evacuation times of the frequency packets. An additional feature of the long laser pulse is that it results in averaging of the laser energy distribution in one dimension.

The cell is constructed from lead molybdate, with an active aperture of 20mm in the acoustic column direction by 6mm across the electrodes. A centre frequency of 150MHz has been used which makes the fractional bandwidth low to facilitate a high efficiency over the band. A drive power of 5W per channel is required which is gated for 400µs around each laser pulse.

The electrodes used are nested (reference 3). This provides a smoother deflection efficiency profile across the electrode and enables overlap of adjacent pixels (figure 4). This overlap, in addition to improving the quality of characters produced, has the added benefit of minimising deflection efficiency troughs between acoustic columns.

Optics

The simple anamorphic optics shown in figure 2 can be configured to obtain a primary image of the correct aspect ratio, but are unwieldy for the multicharacter cell format, and relaying the image to a remote target with an acceptable depth of focus is difficult.

The optical train adopted is shown in figure 5. After the cell, a spherical beam reduction telescope produces a reduced virtual image of the acoustic columns with appropriately magnified deflection angles and divergences. A positive cylindrical lens placed at its focal length from this image collimates the image in the non deflection sense. The power of the telescope and cylindrical lens focal length are chosen to give the correct aspect ratio in the final image.

The final imaging is achieved with a spherical lens with a focal length chosen to give the desired image size. The separation of this lens from the rest of the optical train can be varied without altering the final image. This is of importance to the beam delivery system, which must be tailored to suit individual installations.

Electronics

A block diagram of the electronic drive network for a six character system is shown in figure 6. The sequence of cell frequencies is generated from a single voltage controlled oscillator (VCO) and fed to the seven drive amplifiers via double balanced modulators. The ASCII

character strings from the microcomputer are stored in RAM and con-
verted by the character generation ROM to control the double balanced
modulators in synchronisation with the VCO driver.

Although the system illustrated has been designed for a six
character string, a facility is provided which allows shorter
character strings to be created without energy wastage by appropriate
reduction of the number of frequencies generated.

System Performance

The system produces strings of six alphanumeric characters with a
pixel exposure variation of less than 5%. If all six characters are
generated, the average pixel energy for a 3J laser output is 7mJ.

For low marking threshhold materials, such as inked paper and card,
typical character heights are 2mm whereas for inked and painted metal
surfaces a 1mm character height is generally used. For higher threshhold materials, the character height can be further reduced or the
number of characters marked per pulse may be restricted. Our
experience to date has shown that the majority of plastic materials
can be marked with a six character string of 1mm height or greater.

CARBON DIOXIDE LASER SYSTEM

The obvious choice of laser is a T.E.A. device with a pulse length
in the region of 1μs. Such a pulse length will effectively 'freeze'
the acoustic columns in the cell, implying that, unlike the Nd:YAG
system, all frequencies used must be present in the cell when the
laser fires.

To create a six character string from a cell with seven electrodes
and 30 frequencies now becomes impractical. A 1μs duration at each
frequency would imply a 30μs cell aperture, equivalent to 165mm in
germanium. To obtain sufficient acoustic column control over such a
distance would demand an excessively large electrode size, with pro-
hibitive drive requirements. Reduction to a four character string
does not represent a significant improvement.

The approach adopted is to generate a 2 x 2 matrix of characters
using two groups of 5 acoustic columns and driving each with fourteen
frequencies (figure 7). The resolution requirement for the cell is
thus 15 spots, and this is achieved with a bandwidth of 28MHz at a
centre frequency of 70MHz and a duration at each frequency of 0.5μs.
The germanium cell is 40mm long, with an electrode pitch for each
nested group of 2.5mm. The drive power required is 50W per channel,
which is typical for germanium cells.

System Performance

The overall system efficiency of the 10.6μm character generator is

65% which is comparable to the beam utilisation of a conventional four character stencil mask. The programmable marker therefore offers the same material marking capability as existing systems.

For 'laser friendly' materials the laser energy available in existing systems is considerably greater than that required for a typical four character mark. This enables the laser energy to be split into two channels to give both a conventional stencil mark together with a programmable character mark.

CONCLUSION

The application of multi-channel acousto-optic technology to the field of laser marking has enabled a fully programmable character generation technique to be developed. This technique has been applied to both pulsed Nd.YAG and Carbon Dioxide lasers.

In the Nd.YAG case, high quality six character strings can be marked on a variety of common packaging materials with typical character heights of 1mm. In the Carbon Dioxide version a 2 x 2 character matrix is generated with a similar marking performance to conventional stencil marking systems.

The extention of these techniques to encompass more complex formats such as Japanese characters is currently being studied.

REFERENCES

(1)   Patent Application Filed.
(2)   Patent Application Filed.
(3)   L. BADEMIAN. Proc. SPIE 498, 102 (1984)

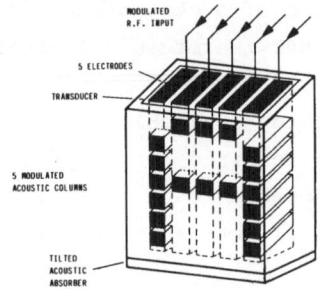

Figure 1. Five Channel
Acousto-Optic Deflector
Single Character Generation

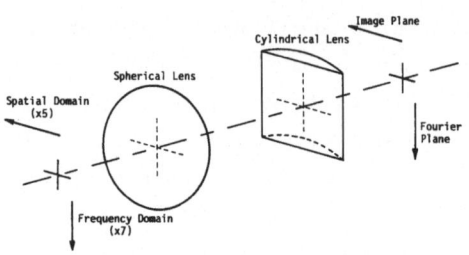

Figure 2. Anamorphic Optics for
Realisation of Character with
Multi-Frequency Technique

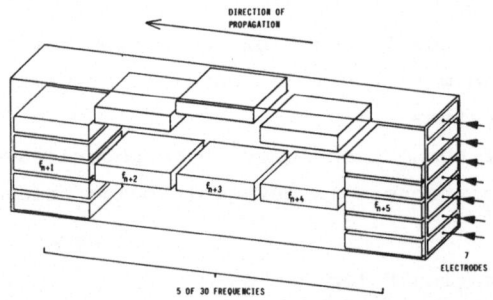

Figure 3. Seven Channel Acousto-Optic Cell for the
Generation of a six Character String at 1.06 Micron

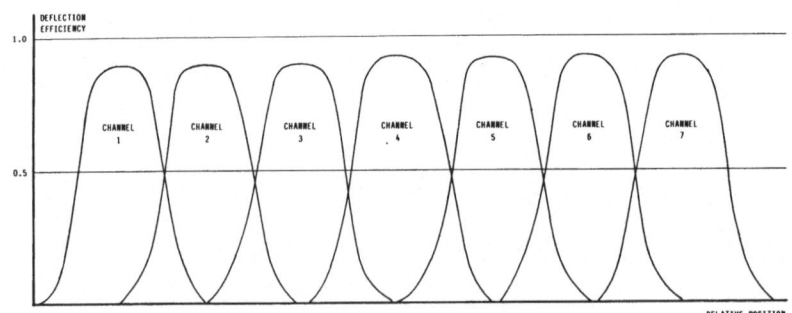

Figure 4. Acoustic Column Deflection Efficiencies
for 1.06 Micron Cell with Nested Electrodes

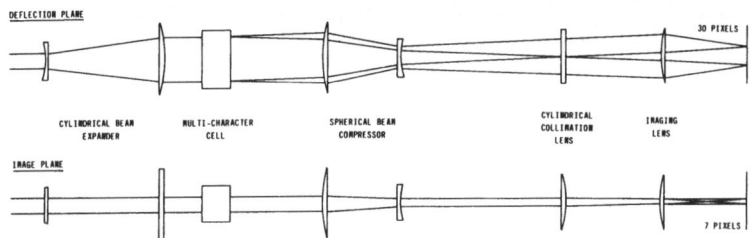

Figure 5. 1.06 Micron Multi-Character Generator Optical Train

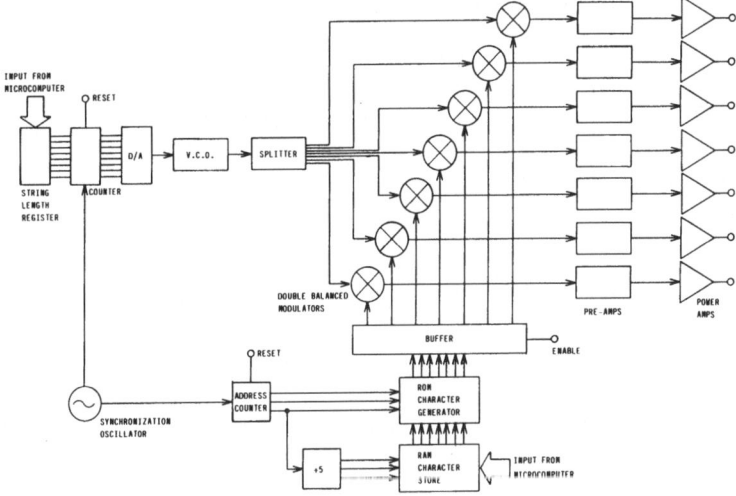

Figure 6. 1.06 Micron Electronic Drive Network

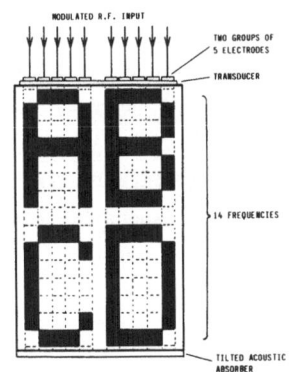

Figure 7. 10.6 Micron Acousto-Optic Cell Configuration for the Generation of a 2 x 2 Character Matrix

Laser Cladding with Multi Elemental Powder Feed

T. Takeda, W.M. Steen and D.R.F. West

Metallurgy Department

Imperial College, London SW7 2BP

Abstract

An investigation is reported on surface cladding with a laser using a mixed powder feed. Success has been achieved in obtaining either homogeneous or non homogeneous layers depending upon the operating conditions. The extent of mixing and the flow structure within the melt pool have been partially observed by the use of copper marker wires embedded in the substrate. Using two powder delivery pipes a 'layered' deposit was achieved.

Introduction

There is a need for a non contact cladding process which can be controlled by automation and which is suitable for robot guided laser beams. Such a process is the blown powder cladding process discussed elsewhere (1,2,3). Discussed here is a process variation in which the blown powder consists of separate elemental or alloy particles which are alloyed in-situ in the laser generated melt pool.

The extent of the melt pool, flow patterns and heat transfer all affect the level of homogeneity of the resultant clad as discussed in this paper which reports some results from a wide ranging set of experiments using this new process.

Experimental

Cladding with mixed powder feed onto an En3 steel substrate (0.2%C) was performed in three different ways:

(a) Using premixed powder in a single hopper with a single feed pipe (fig. 1a)

(b) Multiple hoppers feeding into a single delivery pipe (fig 1b); only three hoppers were used each separately controlled by its own stepping motor and gas delivery system.

(c) Multiple hoppers feeding separate delivery pipes (fig 1c); only two hoppers and two feed pipes were used.

In each experiment, a single track was laid down, using a Control Laser 2kW laser under carefully monitored conditions of laser power, beam diameter mode structure, powder feed rate, traverse speed and powder injection position and angle.

The tracks produced were generally examined for size and shape for microhardness as seen by optical microscopy for chemical composition, as measured by electron microprobe analysis (EPMA) using a spectrometer on a JEOL JSM 35 instrument and for micro-hardness, as measured on a Leitz Miniload 2. Further tracks were produced in which the clad was laid down over a copper marker wire which was embedded in the substrate surface. These tracks were only examined by EPMA with a view to establishing data on the flow patern within the melt pool while cladding.

The alloy system studied is that of Fe/Cr/Ni which is of interest in relation to stainless steels and for which metallurgical data are available with which to compare the results (4,5). In methods a & b powders were fed as elemental powders. In the twin feed system method 'c' only two powders were used. They were nickel and Colmonoy 5 (a nickel based hardfacing alloy).

Results

The single feed pipe techniques (a) and (b) gave similar results. For 1.7kW laser power and beam diameter 7.9mm with a near Gaussian mode and powder feed rate of 0.293g/s an approximately uniform composition was observed using microprobe technique (10 μm spot size) for traverse speeds less than 7mm/s. (fig 2). The observed and expected hardness values are shown in table 1. Above this speed a non uniform clad was produced due to insufficient mixing and diffusion time.

Table 1. Comparison of the literature values (5) and observed hardness for the three compositions examined. (method a).

Cr/Ni Ratio	Structure Indicated by Schaeffler dia.(4)	Hv expected (5)	Hv observed.
13:6	Martensitic	(type 414)* 450 (Hardened) (C% 0.15max)	330 — 380
18:9	Austenite/Martensite Ferrite	(type 301)*400 (work hardened) (C 0.15max)	322 — 386
25:20	Austenite/Ferrite	(type 310)*180 (annealed) (0.25%C max)	158 — 183

* Stainless steel			Composition	(wt%)	
Type	C	Cr	Ni	Si	Mn
414	0.15	11.4-13.5	1.25-2.5	1.0	1.0
301	0.15	16-18	6-8	1.0	2.0
310	0.25	24-26	19-22	1.5	2.0

The twin feed pipe system (c) gave a structural variation band whose location was dependent upon the pointing of the feed system relative to the molten pool. This band was associated with good mixing beneath it and poor mixing above as well as a slightly higher concentration of the upper feed material (Colmonoy 5) in the upper layers and vice versa with the lower feed material (Ni). The copper marker experiments gave data on the melt pool size (fig 4, fig 5). They also indicated the flow pattern and velocities in the melt zone.

Discussion

The process of cladding with mixed powder feed can produce relatively homogeneous layers at slow speeds or layers of variable composition at higher speeds. The concentration peaks seen in a deposit made at higher speeds (7mm/s for 1.7kW and 8mm beam dia. fig. 3) have a typical size of around 100μm similar to the fed particle size. This suggests lack of diffusion or melting. This process depends upon molten time^V can be found from the dimensions of the melt pool as measured by the copper in the marker experiments. Results are shown in table 2.

Table 2. Pool sizes and molten times for various traverse speeds for a laser power of 1.86kW, 4mm beam dia. and 0.315g/s powder feed.

Speed mm/s	Pool Length mm	Molten Time s	Theoretical Molten Time. s
4.4	5	1.13	
7	3.5	0.5	0.3 (cf fig.6).
12	1.8	0.15	
19.5	1.7	0.09	

The centre line molten times would be expected to vary across the section as shown from Weerasinghe's model (6) fig 6. The calculated diffusion distance for a 7mm/s traverse speed 1.7kW, 8mm beam is of the order of 50μm and therefore would explain why this traverse speed and processing condition represents the divide between homogeneity and variable composition. Since the edges were also found to be approximately uniform convection must have played a large role in enhancing diffusion in these regions of reduced molten time.

Concerning hardness values (table 1) comparison is hindered by the higher carbon levels of the commercial alloys compared with the very low carbon values expected with alloying using the pure elements. In the 18:9 insitu alloy the hardness was much higher than the annealed 301 steel but similar to the work hardened material; this contrasts with the 25:20 steel whose hardness corresponded to the annealed state and not the work hardened state. Laser surface melted tracks in 17:11 stainless steel (316) were found by Lamb (7) to have hardness values of only 220Hv i.e. close to the annealed value. The reason for the higher values observed in the insitu 18:9 alloy is the subject of further study as is the analysis of the structure. The banding observed in the twin feed method suggests that there is freezing from the base and so late arriving particles cannot mix throughout the depth. The EDAX scan in fig 7 shows that there is considerable mixing between the layers probably as a consequence of the mixing of the powder in the impinging powder streams. However the change in scale of the dendrites without any large compositional change being observed in the EDAX scans for either Ni or Cr is also suggestive of a variation in cooling rate. The flow structure in forming these deposits is the subject of future work.

Summary and Conclusions

1. Insitu alloys can be clad by blowing mixed elemental powders.

2. Homogeneity depends on having a certain molten time and so varies with traverse speed, location in deposit and impingement point of the powder.

3. There is a lack of homogeneity at clad speeds greater than 7mm/s for a laser power of 1.7kW beam diameter 8mm and powder feed rate of 0.293g/s.

4. Layered structures can be prepared from a layered powder feed if the impingement point of the powder is correctly directed.

5. A range of stainless steels have been prepared by blowing a premixed powder feed.

Acknowledgements

The authors wish to express their thanks to Komatsu Ltd, Tokyo Japan for supporting Mr. T. Takeda's studies at Imperial College and to Quantum Laser Corp. USA for financial support for this work. They would also like to acknowledge the technical help of Mr. G. Briers with the EPMA measurements, Mr. R.Stracey with the operation of the laser and Mr. A. Pace for the experimental work on copper markers.

References

1. V.M. Weerasinghe, W.M. Steen Proc. Conf.'Lasers in Material Processing' Los Angeles 1983 ed. E. Metzbower. Publ. ASM Metals Park, Ohio 44073.

2. R.M. McIntyre Paper 8301-022 pp230-233 1983.

3. V.M. Weerasinghe, W.M. Steen Weld & Met Fabr. Nov. 1983.

4. A. Schaeffler Metal Progress 56 (5) 680 and 680b1949.

5. Metals Handbook vol.1 8th Edition p.414. 1961.

6. V.M. Weerasinghe, W.M. Steen. Proc. Conf. 'Transport phenomena in materials processing'. ASME pub. 345 East 47 St. NY 10017 p.15-23. 1983.

7. Lamb. M., Ph.D. thesis, London University 1985

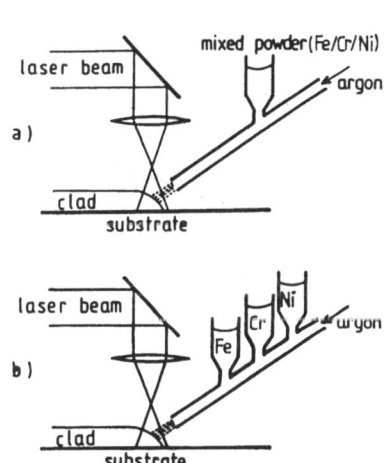

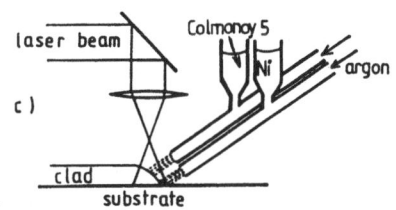

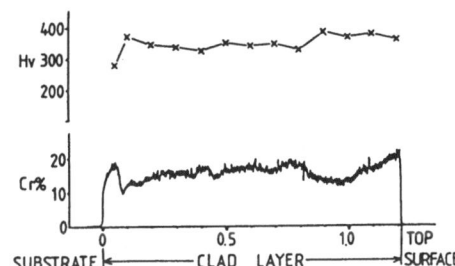

Fig 2.
Variations in hardness and chromium content, measured by EPMA through the centreline of a clad sample from a premixed powder feed of 18:9 Cr:Ni bal. Fe.
Processing conditions were: laser power 1.7kw, beam diameter 8mm (TEMOO), powder feed rate 0.293 g/s, traverse speed 7mm/s

Fig 1. Experimental Arrangement
a) Arrangement for the single hopper single feed system.
b) Arrangement for the triple hopper single feed system.
c) Arrangement for the twin hopper twin feed system

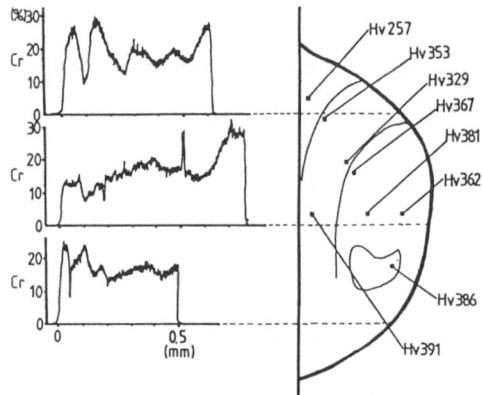

Fig 3.
Chromium % from EPMA scan across
a sample of 18:9:Cr:Ni premixed
feed deposited at 13.6 mm/s.
Laser power 1.5kw, beam diameter
8 mm, powder feed rate 0.293g/s

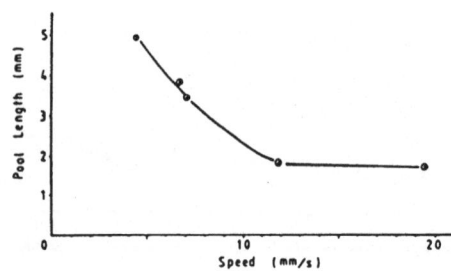

Fig 4.
Variation in melt pool length with
speed

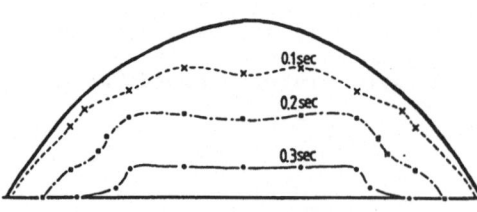

Fig 6.
Theoretical molten time contours
for a laser power of 1.83kw, beam
diam. 5 mm, traverse speed 6.67
mm/s, powder feed rate 0.2g/s

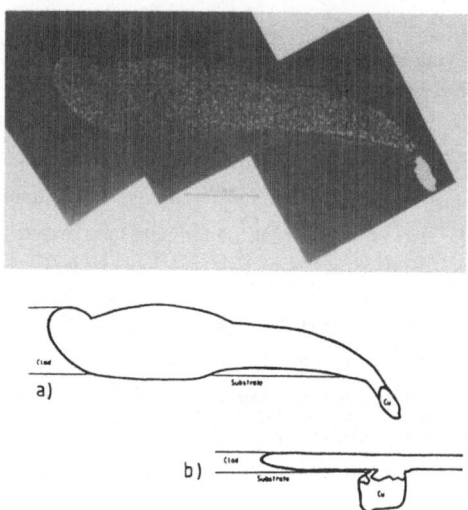

Fig 5.
EDAX scan and diagram of the
location of copper in a longitudinal section of a clad.
The diagram shows the location of
copper in tracks taken at two
speeds: a) 4mm/s b) 12mm/s

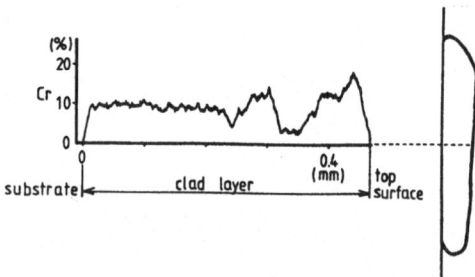

Fig 7.
EDAX scan of a twin feed clad
layer made with Colmonoy 5 above
and nickel feed beneath.
The track was made at 6.7mm/s,
laser power 1.72kw and beam diam.
5mm

Thermochemical Treatment of Titanium Alloys with Lasers (Laser Gas Alloying)

H. W. Bergmann, T. Bell[*], S. Lee
Institut für Werkstoffkunde und Werkstofftechnik, TU Clausthal
Agricolsatr. 2, 3392 Clausthal-Zellerfeld

Abstract

The present paper describes the possibility of thermochemical treatments of Ti and Ti alloys by laser melting of the surface and on
reaction of the melt with suitable gasses like O_2, N_2, CH_4, etc.
forming surface layers containing compounds like TiN, TiO, TiC, etc.

Introduction

The use of lasers in material processing has increased during recent
years. In the beginning mainly writing, cutting and welding applications were used but later the use /1/ of lasers in surface technology
came up /2/. In addition today laser enhanced CVD techniques are known.
In the present paper the application of lasers for thermochemical reaction is demonstrated. The principle of thermochemical treatments is to
anneal a workpiece at a certain temperature together with a reaction
partner in the solid, liquid, gaseous or plasma state and allow for a
concentrational change by a diffusion process /3/. The aim of the process can be either a diffusional zone (carburizing, nitriding) or a
compound layer (γ,ε). In the classical thermochemical treatments the
necessary case depth is obtained by a solid state diffusion process
requiring high temperature and long processing times. A first approximation of the process time can be obtained from a parabolic growth law.

$$x = 2 * \sqrt{D\,t} \quad (1) \qquad\qquad D = D_0 \exp^{Q/RT} \quad (2)$$

---

[*] Dep. of Metallurgy and Material Science, University of Birmingham

High temperatures and long treatment times often cause considerable distortion. This requires remachining and very often lead to a removal of the surface layer. Laser material processing enables a completely new process technique, the significance of which can not yet be fore seen. By local laser melting of the workpiece a liquid gas reaction takes place and the absorbed gas or reaction products of the liquid and the gas are transported into the depth by convection of the melt. This allows for a more rapid transport mechanism compared to solid state diffusion. By this techniques alloys and compounds of the bulk material and natural gases like N_2, O_2 or artifical gases like CH_4 and metal aromates can be obtained. Part of the process is the decomposition of the gas in the laser plasma. The reactions can take place with the basic metal or with the alloying elements of it. Typical examples are the carburizing of steel and the carburizing and nitriding of titanium for the first typ of reaction and the nitriding of chromium steels and the selective internal oxidation of Ag-Cd-type of contact materials.

Titanium is like no other metal predestinated for a thermochemical treatment. On the one hand it has low weight and high strength on the other hand its applications are restricted due to a high coeffcient of friction and unfavourable tribological properties. The results obtained by laser gas alloying of titanium with N_2 and CH_4 have been reported in detail elsewhere /4/. It is the aim of the present paper to point out the manyfold possibilities that can be obtained by laser thermochemical processing.

Experimental

Commercial pure titanium and IMI 318 (TiAl6V4) was cut into 100 cm square pieces , 15 mm thick. Both sides were fine machined and than laser processed in the way which is given in the schematic drawings, fig. 1 and 2. A subsequent melting of the surface under various process gases or mixtures of them was carried out, see table. 1. N_2, O_2 and CH_4 were used as processing gases sometimes diluted in argon in order to avoid hazardeous reactions. The feed rate used was 0.5 m/min, the focal distance was choosen 0.5 cm below the focal point, the pulse frequency was 500 Hz and the applied pulse power was 4,3,2 and 1 kW respectively. By different subsequent treatments the mass transfer obtained and the influence of subsequent treatment on hardness, microstructure and roughness was studied. The aim of these investigations were to optimize between tribological properties, fatigue behaviour and finish of the product.

Results

The results of the thermochemical treatments are given in detail in
table 1. First of all it can be observed that with increasing exposure-
time or by repeating the process the volume fractions of the interme-
tallic phases formed on liquid gas reactions e.g.TiN, TiC on TiO increases and with that fraction the obtained hardness see fig. 3 a,b.
Typical microstructures of the surface layer are given in fig. 4 a-d.
From these it can be realized that the layer consists of a dendriti-
cally solidified compound with an interdendritically solidified remaining melt. For comparison the microstructure obtained for melting in
inert atmosphere is given in fig. 4 e,f. The influence of subsequent
altering the process gas is demonstrated in the hardness values,
table 1, and the macrostructures shown in fig. 5.

Furthermore it can be seen in table 1, how the colour and the finish of
the specimen can be altered by the sequence of multiple treatments, e.g.
a dramatic improve in the tribological behaviour can be achieved together with considerably high hardness. Which of the processes listed
so far in table 1 give the optimum properties in terms of wear, fatigue
and finish can not yet be said and in addition there are further processes to be considered. A first guess can be derived from the fatigue
properties given in fig. 6 and the tensile tests given in table 2.

For the moment it is not yet possible to derive conclusions from the
results obtained so far but the paper demonstrates the enormous capabilities of this type of processes.

Acknowledgement

The autors acknowledge the financial support of the BMFT and the
Commission of the European Community.

References

/1/ 'Metals handbook' 9th edn, Vol. 1, 1978, Metals Park, Ohio,
 ASM.
/2/ B. L. MORDIKE and H. W. BERGMANN: in"Rapidly solidified
 amorphous and crystalline alloys", ed. B. H. Kear et al,
 MRS Symp. Proc. Vol. 8, 463, 1982, New York, Elsevier
/3/ A. STAINES and T. BELL: in"Heat treatment, methods and
 media", 58 - 69, 1979, London, Institution of Metallurgists
/4/ T. BELL, H. W. BERGMANN, J. LANAGAN, P. H. MORTON, A. STAINES
 Surface Engineering 1, (1985), in press

Table 1. Properties of laser gas alloyed Titanium, IMI 318 (TiAl6V4) after processing in various treatment gases

No.	Treatment conditions	Colour	Rough-ness	Micro-hardness	Macro-hardness	Micro-structure	Phases	Comments
0	starting material	silver	7,75	336	296	equiaxed graines of α,β-Ti	α', β-Ti	
1	1 x molten under N_2	yellow golden	38,8	594	879	TiN-dendrites in α-matrix (dendritic zone)	α'-Ti + TiN	dense layer
2	2 x molten under N_2	yellow golden	23,9	652	1033	$\sim 50 - 100$ μm	α'-Ti + TiN	dense layer
3	4 x molten under N_2	yellow golden	21,1	595	1222	heat affected zone ~ 50 μm	α'-Ti + TiN	uniform dense layer
4	8 x molten under N_2	yellow golden	12,9	956	934	dendrite armspacing $5 - 6$ μm	α'-Ti + TiN	uniform dense layer cracks starting at hardness impr
5	1 x molten under CH_4+ argon (7,5% CH_4)	dark grey	12,8	527	528	TiC-precipitation in α'-matrix	α'-Ti + TiC	insufficient layer
6	2 x molten under CH_4 + Argon (7,5 % CH_4)	grey	12,3	519	368	TiC-dendrites in α'-matrix	α'-Ti + TiC	uniform layer

7	4 x molten dark grey under CH_4 + Argon (7,5 % CH_4)	12,5	573	623	molten case 150–200μm heat affected zone ~ 50 μm	α'-Ti + TiC	uniform dense layer
8	8 x molten light grey under CH_4 + Argon (7,5 % CH_4)	12,1	625	770	dendrite arm-spacing 1–2 μm	α'-Ti + TiC	uniform dense layer
9	1 x molten dark grey under O_2 + Argon (7,5 % O_2)	15,9	387	373		TiO + Ti_2O	non uniform dense layer
10	2 x molten dark grey under O_2 + Argon (7,5 % O_2)	12,4 3,8	581 487	467– 387		TiO + Ti_2O	dense layer
11	4 x molten grey under O_2 + Argon (7,5 % O_2)	22,9	460	566	scale consisting of TiO + Ti_2O which sparks off	TiO + TiO_2	wavy surface
12	3 x molten silver under N_2 then 1 x under Argon	20,6	626	435	above the dendritic TiN-layer is a layer with α'-Ti + TiN precipitations (30 μm)	α'-Ti + TiN	large brittle-ness, dense layer

No.	Process	Color				Structure	Phase	Remarks
13	3 x molten under CH_4 + Argon then 1 x under Argon	silver	18,1	469	257	above the dendritic TiC-layer is a layer with α'-Ti + TiC-precipitations (30 μm)	α'-Ti + TiC	dense layer, no brittleness
14	3 x molten under O_2 + 1 x Ar					not evaluable	TiO + Ti_2O + α'-Ti	bad roughness
15	3 x molten under N_2 + 3 x CH_4	grey	12,9	632	526	80 μm coarse TiN-dendrites, fine TiC-dendrites (60 μm)	α'-Ti+TiC after grinding α'-Ti+TiN	dense layer
16	3 x molten under CH_4 + 3 x N_2	golden	9,25	542	701	80 μm fine TiC-dendrites, coarse TiN-dendrites (60 μm)	α'-Ti+TiC after grinding α'-Ti+TiN	dense layer
17	3 x molten under N_2 + 3 x O_2 + Argon	grey	17,8	611	411	80 μm coarse TiN-dendrites, Ti_2O + TiO-dendrites (60 μm)	TiO + Ti_2O TiN after grinding	inhomogenious layer
18	3 x molten under O_2 + Argon, 3 x N_2	golden	14,1	456	796	80 μm Ti_2O + TiO-dendrites, TiN-dendrites (60 μm)	TiN after grind. TiO + Ti_2O	insufficient layer
19	3 x molten with CH_4, then 3 x O_2 + Ar	grey	16,6	475	386	80 μm fine TiC-dendr. TiO + Ti_2O (60 μm)	TiO + Ti_2 after grind. TiC	insufficient layer

No.	Treatment	Colour			Structure	Phase	Layer
20	3 x molten with O$_2$ + Argon, then 3 x CH$_4$ + Ar	light grey	22,3	957	559	80 μm Ti$_2$O + TiO-dendr. TiC-dendr. (60 μm)	TiC after grind. TiO +Ti$_2$O	insufficient layer
21	3 x (1 x N$_2$ + 1 x CH$_4$ + Ar	grey	33,3	608	465	Ti(N,C) dendrites	Ti(N,C)	sufficient layer
22	6 x molten (N$_2$ + CH$_4$ + Argon)	grey	31,3	497	545	Ti(N,C) dendrites	Ti(N,C)	dense layer
23	6 x molten N$_2$ by reduction of energy	dark golden	34	652	1042	see No.3-4	α'-Ti + TiN	very dense layer
24	6 x molten CH$_4$ + Ar reduction of energy	grey	14,5	606	734	see No.7-8	α'-Ti + TiC	very dense layer
25	1 x molten Ar	silver	4,2	328	187	martensite	β-,α'-Ti	
26	1 x Ar + H$_2$	silver	4,6	354	276	martensite	β-,α'-Ti	
27	4 x Ar + H$_2$	silver	4,6	361	255	martensite	β-,α'-Ti	

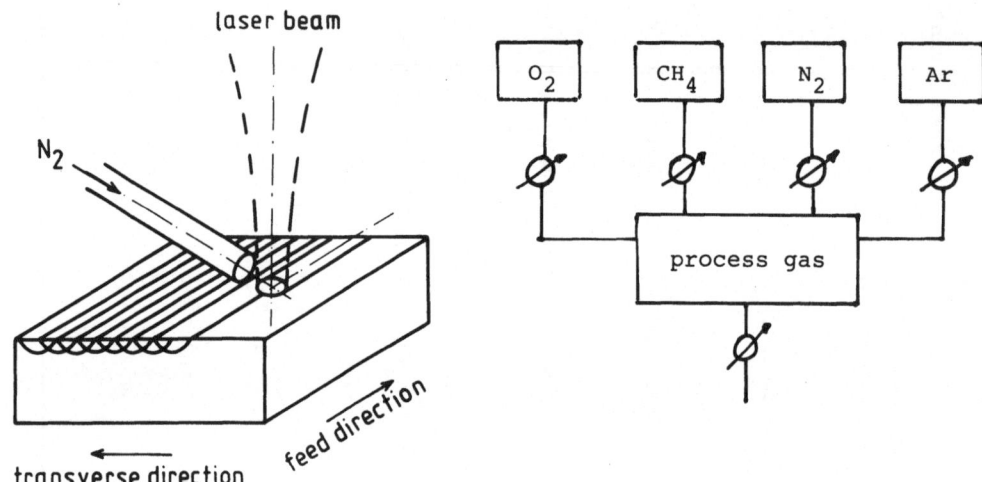

Fig. 1. Schematic drawing of
the laser gas alloying
process

Fig. 2. Schematic drawing of
the equipment used

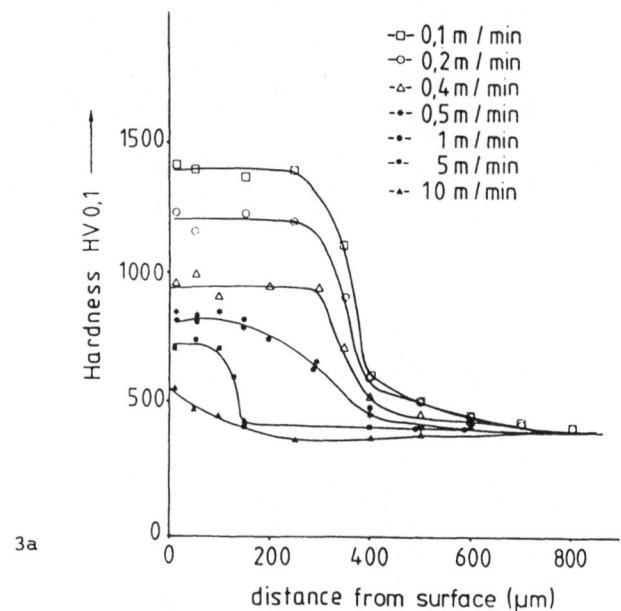

3a

Fig. 3. Hardness profiles for gas alloying of Titanium with different
exposure times
a) nitriding

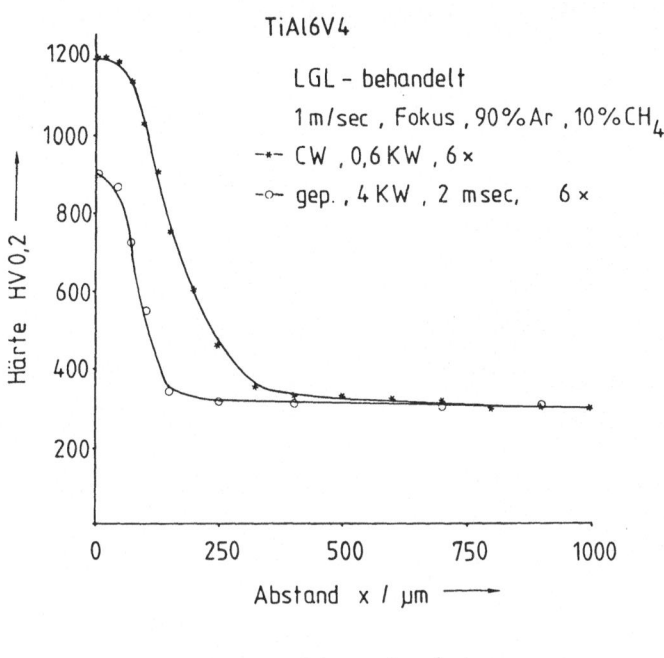

b) carburizing

Table 2. Tensile properties of laser melted IMI Ti 318

Condition	U.T.S. (N/mm^2)	% Elongation
As received	1010	19
Laser melted in He (CW)	955	6
Laser melted in N_2 (CW)	975	4
Laser melted in N_2 (pulsed)	985	6.5

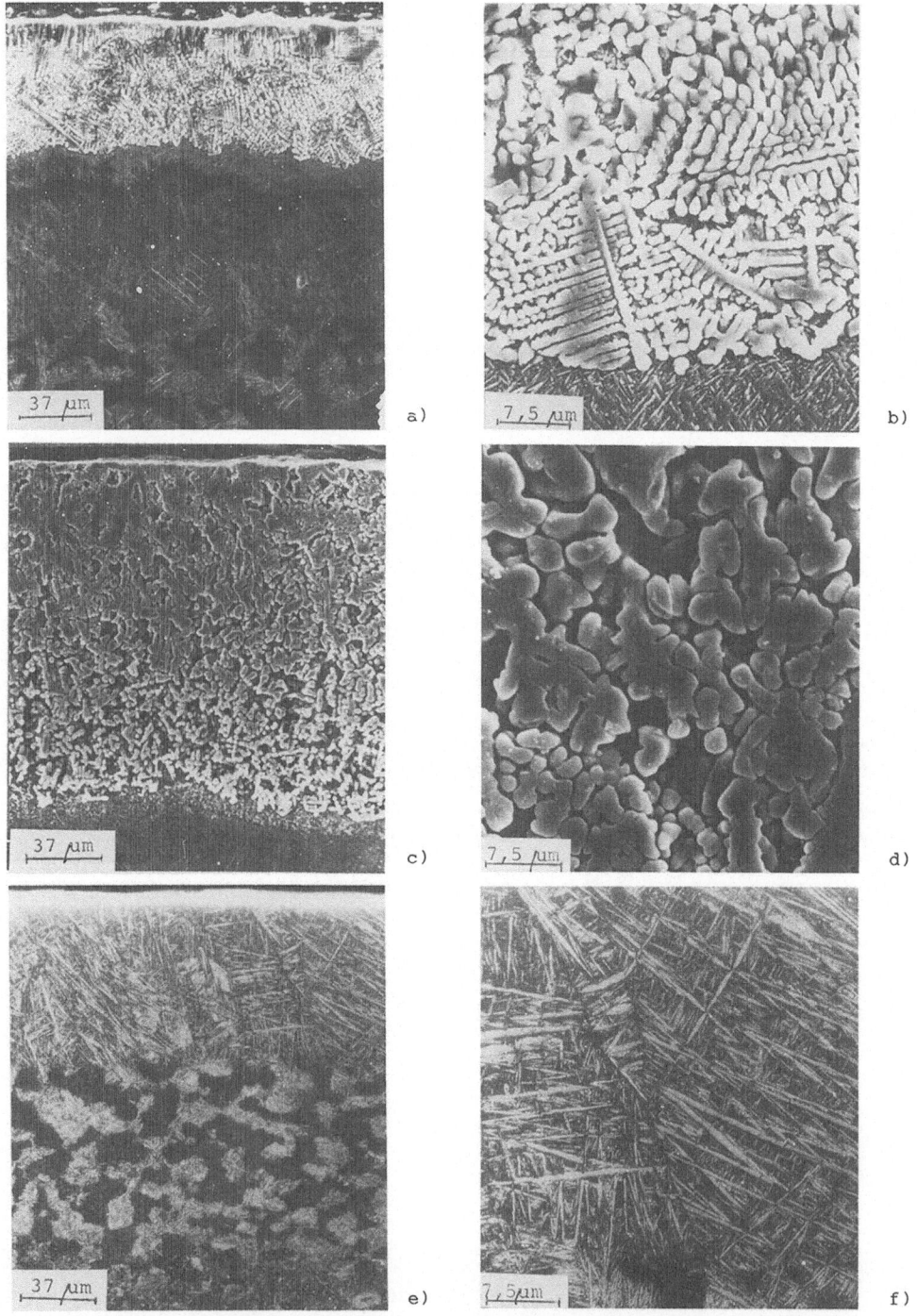

Fig. 4. Microstructures of laser gas alloyed Titanium processed in
a) N_2 b) selected area of a)
c) CH_4 + Argon   d) selected area of c)
e) Argon   f) selected area of e)

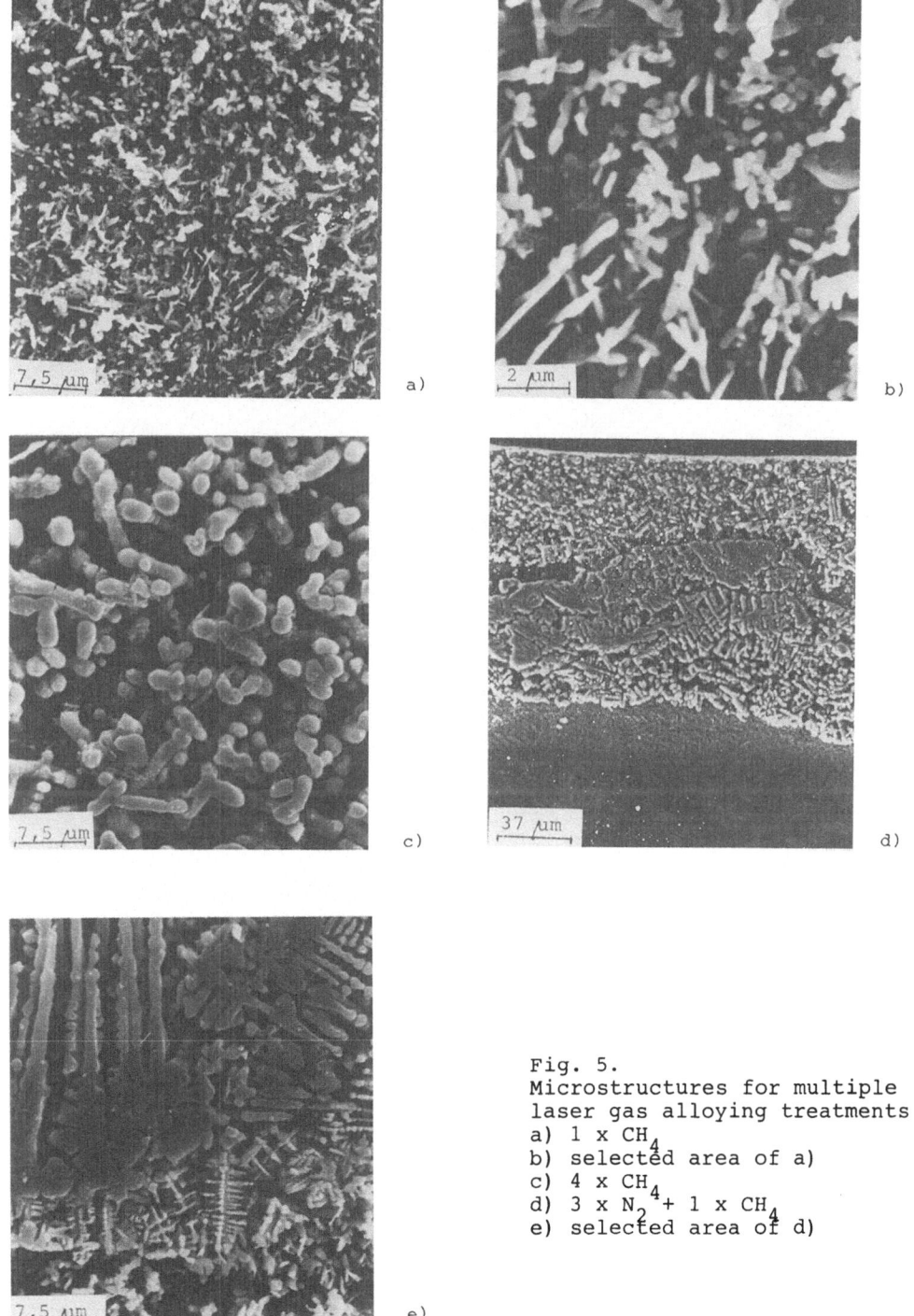

7,5 μm   a)

2 μm   b)

7,5 μm   c)

37 μm   d)

7,5 μm   e)

Fig. 5.
Microstructures for multiple
laser gas alloying treatments
a) 1 x CH_4
b) selected area of a)
c) 4 x CH_4
d) 3 x N_2 + 1 x CH_4
e) selected area of d)

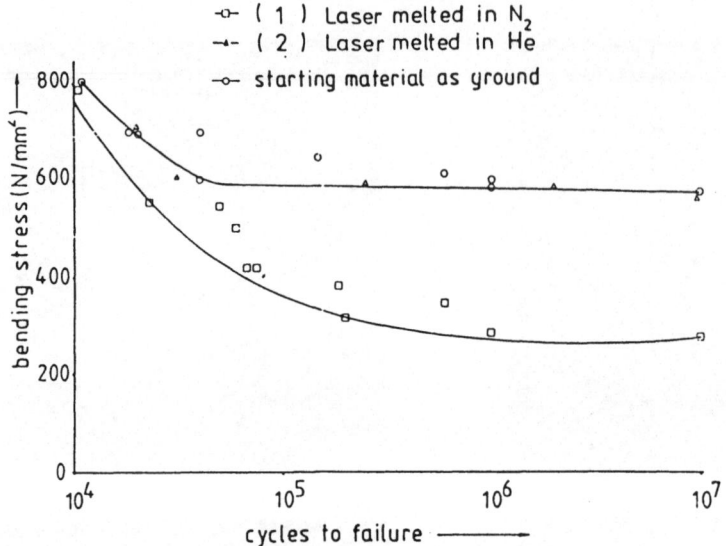

Fig. 6. Fatigue properties of laser gas alloyed Titanium
a) nitriding

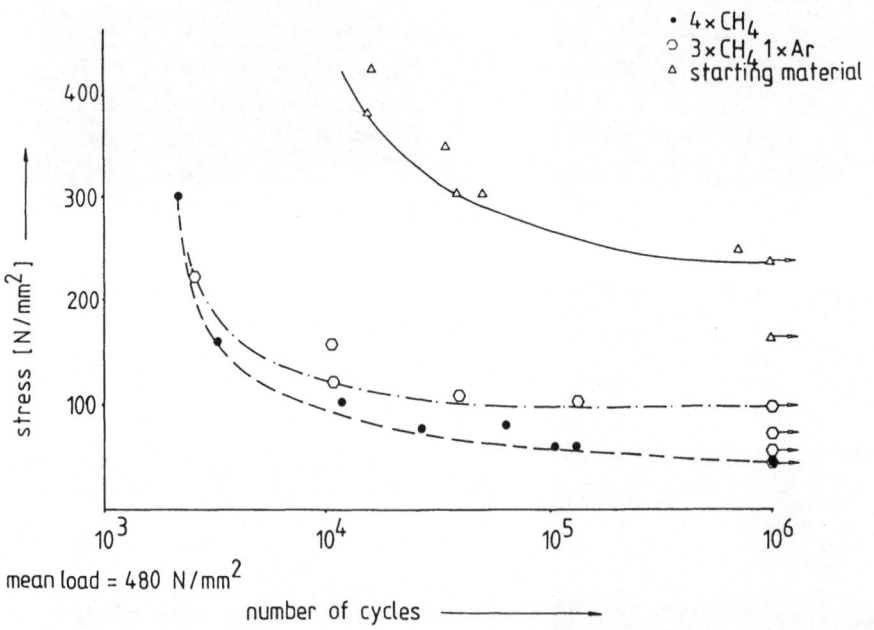

b) carburizing

Laser-Hardening of Machine Components

R. Krumphold, H. Paul, W. Reizenstein, Zentralinstitut für Festkörperphysik und
Werkstoff-Forschung der Akademie der Wissenschaften der DDR, Helmholtzstr. 20
8027 Dresden / DDR

The development and industrial introduction of technologies concerning the laser
surface modification of iron materials present new ways of local heat treatment of
operating faces of machine components. The spatial and temporal extension of the
temperature field caused in the irradiated material by energy absorption determines
the geometry of the structural change in the solid state or via the liquid state,
where the optimum energy input is controlled by laser power, beam cross-section and
the moving speed of the laser rays relative to the workpiece.
Workability results of different steels and cast-iron materials obtained by lasers
in the power range 200 to 1000 W give material-specific and process engineering relationships. In the case of application examples the laser is coupled to a CNC handling system allowing the movement of the table at a high accuracy of positioning
within 4 programmable axes and workpiece loading up to 1.5 t. The multivalent utilization of this laser-machine combination in industrially relevant components and the
advantages reached under industrial conditions are discussed.

Einschmelzlegieren mit Laserstrahlen

A.Gasser, A.Gillner, K.Wissenbach

Fraunhofer-Institut für Lasertechnik, D-5100 Aachen

Einleitung

Durch Oberflächenbehandeln mit Laserstrahlen lassen sich die Verschleißeigenschaften niedriglegierter Stähle deutlich verbessern. Neben der reinen Wärmebehandlung, dem Laserstrahlhärten /1/ bieten die Laserstrahlverfahren die Möglichkeit, durch Einschmelzen von Chrom, Wolfram, Molybdän und ähnlichen Elementen an der Oberfläche die tribologischen Eigenschaften von Kaltarbeitsstählen zu erhalten, während das Grundmaterial die Zähigkeitseigenschaften des weichen Stahls beibehält /2-4/. Zur Behandlung größerer Flächen werden entweder einzelne Schmelzspuren überlappend nebeneinandergelegt oder bei Spurbreiten bis ca. 1 cm Strahlablenksysteme verwendet.

Das Verfahren des Einschmelzlegierens ist in Fig.1 schematisch dargestellt. Der einzulegierende Stoff bzw. die Stoffkombination wird in Form einer Emulsion oder durch ein Plasmasprayverfahren auf die Oberfläche des Werkstücks aufgebracht. Bei der anschließenden Bearbeitung mit dem Laser erhält man je nach Prozeßintensität unterschiedliche Bearbeitungsergebnisse. Unterhalb einer Schwellintensität I_C findet ein reines Aufschmelzen von Legierungselement und Grundwerkstoff statt. In diesem Fall wird die Energieeinkopplung durch die natürliche Absorption von Legierungszusatz und Grundwerkstoff bestimmt. Beim Überschreiten der Schwellintensiät I_C bildet sich ein laserinduziertes Plasma aus. Damit verbunden ist eine deutliche Erhöhung der Absorption und somit eine Vergrößerung der Bearbeitungszone. In beiden Fällen erfolgt die Durchmischung von Legierungselement und Grundmaterial durch die Eigendynamik des Schmelzbades.

Bearbeitungsparameter

Die für das Einschmelzlegieren mit Laserstrahlung im wesentlichen relevanten Prozeßparameter sind

> Laserleistung
> Strahlradius
> Vorschubgeschwindigkeit

sowie bei Verwendung von Strahlablenksystemen, wie z.B. Schwingspiegeln die

> Ablenkfrequenz und
> Ablenkamplitude.

In Fig.2 ist die Schmelztiefe in Abhängigkeit von der Vorschubgeschwindigkeit für 3 verschiedene Leistungen aufgetragen, wobei die Kurven nach einem Wärmeleitungsmodell berechnet sind. Die Schmelz- bzw. Legierungstiefe läßt sich danach durch eine einfache Variation der Bearbeitungsgeschwindigkeit steuern. Die Abhängigkeit der Schmelztiefe vom Strahlradius bei vorgegebener Leistung zeigt Fig.3. Die Fokussierung führt zum Überschreiten der für die Bildung eines laserinduzierten Plasmas erforderlichen Schwellintensität I_C und damit zu einer Erhöhung der Schmelztiefe von ca. 100 µm auf ca. 800 µm.

Der Einfluß der erhöhten Energieeinkopplung durch das Plasma ist in Fig.4 anhand zweier Querschliffe von Schmelzspuren dargestellt. Während in der oberen Abbildung deutliche Spuren des Legierungsstoffes an den Schlieren in der Bearbeitungszone zu erkennen sind, führt die Plasmaunterstützte Bearbeitung neben der vergrößerten Bearbeitungstiefe zu einer gleichmäßigeren Vermischung von Legierungselement und Grundwerkstoff.

Durch die Variation der Bearbeitungsparameter läßt sich somit sowohl die Bearbeitungstiefe als auch die Konzentration des Legierungselementes im Grundwerkstoff einstellen.

Großflächige Bearbeitung

Um zu einer größeren Bearbeitungsbreite zu gelangen, müssen entweder Strahlablenksysteme eingesetzt oder mehrere Schmelzspuren überlappend nebeneinandergelegt werden /5/. Die meisten der für den Hochleistungsbetrieb geeigneten Ablenksysteme basieren auf dem Prinzip der Resonanzscanner, die den Laserstrahl sinusförmig über das Werkstück lenken. Neben der Ablenkfrequenz /6/ ist hierbei die Ablenkbreite von entscheidender Bedeutung für die Qualität der Bearbeitung. In Fig.5 ist der Querschliff einer Legierungsspur dargestellt, bei der die Ablenkbreite zu groß gewählt wurde. Deutlich sind die Auswirkungen der zu großen systembedingten Aufenthaltszeiten in den Umkehrpunkten der Laserstrahlablenkung zu erkennen. Bei korrekter Anpassung ergibt sich ein Querschliff, wie er in Fig. 6a dargestellt ist. Die Legierungszone ist annähernd rechteckig. Fig. 6b und c zeigen die Verteilungen der Elemente Fe und des einlegierten Cr. Der Linescan, gemessen in der Mitte der Legierungspur, verdeutlicht die gleichmäßige Durchmischung von Ligand und Grundmaterial. Die Cr-Konzentration in der aufgeschmolzenen Zone liegt bei etwa 1.5 %, was den Verhältnissen eines Kaltarbeitsstahles entspricht. Durch nachfolgendes Härten werden Oberflächenhärtewerte über 1000 HV10 erreicht, wobei die Härte im Übergangsbereich zwischen Schmelzzone und Grundmaterial kontinuierlich abfällt. Dies gewährleistet eine gute Verbindung zwischen dem Grundmaterial und der veredelten Oberflächenschicht.

Werden Bearbeitungsbreiten größer 1 cm bei entsprechender Schmelztiefe benötigt, so sind mehrere Legierungsspuren überlappend nebeneinanderzulegen. In Fig.7a ist der Querschnitt eines in dieser Weise bearbeiteten Werkstücks dargestellt. Die einlegierte Stoffkombination aus Wolframkarbid und Kobalt wurde hierbei im Plasmasprayverfahren aufgetragen und unter Ausnutzung eines laserinduzierten Plasmas einlegiert. Bei einer Flächendeckungsrate von ca. 2 cm^2/min werden Bearbeitungstiefen von ca. 2.5 mm erreicht. Unvermeidbare Oberflächenrauhigkeiten im 100 µm-Bereich können deshalb ohne Verlust der Legierungsqualität abgetragen werden. In Fig.7b ist der zu Fig.7a korrespondierende Härtverlauf dargestellt. Über einen weiten Bereich werden Härtewerte über 1000 HVO.1 erzielt, wobei der Übergang zum Kernbereich wiederum kontinuierlich stattfindet.

Zusammenfassung

Durch Einlegieren von Zusatzstoffen in Kohlenstoffstähle werden Gefüge erzielt, die denen von Kaltarbeitsstählen entsprechen. Durch Erzeugung eines laserinduzierten Plasmas werden Bearbeitungstiefen größer 2 mm erzielt. Die so erzeugten Oberflächenlegierungen haben bis zum Kernbereich eine gleichmäßig hohe Härte von über 1000 HVO.1 und einen kontinierlichen Übergang zur Kernhärte.

Literatur

/1/ K.Wissenbach ; Dissertation, Darmstadt 1985

/2/ D.S.Gnanamuthu; Appl. of Lasers in Materials Processing
          American Soc. of Metals, 1979

/3/ E.M. Breinan, B.H. Kear; in Laser Materials Processing, ed. by
          M. Bass, North Holland Publishing Company, 1983

/4/ N.Rykalin, A.Uglov, A.Kokora; Laser Welding and Machining
          Pergamon Press, Oxford 1978

/5/ R.Becker, R.Rothe, G.Sepold; Proc. of the 3rd Int. Coll. on
          Welding and Melting by Electrons and Laser Beams
          Lyon 1983

/6/ S.Schiller, S.Panzer; Thin Solid Films, 118 (1984), p.85-92

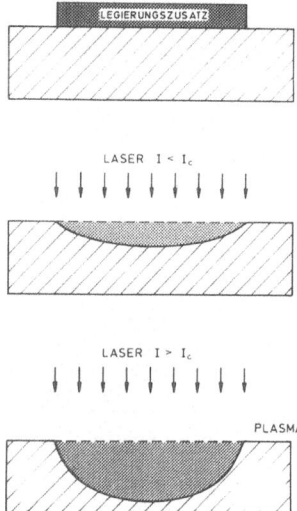

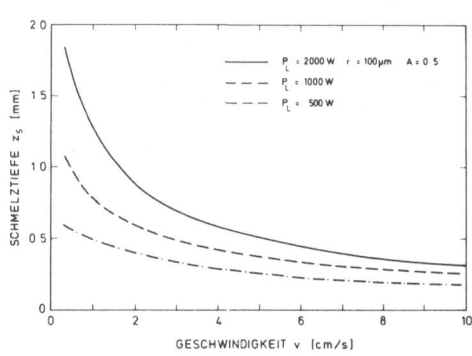

Fig.1. Schematische Darstellung des Einschmelzlegierens. Beim Überschreiten der Schwellintensität IC beim Einschmelzen entsteht ein laserinduziertes Plasma, wodurch die Absorption auf Werte nahe bei Eins erhöht wird

Fig.2. Berechnete Schmelztiefe in Abhängigkeit von der Vorschubgeschwindigkeit v für 3 verschiedene Laserleistungen PL

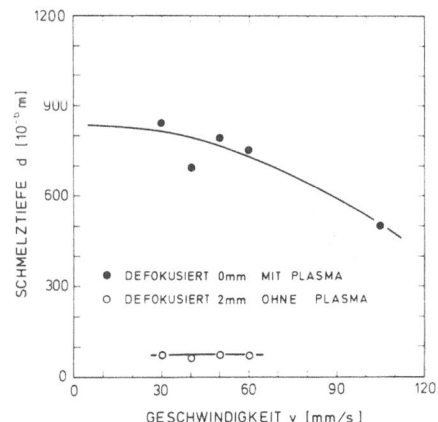

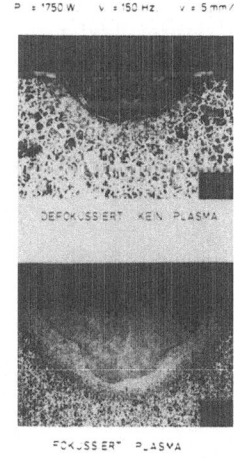

Fig.3. Schmelztiefe in Abhängigkeit von der Vorschubgeschwindigkeit v. Parameter ist der Fokusradius

Fig.4. Querschliffe zweier Einschmelzspuren mit und ohne laserinduziertes Plasma. Die Einschmelztiefe mit Plasma ist etwa einen Faktor 5 größer als ohne Plasma

416

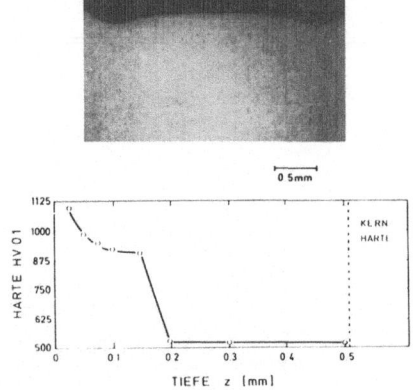

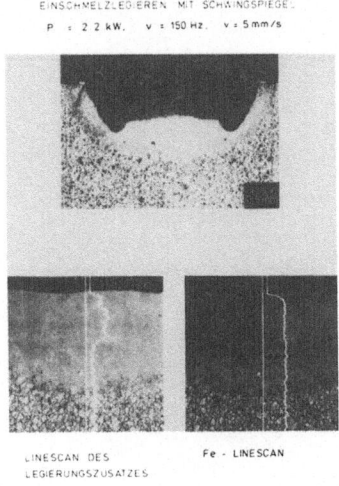

EINSCHMELZLEGIEREN MIT SCHWINGSPIEGEL
P = 2.2 kW, v = 150 Hz, v = 5 mm/s

LINESCAN DES
LEGIERUNGSZUSATZES   Fe - LINESCAN

Fig.5. Querschliff einer Legierungsspur mit Resonanzscanner bei
einer Laserleistung von
P_L = 2.2 kW. Für die eingestellten Parameter ist der Ablenkweg
zu groß. Die Folge davon ist
eine zu geringe Schmelztiefe in
der Mitte der Bearbeitungszone.
Im unteren Teil der Abbildung
ist der entsprechende Härtverlauf dargestellt

<u>Fig.6.</u> Querschliff einer Einschmelz-
spur mit Resonanzscanner

<u>b,c</u> Konzentrationsverlauf der Ele-
mente Chrom und Eisen

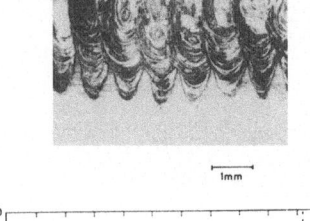

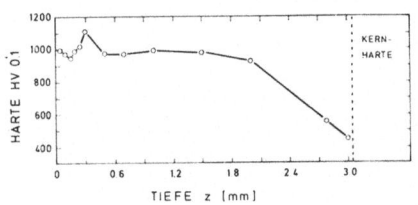

<u>Fig.7a.</u> Querschliff einer Oberflächen-
legierung mit überlappend ne-
beneinandergelegten Schmelzspuren

<u>b.</u> Härteverlauf der Oberflächenle-
gierung; Deutlich ist der kontinuierliche Abfall der Härte
bis zum Kernbereich zu erkennen

Microcrystalline and Amorphous Layers of Laser Melted Boronized Steels

A. Bloyce[*], I. Hancock[*], H. W. Bergmann

Institut für Werkstoffkunde und Werkstofftechnik, TU Clausthal

Abstract

By laser melting of Fe-B-alloys with different CO_2- and Nd:YAG-lasers
a preliminary solidification map was constructed including the various crystalline morphologies and the glass forming region as function of composition and solidification rate.

Introduction

Laser melting of metallic surfaces enables a rapid and directional solidification of the liquid phase /1,2/. Depending on composition and
remelting conditions different solidification morphologies can be obtained /2/. With the increasing quenching rate a finer and finer structure is obtained and finally glassy solidification can be observed /2-
6/. For a given starting concentration the solidification process is
determined by the solidification velocity R, the quenching rate ε and
the temperature gradient G. These parameters depend on the laser parameters - exposer time, power density, etc. - and on material constants e.g. diffusivity /7/. They are interrelated by equation (1)

$$\varepsilon = R * G \tag{1}$$

Increasing quenching rates for constant R/G ratio lead to finer microstructures of equivalent morphology. If for a constant quenching rate
the R/G ratio is varied different morphologies of similar size are obtained. The validity of both relationships has been demonstrated for
Fe-C-Si alloys /2, 3, 6/. For constant composition G-R maps of the mor-

---

[*] Dept. Metallurgy and Materials Science, University Birmingham

phology are obtained and the change of these maps with composition describes the complete solidification behaviour. If on the other hand the temperature gradient is choosen constant R-composition-solidification maps can be generated. If several of them are used, again the complete solidification behaviour can be described. In the present paper a R-concentration map for Fe-B alloys will be generated.

Experimental

Armco-Iron was pack boronized. Compound layers of FeB and Fe_2B with various depths were generated by choosing different temperatures and times. After a vacuum annealing treatment specimen with a Fe_2B mono-layer or eutectic structures like in fig. 1 were obtained depending on the temperature choosen. In melting the surface to a suitable depth alloys with various compositions were obtained by variation of feed rate and power density resulting in different solidification conditions. For vitrification part of the material was melted a second time by a Nd:YAG-laser. Easy glass formation was obtained when boronized samples were vacuum annealed at the eutectic temperature (eutectification) and subsequently melted by a CO_2-laser in order to obtain a fine microstructure before they were finaly laser glazed by a Nd:YAG-laser. The various treatment conditions were investigated by light- and electronmicroscopy and also by x-ray and calorimetric techniques.

Results

Depending on the boron content and the melting- or better solidification parameters various morphologies in the microstructures can be detected. They are given in fig. 2-10. All these microstructures were received by approximately the same solidification conditions. It is a quenching rate of $\epsilon \sim 5 * 10^5$ K/s and a temperature gradient of $G \sim 1 * 10^4$ K/mm and a solidification velocity of $R \sim 50$ mm/s. The morphology changes in fig. 2-10 are therefore simply due to concentrational changes. Fig. 2 shows the planar solidified and subsequently transformed iron microstructure. Even for very high solidification velocities it is known that the maximum solubility of boron in iron is very small for either of the three phases α, γ, and δ. A substantial supersaturation of boron in iron could not be detected (less than 0.5 at% B). An increase of the boron content leads to a cellular structure α-Fe-crystals and intercellular Fe_2B, see fig. 3. A further increase in boron content causes a change in the morphology of the microstructure showing dendritic structures and interdendritic eutectic for 6 and more at% B, see fig. 4. Between 16.8 and 20 at% B, fig. 5-7 three different eutectic morphologies can be found. For the lower boron content a feathery eutectic is observed. Between 18 and 19 at% B an acicular eutectic was detected. Both of them consis of α-Fe and Fe_2B. Near 20 at% B a eutectic is found which contains the cubic $Fe_{23}B_6$ phase. A further increase of the boron content causes a microstructure consisting of primary Fe_2B dendrites (fig. 8) and interdendritic eutectic. Above 30 at% B a cellular microstructure (fig. 9) appears which converts near 33 at% B into a planar solidification mode, fig. 10.

If the solidification behaviour at higher solidification rates is investigated an increase of the region with planar solidification mode is considerable for Fe_2B, however, for α-Fe hardly to be detected. Laser melting leads to a considerable widening of the area with the eutectic solidification behaviour compared to conventional processes. The same tendency can be detected if the solidification velocity is further increased. Near the eutectic composition a considerable high quenching rate causes vitrification on laser melting. For a constant temperature gradient this means that from a critical solidification velocity onward a glassy solidification occurs. In the present paper the vitrification was obtained by Nd:YAG-laser glazing of the surface. Differences in the thickness of the layer and the quality of the glassy region depends on the type of plasma formation created during processing, see fig. 11. Modifications in the plasma formation were obtained by different power densities which could be achieved by defocussing the beam. Nd:YAG-laser glazing of the surface causes a dramatic improvement in the corrosion resistance depending on the plasma formation during processing. Dust particles, e.g. which were sintered onto the surface, altered the corrosion behaviour and acted as heterogenous nuclei for surface crystallisation, see fig. 12. The amorphous surface layer can be identified by x-ray diffractometry. In fig. 13 the amorphous ring in the as-quenched condition is shown together with the crystalline phases obtained after annealing. The morphological results obtained so far are condensed in a preliminary diagram, see fig. 14. The tendency which can be derived from that diagram seems to be correct and is in agreement with theoretical considerations e.g. the extension of the solid solubility and also the extension of the eutectic and the glass forming region with increasing solidification velocity.

420

Conclusion and future work

The present paper confirms the principle ideas on the dependance of
solidificaton morphology on composition and solidification velocity.
There are still too few experimental data to examine the validity of
current solidificaton theories.

Acknowledgement

The authors acknowledge the financial support of the BMFT, SRIC and the
Commission of the European Comunity.

References

/1/ B.L. MORDIKE and H.W. BERGMANN in: Rapidly solidified amorphous
    and crystalline alloys, ed. B.H. Kear et al, MRS Symposium Proc.
    1985, New York, Elsevier
/2/ H.W. BERGMANN, Surface Engineering 1, (1985) 137
/3/ H.W. BERGMANN, B.L. MORDIKE and H.U. FRITSCH, Z. Werstofftechnik
    14, (1983) 237
/4/ H.W. BERGMANN, G. BARTON and J. BETZ, Z. Werkstofftechnik 14,
    (1983) 244
/5/ H.W. BERGMANN and B.L. MORDIKE, Z. Metallkunde 71, (1980) 658
/6/ H. JONES and W. KURZ, Z. Metallkunde 72, (1981) 782
/7/ H.U. FRITSCH and U. LUFT, this volume

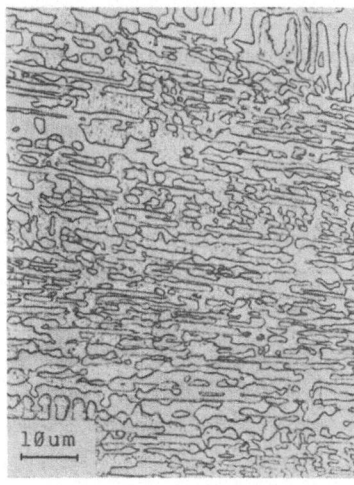

Fig. 1. Microstructure of boro-
        nized Armco-iron after
        vacuum annealing treat-
        ment at the eutectic tem
        perature (eutectification)

Fig. 2. Armco-iron laser mel-
        ted, $R \sim 50$ mm/s, $G \sim$
        $1*10^4$ K/mm, $\varepsilon \sim 5*10^5$ K/s
        planar solidification
        of Fe and subsequent
        transformation

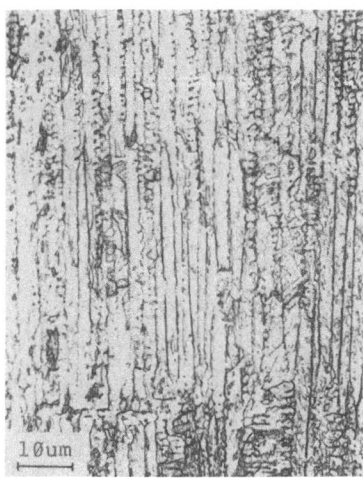

Fig. 3. Armco-iron, boronized, vacuum treated and laser melted C = 1.5 at% B mainly cellular structure of α-Fe and Fe$_2$B

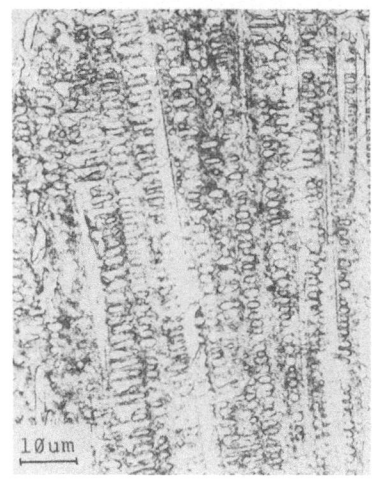

Fig. 4. Armco-iron, boronized, vacuum treated and laser melted C = 6 at% B dendritic solidification of α-Fe and interdendritic Fe$_2$B

Fig. 5. Armco-iron, boronized, vacuum treated and laser melted C = 16.8 at% B feathery eutectic structure of α-Fe and Fe$_2$B

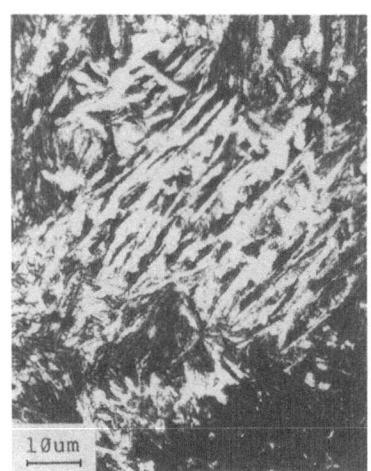

Fig. 6. Armco-iron, boronized, vacuum treated and laser melted C = 18.8 at% B acicular eutectic structure of α-Fe and Fe$_2$B

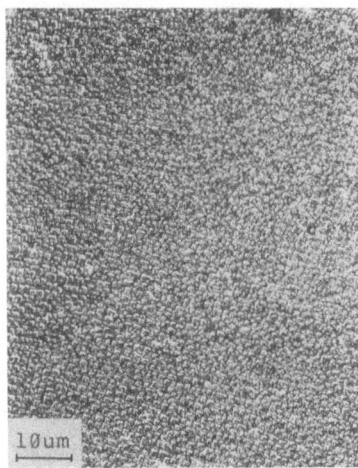

Fig. 7. Armco-iron, boronized,
vacuum treated and laser
melted C = 19.7 at% B eu-
tectic α-Fe and $Fe_{23}B_6$

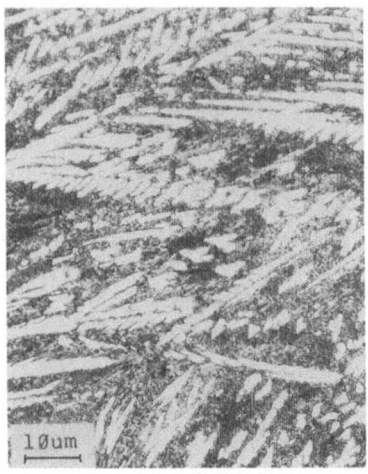

Fig. 8. Armco-iron, boronized,
vacuum treated and laser
melted C = 26.3 at% B
Fe_2B dendrites and
α-Fe - Fe_2B eutectic

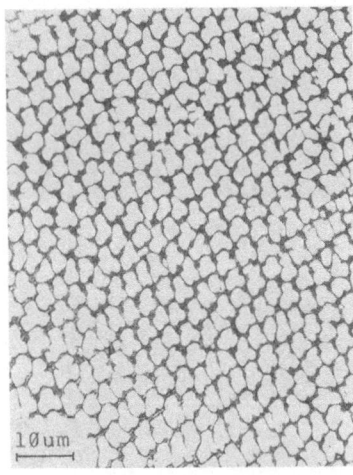

Fig. 9. Armco-iron, boronized,
vacuum treated and laser
melted C = 31 at% B Fe_2B
cells

Fig. 10. Armco-iron, boronized,
vacuum treated and laser
melted C = 33 at% B
planar solidification
of Fe_2B

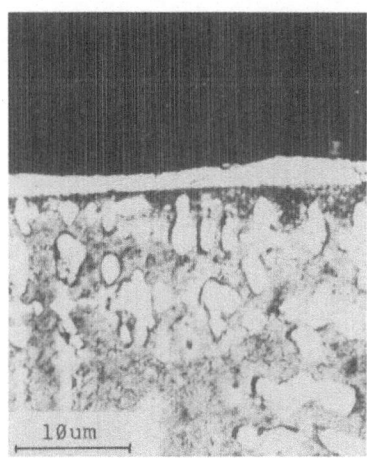

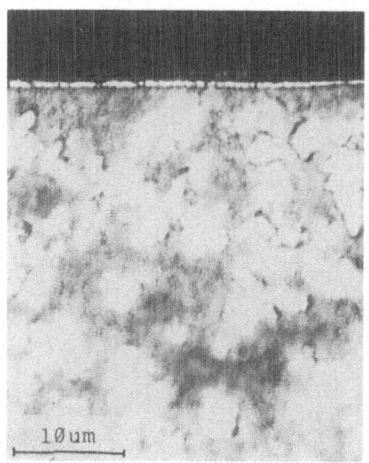

Fig. 11. Micrographs showing the
cross-sections of Armco-
iron, boronized, eutec-
tified and CO_2 laser
melted, prior to Nd:YAG-
laser glazing.
a) maximum power density,
big blue flame visible

b) defocussed, white plasma
flame just visible

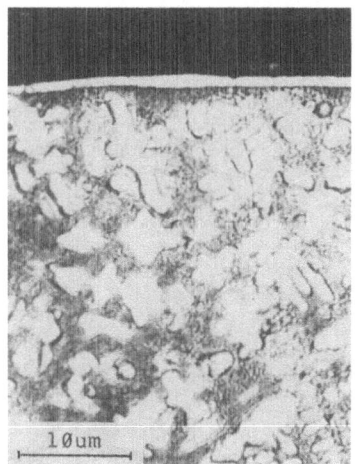

c) intermediate setting, small blue plasma flame

424

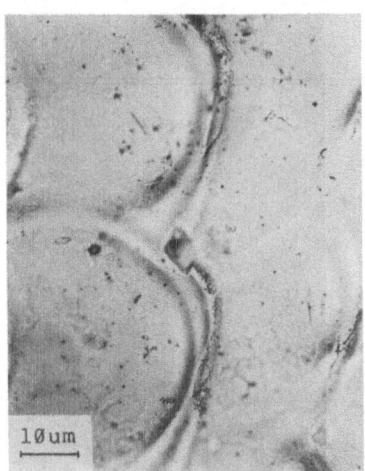

Fig. 12.Surfaces of YAG-laser
melted materials after
long term etching
a) as-quenched,
corresponding to 11 a

b) as-quenched, corresponding
to 11 b

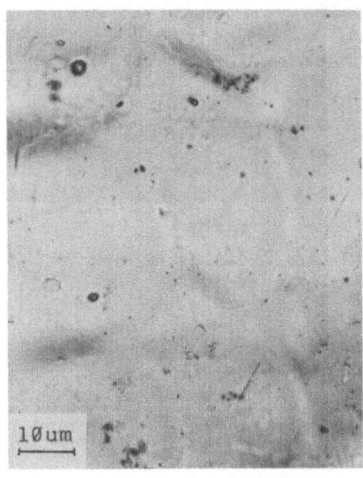

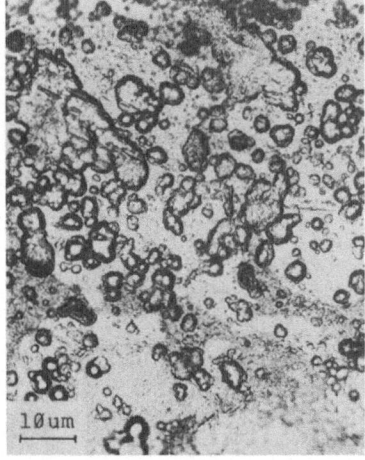

c) as-quenched, corresponding
to 11 c

d) corresponding to 11 c
after crystallisation

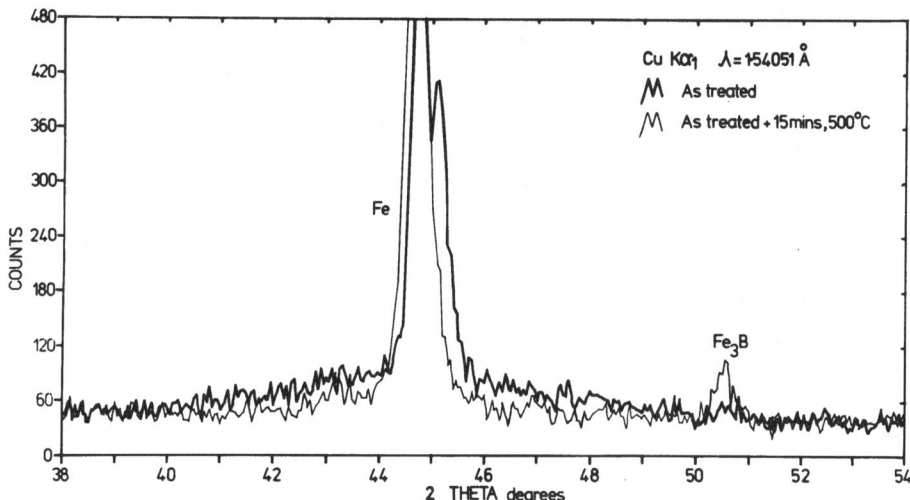

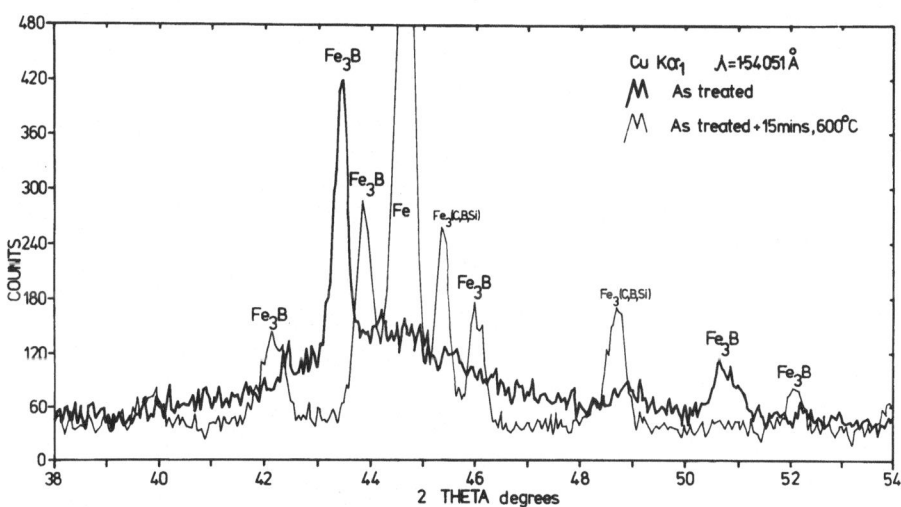

Fig. 13. X-ray diffractometer results showing laser glazed materials
before and after crystallisation
a) Armco-iron substrate
b) steel-substrate 2.1 wt% C 13 wt% Cr

426

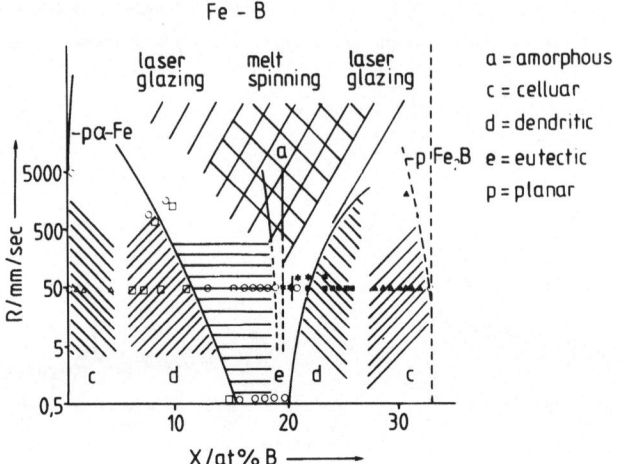

Fig. 14. Solidification map for Fe-B alloys for a fixed temperature gradient of $1 * 10^4$ K/mm

Problemstellungen beim Laserstrahlhärten von Bauteilen

Prof. Dr.-Ing. Dr.h.c. W. König
Dr.-Ing. F.U. Meis
Dipl.-Ing. H. Willerscheid
Dipl.-Ing. Cl. Schmitz-Justen
Fraunhofer-Institut für Produktionstechnologie

Eine Wärmebehandlung ist im Fertigungsablauf von Bauteilen notwendig, einerseits, um spanende oder umformende Arbeitsgänge zu ermöglichen oder zu erleichtern, andererseits, um die erforderlichen werkstoffgebundenen Betriebseigenschaften einzustellen. Gerade im Zusammenhang mit der Wärmebehandlung der Randzone ist die Anwendung der Lasertechnik gegenwärtig von großem Interesse. Dies umfaßt die Prozesse der martensitischen Umwandlungshärtung, des Umschmelzens und des Oberflächenlegierens mittels Hochleistungslasern.

Um das technologische Potential dieser Verfahren - speziell auch im Vergleich zu konventionellen Techniken - charakterisieren und beurteilen zu können, sind zunächst die Zielsetzung einer Randschichthärtung und die daraus resultierenden Anforderungen an das Verfahren zu betrachten.

Zum einen wird eine Leistungssteigerung von Bauteilen durch Erhöhung der Belastbarkeit angestrebt. Beispiele hierfür sind Wellen in Motoren und Antriebssträngen, bei denen Hohlkehlen an den Querschnittsübergängen gehärtet werden. Ähnliche Verhältnisse sind auch bei der Härtung von Verzahnungen bis in die Zahnfußausrundung hinein gegeben. Die zweite Zielsetzung ist die Erhöhung des Widerstandes gegen Verschleiß, Korrosion, Kavitation etc., was beim Härten von Werkzeugen, Führungsbahnen, Lagern und Dichtflächen praktiziert wird. Abbildung 1 stellt am Beispiel einer Getriebewelle diese verschiedenartigen aus der Bauteilfunktion heraus notwendigen Härtezonen exemplarisch dar.

Dem Konstrukteur kommt die Aufgabe zu, die Härtespezifikationen nach Härtewerten, Lage und Ausdehnung sowie Tiefe der zu behandelnden Zonen festzulegen. Die Ausgangsbasis muß dabei der Festigkeitsnachweis sein, solange ausschließlich eine Erhöhung der Tragfähigkeit angestrebt wird. Dieser bestimmt auch die Vorgaben für die Kernfestigkeit des Werkstückes sowie die erforderliche Zähigkeit, die meist durch eine Kombination aus Härten und Anlassen erreicht wird. Im Falle von verschleißhemmenden Härteschichten sind neben dem Endbearbeitungsaufmaß die Anzahl und Abmessungen möglicher Ausbesserungs-, bzw. Reparaturstufen

des Bauteils einzubeziehen. Schließlich muß auch der für die Bauteilfunktion, bzw. -weiterbearbeitung tolerierbare Härteverzug bei der Auslegung einer Wärmebehandlung berücksichtigt werden.

Die neuerliche Festlegung all dieser Faktoren am spezifischen Bauteil
ist insbesondere bei der Untersuchung und Beurteilung einer neuen Technik wie in diesem Fall der Laserhärtung grundsätzlich notwendig, damit
nicht anstatt der tatsächlich funktionsbedingten Anforderungen die mit
konventionellen Prozessen erreichten Werte als Ziel- und Bewertungs-
größen übernommen werden.

Abbildung 2 zeigt eine Gliederung der Randschichtwärmebehandlungsver-
fahren nach ihrem Wirkprinzip. In der Konstruktion und Arbeitsplanung
erfolgt die Auswahl des Werkstoffes, Festlegung der Wärmegrundbehand-
lung, der mechanischen Bearbeitungsfolge, der Randzonenbehandlung und
der darauffolgenden Endbearbeitung. Dabei kommt den einzelnen Ferti-
gungsschritten je nach Werkstoff und Bauteilfunktion unterschiedliche
Bedeutung zu.

Um das bestgeeignete Härteverfahren für den jeweiligen Anwendungsfall
zu bestimmen, müssen die bauteilspezifischen Anforderungen den charakteristischen Merkmalen und Leistungsmöglichkeiten der einzelnen Techniken gegenübergestellt werden. Deshalb sollen hier die Eigenschaften
der Laserstrahlverfahren,die in diesem Zusammenhang allerdings nur die
martensitische Umwandlungshärtung und das Umschmelzen einschließen,
nicht aber die Veränderung der chem. Werkstoffzusammensetzung durch
Laserlegieren, vergleichend zu denen herkömmlicher Prozesse betrachtet
werden. Abbildung 3 gibt einen Überblick über verschiedene dieser Kriterien.

Die Laserstrahlhärtung ist stets ein Verfahren der Randschichtbehand-
lung. Erzielbare Härtetiefen ohne Oberflächenaufschmelzung betragen
maximal 1,5 - 2,0 mm. Dieser Wert wird durch verfahrensspezifische Parameter wie Laserleistung, -leistungsverteilung, Einwirkzeit sowie Ab-
sorptionsmittel und gleichermaßen durch bauteilspezifische Randbedingungen wie Werkstückvolumen, Ausdehnung und Lage der Härtezone, mechanischen und Wärmevorbehandlungszustand, Einstrahlwinkel und Ausgangstemperatur des Werkstücks bestimmt.

Das Laserstrahlhärten ist das von seinen potentiellen Anwendungsmöglichkeiten her am ehesten den konventionellen Techniken der Induktionshärtung, des Nitrierens, Borierens und im weiteren Sinne der Einsatzhär-
tung vergleichbar. Oberflächenbeschichtungen nach den Verfahren des
PVD, bzw. CVD sind wegen der um etwa eine Größenordnung geringeren Härte-
schichtdicke hier nicht berücksichtigt worden. Reduziert man den Anwen-
dungsvergleich auf die praktisch sinnvolle partielle Randschichthärtung

der Bauteiloberflächen, dann steht die Lasertechnik im wesentlichen
der Induktionshärtung und dem Nitrieren gegenüber.

Im Verhältnis zum Nitrieren läßt sich mit der Laserstrahlhärtung eine
größere Einhärtetiefe erzeugen, während die absoluten Härtewerte in
Abhängigkeit des Werkstoffes teilweise niedriger liegen. Betrachtet
man die Behandlungszeit, so ist das Laserhärten bei kleinen Stückzahlen und hoher Fertigungsdringlichkeit als günstiger anzusehen, während
das Nitrieren mit Behandlungszyklen von mehreren Stunden bei großen Losen
oder niedrigerer Priorität, jedenfalls aber bei großflächiger Härtung,
vorteilhaft ist. Der Vergleich mit den Induktionsverfahren hingegen
zeigt, daß die durch Laserstrahlhärten maximal erzielbare Einhärtetiefe
geringer ist. Dies ist auf die höhere örtliche Konzentration der Leistung und die damit verbundene Aufschmelzgrenze zurückzuführen. Insgesamt wird bei der Laserhärtung dem Bauteil jedoch weniger Energie zugeführt und ein kleineres Volumen erwärmt, was sich in den deutlich niedrigeren Verzugswerten von etwa 15 - 25 μm für eine einzelne Härtespur
widerspiegelt. Lediglich das Nitrieren ergibt hierbei noch günstigere
Werte; dies ist auf die verhältnismäßig niedrigen Prozeßtemperaturen
unterhalb von 600°C zurückzuführen.

Betrachtet man die Art und Qualität des entstandenen Härtungsgefüges,
so besitzt der durch die Laserstrahlhärtung erzeugte Martensit aufgrund der prozeßspezifischen Aufheiz- und Abkühlgeschwindigkeit eine
sehr feinnadelige oder -körnige Struktur und zeigt auch einen entsprechend homogenen Härteverlauf. Der Effekt der Selbstabschreckung,
bei dem die Gesamtmasse des Bauteils als "Wärmesenke" die notwendige
kritische Abkühlgeschwindigkeit herbeiführt, ist jedoch nur dann gewährleistet, wenn von der Bauteilgeometrie her genügend kühlendes Volumen im Bereich der zu härtenden Zone vorhanden ist. Anderenfalls kommt
es aufgrund des auftretenden Wärmestaus in Abhängigkeit von der Werkstückdicke zur Durchhärtung oder Überhitzung des Werkstoffes bis hin
zur lokalen Aufschmelzung. Neben unerwünschten Gefügeinhomogenitäten
nehmen dabei auch die Verzugswerte zu. Diese Zusammenhänge sind besonders bei der konstruktiven Auslegung von dünnwandigen Bauteilen zu beachten, die gehärtet werden sollen, beispielsweise rohrförmige oder
stark verrippte Strukturen. Die für eine Selbstabschreckung notwendigen
Voraussetzungen bestimmen im wesentlichen auch die für eine Laserhärtung prädestinierten Werkstoffe. Da zur Erzeugung von Martensit ebenso
wie bei der konventionellen Härtung ein Mindestkohlenstoffgehalt von
0,3% vorhanden sein muß, kommen in erster Linie Vergütungsstähle in
Betracht. Diese umfassen zum einen reine Kohlenstoffstähle; Ergebnisse
zum Laserhärten liegen beispielsweise für den Ck 45 oder den Cf 53 vor.

Zum anderen sind für die Selbstabschreckung jedoch solche Werkstoffe
weitaus besser geeignet, deren Legierungselemente die kritische Abkühlgeschwindigkeit herabsetzen, wie es besonders bei Cr und Mn der
Fall ist. Daneben werden häufig noch festigkeitssteigernde Elemente
wie Mo oder V zulegiert. Die gegenwärtig meist für Laserhärtungsunter-
suchungen verwendeten Stähle sind daher die Sorten 42 Cr Mo 4, 50 Cr V 4,
34 Cr Ni Mo 6. Ebenfalls härtbar sind Gußwerkstoffe mit perlitischer
Matrix, wobei die Graphiteinschlüsse bei der Härtung nicht aufgelöst
werden, was bei der späteren Bauteilbeanspruchung tribologische Vorteile erbringen kann. Beispiele sind GG 25, GGG 60 und legierte Gußsorten.

Besonders bei der Kurzzeitwärmebehandlung, wie sie bei der Laserstrahlhärtung ja vorliegt, ist der Ausgangszustand des Werkstoffes vor der
Erwärmung ein wesentlicher Einflußfaktor auf die erzielbare Qualität
des Härtezonengefüges, weil in der zur Verfügung stehenden Austenitisie-
rungszeit nur kurze Diffusionswege zurückgelegt werden können. Daher
ist eine möglichst gleichmäßige Verteilung aller einzelnen Werkstoffbe-
standteile sehr vorteilhaft. Aus diesem Grunde werden für die genannten
Stahlsorten die besten Ergebnisse bei der Laserstrahlhärtung aus vor-
vergütetem Zustand erreicht.
Wie die hier vorgestellten Überlegungen zeigen, kann das Laserstrahlhärten nur in einem sehr engen Anwendungsgebiet ein direktes Substi-
tutionsverfahren für die konventionellen Härtetechniken darstellen.
Vielmehr bietet es für potentielle industrielle Anwendungen eine Ergänzung in solchen Einsatzfällen, die bisher nicht oder nur unzureichend
abgedeckt werden konnten. Die Schwerpunkte der Laserhärtung dürften dort
liegen, wo partielle Härtezonen auf der Bauteiloberfläche für konventionelle Verfahren nicht zugänglich sind und wo zahlreiche unterschiedliche Varianten dieser Bauteile in relativ kleinen Stückzahlen gefertigt
werden sollen. Die Laserwärmebehandlung ist durch ihre Integrierbarkeit
in den Fertigungsfluß ein Verfahren, welches der angestrebten flexiblen
Automatisierung in der industriellen Fertigung weitgehend entgegenkommt.
Dies gilt umso mehr, als die Lasertechnik sehr breitgespannte Möglichkeiten bietet, dieselbe Strahlquelle gleichermaßen für unterschiedliche
Verfahren zu nutzen. Beispielsweise kann im Verbund einer Gesamtanlage
auf mehreren Stationen getrennt geschweißt, gehärtet und örtlich angelassen werden, ohne daß das Bauteil die Fertigungslinie verlassen müßte.
Die wirtschaftlichen Beurteilungskriterien für den Einsatz des Lasers
liegen also nicht nur in der Gegenüberstellung zu den klassischen Verfahren, sondern weitergehend in einer Optimierung des Fertigungsflusses.

Neben verfahrensspezifischen Merkmalen und den Wirtschaftlichkeitsgrößen

müssen auch die Randbedingungen für die Anwendung der Laserstrahlhärtung am einzelnen Bauteil bekannt sein und beachtet werden. Einen wesentlichen Einfluß besitzt in diesem Zusammenhang der Einfallswinkel,
definiert zwischen Strahlachse und der Normalen auf die Werkstückoberfläche. Abbildung 4 zeigt, daß mit steigendem Einfallswinkel die Einhärtetiefe abfällt und die Härtezone eine asymmetrische Form annimmt.
Dies liegt zum einen an der Zunahme des projizierten Arbeitsfleckdurchmessers und an der einseitig günstigeren Wärmeabfuhr in den Grundwerkstoff. Versuchshärtungen an realen Bauteilen zeigen, daß der Einfallswinkel auf einer ebenen Fläche den Wert von 30° nicht überschreiten sollte, um noch eine ausreichend gleichmäßige Härtezone zu gewährleisten. Dies muß in der quellenexternen Strahlführung und -formung
- gegebenenfalls durch zusätzliche Spiegel - sowie in der Aufteilung
von räumlichen Freiheitsgraden auf Strahl und Werkstück berücksichtigt
werden.

Neben diesen Überlegungen zur Handhabung des Werkstückes ist auch die
Einordnung der Laserstrahlhärtung in den Ablauf der mechanischen Bearbeitung wesentlich. Abbildung 5 stellt den Einfluß der Oberflächenrauheit dar, wie sie für unterschiedliche Fertigungsverfahren vorliegt.
Es ist erkennbar, daß diese Größe bei sonst konstanten Prozeßparametern
eine Differenz der Härtetiefe bis zu ca. 30% hervorruft. Mit fallender
Einhärtetiefe bei glatteren Oberflächen ist auch ein Absinken des Härteverzuges festzustellen. Dieser Zusammenhang kann als Kriterium dafür
gelten, ob nach der Laserhärtung noch eine abschließende mechanische
Feinbearbeitung, z.B. Schleifen, zu erfolgen hat, um die Maßtoleranzen
des Bauteils zu erfüllen.

Maßgeblichen Einfluß auf das Härteergebnis besitzt auch die Ausgangstemperatur des Werkstücks. Abbildung 6 gibt dies für den Werkstoff
42 Cr Mo 4 anhand von Querschliffen wieder. Mit steigender Vorwärmetemperatur bis maximal 200 °C nimmt die Größe der Härtezone zu, was
auf die Austenitisierung eines größeren Werkstoffvolumens durch die
gleiche Energiemenge zurückzuführen ist. Wesentlicher noch ist das
veränderte Übergangsverhalten von der Härtezone zum Grundwerkstoff,
welches bei einigen Werkstoffen zu beobachten ist. Im Gegensatz zu dem
verhältnismäßig schroffen Härteabfall, wie er in vielen Fällen bei der
Laserhärtung aus Raumtemperatur auftritt, ergibt sich hier ein deutlich ausgeprägter Übergangsbereich, der einen besseren Verbund zwischen
gehärtetem und Grundgefüge aufweist. Die bei der Werkstückvorwärmung
auftretenden Erscheinungen werden auch beobachtet, wenn Härtespuren
überlappend unmittelbar nebeneinander gelegt werden. So ist in Abbildung 7 die linke der beiden Spuren, die zuletzt erzeugt wurde, deut

lich tiefer als die erste, rechte Spur. Bei einer größeren Anzahl von
Spuren kann dies ohne weiteres zu einer Aufschmelzung bei konstanten
Prozeßstellgrößen führen, so daß eine Nachführung von Vorschubgeschwindigkeit und / oder Laserleistung notwendig wird. Da mit derzeit industriell verfügbaren Laserquellen im Leistungsbereich bis 5 kW und Lin-
sen-, bzw. Integratoroptiken Einzelspurbreiten von maximal 15 mm erzeugt werden können, ist eine überlappende Härtung - beispielsweise von
Lagersitzen - erforderlich. Typischerweise tritt hierbei neben doppelt
gehärteten Bereichen auch eine Anlaßwirkung der letztgehärteten Spur
auf die vorhergehende ein. Die Zone des entsprechenden Härteabfalls,
der jedoch in der Regel nicht die Werkstoffgrundhärte erreicht, hat
werkstoffabhängig eine Breitenausdehnung bis zu etwa 4 mm. Um breitere
Zonen auch ohne diese Übergänge härten zu können, werden gegenwärtig
Untersuchungen auch mit anderen Strahlformungssystemen, beispielsweise
einer Oszillatoroptik durchgeführt.

Gleiche Probleme wie bei der Spurüberlappung treten auch am Anschluß-
stoß bei der Härtung einer geschlossenen Bahn auf. In Abbildung 8 ist
die entsprechende Zone anhand eines Mikroschliffes und die zugehörige
Härtevermessung in einem oberflächennahen Bereich dargestellt. Bei
zahlreichen Bauteilen ist es allerdings möglich, die Anlaßzone mit reduzierter Härte in einen unbeanspruchten Bereich der Kontur zu legen,
beispielsweise an Steuernocken oder im druckentlasteten Bereich von
Gleitlagern.

Über die gezeigten Zusammenhänge hinaus wird das Härteergebnis bei der
Laserwärmebehandlung durch weitere Faktoren beeinflußt. Hierzu zählen
die Art und Dicke der Oberflächenschicht, die zur Erhöhung der Strahl-
absorption aufgebracht wird sowie die örtliche und zeitliche Veränderung der Leistungsdichteverteilung des Laserstrahles. Um trotz dieses
wechselnden Einflusses eine gleichbleibend hohe Fertigungsqualität und
-sicherheit zu gewährleisten, ist eine online-Aufnahme des Strahlpro-
fils während des Fertigungsverlaufs notwendig. Durch eine hochauflösende Strahlvermessung und Korrelation der Meßdaten mit dem Arbeits-
ergebnis können Kenngrößen gebildet werden, die eine Überwachung der
Laseranlage ermöglichen, Störungen signalisieren und gegebenenfalls den
Prozeß abbrechen.

Unregelmäßigkeiten des Prozeßablaufes an der Bearbeitungsstelle selbst
können durch den Einsatz eines Thermographiesystems berührungslos erfaßt werden. Abbildung 9 zeigt die unterschiedlichen Temperaturen im
Arbeitsfleck bei einer Variation der Vorschubgeschwindigkeit für das
Laserstrahlhärten. Die bei Unterschreiten eines kritischen v_f auftretende Prozeßstörung "Aufschmelzen" wird in der digitalen Bildverarbei-

tung durch eine besondere Farbskalierung sichtbar gemacht. Auf einer solchen Temperaturanalyse aufbauend, ist die Erweiterung der Strahlüberwachung zur Prozeßüberwachung durchaus vorstellbar.

Zusammenfassend bleibt die Feststellung, daß die Prozeßentwicklung der Laserwärmebehandlung und insbesondere der Laserstrahlhärtung dem Bauteilkonstrukteur ein Verfahren an die Hand gibt, welches nicht primär konventionelle Techniken der Randzonenbehandlung substituiert, sondern vielmehr eine technologisch und wirtschaftlich sinnvolle Ergänzung bietet, wie sie beispielhaft in Abbildung 10 dargestellt ist. Wenn auch bei der Laserwärmebehandlung vielfältige Problemstellungen hinsichtlich spezieller Bauteilmerkmale noch zu lösen sind, so werden mit der Strahlüberwachung und Prozeßbeobachtung bereits die Ansätze für ein industriell einsetzbares Fertigungsverfahren hoher Verfügbarkeit geschaffen.

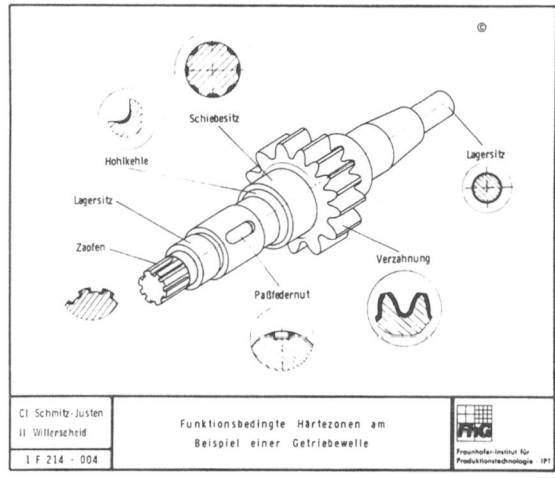

Abbildung 1.

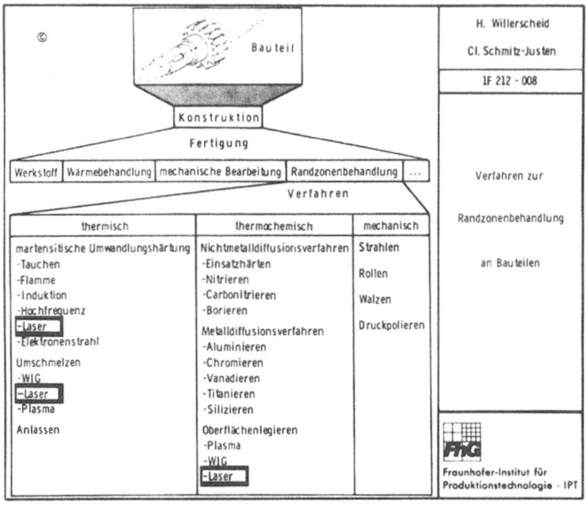

Abbildung 2.

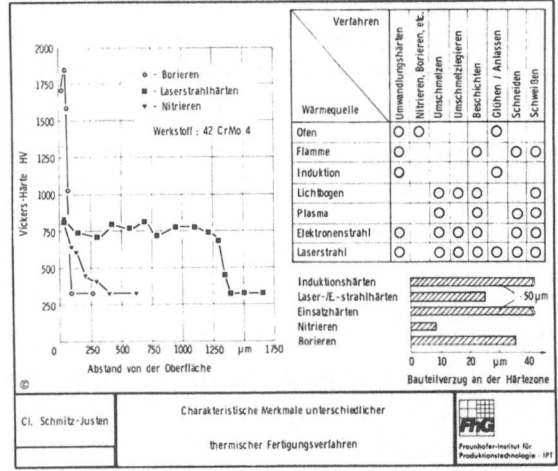

Abbildung 3.

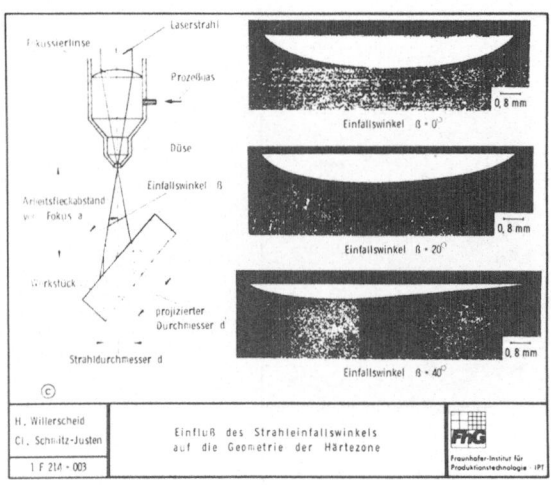

Abbildung 4.

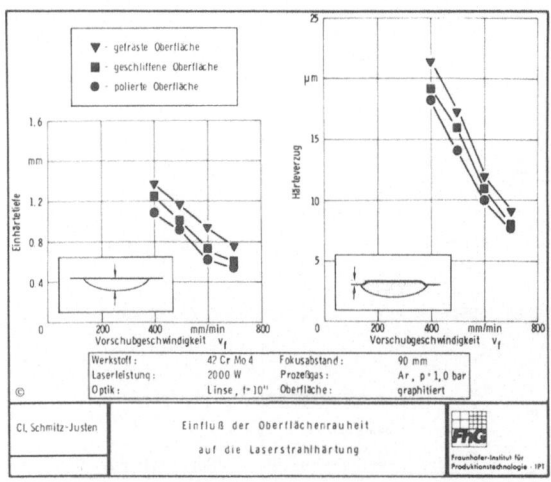

Abbildung 5.

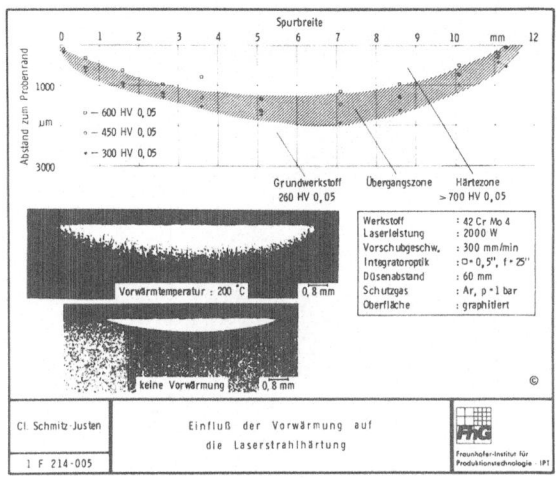

Abbildung 6.

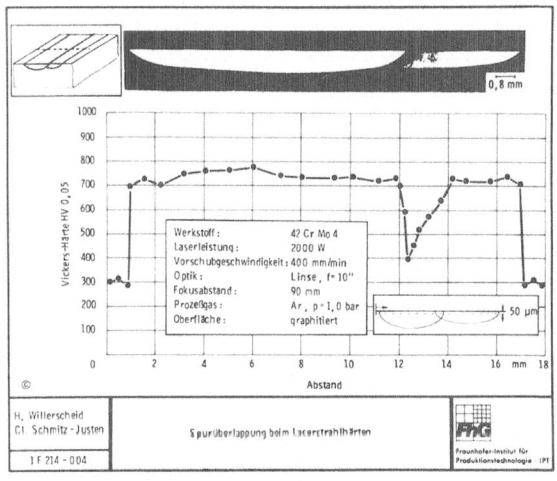

Abbildung 7.

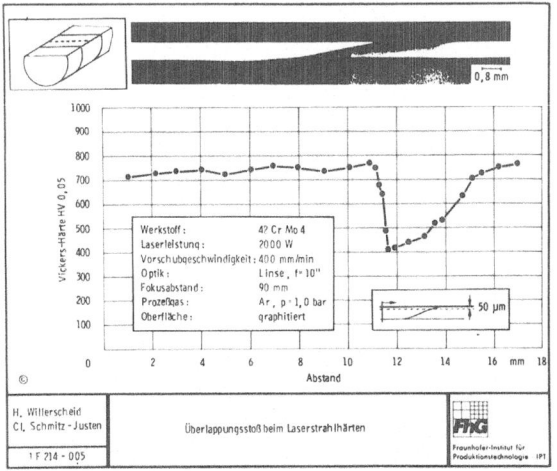

Abbildung 8.

436

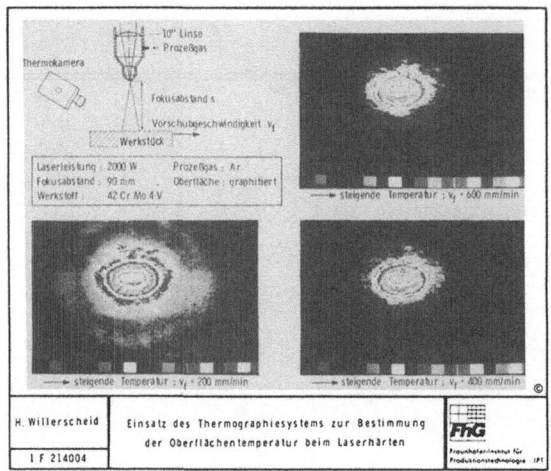

Abbildung 9.

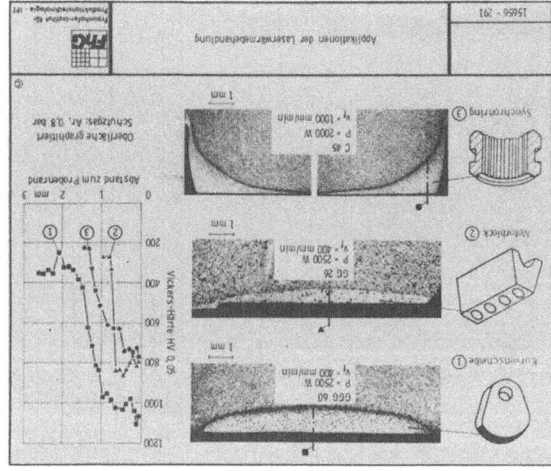

Abbildung 1o.

Laserumschmelzhärten von Schnellarbeitsstählen

H.W. Bergmann, I. Hancock*, T. Bell*, H.U. Fritsch
Institut für Werkstoffkunde und Werkstofftechnik, TU Clausthal
Agricolastr. 2, D-3392 Clausthal-Zellerfeld

Zusammenfassung

Es wird über die Eigenschaften von randschichtumschmelzgehärteten
Schnellarbeitsstählen berichtet, die mit Hilfe eines CO_2-Hochlei-
stungslasers umgeschmolzen wurden. Die beim Erstarren auftretenden
Gefügeänderungen werden als Funktion der Umschmelzparameter diskutiert. Des weiteren werden erste Gebrauchseigenschaften und mögliche technische Anwendungen dargestellt.

Einleitung

Schmilzt man die Oberfläche metallischer Legierungen mit einem Laser
um, so kann es bei der raschen Erstarrung der umgeschmolzenen Schicht
durch die Selbstabschreckung, die das massive Substrat bewirkt, zur
Bildung veränderter Gefügemorphologien und zur Ausbildung metastabiler Phasen kommen /1-4/. Vor allem bei Eisenwerkstoffen sind derartige Vorgänge von wissenschaftlicher und technischer Bedeutung. Von
wissenschaftlicher Bedeutung ist die Tatsache, daß die Veränderungen
von Gefügen und Strukturen es ermöglichen, die beim Eisen auftretenden Umwandlungen δ-γ und γ-α zu beeinflussen /5,6/. Die technische Bedeutung leitet sich aus den vielfältigen Anwendungen von
Eisenbasiswerkstoffen ab. Die heute üblichen Legierungen, die in
technischen Werkstoffen verwirklicht werden, sind in ihrer Zusammensetzung auf bestimmte technische Anwendungen zugeschnitten, um
eine optimale Kombination von Eigenschaften zu ermöglichen. Es ist

---

* Dept. Metallurgy and Materials Science, University of Birmingham

deshalb nicht verwunderlich, daß Laserumschmelzversuche sich an
diesen etablierten Zusammensetzungen orientieren /3-9/.

Die beim Laserumschmelzen auftretenden Gefügeveränderungen lassen
sich grob in drei Typen unterteilen. So bewirkt eine Erhöhung der
Abschreckgeschwindigkeit ein feineres Gefüge /3,4,9/. Bei einem
dendritischen Gefüge bedeutet dies einen geringeren Dendritenarm-
abstand. Andererseits kann eine Zunahme der Erstarrungsgeschwindigkeit eine Veränderung der Morphologie bewirken. So wurde z.B. bei
Fe-Cr-C- und Fe-Si-C-Legierungen eine Abfolge gefunden von a)
Dendriten mit interdendritischem Eutektikum, b) Dendriten mit
interdendritischem, zusammenhängenden Karbidnetz und c) homogenen
Mischkristallen mit Seigerungen /3,4,7,8,9/. Der dritte Typ ist die
Bildung von metastabilen oder glasigen Phasen /1,3,4,9,10/.

Die vorliegenden Resultate zeigen, daß sich auch das Umschmelzverhal-
ten von Schnellarbeitsstählen in diese Arbeiten einreiht. Sie deuten
darauf hin, daß hier möglicherweise ein allgemeingültiger Zusammenhang vorliegt.

Experimentelles

Im Gegensatz zu vielen anderen Fe-Basislegierungen ist das Umschmelzen von Schnellarbeitsstählen des Typs S-6-5-2 unproblematisch, da
in allen untersuchten Fällen rißfreie Oberflächen auftraten, ohne
daß spezielle Vorkehrungen hinsichtlich Fokussierung, Gasführung und
Vorschubgeschwindigkeit getroffen werden mußten. Beide Stähle
konnten durch geeignete Prozeßgase vor Oxidation etc. geschützt
werden. Andererseits war es möglich, sie aufzukohlen oder zu nitrieren, wenn dies gewünscht war. Ein typischer Zusammenhang zwischen
erreichter Einschmelztiefe und Vorschubgeschwindigkeit für verschiedene Prozeßbedingungen ist in Bild 1 dargestellt. Umschmelzungen
wurden an genormten Flach- und Rundproben, Bild 2 und 3, sowie
an Drehmeißeln, Bild 4, und Stanzwerkzeugen durchgeführt.
Die auftretenden Gefüge, Texturen und Härtewerte wurden als Funktion
der Prozeßparameter ermittelt. Zusätzlich wurden Untersuchungen an
wärmebehandelten Proben durchgeführt.

Ergebnisse

Durch Variation von Vorwärmung, Laserleistungsdichte und Vorschubgeschwindigkeit wurden unterschiedliche Abschreck- und Erstarrungsgeschwindigkeiten realisiert. In Bild 5 sind zwei Gefüge für unterschiedliche Abschreckraten dargestellt, erkennbar an den unterschiedlichen sekundären Dendritenarmabständen. Der Einfluß der Erstarrungsgeschwindigkeit auf die Morphologie kommt in Bild 6 zum Ausdruck, wo
durch Variation der Erstarrungsgeschwindigkeit eine Änderung in der
Morphologie erreicht werden konnte, wie sie auch bei anderen Eisenwerkstoffen gefunden wurde. So zeigt z.B. Bild 6a ein dendritisches
Gefüge mit dem interdendritischen Fe-W-C-Eutektikum (Fischgrätenmu-
ster). In Bild 6b ist dagegen ein zusammenhängendes nichteutektisches
Karbidnetz zu sehen und Bild 6c schließlich zeigt den Übergang zu
homogenen, stark übersättigt erstarrten Mischkristallen. Tatsächlich
ist jedoch das Erstarrungsverhalten noch etwas komplizierter, wie
röntgenographische Untersuchungen zeigen. So beobachtet man z.B. bei
geringen Erstarrungsgeschwindigkeiten ein Gefüge, welches aus primä-
rem γ und interdendritischem Eutektikum besteht. Dabei sind die γ-Den-
driten zum Teil martensitisch umgewandelt. Mit zunehmender Erstarrungsgeschwindigkeit kommt es zunächst zu einer Übersättigung des γ
an Kohlenstoff. Mit zunehmender Abschreckgeschwindigkeit, d.h. mit zunehmender Unterkühlung bildet sich primär stark übersättigtes δ-Eisen
mit interdendritischem γ und γ-Fe + Karbid-Eutektikum. Bei gleichzeitig hoher Erstarrungs- und Abschreckgeschwindigkeit bildet sich
ein homogener δ-Mischkristall, der nach einer Wärmebehandlung eine
Vielzahl von Ausscheidungen enthält. Entsprechend sind auch die Här-
tekurven in Bild 7 zu verstehen. Bei geringer Erstarrungsgeschwindigkeit, d.h. γ-Dendriten, erhöht sich die as-quenched-Härte durch
Ausscheidungs- und Umwandlungshärtung, während bei dem übersättigt
erstarrten δ-Eisen (höhere Erstarrungsgeschwindigkeit) ausschließlich
eine Ausscheidungshärtung für eine Härtesteigerung verantwortlich
ist. Über die Frage, ob die Erstarrung bei einem HSS-Stahl der Zusammensetzung S-6-5-2 tatsächlich in Form eines übersättigten δ-
Kristalls erfolgt, hat es eine Reihe von Auseinandersetzungen
gegeben /5,6/. Es scheint jedoch, ausgehend von Texturuntersuchungen,
nunmehr sicher zu sein, daß eine Erstarrung als stark übersättigter
δ-Mischkristall möglich ist. Als Indiz hierfür kann angesehen
werden, daß sich vollständig unterschiedliche Texturen einstellen,
je nachdem, welche Phase zuerst erstarrt ist und ob eine Umwandlung
im festen Zustand stattfindet. So entsprechen die in Bild 8a darge-

440

stellten Texturen einer Erstarrung als γ-Fe, die in die in Bild 8b
dargestellten Texturen bei der martensitischen Umwandlung übergehen.
Bei höherer Erstarrungsgeschwindigkeit bildet sich die typische
Erstarrungstextur für die kubisch raumzentrierte Phase. Der ermittelte Gitterabstand läßt vermuten, daß sich hier eine stark über-
sättigte metastabile δ-Phase bei der Erstarrung gebildet hat, wofür
auch die Härtewerte sprechen.

Technische Eigenschaften und Anwendungen

Die laserumgeschmolzenen Schnellarbeitsstähle zeigen sowohl hinsichtlich der Zerspanversuche als auch in Bezug auf Dauerfestigkeit und
Oxidationsverhalten eine deutliche Verbesserung gegenüber konventionellen Schnellarbeitsstählen, siehe Tabelle 1 und 2. Diese lediglich
durch Umschmelzen hervorgerufene Steigerung kann durch das Umschmelzen unter einer reaktiven Atmosphäre oder durch Einschmelzen
von Hartstoffen weiter erhöht werden.

Hinsichtlich der technischen Anwendung sehen die Autoren die Möglichkeit, einerseits kostengünstiger kompliziertere Geometrien zu
härten, andererseits aber auch Härtungen mit einer höheren Zähigkeit
des gehärteten Materials vorzunehmen. Dies beinhaltet Anwendungen im
Bereich der Stanz-, Schneid- und Prägewerkzeuge.

Danksagungen

Die Autoren danken dem BMFT, der SRIC und der Kommission der EG für
die finanzielle Unterstützung der Arbeiten.

Literatur

/1/ H.W. BERGMANN, Proc. 4th Int. Conf. Rapidly Quenched Metals,
    Sendai, 1981, 197
/2/ B. KEAR et al, Proc. 3rd Int. Conf. Rapidly Quenched Metals,
    Brighton, 1978, 171
/3/ H.W. BERGMANN und B.L. MORDIKE, Mat. Res. Soc. Symp. Proc.,
    Vol. 8, North Holland Publ. Comp., 1982, 463
/4/ H.W. BERGMANN und B.L. MORDIKE, Mat. Res. Soc. Symp. Proc.,
    Vol. 9, North Holland Publ. Comp., 1984
/5/ B. CANTOR et al, Proc. 3rd Int. Conf. Rapidly Quenched Metals,
    Brighton, 1978, 110
/6/ J. WOOD et al, ibid, 94
/7/ H.W. BERGMANN und G. BARTON, Z. Elektrowärme, im Druck
/8/ T. Bell und I. HANCOCK, in Vorbereitung
/9/ H.W. BERGMANN und B.L. MORDIKE, Z. Metallkunde 71, (1980), 658
/10/ H.W. BERGMANN, A.BLOYCE und I. HANCOCK, this volume

Tabelle 1. Zerspanungsverhalten von laserumgeschmolzenen S-6-5-2

Zerspanversuche (im Trockenschnitt):

Schneidmeißel: S-6-5-2

Freiwinkel α = 8°, Schneidwinkel γ = 0°

Radius der Eckenrundung ca. 0,4 mm

A - Konventionelle Wärmebehandlung

B - Konventionelle Wärmebehandlung + gelasert + angelassen, 3 x 560 °C, 2 h

C - Ohne konventionelle Wärmbehandlung, gelasert + angelassen, 3 x 560 °C, 2 h

Prüfwerkstoff: 9S20K Ø 70 mm, Länge 30 cm

1. Versuch: n = 400 U/min

Schnittiefe 0,75 mm, Vorschub 0,1 mm/U

A: 5 min, B: 13 min, C: 11 min

2. Versuch: n = 280 U/min

Schnittiefe: 0,5 mm, Vorschub 0,1 mm/U

1. Durchgang (L = 30 cm) = 6,2 min

Dauer des Versuchs: 75 min (12 Durchgänge)

Ergebnis: Vergleichbare Kolk-Form und -Tiefe für A und B

Tabelle 2. Ermüdungsverhalten von laserumgeschmolzenem S-6-5-2

Dicke mm	Breite mm	Spannung N/mm^2	Lastwechsel	Bemerkung
2,04	20,4	430	10^7	laserumge-
1,95	20,4	430	10^7	schmolzen
2,04	20,4	430	10^7	2 m/min
1,99	20,3	430	10^7	Ausgangsmaterial

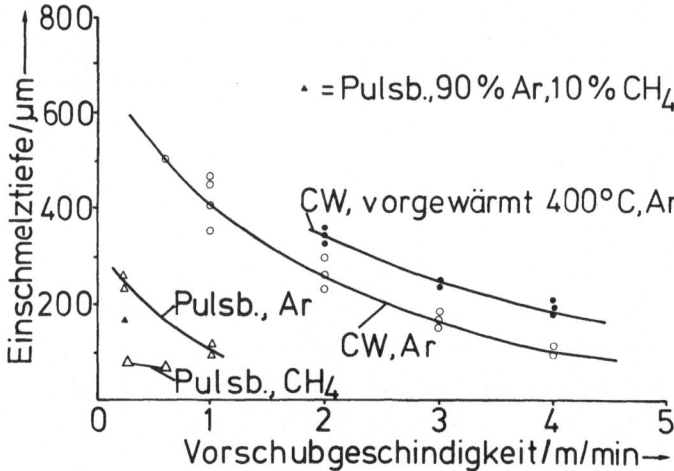

Bild 1. Einschmelztiefe als Funktion der Vorschubgeschwindigkeit
für verschiedene Prozeßparameter. Umgeschmolzen wurde im
Fokus mit 600 Watt im CW-Betrieb und bei 500 Hz und 2 ms
bei 3 kW im Pulsbetrieb

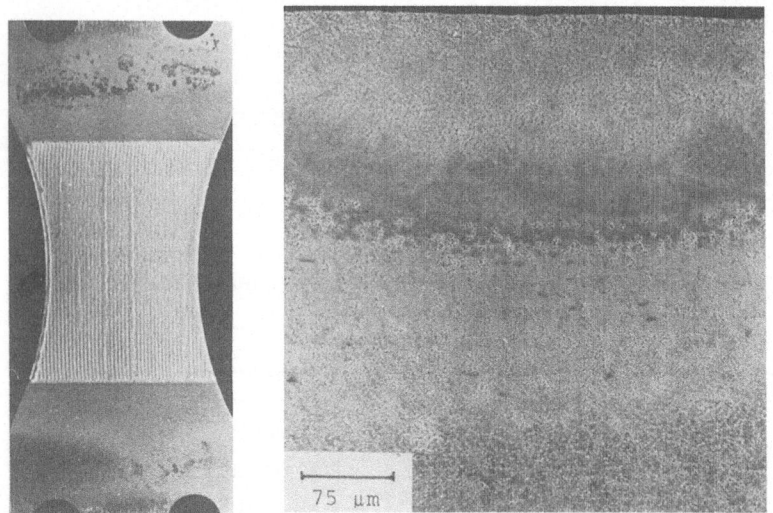

Bild 2. Umgeschmolzene Flachproben aus S-6-5-2 für Ermüdungsversuche.
Die Proben wurden beidseitig umgeschmolzen
a) Makroaufnahme
b) Querschliff

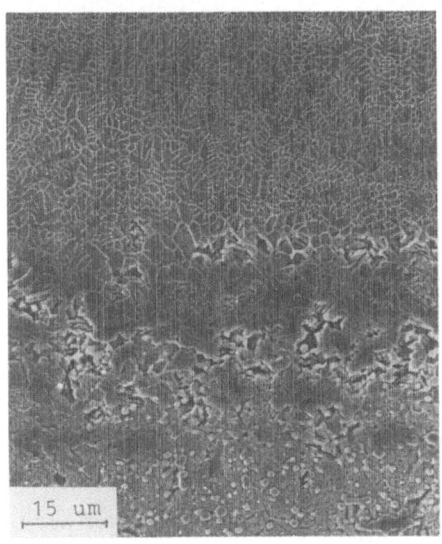 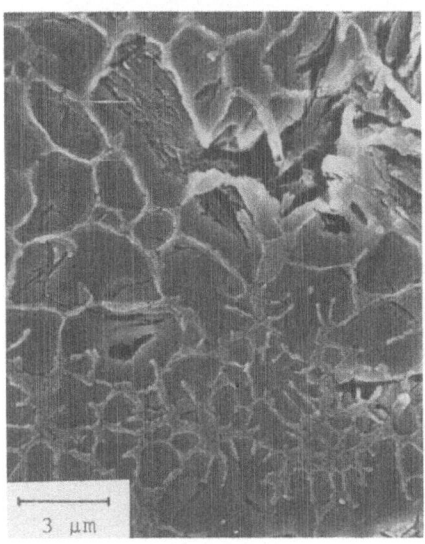

c) Ausschnittsvergrößerung von b)   d) Ausschnittsvergrößerung von c)

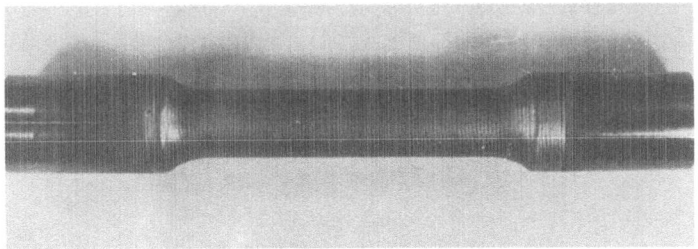

Bild 3. Umgeschmolzene Rundprobe aus S-6-5-2 für Ermüdungsversuche.
        Die Proben wurden beidseitig umgeschmolzen
        a) Makroaufnahme

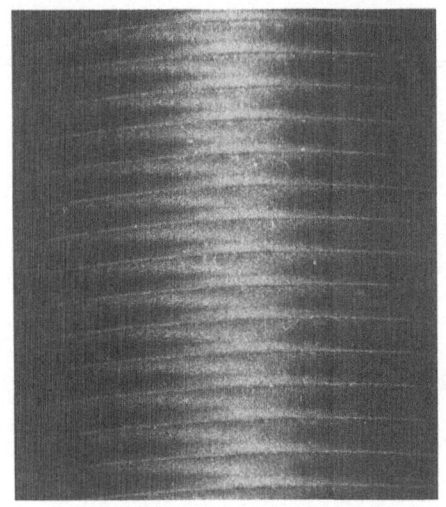

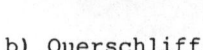

b) Querschliff

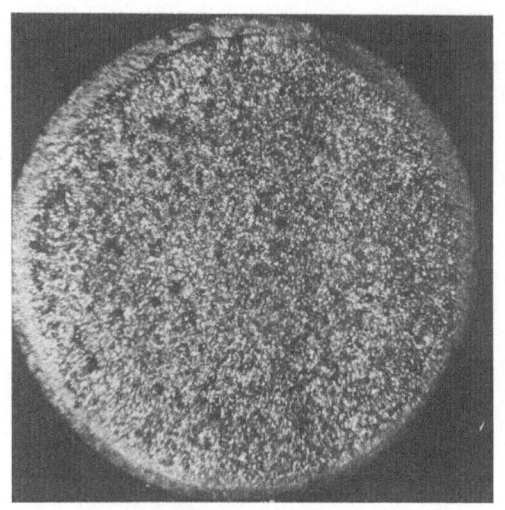

c) Ausschnittsvergrößerung von a)

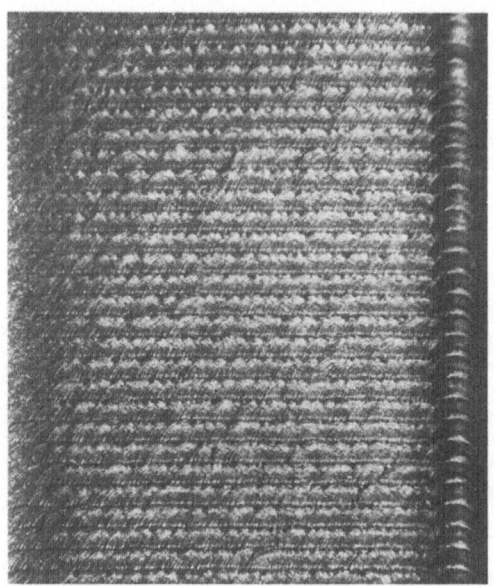

Bild 4. Laserumgeschmolzene Drehmeißel aus S-6-5-2
   a) Makroaufnahme, b) Oberfläche des umgeschmolzenen Bereichs

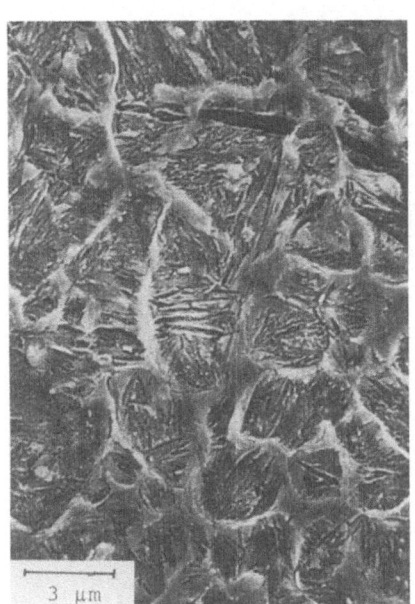 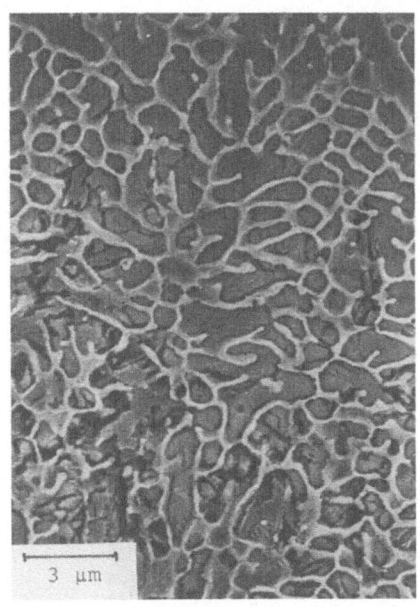

Bild 5. Gefüge eines laserumgeschmolzenen S-6-5-2 für unterschiedliche Abschreckgeschwindigkeiten

a) $\varepsilon \sim 10^4$ K/s   b) $\varepsilon \sim 10^6$ K/s

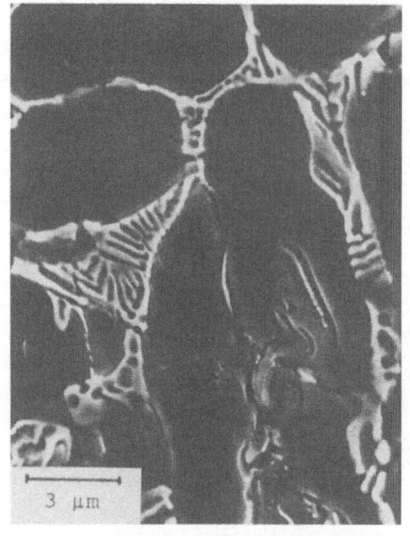

a)

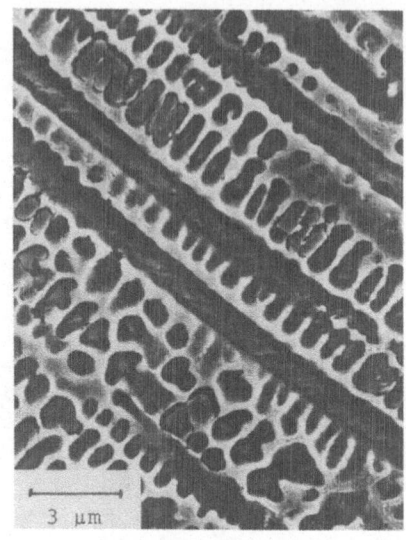

b)

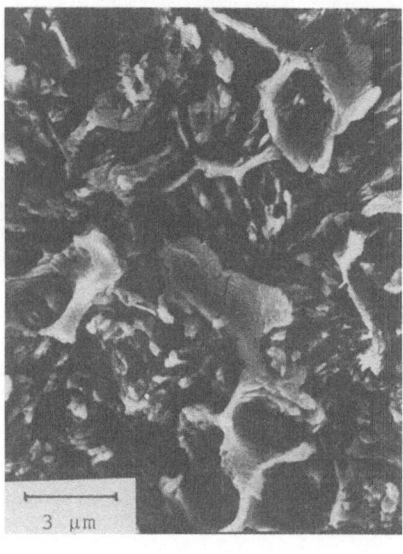

c)

Bild 6.
Gefüge eines laserumgeschmolzenen S-6-5-2 für unterschiedliche Erstarrungsgeschwindigkeiten
a) R = 0,1 m/min Dendriten +
   Eutektikum
b) R = 1   m/min Dendriten +
   Karbide
c) R = 10 m/min Mischkristall
   mit Seigerungen + Ausscheidungen

a)

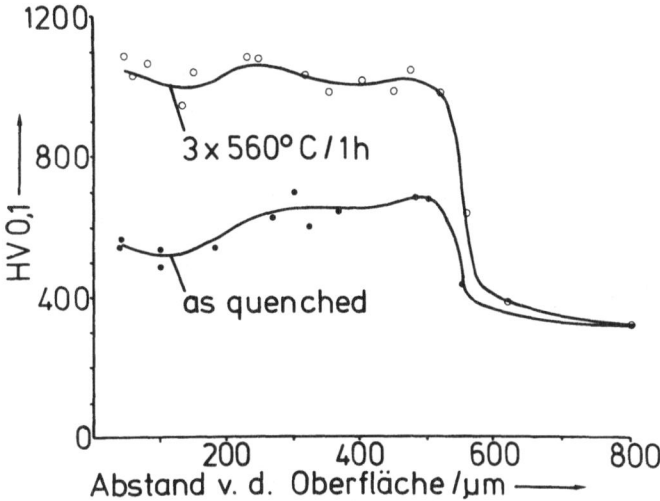

b)

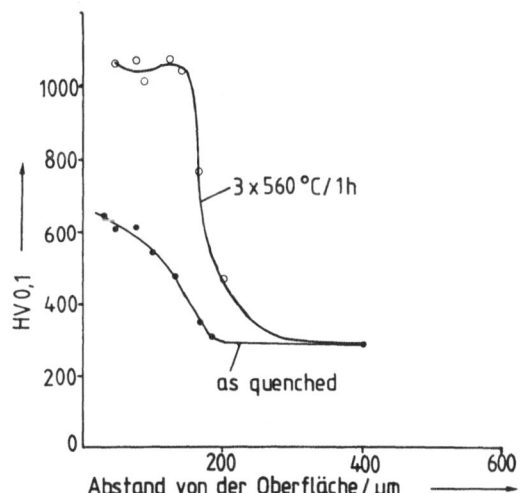

Bild 7. Härteprofile
    a) Geringe Abschreckgeschwindigkeit
    b) Höhere Abschreckgewindigkeit

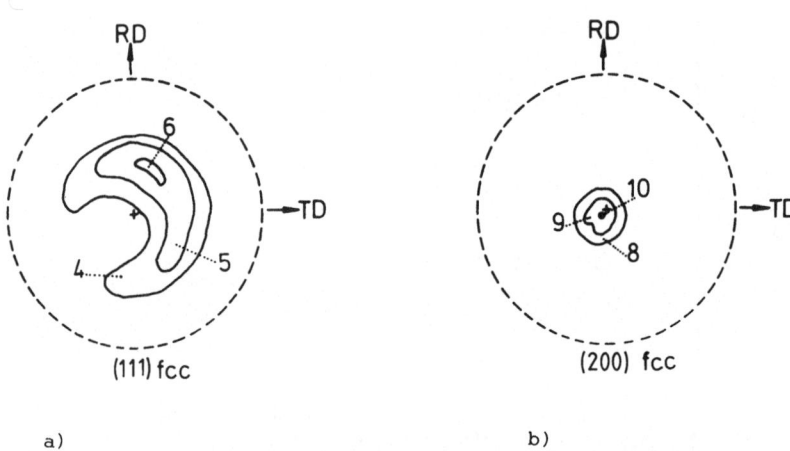

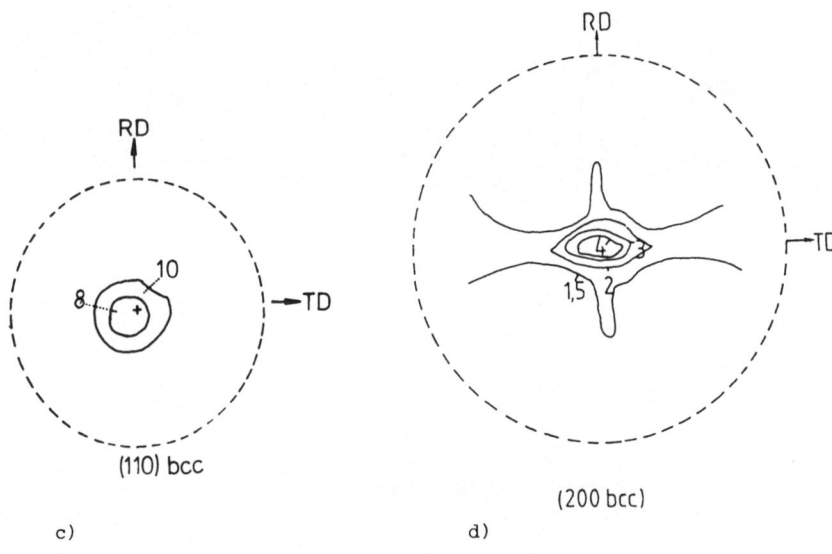

Bild 8. Textur eines laserumgeschmolzenen S-6-5-2
a, b) Erstarrungstextur des γ-Fe
c) Umwandlungstextur des α-Fe aus 9a
d) Erstarrungstextur des δ-Fe

Härten mit CO_2-Laserstrahlung

K.Wissenbach, A.Gillner, E.W.Kreutz,
Fraunhofer-Institut für Lasertechnik, D-51 Aachen

1. Einleitung

Der Laser wird als flexibles Werkzeug in der industriellen Fertigung eingesetzt. Von den verschiedenen Bearbeitungsverfahren gewinnt die Oberflächenveredlung von Werkstoffen zunehmend an Bedeutung. Die verschiedenen Wärmebehandlungsverfahren wie Tempern, Beschichten, Umwandlungshärten, Ein-und Umschmelzlegieren werden in der Regel bei Laserintensitäten $I < I_C$ durchgeführt, wobei I_C die kritische Intensität für das Entstehen eines laserinduzierten Plasmas ist/1-3/. Ausgangspunkt für die Beschreibung dieser Bearbeitungsprozesse ist die dreidimensionale Wärmeleitungsgleichung mit temperaturabhängigen thermophysikalischen Koeffizienten.

Auf Grund der guten Fokussierbarkeit und der Möglichkeiten der Strahlformung und -führung eignet sich der Laser besonders zum Umwandlungshärten kleiner Volumina ($V_H < 3mm^3$) und komplizierter Härtgeometrien. Um die spezifischen Vorteile des Lasers ausnutzen zu können, müssen Laser- und Werkstückparameter an die jeweilige Härtaufgabe angepaßt werden. Im folgenden wird an Hand von Versuchshärtungen und Modellrechnungen (basierend auf Lösungen der Wärmeleitungsgleichung) der Einfluß des Umwandlungsverhaltens verschiedener Stähle, die Temperaturabhängigkeit der Werkstückparameter, verschiedener Intensitätsverteilungen und endlicher Werkstückgeometrien auf das Härtergebnis untersucht.

2. Experimentelle und theoretische Ergebnisse

In Fig.1 ist der Härtvorgang schematisch dargestellt. Der Laserstrahl trifft senkrecht auf das Werkstück, das sich mit der Geschwindigkeit v in x-Richtung bewegt. Über Laserleistung P_L, Strahlradius r_B, Vorschubgeschwindigkeit v und Intensitätsverteilung $I(x,y)$ sind die Laserparameter festgelegt. Die Härtgeometrie, gegeben durch Härtbreite w_H und Härttiefe z_H, hängt von den Laserparametern, den Werkstückeigenschaften (thermische Koeffizienten, Absorptionsverhalten) und dem Umwandlungsverhalten der Stähle ab. Das Absorptionsverhalten von Stählen mit absorbierenden Schichten (Coatings) wird an anderer Stelle /4,5/ ausführlich diskutiert.

Ein Beispiel für das unterschiedliche Umwandlungsverhalten zweier Stähle zeigt Fig.2. Hier ist die Härttiefe in Abhängigkeit vom Strahlparameter Ir_B dargestellt. Die Auftragung über Ir_B wird gewählt, da für eine gaußförmige Intensitätsverteilung die Temperatur im Werkstück $T(x,y,z,t) \sim Ir_B$ ist/4/. Die Härttiefe liegt für den

450

Stahl Cf 53 (Kohlenstoffstahl) über der von 100 Cr6 (Schnellarbeitsstahl mit 1.5%
Cr), da 100 Cr6 umwandlungsträger ist als Cf 53. Dies zeigt sich in den Umwand-
lungsschaubildern (Z-T-A-Diagrammen), wonach sich für Aufheizraten von $2.4 \cdot 10^3$ K/s
die A_{C1}-Temperaturen nach Orlich/6,7/ zu

$$A_{C1} \text{ (Cf 53)} \simeq 800^0C \text{ , } A_{C1} \text{ (100 Cr6)} \simeq 830^0C$$

ergeben. Mit zunehmendem Ir_B steigt die Härttiefe auf Grund zunehmender Wärmelei-
tungsverluste/4/ nicht weiter an, d.h. bei konstanter Laserleistung nehmen die Wär-
meleitungsverluste mit kleiner werdendem Strahlradius zu. Zusätzlich sind die nach
einem dreidimensionalen Wärmeleitungsmodell für "halbunendliche" Werkstückgeomet-
rie/4,8/ berechneten Härttiefen eingetragen. Die Abweichungen zwischen Theorie und
Experiment sind kleiner 10%. Ein Grund für die Abweichungen ist die Nichtberücksich-
tigung der Temperaturabhängigkeit der thermophysikalischen Koeffizienten. Um diesen
Einfluß auf die Temperaturverteitlung zu studieren, wird die nichtlineare, eindimen-
sionale Wärmeleitungsgleichung

$$\rho(T) \ c(T) \ \frac{\partial T}{\partial t} = \frac{\partial}{\partial z} \left[K(T) \ \frac{\partial T}{\partial z} \right] \tag{1}$$

ρ: Dichte
c: spezifische Wärmekapazität
K: Wärmeleitfähigkeit

numerisch mit einem Differenzenverfahren/4/ gelöst. Die Werkstückdicke wird so gewählt, daß das Werkstück in z-Richtung als unendlich ausgedehnt betrachtet werden
kann. Die exakte Temperaturverteilung kann mit analytischen Lösungen mit konstanten
Koeffizienten/9,10/ verglichen werden. Die Ergebnisse dieser Rechnung sind in Fig.3
für den beim Laserhärten typischen Temperaturbereich dargestellt. Der Temperaturverlauf mit konstanten Koeffizienten ist für zwei Fälle eingezeichnet:

- arithmetisch gemittelte Koeffizienten
- Koeffizienten bei Raumtemperatur

Zur Überprüfung des numerischen Verfahrens wird die Temperaurverteilung für verschiedene Schrittweitenverhältnisse (λ-Werte) berechnet. Alle Rechnungen stimmen
überein. Aus beiden Darstellungen kann gefolgert werden, daß bei geeigneter Mittel-
wertbildung der thermophysikalischen Koeffizienten die Abweichungen vom exakten Verlauf für die untersuchten Stähle kleiner 10% sind. Aus diesem Grund wird bei dreidimensionalen Rechnungen mit gemittelten Koeffizienten gearbeitet.
Das in /8/ vorgestellte Modell ist auch auf instationäre Wärmeleitungsvorgänge
anwendbar. Als Beispiel dafür ist in Fig.4 das Einlaufverhalten einer Vorschubhärtung in Form eines Längsschliffs dargestellt. Nach etwa 3.5 mm wird die Härttiefe

konstant, was gleichbedeutend mit dem Erreichen des stationären Zustandes ist. Im unteren Teil des Bildes ist zu sehen, daß Theorie und Experiment gut übereinstimmen. Den Einfluß der Intensitätsverteilung auf die Härtgeometrie wird am Grundmode TEM_{00} und am Ringmode TEM_{01*} , der eine Überlagerung aus TEM_{01} und TEM_{10} darstellt, untersucht. In Fig.5 sind beide Intensitätsverteilungen dargestellt. Da beim Ringmode das Intensitätsmaximum nicht im Strahlzentrum liegt, wird der Wärmetransport in die Umgebung effektiver, d.h. es können größere Flächen gehärtet werden. Dies zeigt Fig.6. Hier sind zwei Querschliffe von Härtungen dargestellt, bei denen die Laserparameter gleich sind. Die Breite, der mit dem Ringmode gehärteten Spur ist etwa 20% größer als die mit dem Grundmode. Im unteren Teil sind die zugehörigen Härtverläufe in Abhängigkeit von der z-Koordinate dargestellt. Die Härtverläufe sind für beide Strahlverteilungen prinzipiell gleich; es werden Maximalhärten von über 900 HV0.5 erzielt. In Fig.7 ist der Vergleich der Härttiefen für beide Verteilungen in Abhängigkeit von Ir_B dargestellt. Im Bereich $Ir_B \geq$ 1100 W/cm ist die Härttiefe für den Ringmode etwa 20% größer als für den Grundmode. Für $Ir_B \leq$ 900 W/cm gleichen sich die Härttiefen an. Auf Grund der höheren Spitzenintensität des TEM_{00} wird bei kleineren Werten von Ir_B noch eine meßbare Härttiefe erzielt, während die Intensität des Ringmodes nicht mehr ausreicht, um die Oberfläche des Werkstücks auf Härttemperatur zu bringen. Zusätzlich eingetragen sind die berechneten Härttiefen. Dabei wird für den Grundmode das Modell nach /8/ verwendet, für den TEM_{01*} wird ein Differenzenverfahren zur Berechnung der Temperaturverteilung benutzt/4/. Fig.7 zeigt, daß Theorie und Experiment im Rahmen der durchgeführten Näherungen übereinstimmen.

Da der Laser besonders zum Härten kleiner Volumina geeignet ist, müssen Wärmeleitungsmodelle für endliche Werkstückgeometrien entwickelt werden. In Fig.8 ist schematisch ein Werkstück dargestellt, das in x-Richtung als unendlich ausgedehnt betrachtet werden kann. Entscheidendes Kriterium, ob die Wärmeleitungsgleichung in endlicher oder in halbunendlicher Geometrie gelöst werden muß, ist das Verhältnis δ_W/D von Wärmeeindringtiefe δ_W gegeben durch

$$\delta_W = \sqrt{4\kappa t_L} \qquad (2)$$

κ: Temperaturleitfähigkeit

t_L: Prozeßdauer

und der Werkstückdimension D in einer Richtung (im Beispiel:z-Richtung). Für $\delta_W \ll$ D kann das Werkstück als unendlich angenommen werden und das Modell aus /8/ kann angewendet werden. Ist die Ungleichung nicht erfüllt, müssen andere Modelle zur Berechnung der Temperaturverteilung herangezogen werden. Als Beispiel wird die Härtgeometrie in dem in Fig.8 dargestellten Werkstück berechnet und mit dem Experiment verglichen. Für stationäre Verhältnisse lautet die Wärmeleitungsgleichung

$$\kappa \, \Delta T = v \, grad \, T \qquad (3)$$

452

Diese Gleichung wird nach folgendem Differenzenverfahren approximiert/4/:

$$\kappa \left[\frac{T_{i,k,j-1} - 2T_{i,k,j} + T_{i,k,j+1}}{(\Delta x)^2} + \frac{T_{i-1,k,j} - 2T_{i,k,j} + T_{i+1,k,j}}{(\Delta y)^2} \right.$$

$$\left. + \frac{T_{i,k-1,j} - 2T_{i,k,j} + T_{i,k+1,j}}{(\Delta z)^2} \right] \qquad (4)$$

$$= v \left[\beta \frac{T_{i,k,j} - T_{i,k,j-1}}{\Delta x} + (1 - \beta) \frac{T_{i,k,j+1} - T_{i,k,j}}{\Delta x} \right]$$

Dabei wird dT/dx in Gl.4 durch eine Kombination von vorwärts und rückwärts ge-
richteten Differenzenquotienten genähert, die mit dem Gewicht β bzw. (1-β) versehen
werden. Im mittleren Teil ist ein Querschliff einer solchen Härtung dargestellt.
Zusätzlich ist die nach dem Modell (Gl.4) berechnete Härtgeometrie eingetragen.
Theorie und Experiment stimmen überein. Wird die Werkstückdicke D sehr viel größer
als die Wärmeeindringtiefe δ_W, so wird der Grenzfall des in z-Richtung unendlich
ausgedehnten Werkstücks realisiert. Ein Querschliff einer solchen Härtung ist im
unteren Teil des Bildes dargestellt. Auch hier stimmt die nach dem Modell/4,8/ berechnete Härtgeometrie mit dem experimentellen Ergebnis überein.

3. Zusammenfassung

Bei Kenntnis der Laserparameter Leistung, Strahlradius, Vorschubgeschwindigkeit bzw.
Prozeßzeit sowie des Umwandlungs-bzw. Absorptionsverhaltens der Stähle liefern die
vorgestellten Wärmeleitungsmodelle befriedigende Ergebnisse in folgenden Punkten:

- Beschreibung stationärer und instationärer Wärmeleitungsvorgänge
- Einfluß verschiedener Intensitätsverteilungen auf die Härtgeometrie
- Beschreibung der Wärmeleitung in endlicher und halbunendlicher
  Werkstückgeometrie

Literatur

/1/ G.Herziger
    Feinwerktechnik und Meßtechnik 91, 156 (1983)
/2/ E.Beyer
    Einfluß des laserinduzierten Plasmas beim Schweißen mit CO_2-Lasern
    Dissertation, TH Darmstadt 1985
/3/ R.Poprawe
    Materialabtragung und Plasmaformation im Strahlungsfeld von UV-Lasern
    Dissertation, TH Darmstadt 1984
/4/ K.Wissenbach
    Umwandlungshärten mit CO_2-Laserstrahlung
    Dissertation, TH Darmstadt 1985

/5/ K.Behler,A.Gillner,G.Herziger,E.W.Kreutz,K.Wissenbach
J. de Physique C6 ,47 (1984)

/6/ J.Orlich
Beschreibung der Austenitisierungsvorgänge unlegierter und legierter Stähle
bei induktiver Schnellerwärmung
Dissertation,TU Berlin (1971)

/7/ J.Orlich,H.-J.Pietrzeniuk
Atlas zur Wärmebehandlung der Stähle
Verlag Stahleisen M.B.H. Düsseldorf (1976)

/8/ K.Wissenbach et al.
Proc. "Laser 83"
Springer Verlag Berlin 1984

/9/ H.S.Carslaw,J.C.Jäger
Conduction of Heat in Solids
University Press , Oxford (1959)

/10/ N.N.Rykalin
Berechnung der Wärmeleitungsvorgänge beim Schweißen
Verlag Technik, Berlin (1957)

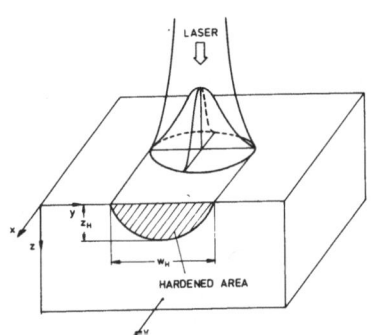

Fig.1. Schematische Darstellung einer Vorschubhärtung mit der Geschwindigkeit v mit den die Härtgeometrie bestimmenden Größen

Fig.2. Härttiefe z_H in Abhängigkeit von (Ir_B) für die Stähle Cf 53 und 100 Cr6. Die Kurven sind nach einem dreidimensionalen Wärmeleitungsmodell/4/ berechnet

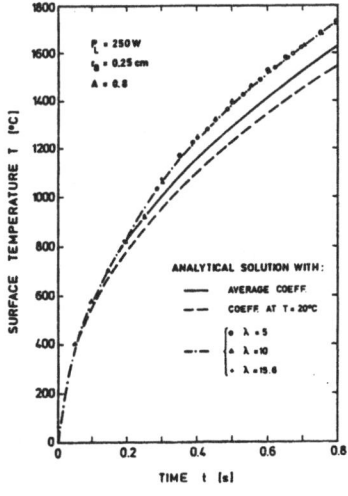

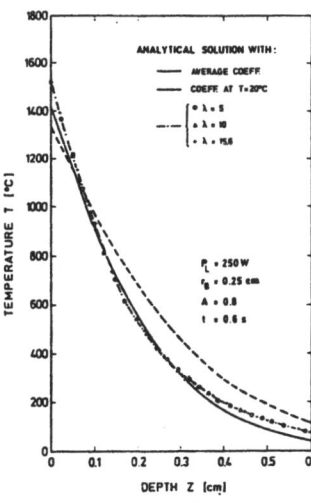

Fig.3. Berechnete Temperaturverteilung als Lösung von Gl.1 in Abhängigkeit von
a) Zeit t,
b) Tiefe z

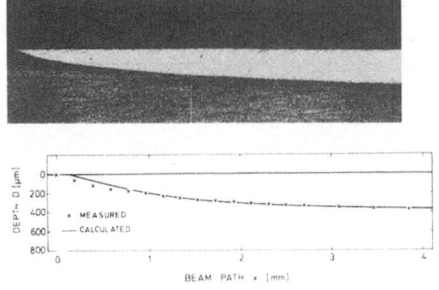

Fig.4. Einlaufverhalten der Härttiefe bei einer Vorschunbhärtung
Oben: Längsschliff in Vorschubrichtung x
Unten: Berechnete und gemessene Härttiefe in Abhängigkeit von der Vorschubrichtung

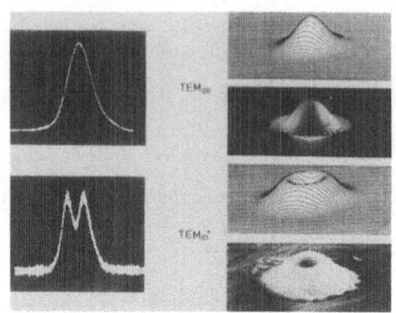

Fig.5. Intensitätsverteilungen TEM_{00} und TEM_{01*} in drei Darstellungen
1) Berechnet
2) Einbrand in Plexiglas
3) Gemessen mit der Steen-Methode

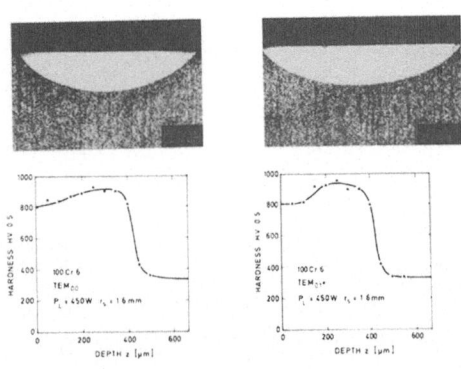

Fig.6. Querschliffe zweier mit den Intensitätsverteilungen TEM_{00} und TEM_{01*} erzeugten Härtspuren mit den dazugehörigen Härtverläufen in Abhängigkeit von der Tiefe z

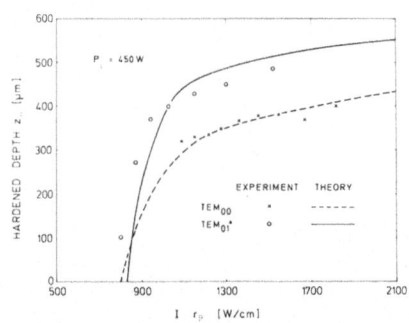

Fig.7. Einfluß der Intensitätsverteilung auf die Härttiefe für eine Vorschubgesschwindigkeit v=0.5 cm/s für den Stahl Cf 53. Die Kurven sind nach dreidimensionalen Wärmeleitungsmodellen /4/ berechnet

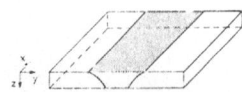

Fig.8. Schematische Darstellung eines Werkstücks und Querschliffe mit berechneten Härtisothermen nach dreidimensionalen Wärmeleitungsmodellen /4/ in endlicher (Mitte) und halbunendlicher (Unten) Werkstückgeometrie

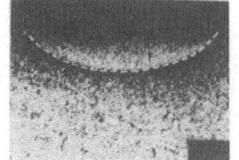

A Kinetic Investigation of Laser-Chemical Deposition of Iron Pentacarbonyl Films

N.A.Kislov, V.I.Dernovsky, I.V.Malikov and V.V.Aristov
Institute of Problems of Microelectronics Technology and
Superpure Materials
USSR Academy of Sciences
142432 Chernogolovka, Moscow District, USSR

The kinetics of LCD films of iron pentacarbonyl $Fe(CO)_5$ vapour have been studied. Fig.1 shows a block diagram of a LCD apparatus. The radiation used was the fourth harmonic of neodymium laser (λ = =266 nm, τ =15 ns, f=25 Hz, average power W=0.01 to 0.3 mW) which was focused onto a substrate as a spot of 150 μm in diameter. By their appearance the deposits were clearly divided into two distinct regions. The central region was formed in the area immediately exposed to the laser beam being of the same shape as the latter. In this

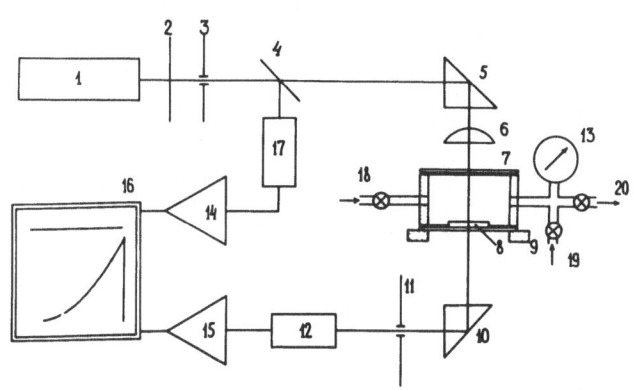

Fig.1. Block diagram of a laser-chemical
deposition apparatus: 1 - laser, 2 -
light filter, 3, 11 - diaphragms,
4 - translucent mirror, 5, 10 -
prisms, 6 - focusing lens, 7 - laser-
chemical cell window, 8 - support,
12, 17 - photocells, 14, 15 - amplifiers, 16 - recorder

region the deposit had a good adhesion and could not be notched with a steel blade. In the second region (beyond the immediately exposed spot) the deposit is also glossy but has a much poorer adhesion (the film is easily separated). From the fact that a film deposit was also formed beyond the exposed region it can be inferred that the deposition occurs as a result of the photolysis of $Fe(CO)_5$ within the laser beam volume.

Substituting the interferometrically determined thicknesses and corresponding optical absorptivities into the Buger-Lambert law we determined the absorptivity of the $Fe(CO)_5$ films as α =(1.7$\pm$0.5) $10^7 m^{-1}$.
Fig.2 shows the deposition rate as a function of the total gas pres-

sure. The deposition rate is seen to decrease rapidly as the total pressure in the cell increases. This may be due to the diffusion mechanism of mass transfer in the process under investigation 1 .

Processing of the experimental data has shown that the relationship between the deposition rate and $Fe(CO)_5$ vapour pressure remained linear within 20-200 Pa under a constant argon pressure of 8×10^4 Pa (Fig.3). The fact that the straight line extrapolated to zero pressures does not pass through the origin may be due to the contribution of the adsorbed $Fe(CO)_5$ layer to the film buildup rate 2 .

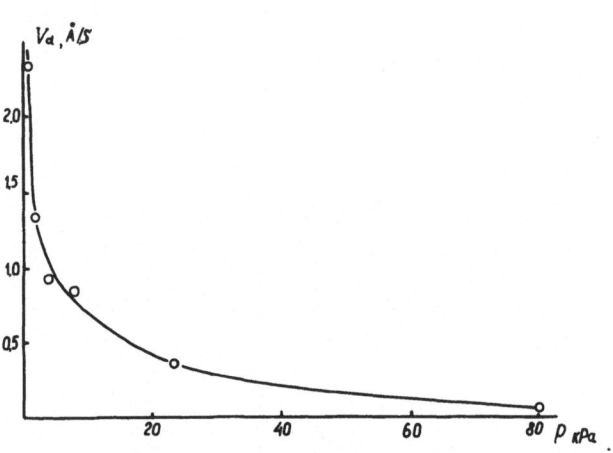

Fig.2. Deposition rate as a function of gas mixture pressure in the cell. $Fe(CO)_5$ pressure 130 Pa; W=0.25 mW

From the kinetic curves for the absorption of the growing film we have also obtained a linear ralationship between the deposition rate and radiated power (Fig.4) which points to the unsaturated single-photon mechanism in $Fe(CO)_5$ dissociation 3 .

The analysis of the correlation between LCD and UV-absorption of $Fe(CO)_5$ must be based on the comparison of the effec-

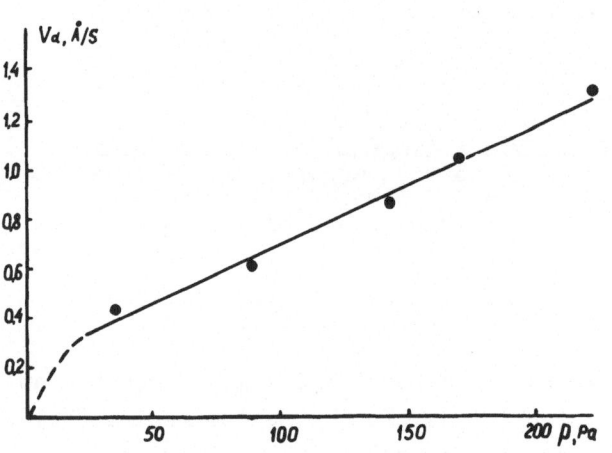

Fig.3. Deposition rate as a function of $Fe(CO)_5$ pressure. Pressure of the $Fe(CO)_5$+Ar mixture 80 KPa; W=0.25 mW

tive cross-sections of the two processes. Obviously, the deposition rate must be related to the effective cross-section of photodeposition δ_0 . Using the equation for the material flux at the beam centre 1 we obtain the following expression for the deposition rate V_d:

$$V_d = WMP\delta_0 / \pi d h \nu \rho R T \qquad (1)$$

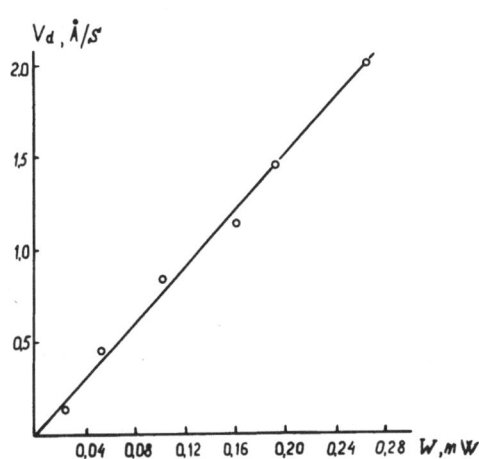

Fig.4. Deposition rate as a function of laser radiation power. Fe(CO)$_5$ pressure 130 Pa; Fe(CO)$_5$+Ar mixture pressure 80 kPa

where W is average power of laser radiation; d is the beam diameter; M, ρ are the molar mass and density of the deposited compound; $h\nu$ is the radiated quantum energy; P is Fe(CO)$_5$ vapour pressure; R is the universal gas constant; T is temperature. For clarity it will be assumed that iron is the component to be deposited. Then, differentiating (1) with respect to P and substituting M=56 kg/kmol; =7.85 10^3 kg/m^3 and dv$_d$/dp found at room temperature from the slope of the curve in Fig.2 we obtain σ_0=(3.0±1.1)x10^{-22}m^2. Under the assumptions used this

value is the upper bound of σ_0 for LCD from Fe(CO)$_5$. On the other hand, the effective cross-section σ_f can be determined from Fe(CO)$_5$ absorptivity data 4

$$\sigma_f = kT\ln(I_e/I)/HP \qquad (2)$$

where k is the Bolzmann constant; I is the intensity of radiation transmitted through the cell containing the absorbing compound; Ie is the same for an evacuated cell; H is the cell height. Fig.5 shows the absorption of radiation by Fe(CO)$_5$ in the cell as a function of their pressure. The slope of the relationship approximated by a straight line represents, according to (2), the UV absorption cross-section that was found to be equal to σ_f =(5.0±0.7) 10^{-22}m^2. Obviously, the quantum yield F of deposition, defined as the ratio of the number of dissociated molecules

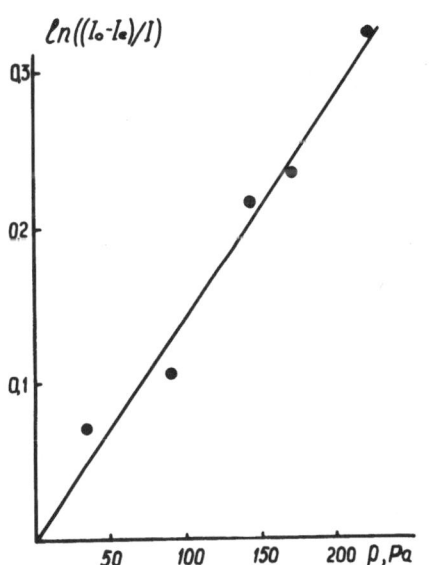

Fig.5. Optical absorption as a function of Fe(CO)$_5$ pressure. Fe(CO)$_5$+Ar mixture pressure 80 kPa

458

to the number of quanta absorbed by the reactant $Fe(CO)_5$ will for an optically thin cell be written as

$$F = \sigma_o / \sigma_f \tag{3}$$

considering the experimental error, $F = 0.60 \pm 0.25$. The fact that $F < 1$ may be due to other photochemical processes apart from photodissociation luminiscence.

To determine likely photolysis products let us compare the quantum energy $E = 445$ kJ/mol with the energy of the Fe-C bond E_b in the $Fe(CO)_5$ molecule, $E_b = 116.9$ kJ/mol 5 . Apparently, the quantum energy is insufficient to break the five Fe-C bonds to liberate pure iron in a photolysis act. Moreover, the quantum energy is smaller than the total energy required to break four Fe-C bonds by a value of $E_a = 22.6$ kJ/mol. It may be assumed that only those molecules will undergo dissociation by the route:

$$Fe(CO)_5 + h\nu \rightarrow Fe(CO) + 4CO \tag{4}$$

which have an energy equal to or higher than E_a prior to quantum absorption. From the Bolzmann particle energy distribution it follows that the concentration of molecules with the energy E_a is:

$$N_a = (P/RT) \exp(-E_a/RT) \tag{5}$$

Substituting the value of P/RT and σ_o in (1) for N_a and $\sigma_r = 10^{-17} m^2$ typical of resonance absorption, we find $V_d = 3.2$ H/S. Hence, the rate of the process involving the participation of thermally excited molecules has the same order of magnitude as the experimental value. The semiquantitative Auger analysis of the deposited films on oxidized silicon supports has shown that iron and carbon are present in approximately equal proportions. The Auger spectra suggest that carbon is unbound there.

It follows from the foregoing that the LCD process occurs mainly via the unsaturated single-photon dissociation of $Fe(CO)_5$ within the laser beam volume. $Fe(CO)$ is the most probable product of the laser flash photolysis ($\lambda = 266$ nm). To dissociate the vapours to pure iron the laser flash with quantum energy close to the total energy of five Fe-C bonds (585 kJ/mol) should be used.

References
1 D.J.Ehrlich, R.M.Ir.Osgood, T.F.Deutsch: J.IEEE J. Quantum Electron. QE-16, 1233 (1980)
2 J.V.Isao, R.A.Becker, D.J.Ehrlich, F.J.Leonberger: J. Appl. Phys. Lett. 42, 559 (1983)
3 D.J.Ehrlich, R.M.Ir.Osgood, T.F.Deutsch: J. Electrochem. Soc. 128, 2039 (1981)
4 A.Hartford, Jr.; J.H.Clark: Chemical and Biochemical Applications of Lasers, ed. C.B.Moore, 5, Academic Press, New York, (1980)
5 G.Nathanson, B.Gilten, A.M.Rosan, J.T.Yardley: J.Chem.Phys. 74, 361 (1981)

Laserschneiden verschiedener metallischer Werkstoffe – Physikalische Einflußgrößen und Prozeßwirkungsgrad

I. Decker und J. Ruge
Institut für Schweißtechnik der TU Braunschweig
Langer Kamp 8, D-3300 Braunschweig

Das Laserschneiden ist in physikalischer Hinsicht ein sehr komplexer
Vorgang, da mehrere Teilprozesse ineinander greifen. Hierzu gehören
die Strahlungsabsorption in der Schnittfuge, das Aufschmelzen und teilweise Verdampfen des Schnittfugenmaterials, die chemische Reaktion mit
und der Schmelzaustrieb durch den Schneidgasstrom sowie der Wärmetransport. In einer früheren Arbeit /1/ wurde ein Schneidmodell vorgestellt,
das für unlegierten Stahl mit den experimentellen Ergebnissen bereits
in befriedigender Weise übereinstimmt. (Ein anderer Zugang wurde in
/2/ gewählt). Im Rahmen theoretischer Arbeiten sind seither einige
Teilvorgänge eingehender untersucht worden /3/. Hier soll an einigen
Beispielen das unterschiedliche Verhalten verschiedener Werkstoffe
dargestellt werden, um so die Wirkung einzelner Einflußgrößen auf den
Schneidprozeß aufzuzeigen.

Laserschneiden als Materialabtrag

Die allein für den Materialabtrag notwendige thermische Energie ist
proportional zum erzeugten Schnittfugenvolumen, ebenso der Gewinn
durch eine anteilige Werkstoffverbrennung im Sauerstoffstrom. Hieraus
ergibt sich der bekannte Zusammenhang zwischen Laserleistung und dem
Produkt aus Schneidgeschwindigkeit und Blechdicke, $L \sim vd$, der zur Abschätzung der maximalen Schneidgeschwindigkeit bei Baustählen in guter
Näherung benutzt werden kann /4/.
Beim Laserbrennschneiden von Titanwerkstoffen liegt eine andere Ge-
setzmäßigkeit vor (Bild 1), da aufgrund der stark exothermen Reaktion
die Schneidgaszufuhr den größeren Einfluß hat /5/: Der Schneidwirkungsgrad vd/L nimmt mit fallender Laserleistung zu. Beim Laserschmelz-
schneiden dagegen wirkt sich eine Zunahme der Leistung Wirkungsgrad-
erhöhend aus, weil sich mit wachsender Schneidgeschwindigkeit die in
das Werkstück durch Wärmeleitung abgeführte Wärmeenergie pro Strecken-
element verringert.
Der geringe Wirkungsgrad bei Aluminium wird mit dem hohen Reflexionsvermögen in Verbindung gebracht. Für sich genommen ist diese Aussage
nicht korrekt, denn unter der Annahme Fresnelscher Absorption kann bei

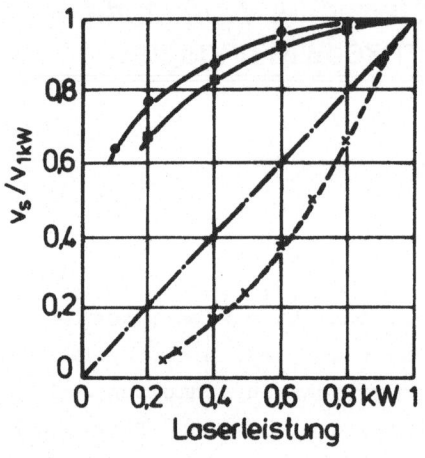

Bild 1. Abhängigkeit der Schneidgeschwindigkeit (jeweils bezogen auf erreichbare Ge-
schwindigkeit bei 1 kW) von
der Laserleistung

        Schneidgas O_2

● TiAl6V4 2.5 mm/0.08 MPa
■ Ti   5   mm/0.05 MPa

        Schneidgas   Ar

✗ TiAl6V4 2.5 mm/0.14 MPa

Aluminium ebensoviel Strahlungsenergie eingekoppelt werden wie bei
Stahl, sofern die Schneidfront steiler verläuft. Dieses läßt sich jedoch mit dem Schneidgasstrom wegen der Oberflächenbeschaffenheit eines
Aluminiumschmelzfilms nicht erreichen, so daß die Strahlungsabsorption
nicht unter optimalen Bedingungen erfolgt. Hinzu kommt die Wärmeableitung in das Werkstück entsprechend der hohen Wärmeleitfähigkeit.

Schneidparameteroptimierung

Bild 2 zeigt für Baustahl die typischerweise auftretende Abhängigkeit
zwischen Riefenbildung und Schneidgeschwindigkeit: Die optimale
Schnittqualität ergibt sich bei etwa 60 bis 80 % der maximalen Schneidgeschwindigkeit. Eine Begründung läßt sich daraus ableiten, daß die
Schneidzonentemperatur mit zunehmender Vorschubgeschwindigkeit an-

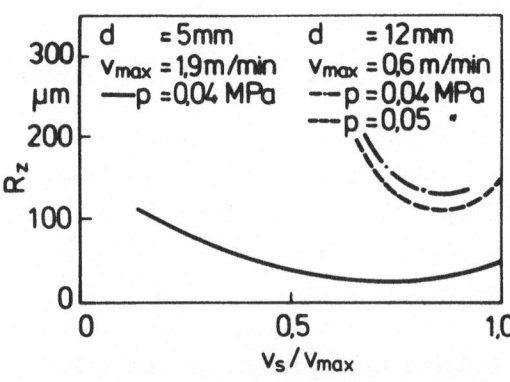

Bild 2. Rauhtiefe in Abhängigkeit
der Schneidgeschwindigkeit
beim Laserbrennschneiden
von Baustahl

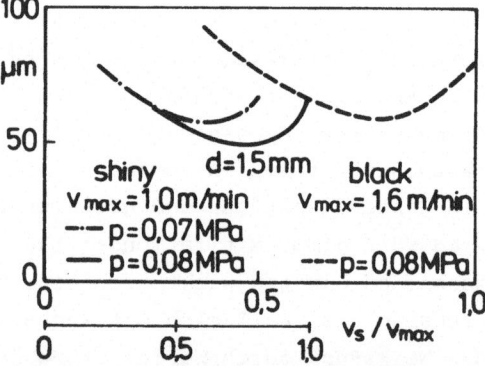

Bild 3. Rauhtiefe beim Laserschneiden von Aluminium
in Abhängigkeit von Ge-
schwindigkeit, Gasdruck
und Oberflächenzustand
(Schneidgas O_2)

steigt, bis es aufgrund der größer gewordenen Schmelzfilmdicke zu verstärkten Schwankungen des Ausfließens kommt. Bei größeren Blechdicken
ist diese Abhängigkeit stärker ausgeprägt.

Aluminium zeigt ein ähnliches Verhalten, wobei in Bild 3 der zusätzliche Einfluß des Gasdruckes und einer Oberflächenschwärzung angedeutet
ist. Für TiAl6V4 sind in Bild 4 die Rauhtiefe und die Unebenheit der
Schnittfläche sowohl für das Laserbrenn- als auch das -schmelzschneiden
aufgetragen: Die Minima der beiden Qualitätskenngrößen liegen bei verschiedenen Schneidparametern.

Einfluß des Schneidgases

Je nach verwendetem Gas handelt es sich um ein Brenn- oder Schmelz-
schneiden. Bei Aluminium läßt sich diese Unterscheidung nicht treffen:
Obwohl Aluminium eine hohe Affinität zu Sauerstoff besitzt, kommt es,
wie chemische Analysen des ausgeblasenen Schnittfugenmaterials gezeigt
haben, beim Laserschneiden mit Sauerstoffunterstützung zu keiner meß-
baren Werkstoffverbrennung. Dennoch wird der Schneidvorgang durch die
Wahl des Gases beeinflußt (vgl. Bild 5): Bei Argon fließt das erschmol-
zene Material an den Schnittflächen unter Zurücklassung einer ausgeprägten Riefenstruktur herunter und bleibt aufgrund der Oberflächenspannung in einem schlanken Grat an der Schnittunterkante hängen. Bei
Sauerstoff hingegen treten in der Schnittfuge explosionsartige Vorgänge
auf, die sich in einer sehr rauhen Schnittflächenstruktur und in einem
seitlich auseinandergetriebenen Schmelzgrat unterhalb des Bleches äu-
ßern.

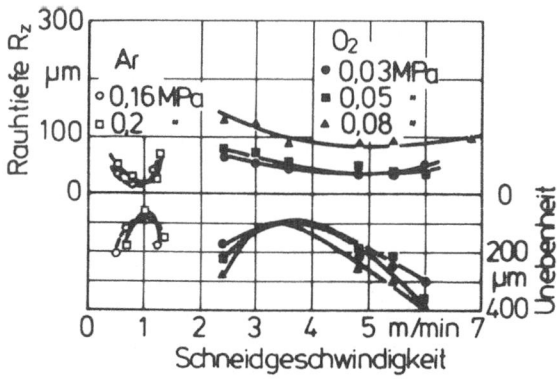

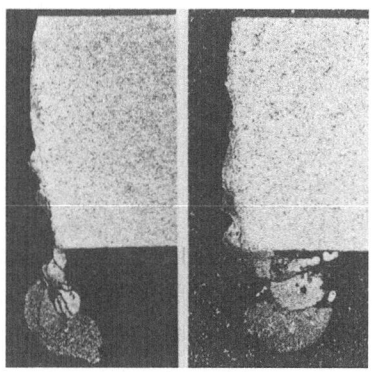

Bild 4. Schnittqualität in Abhängigkeit der Schneidgeschwindigkeit und des Gasdruckes beim
Laserbrenn- und -schmelz-
schneiden von TiAl6V4 der
Dicke 2.5 mm

Bild 5. Gratbildung beim Laserschneiden von Aluminium
mit Argon (links) und
Sauerstoff (rechts)

462

Wärmeverlust im Werkstück

Der den Schneidwirkungsgrad vermindernde Einfluß der Wärmeableitung in
die schnittflächennahen Bereiche des Werkstücks wurde bereits genannt.
Dieser ist jedoch nicht nur von der Wärmeleitfähigkeit des Werkstoffes
und der Schneidgeschwindigkeit abhängig, sondern auch von den hydrody-
namischen Gegebenheiten des Schmelzausflusses. Bild 6 zeigt das mit
der Schneidfront mitbewegte Temperaturprofil beim Laserschmelzschnei-
den von Titan in einem Abstand von etwa 0.5 mm entlang der entstehenden
Schnittfläche, welches sich aus einer numerischen Berechnung der Wärme-
ausbreitung ergibt /6/ (Verlauf Min.). Der zum Vergleich dargestellte
experimentell ermittelte Temperaturverlauf belegt, daß die Wärmeein-
kopplung beim realen Schneidprozeß wesentlich höher ist: Die in den
sich bildenden Riefentälern ausfließende Schmelze führt zu einer zusätzlichen, noch hinter der Schneidfront stattfindenden und damit den
Schneidvorgang nicht fördernden Wärmeübertragung. Hinzu kommt die Ener-
gieaufnahme aus dem sich abkühlenden Schmelzgrat an der Schnittunterkante.

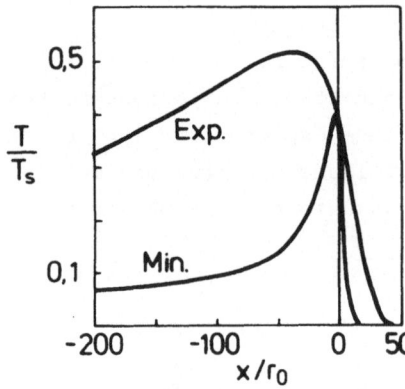

Bild 6. Temperaturprofil entlang
der Schnittfuge im Abstand
0.5 mm von der Schnittfläche beim Laserschmelzschnei-
den von TiAl6V4 der Dicke
2.5 mm: numerisch berechnet
(Min) und experimentell ermittelt (Exp), r_0: Fokusradius, $v = 1.2$ m/min

Die Untersuchungen wurden durch den
Bundesminister für Forschung und
Technologie sowie der Messer Gries-
heim GmbH, Frankfurt, gefördert.

Literatur
(1) I. DECKER, J. RUGE, U. ATZERT: Physical Models and Technological
    Aspects of Laser Gas Cutting. SPIE Vol. 455 (1984) 81-87
(2) D. SCHUÖCKER, W. ABEL: Material removal mechanism of laser cutting.
    SPIE Vol 455 (1984) 88-95
(3) D. BECKER, W. SCHULZ, G. SIMON, H.M. URBASSEK, M. VICANEK, I. DECKER:
    Physikalisches Modell des Laserschneidvorgangs. Beitrag in diesem
    Band
(4) G. CHEN, I. DECKER, G. RAUSCHER, J. RUGE, K. THOMAS: Modellbetrach-
    tungen zum Laserbrennschneiden. "Materialbearbeitung mit CO_2-Hoch-
    leistungslasern", VDI-Technologiezentrum, Berlin 1982
(5) I. DECKER, J. RUGE, Y.-H. HAN: Laserschneiden von Titanwerkstoffen.
    Erscheint in Schweißen und Schneiden 1985
(6) W. SCHULZ: Diplomarbeit, Inst.f.Theor. Physik, TU Braunschweig 1985

Schneiden und Schweißen räumlicher Konturen mit CO_2-Hochleistungslasern – derzeitige Möglichkeiten und Grenzen

Dr. Armin Gukelberger

Spectra-Physics GmbH

Siemensstrasse 20, D-6100 Darmstadt-Kranichstein

Das Laserschneiden ebener Teile mit herkömmlichen X-Y-Tischen sowie das Laser-schweißen von Rundteilen ist heute Stand der Technik. Das Schneiden dreidimensional geformter Teile dagegen wie z.B. Armatur-, Tür- und Teppichauskleidungen von Kraft-fahrzeugen sowie von Blechteilen im Bereich Prototypenbau und das Schweißen dreidi-mensionaler Konturen aus dem Karosserie- und Vorrichtungsbau sind heute oft nur mit Qualitätseinbußen und unter Verwendung aufwendiger Vorrichtungen machbar.

Es soll deshalb über den Aufbau neu entwickelter Systeme berichtet werden, welche die Bearbeitung räumlicher Konturen ohne die erwähnten Einschränkungen ermöglichen. Diese sogenannten Fünf-Achsensysteme stellen sicher, daß der Laserstrahl jeweils senkrecht auf die zu bearbeitende Fläche auftrifft, was bei exakter Führung des Fokuspunktes in bezug auf die Werkstückoberfläche eine gleichbleibende Schneid-bzw. Schweißqualität sicherstellt.

Zwei sich prinzipiell unterscheidende Systemlösungen werden heute angeboten: zum einen handelt es sich um Fünf-Achsen-Schneid-bzw. Schweißsysteme in Portalbauweise, bei der alle Bewegungsachsen (X, Y, Z, Drehen, Kippen) von der Bearbeitungsdüse aus-geführt werden und die einen großen Arbeitsbereich abdecken können. Zum anderen handelt es sich um Laser-Roboter-Systeme, die aus den Komponenten Laser, Roboter und flexibler Strahlführung als Verbindung zwischen Laser und Roboter bestehen. Im Folgenden werden einige derartige Systeme beschrieben und, sofern bereits im Einsatz, deren Anwendungen erläutert.

1. Fünf-Achsensystem in Portalbauweise

Bei den zunächst realisierten Fünf-Achsen-Schneidsystemen wurden lediglich die Be-wegungsachsen Z, Drehen und Kippen mit der Laserdüse ausgeführt, während die Bewe-gungsachsen X- und Y nach wie vor von einem herkömmlichen Tisch übernommen wurden. Die Nachteile dieser Systemauslegung sind eingeschränkte Zugänglichkeit, größerer Platzbedarf und Gewichtsbeschränkung je nach Belastbarkeit des X-Y-Tisches. Diese Nachteile wurden mit der Realisierung von Fünf-Achsen-Systemen ausgeräumt, bei de-nen alle Bewegungsachsen, also auch X und Y, vom Portal ausgeführt werden. Diese Bauweise stellt allerdings wegen der sich stark ändernden Länge des Strahlenganges während der Bearbeitung auch erhöhte Anforderungen an die Qualität des Laser-strahls. So muß sich diese durch eine nur geringfügige Strahldivergenz und eine hohe Richtungsstabilität auszeichnen.

Die Vorteile der Portallösung sind:

* Geringer Platzbedarf
* Freie Zugänglichkeit wenigstens von zwei sich gegenüberliegenden Seiten, was die Verkettung des Lasersystems in einer Transferstraße oder den Einbau einer Wechseltischvorrichtung zur Reduzierung der Stillstandzeiten beim Werkstückwechsel erlaubt.
* Die Werkstücke müssen nicht aufgespannt werden.
* Die zu verfahrenden Massen (hier die Laserdüse) sind unabhängig von der Masse des Werkstücks immer gleich.

Ein solches System (Bild 1) wurde vor etwa 6 Monaten bei einem Zulieferanten für die Automobilindustrie in Betrieb genommen. Es bearbeitet in einer Presse hergestellte Fertigautohimmel, wobei die äußere Kontur sowie eine Reihe von Aussparungen im Autohimmel zu schneiden sind. Dieses System arbeitet mit einer Wechseltischanordnung und zeichnet sich durch eine hohe Flexibilität aus. So kann der Lieferant dieser Autohimmel in kürzester Zeit auf Änderungswünsche der Automobilhersteller eingehen.

Die technischen Daten dieses Systems (Fabrikat Held GmbH Sondermaschinenbau) sind:

Bewegung in X-Richtung:	2600 mm
Bewegung in Y-Richtung:	1600 mm
Bewegung in Z-Richtung:	300 mm
Drehen des Schneidkopfes:	360°
Kippen des Schneidkopfes:	$\pm$ 90°
Maximale Bahngeschwindigkeit:	60 m/min
Laser:	Spectra-Physics Modell 820, 1500 Watt Laserleistung
Steuerung:	Fünf-Achsen-Steuerung, Fabrikat IBH

Für die Bearbeitung metallischer Werkstoffe wird die Schneiddüse zusätzlich mit einem kapazitiv arbeitenden Abstandssensor und einer sechsten Bewegungsachse in der Schneiddüse (Verfahrweg ca. 10 mm) ausgestattet.

Ein ähnlich aufgebautes Portalsystem (Fabrikat MBB) für das Schweißen von Karosserieteilen ist derzeit im Bau.

Die Srahlführung dieser Systeme ist so ausgelegt, daß Laserleistungen bis max. 5kW übertragen werden können. Es ist dehalb zu erwarten, daß diese Portalsysteme auch zum Anbringen von Härtebahnen in schweren und unhandlichen Bauteilen eingesetzt werden.

Bild 1. Fünf-Achsen-Laserschneidsystem für die Bearbeitung räumlich geformter Kunststoffteile

2. Laser-Roboter-Systeme

Es bieten sich zwei Strahlführungssysteme an:
entweder den Strahl über den Knickarm des Roboters zu führen oder, wie im Folgenden erläutert, den Strahl über ein Teleskoprohr direkt zum Bearbeitungskopf des Roboters zu führen (System Laserflex).

Das System Laserflex zeichnet sich durch die folgenden konstruktiven Merkmale aus:

* Wenig Spiegel: Das Laserflex besteht aus 4 Spiegeln. Es ist in der Regel sinnvoll, einen weiteren Strahlumlenker zur Einkopplung und besseren Justiermöglichkeit des Laserstrahls vorzusehen. Strahlführungssysteme, die über den Arm des Roboters gehen, lassen zwar mehr Bewegungsfreiheit zu, arbeiten aber mit 9 bis 12 Spiegeln.
* Sicherheitssensoren überwachen das Teleskoprohr, die Temperatur der Spiegel und die mechanische Belastung der Roboteraufnahme.
* Hohe Genauigkeit durch Verwendung robuster Spiegelgehäuse und Präzisionslager, verbunden mit der Möglichkeit, die Spiegel exakt, einfach und präzise zu justieren.
* Modulare Bauweise, die erlaubt, die Komponenten auch einzeln zu verwenden.

466

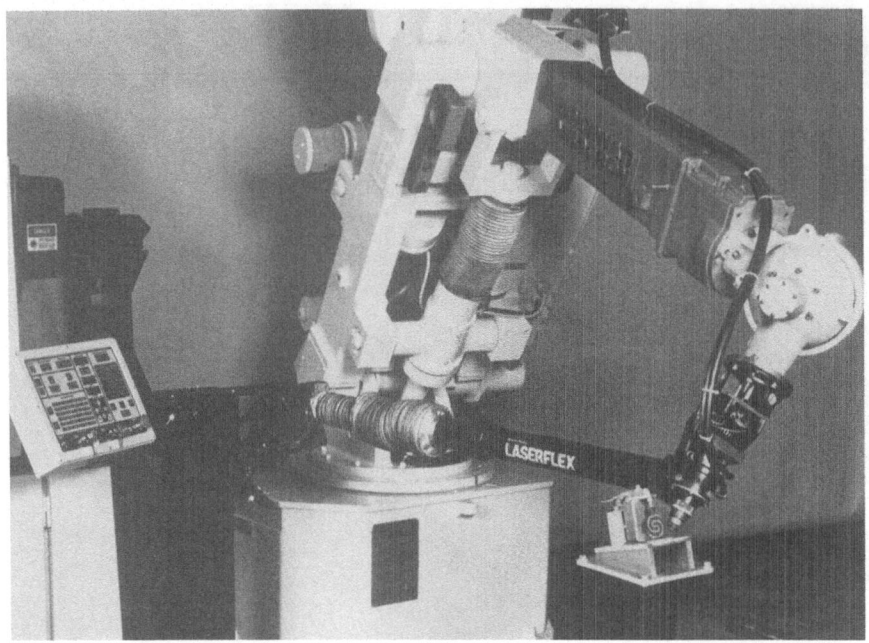

Bild 2.   Flexibles Strahlführungssystem Laserflex zur Verknüpfung von Laser und
           Roboter

Der Autor ist sich bewußt, daß zur Vergrößerung des Einsatzes der Laser-Roboter-
Systeme (dies ist in geringerem Umfang auch für die Portalsysteme gültig) weiterge-
hende Entwicklungsarbeiten auf folgenden Gebieten erforderlich sind:

* Verbesserung des Bewegungsverhaltens des Roboters bei geradliniger Bewegung und
  scharfen Richtungsänderungen
* Verbesserung der Robotergenauigkeit
* Entwicklung von Nahtsuch,- Nahtfolge- und Verbesserung von Abstandssensorsyst-
  emen
* Verbesserung der Programmiermöglichkeiten unter Einbeziehung des Lasers
  (Leistungssteuerung)

Die ersten Lasersysteme zur Bearbeitung dreidimensionaler Bauteile sind bereits im
Einsatz. Hohe Zuwachsraten sind vor allem in der Automobilindustrie zu erwarten.
Dabei wird es sich sowohl um Schneid- als auch um Schweißanwendungen handeln. Durch
die Verbesserung der Genauigkeit und des Ansprechverhaltens der Roboter und die
Weiterentwicklung von Sensorsystemen wird der potentielle Kundenkreis weiter ver-
größert werden.

Influence of the Cutting Method on the Fatigue Behaviour of Steel Sheets

W. Tamaschke*, J. Betz, G. Neuse

Institut für Werkstoffkunde und Werkstofftechnik, TU Clausthal

Agricolastr. 2, 3392 Clausthal-Zellerfeld

Abstract

The present paper gives the fatigue behavior of test specimens which have been obtained by laser cutting, plasma cutting or punching from steel sheets. The paper demonstrates the different damage to the surface introduced by the three techniques. For thin sheets (1.5 mm) laser cut specimens show advantages, for thicker sheets (6 mm) plasma cut sheets exhibit better fatigue properties.

Introduction

Modern CNC controlled sheet handling systems enable high degree of automatisation and economic manufacturing for both cutting and punching techniques /1/. In a number of papers the metallurgical and technical aspects of laser cutting systems have been discussed /2-5/. In the present paper the mechanical and fatigue properties of different steels are investigated in order to elaborate the differences in quality of laser cut, plasma cut and punshed parts.

Experimental

Steel sheets of STW 22 with 1.5, 4 and 6 mm thickness and a stainless steel sheet (X 8 CrNi 18 8) of 1.5 mm thickness have been investigated. In collaboration with various companies the quality of the cut was optimised. Fig. 1 and 2 show macrographs of punshed and cut specimen together with the corresponding roughness profiles.

---

* C. Behrens AG, Hackelmasch 1, 3220 Alfeld

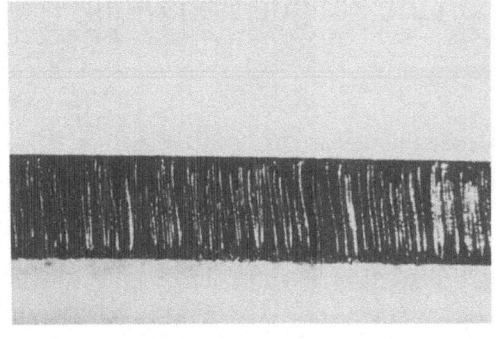

Fig. 1 a) Macrograph of a laser cut
          specimen of V2A, 1.5 mm,
          cutting speed 5 m/min

b) Corresponding roughness
   profile of the cut in
   feed direction

Fig. 2 a) Macrograph of a punshed
          specimen of V2A, 1.5 mm,

b) Corresponding roughness
   profile in longitudinal
   direction of the specimen

Fig. 3 and 4 show the corresponding graphs and profiles for laser and
plasma cut specimens respectively. The thermal cutting methods show
different tolerances on the upper and lower side of the cut specimens.
This can be seen in Fig. 5. The lasered specimen does only exhibit a
very small heat affected zone and no melt eruptions. In order to
ensure reproduciability all specimens were taken with their longitudinal axis parallel to the rolling direction.

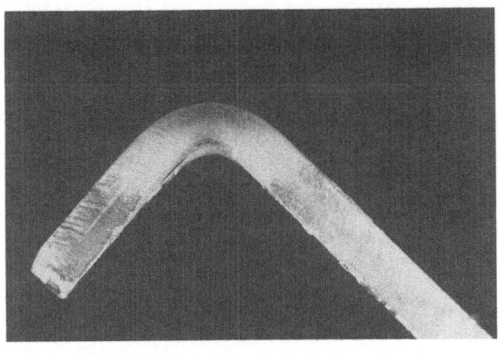
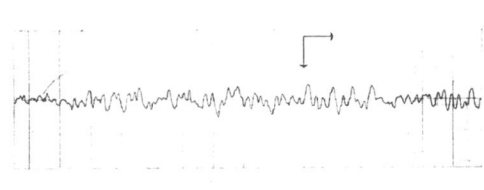

Fig. 3 a) Macrograph of a laser cut   b) Corresponding roughness
         specimen of STW 22, 6 mm,   profile
         cutting speed 2 m/min after
         90° bending

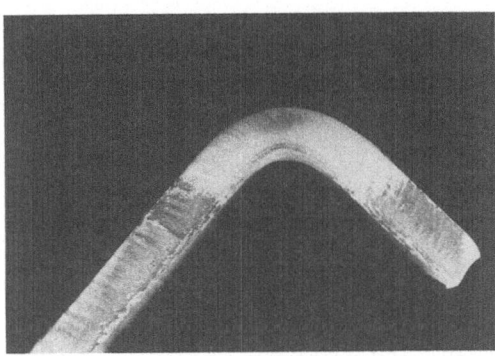
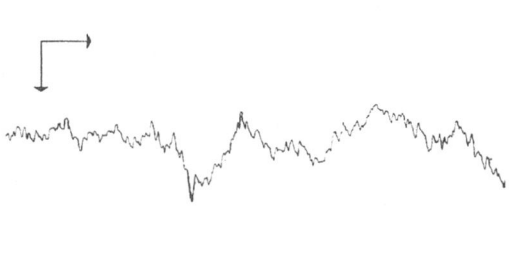

Fig. 4 a) Macrograph of a plasma cut   b) Corresponding roughness
         specimen of STW 22, 6 mm,   profile
         cutting speed 2 m/min after
         90° bending

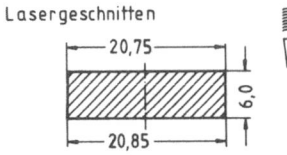
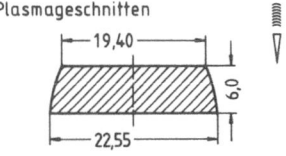

Fig. 5. Schematic drawing of the specimen cross section for laser
        and plasma cutting

The tensile test results for the various materials are given in Table 1.
From these results it is obvious that the three techniques lead to a
different damage of the surfaces and therefore to a different deterioration of the properties. The fatigue properties were measured with an
Instron 1255 in push-push tests with a mean load of 0.75 $R_{p0.2}$. In or-

der to see whether or not local hardening has occurred on thermal cutting metallographic investigations and 90° bending tests were carried out showing that no cracks occur on bending.

Table 1. Mean values of the tensile test properties

Sample treatment	Thickness a_o/mm	UTS R_m/Nmm^{-2}	Yield stress $R_{p0.2}$/Nmm^{-2}	Elongation to fracture A/%
STW 22 punshed	1.50	317.4	233.6	30.6
STW 22 lasercut	1.50	322.4	260.7	33.3
STW 22 punshed	2.00	292.8	207.1	34.4
STW 22 lasercut	2.00	297.0	241.4	36.4
V2A punshed	1.50	533.3	325.0	15.6
V2A lasercut	1.50	643.6	246.6	55.5
STW 22 lasercut	3.80	364.0	287.4	34.5
STW 22 plasmacut	3.80	380.2	328.3	28.8
STW 22 lasercut	6.00	345.4	248.3	37.1
STW 22 plasmacut	6.00	354.6	276.4	33.3

Results

Modifications in the microstructure resulting from the different types of cutting methods are shown in Fig. 6. The punched specimen show a small deformed zone. The laser cut specimen show a thin oxide layer and a thin resolidified zone. Almost no heat affected zone can be detected. The plasma cut specimen show a few mm wide recrystallised zone below the oxide layer.

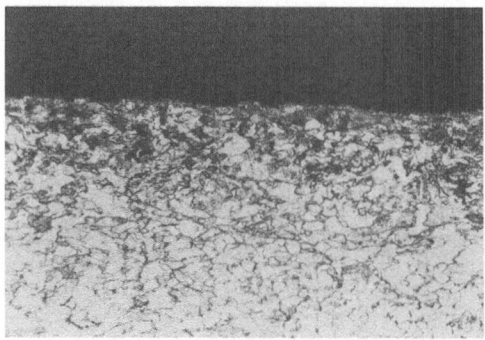

Fig. 6 a) Microstructure of punshed STW 22 at the surface, 1.5 mm

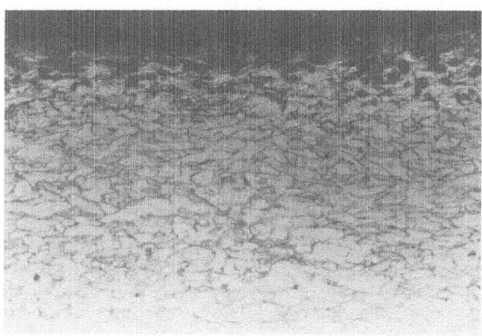 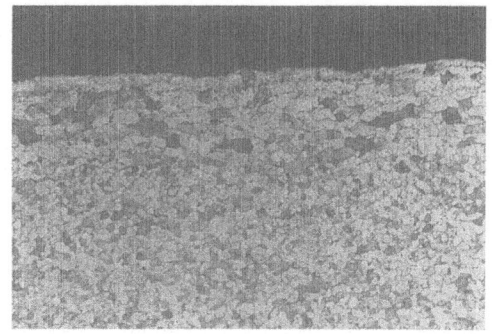

Fig. 6 b) Microstructure of laser cut   c) Microstructure of plasma
          STW 22 at the surface, 6 mm   cut STW 22 at the surface,
                                              6 mm

The fatigue properties are given in Fig. 7 - 10. The 1.5 mm thick STW 22
specimen show equivalent behaviour in the punshed and laser cut state
if the normal statistic variations are taken into account. The
roughness profiles for the laser cut specimens show in general a stronger fluctuation, however, the punshed samples sometimes contain deep
grooves resulting in equivalent fatigue behaviour. The stainless steel
samples show a superiour behaviour if they are laser cut. Tensile
strength and fatigue limit are both higher for the laser cut compared
to the punshed samples. The thicker sheets of 4 and 6 mm thickness show
only small differences between the laser cut and plasma cut state.

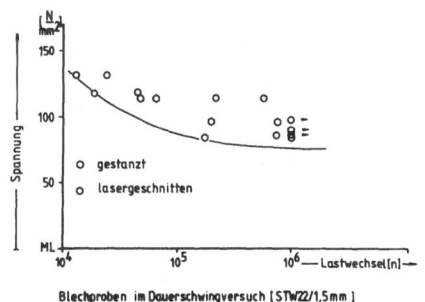 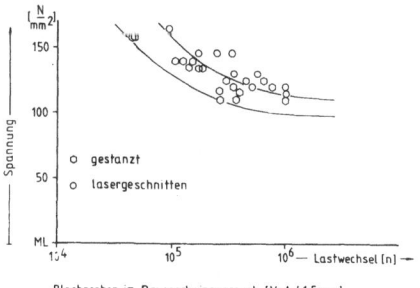

Fig. 7. Fatigue properties of punshed   Fig. 8. Fatigue properties of
        and laser cut STW 22, 1.5 mm   punshed and laser cut
        in pull-pull tests   X10CrNi18/8, 1.5 mm,
                                                    in pull-pull tests

472

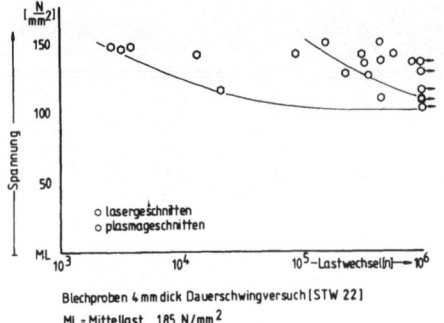

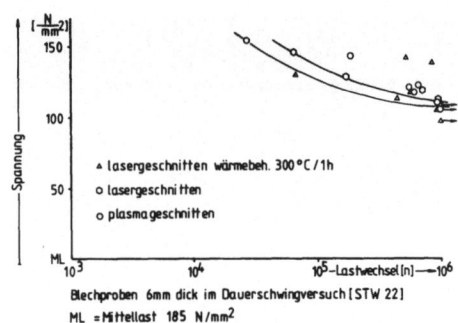

Fig. 9. Fatigue properties of laser and plasma cut STW 22,
4 mm

Fig. 10. Fatigue properties of laser and plasma cut STW 22,
6 mm. For comparison the
behaviour after annealing
is also given

The statistical variations are quite severe arriving from deeper
grooves which occur both in plasma cut and laser cut samples. The
annealing treatment prior to the fatigue test does not alter the
fatigue limit indicating that crack initiation starting from deeper
grooves is responsible for the deterioration in the fatigue properties rather than from a hardening effect caused by the thermal
cutting treatment.

Conclusions

All three techniques allow to use the samples for construction purposes without further treatments as far as strength and fatigue properties are concerned. Deteriorations by a heat affected zone for the
thermal cutting processes can be neglected. The three techniques differ mainly in their ability of maintaining dimensions, in surface
roughness and in production costs.

References

/1/ W. TAMASCHKE, Z. Industrielle Fertigung 6 (1982) 323
/2/ G. BARTON, M. KOSCHLIK, H.W. BERGMANN, Z. Werkstofftechnik 14
(1983) 257
/3/ W.M. STEEN, J.N. KAMALU, in: Laser Materials Processing,
Ed. M. Bass, North Holland Publishing Company (1983) 15
/4/ J. SELLNER, in: Optoelectronik in der Technik, Hrsg.
W. Waidelich, Springer Verlag (1984) 305
/5/ R. JOHNSON, G. THOMAS, M. FENDER, in: Proc. 3rd Int. Conf.
on Welding and Melting by Electrons and Laser Beams,
Ed. M. Contré and M. Kuncevic, Lyon (1983) 671

Potential Uses of Laser Welding and Cutting in the Nuclear Fuel Cycle

M Hill, JHPC Megaw and SJ Osbourn

UKAEA Culham Laboratory, Abingdon Oxon, OX14 3DB, UK

INTRODUCTION

Lasers offer the possibility of remote operation in hostile environments and a range of potential applications of laser cutting and welding in the nuclear industry are now recognised[1]. Many are under investigation internationally, and in some cases, lasers are being used in production and pre-production operations. Within the context of this paper, we first summarise some of these areas and then describe three relevant topics of work carried out in the Laser Applications Group at Culham Laboratory.

The nuclear fuel cycle is illustrated schematically in Figure 1 which shows several potential laser applications. They can be divided into three main areas as follows:-

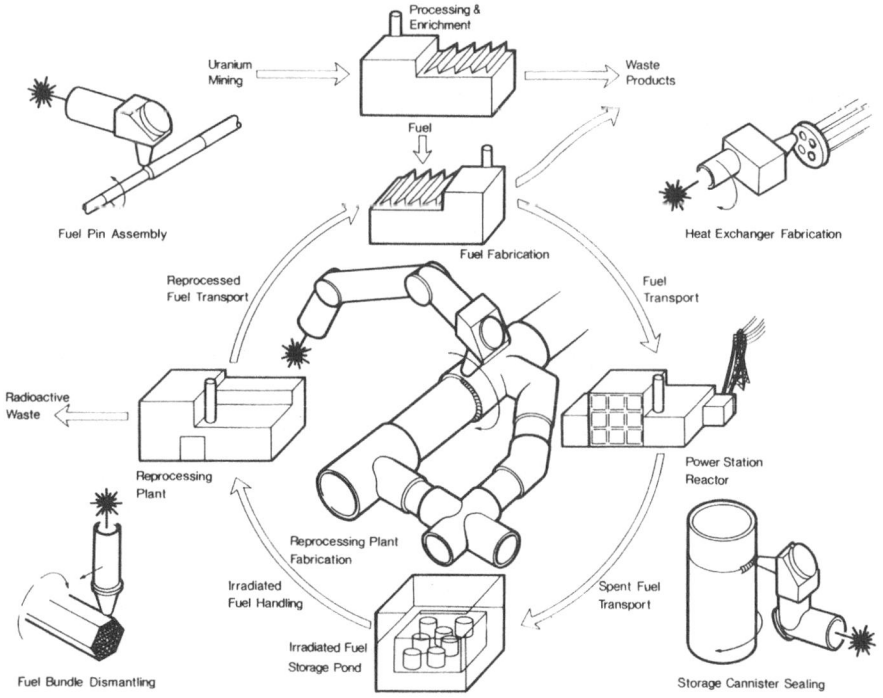

Fig. 1. Potential laser applications in the nuclear fuel cycle

Fuel Fabrication

Fuel is 'canned' in a number of configurations and materials which depend on reactor type. For example, it may be sealed by the welding of end-caps into 'pins' consisting of tubes (~10mm diameter x <1mm wall) of stainless steel or Zircaloy. Subkilowatt pulsed NdYAG lasers and cw CO_2 lasers of ~1kW output appear to offer excellent capabilities for this welding, and a number of accounts have been published including one which describes a comparison of laser, eb and TIG welding[2]. Lasers also offer potential for ancillary operations such as the fixing of wear-resistant pads on to wrappers containing bundles of fuel pins[3].

Uses at the Reactor

Much of the fabrication for nuclear power plant requires the welding of structural, pressure vessel and stainless steels. Industrially rated multi-kilowatt lasers are now capable of welding steels ≈15mm thick in a single pass[4] and multipass operation, with use of filler, can considerably extend single pass penetration[5]. In parallel with assessment of laser-weldabiliy of particular material compositions, studies have been carried out for specific applications such as the sealing of flasks used to transport irradiated fuel from the reactor to the reprocessing plant[6].

Reprocessing

Here, attention has focused on two particular aspects. The first is laser dismantling of wrappers, prior to removal of pins for reprocessing. This technique is well established and is being exploited at the Dounreay Prototype Fast Reactor for fuel post-irradiation examination and reprocessing[7]. The second concerns the concept of re-usable fuel transport flasks, where studies have demonstrated that laser cut edges can be re-welded by laser[8] with, if necessary, the use of filler[9].

EQUIPMENT DEVELOPMENT

Multikilowatt laser welding is a fast, deep penetration process performed by a high power density (>1 x $10^6 Wcm^{-2}$) focal spot which can produce full penetration welds in thick section materials in a single pass. Beam penetration depends on the formation of a capillary, or 'keyhole', filled with hot vapour and plasma which absorbs laser beam power by the inverse bremsstrahlung process. Energy thus absorbed within the keyhole is efficiently coupled to the metal for welding. However, material streaming from the keyhole into the incident beam path promotes extension of the plasma above the workpiece. Beam energy absorbed there is not efficiently used and can result in broadening of the weld crown and loss of penetration. The growth of this plasma may be controlled by directing at the interaction point a jet of inert gas having high

ionisation potential (such as helium), and having flow and pressure
distributions which assist with the formation of the keyhole and the
convective removal of plasma energy. Although this gas jet excludes air
from the interaction point, additional inert shielding must be provided
above and below the welded specimen in order to prevent oxidation of the
solidifying metal.

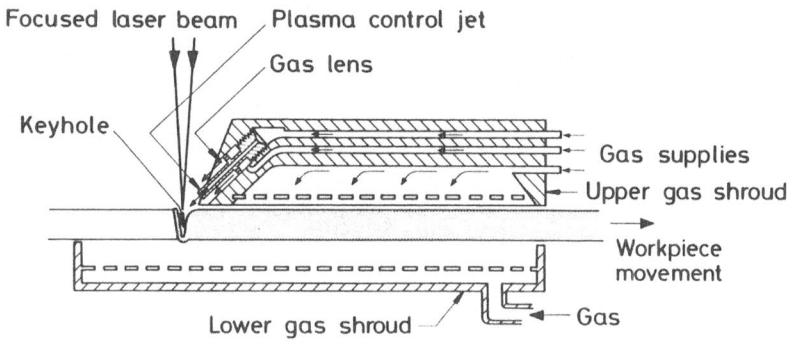

Fig. 2. Plasma control and gas shielding arrangement

During the course of laser welding process development, part of
which has been directed towards nuclear components, the authors have
designed and developed an effective plasma control and gas shielding
arrangement. A longitudinal section view of this device is shown in
Figure 2. A plasma control jet of $\approx$2mm bore is directed at the laser
beam/workpiece interaction point at an angle of 45^0 to the direction of
movement. A gas lens, shown here coaxial with the jet, increases the
extent of inert gas coverage and prevents entrainment of air around the
jet; laminar flow in the lens is arranged by 6 equi-spaced 1mm diameter
holes. Uniform gas distribution and laminar flow in the upper and lower
shrouds is arranged by finely perforated plates as shown, or by gauze
sheets.

This gas control arrangement has been used successfully on a variety
of flat and cylindrical components, for both external and internal welds,
with Culham's multi-kilowatt lasers CL5 and CL10 operating in the power
range <1.0 to >10kW. Welding speeds varying from <5 to >200mms^{-1} have
been used in thicknesses from <1.0 to >20mm. The shroud integrity may be
judged from its ability to produce welds in reactive metals such as
titanium and zirconium without any trace of oxide discolouration.
Typical helium flow rates depend on the laser power, welding speed and
material thickness, and for the jet, lens and each trailing shroud are
5-25, 20-50 and 50-100 slmin^{-1} respectively. Argon, at flow rates
<20slmin^{-1}, may be substituted for helium in each shroud.

A second equipment development relevant to nuclear applications

Fig. 3. Moving optics facility
for horizontal pipe welding

studies is shown in Figure 3. It consists of a lathe-mounted focusing head, with appropriate moving optical path, which permits studies at powers up to 10kW of the girth welding of fixed pipe samples having horizontal axes. Thus effects such as gravity on the weld pool stability can be investigated in a configuration appropriate to process plant fabrication.

WELDING APPLICATIONS

Zircaloy

Zircaloy 2 is widely used as a fuel-cladding material in water-cooled reactors, and is conventionally welded by TIG or eb methods. An investigation was carried out to assess the possibilities of laser fabrication of this material in 4mm thick flat plate. Other earlier work[10] concerned laser welding in thicknesses predominantly below 1mm. In the present study, attention was directed to the following aspects:- (a) Fusion zone geometry, including underside reinforcement

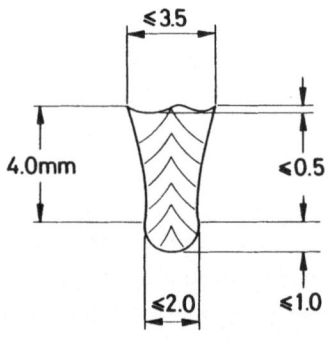

Fig. 4. Specified bead profile
for Zircaloy plate welds

and topside undercutting, and the target specification is shown in Figure 4. (b) Spatter, since adherent material could survive post-weld cleaning but become dislodged in service. (c) Microporosity, which can be prevalent in beam welding of this material, and which may be related to: alloying additions (the vaporisation temperature of the tin in Zircaloy is a little over half that of the zirconium); the different solubilities of gases (O_2, H_2) in α and β phases; possibility of oxides and hydrides being present on incorrectly prepared joints. The target specification was the absence of any pores >0.4mm diameter, and that any weld cross-section should exhibit no more than one micropore.

A parametric study to establish optimum laser welding conditions was performed by making a series of melt runs (bead-on-plate welds) using three different focal length ZnSe lenses and a gas control arrangement as described above. The work was carried out on Culham's 5kW CO_2 laser CL5 and a range of welding speeds and powers at the workpiece were investigated. Three workpiece-to-lens distances (FD) were used for each lens, the middle one in each case being that which gave full penetration at maximum speed. Welding conditions are plotted in the graphs shown in Figure 5. Unacceptable levels of spatter and lack of penetration were established by visual examination of the underside of each plate. A more sensitive indication of spatter was achieved by viewing the underside of the plates through a window in the under shroud during welding. Weld bead widths and the incidence of any underbead undercutting were found by sectioning. Radiography of the complete plates could resolve pores >0.08mm diameter whilst metallography could resolve those >0.01mm.

It was found that acceptable weld profiles were achieved with both the f/6.4 and f/9 lenses, but those made with f/3.6 were generally too narrow. Those made at powers >3kW gave high spatter levels and those made at >50mms^{-1} showed considerable underbead undercutting. Spatter levels were reduced by minimising the plasma control jet helium flow rate. Macro-sections of welds made under optimum conditions are shown in Figure 6. Porosity levels were minimised by careful handling and cleaning of the plates. However, whilst spatter is minimised if the keyhole does not break through, incidence of porosity is then increased due to reduced venting of vapour. Thus, operation below 2.5kW gave lower spatter but greater porosity, whilst the converse was true at 3kW. Further work is required to understand more fully and control these phenomena.

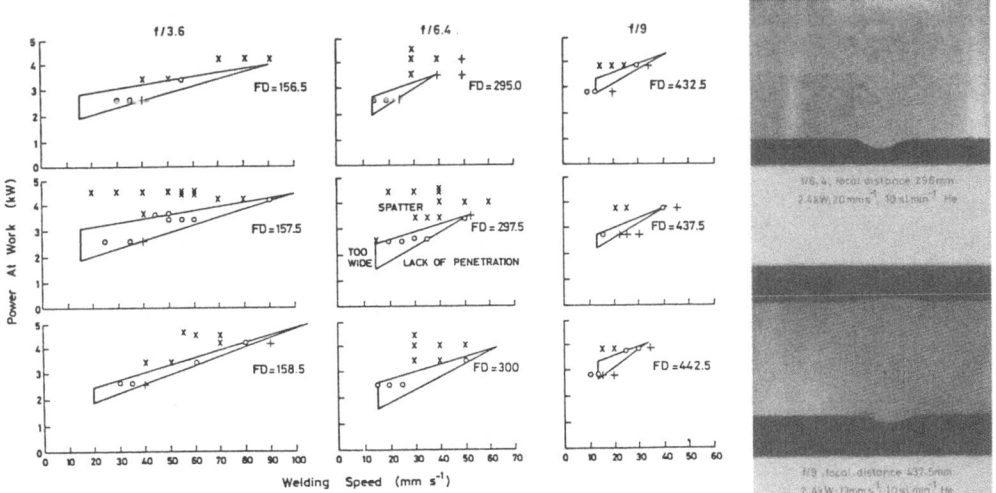

Fig. 5. Laser power vs. welding speed in 4mm Zircaloy: (o) acceptable spatter & penetration; (x) unacceptable spatter; (+) lack of penetration

Fig. 6. Optimum melt run profiles in 4mm Zircaloy

Dissolver Plant

The reprocessing of spent nuclear fuel includes dissolution of the fuel to separate it from its canning. Hot nitric acid is normally used and is contained within a dissolver vessel which may be of materials such as austentic stainless steel, Inconel 690 or zirconium. Typically, the vessels are large and of complex shape so that a joining technique such as laser welding is attractive because it creates minimum distortion and shrinkage. Additionally, the limitation of thermal disturbance is also important; in stainless steel, carbide formation and therefore risk of intergranular attack should be minimised. Although significant effort will be required to develop appropriate beam delivery systems in order to permit laser fabrication of complete plant, some related trials have commenced.

Preliminary work features the creation of test laser welds in coupons of candidate materials. These coupons are currently undergoing long term corrosion tests which will be reported elsewhere. However, the indications are that the weld region behaves at least as well as the parent plate.

Work using the 5kW laser CL5 was carried out on pipe of 100mm outer diameter x 6mm wall. Specimens of 25Cr/20Ni Nitric Acid Grade (NAG) stainless steel were successfully welded with 4.5kW at the work and at 25mms^{-1}. The welding conditions were then repeated on joints between pieces of similar size pipe and cold-formed T-sections made from Sandvic Type 2RE10 stainless steel. An internal diameter mismatch of $\approx$1mm presented no problems. Very smooth, continuous inner beads were achieved, as shown in the longitudinal sections in Figure 7. The laser welds were much narrower than the TIG welds incorporated in the same test piece (Figure 7) and, in contrast, did not require internal dressing of the underbead.

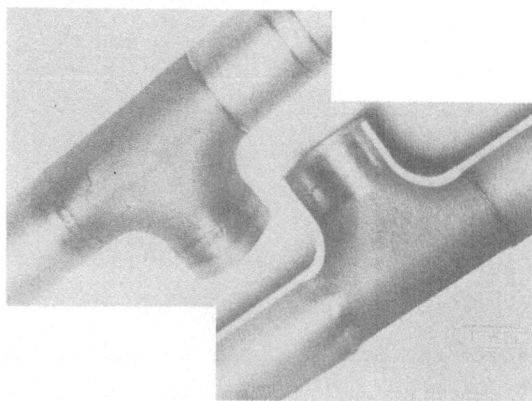

Fig. 7. Comparison between laser and TIG welds in 6mm stainless steel

Work on the 10kW laser CL10 featured butt welding of Type 304L stainless steel pipe, $\approx$250mm O.D. x 10mm wall. These specimens were welded at 25mms^{-1} with 9kW using the lathe-mounted moving optics facility. Narrow, parallel sided welds were produced, but with some underbead undercutting and the occasional pore, as shown in Fig. 8. Very even penetration was achieved around the entire

circumference. A short section
of this material, containing one
laser weld has been incorporated
in the Prototype Batch
Dissolver[11] at Harwell. This
weld is situated near the
proposed liquid/vapour interface,
but has not yet been tested with
nitric acid.

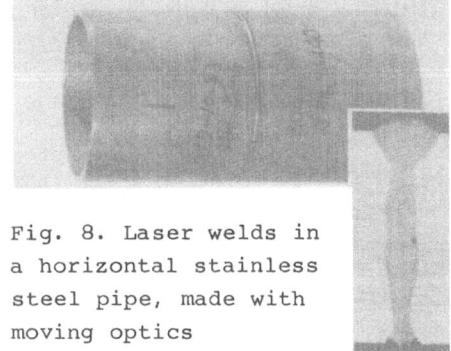

Fig. 8. Laser welds in
a horizontal stainless
steel pipe, made with
moving optics

CONCLUSIONS

We have described equipment developments of gas shielding and beam
manipulation which significantly facilitate and extend the scope for
laser welding investigations. Two welding studies which utilise these
developments have also been described. These have demonstrated that high
integrity welds can be produced using higher speeds and lower heat inputs
than those found in conventional processes. The studies identify a need
for further work in the areas of beam manipulation and microporosity
control, but their results confirm that laser welding offers necessary
and special capabilities which make it of significant interest to the
nuclear industry.

ACKNOWLEDGEMENTS

The authors are grateful for the support of colleagues in the UKAEA,
in particular Dr. T.D. Hodgson and Mr. D. Wood from Harwell. They also
thank Mr. D.S. Manser, Rolls Royce Ltd, for his help and advice.

REFERENCES

1. Spalding IJ, Taylor AF, Kaye AS, Megaw JHPC, Ward BA; ANS, Vol.38,
   702, Miami, June 1981.
2. Goswami GL, Verma R, Prasad GJ, Ghosh JK, Roy PR; 3rd CISFFEL, Vol.
   2, 773, Lyon, September 1983.
3. Johnson R; Laser Welding, Cutting and Surface Treatment, Weld. Inst.,
   43, 1984.
4. Kaye AS, Delph AG, Hanley E, Nicholson CJ; Appl. Phys. Lett., 43(5),
   412, September 1983.
5. Megaw JHPC, Hill M; LIA, Vol. 31, ICALEO, 108, Boston, 1982.
6. Megaw JHPC, Hill M, Bernard J, Moulin M, Vivien J, Geoffroy J,
   Noel JP; 3rd CISFFEL, Vol. 2, 681, Lyon, September 1983.
7. Higginson PR, Campbell DA; PIE Conf., ENES, Grange-over-Sands, May
   1980.
8. Johnson R, Hill M, Megaw JHPC; Inst. Metall. Conf., No. 18, Vol.2, 1,
   April 1981.
9. Megaw JHPC, Hill M, Johnson R; Inst. Metall. Conf., No. 18, Vol. 2,
   146, April 1981.
10. Ram V, Kohn G, Stern A; 3rd CISFFEL, Vol. 2, 653, September 1983.
11. Allardice RH, Hickley HB, Smith GEI, Walker BJ, Ward MD; ANS, Vol. 2,
    214, August 1984.

Influence of Shielding Gas in Laser Beam Welding

E.Beyer, G.Herziger, U.Petschke, W.Sokolowski
Institut für Lasertechnik ILT
Drosselweg 87, D-5100 Aachen

Introduction

During a welding process a shielding gas is used to protect the welding area from the ambient atmosphere. In laser beam welding, the shielding gas has a second function. That is – the laser induced plasma above the workpiece surface can be controlled by various gases. /1-8/

If the shielding gas parameters are not adapted to the processing parameters, a laser induced shielding gas plasma can be formed. Part of the laser power will be absorbed in this plasma, so that the efficiency of the welding process decreases. This can happen in the intensity range from $10^6 < I(W/cm^2) < 10^7$ which is typical for welding with cw CO_2-lasers. This shielding gas plasma is always ignited with the support of a laser induced metallic vapour plasma. After the ignition, this plasma is independed of the welding process. On removing the workpiece a continuous shielding gas plasma remains, especially when using Argon. /8/

1. Shielding Effect

If the absorbed laser energy is higher than the energy losses by heat conduction, evapouration takes place at the interaction area. If the evapouration rate and the laser intensity are sufficiently high, a laser induced plasma will be formed at the workpiece surface /8-10/. This plasma is responsible for a higher absorption of the laser power combined with the development of a 'key hole'. With this increase of power absorption the local vapour density increases too. This vapour flowing in the direction of the incident laser beam will decrease due to rarefraction. Depending of the incident laser intensity and the local vapour density, the beam absorption in the plasma above the workpiece surface can become so high that the laser intensity is insufficient to sustain the evaporation process. The plasma detaches compleatly from the workpiece surface until it becomes transparent again by rarefraction. During this time the welding process will be interrupted as shown in figure 1.

Because the vapour velocity is in the range $2 \cdot 10^5$ (cm/s) to $6 \cdot 10^5$ (cm/s), this effect will occur in times, typically between 10^{-7} s to 10^{-5} s. However, the efficiency of the welding process will decrease. The property of plasma shielding can be influenced particularly by a suitable shielding gas flow. The streak photographes in figure 2 show that using helium as a shielding gas results in the plasma becoming detached from the workpiece surface at higher laser intensities.

2. Plasma Model

The plasma characteristics can be described by the rate equations for electron energy, electron momentum, electron density, and metallic vapour density. For welding with cw-CO_2 lasers it is sufficient to consider times where $t > 10^{-10}$s so that a "Maxwellian distribution" for the plasma electrons can be assumed. In this case, the avarage electron energy can be used in the rate equations. If only the time behaviour of finite volume is considered, the momentum equation will be contained in the diffusion losses in the energy and particle rate equations. Therefor, a typical diffusion length of the order of the plasma dimension must be assumed. However, compared .with the ionisation and recombination rates at the fully developed plasma, the diffusion rate can be neglected.

The vapour density can be obtained by an evapouration model such as that presented by Krokhin /11/ and extended by power losses due to heat conduction. The metallic vapour density is important because the plasma absorption coefficient is a function of vapour density. This is only valid as long as collisions between electrons and atoms exceed those due to electrons and ions.

The three-body recombination rate must split into two terms. It must be distinguished whether the third particle is an electron or a neutral atom. The cross-section for an electron ion collision is greater than the cross-section for an electron-atom collision, but the metallic vapour density is some order of magnitude higher than the electron density, therefore the recombination rate with a neutral atom as third collision particle can be dominant. /8/

SYSTEM OF RATE EQUATION FOR LASER INDUCED PLASMA

ELECTRON DENSITY $n_e(t)$

$$\frac{d\,n_e}{d\,t} = R_i - R_{diff} - R_{ee} - R_{en}$$

R_{ee} : $f(n_e, \bar{\epsilon}) n_e^3$ THREE-BODY RECOMBINATION (ELECTRON-ION-ELECTRON)

R_{en} : $f(n_e, \bar{\epsilon}) \sum_g \frac{ng}{M_g} n_e^2$ THREE-BODY RECOMBINATION (ELECTRON-ION-ATOM)

R_{diff} DIFFUSION RATE

R_i IONIZATION RATE

AVERAGE ELECTRON ENERGY $\bar{\epsilon}(t)$

$$n_e \frac{d\bar{\epsilon}}{d\,t} = \alpha\, I - P_c - P_e - R_i(E_i + \bar{\epsilon}) + R_{ee}(E_i - E_e \cdot \bar{\epsilon})$$

$\alpha(n_M, n_e)$ ABSORPTION COEFFICIENT

I LASER INTENSITY

P_c POWER LOSSES BY ELASTIC COLLISIONS

P_e POWER LOSSES BY EXCITATION

E_i IONIZATION ENERGY

E_e EXCITATION ENERGY

VAPOUR DENSITY $n_M(t)$

$$n_M(t) = const. \left[A\, I - \frac{T_V\, K}{2\sqrt{\varkappa\, t}} \; \frac{1}{\frac{1}{\sqrt{\pi}} - \text{ierfc}\left(\frac{r_F}{2\sqrt{\varkappa\, t}}\right)} \right]$$

A	ABSORPTANCE	$\varkappa$	THERMAL DIFFUSIVITY
T_V	VAPORIZATION TEMPERATURE	r_F	FOCUS RADIUS
L	THERMAL DIFFUSIVITY	K	HEAT CONDUCTIVITY

The numerical solution of the equations system is shown in figure 3. In the metallic vapour density, the avarage electron energy E increases rapidly to a constant value of approximately 1 eV. At this value the ionisation rate in the equation system becomes dominant so that an avalanche ionisation process starts. /8-10/

The electron energy increases again if the absorption coefficient increases as a result of the electron-ion collisions. Dependent on the recombination rate, the electron density reaches a constant value. Using various shielding gases, different constant values of electron densities corresponding to equation (1) can be obtained. The plasma absorption coefficient is a linear function of electron density $\alpha \sim n_e$ so that the plasma shielding above the workpiece goes hand in hand with the electron density. This effect can be seen in figure 4, while figure 5 shows the influence of the gas flow on welding depth.

3. Shielding Gas Plasma

If the shielding gas parameters are not adapted to the processing parameter, a laser induced shielding gas plasma can be formed. Figure 6 shows a typical development of a laserinduced argon plasma. After ignition this plasma is independent of the welding process. On removing the workpiece a continuous argon plasma is visible. Figure 7 shows the calculated threshold intensity for different initial electron densities. An argon plasma, under normal conditions, is produced with laser intensities above 10^8 W/cm^2. With initial electron densities above 10^{12}(cm^{-3}) the required laser intensity is smaller because the diffusion losses decrease by becomming ambipolar./12/ At initial electron densities above 10^{16} (cm^{-3}) the absorption coefficient increases due to electron-ion collisions. The calculations show that with initial electron densities above 10^{16}(cm^{-3}), the formation of a laser induced argon plasma is possible at intensities of about 10^6 W/cm^2. During the welding process such a plasma can be ignited by a laserinduced metallic vapour plasma.

However, formation of a shielding gas plasma always decreases the efficiency of the welding process.

Literature
/1/ F.D. Seaman: The Role of Shielding Gas in High Power CO$_2$ Laser Welding, SME TECHN. PAP. (1977) S.1-11, Paper Nr.: MR 77-982
/2/ M. Banas: High Power Laser Material Processing ,Proc. GCL Gas Flow and Chemical Lasers Köln
/3/ Y.Arata et. al.: Characteristics of High Power CO$_2$ Laser Welding, DVS 63 (1980) S.183
/4/ K.Shinada et.al.: Basic Study of Laser Welding, Int.Laser Conf. Laser Inst. of Am. Anaheim USA, (1981) Nov., S.3.1-3.9
/5/ G. Giersch, H.Hoffmeister: Untersuchungen zum Strahlschweißen mit Hoch-leistungslaser, DVS 52 (1978), S.194-197
/6/ P.F.Awramtschenko et.al.: Schweißen von Staehlen und Titanlegierungen mit Laserleistungen bis 5kW, DVS 63 (1980), S.192-196
/7/ P.F.Avramcenco, I.V. Molcan: Laserschweißen, Automat. Svarka (1983)5, S.68-69

/8/ E.Beyer: Einfluß des Laserinduzierten Plasmas beim Schweißen mit CO_2-Lasern, Thesis TH Darmstadt (1984)

/9/ G.Herziger: Basis Elements of Laser Material Processing, Proc. ÖPG Linz 1983, SPIE *455, P.66*

/10/ E.Beyer et.al.: Development and Optical Absorption Properties of a Laser Induced Plasma during CO_2 Laser Processing, Proc. ÖPG Linz 1983, SPIE *455, P.75*

/11/ O.N.Krokhin: Generation of High Temperatur Vapours and Plasmas by Laser Radiation, Laser Handbook North Holland Publ. Comp. (1972)

/12/ D.C.Smith: Gasbreakdown Initiated by Laser Radiation Interaction with Aerosols and Solid Surface, J.of. Appl. Phys. Vol.48, No.6, June (1972),2217

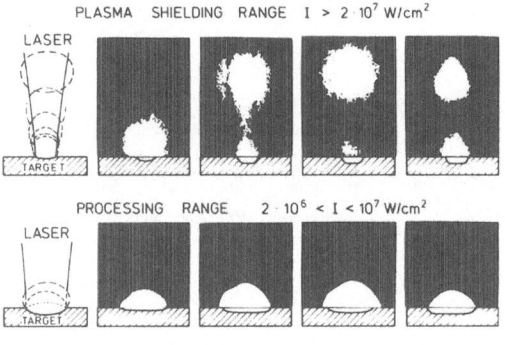

Fig.1. Framing images of the plasma formation above a target at different intensities of a pulsed CO_2 laser (Interframe time 50ns)

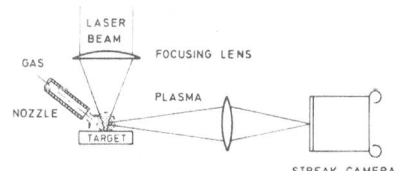

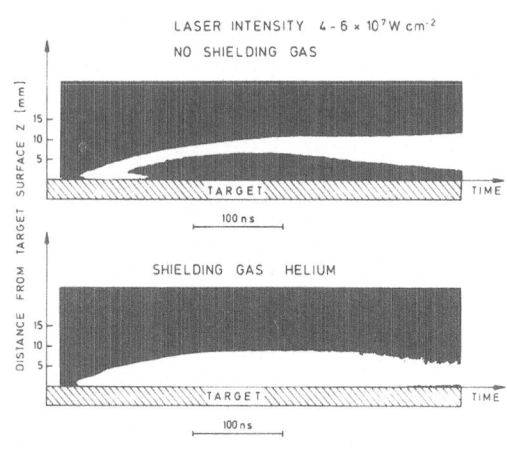

Fig.2. Experimental arrangement and streak images of a laser induced plasma obtained with CO2 laser under various shielding gas conditions

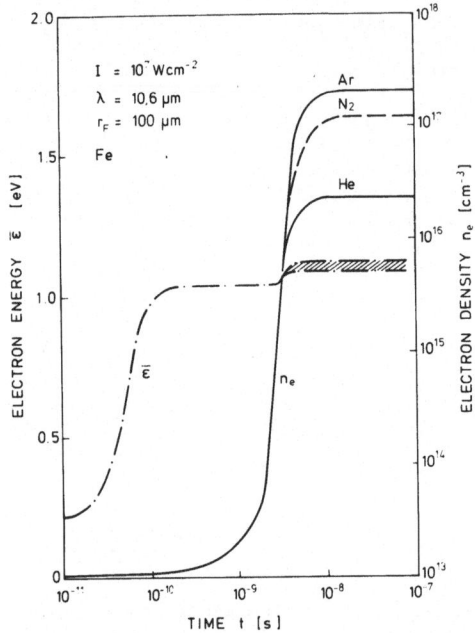

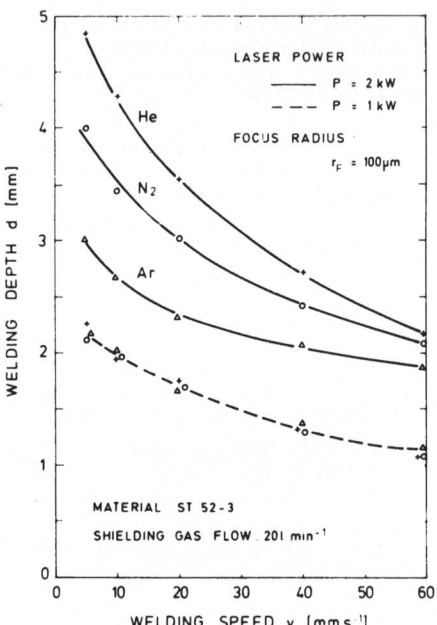

Fig.3. Development of a laser induced plasma calculated with the system of rate equations with different shielding gases

Fig.4. Relation between welding velocity and penetration depth determined in experiments with various shielding gases

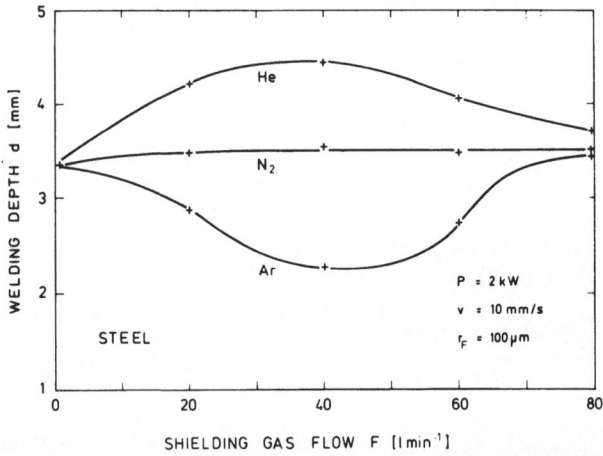

Fig.5. Relation between gas flow and welding depth

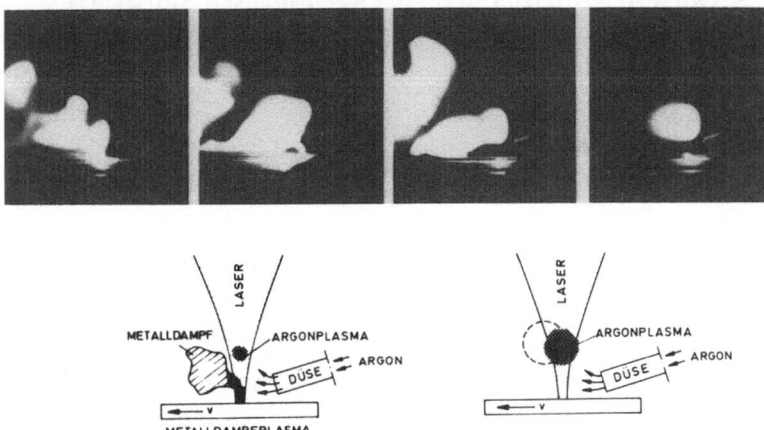

Fig.6. Development of a laser induced argon plasma. The absorption of the laser beam in this plasma interrupts the welding process

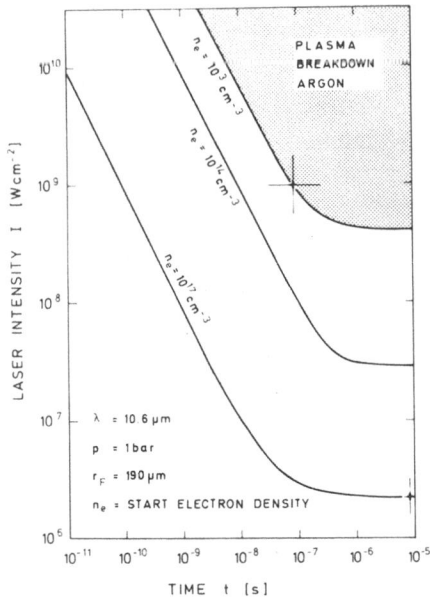

Fig.7. Threshold intensities for the plasma formation in argon calculated with the rate equation system for different initial electron densities

Einfluß der Nahtvorbereitung auf die Schweißqualität im Multi-kW-Bereich

C. Hamann, H.-G. Rosen

Siemens AG, München

1. Einleitung

Aus der konventionellen Schweißtechnik ist bekannt, daß die Nahtvorbereitung einen wesentlichen Einfluß auf die Qualität der Schweißnaht hat. Zur Naht vorbereitung gehört neben der Gestaltung der Fuge die Wahl der geeigneten Oberfläche im Bereich der Schweißnaht.

Untersucht wurden die Auswirkungen unterschiedlich bearbeiteter Oberflächen und verschiedener Korrosionsschutzmittel auf die Porenhäufigkeit beim Schweißen von Blechen im I-Stoß und an Blindschweißungen.

2. Untersuchungen

Die Untersuchungen wurden mit einem Laser von UTRC (United Technologies Research Center) bei einer Leistung von 10 kW am Werkstück durchgeführt. Mit einer Spiegeloptik ergab sich bei einem Öffnungsverhältnis von F 6 ein Fokusdurchmesser von 0,8 mm. Die unterschiedlichen Schweißtiefen im Blech (St 52-3, 10 mm dick) wurden bei konstanter Laserleistung von 10 kW durch Variation der Schweißgeschwindigkeit erreicht. Für die Gruppe der Blindschweißungen wurden sowohl Einschweißungen (30 mm/s) als auch Durchschweißungen (16,7 mm/s) durchgeführt. Um die Aussagekraft der Blindschweißungen in bezug auf reale Fügeaufgaben zu überprüfen, wurden auch Schweißungen mit I-Stoß (16,7 mm/s) in die Untersuchungen einbezogen.

Aus der Vielzahl möglicher Oberflächen in der Metallverarbeitung wurden die folgenden ausgewählt: warmgewalzt, gefräst, geschliffen, gestrahlt.

In Voruntersuchungen wurde ermittelt, daß Stahlkies bei den Strahlverfahren den geringsten Einfluß auf die Schweißqualität hat.

In der Industrie werden verschiedene Korrosionsschutzarten angewendet, die auch zwischen verschiedenen Bearbeitungsschritten (z.B. bei längerer Ein-lagerung) aufgetragen werden. Zu den meist verwendeten Verfahren zählen: lackieren, wachsen, ölen. Sie wurden auf einer gestrahlten als auch auf einer verzunderten Oberfläche untersucht. Beim I-Stoß wurden die Stirnflächen zusätzlich mit Öl und Wachs benetzt.

3. Ergebnisse

Als Bewertungsmaßstab für die Schweißqualität wurde die Porenhäufigkeit über eine größere Länge gemittelt. Für die Auswertung wurden sowohl Durchstrahlungs- bilder als auch Schliffbilder erstellt.

3.1 Einfluß der Bearbeitungsverfahren

Bild 1 zeigt die Ergebnisse der unterschiedlichen Bearbeitungsverfahren.

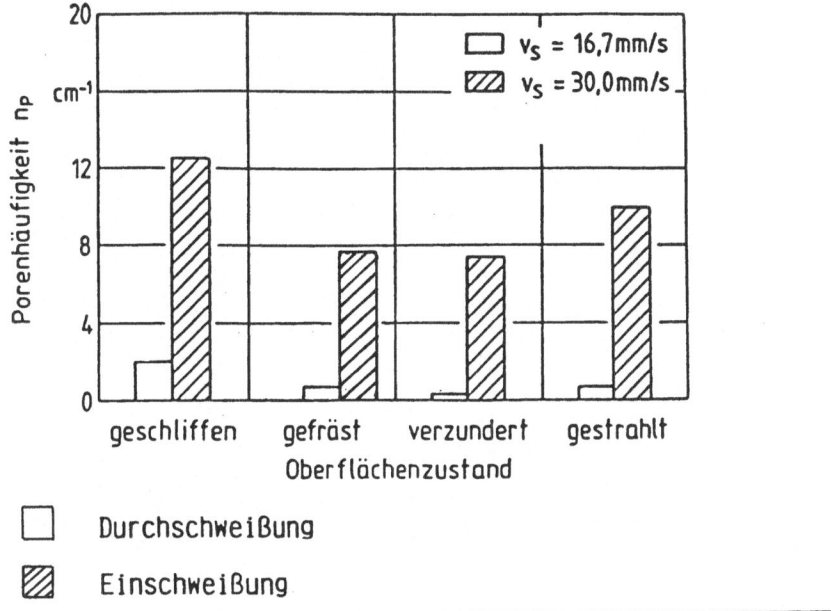

Bild 1. Einfluß des Oberflächenzustands auf die Porenhäufigkeit n_p an Blindschweißungen

Markant ist der Unterschied zwischen Durchschweißung und Einschweißung, wo hingegen sich der Oberflächenzustand nur geringfügig auswirkt. Typische Nahtausbildungen der Einschweißungen zeigen die Längsschliffe in Bild 2. Das ungleichmäßige Aussehen der Unterkante der Schmelzzone ist laserspezi- fisch und auf die dynamische Badbewegung zurückzuführen. Die Porenanhäufung (Porendurchmesser 150 - 250 μm) im unteren Teil der Schmelze hat zwei Ursachen: Zum einen entstehen sie im untersten Bereich der Schmelze, zum anderen werden sie dann durch die rasche Abkühlung in diesem Bereich fixiert. Die nahezu porenfreie Durchschweißung läßt sich damit erklären daß beim Durchtritt des Laserstrahles durch das Blech das Keyhole sich andersartig ausbildet.

Werkstoff: St 52-3
Lasertyp: UTRC TM 34-9
Fokussieroptik: sphärischer Spiegel
                   Öffnungsverhältnis: F 6
Schutzgas: V = 50 l/min He
Schweißparameter: P_L = 10 kW, V_S = 30 mm/s, z = -3mm

Oberflächenzustand

gefräst  verzundert

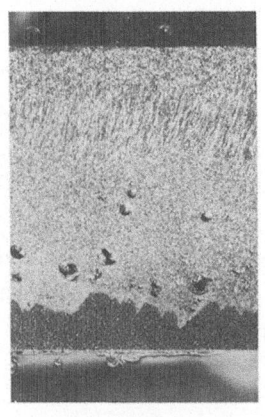 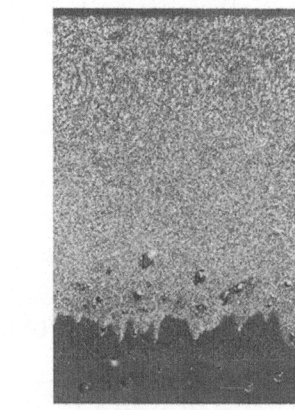

geschliffen   gestrahlt

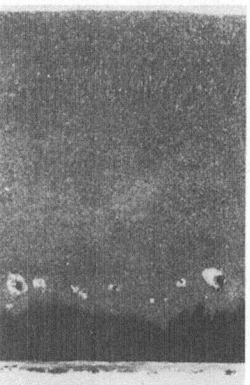 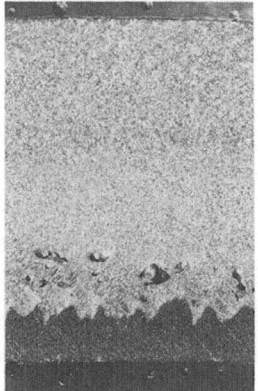

Bild 2.   Oberflächeneinfluß auf die Schweißqualität

3.2 Einfluß der Korrosionsschutzmittel

Als Beispiel für den Einfluß der Korrosionsschutzmittel auf die Qualität der Schweißnaht sind im Bild 3 Längs- und Querschliffe einer gestrahlten Probe mit Ölfilm dargestellt.

Werkstoff:	St 52-3
Laserty:	UTRC TM 34-9
Fokussieroptik:	sphärischer Spiegel, Öffnungsverhältnis F 6
Schutzgas:	V = 50 l/min He
Schweißparameter:	P_L = 10 kw, V_s = 16,7 mm/s, 30 mm/s; Z = - 3mm

Oberflächenzustand: gestrahlt, geölt

I-Stoß

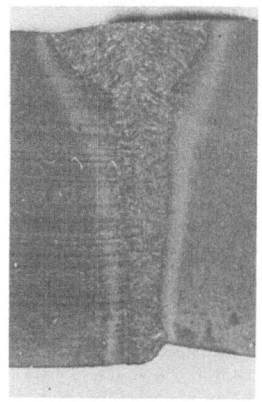 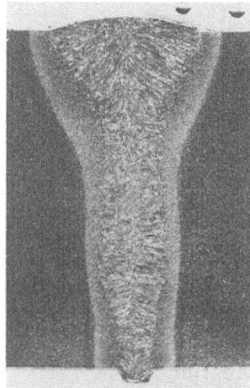

Öl an Stoßkante

16,7 mm/s   16,7 mm/s

Blindnaht

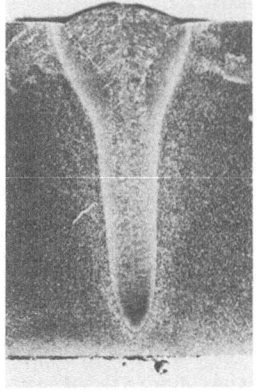 

30 mm/s   30 mm/s

2mm

Bild 3.   Oberflächeneinfluß auf die Schweißqualität

I-Stoß und Durchschweißung zeigen gleich gute Qualität und keinen Unterschied zur ölfreien Probe. Gleichartig verhält sich auch die Einschweißung. Die Querschliffe ergeben, daß sich die Poren überwiegend in der Mitte der Schweißnaht befinden. Einen Überblick über den Einfluß von Öl, Wachs und Farbe auf die Porenhäufigkeit bei gestrahltem Grundwerkstoff wird in Bild 4 gegeben.

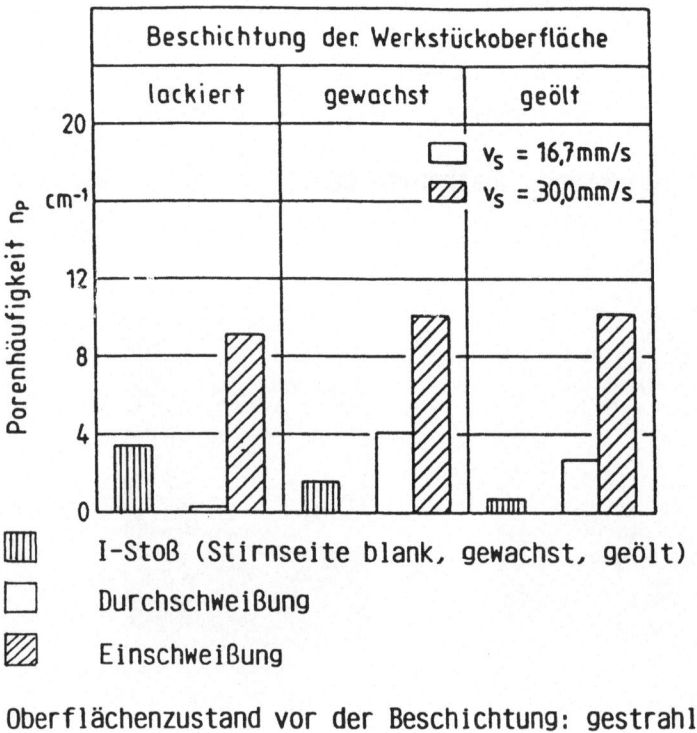

Bild 4. Einfluß der Beschichtung der Werkstückoberfläche auf die Porenhäufigkeit

Die Einschweißung zeigt auch hier erhöhte Porenhäufigkeit. Der geringe Einfluß der Beschichtungen ist darauf zurückzuführen, daß durch die dem Stichloch vorlaufende Wärmefront die leichtflüchtigen Stoffe verdampfen, bevor der eigentliche Schweißvorgang einsetzt. Auf verzundertem Grundwerkstoff aufgebrachte Beschichtungen führten zu gleichen Schweißergebnissen wie bei gestrahlten Untergrund.

Die Ergebnisse dieser Untersuchungen wurden mit einem CW-Laser gewonnen. Eine Übertragbarkeit auf Schweißungen mit gepulsten Laserstrahl ist damit noch nicht gegeben, da die vorlaufende Wärmezone kleiner ist.

Diese Arbeiten entstanden im Rahmen des BMFT-Projektes "Schweißen mit CO_2-Hochleistungslasern".

Laser Welding with Optical Fibres

C.J. Nonhof, K. Schildbach (Nederlandse Philips Bedrijven B.V.),
R. Iffländer (HAAS - Laser)

Introduction

Fibre optics for the transmission of high energy pulses has
greatly expanded the scope of using lasers in material
processing, especially welding. There is a cost reduction factor
if one uses one laser for several welding sites at the same or at
different machines. Also one can remove the laser from the
machine itself and replace it by a fibre and a small optic. This
gives greater accessibility to the machine and allows for a
simpler construction.

To make full use of these new possibilities there are new demands
placed on the lasersystem. If one laser is serving several
asynchronous machines, its power supply should be able to respond
to the random requests from the welding stations for pulses with
different pulse energy and pulse duration. Also the response
from the laser resonator to the flash of the lamp should be
independent of previous pulses. The single physical phenomenon
that stands in the way of the ideal laser is thermal lensing. The
laser rod is heated homogeneously but cooled on the outside.
Because of linear expansion differences the laser rod can be seen
to become a thick lens. Even more important is that the
refractive index of the YAG or glass rod material is temperature
dependent. The thermal relaxation of the rod material ranges from
2 s for a 1/4" YAG rod to 50 s for a 3/8" glass rod, see
table I.

This gives the laser resonator a memory for previous pulses. The
thermal lensing effect gives unpredictable pulse energy and laser
beam parameters as beam waist and divergence angle in a
randomly running laser system.

Table I . Thermal relaxation time $T \sim r^2_0/k$ of laser rod

$2r_0$ / T	YAG (k=4.6 . 10^{-6} m^2/s)	glass (k~5 . 10^{-7} m^2/s)
1/4"	2 s	18 s
3/8"	5,5	50

Near field - far field imaging

The full formula for focusing a laser beam with a beam waist w_1, and divergence angle θ_1 to a spot radius w_2, see figure 1, reads.

$$\frac{1}{w_2^2} = \frac{1}{w_1^2} \left(1 - \frac{s_1}{f}\right)^2 + \frac{1}{\theta_1^2 f^2} . \qquad (1)$$

Where s_1 is the distance between the beam waist w_1 and the lens and f is the focal length of the lens. The formula reduces to

$$w_2 = \theta_1 f \text{ when } s_1 = f. \qquad (1a)$$

This is the so-called far field imaging limit. The near field imaging limit is reached for s_1 "large" (usually $\geq$ 2 m), then

$$w_2 = \frac{f}{d_1 - f} w_1. \qquad (1b)$$

For lasers with a simple resonator design consisting of two planar mirrors symmetrically around the laser rod, the divergence angle θ_1 is strongly dependent on the thermal lensing. The beam waist is, however, constant to high power levels. With these resonators near field imaging should be used: $s_1 \geq$ 2 m in formula (1b). For lasers which have an optimized resonator for a constant divergence angle θ_1 far field imaging may be used allowing a compact laser system.

Beam quality factor

Formula (1) together with the propagation formula for a laser beam

$$w(z) = w \sqrt{1 + (z/\theta f)^2} \qquad (2)$$

gives us that the property

$$w_1 \theta_1 = w_2 \theta_2 \qquad (3)$$

is conserved when the laser beam passes through lens systems. We call $w\theta$ the beam quality factor. It can also be written in a different form

$$w\theta = C_{nm}^2 \lambda/\pi \quad \text{with } C_{nm} \geq C_{oo} = 1 \qquad (4)$$

where λ is the wavelength of the laser radiation and C_{nm} is dependent on the mode number. In a multimode laser we use an effective C_{nm}. No laser beam can have a better beam quality factor than the TEM_{oo} or Gaussian mode. The multimode lasers used in welding are up to a hundred times worse than TEM_{oo}.

Fibre optics

In focusing the laser beam into a step index fibre with a core of 600 μm the beam quality is conserved in focusing, see figure 1. In the fibre the focusing angle θ_2 is conserved under certain circumstances, see figure 2. The output beam waist w_3 reflects the core diameter and is irrespective of the input beam waist w_2. Here we have the possible danger of loosing beam quality. Therefore w_2 should be matched to the core diameter of the fibre for all power levels of the laser. When we fail in this respect, we must make the final focusing optics larger in diameter than is necessary. Small final focusing optics and reasonable working distances are obtained only for small beam quality factors.

494

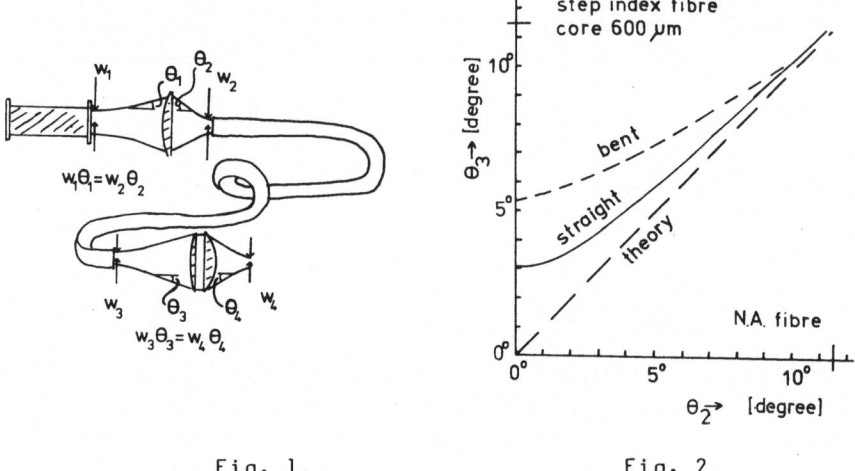

Fig. 1.   Fig. 2.

Note

Also in straight welding and cutting applications a constant spot size $2w_2$ is of the utmost importance. Remarks as: "This high power laser actually works better at low output power," can quite often be traced back to using far field imaging and a simple resonator design.

Wear and Fatigue Properties of Laser Melted Cast Iron

H. W. Bergmann, B. L. Mordike, F. Reinke*, J. Betz
Institut für Werkstoffkunde und Werkstofftechnik
Technische Universität Clausthal
Agricolastr. 2, D-3392 Clausthal-Zellerfeld, FRG
*AEG-Elotherm, D-5630 Remscheid, FRG

Abstract

This paper gives some examples of laser melted cast irons including microstructures and resultant properties. Possible applications of laser melted components are also discussed.

Introduction

Various techniques have been used to modify the surface structure of grey cast iron in order to achieve a white, ledeburitic microstructure whilst retaining the grey core. The most prominent techniques are chilled surface casting and Tungsten Inert Gas Welding (TIG). Today, both methods are widely accepted in industry, for instance in camshaft production. The advantage of such treatments is the combination of the core ductility, due to stable Fe-C solidification resulting in an iron-graphite microstructure, and the wear resistance of the surface due to metastable solidification giving a $Fe-Fe_3C$ layer. Recently, the availability of high-powered lasers has enabled high speed surface melting and large scale automation. This paper describes the laser melting process, the microstructures obtained and a variety of resultant properties.

Experimental

SG iron with ferritic, pearlitic and bainitic matrix was surface melted using a 500 W CO_2 laser focussed beam with TEM 00 mode. CO_2 or He was used to increase the amount of incident energy absorbed by forming a plasma. The melted layer was between 0.3 and 0.5 mm thick depending on the feed rate. The surface was perfectly uniform, crack free and untarnished. The surface roughness was less then $\pm$ 8 µm in the lasered condition. The laser treatment caused negligible distortion. A typical picture is given in Fig.1.

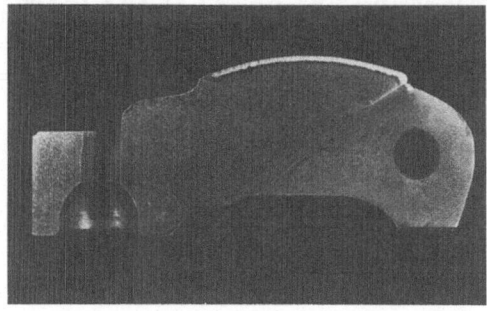

Fig. 1. Laser melted SG iron rocker arm

Results and discussion

Laser melting of cast irons may lead to surface hardness improvements
by directional rapid solidification. Rapid solidification results in
grain refinement, metastable phase formation, e.g. conversion of graphite to Fe_3C, and supersaturation which offers the possibilities of
further transformation and/or precipitation hardening. A typical microstructure is given in Fig. 2 showing the ledeburitic case and the
ferritic SG iron core. There are further improvements possible by laser alloying which are discussed elsewhere (1).
The mechanical properties of laser melted S.G. irons depend on both
the area ratio of melt depth to substrate and HAZ to substrate. For
S.G. iron the HAZ is small when compared to the melt depth and may be
neglected in a first approximation. Fig. 3 shows a typical stress-
strain curve for a laser melted test sample.

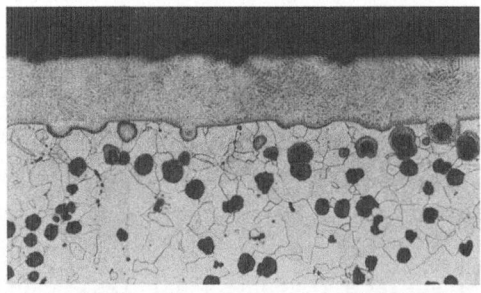

Fig. 2. Microstructure of laser melted SG iron

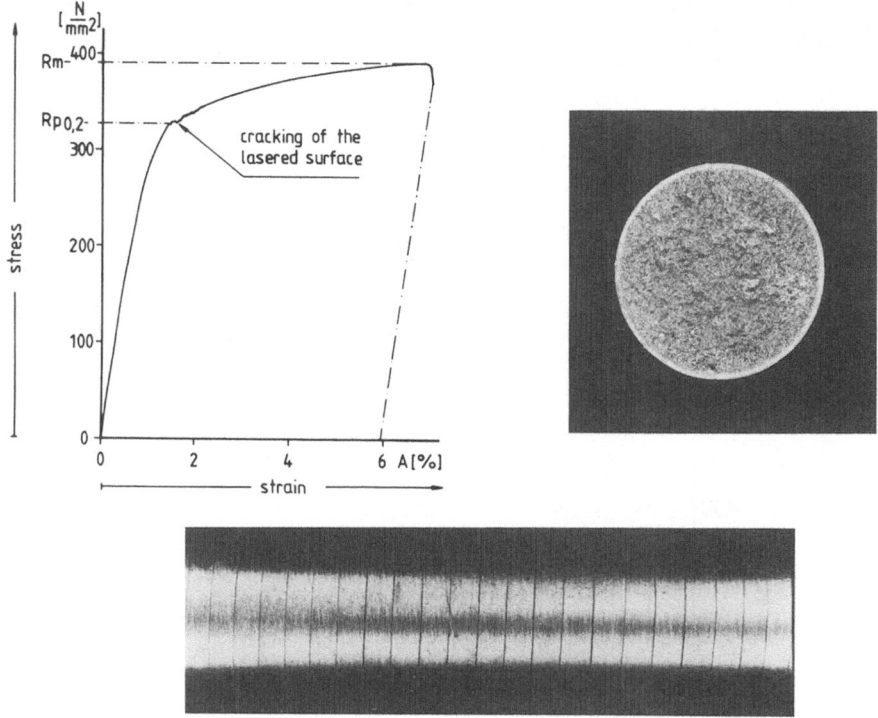

Fig. 3. Stress-strain curve of laser melted SG iron
and fractured tensile test specimens

The stress-strain curves for untreated and laser melted test samples
are similar, however, above the yield point, the case cracks circum-
ferentially giving a feature on the curve not dissimilar to disconti-
nuous yielding and this is accompanied by an acoustic emission. At
least 1 % elongation occurs before the case cracks. Fractographs
show a ductile fracture of the core and a more brittle fracture of
the outer layer. Cracks propagate through the case mainly in the
interdendritic eutectic phase, where microporosity may also be
present. A unique feature of this material after tensile testing is
the occurrence of equidistant circumferential cracks along the gauge
length. These cracks do not penetrate further than the case core interface because of workhardening of the ductile core. This effect is
not influenced by metallurgical inhomogeneities between laser tracks
as there are none and it occurs with longitudinal tracks as well as
for spiral melting. This behaviour is observed for ferritic, pearlitic and bainitic core structures which have a higher toughness than
the case. When a martensitic matrix is used, the initial case crack
continues through the brittle core.

Laser surface melting results in internal stresses on solidification, see Fig. 4. Several factors contribute to the final stress situations, for example, the volume difference on solidification, differences in shrinkage in the solid state for core and surface due to both temperature differences and different coefficients of thermal expansion. However, the various transformations in both case and core have an effect. For cast iron with pearlitic matrix and flaky graphite (2) it was found that in the surface, compressive stresses were present which change to tensile stresses at some distance below the surface. Depending on the case depth, the sign of the stress changes in either the core or the surface layer.

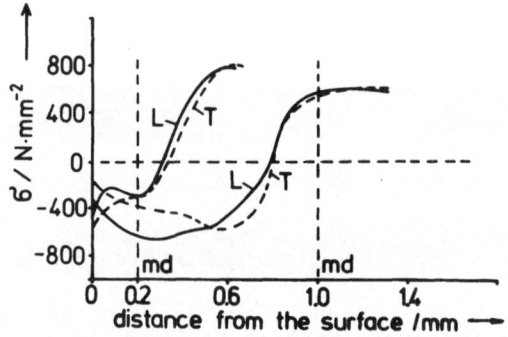

Fig. 4. Internal stresses in laser melted SG iron (md = case depth, L = stresses in longitudinal direction, T = stresses in transversal direction)

The fatigue properties of ferritic S.G. iron are given in Fig. 5. Untreated specimens were ground after machining. As no relevant data are available in the literature, optimised processing parameters for laser treatment had to be defined. This was done by keeping the case depth constant at about 10 % of the cross-section. Compared to the untreated S.G. iron, a decrease in fatigue limit was found in the as lasered condition. This was more pronounced when N_2, CO_2 or Ar was used as protective gas, but less than with He.

In addition, a spiral laser track gives better results than a longitudinal series of tracks. Grinding the surface, i.e., smoothing small amounts of roughness produced by the laser treatment leads to a small decrease in fatigue limit as compared with untreated specimen. The most favourable values were found when the specimens were annealed at 240°C for 2 hrs after laser treatment with helium and subsequently ground.

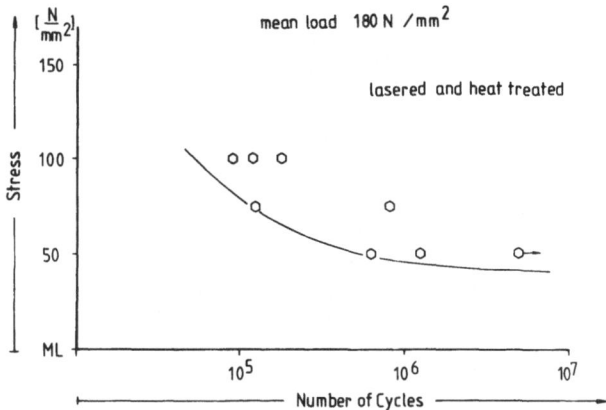

Fig. 5. Fatigue Properties of laser melted SG iron
a) 15 mm diameter, 0.3 mm case depth

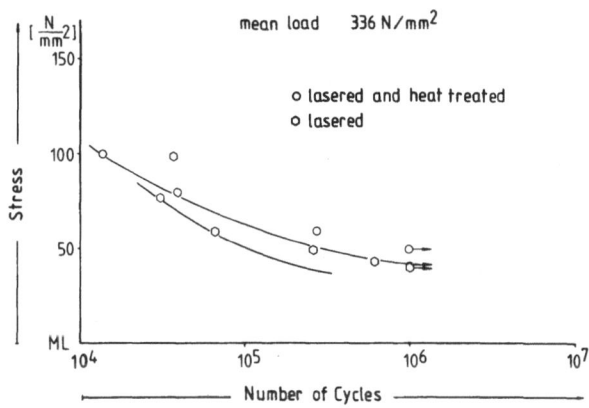

b) 8 mm diameter, 1 mm case depth

The excellent wear properties of ledeburitic surface layers on cast
iron has been demonstrated over recent years. Most of the work has
been carried out on TIG melted surfaces, while for the new technique
of laser melting, only a limited amount of data is available (3,4).
For lasered grey iron with pearlitic matrix and flaky graphite, Bell
and co-workers demonstrated that the wear properties in dry pin on
disc tests are increased with laser melting by one order of magnitude
compared to the untreated situation. There was still a significant
increase compared to fully martensitic hardened microstructures.
The advantage of laser melted S.G. iron was demonstrated for rolls
which run dry against each other with a definite relative slip, one
wheel being driven and the other partially braked (Fig. 6). Two loads
were applied (500 and 1000 Nmm^{-2}) and the humidity was controlled.

500

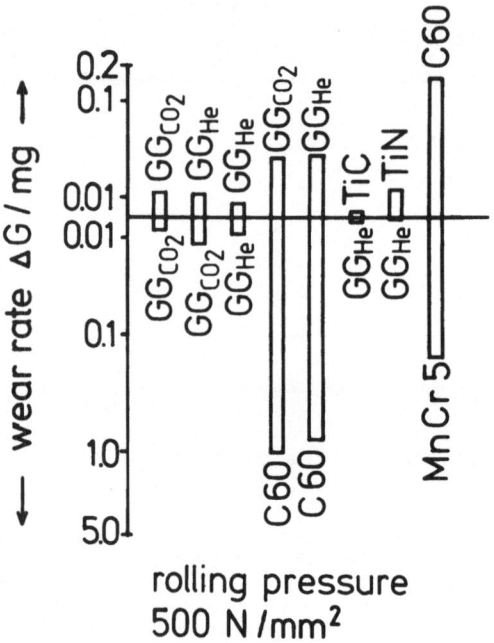

Fig. 6. Wear properties of laser melted SG iron

For a slip of approximately 3 - 5 % and the higher load, melting
occurred after a couple of minutes when steel rolls were used, but
the cast irons showed no evidence of any deformation. For comparison,
1 % slip was used. It is obvious that the wear properties of lasered
cast iron are superior to those of hardened, conventional steels.
This improvement is less pronounced if the steels have been hard
faced or nitrided. Combinations of lasered irons and steels must be
avoided as massive deformation of the steel occurs. When laser
treatments were carried out under He, better results were found than
with other gases. Excellent results were obtained when laser melted
S.G. irons were used in combination with TiN or TiC coated hardened
steels, the TiN/TiC surface suffering the wear. After the tests, no
deformation is visible on the ledeburitic surface.

General comments and possible applications

Apart from the fact that laser melting enhances certain desirable
properties mentioned above, there are additional beneficial features
obtained with this technique. The melting process can be carried out

on an almost finished component as the distortion associated with the process is negligible and surface roughness is in the order of 10 μm, which can easily be improved by grinding if necessary. There is no need for a dedicated handling system as a uniform and homogeneous case can be achieved if lasered within the focal distance, i.e., $\pm$ 1.5 cm for a 25.4 cm focal length lens. Laser melting is a hardening process which can be fully automated. It is also a fast and cheap process.

Achnowledgements

This work was supported by the VW-Stiftung.

References

1. H.W. BERGMANN and B.L. MORDIKE, Z. Werkstofftechnik <u>14</u>, 228, (1983).
2. H.W. BERGMANN, B.L. MORDIKE and T. BELL, Laser 83, München, Springer Verlag Berlin, (1985).
3. D.N.H. TRAFFORD, T. BELL, J.H.P.C. MEGAW and A.S. BRANSDEN, Proc. Conf. Heat Treatment, (1981), The Metals Society, London.
4. H.W. BERGMANN and W. WITZEL, Textures in Rapidly Solidified Metals, Proc. ICOTOM 7, Holland, (1984).

Anpassung der Verfahrensparameter beim Schweißen verschiedener Metalle mit dem Nd:YAG-Laser

I. Decker, G. Chen und J. Ruge
Institut für Schweißtechnik der TU Braunschweig
Langer Kamp 8, D-3300 Braunschweig

Die Anwendung des im Pulsbetrieb arbeitenden Nd-Festkörperlasers zum Punkt- und Nahtschweißen an kleinen Bauteilen ist bereits Stand der Technik. Unser Interesse gilt dem Einfluß der Werkstoffeigenschaften auf den Schweißvorgang /1/. Die Schweißversuche wurden mit einem handelsüblichen multimoden-Nd:YAG-Laser durchgeführt.

Versuchsbedingungen

Die Schweißung wird durch verschiedene Parameter beeinflußt. An der Anlage einzustellen sind die Pulsenergie, die Pulsdauer und die Lage des Strahlfokus in bezug auf die Werkstückoberfläche. Festgelegt sind dann die Intensitätsverteilung im Strahl aufgrund der Resonatoranordnung (Bild 1) und der Brennfleckdurchmesser aufgrund der verwendeten Arbeitsoptik. Der zeitliche Verlauf eines Pulses kann in einem bestimmten Umfang variiert werden (Bild 2). Es wurde stets ein trapezförmiger Pulsverlauf gewählt (Verlauf c), da hierbei das Verhältnis zwischen Pulsenergie und Strahlungsleistung fest vorgegeben ist. Das experimentell ermittelte Intensitätsprofil des multimode-Strahls ließ sich durch eine Gaußverteilung bei jedoch gegenüber dem Grundmode vergrößerter

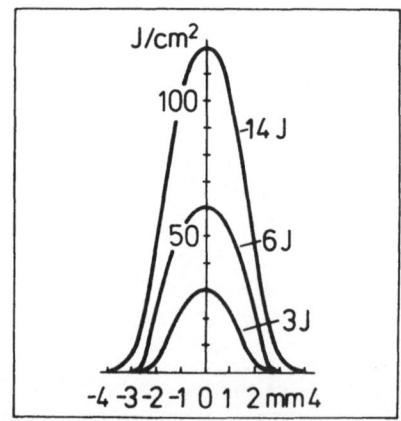

Bild 1. Pulsenergieverteilung im Laserstrahl für verschiedene Pulsenergien

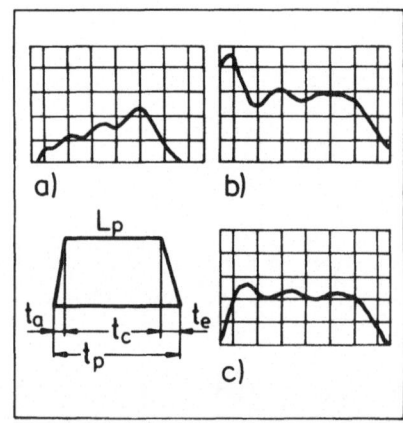

Bild 2. Zeitliche Verläufe des Laserpulses

Strahldivergenz annähern. Daher
konnte auch die Intensitätsverteilung im Brennfleck als gaußförmig
mit entsprechend angepaßtem Fleckradius angenommen werden. Rechnerisch ergab sich für die maximale
Intensität im Brennfleckzentrum
während der Dauer eines Pulses die
in Bild 3 gezeigte Abhängigkeit.

(Bei dieser vereinfachenden Betrachtung werden die im Brennfleck eines
multimoden-Laserstrahls momentan und
lokal auftretenden Intensitätsspit-
zen nicht berücksichtigt!)

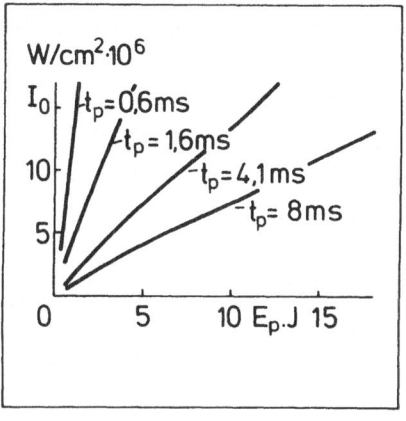

Bild 3. Maximale Intensität im
Brennfleck in Abhängig-
keit von Pulsenergie
und -dauer

Schweißvorgang, Schwellenverhalten

Bei niedriger Pulsenergie entstehen flache Schweißnähte, es handelt
sich um ein Wärmeleitungsschweißen. Oberhalb einer Energieschwelle
führt der Tiefschweißeffekt zu einer Naht mit einem Tiefe-zu-Breite-
Verhältnis größer als Eins. Wie in Bild 4 gezeigt ist, liegt dazwischen ein Bereich mit verstärktem Werkstoffabtrag und deshalb unbe-
friedigendem Schweißergebnis. Bei sehr hohen Pulsenergien geht der
Tiefschweißvorgang aufgrund der Verdampfung in das Laserbohren über.
Diese vier Energiebereiche konnten bei allen untersuchten Werkstoffen
(un- und hochlegierte Stähle, Titan- und Aluminiumwerkstoffe, Nickel,
Molybdän, Kupfer, Gold und Silber) deutlich gegeneinander abgegrenzt
werden.

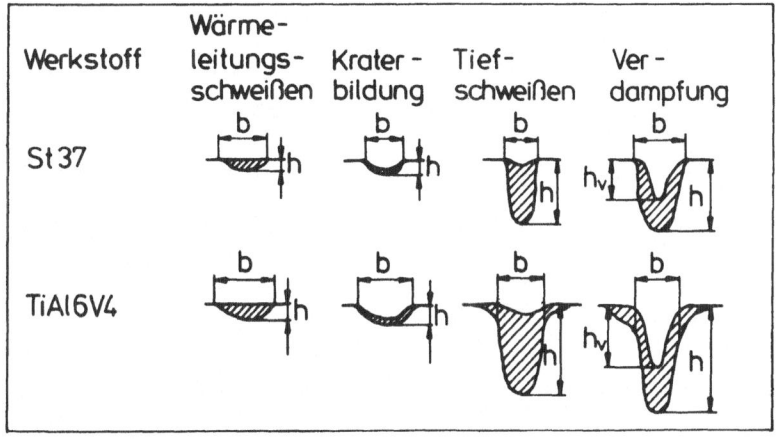

Bild 4. Nahtaussehen für verschiedene Pulsenergien

Einige zahlenmäßige Ergebnisse sind in Bild 5 dargestellt. Der Bereich
der Kraterbildung ist am relativen Maximum des Verhältnisses von ver-
dampftem zu insgesamt aufgeschmolzenem Werkstoffvolumen zu erkennen,
der Bereich des Tiefschweißens am relativen Minimum. Der durch kalori-
sche Messung ermittelte Absorptionsgrad steigt im Übergangsbereich von
dem durch das metallische Reflexionsvermögen bedingten niedrigen Wert
auf nahezu 100 % an. Für das Tiefschweißen existieren somit günstige
Pulsenergiebereiche, deren Lage und Breite von den anderen Parametern
abhängig sind (Bild 6): Mit zunehmender Pulsdauer muß eine höhere Pulsenergie gewählt werden. Die Tiefschweißbereiche für hochreflektierende
Werkstoffe liegen bei hohen Energien und besitzen eine kleinere relative Breite, so daß die Schweißparametereinstellung hier mit erhöhter
Genauigkeit vorzunehmen ist.

Physikalische Bedeutung der Schwellintensität

Die für das Tiefschweißen notwendige Intensität im Zentrum des Brenn-
flecks ist in Bild 7 als Funktion der Wärmeleitfähigkeit der einzelnen
untersuchten Werkstoffe aufgetragen. Sie wird verglichen mit der Intensität, die zum Erhitzen eines Oberflächenpunktes auf Verdampfungstem-
peratur unter Berücksichtigung der Reflexionsverluste notwendig ist,
sowie mit der Schwellintensität für das Zünden eines oberflächennahen
Metalldampfplasmas /2/. Im Bereich hohen Wärmeleitvermögens vollzieht

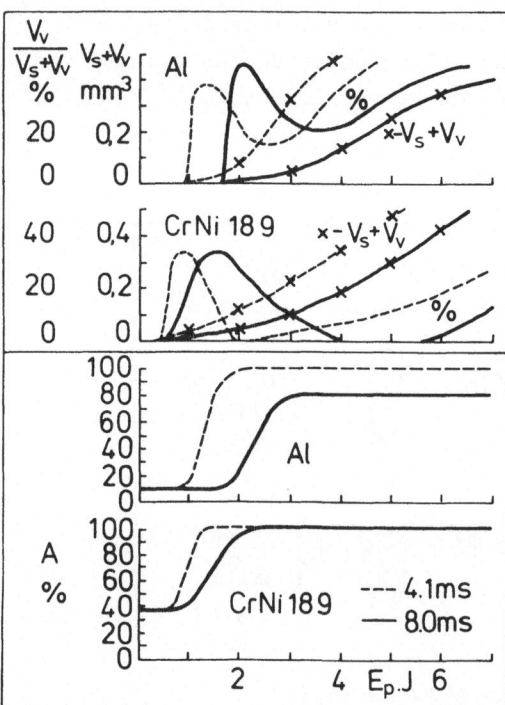

Bild 5. Aufgeschmolzenes Volumen,
dabei verdampfter Anteil
und absorbierte Strah-
lungsenergie in Abhängig-
keit der Pulsenergie

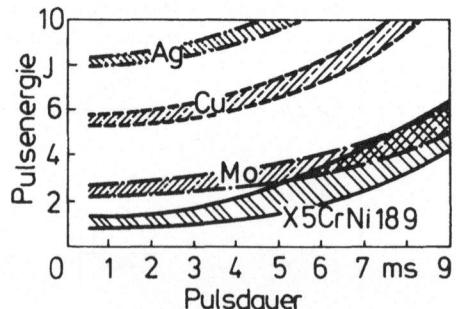

Bild 6. Pulsenergiebereiche für
das Tiefschweißen bei
verschiedenen Werkstoffen

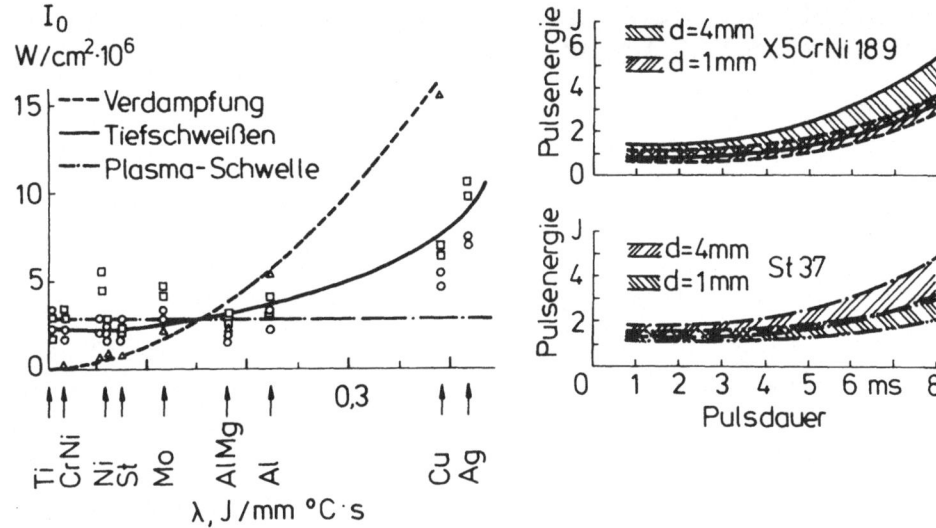

Bild 7. Schwellintensität (auf die Tief-
schweiß-Pulsenergie bezogene
Mittelwerte) in Abhängigkeit
der Wärmeleitfähigkeit

Bild 8. Pulsenergiebereiche
für das Tief-
schweißen bei ver-
schiedenen Blech-
dicken

sich das Tiefschweißen oberhalb der Plasmaschwelle und unterhalb der
Verdampfungsintensität. Deshalb läuft die plasmakontrollierte anomale
Strahlungsabsorption nicht in dynamisch stabiler Form ab. Es kommt zu
einem fluktuierenden Tiefschweißeffekt, der mit der beobachteten stär-
keren Werkstoffverdampfung und einer unruhigen Schmelzbadbewegung (ausgeprägte Spritzerbildung) verbunden ist.

Einfluß der Werkstückgeometrie

Die Lage der Pulsenergiebereiche für das Tiefschweißen hängt neben den
genannten Parametern auch von der Werkstückbeschaffenheit ab: Oberflächenzustand, Temperatur, Lage der Schweißstelle und geometrische Form
des Bauteils. Bild 8 zeigt den Einfluß der Blechdicke: Bei den 1 mm
dicken Blechen handelt es sich um eine Durchschweißung, während bei
den 4 mm-Blechen eine Einschweißung vorgenommen worden ist.

Literatur
(1) G. CHEN: Verhalten verschiedener Werkstoffe beim Schweißen mit dem
    Nd:YAG-Laser. Dissertation, TU Braunschweig 1984
(2) E. BEYER, G. HERZIGER u.a.: Formation and Influence of Laser In-
    duced Plasma during Laser Welding. LASER 83, ed. W. Waidelich,
    Springer-Verlag Berlin Heidelberg 1984, p. 367-372

Feinschweißen, -Trennen und Bohren mit dem gepulsten Nd:YAG Laser

P.K. Affolter und H.P. Schwob, Lasag AG, CH-3600 Thun

1. Anforderungen an Fertigungssysteme für die Feintechnik

In den vergangenen Jahren stieg die Nachfrage von anpassungsfähigen, flexiblen Fertigungssystemen, insbesondere auch bei "High Technology" Produktionsanlagen in der Feinwerktechnik. Bei hochwertigen Laserwerkzeugmaschinen ist nebst der optimalen Prozessbeherrschung die Betriebswirtschaftlichkeit (einfache Handhabung, kurze Einricht- und Umrüstzeiten, die Zuverlässigkeit und niedrige Betriebskosten) immer mehr d a s entscheidende Kriterium für den erfolgreichen Einsatz.

2. Anforderungen an ein flexibles, wirtschaftliches Laserfertigunssystem

Optimale Feinschweiss-, Schneid- und Bohrprozesse erfordern:

1. Eine anpassungsfähige Strahlquelle
2. Bearbeitungsspezifisch konzipierte Strahlführungs- und Fokussierelemente
3. Angepasste Schutz- und Schneidgaszuführelemente
4. Werkteilpositionier-, Steuerungs-, Ueberwachungs- und Sicherheitselemente, welche dem Standard von hochwertigen Werkzeugmaschinen entsprechen

Dabei wird heute auch bei Laserwerkzeugmaschinen besonders darauf geachtet, dass die zugeführte Leistung optimal für die geforderten Bearbeitungsprozesse umgesetzt wird.

2.1 Eine anpassungsfähige, multifunktionale Strahlquelle

Die elektrischen Versorgungs- und Steuergeräte sind dabei so konzipiert, dass für Bohr- und Schneidaufgaben mit hohen Impulsleistungen (bis zu 30 kW) bei maximalen mittleren Leistungen bearbeitet werden kann.
Für Schweissprozesse sind Impulsenergien bis zu 60 Joules und stufenlos programmierbare Schweissimpulslängen bis zu 20 ms wählbar.

Energiehochlauf- und Abschlussrampen sind für viele Werkstoffe und Werkteilgeometrien notwendig und können vorprogrammiert werden.

Durch theoretische Untersuchungen, Rechenprogramme und experimentelle Arbeiten wurden Resonatoren entwickelt, welche durch einfache Vorkehrungen in einem grossen Bereich fürs Schweissen, Schneiden oder Bohren gezielt angepasst werden.

Dabei kommen grundsätzlich zwei Anordnungen der optischen Resonator-Elemente zur Anwendung:

a) Durch Auswechseln von fest einjustierten Spiegeln (s1-s4) mit definierten Krümmungsradien wird das günstigste Strahlprofil für jeden geforderten Bearbeitungseffekt gewählt (Abb.1).

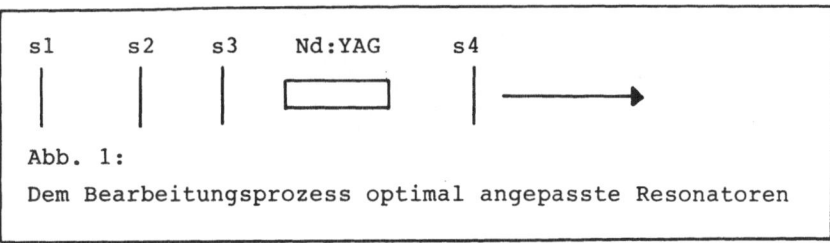

Abb. 1:
Dem Bearbeitungsprozess optimal angepasste Resonatoren

b) Durch ein variables Element (V) wird der Laserresonator der gewünschten Bearbeitungsart optimal angepasst (Abb. 2).

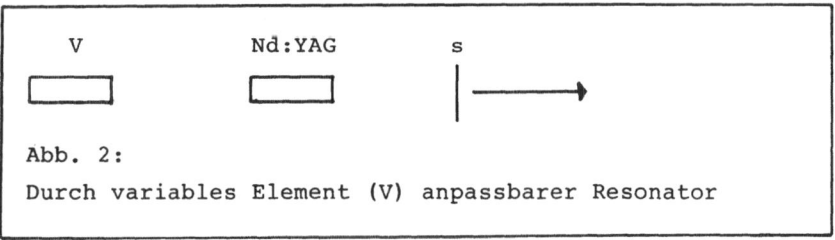

Abb. 2:
Durch variables Element (V) anpassbarer Resonator

2.2 Bearbeitungsspezifisch konzipierte Strahlführungs- und Fokussierelemente

Der Laserstrahl wird über feste, ansteuerbare Spiegel oder Lichtleiter (Abb. 3) zur Fokussieroptik geführt, welche für die geforderte Bearbeitung speziell ausgelegt und optisch korrigiert ist. Reproduzierbar vorwählbare Brennfleckdurchmesser werden durch motorisch angetriebene variable Strahlaufweiter erzielt.

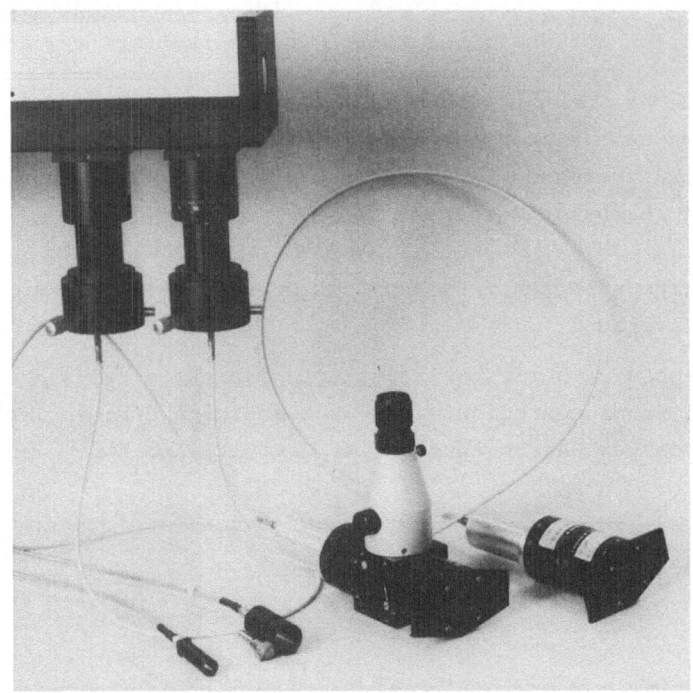

Abb. 3.   Komponenten für Lichtleitertechnik:
Einkopplungselemente, Lichtleiter, Bearbeitungsköpfe für
Punkt-, Nahtschweissen und Feinstschneiden

2.3 Angepasste Schutz- und Schneidgaszuführelemente

Fürs Schweissen, Schneiden und Bohren ist die Konstruktion der Gas-
zuführelemente, Gasart, -Druck, -Menge und -Richtung oft ebenso wichtig
wie die Laserparameter.
Beim Schweissen ermöglichen spezielle Zuführrohre ein laminares Ausströmen der Gase oder Gasgemische und bewirken einen optimalen Schutz
der Schweisszone.

Die Schneid- und Bohrdüsen sind für qualitativ gute Schnitte und
Bohrungen und grösste Bearbeitungsgeschwindigkeiten von entscheidender
Bedeutung.
Aufgrund von Effekten bei bekannten thermischen Trennverfahren und
Experimenten sind fürs Schneiden mit dem Nd:YAG Laser, Düsen entwickelt
worden, die sich auch bei extremen Bearbeitungsaufgaben bewähren: durch
Optimierung des Schneid-/Bohrgasjets können selbst stark gewölbte
Bauteile ohne Nachführung der Fokussieroptik scharfkantig und recht-
winklig geschnitten werden (Abb.4).

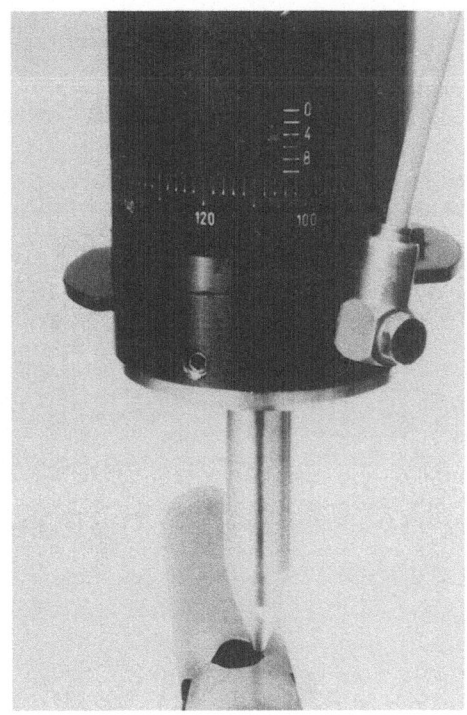

Abb. 4. Schneiddüse für Schnitte an gewölbten Hohlkörpern

3. Richtiges Zusammenwirken der Bearbeitungsgrössen: Typische Bearbeitungsbeispiele und Bearbeitungsdaten

Eine wirtschaftliche Fertigung erfordert heute und in Zukunft komplette Technologiedaten für den Benützer von Laserwerkzeugmaschinen, da nur so innert nützlicher Frist beste Schweissungen, Schnitte und Bohrungen erzielbar sind.

3.1 Schweissen

Das verzugsfreie, kraftschlüssige Verbinden von Trafo-Kernblechpaketen (Abb. 5) bei hohen Bearbeitungsgeschwindigkeiten erfordert grosse mittlere Laserleistungen bei geringen Strahldivergenzen. Der Laser-resonator wird dazu der erforderlichen Schweissbadgeometrie angepasst.

Durch geschickte Wahl des Schweissbades und der Lage der Schweissnähte am Transformator wird der magnetische Fluss im Bauteil nicht gestört.

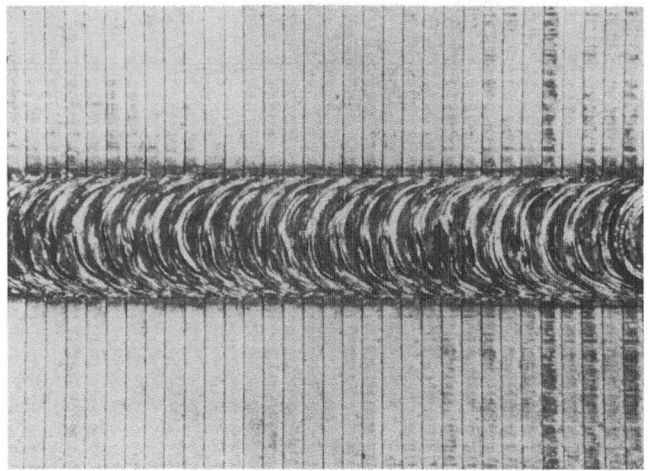

Abb. 5.   Verzugsfreies Nahtschweissen von Kernblechen

3.2 Schneiden

Feinste Schnitte in harte, spröde Werkstoffe wie z.B. in polykristallinen Diamant (Abb. 6) werden durch kleinste Strahldivergenzen bei
grossen mittleren Laserleistungen wirtschaftlich realisiert.
Kundenspezifische Feinschnitte für spanabhebende Formwerkzeuge lassen
sich mit schmaler Schnittfuge (0.15 mm) und der notwendigen Ober-
flächengüte herstellen.

Abb. 6.   Feinschnitt in polykristallinen Diamant

3.3 Bohren

Die für Bohrprozesse eingestellte Strahlquelle (für hohe Impuls-
leistungen bei grösseren Impulsraten), eine rotierende Bearbeitungs-
optik und eine spezielle Bohrdüse ermöglichen auch äusserst extreme
Bohrungen in Superlegierungen (Abb. 7).
Kühllöcher für Triebwerkteile können dabei sogar gegen die Strahl-/-
Jetaustrittsseite hin konisch gebohrt werden.

Typische Technologiedaten für Schweiss-, Schneid- und Bohrprozesse

Bearbeitungsart	Schweissen	Schneiden	Bohren
Werkstoffe	4301 Edelstahl	polykristal-liner Diamant	Hastelloy X
Wandstärke	1.5 mm	3 mm	6.5 mm
Anforderungen:			
Einschweisstiefe	1.5 mm	---	---
Schnittbreite	---	0.15 mm	0.3 mm
Oberflächengüte	spritzerfrei	Rauhtiefe = 6 μm	recast layer = 20 μm
Bearbeitungsge-schwindigkeit	400 mm/Min	80 mm/Min	60 mm/Min
Bearbeitungsparameter:			
Impulsenergie (J)	15 J	4 J	7 J
Impulslänge (ms)	6 ms	0.4 ms	0.6 ms
Impulsfrequenz (Hz)	20 Hz	50 Hz	30 Hz
Resonatorart / Ein-stellung	S 1	S n 2	Bo 3
Düsenart	Rohr A	Nr. 3	Nr. 4
Gas	Argon	---	O_2
Gasmenge (Druck)	5 l/Min	---	6 bar

512

Abb. 7. Schliff durch eine negativ konische Bohrung in Hastelloy X

Zusammenfassung

Durch prozesszentrierte Untersuchungen von Laserparametern und anderen wichtigen Einflussgrössen, werden mit der notwendigen Laserleistung die geforderte Bearbeitungsqualität und -Geschwindigkeit erreicht und für den Fertigungsfachmann praxisbezogen dargestellt. Anhand einiger Beispiele wird gezeigt, wie mit einer anpassungsfähigen multi-funktionalen Laserbearbeitungsmaschine über einen grossen Bereich optimale Bearbeitungseffekte erzielt werden.

Optoelektronische Signalübertragung
Optoelectronic Signal Transmission

Entwicklungstendenzen in der optischen Breitbandkommunikation

Clemens Baack

Heinrich-Hertz-Institut für Nachrichtentechnik Berlin GmbH,

Einsteinufer 37, .D-1000 Berlin 10

Basistechnologien zukünftiger Kommunikationssysteme sind

- die Mikroelektronik, durch die eine kostengünstige und außerordentlich vielseitige Verarbeitung digitaler Signale ermöglicht wird,
- die Optische Nachrichtentechnik, die die Übertragung großer Informationsmengen in digitaler Form gestattet,
- die Integrierte Optik, die langfristig in der Optischen Nachrichtentechnik eine kostengünstige Massenproduktion optoelektronischer Systeme gewährleisten soll.

Die Bedeutung der Mikroelektronik für zukünftige Kommunikationssysteme ist hinreichend bekannt und soll hier nicht näher erläutert werden. Diese Aussage gilt auch für die Optische Nachrichtentechnik. In diesem Beitrag soll demnach nicht die heutige, "konventionelle" Optische Nachrichtentechnik behandelt werden, vielmehr gilt es, Einsatzmöglichkeiten der nächsten Generation der Optischen Nachrichtentechnik, der sog. Kohärenten Optischen Nachrichtentechnik, in zukünftigen Kommunikationssystemen aufzuzeigen. Schließlich gilt es, den Begriff Integrierte Optik zu klären und die Bedeutung der Integrierten Optik in Verbindung mit der Kohärenten Optischen Nachrichtentechnik für die zukünftige Nachrichtentechnik darzustellen.

1. Kohärente Optische Nachrichtentechnik (KONT)

Statt der heute üblichen Optischen Nachrichtentechnik, die nach Bild 1a mit optischen Direktempfängern, bestehend aus Photodioden mit nachgeschalteten Verstärkern, arbeiten, wird man in der KONT optische Überlagerungsempfänger einsetzen (Bild 1b). Analog zum Überlagerungsprinzip in der Rundfunktechnik wird in einem optischen Überlagerungsempfänger das Empfangssignal in einem Richtkoppler mit dem Licht eines lokalen Lasers überlagert. Am Ausgang der Photodiode entsteht ein Mikrowellensignal mit einer Zwischenfrequenz, die der Differenzfrequenz der beiden Lichtsignale entspricht. Der optische Überlagerungsempfänger ist hoch selektiv und wesentlich empfindlicher als der Direktempfänger. Es wird nun möglich, sehr viele Lichtträger in sehr geringem Abstand (z.B. einige GHz) über eine Faser zu übertragen. Empfangen wird nur die Information des Trägers, der mit dem Licht des lokalen Lasers eine dem ZF-Verstärker entsprechende Zwischenfrequenz bildet /1/.

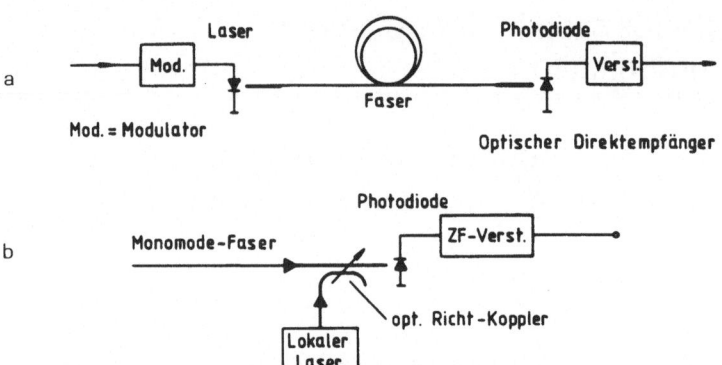

Bild 1. Optischer Direktempfang   (a)

          Optischer Überlagerungsempfang (b)

Zur Zeit ist man bestrebt, die Übertragungskapazität der Faser durch die λ -Multi-
plextechnik besser auszunutzen,   indem mehrere Lichtträger unterschiedlicher Wellenlänge gleichzeitig über die Faser übertragen und am Faserausgang durch optische
Filter wieder getrennt werden. Beim heutigen Stand der Filter- und Lasertechnik
müssen die Lichtträger wenigstens einige zehn Nanometer auseinanderliegen, um eine
sichere Trennung zu ermöglichen. Wählt man einen Wellenlängenabstand von z.B. $\Delta\lambda$ =
20 nm (entsprechend etwa 10 THz), so lassen sich in dem dämpfungsarmen Wellenlängenbereich von 0,7 µm bis 1,8 µm einer modernen Monomodefaser etwa 40 Lichtträger
unterbringen (Bild 2a). Bei Verwendung eines optischen Überlagerungsempfängers läßt

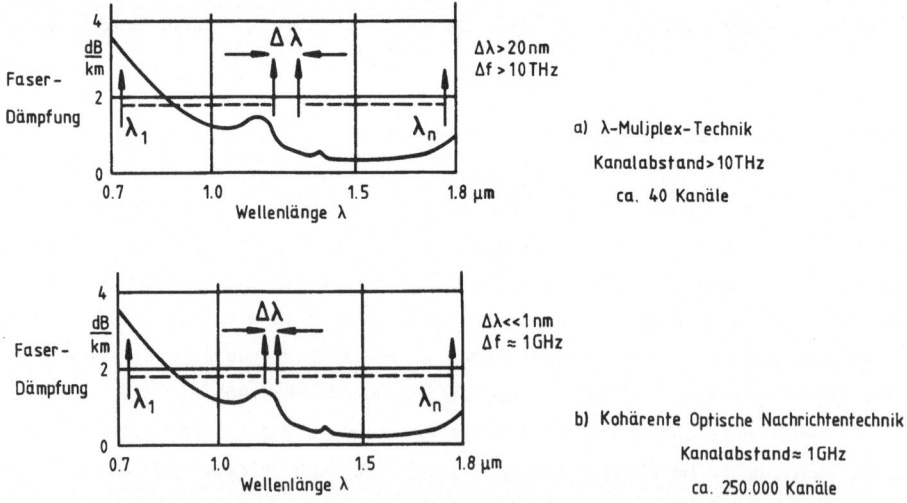

Bild 2. Übertragung eines Lichtfrequenzvielfaches über eine Monomodefaser mit λ -Mul-
tiplextechnik und mit Kohärenter Optischer Nachrichtentechnik

sich ein Kanalabstand von z.B. 1 GHz realisieren (Bild 2b). Damit wird prinzipiell der Übertragung von ca. 250.000 Lichtträgern über eine Monomodefaser möglich; die praktischen Grenzen werden durch die Nichtlinearitäten in der Faser bestimmt.

2. Integrierte Optik (IO)

Die IO hat zum Ziel, optische aktive und passive sowie elektronische Komponenten auf einem Subtrat monolithisch zu integrieren. Optische aktive Komponenten, wie Laser und Photodioden, dienen zur elektronisch-optischen bzw. optisch-elektronischen Wandlung. Optische passive Komponenten, wie z.B. Wellenleiter, Gitter, Linsen, Filter und Koppler, dienen zur Führung, Beugung, Abbildung, spektralen Zerlegen und zum Schalten des Lichts. Elektronische Komponenten schließlich, wie z.B. Transistoren, Widerstände, Kondensatoren, dienen zur Ansteuerung optischer Komponenten und zur elektronischen Signalverarbeitung. Diese Opto Electronic Integrated Circuits (OEIC) werden zur optoelektronischen Signalverarbeitung eingesetzt /2/.

Die IO hat Parallelen zur Mikroelektronik. In der Mikroelektronik werden hochkomplexe elektronische Systeme auf einem kleinen, robusten, zuverlässigen und kostengünstig in großer Stückzahl zu fertigenden Silizium-Chip untergebracht. Die IO hat das Ziel, opto-elektronische Systeme als kleine, robuste, zuverlässig und kostengünstig in großer Stückzahl zu fertigende Halbleiterchips zu entwickeln. Im Aufbau unterscheiden sich die Chips der Mikroelektronik (Electronic Integrated Circuits-EIC) wesentlich von den OEICs der IO. Die einzelnen Komponenten der EICs haben µm-Abmessungen, rechtwinklige Konturen und ein Längen- zu Breitenverhältnis von etwa 1:1. Infolgedessen lassen sich sehr hohe Packungsdichten von einigen hunderttausend Komponenten je Chip erzielen. Die optischen Komponenten der OEICs haben i.allg. krummlinige Konturen, und das Längen- zu Breitenverhältnis kann einige Größenordnungen umfassen; ein optischer Richtkoppler z.B. besteht aus zwei Wellenleitern, die einige µm breit sind und die über einige Millimeter hinweg im Abstand von einigen µm parallel geführt werden müssen. Daraus ergeben sich ganz unterschiedliche Architekturen für EICs und OEICs und wesentlich geringere Packungsdichten für OEICs. Trotz der vergleichsweise sehr geringen Komponentenzahl sind die OEICs außerordentlich leistungsfähig. Beim Vergleich von integrierten Schaltungen der Mikroelektronik und der IO ist nicht die Anzahl der integrierten Komponenten maßgebend, entscheidend sind die völlig unterschiedlichen physikalischen Funktionen der elektronischen und optischen Komponenten.

Als Halbleitermaterial für OEICs kommt nicht das technologisch gut beherrschte Silizium in Frage. Die indirekte Bandstruktur des Silizium verbietet die Herstellung von Lichtquellen, wie z.B. Laser. Für die Entwicklung von OIECs kommen Verbindungshalbleiter des III-IV-Materialsystems zum Einsatz. OEICs auf InP-Basis sind auf den günstigsten Übertragungsbereich von Glasfasern (λ = 1,3 ... 1,6 µm) abgestimmt und sind somit für den Einsatz in allen Ebenen öffentlicher Nachrichtennetze geeignet.

3. Kohärente Optische Nachrichtentechnik und Integrierte Optik in zukünftigen Kommunikationssystemen

Nachfolgend werden Beispiele für den möglichen Einsatz von KONT und IO in der Fern- und Teilnehmerebene zukünftiger öffentlicher Nachrichtennetze sowie in zukünftigen lokalen Netzen genannt. KONT und IO erfordern als Übertragungsmedium Monomodefasern. Der Einsatz dieser Techniken führt somit zur Verlegung von Monomodefasern nicht nur in der Fernebene, das gilt heute als selbstverständlich, sondern langfristig auch in der Teilnehmerebene des öffentlichen Netzes und sogar in lokalen Netzen.

3.1 Fernebene des öffentlichen Netzes

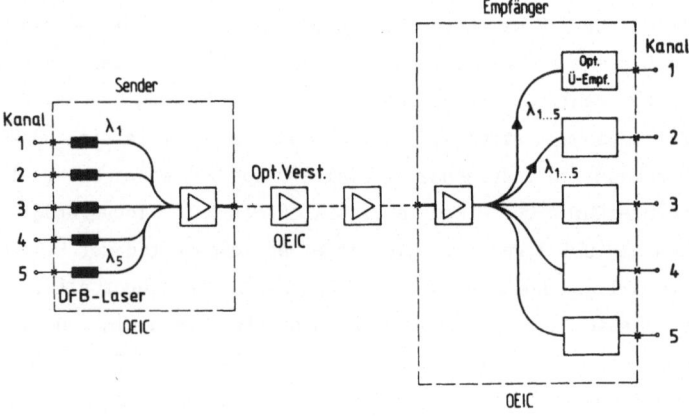

Bild 3. Ferntrasse mit optischen Verstärkern und optischen Überlagerungsempfängern

Zukünftige Breitband-Kommunikationsnetze werden den Dienst Bildtelefon mit hoher Bildqualität anbieten, was zu außerordentlich hohen Informationsströmen in den Fernebenen führen wird. Die Ferntrassen werden aus zahlreichen parallelen Monomodefasern bestehen, jede Faser wird durch λ-Multiplextechnik mehrfach genutzt werden und jeder Lichtträger wird einen hochratigen Informationsstrom übertragen. Die Vielzahl der Fasern, die λ-Multiplextechnik und die hochratige Übertragungstechnik werden zu sehr komplexen Repeaterstationen führen. Diese Probleme lassen sich durch den Einsatz von KONT in Verbindung mit IO drastisch reduzieren. Nach Bild 3 werden z.B. fünf Lichtträger über eine Monomodefaser übertragen, der Abstand zwischen den Trägern beträgt jedoch, im Gegensatz zur λ-Multiplextechnik, nur wenige GHz, so daß alle fünf Träger von einem optischen Verstärker verstärkt werden können.

Die Repeaterstationen enthalten weder λ-Multiplexer/Demultiplexer noch optoelektronische bzw. elektrooptische Wandler. Die optischen Signale in der Faser werden durch
die optischen Verstärker lediglich analog verstärkt, jedoch nicht, wie in einem
Repeater, regeneriert. Die Gesamtübertragungsreichweite der Ferntrasse ist demnach
nicht beliebig groß. Am Ausgang der Ferntrasse werden die verschiedenen Träger durch
optische Überlagerungsempfänger selektiert und detektiert.

Der Einsatz solcher Ferntrassen, die an Stelle der heute üblichen elektronischen
Repeater mit optischen Verstärkern (Halbleiterlaser mit entspiegelten Endflächen)
arbeiten, ist erst denkbar, wenn es gelingt, die Sendeeinheiten, die optischen
Verstärker und die Empfangseinheiten als OEICs zu realisieren.

3.2 Teilnehmerebene des öffentlichen Netzes

Wie eingangs erwähnt, wird bei Einsatz von IO und KONT im Teilnehmerbereich die
Verlegung von Monomodefasern im Teilnehmerbereich erforderlich. Entscheidend ist
hier die Frage nach der Spleiß- und Steckertechnik. Das Spleiß- und Steckerproblem
für Monomodefasern darf in absehbarer Zeit als zufriedenstellend gelöst betrachtet
werden, immerhin sind bereits heute die ersten automatischen Spleißgeräte sowie
dämpfungsarme Stecker mit hoher Zuverlässigkeit käuflich.

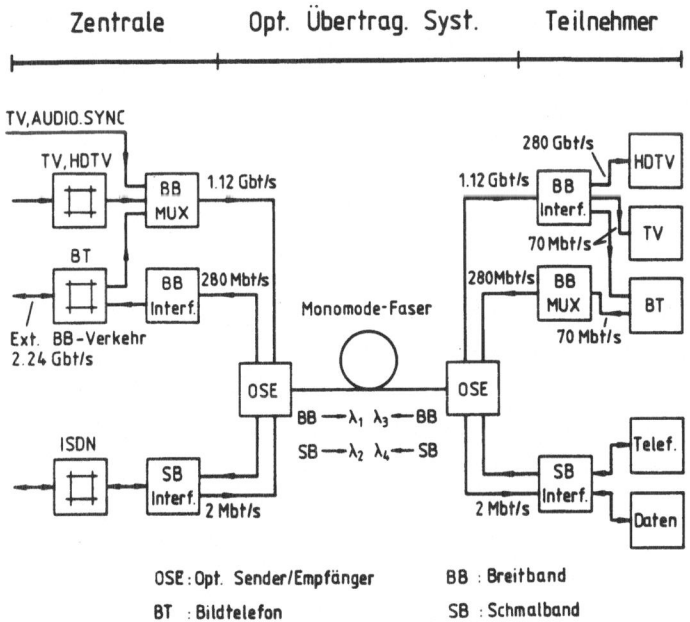

Bild 4. Experimentalsystem eines Teilnehmernetzes mit Monomodefasern

Außerdem kann man im Teilnehmerbereich auf verfügbare Stecker zurückgreifen, die infolge einer Strahlaufweitung mit wesentlich geringeren Toleranzanforderungen arbeiten; die höheren Verluste dieser Stecker sind wegen der geringeren Entfernungen im Teilnehmerbereich tolerierbar.

Zur Zeit wird im HHI ein Experimentalsystem /3/ (Bild 4) entwickelt, das u.a. zur Klärung der Frage beitragen soll, ob von der Monomodefaser im Teilnehmerbereich unmittelbar Vorteile zu erwarten sind, also nicht erst durch den Einsatz von KONT und IO, beide Techniken werden erst langfristig verfügbar sein.

Die monomodale Teilnehmeranschlußleitung (Bild 4) erlaubt die Übertragung hoher Bitraten (1,12 Gbit/s) zum Teilnehmer und ermöglicht damit die gleichzeitige Übertragung von z.B. 16 digitalen TV-Signalen im heutigen Standard (mit 70 Mbit/s je Signal) oder von mehreren TV-Signalen in einem zukünftigen High Definition Television (HDTV)-Standard (280 Mbit/s) /4/. Die hochratige Übertragung über Monomodefasern ist heute nicht mehr Thema der Forschung, hochratige Ferntrassen werden z.Z. von der Industrie entwickelt. Entscheidend für den Einsatz hochratiger Teilnehmeranschlußleitungen ist die Elektronik. Die sendeseitige und empfangsseitige hochratige Elektronik muß jeweils in einem Chip mit vertretbarer Verlustleistung realisierbar sein. Der sendeseitige hochratige Baustein (BB-Mux, 280/1120 Mbit/s in Bild 4) liegt inzwischen als kundenspezifischer Si-Schaltkreis vor /5/. Der empfangsseitige hochratige Baustein (BB-Interf.) liegt als Dickfilmschaltung (2" x 1"; 1,5 W) vor /6/, die monolithische Integration steht bevor. Ein Ziel dieses Projekts ist es, die Entwicklung hochratiger integrierter Schaltkreise (HSICs) für die Nachrichtentechnik zu forcieren.

Die Schmalbanddienste werden in diesem System über getrennte Wellenlängen mit 2 Mbit/s abgewickelt (Notstromversorgung des Telefons!). Die Verbindung zur Fernebene erfolgt über eine 2,24 Gbit/s-Trasse.

Der zentralseitige optische Sender/Empfänger (OSE, Bild 4) ist in Bild 5 dargestellt. Derartige Sende/Empfänger sind heute aus einzelnen Komponenten zusammengesetzt. Der optische Multiplexer/Demultiplexer (λ-Muldex) ist ein mikro- oder faseroptischer und damit ein feinmechanisch anspruchsvoller Baustein, mit dem die Laser und Photodioden über Fasern verbunden sind. Die Emissionswellenlängen der Laser sind sorgfältig auf die Durchlaßbereiche der Multiplexer abzustimmen. Schließlich sind die Laser mit den Modulatoren und die Photodioden mit den Verstärkern zu verbinden.

Langfristig muß es gelingen, die optischen Sender/Empfänger, die in heutiger diskreter Bautechnik keineswegs für eine kostengünstige Massenproduktion geeignet sind, als OEIC zu realisieren; dies wird eine der dringlichsten Aufgaben der IO sein.

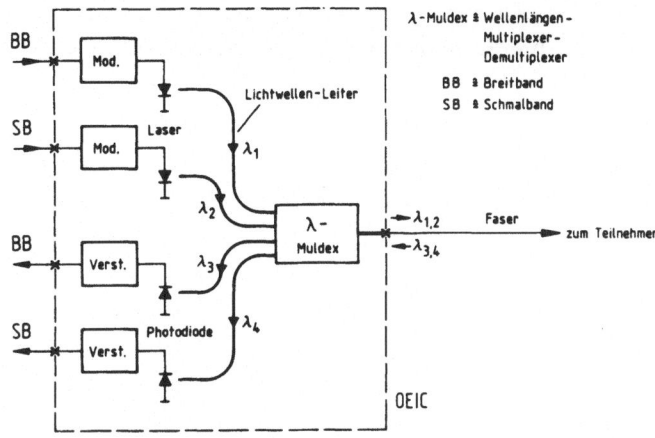

Bild 5. Optischer Sender/Empfänger

Die KONT bietet attraktive Möglichkeiten zur Verteilung von Breitbanddiensten im Teilnehmerbereich. Will man allen Teilnehmern z.B. 100 TV-Programme anbieten, so ordnet man jedem Programm einen Lichtträger zu, alle 100 Träger ($\lambda_1 \ldots \lambda_{100}$ in Bild 6) werden über die Teilnehmeranschlußleitungen übertragen. Jeder Teilnehmer kann

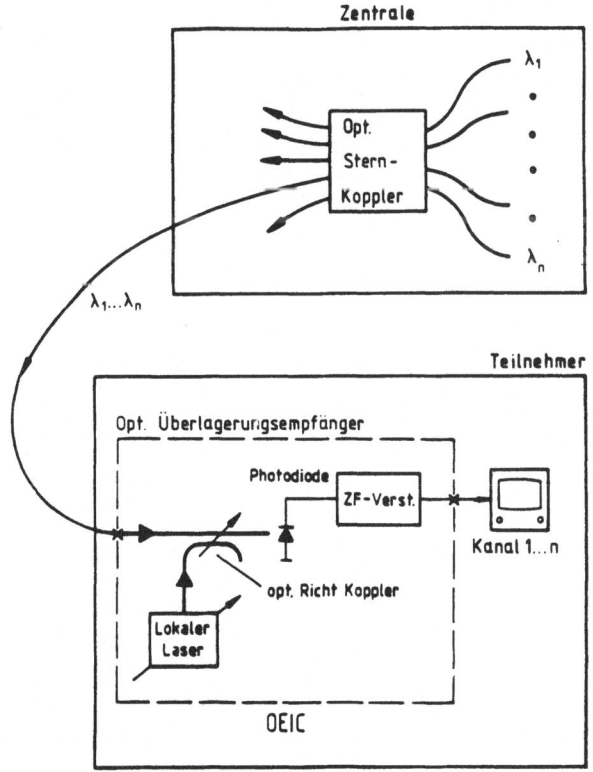

Bild 6. Breitband-Verteilnetz

mit Hilfe eines durchstimmbaren optischen Überlagerungsempfängers das gewünschte Programm auswählen /1/. Diese interessante Lösung eines optischen Verteilnetzes ist erst denkbar, wenn es mit der IO gelingt, den durchstimmbaren optischen Überlagerungsempfänger als OEIC zu realisieren.

In einem Forschungsvorhaben, das im Auftrage der DBP im HHI durchgeführt wird, gelang inzwischen die Übertragung von drei Lichtträgern im Abstand von ca. 5 GHz über eine Monomodefaser. Am Faserausgang werden beide Träger mit einem durchstimmbaren optischen Überlagerungsempfänger selektiert und detektiert.

Langfristig könnten die 100 dichtgepackten TV-Signale (Bild 6) z.B. über den λ_1-Multiplexkanal des Systems in Bild 4 verteilt werden. Die mittelfristig realistische Zeitmultiplextechnik, die in Bild 4 zur Verteilung von Breitbandsignalen eingesetzt wird, könnte in ferner Zukunft durch eine optische Frequenzmultiplextechnik ersetzt werden.

Die Monomodefaser in der Teilnehmerebene bietet somit folgende Vorteile:
- große Systemflexibilität bei der Verteilung von TV- und HDTV-Programmen durch hochratige Zeitmultiplextechnik
- Einsatz von integrierten optoelektronischen Komponenten, die kostengünstig in großer Stückzahl gefertigt werden können (langfristig)
- große Systemflexibilität bei der Verteilung von TV- und HDTV-Programmen durch optische Frequenzmultiplextechnik hoher Kanalzahl mit Hilfe der Kohärenten Optischen Nachrichtentechnik und der Integrierten Optik (langfristig)
- Einheitsfaser für alle Netzebenen.

3.3 Lokale Netze

Lokale Netze werden in Zukunft für die Industrie- und Büroautomation sowie für den Rechnerverbund eine sehr wichtige Rolle spielen. Die Glasfaser ist das geeignete Übertragungsmedium, sie ist breitbandig, unempfindlich gegenüber elektromagnetischen Störfeldern und frei von Erdschleifen.

Besonders von japanischen Firmen wird eine Vielfalt solcher optischer Netze angeboten. Das Spektrum möglicher OEICs in diesem Bereich ist unüberschaubar, und das in Japan vom MITI gesteuerte Opto-Projekt zur Förderung der Integrierten Optik hat die Aufgabe, OEICs für den Einsatz in lokalen Netzen zu entwickeln. Die Teilnehmerzahl lokaler Netze ist im Vergleich zum öffentlichen Netz gering, so daß hier neben der optischen Nachrichtenübertragung auch die optische Nachrichtenvermittlung mit Hilfe der Integrierten Optik aussichtsreich erscheint.

Das in Bild 7 dargestellte Experimentalsystem hat u.a. das Ziel, Erfahrungen auf dem Gebiet der optischen Vermittlungstechnik zu sammeln /7/. Das System soll zur Übermittlung einer Vielzahl von ISDN-spezifischen Schmalbanddiensten, von LAN-spezifischen Mittelbanddiensten sowie zur Übermittlung und Verteilung von Breitbanddiensten (Bildtelefon, TV-Studiosignale heutigen Standards und eines zukünftigen HDTV-Standards, Medizintechnik usw.) dienen. Die Übertragung der Schmal- und Mittelbanddienste wird über die Wellenlänge λ_1 (λ-Multiplextechnik) und die Übertragung der Breitbanddienste über die Wellenlänge λ_2 erfolgen.

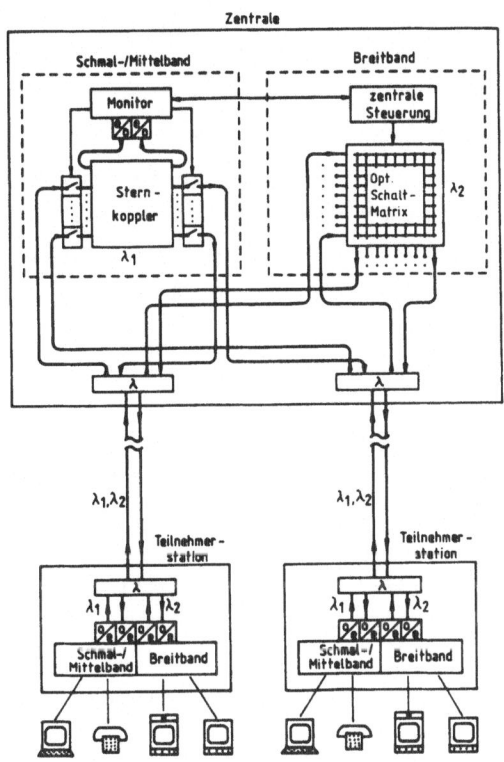

Bild 7. Lokales Breitbandnetz mit Optischer Vermittlungstechnik

Der Sternkoppler (64 x 64) in Bild 7 dient zur dezentralen optischen Vermittlung der Schmal- und Mittelbanddienste. Die Basiselemente monomodaler Sternkoppler sind 3 dB-Richtkoppler /8/, die zu 4 x 4 - oder 8 x 8-Substernkopplern integriert werden, der gesamte Koppler entsteht durch Kaskadierung dieser Substernkoppler. Der 64 x 64 monomodale Sternkoppler wurde sowohl aus käuflichen 3 dB-Faserrichtkopplern als auch aus integrierten 8 x 8 Sternkopplern ($LiNbO_3$) realisiert /9/. Statt der Teilerverluste von 18 dB wurde in beiden Fällen eine Einfügungsdämpfung von ca. 30 dB gemessen, diese Dämpfung ist mit käuflichen Sende- und Empfangselementen problemlos zu überwinden. Die üblichen Nachteile der dezentralen Vermittlung, wie z.B. unzureichende

Abhörsicherheit oder Störung des Netzes durch einzelne Teilnehmer /10/, werden nach Bild 7 durch zusätzlich integrierte optische Schalter vermieden.

Die optische Schaltmatrix (64 x 64) in Bild 7 dient zur zentralen Vermittlung der Breitbanddienste. Die Schaltmatrix entsteht durch Kaskadierung von 4 x 4- und 3 x 5-Subschaltmatrizen, die z.Z. in $LiNbO_3$ entwickelt werden. Die dezentralen und zentralen Vermittlungseinrichtungen in Bild 7 werden z.Z. auf $LinBO_3$-Basis integriert, um Systemerfahrungen für eine spätere Integration auf InP-Basis sammeln zu können. Wie eingangs erwähnt, ist für den Einsatz integriertoptischer Komponenten in lokalen Netzen die Monomodefaser als Übertragungsmedium unabdingbar.

4. Schlußbemerkung

Der Beitrag versucht, an Hand von Systemvorschlägen die Bedeutung der Integrierten Optik in Verbindung mit der Kohärenten Optischen Nachrichtentechnik für die zukünftige Kommunikationstechnik darzustellen. Erst die Integrierte Optik wird eine kostengünstige Massenproduktion optoelektronischer Bausteine für den Einsatz in der Optischen Nachrichtentechnik gestatten. Die Kohärente Optische Nachrichtentechnik ermöglicht die Übertragung eines optischen Frequenzvielfaches mit sehr geringem Kanalabstand über eine Glasfaser, die Selektion der einzelnen Kanäle erfolgt mit optischen Überlagerungsempfängern; die Technik bietet ganz neuartige Systemkonzepte. Durch Integrierte Optik und Kohärente Optische Nachrichtentechnik wird ein Trend zur Monomodefaser in allen Ebenen öffentlicher Netze und in lokalen Netzen erkennbar.

Im HHI wird z.Z. aus Mitteln des BMFT und des Landes Berlin ein Bereich errichtet, der die monolithische Integration von optischen und elektronischen Komponenten auf InP-Basis zum Ziel hat. Die vorgestellten Systemvorschläge werden im HHI in Projekten, die aus Mitteln des BMFT und der DBP gefördert werden, untersucht.

Literatur

/1/ Baack,C.; Bachus,E.-J.; Strebel,B: ntz, Bd. 35 (1982), Heft 11
/2/ Baack,C.: ntz, Bd. 37 (1984), Heft 6
/3/ Heydt,G.; Teich,G.; Walf,G.: ISSLS 84, Oct. 1984
/4/ Kummerow,Th.: Frequenz, Bd. 37, Nr. 11-12 (Nov./Dez. 83), S. 278-285
/5/ Rein,H.-M.; Daniel,D.:IEEE Journal of Solid-State Circuits, sc-19, no.3, June 84
/6/ Teich,G.: ntz Archiv, Bd.6 (1984), H.5
/7/ Hermes,Th.; Saniter,J.; Schmidt,F.; Werner,W.: OFC´85, Febr. 1985, San Diego
/8/ Hermes,Th.; Saniter,J.; Schmidt,F.: ntz, Bd.37 (1984), Heft 10
/9/ Werner,W.; Doeldissen,W.; Heidrich,H.; Hermes,Th.; Hoffmann,D.; Saniter,J.;
     Schmidt,F.: ECOC´85, Okt. 1985, Venedig
/10/ Hermes,Th.: ntz, Band 38, (1985), Heft 2

Sende- und Empfangselemente

H. P. Vollmer

AEG Forschungsinstitut Ulm
Postfach 1730, D - 7900 Ulm

Der für die optische Signalübertragung verwendete Wellenlängenbereich hat sich in letzter Zeit mehr und mehr in den langwelligen Spektralbereich zwischen 1,0 und 1,6 µm verlagert. Das ist auf die besseren Eigenschaften der Lichtwellenleiter bezüglich Dämpfung und Dispersion in diesem Bereich im Vergleich zu dem früher genutzten um 850 nm zurückzuführen. Als aktives Halbleitermaterial für die elektro-optischen Komponenten (Sender und Empfänger) wird im langwelligen Bereich meist die ternäre/quaternäre III/V-Verbindung InGaAs(P) verwendet. Durch geeignete Wahl der Materialzusammensetzung kann die Emissionswellenlänge bzw. Absorptionskante einge-stellt werden, wobei gleichzeitig Gitteranpassung an das als Substrat verwendete InP erreicht wird. Im folgenden wird über Strukturen und Eigenschaften der Sende-elemente Laser und Lumineszenzdiode (LED) und der Empfangselemente PIN- und Lawinenphotodiode (APD) berichtet und auf sich abzeichnende Trends eingegangen.

Ein wesentlicher Unterschied der InGaAsP/InP-Laser gegenüber dem GaAs/GaAlAs-Laser ist die größere Abhängigkeit des Schwellstroms von der Temperatur, eine dem quaternären Material eigene Eigenschaft /1/. Dadurch bedingt sind niedrige Schwell-ströme möglichst unter 30 mA Voraussetzung, um auch kontinuierlichen Betrieb bei Temperaturen von 40 - 60 °C zu ermöglichen, wie sie bei Systemanwendungen auftreten können. Diese Schwellströme können nur mit indexgeführten Strukturen erreicht werden. Von der Vielzahl der vorgeschlagenen Strukturen /2 - 5/ ist als Beispiel der DC-PBH-Laser /6/ schematisch in Bild 1 dargestellt. Allen Strukturen gemeinsam ist eine aufwendige Technologie, sei es, daß zu ihrer Herstellung zwei Epitaxie-schritte notwendig sind, sei es, daß eine größere Zahl von teilweise sehr dünnen Schichten gewachsen werden müssen. Es besteht daher der Trend zu weniger komplexen Strukturen, z.B. dem Pilzlaser /7/ und dem daraus abgeleiteten Mass-Transport-Laser /8/, allerdings ist die bei ihnen erreichte Zuverlässigkeit bisher noch nicht aus-reichend.

Typische Puls- und Dauerstrichkennlinien, hier für einen bei uns hergestellten DC-PBH-Laser, sind in Bild 2 gezeigt. Der Schwellstrom liegt bei 25 °C zwischen 10 und 30 mA, die maximale Temperatur für kontinuierlichen Betrieb bei 70 - 100 °C. In Systemen werden die Laser üblicherweise bei einer optischen Ausgangsleistung von 5 mW betrieben, von denen 2 - 3 mW sowohl in eine Standardgradientenfaser als auch in eine Monomodefaser gekoppelt werden können. Die Laser können bis zu mehr als einem Gbit/s moduliert werden, in Einzelfällen bis zu 10 Gbit/s.

Das optische Spektrum dieser Laser ist bereits im kontinuierlichen Betrieb meist multimodig und wird bei Modulation noch verbreitert. Dadurch werden die erreichbaren Übertragungsraten bzw. -längen begrenzt. Es wurden deshalb erhebliche Anstrengungen unternommen, Verfahren bzw. Laserstrukturen zu entwickeln, bei denen monomodiger Betrieb auch bei Modulation gewährleistet ist (DSM- = Dynamic Single Mode). Dazu wird der Laser stabilisiert entweder durch Verkopplung mit einem geeigneten Resonator, passiv beim externen Resonator /9/, aktiv beim C3- (Cleaved-Coupled-Cavity) Laser /10/ oder durch nebeneinander angeordnete Streifen /11/, durch kohärente Einstrahlung oder indem die Resonatorspiegel durch Gitterstrukturen ersetzt werden wie beim DFB- /12/ oder DBR-Laser /13/. In Bild 3 sind die Anordnungen der wesentlichen Verfahren zusammengestellt.

Hinzuweisen ist schließlich auf einen Trend in der Technologie, wo neben der Flüssigphasenepitaxie die Molekularstrahlepitaxie (MBE = Molecular Beam Epitaxy) und die Gasphasenepitaxie unter Verwendung von metallorganischen Verbindungen (MO-VPE = Metal Organic Vapour Phase Epitaxy) zunehmend an Bedeutung gewinnen. Bei diesen Epitaxien können sehr dünne Schichten im Bereich einiger Atomlagen homogen über eine Fläche von mehr als 10 cm² abgeschieden werden. Damit ist auf der einen Seite eine Erhöhung der Ausbeute verbunden, auf der anderen Seite sind völlig neue Strukturen wie z.B. der Multiquantum-Well-Laser /14/ herstellbar.

Lumineszenzdioden (LED) haben gegenüber Lasern den Vorteil, daß mit der Kohärenz zusammenhängende Probleme wie Pulsationen, Rückwirkungsempfindlichkeit, Moden- und Modenverteilungsrauschen bei ihnen nicht auftreten, daß ihre Temperaturempfindlichkeit geringer ist und daß schließlich der Einfluß der Technologie auf ihre Eigenschaften unkritischer ist und sie dadurch mit besserer Ausbeute hergestellt werden können. Auf der anderen Seite sind die in Lichtwellenleiter koppelbaren Leistungen und die erreichbare Modulationsrate niedriger, das optische Spektrum ist breiter, so daß sie nur für Systeme mit begrenzter Übertragungsrate und -länge geeignet sind.

Man unterscheidet zwei wesentliche LED-Typen. Beim Oberflächenemitter (Bild 4) wird das senkrecht zur aktiven Zone abgestrahlte Licht genutzt /15/. Für eine effektive Kopplung ist der Durchmesser des aktiven Flecks möglichst klein zu halten, ohne daß die Erwärmung durch die Stromdichte zu hoch wird. Optimale Fleckdurchmesser liegen bei 30 - 50 µm. - Der Kantenemitter /16/ ähnelt im Aufbau einem Streifenlaser. Allerdings wird die rücklaufende Welle in dem ungepumpten Bereich hinter dem gepumpten Streifen absorbiert, der Resonator ist unwirksam. Abstrahlung erfolgt nur in Vorwärtsrichtung parallel zur aktiven Zone. In Bild 5 ist als Beispiel die bei uns entwickelte InGaAsP/InP-V-Nut-LED gezeigt. Daten für beide Typen sind in Tabelle 1 zusammengetellt.

Seit kurzem wird die Anwendung von Kantenemitter-LEDs im Ortsnetz in Verbindung mit Monomodefasern diskutiert /17/. Mit den erreichten eingekoppelten optischen Leistungen zwischen 5 und 10 µW lassen sich bei einer Übertragungsrate von 140 Mbit/s mehr als 20km überbrücken. - Durch hohe p^+-Dotierung der aktiven Zone kann schließlich die Modulationsrate auf über 1 Gbit/s gesteigert werden /18/, allerdings muß eine deutliche Abnahme der verfügbaren optischen Leistung in Kauf genommen werden.

<u>Photodioden</u> werden heute meist als Mesadioden aus dem Materialsystem InGaAs/InP (Bild 6) aufgebaut. Die InGaAs-Absorptionszone hat zur Folge, daß diese Dioden für den gesamten interessierenden Spektralbereich von unter 1 µm bis über 1,6 µm empfindlich sind. Für niedrige Kapazitäten und Dunkelströme sind die aktiven Flächen klein zu halten. Die Einstrahlung erfolgt durch das Substrat, wozu ein spezieller Aufbau auf einem Keramikträger mit Schlitz über die Übertragungsfaser gewählt wurde. Die Empfindlichkeit liegt typischerweise zwischen 0,7 und 0,9 A/W.

Da aufgrund des geringen Bandabstands im ternären InGaAs der Tunnelstrom stark ansteigt, bevor bei steigender Spannung Ladungsträgerverstärkung einsetzt, wurde für InGaAs-Lawinenphotodioden eine spezielle Struktur vorgeschlagen (Bild 7) /20/. Bei dieser SAGM- (Separated Absorption and Graded Multiplication)-Struktur ist der Absorptionsbereich vom Verstärkungsbereich getrennt. Zwischen beide Bereiche ist eine dünne InGaAsP-Schicht eingefügt, die den Übergang zwischen den Bandabständen besser abstuft, um Ladungsträgerspeichereffekte zu vermeiden, die die Schnelligkeit erheblich herabsetzen. Typische Daten für beide Dioden sind in Tabelle 2 zusammengestellt.

In jüngster Zeit wird verstärkt an planaren Strukturen gearbeitet /21/, außerdem wird versucht, Photodiode und einen ersten Transistor oder sogar eine erste Verstärkerstufe monolithisch herzustellen.

Die diesem Bericht zugrunde liegenden Arbeiten wurden mit Mitteln des Bundesministeriums für Forschung und Technologie gefördert.

Literatur

/1/ M. ASADA et al; IEEE J. Quant. Electron., QE-19 (1983) 917
/2/ E. OOMURA et al; IEEE J. Quant. Electr. QE-20 (1984) 866
/3/ H. IMANAKA et al; Appl. Phys. Lett. 45 (1984) 282
/4/ M.C. AMANN, Electr. Lett. 15 (1979) 441
/5/ M. HIRAO et al; J. Appl. Phys. 51 (1980) 4539
/6/ I. Mito et al; J. Lightwave Techn. LT 1 (1983) 195
/7/ H. BURKHARD et al; Jpn. J. Appl. Phys. 22 (1983) L 721
/8/ z. B. Z. LIAN et al; IEEE J. Quant. Electr. QE-20 (1984) 855
/9/ J. VAN DER ZIEL et al; IEEE J. Quant. Electr. QE-20 (1984) 223
/10/ W.T. TSANG et al; Appl. Phys. Lett. 43 (1983) 1003
/11/ F. KAPPELER, Electr. Lett. 20 (1984) 1040
/12/ M. KITAMURA et al; J. Lightwave Techn. LT 2 (1984) 363
/13/ F. KOYAMA et al; Electr. Lett. 19 (1983) 325
/14/ W.T. TSANG, Appl. Phys. Lett. 38 (1981) 204
/15/ H. TEMKIN et al; Bell Syst. Techn. Rep. 62 (1983) 1
/16/ G. ARNOLD et al; Proc. 10th ECOC, Stuttgart (1984) 154
/17/ G. ARNOLD et al; Electr. Lett. 21 (1985) 390
/18/ D. GLOGE et al; Bell Syst. Techn. J. 59 (1980) 1365
/19/ E.A. SCHURR et al; Optoelektronik in der Technik - Herausgeber W. Waidelich, Springer-Verlag Berlin, Heidelberg, New York, Tokyo (1984) 439
/20/ Y. TAKANASHI et al; Jap. J.Appl. Phys. 7 (1981) 1271
/21/ T. SHIRAI et al; Electr. Lett. 17 (1981) 22

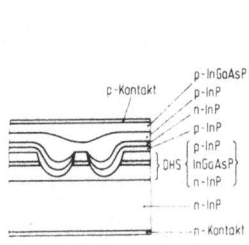

Bild 1: InGaAsP/InP DC-PBH-Laser

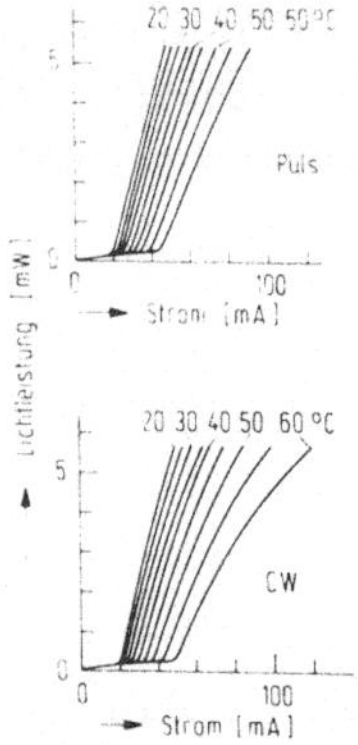

Bild 2: Optische Kennlinien vom DC-PBH-Laser

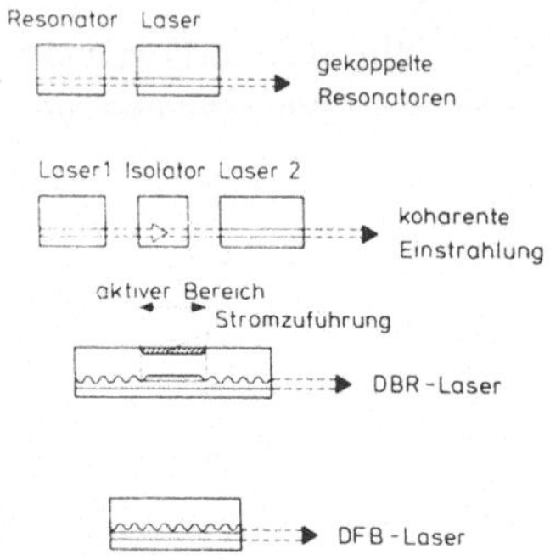

Bild 3: Dynamisch einwellige Laser (DSM)

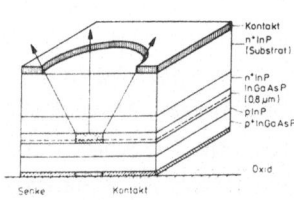

Bild 4: InGaAsP/InP-Oberflächenemitter (LED)

Bild 5: InGaAsP/InP-V-Nut-Kantenemitter (LED)

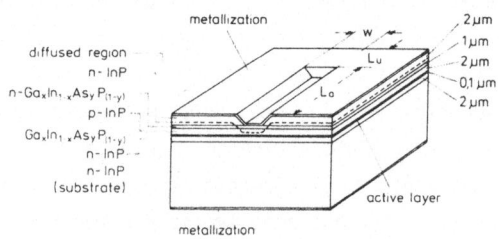

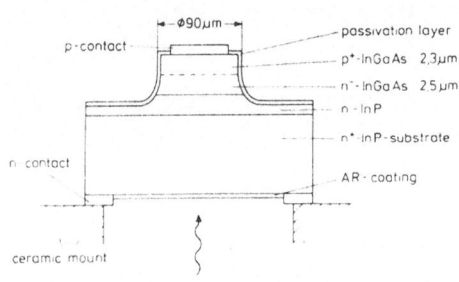

Bild 6: InGaAs/InP-PIN-Photodiode

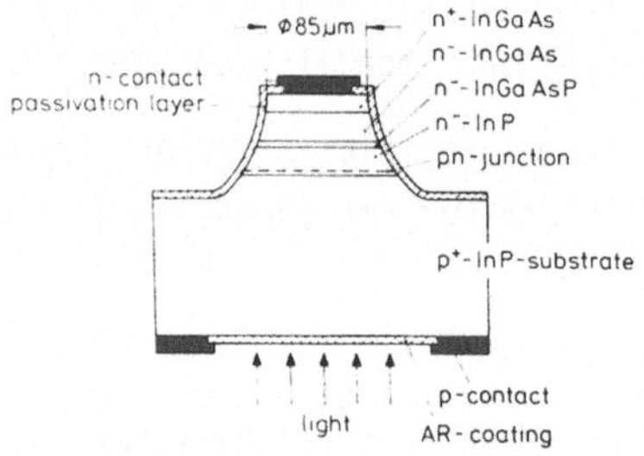

Bild 7: InGaAs/InP-SAGM-Lawinenphotodiode (APD)

	Oberflächen-emitter	Kanten
gekoppelte Leistung P (Gradientenfaser; 100 mA)	20 – 40 μW	40 – 70 μW
Spektralbreite	100 – 150 nm	80 – 100 nm
Anstiegs/Abfallzeit	~5 ns	~2 ns
Modulation	< 100 Mbit/s	> 140 Mbit/s

Tabelle 1: Daten für InGaAsP/InP-LEDs

PIN-Photodiode

Dunkelstrom I_D	bei –10 V	1 – 10 nA
Kapazität C	bei –10 V	einige 0,1 pF
Empfindlichkeit S		0,7 – 0,9 A/W
Demodulation		Pulse von 100 ps

Lawinenphotodiode (APD)

Durchbruchspannung U_D	– (50 – 150) V
Dunkelstrom bei Verstärkung 10	< 10 nA
Kapazität C	einige 0,1 pF
Empfindlichkeit S	0,7 – 0,9 A/W
Verstärkung M	> 100
Verstärkung Bandbreite	~ 30 GHz

Tabelle 2: Daten für InGaAs/InP-Photodioden

Lichtwellenleiter

H. P. Huber
AEG Forschungsinstitut Ulm
Postfach 1730, D - 7900 Ulm

Die Vorteile der optischen Nachrichtentechnik basieren weitgehend auf den Eigenschaften des Übertragungsmediums Lichtwellenleiter. Gegenüber der Kupferkabeltechnik sind insbesondere die wesentlich höhere Übertragungsbandbreite und die niedrigeren Verluste zu nennen. Dadurch werden größere Repeaterabstände bei der Fernübertragung und höhere Übertragungsraten möglich. Weitere Vorteile sind das geringere Gewicht und Volumen eines LWL-Kabels, die Unempfindlichkeit gegenüber elektromagnetischen Störfeldern sowie die erwarteten geringeren Kosten.

Übertragungsbandbreite

Die Übertragungsbandbreite von Lichtwellenleitern wird durch ein zeitliches Auseinanderlaufen eines Lichtimpulses bei der Übertragung begrenzt. Diese Pulsverbreiterung ist für verschiedene Fasertypen sehr unterschiedlich. In Multimode-Stufenprofilfasern kommt die Pulsverbreiterung in erster Linie durch Laufzeitunterschiede einzelner Moden zustande und beträgt für typische Fasern um 14 ns/km. Die Übertragungsbandbreite ist dann auf ungefähr 20 MHz km begrenzt. In Multimode-Gradientenprofilfasern werden die Modenlaufzeitunterschiede durch ein nahezu parabolisches Brechzahlprofil im Faserkern in erster Näherung ausgeglichen. Theoretisch kann die Modendispersion dadurch auf rund 0,02 ns/km verringert werden. Neben der Modendispersion ist in Gradientenprofilfasern die Materialdispersion zu berücksichtigen. Sie beträgt in Quarzglasfasern bei 850 nm Wellenlänge ungefähr 0,1 ns/km nm und nimmt zu größeren Wellenlängen hin ab, wird bei 1,3 µm zu Null und nimmt dann wieder zu. Bei 1,3 µm Wellenlänge ist daher mit spektral schmalen Sendern theoretisch eine nur durch Modendispersion begrenzte Bandbreite um 10 GHz km möglich. In der Praxis werden allerdings wegen nicht zu vermeidender Profilfehler nur Werte um 1 bis 3 GHz km erreicht.

Noch höhere Übertragungsbandbreiten sind mit Einmodenfasern möglich. Hier wird die Pulsverbreiterung durch die Summe aus der Material- und der bei Multimodefasern unbedeutenden Wellenleiterdispersion bestimmt. Ihr Verlauf als Funktion der Wellenlänge ist in Bild 1 dargestellt. Bei normalen Einmodenfasern ist die Gesamtdispersion bei einer Wellenlänge um 1,3 µm Null. Mit spektral schmalen Lichtquellen (Halbleiterlaser, $\Delta \lambda \approx 4$ nm) beträgt die Bandbreite mehr als 1000 GHz km. In neueren Faserdesigns mit speziellen Brechzahlprofilen (Bild 2) wird durch Verändern der

Wellenleiterdispersion die Wellenlänge verschwindender Gesamtdispersion bis in den Wellenlängenbereich um 1,55 µm verschoben, um neben der hohen Bandbreite das dort liegende Dämpfungsminimum in Quarzglasfasern nutzen zu können /1/. Durch mehrfach gestufte Kern- oder Mantelbrechzahlprofile wird auch versucht, die Gesamtdispersion in einem größeren Wellenlängenbereich von 1,3 µm bis 1,6 µm gering zu halten /2/. Damit wäre eine hochbitratige Übertragung über einen größeren Spektralbereich auch mit nicht extrem schmalbandigen Sendern möglich.

Verluste

Für die optische Nachrichtenübertragung werden derzeit nahezu ausschließlich schwach dotierte Quarzglasfasern verwendet. Im Spektralbereich von 0,7 µm bis 1,7µm sind die Verluste heutiger Fasern hauptsächlich durch die Rayleighstreuung, die mit λ^{-4} zu größeren Wellenlängen hin abnimmt, begrenzt. Die Eigenabsorption ist in diesem Wellenlängenbereich vernachlässigbar klein, und eine nennenswerte Fremdabsorption durch Verunreinigungen ist nur in Form von zwei schwer vermeidbaren OH Absorptionsbanden bei 1,39 µm und 1,24 µm vorhanden. Bild 3 zeigt als Beispiel die spektralen Verluste einer Einmodenfaser. Bestwerte liegen um 0,35 dB/km für 1,3 µm Wellenlänge und um 0,16 dB/km für 1,56 µm. Neben den Materialverlusten können noch Wellenleiterverluste, die durch Krümmungen und Mikrokrümmungen verursacht werden, auftreten.

Lichtwellenleiterherstellung

Für die Herstellung dämpfungsarmer Quarzglasfasern sind in den letzten Jahren mehrere verschiedene Verfahren bis zur Fertigungsreife entwickelt worden. Allen liegt die Erzeugung von Quarzglas höchster Reinheit durch Reaktion von Siliziumtetrachlorid und Sauerstoff aus der Gasphase zugrunde. Zur Brechzahländerung werden geringe Dotierstoffmengen wie Germaniumoxid, Phosphoroxid oder Fluor benutzt, die bereits in die Gasphasenreaktion als Halogenide eingebracht werden. Durch Abscheidung von dotiertem und undotiertem Quarzglas auf einen Träger entsteht ein 'Vorform' genannter Glasstab, aus dem schließlich die Faser gezogen wird. Die wichtigsten Verfahren zur Vorformherstellung sind in Bild 4 dargestellt. Beim MCVD-Verfahren (modified chemical vapor deposition) werden die Reaktionsgase in ein Quarzglasrohr eingeleitet und durch Erhitzen von außen zur Reaktion gebracht /3/. Die Abscheidung erfolgt schichtweise auf der Innenwand, anschließend wird das Vorformrohr zum Stab kollabiert. Das Quarzglasrohr wird Bestandteil der Vorform und bildet in der Faser den äußeren Mantel. Das PCVD-Verfahren (plasma activated chemical vapor deposition) ist ebenfalls ein Innenbeschichtungsverfahren, bei dem die Reaktion jedoch in einem Niederdruckplasma stattfindet /4/. Beim OVPO /5/ (outside vapor phase oxidation) erfolgt die Reaktion in einer Knallgasflamme und das entstehende Quarzglaspulver wird auf einem Trägerstab niedergeschlagen. Danach wird der Träger entfernt, der Sinterkörper zunächst in einem Chlorgasstrom getrocknet und anschließend zur Vorform verglast und kollabiert. Das VAD-Verfahren (vapor axial deposition) arbeitet ebenfalls mit einer Flammenreaktion /6/. Die Abscheidung wird an einem Hilfsstab begonnen und schreitet dann in axialer Richtung des entstehenden Vorformstabes fort. Die Weiterverarbeitung entspricht weitgehend dem OVPO-Verfahren. Alle Verfahren sind sehr leistungsfähig mit gewissen Vor- und Nachteilen und stehen derzeit miteinander in Konkurrenz. Die Abscheideraten liegen zwischen 1 und 5 g/min. Durch Übermanteln der Vorformen mit Quarzglasrohren als äußerem Fasermantel werden Faserfertigungsraten von mehr als 5 km/Std erreicht. Einzelne Vorformgrößen für Faserlängen von über 100 km sind realisiert worden, übliche Größen liegen zwischen 15 und 30 km.

Für die Verkabelung und Verlegung müssen die Lichtwellenleiter eine ausreichende, langzeitstabile Festigkeit haben. Fehlerfreie Quarzglasfasern haben Festigkeitswerte von über 5500 N/mm². Um die hohe Festigkeit zu erhalten, werden die Fasern beim Ziehprozeß mit einem Schutzlack für die Oberfläche überzogen. Weiter muß die Kabelkonstruktion für den Feldeinsatz eine möglichst spannungsfreie Verlegung gewährleisten sowie Zusatzverluste durch Mikrokrümmungen, die durch Druck auf die Faser von außen oder durch temperaturbedingte Längenänderungen hervorgerufen werden könnten, vermeiden.

Literatur
/1/ V.A. BHAGAVATULA, M.S. SPOTE AND W.F. LOVE: Optics Lett. 9, 186-188 (May 1984)
/2/ L.G. COHEN, W.L. MAMMEL and S.J.JONG, Electr. Lett. 18, 1022-1023 (Nov. 1982)
/3/ S.R. NAGEL, J.B. MAC CHESNEY, and K.L. WALKER, IEEE J. Quant.El., vol. QE-18, no. 4, 459 - 476 (April 1982)
/4/ P. GEITTNER, D. KÜPPERS, and H. LYDTIN, Appl. Phys. Lett., vol. 28, no. 11, 645 - 646 (June 1976)
/5/ M.C. BLANKENSHIP and CH. DENEKA, IEEE J. Quant. El., vol. QE-18, no. 10, 1418-1423 (Oct. 1982)

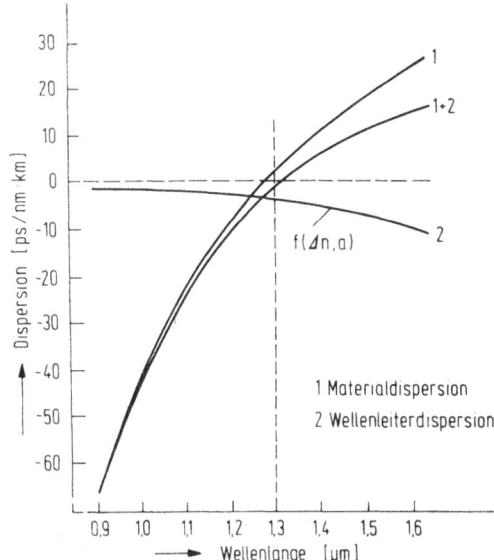

Bild 1.

Materialdispersion (1), Wellenleiterdispersion (2) und Gesamtdispersion einer Einmodenfaser

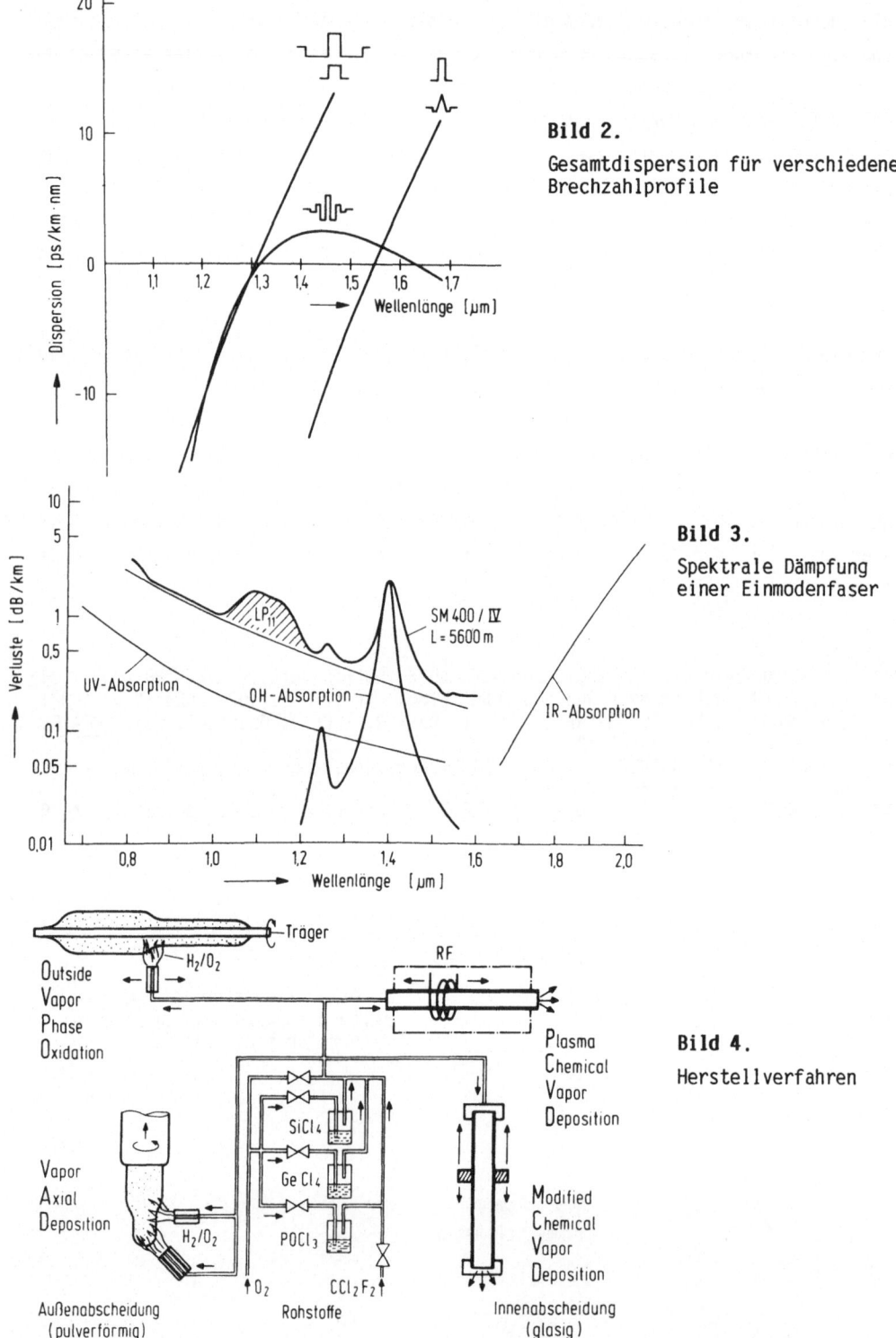

Bild 2.

Gesamtdispersion für verschiedene Brechzahlprofile

Bild 3.

Spektrale Dämpfung einer Einmodenfaser

Bild 4.

Herstellverfahren

Wellenlängenmultiplextechnik

E. Weidel

AEG, Forschungsinstitut Ulm

Sedanstr. 10, D-7900 Ulm

Einleitung

Glasfaser-Lichtwellenleiter weisen in einem weiten Spektralbereich hervorragende
Übertragungseigenschaften auf: Die Dämpfung ist für Wellenlängen zwischen 1,2 µm
und 1,6 µm kleiner als 0,5 dB/km und auch das Basisband ist über diesen weiten Spektralbereich bei richtig geformtem Brechzahlprofil groß. Dies gilt sowohl für Gradientenprofilfasern als auch für Monomodefasern. Da Sende- und Empfangselemente zwischen
1,1 µm und 1,6 µm prinzipiell zur Verfügung stehen, erscheint die gleichzeitige Übertragung einer Vielzahl von Kanälen mit unterschiedlichen Wellenlängen über eine Faser
attraktiv - die Wellenlängenmultiplextechnik.

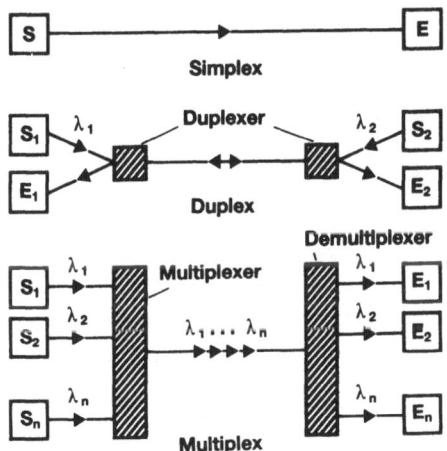

Bild 1. Betriebsarten

Bild 1 zeigt drei Übertragungssysteme. Neben der einfachen Punkt-zu-Punkt-Verbindung
(Simplexbetrieb) sind zwei Multiplexsysteme skizziert, der Duplex- und der Wellen-
längen-Multiplexbetrieb. Beim Duplexbetrieb werden Informationen in beiden Richtungen
auf einer Glasfaser mit Hilfe zweier unterschiedlicher Lichwellenlängen übertragen.
Diese sogenannte bidirektionale Übertragung schafft die Voraussetzungen für den optischen Teilnehmeranschluß. Beim Multiplexbetrieb hingegen erfolgt die Übertragung
nur in einer Richtung. Die Übertragungskapazität der Leitung wird dabei um die Zahl
der Lichtwellenlängen vervielfacht.
Zum Aufbau derartiger Systeme werden Koppler benötigt: Duplexer, Multiplexer und Demultiplexer. Aus der großen Anzahl bereits realisierter Koppler /1, 2/ werden einige
technisch besonders interessante vorgestellt:

- Koppler mit Interferenzfiltern, in faseroptischer oder mikrooptischer Technik
- Wellenlängenabhängige Richtkoppler
- Koppler mit Beugungsgittern.

Koppler mit Interferenzfilter

Für Koppler mit Interferenzfiltern (Bild 2 bis 4) werden Kanten- oder Bandfilter verwendet, die so aufgebaut sind, daß sie für einen Wellenlängenbereich möglichst transparent und für einen zweiten hoch reflektierend sind. Die räumliche Trennung des reflektierten vom einfallenden Licht wird durch schräges Auftreffen des Lichtbündels auf die Filterschicht erreicht. Mit einem Filter können somit zwei Wellenlängenbereiche getrennt (oder zusammengeführt) werden.

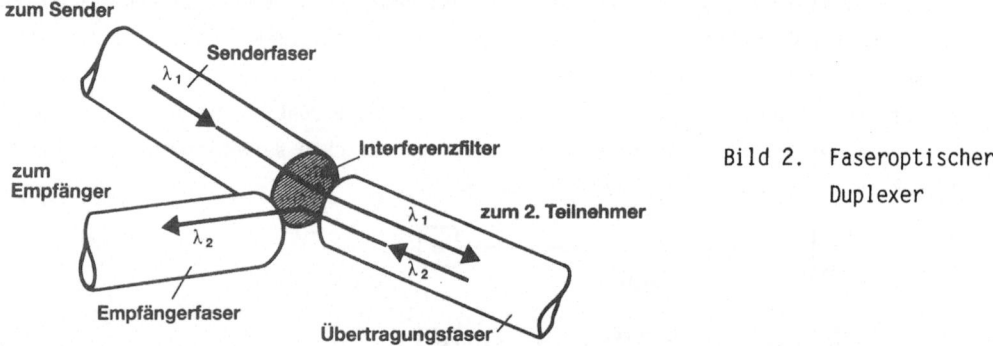

Bild 2. Faseroptischer Duplexer

Bild 2 zeigt einen faseroptisch aufgebauten Koppler. Sein Kernstück ist eine Anordnung von drei Fasern unter bestimmten Winkeln in einer Ebene. Eine Faser stellt die Verbindung zur Übertragungsleitung her (Übertragungsfaser), eine zweite führt zum Sender (Senderfaser) und eine dritte zum Empfänger (Empfängerfaser). Das Interferenzfilter zur Trennung von λ_1 und λ_2 ist auf dem schrägen, planpolierten Ende der Senderfaser aufgebracht. Um die Verluste durch Aufweitung des Lichtbündels zwischen den Fasern zu vermindern, sind die Enden der Empfängerfaser und der Übertragungsfaser rund geschmolzen.

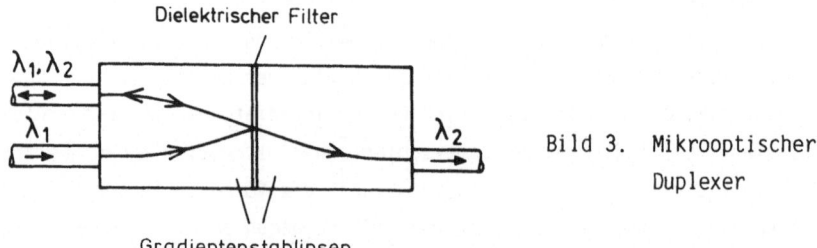

Bild 3. Mikrooptischer Duplexer

Beim mikrooptischen Aufbau nach Bild 3 werden Gradientenstablinsen benutzt, um das divergente Lichtbündel aus der Faser in ein paralleles Bündel zu transformieren. Nach Transmission oder Reflexion am Filter wird das Lichtbündel auf die Endflächen der Ausgangsfasern fokussiert. Der mikrooptische Koppler ist aufwendiger in der Herstellung verglichen mit dem faseroptischen; sein Vorteil liegt darin, daß das Licht parallel und nur wenige Grad geneigt zur Normalen auf das Filter auftrifft und deshalb steilere Filterkanten realisiert werden können.

Für die Koppler nach Bild 2 und 3 können Verluste von 1 bis 2 dB und eine Übersprechdämpfung von 40-60 dB erreicht werden.

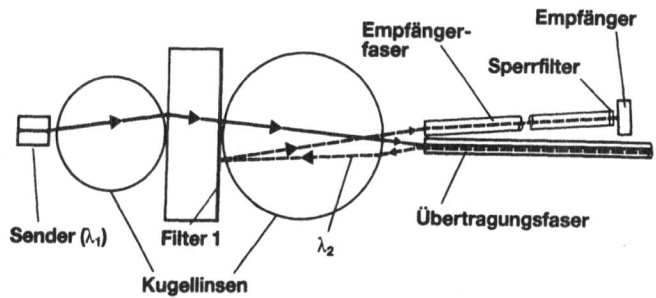

Bild 4. Duplexer mit hybrid integrierten aktiven Komponenten

Bild 4 zeigt ein weiterführendes Konzept, den "Duplexer mit hybrid integrierten aktiven Komponenten" /3/. Hier sind die Interferenzfilter (Filter 1 und Sperrfilter) in einem Gehäuse mit der Sendediode und der Empfangsdiode untergebracht.

Die Ausgangsstrahlung mit der Wellenlänge λ_1 des optischen Senders (Laserdiode, Kantenemitter-LED) wird dabei über zwei Kugellinsen in die Übertragungsfaser eingekoppelt. Als Übertragungsfaser wurde bisher eine Gradientenfaser mit 50 μm Kern- und 125 μm Außendurchmesser verwendet; der Aufbau ist jedoch auch für eine Monomodefaser geeignet. Das in dieser Faser vom fernen Teilnehmer ankommende Lichtsignal mit der Wellenlänge λ_2 trifft über die zweite Kugellinse auf das Interferenzfilter des Glasplättchens zwischen den beiden Kugellinsen, wird dort reflektiert und gelangt über ein kurzes Faserstück, das als Blende für Streulicht wirkt, auf den Detektor. Das komplette Bauelement ist sehr klein und weist nur einen optischen sowie die elektrischen Anschlüsse für Sendediode und Detektor auf. Der Koppelwirkungsgrad vom Sender in die Übertragungsfaser ist abhängig von der Abstrahlcharakteristik der Sendediode: Für Laserdioden wurden Werte zwischen 50 % und 60 % gemessen, für LEDs zwischen 10 % und 20 %. Diese Werte sind vergleichbar mit konventionellen Laser-Faser-Kopplungen. Einfügungsverluste treten bei diesem Duplexer also nur zwischen Übertragungsfaser und Detektor auf. Sie sind im wesentlichen durch die Filterverluste gegeben und liegen für die verwendeten Filter bei -2 dB. Die Übersprechdämpfung beträgt für Laserdioden -40 dB und für LEDs -37 dB bezogen auf die Leistung in der Übertragungsfaser.

Wellenlängenabhängiger Richtkoppler

Für Monomodefasern kann zur Trennung von Wellenlängen auch ein wellenlängenabhängiger Richtkoppler verwendet werden. Führt man zwei Wellenleiter parallel in geringem Abstand zueinander, so findet eine vollständige Überkopplung des Lichts aus einem Wellenleiter in den anderen statt, wenn die Ausbreitungskonstanten gleich sind. Die Koppellänge für vollständige Kopplung ist abhängig von der Wellenlänge; bei richtiger Wahl von Abstand und Länge können zwei verschiedene Wellenlängen getrennt (oder zusammengeführt) werden. Verluste von 0,1 dB wurden erreicht.

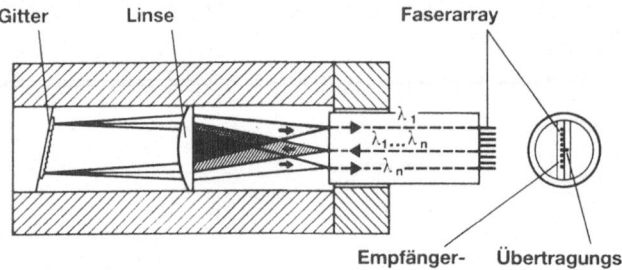

Bild 5. Gitterdemultiplexer

Koppler mit Beugungsgitter

Bei Multiplexern und Demultiplexern für viele Kanäle sind Koppler mit Beugungsgittern optimal, da hier die Trennung einer großen Anzahl von Kanälen in einem Schritt vorgenommen wird. Bild 5 zeigt die optische Anordnung für einen Demultiplexer. Das in der Übertragungsfaser ankommende Licht wird durch die Linse parallel gerichtet. Das Gitter erzeugt eine wellenlängenabhängige Winkelablenkung, die durch die Linse in eine räumliche Trennung umgesetzt wird. Kehrt man die Richtung des Lichts um, erreicht man statt der Trennung (Demultiplexer) die Zusammenführung (Multiplexer) unterschiedlicher Wellenlängen.

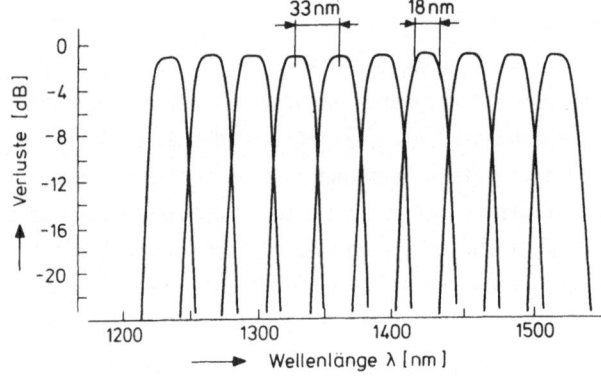

Bild 6. Verluste eines
Demultiplexers

Die Verluste eines Demultiplexers mit 10 Kanälen sind in Bild 6 dargestellt. Bei einem Kanalabstand von 33 nm sind die minimalen Verluste 1 dB für alle Kanäle.

Systemversuche

Eine Anzahl von Systemversuchen wurde mit λ-selektiven Kopplern durchgeführt; Beispiele sind in /4-8/ angegeben. Bei Bitraten von 6,8 Mbit/sec bis 2 Gbit/sec wurden Streckenlängen von 6,5 bis 68 km realisiert.

Literatur
(1) H. ISHIO, J. MINOWA, AND K. NOSU: J. Lightw. Techn. 2, 448 (1984)
(2) G. WINZER: Proc. 8th ECOC, 305 (1982)
(3) B. HILLERICH, M. RODE, E. WEIDEL: Proc. 10th ECOC, 166 (1984)
(4) W. BAMBACH, W. SCHMIDT, H. G. ZIELINSKI: Proc. 10th ECOC, 266 (1984)
(5) T. KANADA, Y. OKANO, K. AOYAMA, I. KITAMI: IEEE Trans. Com. 31, 1095
(6) J. D. SPALINK, R. J. S. BATES, S. J. BUTTERFIELD: Proc. OFC, post-deadline p.,1('83)
(7) H. TRIMMEL, H. F. MAHLEIN, A. REICHELT, B. STUMMER: Proc. 10th ECOC, 262 (1984)
(8) N. A. OLSSON et al.: Proc. OFC '85, 88 (1985)

Lokale Netze mit Lichtwellenleitern

R. Portscht
AEG-Telefunken, Forschungsinstitut
Sedanstr. 10, D - 7900 Ulm

"Ein Lokales Netz (LAN, Local Area Network) ist ein Netz für bitserielle Übertragung
von Information zwischen untereinander verbundenen unabhängigen Geräten. Es unterliegt
vollständig der Zuständigkeit des Anwenders und ist auf dessen Grundstück beschränkt"
/1/. Unter "Geräten" sind beispielsweise Rechner, Terminals, Massenspeicher, Drucker,
Fernkopierer, Telex-, Teletexmaschinen, Telefonapparate, Monitor- und Kontrollgeräte
und Gateways zu öffentlichen oder anderen Lokalen Netzen zu verstehen. Einzelne oder
mehrere Geräte können mit einer Netzstation verbunden werden, die ihrerseits für den
Zugang zum Übertragungsmedium sorgt. Die Netzstationen können wenige Meter bis zu
wenigen Kilometern voneinander entfernt sein, typische Datenübertragungsraten realisierter kommerzieller Lokaler Netze liegen zwischen 100 kbit/s und 20 Mbit/s.
Im Gegensatz zu den zentral gesteuerten privaten Nebenstellenanlagen sind Lokale Netze
dezentral organisiert und sind dadurch für einen schnellen burstartig auftretenden In-
formationsaustausch zwischen wechselnden Partnern besonders gut geeignet. Für diese
Anwendung als Datenpaketvermittlung sind Nebenstellenanlagen bisher nicht ausgelegt.

Während Nebenstellenanlagen eine sternförmige Topologie besitzen, gibt es bei Lokalen
Netzen eine Vielfalt von Topologien: Neben einfachen Punkt-zu-Punkt-Verbindungen und
maschenartigen Strukturen sind vorallem Bus-, Baum-, Ring-, Schleifen- und Doppelring-
strukturen verbreitet. Zudem ist eine große Zahl von Steuerungs- und Zugriffsverfahren
bekannt. Aber nur wenige Kombinationen von Übertragungsmedium, Zugriffsverfahren und
Topologie haben sich als sinnvoll erwiesen und in der Praxis durchgesetzt.

Die zur Zeit am meisten verbreiteten Lokalen Netze gehören zur Familie des Ethernet
/2/, das eine Bus- oder Baumstruktur mit Koaxialkabeln als Übertragungsmedium besitzt. Als Zugriffsverfahren dient CSMA/CD (Carrier Sense Multiple Access with Collision Detection), ein Vielfachzugriffsverfahren aus der Gruppe der Wettbewerbsver-
fahren mit Trägerüberwachung und Kollisionsentdeckung.
Ein zweites bedeutendes Steuerungsverfahren ist der Tokenzugriff, der zur Gruppe der
deterministischen Reservierungsverfahren gehört und sowohl bei Ring- als auch bei Bussystemen angewendet wird /3,4/. Im Token-Ring wird ein Senderecht (der Token) von
Station zu Station weitergereicht. Eine Station kann nur senden, wenn sie das Senderecht erhält, alle anderen aktiven Stationen sind empfangsbereit und können an sie
adressierte Datenblöcke empfangen.

Neben den bisher betrachteten Lokalen Netzen mit Kupferleitern (Koaxialkabeln oder verdrillten bzw. abgeschirmten Kupferleitern) gewinnen in jüngster Zeit immer mehr Lo kale Netze mit Lichtwellenleitern an Bedeutung /5 bis 8/. Die Lichtwellenleiter-Übertragungstechnik hat sich bisher in der Weitverkehrstechnik ("Ferntrasse") und bei Rechner-Rechner-Kommunikation bewährt, wo Daten mit hoher Geschwindigkeit und über große Entfernungen zu übertragen sind. Die Eigenschaften der Glasfasern

* geringe, praktisch frequenzunabhängige Dämpfung und
* großes Bandbreite-Entfernungs-Produkt

werden aber bisher in Lokalen Netzen noch nicht genutzt, weil keine großen Entfernungen und in der Regel keine hohen Datenströme auftreten. Vielmehr wird Glasfasern auf grund der Eigenschaften

* Unempfindlichkeit gegen elektromagnetische Störungen,
* geringes Gewicht, kleine Dimensionen und
* keine Abstrahlung elektromagnetischer Schwingungen, d.h. kein Nebensprechen und hohe Abhörsicherheit

in Lokalen Netzen der Vorzug gegenüber Kupferleitern gegeben. Als typische Anwendungeı seien Prozeßsteuerungen und Datenerfassungssysteme in Industrieanlagen, Bürokommunikation, Kommunikation in Flugzeugen, Fahrzeugen und Schiffen und in militärischen Geräten und Anlagen genannt.

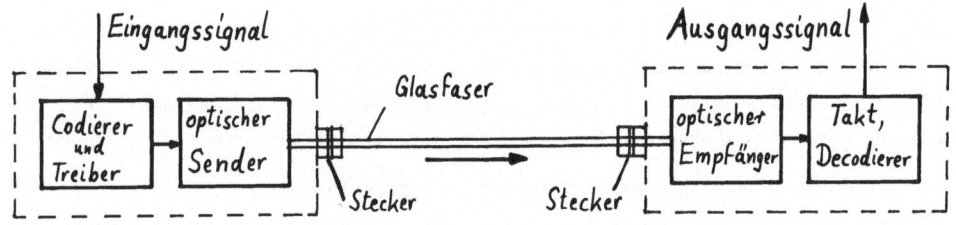

Abb. 1. Gerichtete Punkt-zu-Punkt-Übertragung

Die einfachste Art der Übertragung von Information über Lichtwellenleiter ist eine gerichtete Punkt-zu-Punkt-Übertragung. In Abb. 1 sind die erforderlichen Komponenten eingezeichnet. Das zu übertragende Eingangssignal wird codiert, elektrisch verstärkt und in einem optischen Sender (OS) in Lichtsignale umgesetzt. Die Lichtsignale werden über eine Glasfaser zu einem optischen Empfänger übertragen, es wird der Takt zurückgewonnen und das Ausgangssignal decodiert. Wegen der nicht allzu hohen Anforderungen an das Bandbreite-Entfernungs-Produkt reicht es aus, als Lichtwellenleiter Multimodefasern zu verwenden. Als optische Sender kommen LEDs, LASER- Dioden oder LASER und als Empfänger PIN-Fotodioden bei den Wellenlängen λ= 0,8 μm oder λ= 1,3 μm in Betracht. Um eine flexible Installation der Systeme zu erreichen, ist es empfehlenswert, Steckverbindungen anstelle von Spleißverbindungen vorzusehen. Im genannten Wellenlängenbereich sind erprobte und zuverlässige optische Komponenten (wie Sende- und Empfangsdioden, Glasfasern und Stecker) auf dem Markt erhältlich, mit denen sich einfache Lichtwellenleiter-Übertragungssysteme entsprechend Abb.1 aufbauen lassen /6/.

Bei Anwendung der Lichtwellenleiter-Übertragungstechnik in Lokalen Netzen tritt jedoch
zusätzlich das Problem der Aus- und Einkopplung des Lichts in den Netzanschlußpunkten
auf. Bei rein passiver Auskopplung (s. Abb.2) wird die Lichtleistung in die Richtungen
1 und 2 aufgeteilt, was zu merklichen oder gar beträchtlichen Verlusten in Hauptüber-
tragungsrichtung (Richtung 1 in Abb.2) führt. Man kann leicht abschätzen, daß durch
die Verluste in den Steckern und an den Koppelstellen nach ca. 10 Anschlußpunkten die
Lichtleistung bis in den Bereich des Rauschens der Empfangsdioden gedämpft ist.
Obwohl es verschiedene Lösungen für T-Koppler gibt (Stirnflächen-, Oberflächen-, V-
Nut-Koppler /9/, T-Koppler mit Linsen und Strahlteilern und Taper-Koppler /7/), wird
man - von Sonderfällen abgesehen - in Lokalen Netzen nicht auf aktive Koppler verzichten können. Ein Beispiel für einen aktiven Koppler mit gerichteter Übertragung (bei-
spielsweise in einem Ringnetz) zeigt Abb.3. Die empfangene Lichtleistung wird im opti-
schen Empfänger (OE) in ein elektrisches Signal gewandelt, verstärkt, decodiert und
zum Empfänger (R = Receiver) ausgekoppelt. Die Daten können verarbeitet oder über den
Sender (T = Transceiver) wieder ins Netz eingespeist werden (OS = Optischer Sender).
Zusätzlich ist ein Ringschalter RS in Abb. 3 eingezeichnet, der bei Tokenzugriff in
Ringnetzen vorteilhaft ist und über den Anschluß A angesteuert wird /3, 4/.
Weiterhin ist in Abb.3 ein optischer Bypass eingezeichnet, der mit oder ohne optisches
Relais (OR) betrieben werden kann. Durch diesen Bypass kann erreicht werden, daß bei
Abschalten oder Ausfall der Stromversorgung das Ringnetz funktionsfähig bleibt.

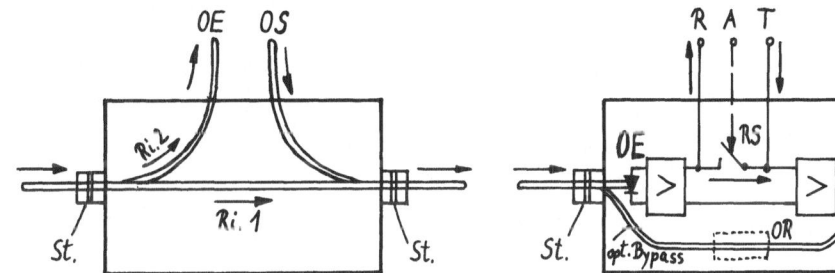

Abb. 2. Passive Ankopplung   Abb. 3. Aktive Ankopplung
     bei unidirektionaler Übertragung   bei unidirektionaler Übertragung

Noch komplizierter und aufwendiger als in Ringsystemen ist das Ein- und Auskoppeln
der Lichtleistung an den Netzanschlußpunkten in Bussystemen, weil hier eine bidirektionale Übertragung erforderlich ist. Ein Beispiel für eine passive Ankopplung ist in
Abb.4 dargestellt (durchgezogene Pfeile: Einkopplung der Lichtleistung von der Geräte-
seite (OS); gestrichelte Pfeile: Auskopplung von Lichtleistung aus Übertragungsrichtung
1, strichpunktierte Pfeile: Auskopplung von Lichtleistung aus Übertragungsrichtung 2).
Auch hier ist die Zahl der passiven Koppelstellen (wie bei Abb.2) auf wenige Anschlußpunkte beschränkt.

540

Zwei Beispiele für eine aktive Ankopplung bei bidirektionaler Übertragung sind in
Abb.5 (mit 2 Glasfasern) und Abb.6 (mit Wellenlängenduplex) dargestellt. Wenn man diesen beträchtlichen Aufwand betreibt, erscheint es vorteilhafter, anstelle einer Bus-
eine Doppelringstruktur zu verwenden, wobei es möglich ist, ein fehlertolerantes System zu realisieren /10/.

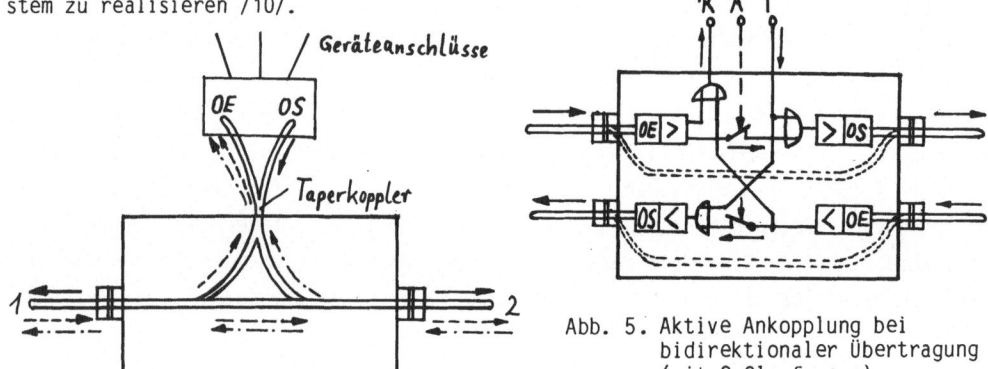

Abb. 5. Aktive Ankopplung bei
bidirektionaler Übertragung
(mit 2 Glasfasern)

Abb. 4. Passive Ankopplung
bei bidirektionaler ·Übertragung

Abb. 6. Aktive Ankopplung bei
bidirektionaler Übertragung
(Wellenlängenduplex)

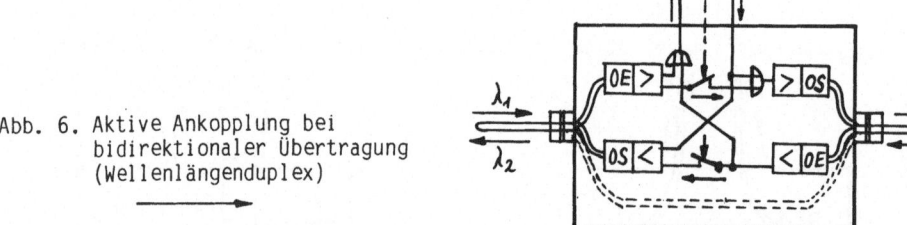

Es gibt auch Realisierungsbeispiele mit sternförmiger Topologie (mit passivem oder ak-
tivem Sternkoppler) /5/. Für den passiven Sternkoppler gelten ähnliche Einschränkungen
in bezug auf die Zahl der Anschlußpunkte wie bei passiven Ring- und Bussystemen (Abb.2
und Abb.4). Bei aktiven Sternkopplern ist ein zentralisiertes Zugriffsverfahren dem
sonst in lokalen Netzen üblichen dezentralisierten Zugriffsverfahren vorzuziehen.

Aufgrund dieser Ausführungen ist es klar, daß eine einfache Ringstruktur mit Tokenzugriff für Lokale Netze mit Lichtwellenleitern die naheliegendste und wirtschaftlichste
Lösung darstellt.

Literatur
(1) I. FROMM: NTZ 35, 634 (1982)
(2) R.M. METCALFE, D.R. BOGGS: Comm. ACM 19.7, 395 (1976)
(3) R. PORTSCHT, J. SWOBODA: NTG-Fachberichte 88, 254 (1985)
(4) W.BUX u.a.: IEEE J. SAC 1.5, 756 (1983)
(5) M.R. FINLEY: IEEE Comm. Magaz. 22.8, 22 (1984)
(6) R. MACIEJKO: Telesis 2, 15 (1984)
(7) H. WITTE: Elektronik 33.26, 80 (1984)
(8) R. ALLAN: El. Design 31, 97 (23.6.1983)
(9) O. KRUMPHOLZ: Elektronik 61.20, 14 (1979)
(10) W. SCHRÖCK: Informatik-Fachberichte 84, 265 (1984)

Technologie der Integrierten Optik

J. Krauser, N. Grote

Heinrich-Hertz-Institut für Nachrichtentechnik Berlin GmbH, Einsteinufer 37,
D-1000 Berlin 10, FRG

1. Zielsetzung der Integrierten Optik

Der Einsatz der optischen Nachrichtentechnik im Teilnehmerbereich zukünftiger öffentlicher Breitbandkommunikationssysteme wird durch die hohen Kosten der optischen
Sender/Empfänger erschwert werden. Abb. 1 zeigt ein Beispiel eines optischen Sen-
ders/Empfängers /1/:

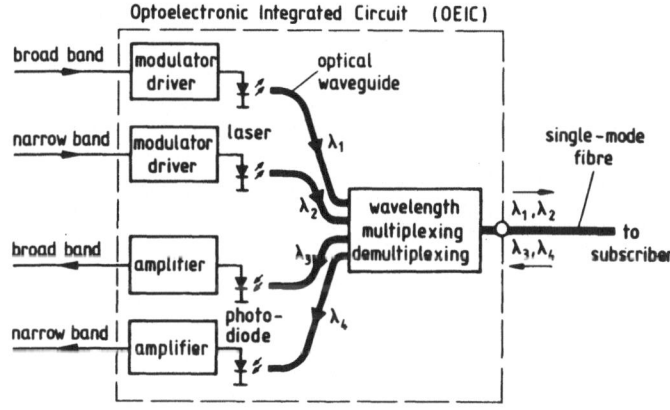

Abb. 1.   Optischer Sende/Empfänger-Modul

Über die Wellenlängen λ_1 und λ_2 werden die Breitband- bzw. Schmalbanddienste von
der Ortsvermittlungsstelle zum Teilnehmer hin und über die Wellenlängen λ_3 und
λ_4 in umgekehrter Richtung übertragen. Derartige optische Sender/Empfänger bestehen heute aus diskreten Komponenten; eine kostengünstige Massenproduktion - und
das ist gerade für den Teilnehmerbereich unerläßlich - ist in dieser Bauweise nicht
möglich. Das vordringliche Ziel der Integrierten Optik ist somit die Entwicklung
von optischen Sendern/Empfängern als monolithisch integrierte Schaltungen, die
- vergleichbar den Chips der Mikroelektronik - in großer Stückzahl, kostengünstig
produziert werden können. Derartige Opto-Electronic Integrated Circuits (OEIC) /2/
sind Schlüsselkomponenten der zukünftigen Kommunikationstechnik. Die Vorteile der

monolithisch integrierten Optik gegenüber einer Hybridtechnik, bei der mehrere Substratmaterialien verwendet werden, sind:

1. Hohe opto-mechanische Stabilität

2. Reduzierung von optischen Reflexionen

3. Aufwendige Justierungen werden durch hochauflösende lithographische Verfahren ersetzt.

4. Kleine Abmessungen

5. Wirtschaftliche Bedeutung wegen der zu erwartenden Massenproduktion

2. Funktionselemente der Integrierten Optik

Ein OEIC besteht aus elektro-optischen, opto-elektronischen, optischen und elektronischen Komponenten (Tab. 1).

Bauelementklasse	Funktionselemente	Verbindungen	Verkopplung
optische Komponenten	Wellenleiter, Spiegel, Y-Abzweige, Polarisatoren, Filter	Wellenleiter	optisch
opto-elektronische, elektro-optische Komponenten	Laser, opt. Verstärker, Photodetektor, Phasenmodulator, -Richtkopplermodulator, Interferometermodulator, opt. Schalter, Modenkonverter, opt. Isolator	1. Wellenleiter 2. metallische Leiterbahnen	1. optisch 2. elektrisch
elektronische Komponenten	Widerstand, Kapazität Transistor	metallische Leiterbahnen	elektrisch

Tab. 1. Zusammenstellung der Bauelementklassen, Art der Verbindungen und Verkopplungen

Die Verbindungen der Bauelemente untereinander werden durch elektrische Leiterbahnen und optische Wellenleiter hergestellt. Die Verkopplung zwischen Verbindung und Bauelement ist entweder elektrisch oder optisch. Eines der schwierigsten Probleme der integrierten Optik ist die optisch-optische Verkopplung verschiedener Bauelemente unterschiedlicher Materialzusammensetzungen, da hierbei Brechungsindexsprünge auftreten können. Um derartige Reflexionen zu vermeiden, muß ein langsamer Übergang (Taper) zwischen Gebieten mit unterschiedlichen Brechungsindices hergestellt werden.

Allgemein lassen sich die Funktionselemente der optischen Integration in folgende
Gruppen einteilen:

Funktion	physikalische Eigenschaften
1. Licht erzeugen, verstärken	Lumineszenz
2. Licht führen	Transparenz, Totalreflexion
Licht schalten/modulieren	elektro-optische Effekte
3. Licht detektieren	Absorption

Für jede dieser Gruppen muß eine andere Materialzusammensetzung gewählt werden.

3. Materialsysteme

Für die Auswahl des Materialsystems für die integrierte Optik ist der gewünschte
Wellenlängenbereich entscheidend. In Abb. 2 wird eine Übersicht über die geeigneten
Materialsysteme gegeben. Für die binären Verbindungen (Eckpunkte in Abb. 3) sind
Bandabstand und Gitterkonstante von der Natur vorgegebene Größen, während für qua-
ternäre Verbindungen Bandabstand und Gitterkonstante unabhängig voneinander gewählt
werden können. Aus praktischen Erwägungen muß für die erforderlichen Schichten ein
Trägermaterial gewählt werden, das in genügender Menge und hinreichender

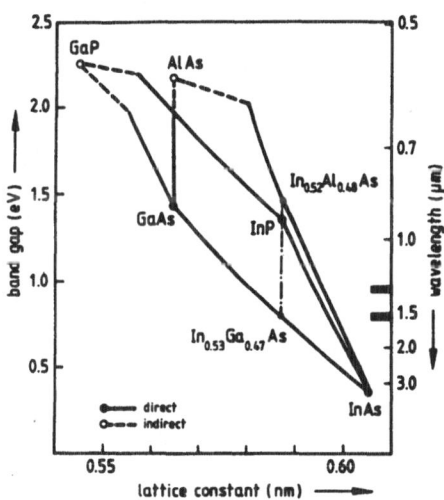

Abb. 2. Bandabstand - Gitterkon-
stantendiagramm für InGaAsP und
InGaAlAs

Qualität zur Verfügung steht. Geht man außerdem vom Bereich geringer Absorption der
Glasfaser aus, so kommen auf "GaAs" und "InP" basierende Materialien zur Anwendung,
wobei die "InP-Systeme" (InGaAsP und InGaAlAs) insbesondere das für den Weitver-
kehr vorzuziehende Spektralgebiet um 1,3 μm und 1,55 μm abdecken. Es können also
durch Variation der Zusammensetzung von InGaAsP bzw. InGaAlAs gitterangepaßte Epitaxieschichten hergestellt werden, je nach Anwendung - abgestimmt auf die Arbeits-
wellenlänge - emittierend, transparent oder absorbierend. Da der Brechungsindex
eine Funktion des Bandabstandes und damit der Zusammensetzung des Mischkristalls

ist, können wellenführende Schichten hergestellt werden. Außerdem lassen sich in diesem Materialsystem auch elektronische Bauelemente realisieren. Damit sind prinzipiell die Voraussetzungen für eine monolithische Integration im InP-System erfüllt. Ungelöst ist noch die Herstellung nicht-reziproker Komponenten (opt. Isolator).

Aus der oben beschriebenen Materialwahl folgt, daß die Technolgie der integrierten Optik eine Halbleitertechnologie ist. Sie umfaßt aufwendige Verfahren von der Epitaxie über Lithographie, mit der Strukturen bis in den Submikronbereich beherrscht werden müssen, bis zu einer Reihe weiterer komplexer Prozeßverfahren zur Herstellung von integrierten Bauelementen. Zu erwähnen sind hier namentlich Trockenätzverfahren für die dimensionsgetreue Strukturübertragung. Darüberhinaus ist eine umfangreiche Meßtechnik sowohl zur Material- als auch zur Bauelementecharakterisierung erforderlich. Die zur Herstellung von opto-elektronisch integrierten Schaltungen notwendigen Verfahren befinden sich teilweise noch im Entwicklungsstadium. Der Fortschritt der "Integrierten Optik" wird auch von der Entwicklung neuartiger Apparaturen/Verfahren abhängen. Zu nennen sind hier besonders "schreibende" Verfahren, die es gestatten, Epitaxieschichten unterschiedlicher Materialzusammensetzungen lateral begrenzt herzustellen. Hierbei könnten maskierte Molekularstrahlepitaxie, photoinduzierte Depositions- und Ätzverfahren sowie mit fokussierten Ionenstrahl arbeitende Ätzmethoden zur Anwendung kommen.

4. Vergleich zwischen elektronischer und optischer Integration

Der Aufbau von opto-elektronisch (OEIC) und rein elektronisch integrierten Schaltungen (EIC), sei es auf Si- oder GaAs-Basis, weist erhebliche Unterschiede auf:

1. Die Funktionen und die Verkopplung elektronischer und optischer Komopnenten sind völlig unterschiedlich.
2. Für die einzelnen optischen Funktionselemente werden aus prinzipiellen Gründen unterschiedliche Materialzusammensetzungen, für EICs nur eine Materialstruktur benötigt.
3. EICs bestehen aus wenigen Bauelementetypen (Widerstände, Kapazitäten, Transistoren); OEICs dagegen aus einer Vielzahl verschiedenartiger Strukturen.
4. Das Längen-/Seitenverhältnis beträgt bei EICs etwa 1:1, bei OEICs bis zu 1000:1.
5. Die Packungsdichte in EICs kann sehr hoch sein; in OEICs ist sie gering.

5. Beispiele einzelner Grundkomponenten der integrierten Optik

In Abb. 3 ist schematisch die monolithische Integration von Laser, Detektor, Modulator und Wellenleiter als Kopplerstruktur in "Hucke-Pack-Anordnung" dargestellt. Die erste Ebene stellt den über den gesamten OEIC laufenden passiven Wellenleiter dar, der am Ausgang des OEIC als Modulator ausgebildet ist.

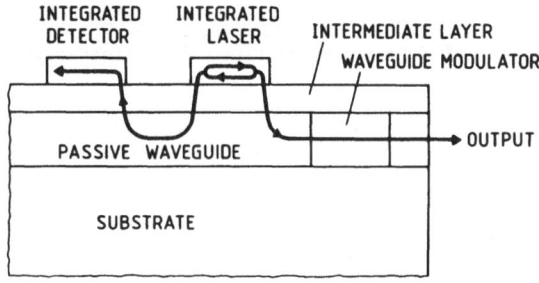

Abb. 3. Schematische Darstellung der monolithischen Integration von Laser, Detektor, Modulator und Wellenleiter

In der nächsten Ebene befindet sich der integrierte Laser, dessen emittiertes Licht in der einen Richtung in den passiven Wellenleiter eingekoppelt wird und in der anderen in den integrierten Detektor gelangt. Eine Ausführungsform des passiven Wellenleiters, der einen Grundbaustein der integrierten Optik darstellt, ist in Abb. 4 dargestellt. Die quaternäre Schicht wird epitaktisch auf InP-Substrat aufgebracht. Da der Brechungsindex der Q-Schicht größer ist als von InP, stellt sie einen Wellenleiter dar. Durch Ätzen einer kleinen Stufe wird ein laterales Confinement erreicht. Der Bestwert der Dämpfung für dieses Material liegt z.Z. bei 2,8 dB/cm für einen 14 mm langen Wellenleiter /4/.

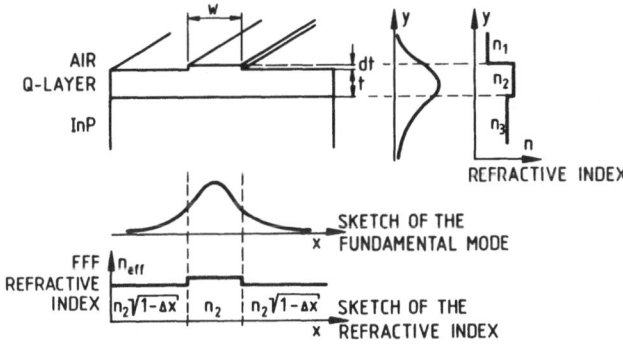

Abb. 4. Stufenwellenleiter

Abb. 5 zeigt einen Phasenmodulator, bei dem durch ein äußeres elektrisches Feld der Brechungsindex und damit die Phase des Lichtes beeinflußt wird /4/. Die Phasenänderung ist proportional dem angelegten elektrischen Feld und der Länge. Durch die spezielle Anordnung der versenkten Elektrode konnten 8°/V mm Phasendrehung erreicht werden /5/.

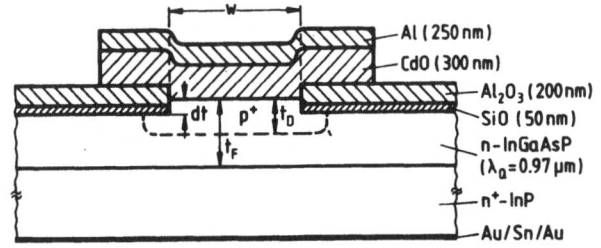

Abb. 5. Phasenmodulator

In Abb. 6 ist eine optisch-optische Mehrfachintegration bestehend aus einem Laser mit integriertem Filter als Resonator"spiegel" und einem passiven Wellenleiter dargestellt. Außerdem ist dies ein Beispiel für die optische Verkopplung /6/.

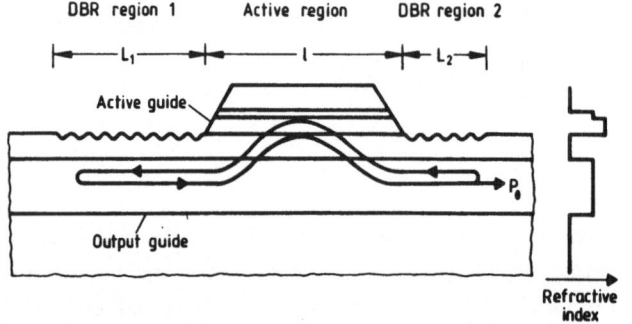

Abb. 6. Distributed Bragg Reflected-Integrated Twin-Guide (DBR – ITG) Laser

Abschließend wird eine komplexere Schaltung (Abb. 7), bei der drei Transistoren als Treiberstufe mit einem Laser monolithisch integriert sind, dargestellt /7/.

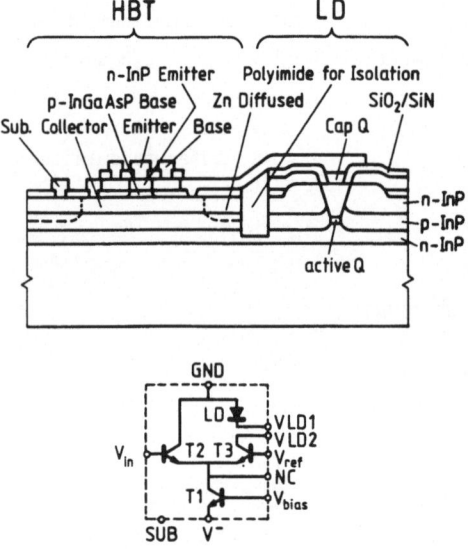

Abb. 7. Monolithische Integration eines Lasers mit einer Treiberstufe

6. Zusammenfassung

Anhand eines Beispiels aus der optischen Nachrichtentechnik wurde die Notwendigkeit der Entwicklung einer monolithisch integrierten Optik dargelegt. Nicht nur für die optische Nachrichtentechnik, sondern auch für die optische Signalverarbeitung und Sensorik wird eine völlig neue Perspektive eröffnet.

References

/1/ C. Baack: "Entwicklungstendenzen in der optische Breitbandkommunikation", in diesem Band

/2/ F.K. Reinhart: Proceedings of SPIE, 272, 66-75 (1981)

/3/ C. Bornholdt, W. Döldissen, D. Franke, N. Grote, J. Krauser, U. Niggebrügge, H.P. Nolting, M. Schlak, I. Tiedke: Electron. Lett. 19, 3, 81-82 (1983)

/4/ H.G. Bach, J. Krauser, H.P. Nolting, R.A. Logan, F.K. Reinhart: Appl. Phys. Lett.: 42,8 (1984)

/5/ C. Bornholdt, W. Döldissen, D. Franke, J. Krauser, U. Niggebrügge, H.P. Nolting, F. Schmitt: Proceedings of the 3rd Euro. Conf., ECIO'85, 121-126 (1985)

/6/ K. Utaka, K.I. Kobayashi, Y. Suematsu: IEEE J. Quantum Electron., QE 17, 651-658 (1981)

/7/ J. Shibata, I. Nakao, Y. Sasai, S. Kimura, N. Hase, H. Serizawa: Appl. Phys. Lett. 45, 3, 191-193 (1984)

Kohärente Optische Nachrichtentechnik

Ernst-Jürgen Bachus

Heinrich-Hertz-Institut für Nachrichtentechnik Berlin GmbH

Die optische Nachrichtentechnik mit Halbleiterlasern und Glasfaserstrecken hat zwar den Nachteil einer geringen Systemdynamik (ca. 40 dB) aufgrund kleiner Sendeleistungen (einige mW) und hohen Empfängerrauschens (Quantennatur des Lichtes), bietet dafür aber eine extrem große Frequenzbandbreite (Terahertz). Es ist schon lange bekannt, daß ein Überlagerungsempfang (wie im Rundfunkempfänger) auch im optischen einen Gewinn an Empfindlichkeit (10-25 dB) bringt /18, 19/, und überhaupt erst die gesamte Bandbreite im Bereich der optischen Wellenlängen mit Hilfe von Trägerfrequenzsystemen auszunutzen erlaubt.

Parallel zur Entwicklung von spektral reinen (Single-Mode) Halbleiterlasern (in Verbindung mit Monomode-Fasern) wurde daher auch folgerichtig mit digitalen Übertragungsexperimenten (etwa 1980 in Japan) mit optischen Heterodynempfang begonnen.

Im folgenden Übersichtsbeitrag sollen Grundlagen, Möglichkeiten und Grenzen dieser zukünftigen Technik skizziert werden.

Der Heterodynempfänger enthält einen Lokallaser, dessen Lichtleistung P_L zum über die Faserstrecke ankommenden Signallicht $P_S(t)$ addiert wird (Bild 1). Der nur auf Lichtleistung reagierende (farbenblinde) Photodetektor sieht eine zeitliche Interferenz, sein Photostrom schwankt mit der Differenzfrequenz von Sende- und Lokallaser.

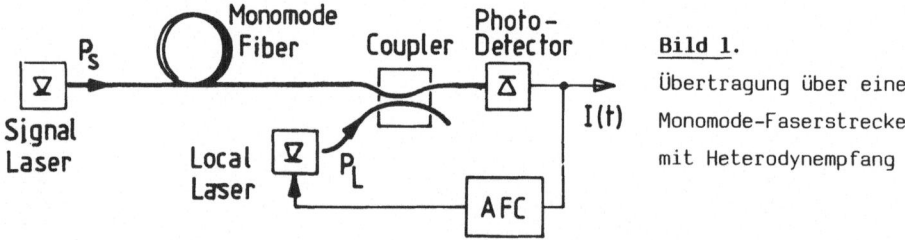

Bild 1.
Übertragung über eine
Monomode-Faserstrecke
mit Heterodynempfang

Diese Zwischenfrequenz kann konstant gehalten werden, wenn der Lokallaser dem Sendelaser frequenzmäßig nachgeregelt wird. Der Photostrom enthält 3 Anteile:

$$I(t)/(0,5\ \eta e/hf) = P_S(t)+P_L+2\sqrt{P_S(t)P_L}\ \cos\left[(\omega_S(t)-\omega_L)t+\psi_S(t)-\psi_L(t)\right] \qquad (1)$$

(I(t) = Photostrom, Faktor 0,5 infolge idealem 3 dB-Koppler, Quantenwirkungsgrad $\eta \approx 0,8$, Elektronenladung $e = 1,601 \cdot 10^{-19}$ As, Planck'sches Wirkungsquantum

h = 6,624·10^{-34} Ws/Hz, f = optische Frequenz, hf = Energie eines einzelnen Photons)
Nur der Zwischenfrequenzanteil (dritter Term) wird ausgefiltert und demoduliert. Sowohl Sendeleistung, -frequenz und -phase (P_S, ω_S, ψ_S) können moduliert werden (ASK, FSK, PSK). Bei kleinen Signalleistungen wird mit großen Lokallaserleistungen eine (scheinbar beliebige) Signalverstärkung erreicht:

$$ZF\text{-Trägerleistung} \sim 2\ P_S P_L \tag{2}$$

bzw. 4 $P_S P_L$, wenn die Zwischenfrequenz gleich Null ist (Homodyn-Empfang).

Heterodyn-Empfang setzt zeitliche Kohärenz voraus. Als Beispiel hat das Zwischenfrequenzspektrum zweier überlagerter CSP-Singlemode-Laser (Hitachi HLP 1400, λ = 830 nm) das (willkürlich) auf 1,5 GHz geregelt wurde, eine Halbwertsbreite von 25 MHz, dem entspricht etwa eine Kohärenzzeit von 40 ns (Bild 2). Der Zeitverlauf auf dem Sampling-Oszillografen zeigt einen fortschreitend sich verwischenden Sinuszug. Wird die Frequenz oder die Phase des Signallichtes als Informationsträger benutzt, so muß die Kohärenzzeit wesentlich höher sein als die Zeit, in der die bestimmte Frequenz oder Phase sicher erkannt werden soll, i.a. also höher als die Bitzeit. Von daher ist also eine hochratige Modulation günstig.

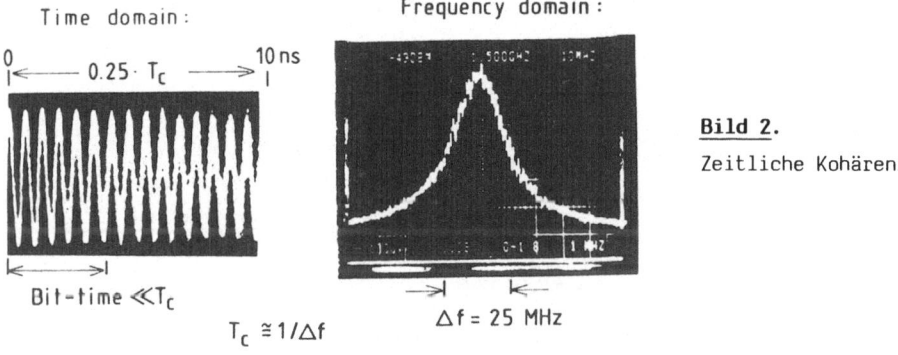

Bild 2.
Zeitliche Kohärenz

Die zweite Bedingung für Heterodyn-Empfang ist die notwendige räumliche Kohärenz.

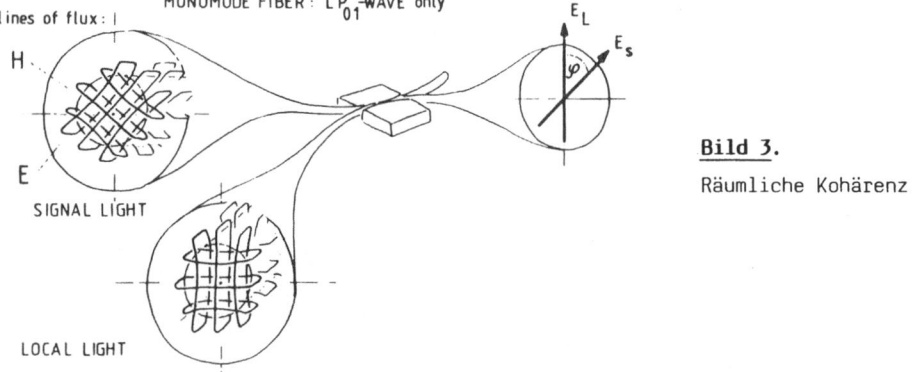

Bild 3.
Räumliche Kohärenz

Sie wird durch die Monomode-Faser per se erfüllt: nur der Grundwellentyp mit geordneter, stets gleicher (und damit interferenzfähiger) Feldstruktur ist ausbreitungsfähig, allerdings mit beliebiger Polarisationsrichtung (der transversalen elektrischen Feldstärke),(Bild 3).Eine orthogonale Polarisation von Sende- und Lokallicht hinter dem Koppler verhindert Interferenz und löscht den Zwischenfrequenzterm in (1) aus. Um dies zu vermeiden, muß z.B. die Polarisation des Lokal-Lichtes nachgeregelt werden. Oder man verwendet polarisationserhaltende Monomode-Fasern.

Ein praktisches Problem ist die Frequenzkonstanz der Halbleiterlaser. Eine Temperaturerhöhung um 1° C läßt z.B. bei den erwähnten CSP-Lasern die optische Frequenz um 40 GHz absinken. Die Temperatur sollte auf weniger als 0,01° C konstant regelbar sein, schnelle Änderungen werden über den Injektionsstrom ausgeregelt (-4 GHz bei $\Delta I = +1$ mA).

Mit welcher kleinsten Lichtsignalleistung kommt nun ein Heterodyn-Empfänger aus? Das hängt vom erforderlichen Signal-Rauschabstand und damit von den vorhandenen Rauschanteilen ab.

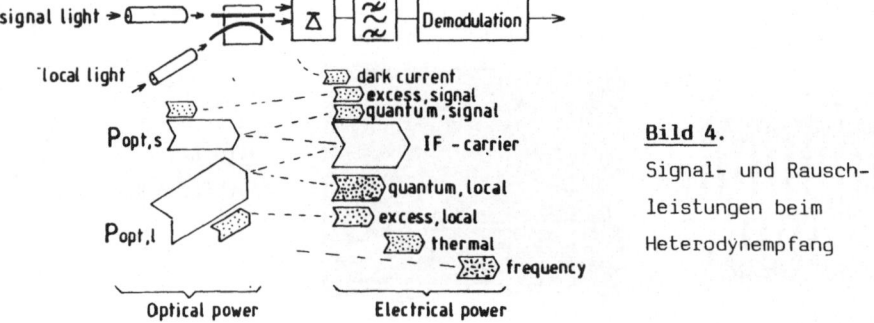

Bild 4.

Signal- und Rauschleistungen beim Heterodynempfang

Im Bild 4 sind links die optischen, rechts die elektrischen Leistungen nach der Detektion dargestellt. Die Rauschleistungsanteile (durch Punkte markiert) sind: Dunkelstrom der Photodiode, Leistungsschwankung des Signallichts (excess noise), beim Detektionsprozess entstehendes Quantenrauschen infolge Sendelicht, infolge Lokallicht, Leistungsschwankungen des Lokallichts, thermisches Rauschen der Mikrowellenstufe (bei Direkt-Detektion meist der größte Rauschanteil) und schließlich das von Sende- und Lokallicht herstammende, nach der Demodulation sichtbare Frequenzrauschen. Hier wird man wegen Gl.(2) eine hohe Lokalleistung anstreben, dann wird das Quantenrauschen infolge des Lokallichtes der größte Rauschanteil und das Signal/Rauschverhältnis geht gegen den Wert:

C/N = ZF-Leistung/Summe aller Rauschleistungen

$$\approx P_{ZF}/N_{QL} = \eta \cdot P_{opt,s}/hf \cdot \Delta T = \eta \cdot \text{Zahl der Signalphotonen in einem Bit} \qquad (3)$$

Dies ist die sog. Quantengrenze. Letzlich darf also eine gewisse Signalleistung

nicht unterschritten werden. Diese hängt auch ab von den verwendeten Modulationsver-
fahren, die im Bild 5 hinsichtlich ihrer Empfindlichkeit miteinander verglichen
werden.

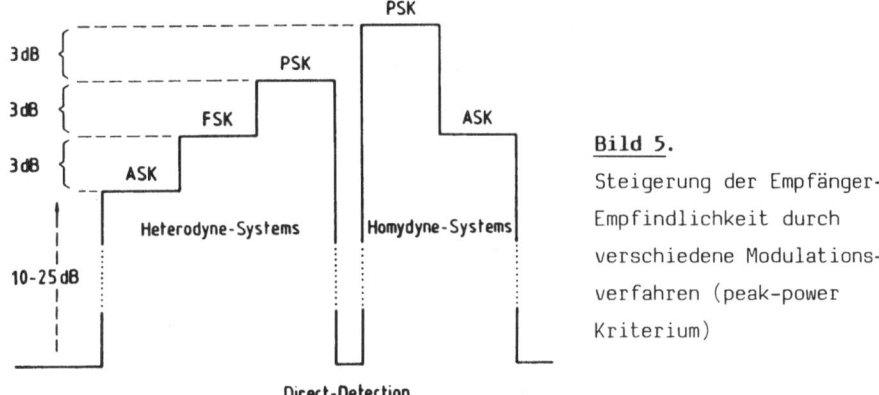

Bild 5.
Steigerung der Empfänger-
Empfindlichkeit durch
verschiedene Modulations-
verfahren (peak-power
Kriterium)

Nun zum anderen Aspekt des optischen Überlagerungsempfangs. Die Möglichkeit, sehr
selektiv eine Lichtsendefrequenz durch Heterodynempfang auszufiltern kann in einem
Lichtfrequenzmultiplex-System ausgenutzt werden (Bild 6).

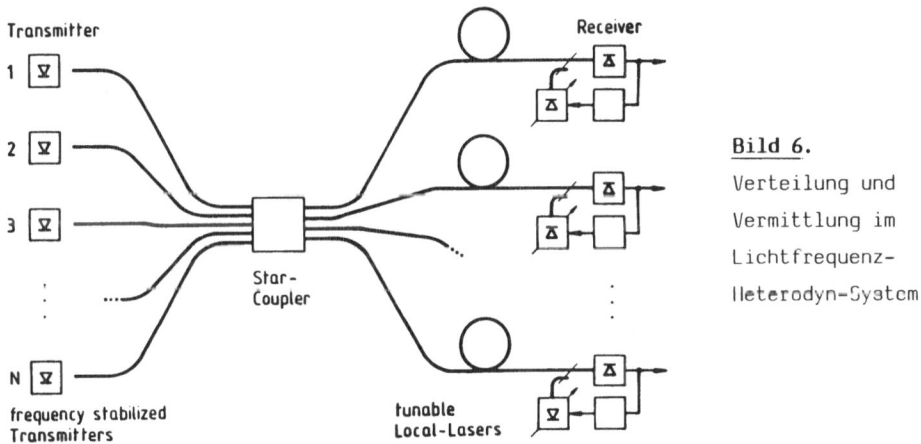

Bild 6.
Verteilung und
Vermittlung im
Lichtfrequenz-
Heterodyn-System

Im Vergleich zum λ-Multiplex (in /15/ ist z.B. ein Kanalabstand von Δλ = 1,35 nm,
entsprechend 180 GHz realisiert worden) können jetzt die Kanäle sehr dicht nebeneinandergelegt werden und damit von einem Lokallaser überstrichen werden. (Bei unseren
Experimenten wurden bisher 3 Kanäle und ein Rückkanal im Abstand von nur 5 GHz bei
70 Mbit/s und einer Fehlerrate von BER < 10^{-9} betrieben).

Der Sternkoppler in Bild 6 ist ein optisches Netzwerk, das jeden Eingang gleichmäßig
auf jeden Ausgang koppelt. Jeder Empfänger kann durch Abstimmen seines Lokallasers
jeden Sender empfangen.

Wie hoch kann nun die maximale Kanalzahl auf einer Monomodefaser sein? Da jedes Sendesignal beim Empfänger nur mit einem Bruchteil der Eingangsleistung (entsprechend der Kanalzahl) vorliegt (bei 1000 Kanälen mindestens 30 dB unter dem Eingangspegel) ist die Obergrenze durch die Empfängerempfindlichkeit vorgegeben. In Bild 7 wurde hierfür eine gemittelte FSK-Empfindlichkeit (aus Bild 8) angenommen. Das Hintergrundlicht aller anderen optischen Signale erzeugt aber ein hohes Quantenrauschen und setzt die mögliche Kanalzahl weiter herab.

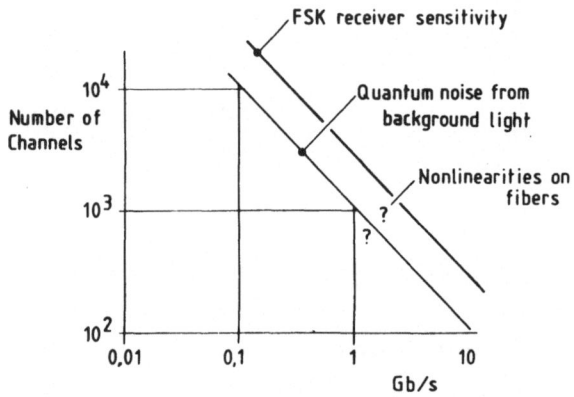

Bild 7.

Abschätzung einer maximalen Kanalzahl (Eingangsleistung zu 1 mW pro Kanal angenommen)

Weiterhin kann durch Kreuzmodulation (hervorgerufen durch Nichtlinearitäten auf der Monomodefaser) die mögliche maximale Kanalzahl noch niedriger werden. Erste Experimente hierzu werden z.Z. durchgeführt.

Zum Abschluß bringt Bild 8 ein Vergleich zwischen bisher realisierten Direkt-Detektions- und Heterodyn/Homodyn-Experimentalsystemen. Als Kriterium dient die erreichte Empfängerempfindlichkeit über der Bitrate (meist mit einer Fehlerrate $< 10^{-9}$). (Schwarze Punkte = Direkt Detektion, Heterodyn/Homodyn-System: Striche = ASK, gestrichelte = FSK, gepunktet = PSK. Ausgezogene Kreise = Faserübertragung, gestrichelte Kreise = Simulationsexperimente). Ein Vergleich von z.B. 10, 11, 9 und 4 zeigt, daß die Empfindlichkeitssteigerung nach Bild 5 auch experimentell bestätigt ist. 1, 5, 6, 7, 10 leiden unter dem starken Frequenzrauschen heutiger Fabry-Perot-Laser. In 9 wurden mit extern spektral-verbesserten Lasern 199 km mit FSK-Modulation überbrückt. Im Vergleich mit Direkt-Systemen (bis 4 Gbit/s) sind bisherige Heterodyn-Systeme noch langsam. Kohärente Systeme mit Gbit/s-Modulation werden sicher in naher Zukunft erprobt.

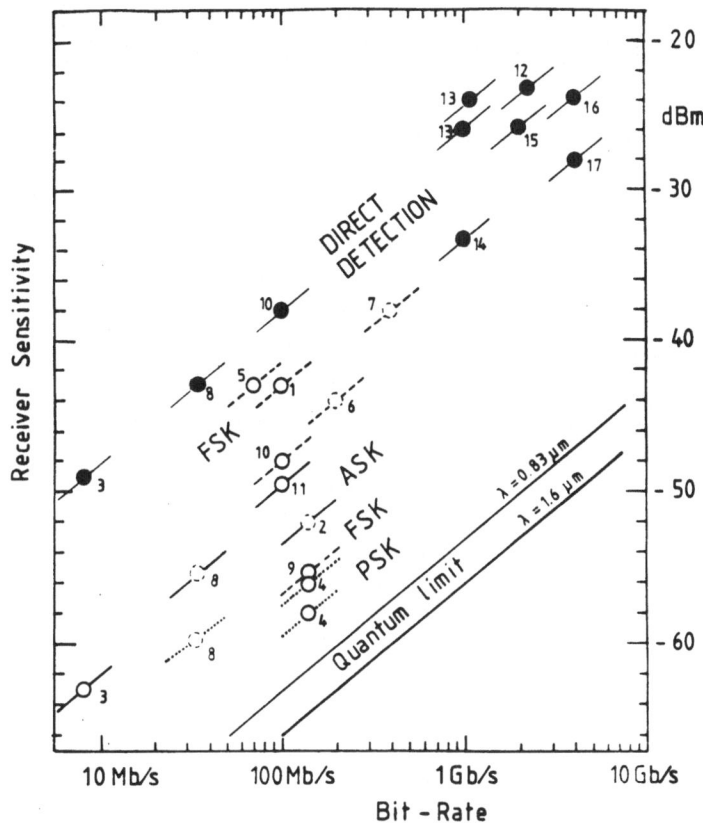

Bild 8. Systemvergleich: Direkt-Detektion mit Heterodyn/Homodyn-Detektionssystemen
(die Ziffern beziehen sich auf die Literaturangaben)

Literatur

/1/ S.Saito et al.: El.Lett. 18, 11 (1982)
/2/ T.G.Hodgkinson et al.: El.Lett. 18, 12 (1982)
/3/ T.G.Hodgkinson et al.: El.Lett. 18, 21 (1982)
/4/ R.Wyatt et al.: El.Lett. 19, 14 (1983)
/5/ E.J.Bachus et al.: IEEE J.Lightwave Tech. LT-2, 4 (1984)
/6/, /7/ S.Saito et al.: IOOC 1983
/8/ M.Shikada et al.: Trans.IECE, Japan, 67, 6 (1984)
/9/ R.Wyatt et al.: El.Lett. 20, 22 (1984)
/10/ K.Emura et al.: El.Lett. 20, 24 (1984)
/11/ M.Shikada et al.: El.Lett. 20, 4 (1984)
/12/ C.Baack: Optical Wideband Transmission, CRC Press, to be published
/13/ R.A.Linke et al.: El.Lett. 19, 19 (1983)
/14/ M.C.Brain et al.: El.Lett. 20, 21 (1984)
/15/ N.A.Olsson et al.: OFC/OFS San Diego (1985)
/16/ S.K.Korotky et al.: OFC/OFS San Diego (1985)
/17/ A.H.Gnauk et al.: OFC/OFS San Diego (1985)
/18/ F.E.Goodwin et al.: IEEE J.Q.E., QE-H (1968)
/19/ O.E.Delange: Proc. IEEE 58, 10 (1970)

Laser und Optoelektronik in der
Weltraumtechnik
Lasers and Optoelectronics in
Space Technology

Lasers and Optoelectronics for Earth Observation

E. V. Browell and A. F. Carter
Atmospheric Sciences Division
NASA Langley Research Center
Hampton, VA 23665

1. Introduction

Lasers have been used extensively since the early 1960's for making measurements of various properties of the Earth's atmosphere. These include molecular and aerosol backscattering, gas concentration profiles, wind velocities, and atmospheric waves. The laser is used as a transmitter of pulsed radiation and needs an appropriate receiver to detect the backscattered signals. This combination is called a lidar (Light Detection and Ranging) system. Lidar development in the early years was primarily concerned with ground-based/laboratory systems, with a few airborne systems making measurements in the late 1960's and early 1970's. By 1975, the lidar technique was well proven and consideration was being given to developing a shuttle lidar system. A spaceborne system is now under-going development, brought about by the rapid maturing of the laser as a transmitter source over the last decade.

There are basically four main atmospheric lidar processes or techniques: elastic scattering, Raman scattering, resonance fluorescence, and differential absorption. Lidar measurements utilizing elastic backscattering have focused primarily on investigations of molecular, aerosol, and cloud properties (Refs. 1-2). Lidar systems detecting Raman shifted frequencies have generally been limited to measurements of gases having high mixing ratios, such as water vapor, at relatively short ranges of 1 to 4 km at night (Ref. 3). Resonance fluorescence has been an important technique used to measure sodium, potassium, and atmospheric waves in the upper atmosphere (Ref. 4-6). The Differential Absorption Lidar (DIAL) technique focuses on measurements of concentration profiles of atmospheric gases. This technique requires the near simultaneous transmission of two wavelengths, usually by two tunable lasers. One of the lasers is tuned "on" the peak of an absorption line of the species to be measured, and the other is tuned "off" the absorption peak to a nearby wavelength. The average gas concentration n between range R_1 and R_2 can be calculated using the approximation (Ref. 7)

$$n = \frac{1}{2 \ (R_2 - R_1) \ (\sigma_{on} - \sigma_{off})} \quad \ell n \quad \frac{P_{off}(R_2) \ P_{on}(R_1)}{P_{off}(R_1) \ P_{on}(R_2)}$$

where $\sigma_{on} - \sigma_{off}$ is the absorption cross-section difference between the on-and-off wavelengths, and $P_{off}(R)$ and $P_{on}(R)$ are the detected backscattering signals received from range R for the on- and off-line wavelengths, respectively.

NASA Langley Research Center has been involved with the development and application of lidar technology for over two decades. Early ground based systems (Refs. 8-10) measured atmospheric aerosols and H_2O and SO_2 in power plant stack plumes. Recently, an advanced airborne DIAL system was developed and has been used to measure O_3, H_2O, and aerosols in the troposphere and lower stratosphere (Ref. 11). This paper will discuss the present state of lidar technology, as well as present and proposed airborne and spaceborne systems. Examples of measurements made with the present NASA Langley Research Center advanced airborne DIAL system will be shown to illustrate the advantages of using lidar systems in conducting scientific investigations of our atmosphere.

2. Advanced Airborne DIAL System

The schematic configuration of the airborne DIAL system mounted in the NASA Electra aircraft is shown in figure 1. Two frequency doubled Nd:YAG lasers are used to pump two high conversion efficiency dye lasers. The on and off wavelengths are produced in sequential laser pulses with less than 100 µs separation. Backscattered lidar returns at the two wavelengths are sequentially detected by a photomultiplier tube, digitized, and stored on high-speed magnetic tape. Gas concentrations and aerosol backscattering profiles are calculated for each measurement in real time by a minicomputer. The selection of the lasers was dictated by three basic requirements: tunability, high power, and reliability. The dye laser is easily tuned over a wide selection of wavelengths, and the output can be doubled into the UV. The Nd:YAG lasers chosen to pump the dye lasers generate very high energy per pulse at reasonably high pulse rates. When this system was assembled in 1979, the Nd:YAG laser had

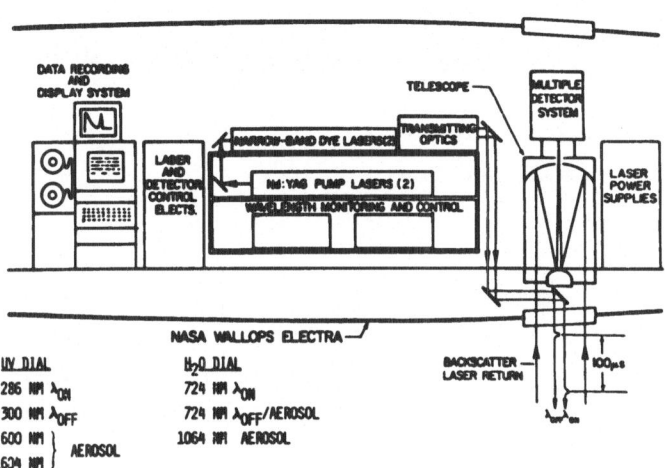

Figure 1. NASA LaRC airborne DIAL system schematic (Ref. 11)

reached a level of technological maturity which resulted in a high
reliability.   This combination of pump and dye laser has given
outstanding performance over the last 6 years, and loss of data due
to laser failure has been rare.   The performance parameters of the
system for operation in the UV and near-IR are listed in Table 1.

Some examples of our airborne DIAL data, shown in figure 2,
illustrate some of the capabilities of the system.   Signal returns
from the UV (286 and 300 nm) and the visible (600 nm) channels are
displayed at the top of the figure to show the contrast due to
increased aerosol backscattering in the atmospheric boundary layer.
The range-corrected visible channel is shown at the lower left, and a
gray scale picture made using these data is shown at the lower
right.   Each vertical line of the gray scale represents a laser shot,
and the brighter regions represent greater atmospheric backscattering, or aerosol concentration.   This gray scale clearly defines
the boundary layer height and presence of clouds (bright regions
followed by absence of return signal from lower altitudes).   This
type of aerosol data can provide information on widely varying
conditions including aerosol layering and transport, clouds,
condensation levels, and topographic features.   Figure 3 shows an
intercomparison of DIAL O_3 measurements with in situ measurements
made onboard a small aircraft spiraling in the vicinity of the
Electra.   The intercomparison indicates that: (1) remote lidar
measurements and in situ measurements are in excellent agreement, (2)
DIAL measurements have 10 percent or less measurement uncertainty for

Table 1.

AIRBORNE DIAL SYSTEM CHARACTERISTICS

Transmitter:

    Two Pump Lasers -- Quantel Model 482

      Pulse Separation -- 100 µs

      Pulse Energy -- 350 mJ at 532 nm

      Repetition Rate -- 10 Hz

      Pulse Length -- 15 nm

    Two Dye Lasers -- Jobin Yvon Model HP-HR

	UV (near 300 nm)	Near-IR (near 720 nm)
Fundamental Dye Output Energy	157 mJ/pulse near 600 nm	63 mJ/pulse
Doubled Dye Output Energy	47 mJ/pulse near 300 nm	--
Transmitted Laser Energy	40 mJ/pulse near 300 nm and 80 mJ/pulse near 600 nm	50 mJ/pulse
Laser linewidth	<4 pm	<2 pm

Receiver:

Area of Receiver	0.086 m^2	0.086 m^2
Receiver Efficiency to PMT	28%	29%
PMT Quantum Efficiency	29%	4.8%
Total Receiver Efficiency	8.1%	1.4%
Receiver Field of View	2 mrad	2 mrad

the resolution shown, and (3) DIAL measurements can be made with high vertical (210 m) and horizontal (6 km) resolution.

Airborne DIAL measurements have been used to study the correlation between spatial distributions of O_3 and aerosols in the troposphere and lower stratosphere. Figure 4 shows O_3 measurements by the DIAL system in the zenith-viewing mode compared with in situ observations. Profiles of potential temperature and ozone measured by an ozonesonde are presented in the figure along with DIAL O_3 data. The location of a change in slope of the potential temperature profile ($\approx$11 km) defines the height of the tropopause. The O_3 profiles show layering of O_3 in the lower stratospheric regions, and there is good agreement between the DIAL and in situ measurements. Further, other DIAL measurements show that this O_3 layering persisted along the flight path of the aircraft. This indicates the ability of an airborne DIAL system to map O_3 layers in the troposphere and lower stratosphere and to study stratosphere and troposphere exchanges.

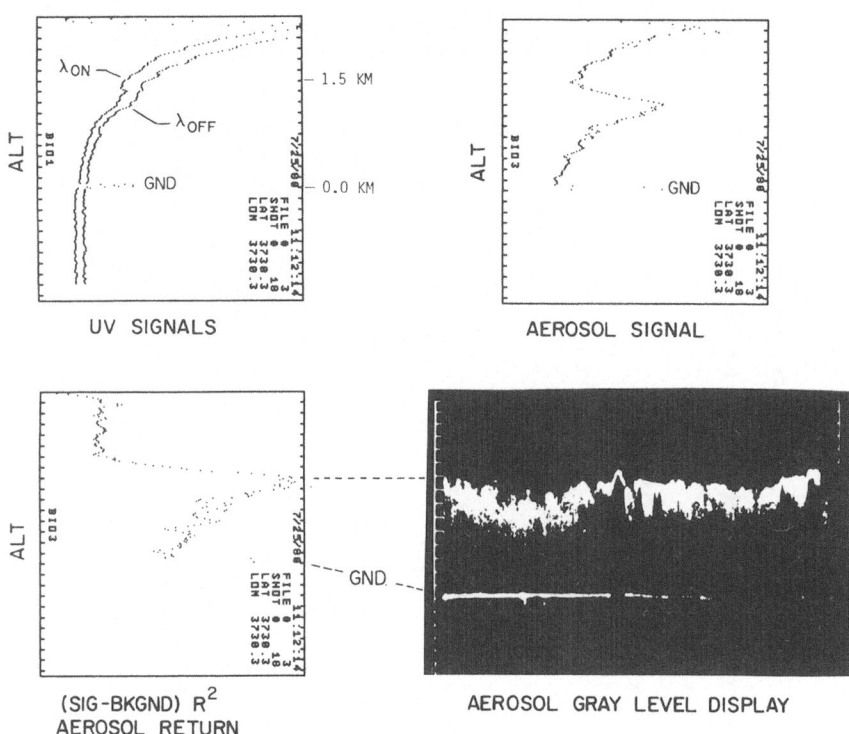

Figure 2. Airborne UV DIAL and aerosol returns; aircraft altitude
was 3 km

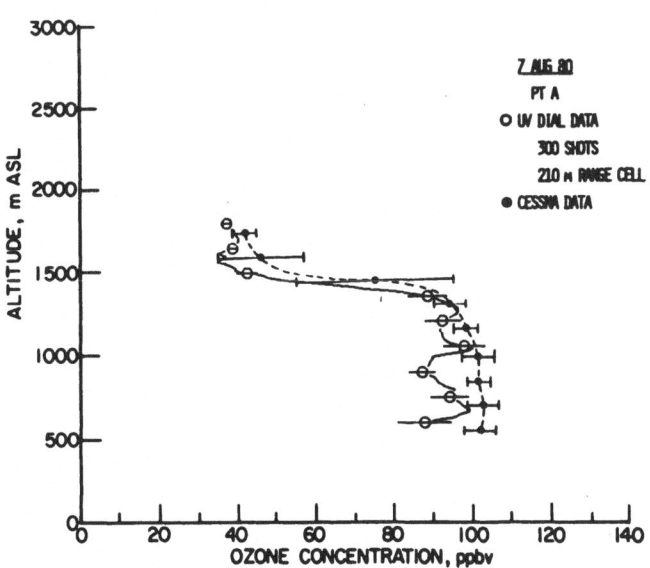

Figure 3. DIAL and in situ ozone measurement comparison (Ref. 12)

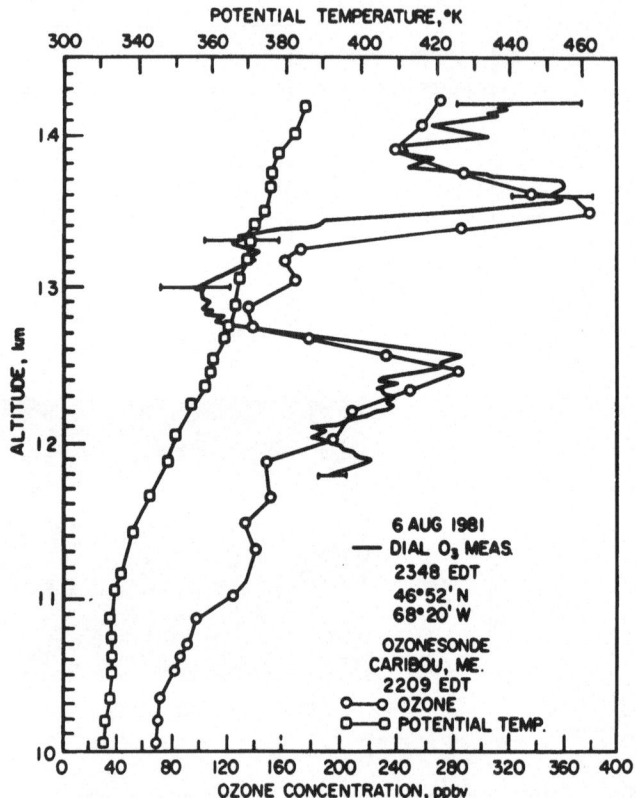

Figure 4. Comparison of airborne DIAL and in situ measurements of stratospheric ozone (Ref. 12)

When the airborne DIAL system is used to study H_2O profiles, the on-line laser wavelength is tuned to coincide with the peak of a H_2O absorption line in the 720-nm region. An intercomparison between the DIAL H_2O measurements and in situ observations is shown in figure 5. The good comparison between these two observations shows the ability of the DIAL system to make H_2O measurements in the lower troposphere.

A list of the demonstrated advanced airborne lidar techniques for O_3, H_2O, aerosol and wind measurements using various types of lasers is given in Table 2 (Refs. 11-15). These experiments have potential for spaceborne lidar applications. Table 2 also contains a list of lidar systems under development that will have an important impact on future lidar applications.

3. ER-2 DIAL SYSTEM

A high altitude (16-21 km), autonomous DIAL system for operation on a NASA ER-2 (Extended Range U-2) aircraft is being developed by NASA. This system is called LASE, for Lidar Atmospheric Sensing

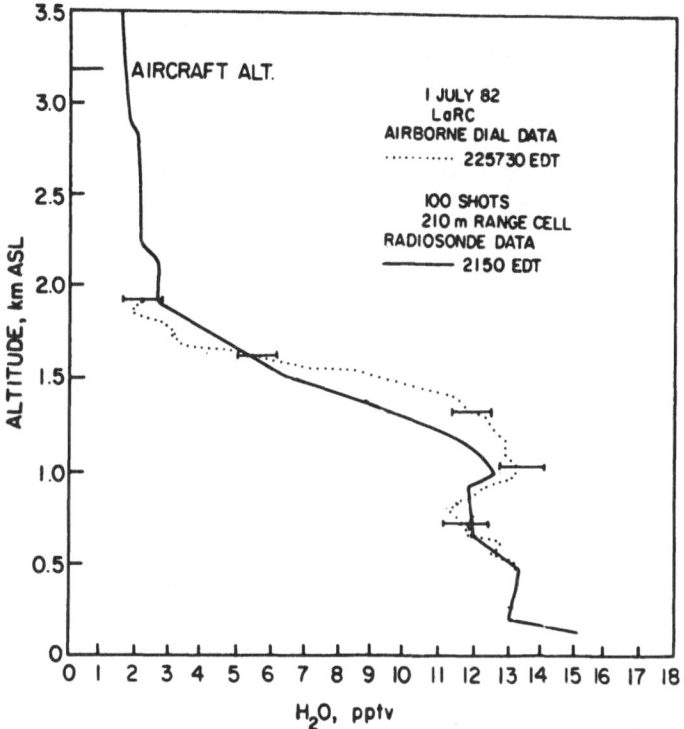

Figure 5. Comparison of DIAL and radiosonde water vapor measurement

Table 2.

AIRBORNE LIDAR TROPOSPHERIC MEASUREMENTS

DEMONSTRATED	NEAR FUTURE (≤ 2 YEARS)	FUTURE (> 2 YEARS)
O_3 (Nd:YAG-DYE[*]; CW-CO_2[+])	O_3 (EXIMER)	O_3 PROFILES (10 μm)
H_2O (Nd:YAG-DYE[**]; CW-CO_2[++])	H_2O (ALEX)	H_2O (1.14 μm)
AEROSOL + CLOUD DIST. (Nd:YAG[*])	P CW-λ Nd:YAG; Nd:YAG-DYE/ALEX.)	CO (4.6 μm)
WINDS (CO_2 DOPPLER[*])	T (Nd:YAG-DYE/ALEX.)	NH_3 (10.3 μm)
SF_6 TRACER (CW-CO_2[++])	AEROSOL DISCRIM (3-λ Nd:YAG-DYE)	CH_4 (3.3 μm)
	ATM. DENSITY AND TEMP. (2-λ Nd:YAG)	N_2O (3.9 μm)
	SO_2 (EXIMER)	CO_2 (2.7 μm)

*Ref. 11; **Ref. 12; +Ref. 13; ++Ref. 14; ψRef. 15

Experiment. It will initially operate near the 720-nm region to
measure H_2O profiles in the lower troposphere, and at a later stage

it will be used near the 940-nm region to measure H_2O profiles in the upper troposphere. In addition, atmospheric pressure and temperature determinations will be made using DIAL O_2 measurements in the 760-nm wavelength region. Figure 6 shows the results of our simulations to determine the random errors associated with H_2O measurements with the LASE system. The random error at a specific altitude depends upon

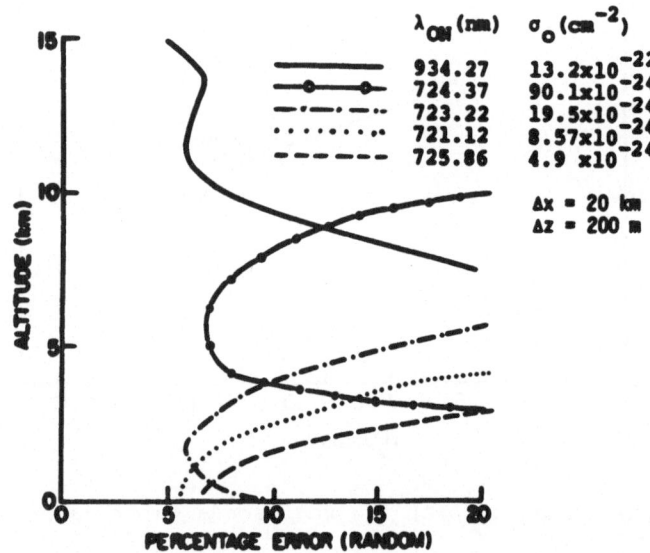

Figure 6. Water vapor DIAL night simulations for the ER-2 lidar system. An aircraft altitude of 16 km and a mid-latitude summer H_2O profile was assumed

the absorption strength of the H_2O line being used, with the stronger lines being more suitable for higher altitude regions. Clearly the random errors for night operation and for a horizontal resolution of 20 km and a vertical resolution of 200 m would permit a 5- to 10-percent measurement uncertainty in a specific altitude region. Our analysis shows that systematic errors could contribute a significant amount to the uncertainty of these measurements. The systematic errors can be caused by: (1) interaction of the laser linewidth with the altitude-dependent H_2O linewidth, (2) Doppler broadening of backscattered light by air molecules, (3) accuracy in the positioning of the laser line with respect to the H_2O line, (4) uncertainty in knowledge of the laser wavelength, and (5) the broadband laser energy component outside the specified on-line region. Our simulations indicate that with proper controls on the laser output and by folding in model atmospheric temperature and pressure information, the systematic errors could be limited to a few percent (Ref. 16). The LASE system would be capable of making measurements on long-range flights (> 4500 km) during both day and night background conditions, and because of its unique autonomous mode of operation, the LASE system would be a precursor to the development of a spaceborne lidar system.

4. SPACEBORNE LIDAR TECHNOLOGY EVOLUTION

The development of a lidar system for operation on a space platform is the next planned step by NASA after development of the ER-2 DIAL system. Spaceborne lidar systems would enhance the ability to conduct atmospheric investigations over wide regions of latitude and longitude, with unique emphasis on the troposphere. In addition, regions of the upper atmosphere that are difficult to study from the ground and with passive remote sensing techniques, like the iono-sphere, would be easily accessible. To better define the role and capabilities of spaceborne lidar in conducting atmospheric investiga-tions, several studies have been conducted by the European space agencies and by NASA during the past decade. The studies conducted by NASA include:

- Atmosphere, Magnetosphere and Plasmas in Space (AMPS) Study (1975)
- Stanford Research Institute Study (NASA CR-132724, 1975)
- Lidar Proposals for Spacelab 1 (June 1976)
- Atmospheric Lidar Working Group Study (NASA SP-433, 1979)
- General Electric Shuttle Lidar System Definition Study (NASA CR-3303, 1980)
- Laser Atmospheric Sounder and Altimeter (LASA) Study for the NASA Earth Observing System (currently in progress)

The study conducted by the Atmospheric Lidar Working Group (Ref. 17) concluded that a spaceborne lidar can significantly con-tribute to major science and application objectives. The Lidar Working Group also identified 26 classes of experiments that could address these objectives. Out of these 26 classes of experiments, which are listed in Table 3, 17 (identified by closed boxes) are measurements where lidar can make a unique contribution. This study concluded that a shuttle lidar development would contribute to an understanding of the processes governing the Earth's atmosphere and would also provide an evaluation of atmospheric susceptibility to manmade and natural perturbations. The General Electric Company conducted a shuttle lidar system definition study (Ref. 18) on the basis of the objectives and experiments identified by the Lidar Working Group. This study concluded that: (1) most of the science

Table 3

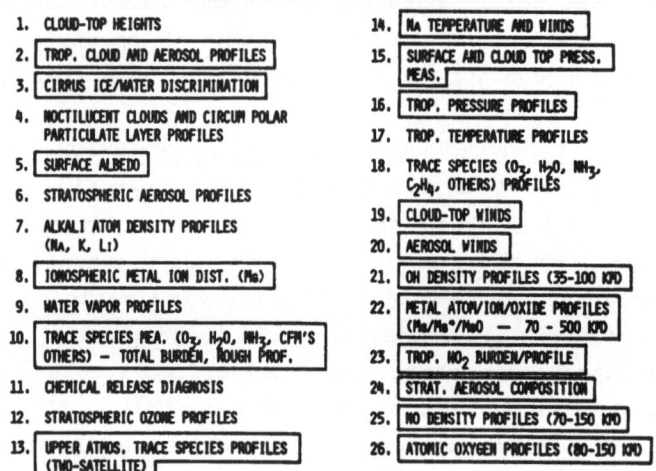

REPRESENTATIVE EXPERIMENT CLASSES

1. CLOUD-TOP HEIGHTS
2. TROP. CLOUD AND AEROSOL PROFILES
3. CIRRUS ICE/WATER DISCRIMINATION
4. NOCTILUCENT CLOUDS AND CIRCUM POLAR PARTICULATE LAYER PROFILES
5. SURFACE ALBEDO
6. STRATOSPHERIC AEROSOL PROFILES
7. ALKALI ATOM DENSITY PROFILES (NA, K, LI)
8. IONOSPHERIC METAL ION DIST. (Ma)
9. WATER VAPOR PROFILES
10. TRACE SPECIES MEA. (O_3, H_2O, NH_3, CFM'S OTHERS) — TOTAL BURDEN, ROUGH PROF.
11. CHEMICAL RELEASE DIAGNOSIS
12. STRATOSPHERIC OZONE PROFILES
13. UPPER ATMOS. TRACE SPECIES PROFILES (TWO-SATELLITE)

14. Na TEMPERATURE AND WINDS
15. SURFACE AND CLOUD TOP PRESS. MEAS.
16. TROP. PRESSURE PROFILES
17. TROP. TEMPERATURE PROFILES
18. TRACE SPECIES (O_3, H_2O, NH_3, C_2H_4, OTHERS) PROFILES
19. CLOUD-TOP WINDS
20. AEROSOL WINDS
21. OH DENSITY PROFILES (35-100 KM)
22. METAL ATOM/ION/OXIDE PROFILES ($Ma/Ma°/MaO$ — 70 - 500 KM)
23. TROP. NO_2 BURDEN/PROFILE
24. STRAT. AEROSOL COMPOSITION
25. NO DENSITY PROFILES (70-150 KM)
26. ATOMIC OXYGEN PROFILES (80-150 KM)

objectives defined by the Working Group can be accomplished using an evolutionary lidar system, (2) the space shuttle can accommodate an evolutionary lidar instrument, and (3) that the lidar technology exists for the initial facility concept. Such studies have shown that there exists a sound scientific and technological basis for the development of a spaceborne lidar system.

The development of spaceborne lidar systems is contingent on the ability of lasers to operate in a space environment. NASA is taking a step in the direction of developing an operational laser capability in space with the Lidar In-Space Technology Experiment, or LITE. The objective of LITE is to develop the technology base and measurement techniques necessary to operate a solid-state laser lidar system from a spaceborne platform. The approach being used is to space-harden existing lidar components technology with minimum change. The initial experiment will utilize an Nd:YAG laser in a lidar system with multimode, single wavelength operation. The construction will incorporate easily modified modular concepts, and the experiments will focus on aerosol and cloud measurements. LITE is scheduled to fly on the space shuttle in 1988.

The next program following LITE to incorporate a lidar system in space will most likely be the Earth Observing System (EOS) Program scheduled for the middle 1990's time period. A lidar system for EOS is being proposed, and it is called the Lidar Atmospheric Sounder and Altimeter (LASA). The science objectives for LASA cover many disciplines, including hydrology, altimetry, geodynamics, aerosol, and cloud studies. The list of lidar measurements proposed for LASA is shown in Table 4. The 14 listed items have all been demonstrated in the laboratory, and most have been demonstrated in both ground based and low altitude airborne systems. LASE data from the ER-2

Table 4.

DEMONSTRATED LIDAR MEAUSREMENTS FOR LASA

		LAB	GROUND-BASED	LOW ALT. AIRBORNE	HIGH ALT. AIRBORNE	LASA SCI. OBJECTIVES
1.	ALTIMETRY	X	X	X		2
2.	RETRO-RANGING	X	X			3
3.	CLOUD TOP HEIGHT	X	X	X	X	1
4.	PBL HEIGHT	X	X	X	X	1
5.	STRAT. AERO.	X	X	X		4
6.	CLOUD PROP.	X	X	X	X	4
7.	TROP. AERO.	X	X	X	X	4
8.	H_2O COLUMN	X	X	X	A	1
9.	SURFACE PRES.	X				5
10.	O_3 COLUMN	X	X	X		5
11.	H_2O PROFILE	X	X	X	A	1
12.	PRES. PROFILE	X	X			5
13.	TEMP. PROFILE	X				5
14.	O_3 PROFILE	X	X	X		5

LASA SCIENCE OBJECTIVES: 1 HYDROLOGY; 2 ALTIMETRY; 3 GEODYNAMICS;
4 AEROSOLS AND CLOUDS; 5 OTHER

A: LASE MEASUREMENTS (1988)

aircraft will further demonstrate high altitude lidar measurement capability. Using the technology base developed from the LASE and LITE programs, LASA should be successful in gathering large quantities of data on a global basis to make a significant step forward in the scientific understanding of our atmosphere.

Beyond LASA, spaceborne lidar techniques will require laser wavelengths that range from about 0.2 to 11 µm to make measurements from the thermosphere to the ground. Specific wavelength requirements for future spaceborne lidar systems are listed in Table 5. Generally, the UV wavelengths are more applicable for the upper atmosphere, and the visible and IR wavelengths are more useful for

Table 5.

LASER WAVELENGTH REQUIREMENTS FOR FUTURE SPACEBORNE LIDAR APPLICATIONS

LASER WAVELENGTH REGION, nm	PARAMETER (WAVELENGTH, nm)
200-350	NO (215) O₃ (226.6) Na⁺ (279.6) Mg (285.2) OH (300) O₃ (310-340)
350-400	T & DENSITY*
400-450	NO₂ (440)
500-600	AEROSOLS* & CLOUDS* Na, T & WINDS (589)
700-800	H₂O (724) O₂ FOR P&T (760) K (770)
900-950	H₂O (940)
1.0-1.15 μm	AEROSOLS*, CLOUDS*, & DOPPLER WINDS* H₂O (1140)
2.0-5.0 μm	CO₂ (2.7 μm) CH₄ (3.3 μm) HCl (3.5 μm) H₂O (5.9 μm) SO₂ (4.0 μm) CO (4.6 μm)
9.0-11.0 μm	O₃ (10 μm) NH₃ (10.3 μm), NWH DOPPLER WINDS*

* NON-SPECIFIC WAVELENGTH

the troposphere. The general characteristics needed for spaceborne lasers and a list of tunable lasers for potential space lidar applications are given in Table 6. Most spaceborne lidar applications require tunable laser wavelengths with high average power. The characteristics given in Table 6 are for a free-flying satellite that may only be visited every 8 to 12 months. The types of tunable lasers given in the same table are examples of general laser types that are currently under development and are potential candidates for future spaceborne lidar missions.

Table 6.

SPACEBORNE LASERS

GENERAL CHARACTERISTICS	TUNABLE LASER TYPES
• HIGH POWER (1 - 20 W)	• ALEXANDERITE (720 -770 nm)
• WAVELENGTH TUNABLE (10 - 40 cm⁻¹)	• Co:MgF₂* 1.6 -2.3 μm
• NARROW LINEWIDTH (< 0.02 cm⁻¹)	• Ti:SAPPHIRE (700 -850 nm)
• LONG LIFETIME (4 - 12 MO)	• EMERALD (751 -759 nm)
• RUGGEDIZED (SHUTTLE LAUNCH, SPACE ENVIRON.)	• Nd:GLASS*
• REDUCED COMPLEXITY AND MODULAR	• EXIMER*
• CAPABLE OF MEETING EYE SAFTY CRITERIA	• CO₂ (LINE TUNABLE AND ISOTOPE BROADENED)

*RAMAN SHIFTED

This paper has discussed the evolution of lidar systems for Earth observations. Measurements of aerosols, O_3, and H_2O made with the NASA LaRC airborne DIAL system were presented as examples of the types of data that could be obtained from a spaceborne lidar system. An autonomous DIAL system under development by NASA for the high-altitude ER-2 aircraft was described. Future spaceborne lidar systems, LITE and LASA, were discussed, as well as the requirements for future spaceborne laser systems. It is clear that the successful development of spaceborne lidar systems depends critically on solving the laser technology challenges of improved laser efficiency and long lifetimes. Adequate power and narrow linewidth with tunability are also necessary for the full utilization of the laser's capability. New thrusts to improve optical detector sensitivity and quantum efficiency can be helpful by reducing power requirements. However, new technology developments must emerge for lidar to fulfill its vast potential. New research thrusts in solid-state laser pumping and solid-state, tunable, efficient, lasing materials are being made by NASA to open these space application frontiers within the next decade.

REFERENCES

1. Fiocco R. T. H. and Grams G. 1966, Observations of the Aerosol Layer at 20 km by Optical Radar, J. Atmos. Sci. 21(3), 323.
2. McCormick M. P. 1975, The Use of Lidar for Atmospheric Measurements, Remote Sensing Energy Related Studies, Vezeroghu T. N. ed., Hemisphere Press, Wahsington, 113.
3. Cooney, J. A. 1971, Comparisons of Water Vapor Profiles Obtained by Radiosonde and Laser Backscatter, J. Appl. Meteo., 10, 301.
4. Bowman M. R. et al. 1969, Atmospheric Sodium Measured by a Tuned Laser Radar, Nature, 221, 456.
5. Megie G. et al. 1978, Simultaneous Nighttime Lidar Measurements of Atmospheric Sodium and Potassium, Planet Space Sci. 26, 27.
6. Rowlett J. R. et al. 1978, Lidar Observations of Wave-like Structures in the Atmospheric Sodium Layer, Geophys. Res. Lett. 5(8), 683.
7. Schotland R. M. 1974, Errors in the Lidar Measurement of Atmospheric Gases by Differential Absorption, J. Appl. Meteor. 13, 71.
8. Browell E.V. et al. 1979, Water Vapor Differential Absorption Lidar Development and Evaluation, Appl. Opt. 18, 3474.
9. Pelon J. and Megie G. 1982, Ozone Monitoring in the Troposphere and Lower Stratosphere: Evaluation and Operation of a Ground Based Lidar Station, J. Geophys. Res. 87(C7), 4947.
10. Cahen C. et al. 1982, Lidar Monitoring of the Water Vapor Cycle in the Troposphere, J. Appl. Meteo., 21, 1506.
11. Browell E. V. et al. 1983, Airborne DIAL System and Measurements of Ozone and Aerosol Profiles, Appl. Opt. 22, 522.
12. Browell, E. V. 1983, Remote Sensing of Tropospheric Gases and Aerosols with an Airborne DIAL system, in Optical and Laser Remote Sensing, D. K. Killinger and A. Moorodian, eds., Springer-Verlag, pp. 138-147.

13. Shumate M. S. et al. 1981, Laser Absorption Spectrometer: Remote Measurement of Tropospheric Ozone, Appl. Opt. 20, 545.
14. Englisch W. et al. 1983, Laser Remote Sensing Measurements of Atmospheric Species and Natural Target Reflectivities in Optical and Laser Remote Sensing, Killinger D. K. and Mooradian A. eds, Springer-Verlag, 38.
15. Bilbro J. W. 1983, Coherent CO_2 Lidar Systems for Remote Atmospheric Measurements, In Optical and Laser Remote Sensing, Killinger D. K. and Mooradian A. eds., Springer-Verlag, 356.
16. Ismail, S. et al. 1984, Sensitivities in DIAL Measurements from Airborne and Spaceborne Platforms, Conf. Abs., Twelfth International Laser Radar Conference, Aix-en-Provence, France, August 13-17, 1985.
17. Browell E. V. 1979, Ed., Shuttle Atmospheric Lidar Research Program-Final Report of Atmospheric Lidar Working Group, NASA, Spec Publ.- 433.
18. Greco R. V. 1980, Ed., Atmospheric Lidar Multi-User Instrument System Definition Study, NASA Contract Rep 3303.

Halbleiterlaser für kohärente optische Nachrichtensysteme

A. Ullrich, Y. Wang
Institut für Nachrichtentechnik, Technische Universität Wien
Gußhausstraße 25, A-1040 Wien

1. Einleitung

Für die Datenübertragung im Weltraum zwischen Satelliten werden von der
Europäischen Weltraumbehörde (ESA) Empfänger für Laserstrahlung im
Wellenlängenbereich des nahen Infrarot untersucht. Aus der herkömmlichen Nachrichtentechnik ist bekannt, daß das Homodyn-Empfangsverfahren
- das ist ein Überlagerungsempfang mit einer Zwischenfrequenz von Null
Hertz - in Verbindung mit Phasenumtastung (PSK) das empfindlichste
System darstellt. Im folgenden wird untersucht, ob ein solches System
auch in der optischen Nachrichtentechnik mit Halbleiterlasern realisierbar ist.

Allgemein sind für kohärente Empfangssysteme - hiezu zählt auch der
Heterodyn-Empfang - folgende Eigenschaften an die optischen Signalquellen zu stellen: Monomodebetrieb, gute Frequenzkonstanz und geringes
Phasenrauschen. Für die Freiraumausbreitung erscheinen GaAlAs-Laserdioden, die bei einer Wellenlänge von etwa 850 nm emittieren, attraktiv,
da dadurch in Empfängern empfindliche und rauscharme Si-Photodioden
eingesetzt werden können. Allerdings bringt der Einsatz derartiger Laser
in Überlagerungsempfängern auch Probleme mit sich: GaAlAs-Laserdioden
zeigen eine starke Abhängigkeit der Emissionsfrequenz von Injektionsstrom (ca. -3 GHz/mA) und Temperatur (ca. -20 GHz/K), außerdem ist das
ausgesandte Signal in seiner Phase stark verrauscht. Dieses Phasenrauschen äußerst sich im Spektrum als instantane Linienbreite. Bei
GaAlAs-Laserdioden beträgt die volle Halbwertsbreite Δf des Lorentz'-
schen Linienprofils einige MHz.

2. Rauschverhalten einer PLL

Eine Voraussetzung für die Realisierung eines Homodyn-Empfängers ist
die Synchronisation des lokalen Oszillators mit dem Eingangssignal in
Frequenz und Phase. Diese Anforderung wird durch den Einsatz einer
Phasenregelschleife (PLL) erfüllt. Das Prinzip eines optischen Homodyn-
Empfängers zeigt Abb.1. Das Eingangssignal und das Signal des lokalen
Oszillators werden einem Mischer zugeführt. Stimmt die Frequenz überein,
so ist der Ausgang des Mischers der Phasendifferenz proportional. Dieses

Fehlersignal wird über ein Schleifenfilter dem steuerbaren lokalen
Oszillator (VCO) zugeführt um den Zustand der Phasensynchronisation
zu gewährleisten. Ist das Eingangssignal phasenmoduliert, so lassen
sich die Daten direkt am Ausgang des Phasendetektors rückgewinnen, sofern die Daten einen spektralen Bereich oberhalb der Eigenfrequenz der
Phasenregelschleife einnehmen.

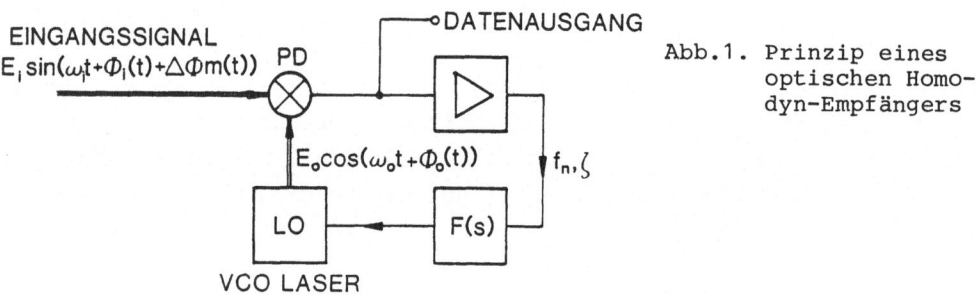

Abb.1. Prinzip eines
optischen Homo-
dyn-Empfängers

In Abb.2 ist das Blockschaltbild einer linearisierten PLL für den Frequenzbereich dargestellt, wobei Frequenzgleichheit zwischen Eingangssignal und lokalem Signal angenommen wurde. In einer optischen PLL ist
der Phasendetektor eine Photodiode, die vor allem auf Grund der mittleren Leistung des lokalen Oszillators neben dem Nutzsignal auch einen
Gleichanteil liefern wird, der ein Rauschsignal zufolge Schrotrauschens
hervorrufen wird. Als Störquellen werden das Schrotrauschen des Phasendetektors und die Linienbreiten (Δf_i, Δf_o) der optischen Signale berück
sichtigt. Auf Grund dieser Störungen ergibt sich ein zeitlich veränderlicher Phasenfehler, der das Signal-Geräusch-Verhältnis im Datenkanal
bestimmen wird. Eine wichtige Kenngröße dieses stochastischen Signals,
die Varianz, läßt sich aus den Leistungsdichtespektren der Störungen
und den Übertragungsfunktionen der geschlossenen Schleife berechnen.

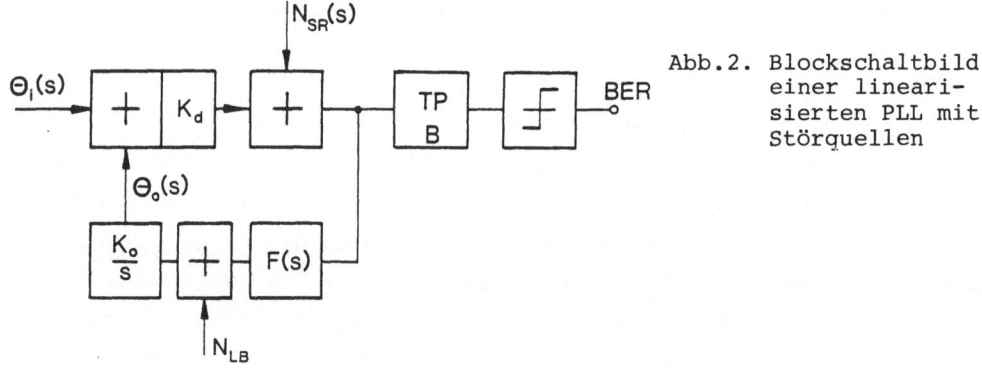

Abb.2. Blockschaltbild
einer lineari-
sierten PLL mit
Störquellen

Für eine PLL 2. Ordnung mit einem aktiven Schleifenfilter F(s) ergibt
sich im Datenkanal ein Phasenfehler zufolge der Linienbreite und des
Schrotrauschens (SR) des Phasendetektors mit einer Varianz von

$$\sigma^2_{\Delta f} = \frac{1}{4\zeta} \frac{\Delta f_i + \Delta f_o}{f_n} \qquad \sigma^2_{SR} = \frac{N_o}{K_d^2} . 2B . X(f_n, \zeta) .$$

Darin bezeichnet f_n die Eigenfrequenz und ζ den Dämpfungsfaktor der
Schleife, K_d den Verstärkungsfaktor des Phasendetektors, N_o die Rausch-
leistungsdichte des Schrotrauschens des Phasendetektors, B die Bandbreite des Tiefpaßfilters im Datenweg und X eine Korrekturfunktion,
die von den Schleifenparametern abhängt. Die gesamte Varianz des Phasenfehlers ergibt sich auf Grund der statistischen Unabhängigkeit der Störquellen als Summe der obengenannten Terme.

3. Bitfehlerrate in PSK-Homodyn-Systemen

Aus der Varianz des Phasenfehlers kann die Bitfehlerrate (BER) für ein
PSK-Homodyn-System nach /1/ zu

$$BER = \frac{1}{2} \, erfc \sqrt{\frac{\sin^2 \Delta\phi}{2(\sigma^2_{\Delta f} + \sigma^2_{SR})}}$$

berechnet werden, worin $\Delta\phi$ den Phasenhub bei bipolarer Modulation und
erfc die komplementäre Fehlerfunktion bezeichnet. Für vernachlässigbare
Linienbreiten ergibt sich ein Minimum für die BER bei einem Modulations-
hub von $\overset{+}{-}45^o$. Abbildung 3a zeigt den Zusammenhang von BER und Eingangsleistung eines Homodyn-Empfängers mit der Summe der Linienbreiten von
Eingangssignal und lokalem Signal als Parameter. Die Systemdegradation
durch die Linienbreiten der Signalquellen ist in Abb.3b dargestellt.
Homodyn-PSK-Systeme mit phasenverrauschten Trägersignalen können, wie
aus Abb.3b hervorgeht, entweder durch Verringerung der Linienbreiten
oder durch Erhöhung der Eigenfrequenz der Phasenregelschleife näher
an die Empfindlichkeitsgrenze, die sich aus dem Schrotrauschen der
Photodiode ergibt, herangeführt werden. Eine Vergrößerung der natürlichen Frequenz der PLL wird durch die technische Realisierbarkeit von
breitbandigen, gleichstromgekoppelten Verstärkern mit linearem Phasen-
gang, wie sie eine PLL benötigt, begrenzt. Eine Reduktion der Linienbreite von Halbleiterlasern läßt sich durch Hinzufügen eines dritten
Spiegels zum Laserresonator erzielen /2/. Mit diesem Verfahren sind
Linienbreiten von etwa 100 kHz zu erreichen. Durch zusätzliche Anti-
reflexbeschichtung einer Austrittfläche des Halbleiterlasers wurde

574

eine Linienbreite bei einem InGaAsP-Laser von 20 kHz erzielt /3/.

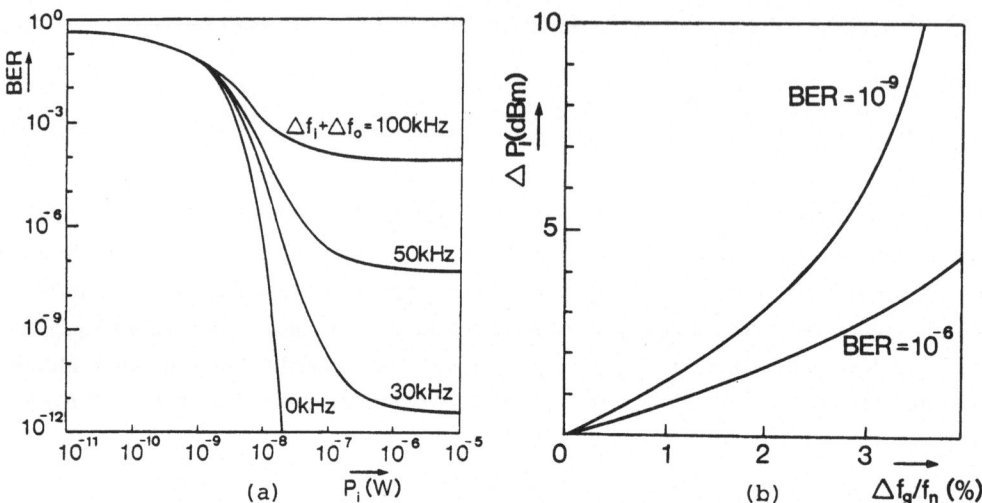

(a)   (b)

Abb.3 (a). Bitfehlerrate für optischen Homodyn-Empfänger. (b) Systemdegra
dation durch Linienbreite der Quellen. (λ = 830 nm, $\Delta\phi$ = 45°, Date
rate 140 Mbit/s; f_n = 1 MHz, ζ = 0.707; Parameter der Photodiode:
Rauschzahl 10, Quantenwirkungsgrad 0.8)

4. Ausblick

Um ein empfindliches optisches PSK-Homodyn-System mit Halbleiterlasern
realisieren zu können, werden vor allem auf dem Gebiet der Linienver-
schmälerung weitere Anstrengungen notwendig sein. Eine Verbesserung de
Empfindlichkeit gegenüber dem nichtkohärenten Direktempfang erscheint
bei den derzeit erreichbaren Linienbreiten durch den Einsatz des kohärenten Systems der Differenzphasentastung in Verbindung mit Heterodyn-
Empfang möglich zu sein, bei dem die Anforderungen an die Linienbreite
der Signalquellen geringer sind /4/.

Literatur

/1/ H.D. LÜKE, "Signalübertragung", Springer 1979.
/2/ S. SAITO, Y. YAMAMOTO, "Direct observation of Lorentzian lineshape
    of semiconductor laser and linewidth reduction with external gra-
    ting feedback," Electron. Lett., vol. 17, pp. 325-327, April 1981.
/3/ M.R. MATTHEWS et al., "Packaged frequency-stable tunable 20 kHz
    linewidth 1.5 µm InGaAsP external cavity laser," Electron. Lett.,
    vol. 21, pp. 113-115, Jan. 1985.
/4/ G. NICHOLSON, "Probability of error for optical heterodyne DPSK
    system with quantum phase noise," Electron. Lett., vol. 20, pp.
    1005-1007, Nov. 1984.

Development of a CO_2-Waveguide Laser with Wide Tuning Range

M.Endemann, E.Golusda

Battelle-Institut e.V.

Am Römerhof 35, D-6000 Frankfurt a.M./D

Design considerations and the actual hardware development of a cw-CO_2-waveguide laser with a wide tuning range of about 1.4 GHz are presented. The laser is developed as local oscillator source for the Intersatellite Laser Link ISL[2] /1/. However, the laser can be used in many different applications as well, like master-oscillator for injection-locked single-mode TEA-lasers, spectroscopy, or local oscillator in heterodyne radiometers and coherent lidars.

The laser is designed to meet the following specifications:
- Tunable over a 1.4 GHz wide range around line center ($\pm$ 700 MHz)
- Line-tunable
- Long lifetime
- Linearly polarized output in fundamental transverse mode
- Good short term and long term frequency stability

It has been shown /2/ that the tuning range df of a homogeneously broadened laser is approximately given by

$$(1) \qquad df = \beta(p) \cdot (g_0 l_a/(a+t) - 1)^{1/2}.$$

$\beta(p)$ is the line width of the transition which is a function of the gas pressure p,

$g_0(p)$ is the small signal gain,

l_a is the length of the active medium ($g_0 l_a$ is the single-pass gain),

t is the transmission of the output-coupler,

a are the residual resonator losses (a+t are the total resonator losses).

The line width $\beta(p)$ can be expressed as a function of the partial pressures of the different gas components of the filling, and of the average temperature T of the discharge /3/:

$$(2) \qquad \beta(p) = 5.69 \text{ MHz/hPa} \cdot (p_{CO2} + 0.64 p_{He}) \cdot ((300 \text{ K})/T)^{1/2}$$

Measurements with waveguide-lasers indicate that the partial pressure of carbon-dioxyde remains rather constant at 60 hPa and only the pressure of helium is adjusted to set the desired filling pressure /4/. The average gas temperature in the discharge has been measured to be near 400 K for a tap-water cooled waveguide /4/. Thus the pressure broadened line width for gas pressures near 300 hPa is approximately

$$(3) \qquad \beta(p) = 3.5 \text{ MHz/hPa} \cdot p = B \cdot p.$$

The gain coefficient g_0 depends strongly on the total gas pressure p, on the gas composistion, and also on the diameter d of the waveguide. An empirical relation for the gain for optimum gas mixture is /5/

$$(4) \qquad g_0 = G/(d \cdot p).$$

The constant G was determined as 0.23 hPa using own measurement data as well as data from ref.6. The dependence on the waveguide diameter d is derived with the scaling-laws for gas discharges /5/. It holds for diameters down to 1 mm. For smaller waveguides wall depletion effects become important and reduce the available small signal gain. Thus for maximum gain at high filling pressures, the selection of a waveguide diameter near one millimeter is optimum.

The length l_a of the active medium is limited by the resonator length l_r which in turn is limited by the condition that the desired frequency spacing of the axial modes $c/2l_r$ is larger than the desired tuning range df; otherwise extra mode-selecting elements must be introduced at the cost of increased complexity of the design. For a tuning range of 1.4 GHz, the maximum resonator length l_r is only 107 mm.

The resonator losses (a+t) must be minimized for a wide tuning range. Thus the transmission t of the output coupler should be as small as possible; it is limited by the resulting high intra-cavity intensities that can damage optical components. A realistic transmission value is about 1 %.

The remaining losses result from the line selecting element (usually a grating), guiding losses of the waveguide section, and the coupling losses between waveguide and resonator mirrors. Grating-losses are at least 5 % and thus form a major contribution to the overall resonator losses. They can be reduced by an additional partially transmitting mirror between grating and waveguide. The losses of this 'enhanced

grating' are predicted to be as low as 0.2 % for large beams /7/. Own
measurements indicate that the losses stay below 1 % even for small beam
diameters.

Guiding losses should remain well below 1 % /8/; however, actual measurements indicate losses above 1 %, probably due to thermal lensing
effects of the gas discharge.

Coupling losses stay below 1 %, as long as flat mirrors are positioned
within 1/10 of a Rayleigh-range behind the ends of the waveguide.

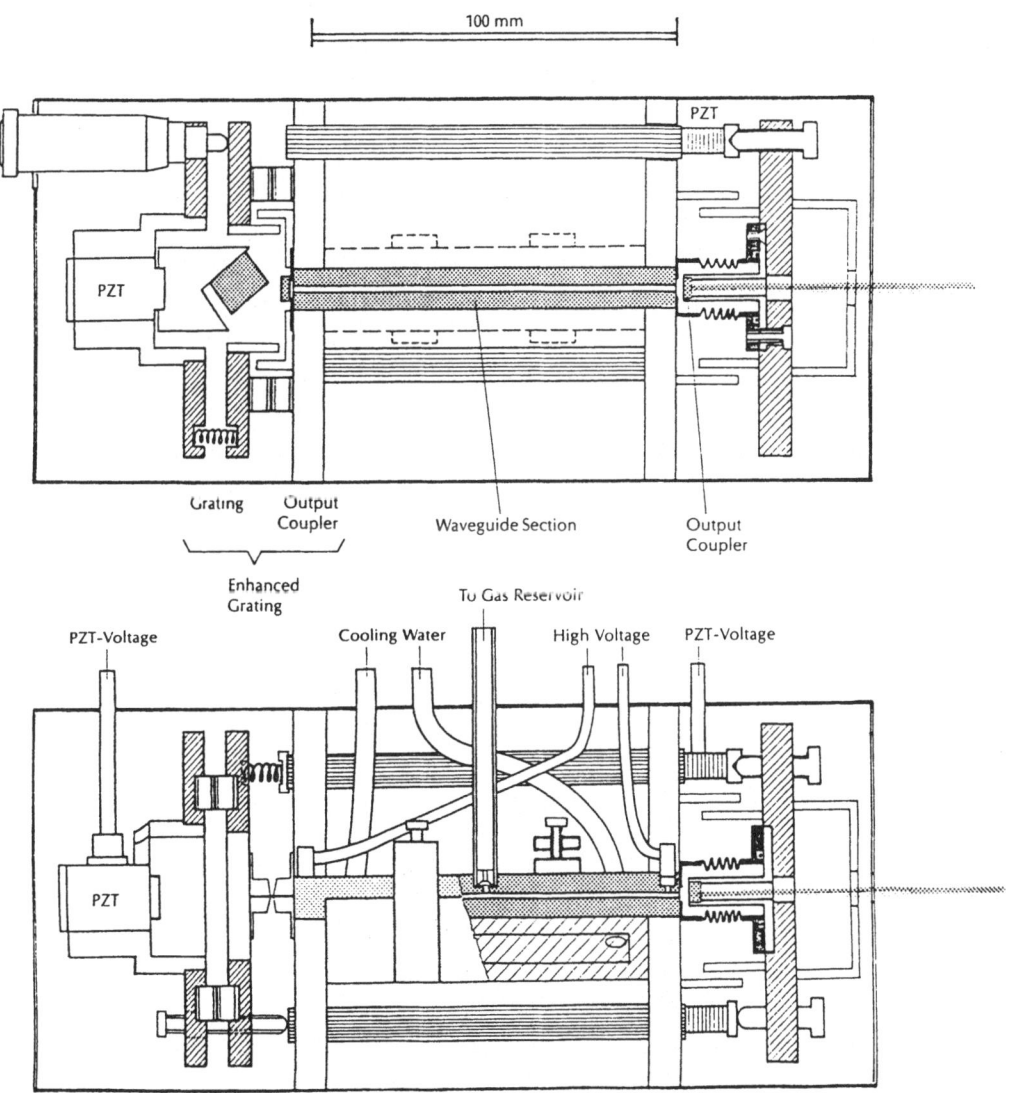

Top- and sideview of the tunable CO_2-waveguide laser

Using the pressure dependence of line width $\beta(p) = B \cdot p$ and the small signal gain $g_0(p) = G/(d \cdot p)$, the tuning range df can be expressed as

$$(5) \qquad df = B \cdot (2Gl_a/(d \cdot (a+t)) - p^2)^{1/2},$$

and it is easily shown that the widest tuning range df_{opt} is equal to the line width $B \cdot p_{opt}$ at optimum filling pressure p_{opt}, which is given by

$$(6) \qquad p_{opt} = Gl_a/(d \cdot (a+t)).$$

For an active length l_a = 98 mm, a waveguide diameter d = 1.5 mm, resonator and guiding losses a+t = 3.5 %, and the gain constant G = 0.23 hPa, the optimum gas pressure is 375 hPa, and the maximum tuning range is 1500 GHz. Thus the desired tuning range can be achieved when the resonator losses are kept very low.

The Figure shows a drawing of the tunable waveguide laser that was designed to fulfill the above specifications. Its gas-envelope is hard-sealed to provide a long operational lifetime. An external resonator structure is used for a good passive frequency stability. The design of the enhanced grating is inherently complex as independent adjustment of grating and partially transparent mirror must be provided. The resonator length and frequency offset is controlled with piezo-stacks at the output-coupler end of the waveguide laser.

References
(1) W.REILAND, M.ENDEMANN, W.ENGLISCH, B.DÖRBRAND: Intersatellite CO_2 Laser Links: Transceiver System Optical Design, Laser 85, München
(2) J.J.DEGNAN: Journ.Appl.Phys. 45, 257 (1973)
(3) R.L.ABRAMS: Appl.Phys.Lett. 25, 1411 (1974)
(4) M.ENDEMANN, W.ENGLISCH: Beam Pointing and Coherent Detection for Laser Data Links, Final Report for ESTEC Contract No.4673/81/NL/HP (1984)
(5) M.ENDEMANN, W.ENGLISCH: Development of a receiver Unit for Laser Data Links - Local Oscillator Laser -, Final Report for ESTEC Contract No.4324/80/NL/HP (1982)
(6) R.L.ABRAMS, W.B.BRIDGES: IEEE Journ.Quant.Electron. 9, 940 (1973)
(7) J.E.BJORKHOLM, T.C.DAMEN, J.SHAH: Opt.Comm. 4, 283 (1971)
(8) E.A.MARCATILI, R.A.SCHMELTZER: Bell Syst.Tech.Journ. 43, 1783 (1964)

Gekühlte Detektor-Vorverstärker-Einheit für einen 10,6 μm Homodyn-Empfänger

R. Flatscher

Technische Universität Wien, Institut für Nachrichtentechnik
Gußhausstraße 25, A-1040 Wien

Für breitbandige Datenübertragung zwischen Satelliten wird im Auftrag
der Europäischen Weltraumbehörde ESA das Labormodell einer Sende-
Empfangseinheit bei einer Wellenlänge von 10,6 μm aufgebaut. Bei einer
vorläufigen Datenrate von 140 Mbit/s soll ein digital phasenmodulier-
tes Eingangssignal mit einer Leistung von 1 nW detektiert werden /1/.
Für den vorliegenden Anwendungsfall zeigt der Homodyn-Empfänger gegenüber anderen Empfangsverfahren deutliche Vorteile /2/. Ein wesentlicher
Teil dieses Empfängers ist eine auf ca. 80 K gekühlte Detektor-Vorver-
stärker-Kombination.

Als Alternative zur Strahlungskühlung ist ein mechanischer Kühler vorgesehen, mit dem Vorteil der Erzielbarkeit von tieferen Temperaturen
und einer weitgehend freien Wahl der Anordnung im Satelliten. Ein
Labormodell wurde mit einem kommerziell erhältlichen Stirling-Kühler
aufgebaut. Der abgesetzte Kaltteil befindet sich in einem Vakuumgefäß.
An ihm sind die Detektordiode und der Vorverstärker befestigt. Der
Kühler hat bei einer Eingangsleistung von ca. 40 W eine Kühlkapazität
von 1 W bei einer Temperatur von 80 K. Die vergleichsweise geringe
Lebensdauer von 2500 Std. und das Auftreten von Vibrationen bieten
Raum für Verbesserungen.

Eine Temperaturstabilisierung kompensiert die Strahlungsverluste zufolge der Leckrate des Vakuumgefäßes und hält die wesentlichen Detek-
toreigenschaften konstant. Beim Homodyn-Empfänger muß vor allem der
Dunkelstrom konstant bleiben,um in der Phasenregelschleife zur Ansteuerung des lokalen Laseroszillators keinen Phasenfehler vorzu-
täuschen. Die Temperatur läßt sich von 80 - 120 K einstellen. Die Abweichung von der Solltemperatur beträgt maximal 5 mK. Die tatsächliche
Temperatur wird mit einer Auflösung von 0,1 K digital angezeigt.

Eine kommerziell erhältliche HgCdTe-Photodiode dient als Detektor. Die
aktive Fläche beträgt 200 μm x 200 μm, die empfohlene Arbeitstemperatur
ca. 80 K. Die gemessene Strom-Spannungs-Kennlinie in Abhängigkeit der

optischen Leistung zeigt Bild 1a. Aus diesem Kennlinienfeld kann der Photostrom als Funktion der optischen Leistung gewonnen werden (siehe Bild 1b). Parameter ist hier die Vorspannung des Elementes. Die Steigung der Kurve entspricht dem Quantenwirkungsgrad, der hier 65% erreicht.

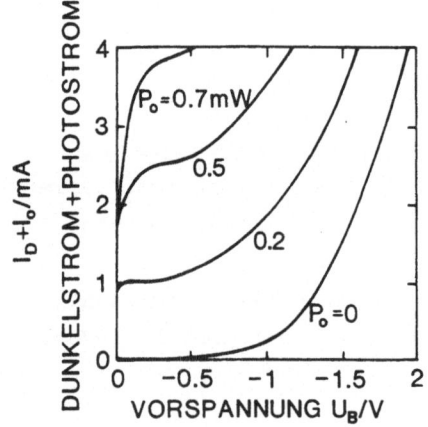

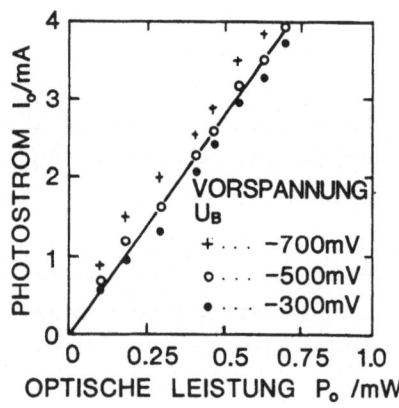

Bild 1a. Strom-Spannungs-Kenn-
        linie, Parameter: opti-
        sche Leistung

Bild 1b. Photostrom über opti-
        scher Leistung , Para-
        meter: Vorspannung

Bei höheren Frequenzen wird die Güte des Detektors durch seine Impedanz charakterisiert. Die entsprechenden Meßergebnisse sind im Bild 2a im Smith-Diagramm dargestellt. Diese Impedanz kann durch das Ersatzschaltbild in Bild 2b angenähert werden. Mit den Werten der Schaltelemente kann das theoretische Signal-Geräusch-Verhältnis errechnet werden.

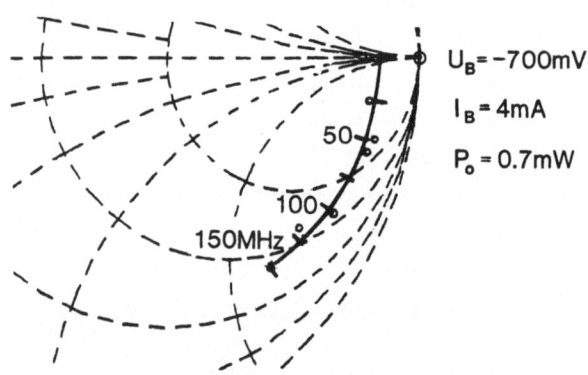

Bild 2a. Detektorimpedanz im Smith-Diagramm

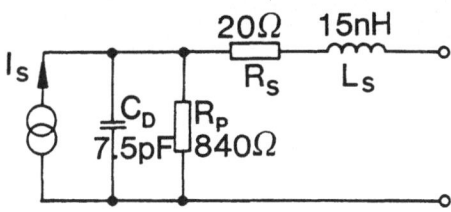

I_S ... Photostrom

C_D ... Parallelkapazität

R_P ... Parallelwiderstand

R_S ... Serienwiderstand

L_S ... Serieninduktivität

Bild 2b. Hochfrequenz-Ersatzschaltbild

Bei großer Leistung des lokalen Oszillators dominiert das Schrotrauschen der Photodiode, und der Signal-Geräusch-Abstand erreicht sein Maximum. Die zulässige optische Leistung des Detektors ist aber mit etwa 0,7 mW begrenzt. Die thermischen Rauschbeiträge der Photodiode und vor allem des Vorverstärkers machen sich deshalb bemerkbar und müssen minimiert werden. Streukapazitäten und Induktivitäten in der Verbindung des De-tektors zum Vorverstärker verschlechtern die Rauscheigenschaften bei hohen Frequenzen entscheidend. Daher muß der Verstärker in räumlicher Nähe des Detektors angebracht und somit zwangsweise gekühlt werden.

Bei Temperaturen um 80 K sind nur Feldeffekttransistoren (FETs) als aktive Bauelemente geeignet. MOSFETs haben eine wesentlich größere Steilheit als bipolare FETs und sind bis etwa 200 MHz, GaAs-FETs bei höheren Frequenzen vorteilhaft verwendbar. Bei allen FETs verbessert sich durch Kühlung sowohl das Rauschverhalten als auch die elektrischen Eigenschaften. Empfindliche optische Breitbandempfänger werden mittels Transimpedanz-Verstärker realisiert. Am Verstärkereingang wird dabei die durch die Verstärkung dividierte Gegenkopplungsimpedanz wirksam. Die relativ kleine Eingangsimpedanz ergibt eine große Bandbreite und gutes dynamisches Verhalten. Bei verschwindendem Rückkopplungsnetzwerk er-hält man einen Verstärker mit großer Eingangsimpedanz. Da der Rausch-beitrag dieses Netzwerkes fehlt, ist diese Variante der empfindlichste Verstärker. Bild 3a zeigt die Schaltung eines solchen Verstärkers mit MOSFETs, der in Chiptechnologie aufgebaut und mit dem mechanischen Küh-ler bei einer Temperatur von 80 K getestet wurde (siehe Bild 3b).

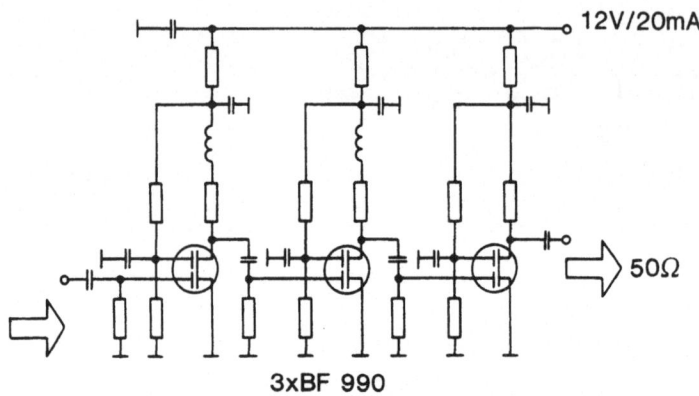

Bild 3a. Schaltbild eines hochohmigen Verstärkers

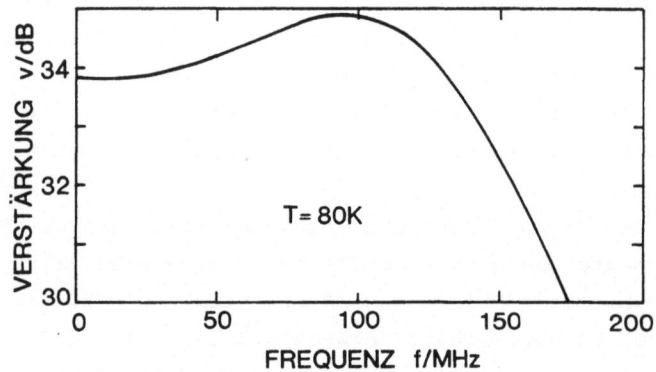

Bild 3b. Frequenzgang der Verstärkung

Die ersten beiden Stufen erzeugen die gewünschte Verstärkung, die letzte
dient der Impedanzanpassung an 50 Ω. Mit einer 3 dB-Bandbreite von 165
MHz ist dieser Verstärker ausgezeichnet für das 140 Mbit/s-System geeignet. Durch die hohe Eingangsimpedanz wird das Eingangssignal integriert. In einer anschließenden Stufe wird die Entzerrung und Bandbe-
grenzung vorgenommen.

Literatur

/1/ W.R. LEEB et al.: Phase-locked Loops for Optical Homodyne Detection,
    Final report ESTEC contract No. 5083/82, Dec. 1983
/2/ W. ENGLISCH et al.: Laser 83 Optoelektronik, Springer-Verlag Heidel-
    berg, S. 470-475, 1984

Intersatellite CO_2 Laser Links: Transceiver System Optical Design

W. REILAND, M. ENDEMANN, W. ENGLISCH
Battelle-Institut e.V.
Am Römerhof 35, D-6000 Frankfurt am Main 90

B. DÖRBAND
Institut für Technische Optik, Universität Stuttgart
Pfaffenwaldring 9, D-7000 Stuttgart 80

INTRODUCTION

The benefits of optical carriers in data transmission and the availability of powerful laser sources have turned first development efforts into a practical technology for communications. The combination of high bandwidth capability with the possibility of obtaining tremendous gain with small size antennae has made optical communication also very attractive for space applications.

From present status of technology, long range bidirectional intersatellite laser links with data transmission rates in the (multi) Gbit/s regime seem to be most feasible by using CO_2 laser technology /1/.

To efficiently accomplish the functions required in an optical communication system, high quality optical components have to be used. In order to achieve the high antenna gains the optics must approach diffraction limited performance. This, together with the requirements for optical homodyning, has to be taken into account in optical subsystem design which is presented in this paper. Its overall performance in checked against by optical ray tracing calculations.

OVERALL SYSTEM CONCEPT

To meat the given mission requirements and technical objectives of e.g. GEO-GEO and GEO-LEO links (GEO = GEOsynchronous satellite; LEO = Low Earth Orbiting satellite) /2/ the most promising concept proofed to be a transceiver package built up of an external electro-optically phase-modulated CO_2 laser transmitter /3, 4/, a coherent optical (homodyne) receiver /5-7/, diffraction limited optics /8/ common to both the transmit and receive channels, and a high performance acquisition and tracking subsystem /9/. Thereby, the optical subsystem acts as an interface and (polarization sensitive) switch between the different transceiver subsystems as schematically shown in fig. 1.

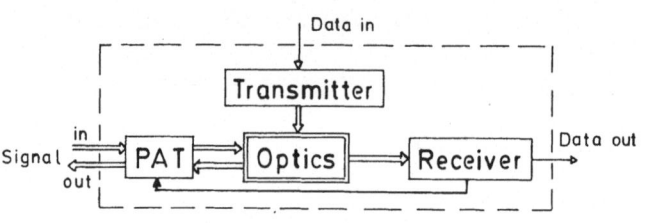

Fig. 1. Schematic representation of the transceiver subsystems mutual optical (⟹) and electronical (⟶) interaction

The optics have to handle both the received nanowatts IR-signal and the data-modulated transmit beam in the one-watt range, simultaneously. A detailed description of the conceptual design of the optical subsystem is given in the following section.

OPTICAL SUBSYSTEM DESIGN

Fig. 2 shows technical drawings (side and top view) of the CO_2 laser

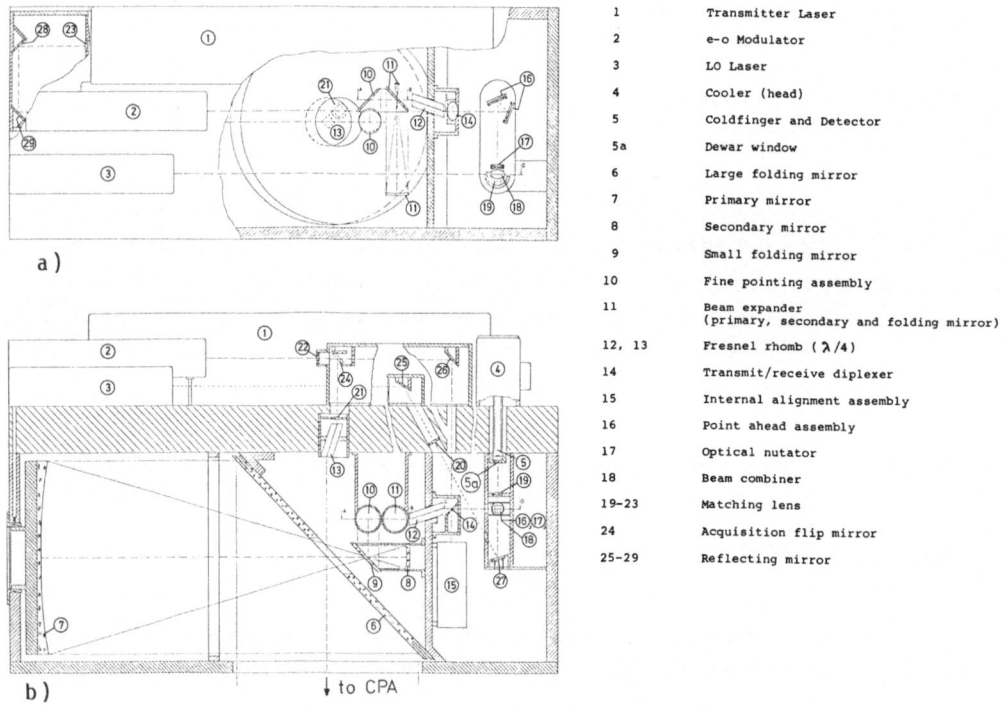

1	Transmitter Laser
2	e-o Modulator
3	LO Laser
4	Cooler (head)
5	Coldfinger and Detector
5a	Dewar window
6	Large folding mirror
7	Primary mirror
8	Secondary mirror
9	Small folding mirror
10	Fine pointing assembly
11	Beam expander (primary, secondary and folding mirror)
12, 13	Fresnel rhomb ($\lambda/4$)
14	Transmit/receive diplexer
15	Internal alignment assembly
16	Point ahead assembly
17	Optical nutator
18	Beam combiner
19-23	Matching lens
24	Acquisition flip mirror
25-29	Reflecting mirror

Fig. 2. CO_2 Laser Transceiver Package (top (a) and side view (b)).

transceiver package, with special emphasis on arrangement and interfacing of the optical components. Mechanical details are largely simplified.

Transmit channel. The laser transmitter (1) emits an unmodulated linear polarized Gaussian beam in TEM$_{OO}$ mode which is suitably matched (23) to the travelling-wave-type electro-optic light pipe modulator (2). Application of a modulating voltage (carrying the information signal) to the CdTe modulator crystals changes the phase state of the light wave thus estabilishing the desired binary PSK (= Phase Shift Keying) modulation format.

The modulated beam, still linearly polarized, then passes another matching lens (22) and is directed (26) into the transmit/receive diplexing assembly where it is efficiently reflected off (R=0.99) the wire-grid polarizer (14) and enters that part of the opto-mechanical subsystem which is common to both the transmit and the receive channel.

Passing the quarter wave phase retarder (Fresnel rhomb (12)) the laser light becomes circularly polarized and enters the intermediate beam expander (11) which enlarges the beam diameter and, correspondingly, reduces the beam divergence by a factor of 6. The enlarged beam is reflected off the fine pointing mirrors (10) which are movable around two othogonal axes and provide the fine pointing capability. They are located as close as possible to the exit pupil of the large telescope in order to minimize vignetting effects. The telescope (6-9) is of folded Gregorian type. It operates in the beam expanding mode and reduces the beam divergence further to final value (another factor of 8.5). The large folding miror (6) of the telescope directs the beam onto the coarse pointing mirrors (optical surface effective diameter = 25 cm each) which are movable around two orthogonal axes (Coudé-type arrangement) and provide the coarse pointing capability with more than hemispherical coverage.

During acquisition the transmitting terminal must provide a widened beacon beam with a beam divergense of about 3.7 mrad (half cone angle) which is about a factor of 100 wider than the communication beam width.

The full beacon beam is extracted from the ordinary beam path via the actuated mirror (24) of the acquisition beam assembly. A quarter wave phase retarder (Fresnel rhomb (13)) generates the required circular polarization state while an appropriate matching lens (21) diverges the beam to the desired value.

The beam circumvents the total front-end of the optical subsystem and passes through the central bore of the telescope's large folding mirror and is directed towards the opposite terminal via the coarse pointing mirrors.

Receive channel. The received laser light originating from the opposite terminal is circularly polarized. It follows the common optical beam path in the vice versa direction. The dimensions of the received light bundle are determined by apertures (25 cm ∅) and field stops which reject unwanted background radiation. They are governed by the laws of diffraction. At the transmit/receive diplexer the Fresnel rhomb (12) converts the received circular polarized light into linear polarized light orthogonal to that of the transmit beam. Consequently the signal beam (about 5 mm in diameter) is transmitted efficiently (0.96) by the wire-grid polarizer (14) and enters the detection optical beam path. It is reflected off the point-ahead mirrors (16) which compensate for the light "aberration" due to the transverse relative velocity of the orbiting terminals. The received light passes further the rotating wedge (17) of the optical nutator which creates a slight spatial modulation required to generate the spatial tracking error signal. The local oscillator (LO,3) beam is appropriately matched (25, 20, 27) and inserted colinearly to the signal beam via the LO beam combiner (18). Both beams are then focused (19) onto the cooled mixer element (5) with the focused spot sizes matched for optimum homodyne detection. The resulting electrical signal is amplified and processed according to the needs of data demodulation, frequency acquisition and tracking, spatial acquisition and tracking, and internal alignment (15).

RAY TRACING RESULTS FOR TOTAL RECEIVER CHANNEL

Completing the optical subsystem design and putting emphasis on the receiver channel for reasons of optimum coherent detection, calculations have been carried out using ray tracing software which is able to handle aspheric surfaces /10/ with individual decentering and tilt and to investigate vignetting performance with central bored surfaces (e.g. elemements 6, 9, and 11). Also waveaberrations, transverse aberrations and tolerances for misalignment and misshape have been calculated for phase front distortions $\leq \lambda/20$ within the detector plane.

<u>Focusing lens (19)</u>. Due to the large F-number of 5.9 and the long wavelength of 10.6 µm used, large tolerances in the vertex radius r and the conic coefficient e can be tolerated while keeping waveaberrations smaller than λ/20. Thus an aspheric surface is not really needed and can be replaced by a spherical element.

Calculating the back focal length of the focusing lens/dewar window (5a) assembly with the given elements specifications, a focus shift of up to 0.14 mm equivalent to a change in thickness of 0.33 mm for the lens or 0.24 mm for the dewar window can be tolerated for wave-aberrations $\leq \lambda$/20.

<u>Vignetting</u>. In order to estimate the effect of vignetting on the performance of the total receiving path, the vignetting factor, i.e. the ratio of transmitted light power for a given field angle to that at field angle 0 has been calculated. Thefine pointing mirrors have been thought to be adjusted to compensate for receiver field angle 0° (signal beam parallel to receiver optical axis). The results show as expected, that vignetting is mainly due to the small folding mirror of the telescope (9) with central (ellipt.) bore of 5 x 3.5 mm and to the two-sided mirror of the beam expander (11) with central elliptical bore of 1,7 x 1,2 mm. Although the vignetting factor is equal to 1 up to about an angle of 0.05°, it drops to about 0.66 for a field angle of 0.15° (fig. 3), which is equal to

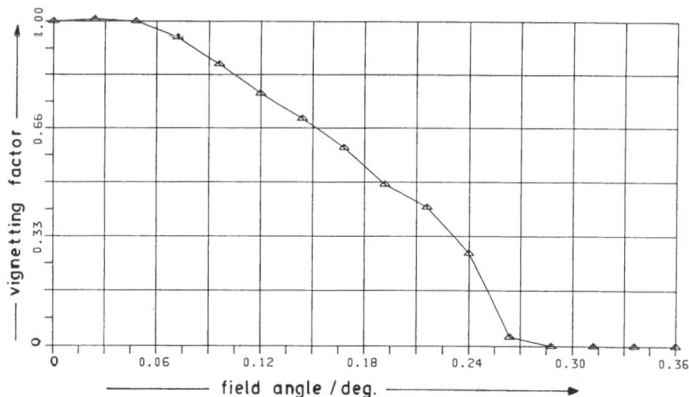

<u>Fig. 3.</u> Vignetting factor as a function of field angle (fine pointing mirrors adjusted)

the acquisition beam pointing uncertainly assumed for a GEO–GEO/LEO link. This result shows that central bore diameters have to be optimized, which isn't as critical however, due to large margin in acquisition link budget.

Spot diagrams and waveaberrations. For the case of maximum field angle of 0.15° and fine pointing mirrors adjusted appropriately, spot diagramas (SD), waveaberrations (WAB) and transverse aberrations (TAB) have been calculated. The SD and WAB results are shown in fig. 4 and are both corrected for zero point offset and

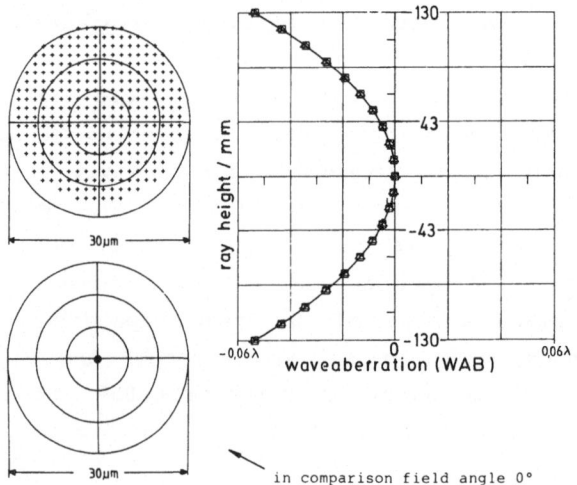

in comparison field angle 0°

Fig. 4. Spot diagram and waveaberration (WAB) for total receiver channel (field angle 0.15°)

linear terms (caused by optical nutator wedge), respectively. Taking into account the quadratic curvature of the WAB as a function of optical ray heights, the spreading of the spot to a diameter of up to 30 µm has to be interpreted as a defocusing term due to the large field angle of 0.15°. Comparing this geometrical spot diameter to the calculated (diffraction limited) Airy disk of 153 µm and realizing that major aberrations result from outer rays, we may conclude for a perfectly aligned system, that aberrations will have only a small effect on overall system performance.

Tolerances of shape and adjustment. Assuming a perfectly aligned system, the tolerances in shape and adjustment have been calculated which would lead to waveaberrations of $\lambda/20$.

Element	Δr (mm)	Δe	Δn	Δd (mm)	Δdef (mm)	Δdec (mm)	Δtilt (°)
prim. mirror (telescope)	0.030	0.006			0.013	0.047	0.004
sec. mirror (telescope)	0.030	0.060			0.013	0.047	0.033
prim. mirror (beam exp.)	0.126	0.043			0.065	0.237	0.071
sec. mirror (beam exp.)	0.126	3.000			0.065	0.237	0.443
focusing lens (front surface)	0.194	4.850				2.1	6.9
focusing lens (back surface)			0.00684	0.332			4.6
dewar window			0.26700	0.240			
detector plane					0.139		

Table 1. Tolerances for maximum waveaberrations of $\lambda/20$ (r vertex radius, e conic coefficient, n refractive index, d thickness, def defocus, dec decentering).

As may be seen from table 1 some critical elements (e.g. tilt angles of primary and secondary telescope mirrors) can be identified. However, one should take into account, that the ray tracing calculations do not include diffraction effects (e.g. compare diameter of spot diagram of fig. 4 with the actual diffraction limited focus of about 150 µm) and overestimate effects resulting from far off-axis rays.

Nevertheless, the presented optical subsystem design proofed to meet the general requirements. It shows high optical throughput, diffraction limited performance and is suitable for coherent optical detection, i.e. small wave front distortions ($\leq \lambda/20$) may be guaranteed if state of the art optics and opto-mechanics are used.

OUTLOOK

Summarizing overall CO_2 laser transceiver package technolgy development activities (under contract from ESA and the German Space Authority) present-day results look very promizing. The progress in hardware realization for near future laboratory breadboard model integration shows that also experimental inflight testing in the early 1990's appears to be feasible.

Thereby, following the rapidly increasing demand on high performance (tele-) communication (intersatellite) links, optical carrier systems grow into a potentially attractive alternative to convential radio- and microwave systems.

REFERENCES

/1/   Proceedings of an ESA workshop on SPLAT,
      Les Diablerets, Switzerland, March 1984 (ESA-SP-202,
      May 84, e.g. "Prospects of Laser Communication in Space"
      W.R. LEEB, pp. P3

/2/   "CO_2 Laser Transceiver Package with Gbit/s capability for
      Space-to-Space Communications"
      W. ENGLISCH, M. ENDEMANN, W. DIEHL, W. WIESEMANN, pp. P99, ibid

/3/   e.g. "Electrooptic gigahertz light-pipe modulator for CO_2 laser
      wavelength"
      A.L. SCHOLTZ, W.R. LEEB, E. BONEK
      IEEE J. Qantum Electronics QE-18 (1982), pp. 14

/4/   "Raumflugtauglicher CO_2-Laser"
      W. WIESEMANN, W. DIEHL, M. ROTHER, M. ENDEMANN, W. ENGLISCH
      Proc. Laser 83 Opto-Electronics (Munich, June 1983)
      Springer-Verlag, Heidelberg, 1984, pp. 470

/5/   "Receiver Concepts for data transmission at 10 microns"
      A.L. SCHOLTZ, K.H. PHILIPP, W.R. LEEB
      Proc. ESA Workshop on SPLAT, March 1984
      (ESA SP-202, May 84) pp. P 107

/6/   "The Concept of the 'low intermediate frequency translation
      loop' LIFTL and its implication on a CO_2 laser intersatellite
      data link system"
      M. WITTIG, pp. P 117, ibid.

/7/   "Development of a CO_2 waveguide laser with wide tuning
      range",
      M. ENDEMANN, this conference

/8/   "Design of a CO_2 laser transceiver package for space-to-space
      optical communication"
      W. ENGLISCH, final report, ESTEC contract no. 4006/79

/9/   "Pointing, Acquisition and Tracking for intersatellite optical
      data links"
      W. AUER, Proc. ESA Workshop on SPLAT, March 1984
      (ESA SP-202, May 1984), pp. P 131

/10/ "Auslegung von Kompensationssystemen zur interferometrischen
      Prüfung asphärischer Flächen"
      B. DÖRBAND, H.J. TIZIANI
      Optik 67, 1984, pp 1

A Pointing, Acquisition and Tracking System (PATS) for Inter-Satellite Laser Links

R. Kern

Teldix GmbH

Grenzhöfer Weg 36, D - 6900 Heidelberg

1. Introduction

For more than 10 years, there have been various activities in laser communications for space applications, e.g. see (1 - 8). In contrast to the already existing micro-wave communication systems, the laser communication offers, among other things, the following advantages: very high data rates, small antennas and privacy. Yet, because of very narrow beams, PAT Systems,offering extremly high accuracies,are required. In the following, the operation of such a System is described and some details of the PATS control are illustrated.

2. Ray Traces and Operational Modes

In Fig. 1, the complete transceiver package is shown which is explained in (1,3). The mechanical assemblies of the PATS, which are the Coarse Pointing Assembly (CPA), the Fine Pointing Assembly (FPA), the Point-Ahead Assembly (PAA) and the Optical Nutator Assembly (ONA), are shaded in Fig. 1 and the ray traces are shown. For testing the PATS breadboard model, additional components such as telescope, beam expander, photodetector and drive and control electronics are required. The received beam is led through the CPA which has a clear aperture of 25 cm and reaches the mirror telescope. It reduces the beam to 3 cm and enlarges the angular relationships. Leaving the tele-scope the beam passes through the FPA and the succeeding beam expander which reduces the beam diameter by 1/6. Up to now, the same trace has been passed through in the reverse direction by the transmitted beam reflected into the trace at this point. Leaving the beam expander the beam received passes through the ONA and a lens fo-cussing it onto the photodetector. The pointing, acquisition and tracking procedures are as follows: Pointing: First, the CPAs of both terminals are directed to each other. For this purpose, the data of the not exactly known satellite positions are processed. Acquisition: One of the satellites illuminates the opposite one with a broadened beam. Now, the receiver of the illuminated satellite scans with a narrow field of view (FOV) of about 20 µrad a region of 0.3° by means of the FPA. The scan pattern is a spiral of Archimedes. Tracking: The tracking procedure starts if the line of sight (LOS) is detected. Then, the opposite satellite starts the acquisition proce-dure to find the LOS and to switch over to the tracking mode.

3. PATS Control

In the block diagram of Fig. 2, the optical and electrical connections of the assemblies are shown. In the acquisition mode, the computer-activated switches are in position A and in the tracking mode in position T. The pointing mode is not explicitly illustrated. In this mode, the positioning of the CPA is executed by the computer and the electronics of the CPA transformation block. All blocks showing a terminal C are connected to the computer. The optical signal paths are marked by a double-line arrow, the electrical ones by a single-line arrow. The mechanical assemblies have motors (MO) or actuators (AC) and pickoffs (PO). The control signals are fed to the CPA's drive units (DU) and to the FPA's mirror units (MU). For the sake of simplification, the PAA is not taken into account because it has only electrical connections to the computer.

In the acquisition mode, the acquisition logic starts the wave generator. This device controls the FPA mirror units in such a way that the optical input of the FPA or the object space is scanned with the pattern of a spiral. If the FPA mirrors reach a position in which the skew-whiff input beam of the FPA leaves it in the direction of the optical axis, the photodetector generates a pulse which is detected by the peak detector. The target, i.e. the direction of the opposite satellite, is detected. Now, the peak detector triggers the acquisition logic which starts with its specific task namely, to switch over from the acquisition to the tracking mode without losing the detected signal. If the FPA mirrors were inertia-free the task would be solved by a simple switching from A to T. However, because of the mirrors' inertia, an ingenious search programme has to be involved to ensure the required switch-over from A to T. Thereafter, the FPA mirror units are controlled by the demodulator such as to illuminate a maximum area of the photodetector. Details of the tracking control loops are shown in Fig. 3 and pointed out in the following.

In a satellite-fixed coordinate system CS, see Fig. 1, the azimuth angle is designated by ϑ and the elevation angle by φ. In analogy with this, the rotation angles of the CPA drive units are designated by ϑ_M and φ_M. ϑ_M describes a rotation around the z-axis whereas the rotation axis of φ_M obviously depends on the actual angle ϑ_M. To receive a beam with its angles ϑ and φ at the time t = 0 the CPA motors must be controlled by the pointing law

$$\vartheta_M(0) \quad = \quad 90° + \vartheta \tag{1}$$

$$\varphi_M(0) \quad = \quad 90° - \varphi . \tag{2}$$

Now, only small deviations from the optical input axis of the FPA are considered. Then, these deviations of a unit vector can be described by the direction cosines ψ, ξ

corresponding to the CS axes x, y respectively. If the CPA is not exactly aligned by the relations of (1) and (2), for example due to the fact that there are only approximation values of θ and φ given in the form

$$\theta_0 = \theta - \Delta\theta , \tag{3}$$
$$\varphi_0 = \varphi - \Delta\varphi , \tag{4}$$

then the following transformation is valid:

$$\begin{bmatrix} \psi \\ \xi \end{bmatrix} = 8.5 \cdot \begin{bmatrix} \cos\varphi_0 \cdot \cos(\varphi_0 + \theta_0) & -\sin(\varphi_0 + \theta_0) \\ \cos\varphi_0 \cdot \sin(\varphi_0 + \theta_0) & \cos(\varphi_0 + \theta_0) \end{bmatrix} \cdot \begin{bmatrix} \Delta\theta \\ \Delta\varphi \end{bmatrix} \tag{5}$$

The deviation from the optical input axis of the FPA causes a deviation from the optical output axis which in the first order approximation is given by (6)

$$\Delta x_1 = (\psi + 2\beta) \cdot f , \tag{6}$$
$$\Delta x_2 = (\xi - 2\alpha) \cdot f , \tag{7}$$

where f is the expansion ratio of the beam expander, and α, β are the rotation angles of the FPA mirrors. By means of the ONA and the photodetector, the error angles Δx_1 and Δx_2 are processed by applying the Conical Scan procedure, and the resulting signal is fed to the demodulator input. The demodulator generates the control signals for the FPA mirror units in the way that $\Delta x_1 = \Delta x_2 = 0$. If in this connection the angles α or β reach 2/3 of their maximum then the CPA drive units are rotated by the angles $\Delta\theta_M$ and $\Delta\varphi_M$ which are determined by the tracking law.

$$\begin{bmatrix} \Delta\theta_M \\ \Delta\varphi_M \end{bmatrix} = 2 \cdot \begin{bmatrix} \dfrac{\cos(\varphi_0 + \theta_0)}{\cos\varphi_0} & \dfrac{\sin(\varphi_0 + \theta_0)}{\cos\varphi_0} \\ -\sin(\varphi_0 + \theta_0) & \cos(\varphi_0 + \theta_0) \end{bmatrix} \cdot \begin{bmatrix} \alpha \\ \beta \end{bmatrix} \tag{8}$$

The discrete instants at which the drive units are moved by $\Delta\theta_M$, $\Delta\varphi_M$ are designated by t_k, k = 0, 1, 2 ..... For the sake of a simple notation, k is used instead of t_k. Now, the time-dependent angles can be written in the following form:

$$\theta_M(k + 1) = \theta_M(k) + \Delta\theta_M(k) \tag{9}$$
$$\varphi_M(k + 1) = \varphi_M(k) + \Delta\varphi_M(k) \tag{10}$$
$$\theta_0(k + 1) = \theta_0(k) + \Delta\theta_M(k) \tag{11}$$
$$\varphi_0(k + 1) = \varphi_0(k) + \Delta\varphi_M(k) \tag{12}$$

From Equ. 8 the singularity of the CPA can be seen which is caused by the elevation angle $\varphi_0 = 90°$. In this case no definite azimuth angle φ_M exists.

4. References

(1) W. ENGLISCH: Design of a CO_2 Laser Transceiver Package for Space-to-Space Optical Communication, Final Report ESTEC Contract No. 4006/79 (1983)

(2) W. AUER: Pointing, Acquisition and Tracking for Inter-Satellite Optical Data Links, ESA-Workshop SPLAT (1984)

(3) E. BONEK, H. LUTZ: CO_2 Laser Communication Technology for Intersatellite Data Links, ESA Journal, Vol. 5 (1981)

(4) F. E. GOODWIN et al: Experiment Definition Phase Shuttle Laboratory, Final Report NASA Contract NAS 5-20018 (1976)

(5) L. LIEBING, F. KUENSTLER: Laser Tracking and Acquisition for Satellite-Communication, 26 Congresso Scientifico Internationale per L'Elettronica (1979)

(6) R. KERN, G. NIEDER, K. KIERSCHKE: Study Note on Conceptual PAT Design, ESTEC Contract No. 5522/83/NL/GM(SC), (1984)

(7) W. ENGLISCH, M. ENDEMANN: Conical Scan-Tracking-Technik mit kohärentem optischem Empfang bei 10 μm, Proceedings of the 6th International Congress Laser 83 Optoelektronik, Springer-Verlag (1984)

(8) M. ROSS et al: Space Optical Communications with the Nd : YAG Laser, Proc. IEEE, Vol. 66, No. 3 (1978)

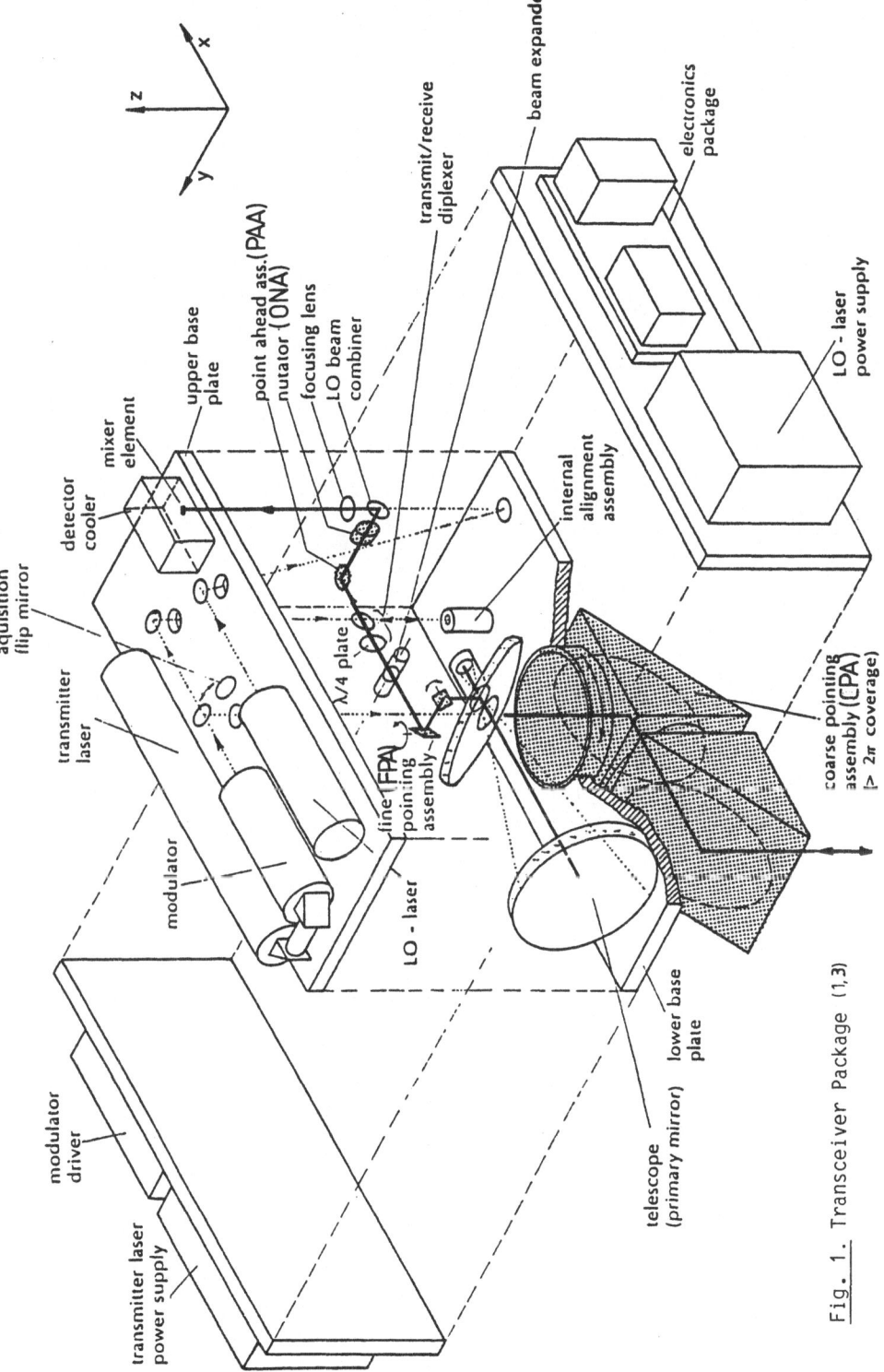

z

x

y

beam expander

transmit/receive diplexer

electronics package

LO - laser power supply

point ahead ass.(PAA)
nutator (ONA)
focusing lens
LO beam combiner

upper base plate

mixer element

detector cooler

internal alignment assembly

aquisition flip mirror

transmitter laser

λ/4 plate

modulator

fine pointing assembly (FPA)

coarse pointing assembly (CPA)
(> 2π coverage)

LO - laser

telescope (primary mirror)

lower base plate

modulator driver

transmitter laser power supply

Fig. 1. Transceiver Package (1,3)

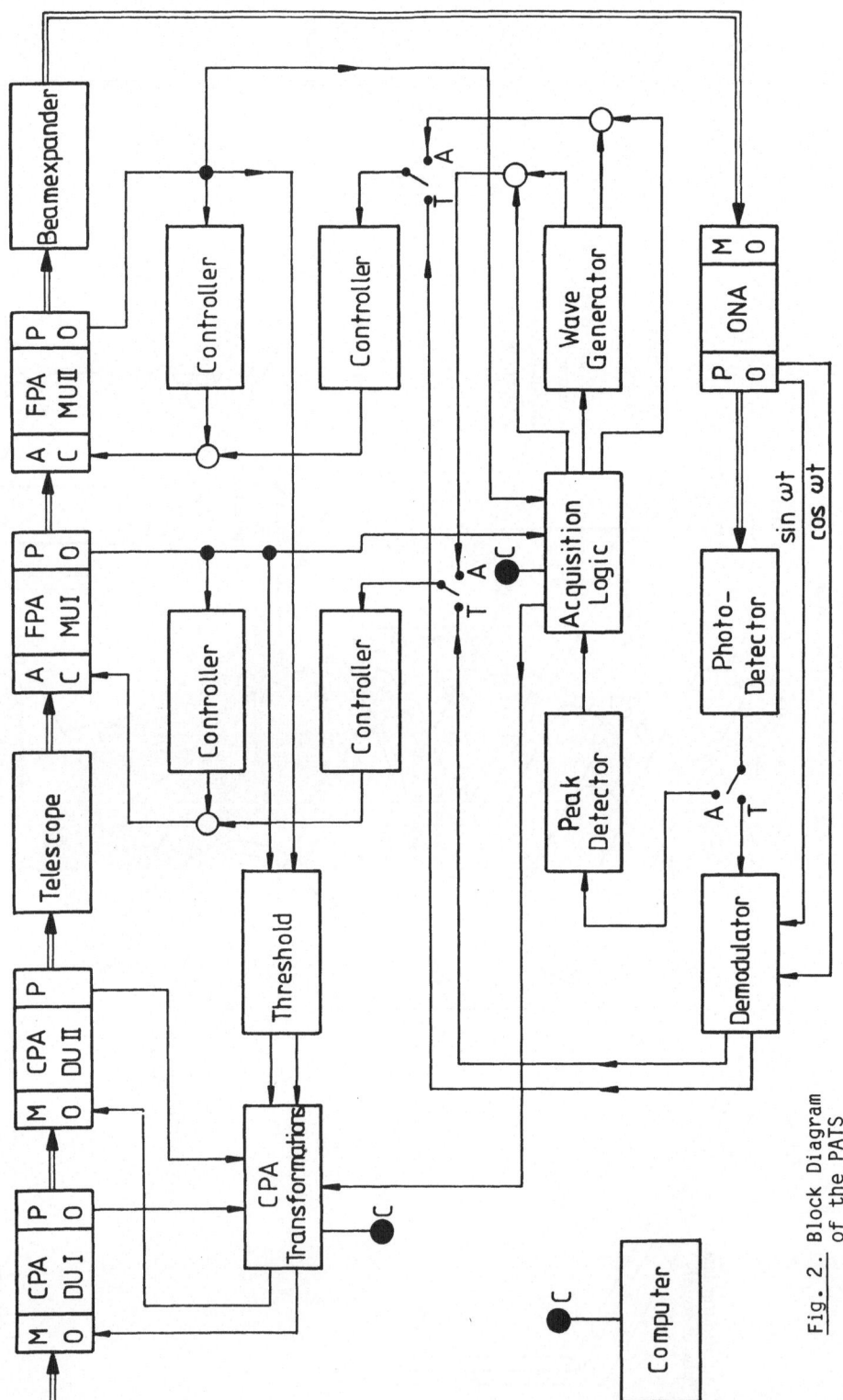

Fig. 2. Block Diagram of the PATS

597

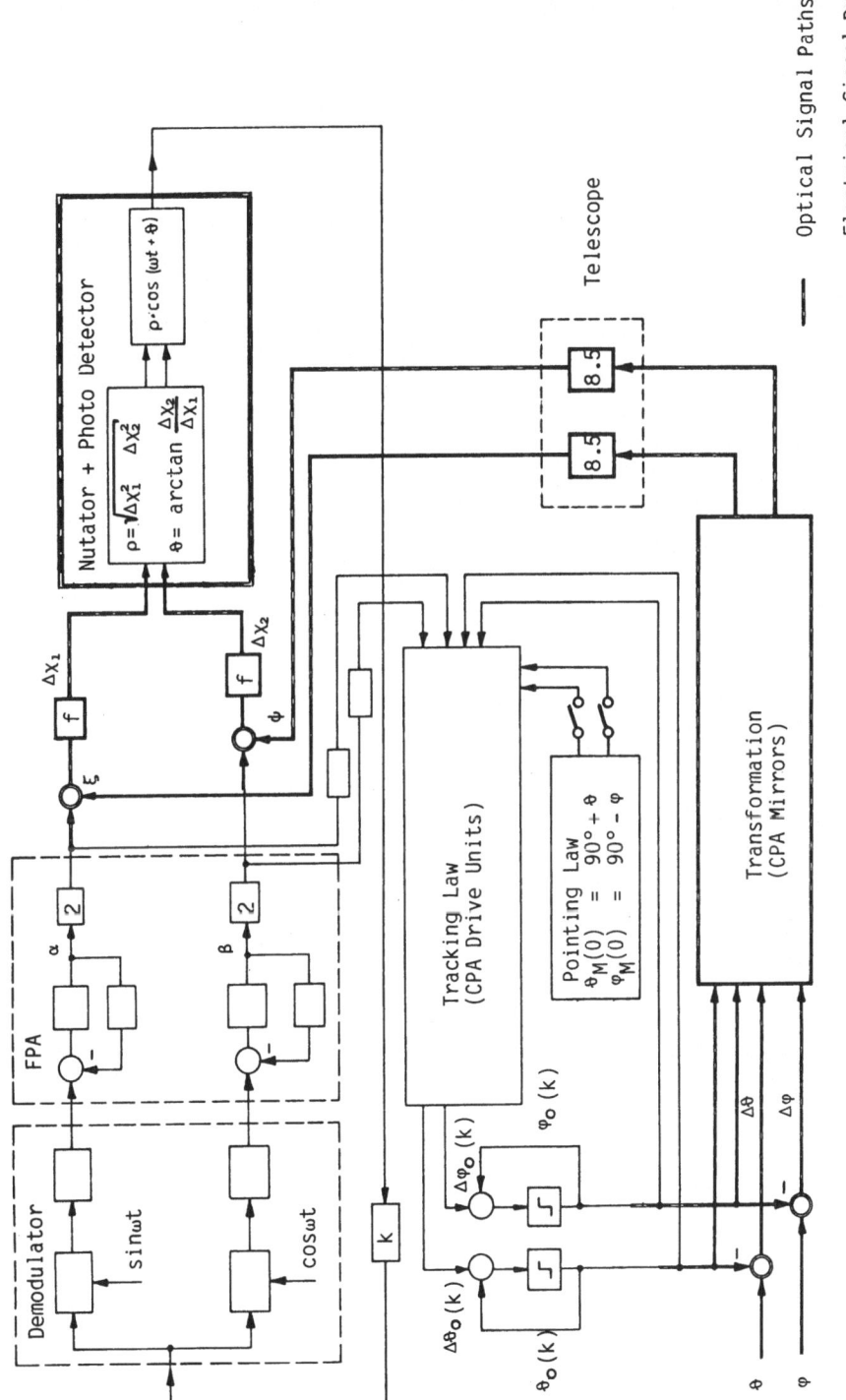

Fig. 3. Block Diagram of the tracking control system

Realization of a Microprocessor Controlled Laser Ranging and Target Tracking System

S.Manhart and G.Barthel

Messerschmitt-Bölkow-Blohm GmbH

D - 8012 Ottobrunn

Under ESTEC contract an autonomously operating laser ranging and tracking system was developed. Consisting of a pulsed diode laser rangefinder, a two-axis beam steering unit, and a microprocessor control unit the realized demonstration model is capable of:

    a) acquiring and discriminating several reflector targets
       within a 30° x 30° field-of-view;

    b) automatically tracking slowly moving targets;

    c) measuring the angular X- and Y-positions to better
       than 1 arcmin.;

    d) determining the range to millimeter accuracy at short
       target distance (0...50 m) and to centimeter accuracy
       at medium target distance (up to 1 km).

Due to its modular design, the demonstration model is very flexible and may easily be adapted to particular requirements of future space missions.

1. Introduction

Because of their small size, ease of operation, and adequate lifetime performance semiconductor lasers are exceptionally suited for a variety of short and medium range space mission requirements. Accompanying technologies, e.g. fast signal processing, beam steering, or microprocessor technology, have proven to be space qualified. Therefore laser ranging and target tracking using diode lasers can be regarded to be mature for space application.

The demonstration model of the laser diode rangefinder and target tracking system (LDRF-DeMo) was jointly developed by the Inst. für Nachrichtenverarbeitung/Universität Siegen, and by MBB/Space Division.

2. Overall LDRF-DeMo Design

Fig. 1 shows the laser diode rangefinder demonstration model with:

a) the terminal, keyboard and display;

b) the microprocessor control unit;

c) the laser rangefinder electronics;

d) the optics head, arranged on a rotation-translation mount.

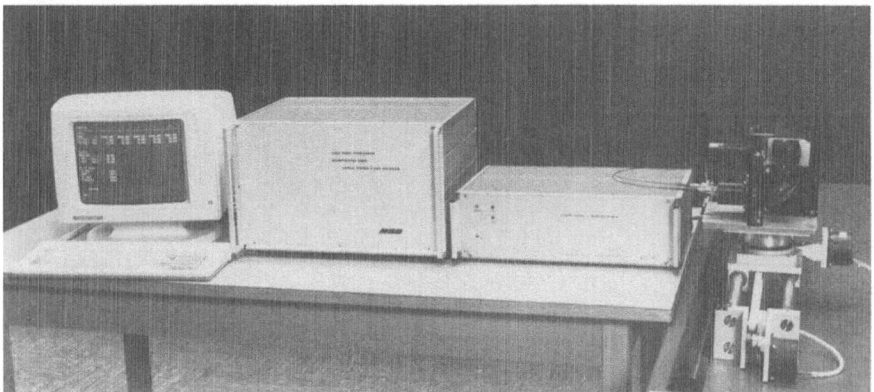

Fig. 1. Setup of the laser diode ranging and tracking system

The control electronics and the rangefinder electronics are contained in standardized 19" housings. The connection between rangefinder electronics and optics head is accomplished by glass fibres. Thereby a maximum of flexibility has been achieved, allowing modifications and replacement of subunits or circuit blocks without affecting the rest of the system. The complete system can be regarded as a working model, which enables verification of ranging and tracking performances with respect to specific future mission requirements.

3. Target Acquisition and Tracking

A sinusoidal raster scan pattern is used for target acquisition (LAS). The total field-of-view of 30° x 30° is scanned in a time frame of 10 s (Fig.2). Because of the angular overlap of subsequent laser shots there will be at least 3 target returns from a single target. Individual targets are discriminated during the LAS, the range and angular position values are displayed after each scan frame. The small angle scan (SAS) is performed in 1 s. A large number of target returns is available from a single reflector allowing the precise determination of the angular position by averaging the individual values. The reproduceability of the angular target position is approximately

0,01°. In the measurement (MSM) mode, the laser beam is directed at the target with the scanners remaining at fixed positions. Tracking of moving targets is achieved by continuously repeating the LAS mode, or by repetitively switching between LAS and MSM. The individual modes are controlled either directly via the terminal or automatically by the microprocessor. The strategy of automatic acquisition and tracking is shown in Fig. 3.

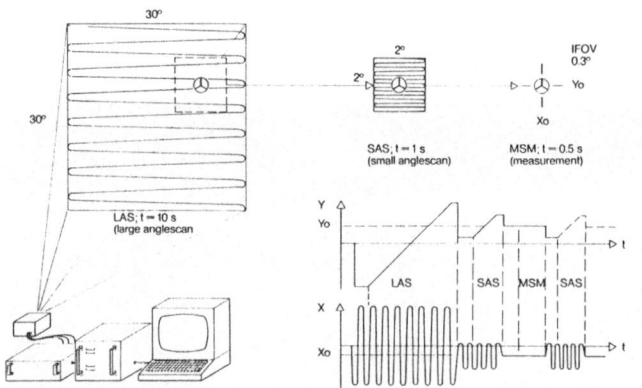

Fig.2.

Scan patterns during acquisition and tracking

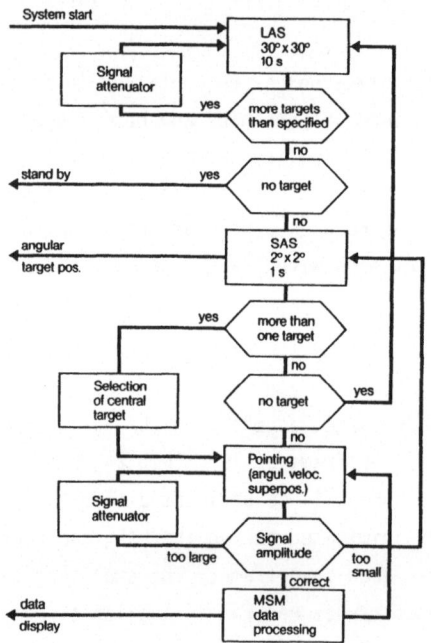

Fig. 3.

Automatic acquisition and tracking

If more targets than originally specified were found, automatic signal attenuation is initiated during LAS. This results in eliminating diffuse reflections from close objects or erroneous glint signals. In the MSM mode signal attenuation is performed to keep the target signal amplitude at a proper level for further processing.

4. Microprocessor Control Unit

The central control and data processing unit is based on an Intel 8086 microprocessor. The microcomputer board, the A/D and D/A converters, the interface electronics, the scan pattern generator circuit and the closed-loop control electronics are contained in the 19" housing.

The software is written in PLM, with the exception of the time critical subroutines, which are established in assembler language to yield faster processor speed. The complete software is stored in an EPROM on the microcomputer board.

5. Scanner

Gold coated deflection mirrors, are used for beam steering. Both mirrors are slightly wedge shaped to reduce mirror inerta. The mirror mounts are made of titanium, because of its low weight and its thermal expansion coefficient close to that of the mirrors. The mirror drives are moving iron galvanometer motors (General Scanning Inc., type G 300 PD). The built-in position transducers are used for acquisition of the actual scan mirror positions, and for speeding the scanner using a PID control circuit. The configuration of the scanner and the beam collimation optics is shown in Fig. 4.

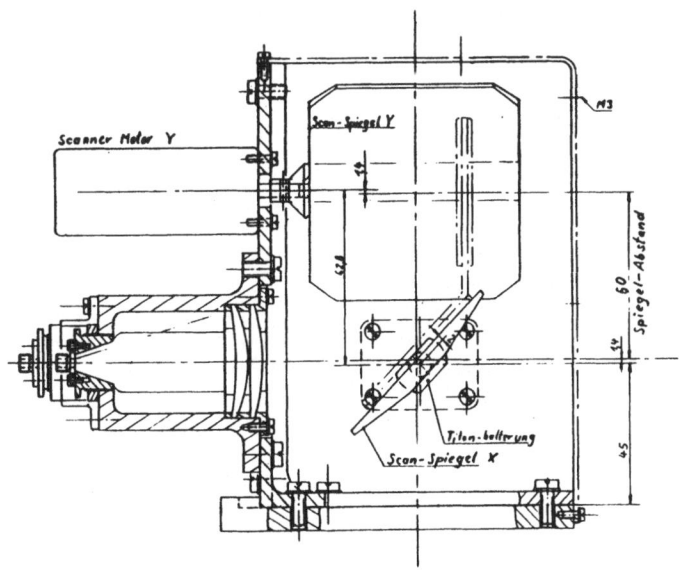

Fig. 4 .

Optics head, consisting of deflection mirrors, galvanometer motors, transmitter and receiver optics

6. Optics and Fibre Coupling

The fibre coupling scheme for the transmitter and receiver signal is shown schematically in Fig. 5.

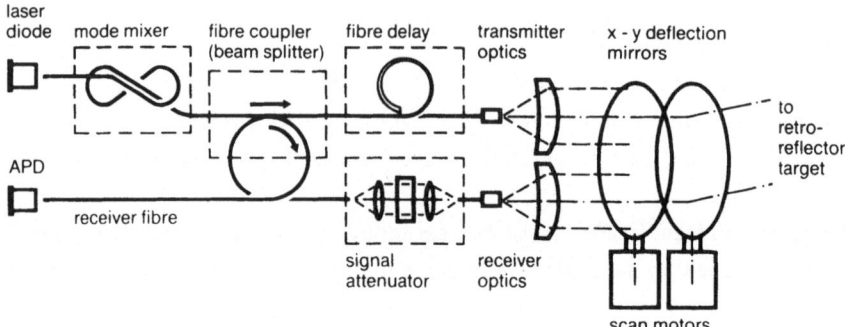

Fig. 5 . Fibre coupling scheme

Near and far field inhomogeneities of the laser diode emission are greatly reduced by the mode mixer. Part of the laser signal is fed directly to the APD to yield a time reference for the target signal. The additional fibre delay avoids collision between reference signal and target signal at close target distance. The attenuator unit makes use of neutral density filters to confine the target signals to adequate amplitude levels for further processing.

The transmitter and receiver optics are part of the same optical housing, but are separated by a wall to prevent laser light from getting directly scattered to the receiver. Additionally the diffusely scattered light from closely spaced objects is significantly reduced by separating the optics. The retroreflector signal, however, is unaffected because of the lateral beam offset (Fig. 6).

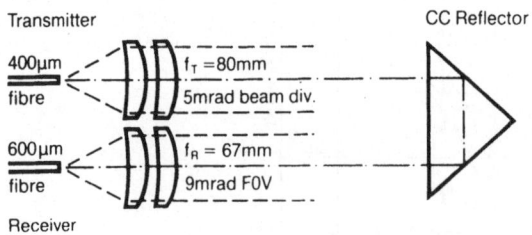

Fig. 6.

Transmitter and receiver optics configuration and lateral beam offset by a CC reflector

The ranging system was designed to acquire targets up to 20 km distance. To compensate for the scan mirror movement during the propagation time of the laser pulse, the receiver field-of-view was chosen to be 9 mrad, corresponding to the 5 mrad laser beam divergence.

scan amplitude : X_o = 15°

scan frequency : f = 10 Hz

max. signal
propagation time : T = 2 R/c = 133 µs

angular movement of the scan mirror
during the signal propagation time:

$$X_{max} = 2 \pi f \cdot X_o \cdot T$$

$$= 0,125° \triangleq 2.2 \text{ mrad}$$

7. Laser Transmitter and Receiver /1/

Transmitter : GaAs single heterostructure diode
M/A Comp. LD 65

peak power 8W

pulse width 9 ns

rise time 500 ps

pulse rep.
freq. 7,5 kHz

Receiver : Si-APD ; RCA C30902E

transimpedance amplifier

of 500 MHz bandwidth

8. Time-of-Flight Measurement

Precise trigger time evaluation is accomplished by a constant fraction discriminator which makes use of the leading edge of the laser pulse. The signal propagation time is measured by counting the pulses of a 48 MHz quartz clock and by precisely determining the remaining interval between arrival of the target signal and the next clock pulse, using a dual-slope time-to-time conversion technique. Thereby the critical time interval is stretched by charging a capacitor at constant current I_1 and discharging it at constant current I_2, yielding a stretching factor of k = I_1/I_2.

9. System Performance

The demonstration model was tested using different kinds of retroreflector targets. Short ranges could be verified in a 53 m tunnel, medium ranges were simulated by inserting a 860 m gradient index glass fibre delay into the propagation path (Fig. 7). Target movements were simulated with the optics head on a motor driven translation-rotation mount. The preliminary test results are:

- acquisition and discrimination of up to 5 retroreflector targets within a FOV of 30° x 30° during 10 s;
- tracking speed   : 0 ... 1 deg/s;
- angular resolution of single targets: 0,01 deg;
- range resolution : 1 ... 3 mm, at short target distance (0 ... 50 m);
  3 ...10 mm, at medium target distance(50 m ... 1 km).

Further testing will be performed using a computer controlled test bench and a statistics evaluation program:

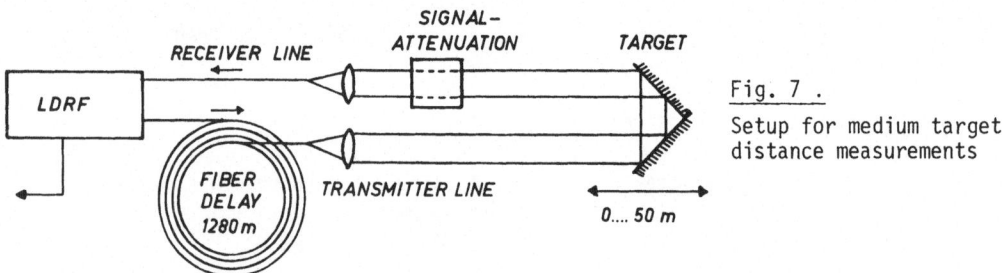

Fig. 7 .

Setup for medium target distance measurements

10. Conclusions

The performances of the diode laser ranging and tracking system have demonstrated that this technology is well suited for a variety of space operations. Since no particular difficulties are expected to convert the existing model into a space qualified unit, the system offers a promising new concept for improved proximity sensors and target tracking devices.

Acknowledgement

We would like to thank Prof. Schwarte, INV, Universität Siegen, and his team for their excellent work on the laser rangefinder. We also appreciate the work on the microprocessor control, done by R.Marquardt and W.Schattmann at MBB.

Literature

/1/ R.SCHWARTE - "Performance Capabilities of Laser Ranging Sensors" Proc.ESA Workshop on SPLAT, ESA SP-202, May 1984, P61;
/2/ Laser Technology for Optical Ranging, Vol.III: Laser Diode Rangefinder - Demonstration Model - ESA Contr. No. 5159/82/NL/HP Final Report, March 1985

The Monocular Electro-Optical Stereo Scanner (MEOSS) Space Experiment

F. Lanzl

DFVLR, D-8031 Oberpfaffenhofen, Germany

The Indian Space Research Organisation ISRO offered DFVLR to fly a space experiment on a Streched ROHINI Satellite (SROSS) with the following

Missions Parameters

circular orbit	450 km
inclination	45.56° (48.2°)
orbital period	92.8 minutes
lifetime	6 - 12 months
launch	1987
telemetry	S-band
data rate	10.4 Mbit per second
payload weight	10 kg
payload power	25 Watt for 20 minutes/day

DFVLR will provide for this mission a camera experiment which will be the first along track threefold stereo scanner in space (1,2). The main characteristics of the proposed threefold stereoscopic CCD line scan camera are:
- a single objective (monocular) imaging system for reduction of weight and adjustment efforts
- no moving mechanism
- passive cooling
- three relatively long CCD lines in one common focal plane

with the following

MEOSS Payload Parameters

numer of elements of linear CCD array	3456
line frequency	131 Hz
signal quantization	8 bit
amplification levels	1, 2, 4
numer of parallel lines	3

focal length of objective	61.1 mm
f number	8
angular distance between CCD lines	± 23.6°
spectral band	570 nm to 700 nm
swath width	256 km
stereo base	2 x 198 km
ground resolution	
along track	52 m
across track	79 m
height resolution	55 m

The <u>principle</u> of monocular stereorecording is indicated in Fig. 1. From there we can conclude that the same area on ground is imaged under different angles in time intervals of 30 seconds.

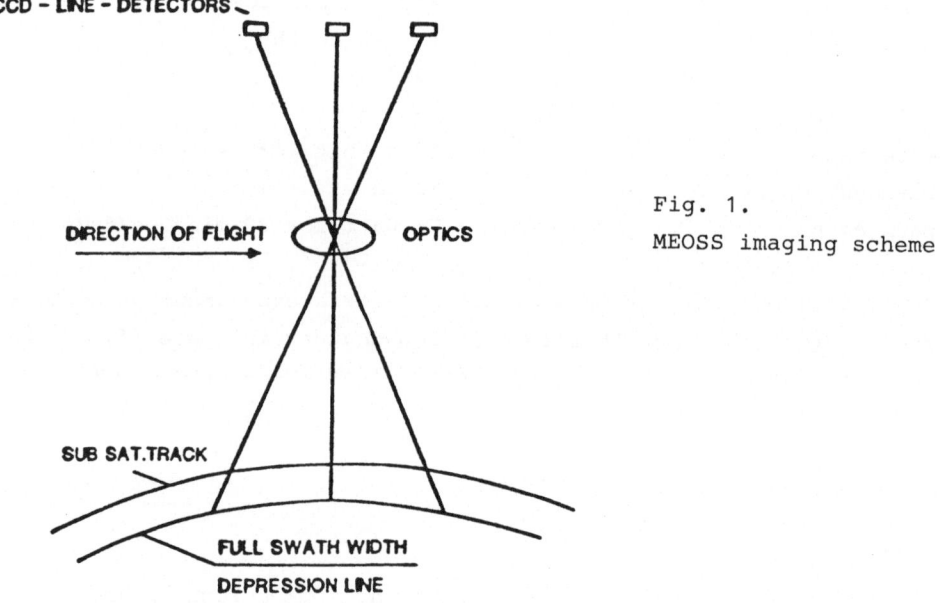

Fig. 1.
MEOSS imaging scheme

The <u>optomechanical design</u> contains a central mechanically independent imaging unit with a space modification of a Zeiss Biogon objective, a baffle acting also as passive cooler and an adapter connecting the camera with the SROSS payload deck. Fig. 2 shows the MEOSS structural model with the observing direction towards left.

The properties of the CCD's are described in a separate presentation of this congress (3).

Fig. 2.
MEOSS structural
model

For the MEOSS mission goal besides the first recording of stereo data
by a linescanner from space, the methods to evaluate these data are of
highest interest. The principle of the correction of a scene necessary
with respect to the attitude variations of the satellite is indicated
in Fig. 3. As one source of data out of the redundant threefold stereo
image data characteristic points forming a wide grid model are selected.
On the other hand satellite orbit and attitude data together with an
dynamic satellite model enter. Both kinds of data enter the linearized
image equations. After several iterations controlled by least square
measures, on the one side a corrected wide grid terrain model and on
the other side improved satellite orbit and attitude data result.
This twofold result allow to
- correct all image data within the improved wide grid terrain model
- improve navigation data by the use of image data.

This expected results of the MEOSS mission will be a sound basis for
future stereo linescan missions with higher resolution in the region of
10 m. Here the attitude stability conditions are rather severe and are
not met by present stabilization systems used in remote sensing of the
earth surface. Simular schemes as indicated above will be used for data
correction.

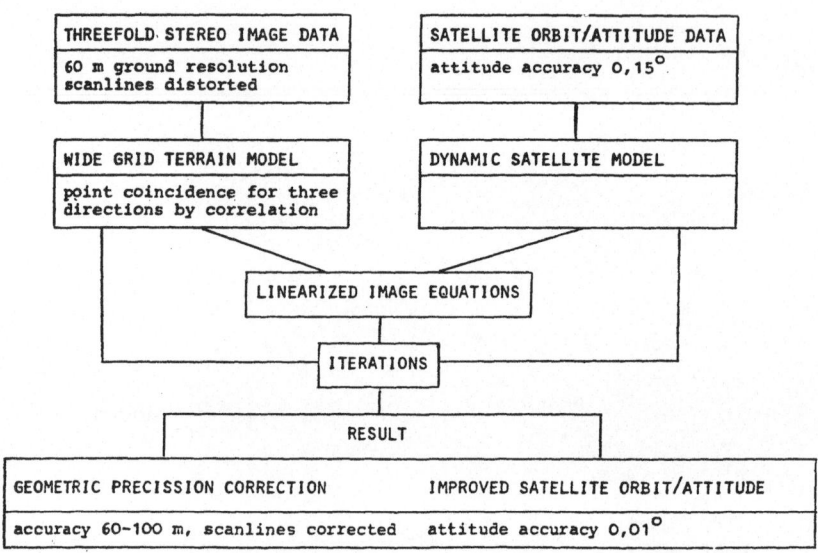

Fig. 3. MEOSS high precission correction scheme

The image data reception is performed by the ISRO-NRSA LANDSAT ground station near Hyderabad and by the DFVLR groundstation Weilheim and is open to other groundstations within the mission constrains.

Literature
(1) A. DRESCHER: "MEOSS an Experimental CCD-Camera for Remote Sensing"
              DFVLR-Mitt. 83-03 p. 35
(2) F. LANZL: "The Monocular Electro-Optical Stereo Scanner Experiment"
              DFVLR-Mitt. 84-12 p. 31
(3) G. RESS, P. SEIGE: "Preliminary Results of the TI TC-104 CCD
              Investigation for Project MEOSS"
              Proc. of this Congress

Preliminary Results of the TI TC-104 CCD Investigation for Project MEOSS

G. Ress, P. Seige

DFVLR, 8031 Oberpfaffenhofen, Germany

Abstract

The Texas Instruments TC-104 linear CCD array with 3456 elements per line has been investigated especially with respect to applications in speaceborne imaging systems where parameters such as linearity, pixel non uniformity dynamic range and decalibratability were of special importance.

1 Introduction

Modern imaging systems for remote sensing applications require high spatial resolution as well as wide coverage. Therefore there is an increasing interest in long arrays presently available with 3456 pixels per line. For the MEOSS (Monocular Electro Optical Stereo Scanner) project the T.I. array with its modern virtual phase technology was selected.

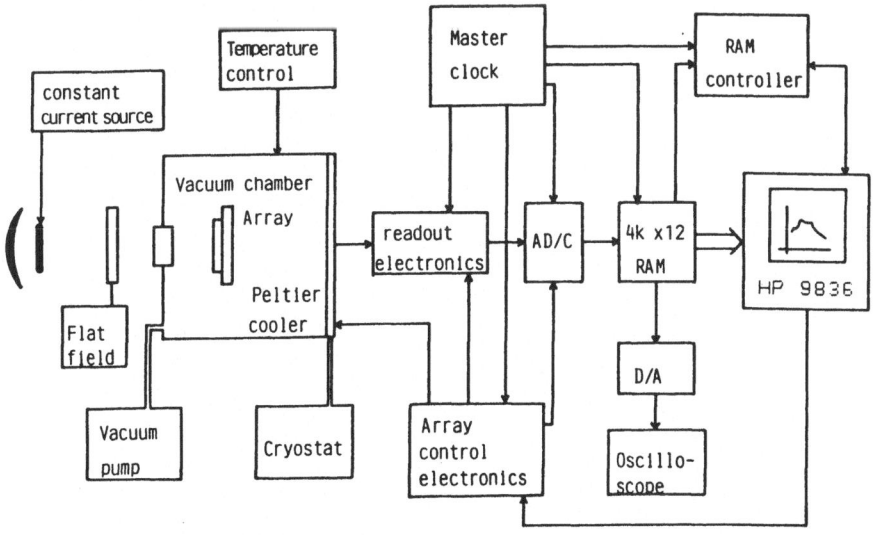

Fig. 1.   CCD Test Facility

In order to be able to properly investigate the TI TC-104 array a
special test facility was set up (Fig. 1). It consists of an optical
set up, the proper CCD drive and readout electronics, a computer
interface as well as a computer facility for system control and data
evaluation.

The CCD is housed in a small vacuum chamber together with a peltier
cooler for operating the devices in a controlled environment between -
45 °C and +30 °C. Included in the chamber is the first stage of the
readout electronics as well. The
array is illuminated through an optical window by means of a flat field
controlled by a constant light source, where a set of neutral
density/colour filters can be added. For the electronics special attention was given to low noise performance.

3 Results

All results shown in Fig. 2 to 7 were obtained at 500 kHz pixel rate
and + 20 °C. Fig. 2 shows a flat field exposure at about 50 % of
saturation. Obvious is the relatively small pixel non uniformity of
about 3 % and the odd-even effect. Corresponding to this effect is the
histogram shown in Fig. 3. Fig. 4 shows the variation of the odd-even
separation with signal level. Since this changes proportional to the
signal level along a straight line a very good decalibration of the
device is possible. Fig. 5 displays the relation between signal vs.
noise showing that the plot follows down to about 25 % of saturation
almost ideally the 1/2 slope. Only then readout noise gets dominant. As
seen in Fig. 6 the linearity is very good over three decades of
illumination levels. Fig. 7 finally shows the surface flatness of the
array an important factor for optical systems with a small depth of
focus.

4 Conclusion

The TI TC-104 is a very good CCD array for high performance imaging
applications with low pixel non uniformity, good linearity and a high
dynamic range related to RMS noise of many times more than 4000:1.

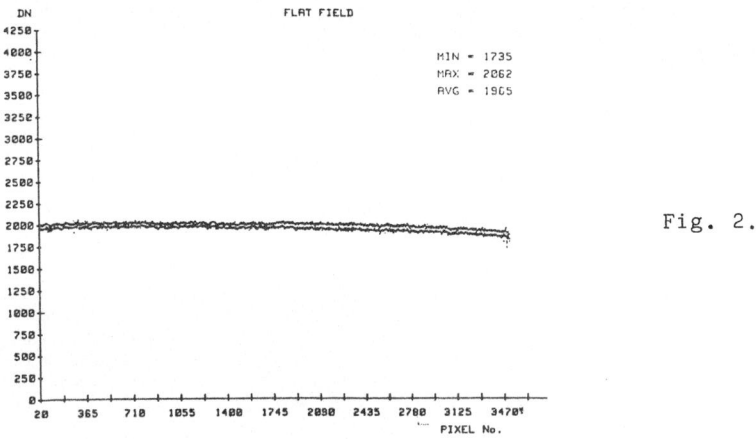

Fig. 2.

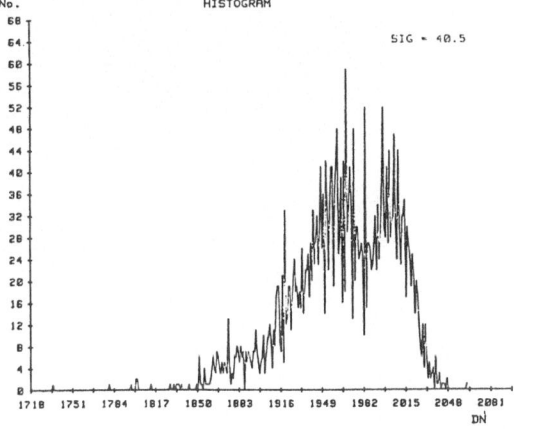

Fig. 3.

Fig. 4.

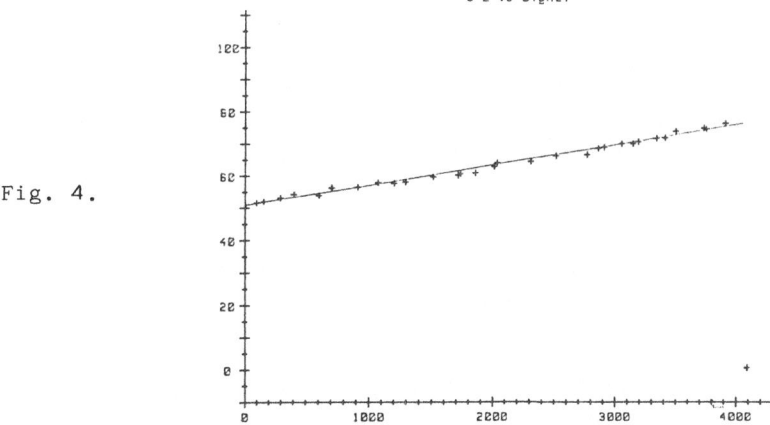

A Small Spaceborne Lidar for Atmospheric Backscatter Measurements

M.Endemann

Battelle-Institut e.V.

Am Römerhof 35, D-6000 Frankfurt a.M./D

Two years ago at the Laser'83-conference, the conceptual design of a spaceborne lidar for measurements from future operational earth-observation satellites was presented /1/. Here we report on the progress of this development into an instrument called Small Spaceborne Lidar for Atmospheric Backscatter-measurements (S^2LAB). This instrument is designed to be part of some sensor package of future operational weather-observation satellites in an 800-km orbit. Fig.1 shows an artist's view of such an instrument on board of some advanced remote-sensing satellite.

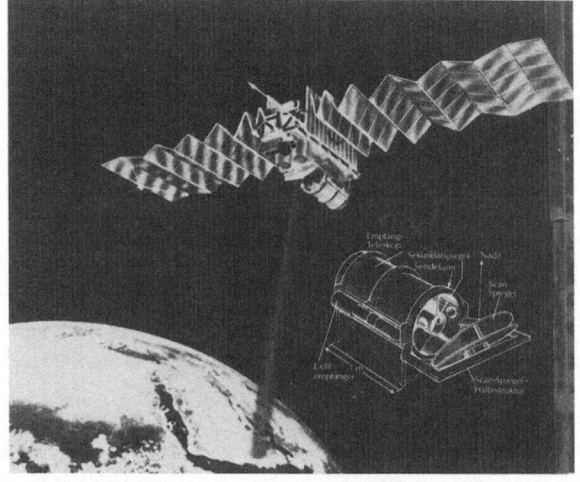

Fig.1. Artist's view of an S^2LAB-instrument on an advanced weather-observation satellite

The data from S^2LAB can be used to improve the vertical resolution and the accuracy of measurements from other passive sensors, and to provide important data on hight and density of the aerosol-layers in troposphere and stratosphere. Tab.1 tries to summarise potential applications of S^2LAB-data for weather forecasting and climatology /2/.

In addition, further development stages of the basic S^2LAB-instrument will allow to perform DIAL-measurements of atmospheric water vapor or temperature from a 250-km orbit in addition to the above parameters. The vertical resolution of these DIAL measurements is much better than that obtainable with passive sensors.

Tab.1. Measurement quantities of the backscatter lidar and derived
       atmospheric parameters

Measured Quantity	Derived Meteorological Parameter	Remarks
Cloud-top height	a) Temperature at cloud-top b) Temperature above cloud-top c) Humidity at cloud-top d) Wind-velocity at cloud-top	Using data from passive sensors
Subvisible clouds (detection and height assignment)	a) Improvement of entire tempera- ture profile b) Improvement of humidity data	Providing correc- tion factors for passive sensors
Tropospheric and stratospheric aerosol layers	a) Net radiation budget b) Global radiation at the ground c) Tracking of aerosol clouds	
Height of plane- tary boundary layer	a) Stability of boundary layer b) Sensible heat flux from ocean	Using jump of aerosol concen- tration
Tropopause height	Improving temperature profiles from passive sensors below and above	Using jump of aerosol concen- tration

Design Objectives

The S^2LAB instrument is designed to improve the measurement resolution
and accuracy on some atmospheric parameters that are used for weather
forecasting and climatology research. For this purpose its data will be
combined with those from other sensors to improve their vertical resolution and accuracy. This calls for the following performance demands:

- Good vertical resolution; a vertical resolution near 150 m can
  be achieved without difficulties with the backscatter lidar.
- Wide measurement track to give a good area coverage every day
  and thus allow easier combination of the data sets from different sensors. This demand requires scanning capabilities of
  the S^2LAB-instrument. A swath-width of 500 km allows a semi-
  global coverage per day.
- Good horizontal resolution of cloud-top height data to provide
  sufficient samples within a 150·150 km^2 large area (as used in
  weather forecasting calculations) for a good assessment of the
  average cloud height. A value of 30·30 km^2 is selected as a
  compromise between the desired high sampling rate and a low
  pulse frequency of the laser (3.7 Hz for a track width of

500 km). Data of aerosol layers in troposphere and stratosphere require averaging over adjacent samples and thus have a
lower horizontal resolution.
- Small size, low weight, low power consumption to allow the
  integration into future sensor packages.
- Long lifetime of about 1/2 year of continuous operation. For
  the measurement frequency of 3.7 Hz, this corresponds to some
  $60 \cdot 10^6$ measurements.

Design Options

The most demanding component of the S^2LAB-instrument is the transmitter
laser. It must provide high reliability and very efficient operation in
a wavelength region where photo-multipliers with good quantum-efficiency
are available. The new Cr-doped vibronic laser materials like Alexandrite or Cr:GSGG are selected as prime candidates for the transmitter
laser. They operate in the 700-800-nm-region where sensitive photomultipliers are still available (GaAs-photocathodes with up to 18 %
quantum efficiency). This region is advantageous as it provides the
possibility for later DIAL-measurements to obtain profiles of water-
vapor and temperature. Furthermore it can be expected that vibronic
lasers can operate with an efficiency near 1.5 % and give pulse energies
near 1 J in q-switched operation /3/. A lifetime of 60-million pulses
can be achieved with conventional flashlamp-pumed devices. The aspect of
tunability becomes important for the later DIAL-measurements.

Scanning of the lidar can be achieved by various means. Originally it
was proposed to use a 45°-scan mirror that reflects the transmitted
laser pulse and the backscattered light. This version is indicated in
the artist's view in Fig.1. However, this arrangement requires an additional large mirror and thus is rather inefficient with respect to size
and weight.

Alternative schemes involving a scanning telescope were examined. Fig.2
shows the design of a small backscatter lidar with a scanning Newtonian
telescope. This design is considerably smaller and lighter than the
alternative with scanning mirror: it fits into a cube of less than
0.6 m^3 volume, and is estimated to weigh less then 80 kg. The scanning
is achieve by swinging the telescope with the secondary mirror (which is
also the steering mirror for the transmitted light pulse within) over an
angle of $\pm 17°$ with a period of 9.4 s (6.4 sweeps per minute). This
rather slow harmonic sweep causes negligible attitude changes to the
sattelite on which it is mounted.

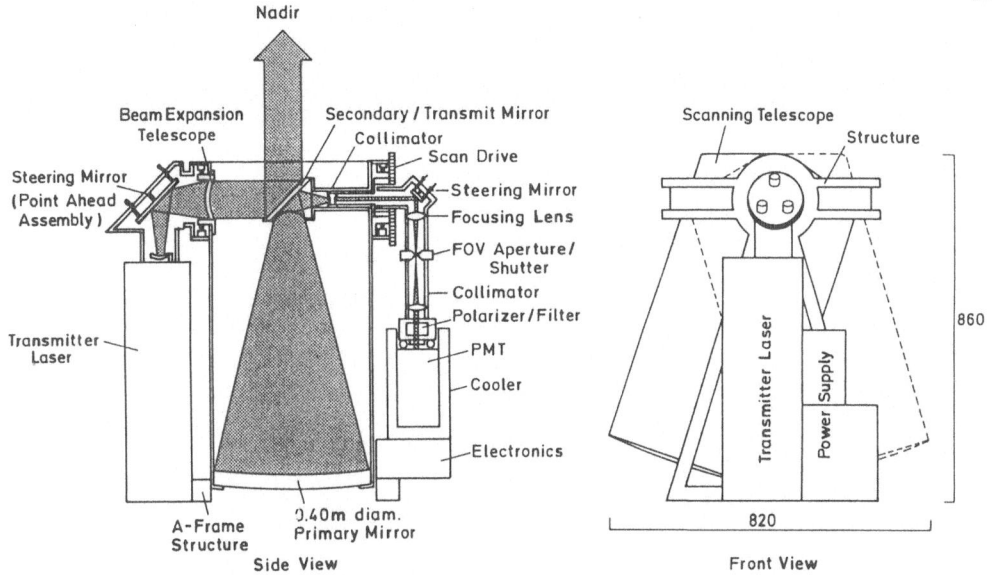

Fig.2. S²LAB design using a swinging Newtonian telescope

The power consumption of the lidar depends to a large extend on the efficiency of the transmitter laser. When an overall efficiency of 1.5 % can be achieved, the 1-J laser consumes some 250 W; receiver electronics, detector, scanner and temperature controls are expected to required additional 90 W, resulting in a total power consumption of 340 W.

An even simpler lidar design using a fibre-optical coupler to connect the focal plane for the receiving telescope with the detector is currently under investigation. It can be expected that this arrangement will allow a further reduction of size and weight and will simplify its mechnical design. An experimental lidar to verify the design assumptions and to assess the performance potential different components is currently under development at Battelle.

References

(1) M.ENDEMANN, W.ENGLISCH: Spaceborne Lidars for Global Meteorological Measurements, Proc. Conf. Laser '83, München, (1983), 505

(2) H.GRASSL: Potential Applications of a Space-Borne LIDAR to Meteorology and Climatological Research, 12 Int.Laser Radar Conference, Aix en Provence (1984)

(3) M.ENDEMANN: Orbiting Lidars for Atmospheric Sounding, Volume 2: Development of the Transmitter Laser, Final Report for ESTEC Contract No.5186/82 (1985)

Laser und Optoelektronik in der
Umweltmeßtechnik
Lasers and Optoelectronics in
Environmental Measurement Techniques

Remote Sensing of Environmental Pollution and Gas Dispersal Using Low Loss Optical Fiber Network System

Humio Inaba

Research Institute of Electrical Communication

Tohoku University

Katahira 2-1-1, Sendai 980, Japan

I. Introduction

Optical fiber sensor technology is a subject of considerable interest and various techniques have been implemented for the measurements of a number of physical and chemical parameters.[1] Based on the author's earlier idea in 1979[2,3] for a new capability of this technology, an optical network system employing low-loss optical fibers has been analyzed and examined for remote sensing of environmental pollution and spilled dispersals of combustible, explosive and toxic gases / vapors by the spectroscopic absorption method, in various industrial and mining complexes as well as in urban and residential area.[3-17]

This regime of fiber-optic gas remote sensing has practically useful features because the optical energy can be concentrated and transmitted in low-loss optical fibers even for long distance over several tens km,[10,12,14] instead of the open atmosphere.[18] Therefore, a fully optical, reliable, sensitive, low-cost (with low-power lasers or even conventional nonlaser sources), feasible, real-time, nonhazardous, e.g., eye-safe and explosion-free technique can be realized for various stressing environments and severe conditions. This method also has the capability of little optical interference, and of continuous surveillance with easy calibration, wide selectivity, and no electrical induction.

The present paper summarizes and discusses mainly the basic subjects and experimental results of a low-loss optical fiber-based remote sensing system of various dangerous and polluting gases which have been investigated in my laboratory, with specific emphasis on the use of near-infrared wavelengths ranging from 1.0 μm to 1.8 μm for this purpose. Measured results of fully optical remote detection achieved in a diameter up to 20 km utilizing presently available very low-loss silica optical fiber links are also presented for hydrocarbons including CH_4, C_3H_8 and C_2H_4 in this wavelength region.

II. Spectroscopic Studies on Molecular Absorption in the Near-Infrared Region

Currently, ultralow-loss silica optical fibers exhibiting transmission loss as low as 1 dB/km or less over the range of about 1.05 - 1.70 μm [19] and high quality optical sources and detectors have been developed primarily for long-distance, high bit-rate optical communications as shown in Fig. 1.[5,10,12] Typical absorption wavelengths of various molecular species present in the atmosphere and environment are also depicted. Consequently, it becomes fairly apparent that the fiber-optic remote gas sensor system fully based on the optical version should be substantially established in this near-infrared region to achieve an excellent capability and a wide area coverage such as, for instance, a few tens km in diameter.[2,3,5]

From this point of view, we are pursuing spectroscopic measurements and analyses of combination and overtone bands of a number of molecules such as CH_4, C_3H_8, C_2H_4, C_2H_2, C_2H_6, C_4H_{10}, NH_3, H_2O, those contained in LNG, LPG, and city gases, and others in this near-infrared region. As the light source with continuous coverage, InGaAsP and InGaAs laser diodes (LD's) are chosen and operated under threshold current level as high-radiant light emitting

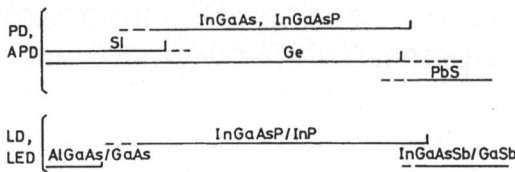

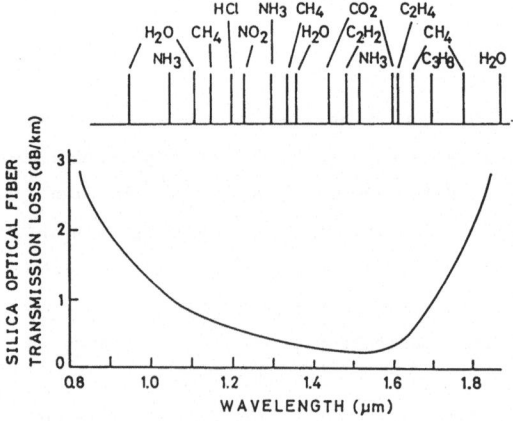

Fig. 1   Currently available, high-quality semiconductor light sources (LD and LED) and detectors (PD and APD) in conjunction with ultralow-loss silica optical fibers in the near-infrared region, and typical absorption wavelengths of various molecular species present in the atmosphere and environment.

diodes (LED's). In the following, we present and briefly discuss the measured spectra as well as their assignments of C_2H_4, CH_4, and C_3H_8 absorption in 1.1 - 1.7 µm region for their applications to fiber-optic gas remote sensing.

Figure 2 shows the measured absorption spectra of C_2H_4 gas around 1.6 - 1.7 µm range. Pure (>99.9%) C_2H_4 gas was introduced into a 50-cm long absorption cell at a pressure of 650 Torr at room temperature. The spectral resolution was 1 nm. This result indicates the existence of a separate absorption band around 1.62 µm and superposition of a few overlapping bands in 1.64 - 1.72 µm. According to the selection rules,[20] the possible assignment will be the binary combination band of $\nu_5 + \nu_9$ for the former and the $\nu_1 + \nu_9$, $\nu_1 + \nu_{11}$ and $\nu_5 + \nu_{11}$ bands for the latter. Comparatively strong absorption spectra are observed in the $\nu_5 + \nu_9$ band, whose detailed measurement is shown in Fig. 3. The assignment of the observed spectra in this figure involving P, Q and R branches can be found in our recent report.[21] Strongest absorption was found at 1.6246 µm, i.e., within the Q branch. Its absorption coefficient was measured to be 18.3 x 10^{-6}/ppm·m.[21] This portion of spectra seems to be suitable for the application of optical fiber remote detection of C_2H_4 gas.

The near-infrared absorption spectrum of CH_4 was measured and studied to find

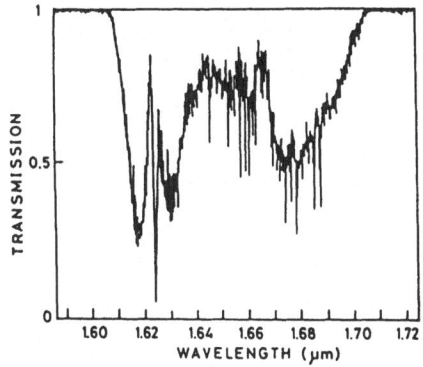

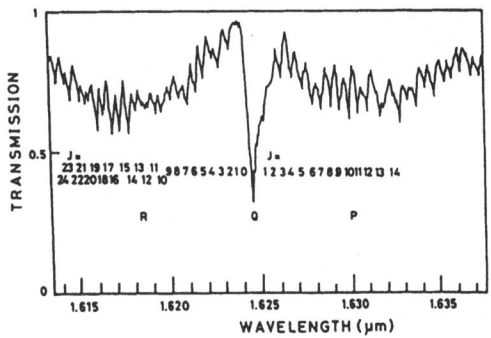

Fig. 2 Measured absorption spectra of C_2H_4 molecules in the 1.6 - 1.7 μm region with the spectral resolution of 0.1 nm, using InGaAsP and InGaAs LED's. C_2H_4 gas pressure in a 50-cm long absorption cell was 650 Torr.

Fig. 3 Measured absorption spectra and their assignment of the $\nu_5+\nu_9$ combination band of C_2H_4 molecules in the 1.61 - 1.64 μm region with a spectral resolution of 0.08 nm. C_2H_4 gas pressure was 100 Torr and absorption cell length was 50 cm.

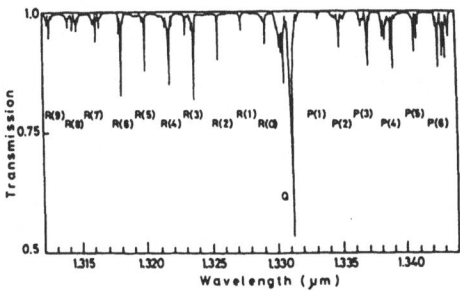

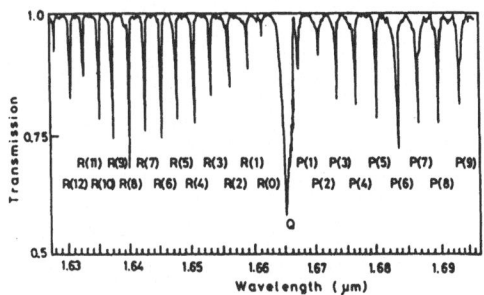

Fig. 4 Measured absorption spectra and their assignment of the $\nu_2+2\nu_3$ combination band of CH_4 molecules in the 1.31 - 1.35 μm region with a spectral resolution of 0.05 nm. CH_4 gas pressure was 300 Torr and absorption cell length was 50 cm.

Fig. 5 Measured absorption spectra and their assignment of the $2\nu_3$ overtone band of CH_4 molecules in the 1.63 - 1.70 μm region with a spectral resolution of 0.3 nm. CH_4 gas pressure was 60 Torr and absorption cell length was 50 cm.

the combination band of $\nu_2+2\nu_3$ around 1.33 μm[5-8,10,12,14,15] and the overtone band of $2\nu_3$ around 1.66 μm.[8-10,12-15] Figures 4 and 5 show their measured absorption spectra and assignments in 1.31 - 1.34 μm and 1.63 - 1.69 μm with the spectral resolution of 0.05 and 0.3 nm, respectively. Pure (99.9%) CH_4 gas was contained in a 50-cm long sample cell with a pressure of 300 Torr for Fig. 4 and 60 Torr for Fig. 5. There are relatively strong absorption spectra around 1.331 μm and 1.666 μm, and the absorption spectra of $2\nu_3$ band are always stronger than that of the $\nu_2+2\nu_3$ band. The strongest absorption exists at 1.6654 μm. Its absorption coefficient was measured to be 9.3 x 10^{-10}/ppm·m[9] which is about twice of that at 1.312 μm at the same spectral resolution of 0.3 nm.[6]

In Fig. 6, a part of the measured absorption spectra of C_3H_8 gas in 1.67 - 1.72 μm region with the spectral resolution of 0.6 nm is illustrated. Pure (99.9%) C_3H_8 gas was contained in a 50-cm long absorption cell with a pressure of 200 Torr. These spectra, believed to be the superposition of many combination and overtone bands,

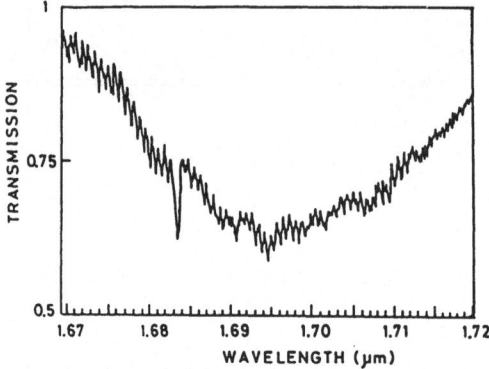

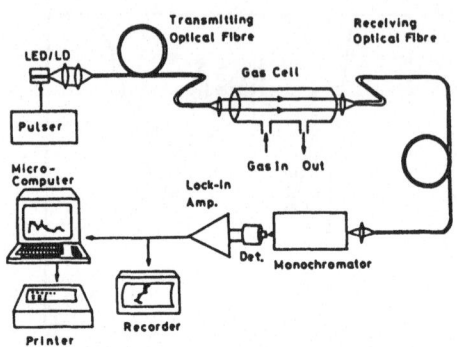

Fig. 6 Measured absorption spectra of C_3H_8 molecules in the 1.67 - 1.72 µm region with a spectral resolution of 0.6 nm. C_3H_8 gas pressure was 200 Torr and absorption cell length was 50 cm.

Fig. 7 Block diagram of an all - optical fiber-based remote sensing system for absorption measurement of low-level combustible, explosive, polluting and toxic gases in an absorption cell as the gas sensor head using a LED or LD operated in the near-infrared region.

appear to be very complicated and difficult to resolve. Comparatively strong absorption is located at 1.6837 µm, whose absorption coefficient was measured to be 4.5×10^{-6} /ppm·m at a resolution of 0.3 nm.[11]

III. Fully Optical Remote Measurement of C_2H_4 and C_3H_8 Gases Using Ultralow-Loss Silica Optical Fiber Links

On the basis of the above-mentioned spectral information, we have performed their remote absorption measurements using InGaAsP and InGaAs LED's in conjunction with long distance, ultralow-loss silica optical fibers. Figure 7 shows the schematic diagram of a fully optical remote gas sensing system using a 50-cm long absorption cell as the gas sensor head. Near-infrared InGaAsP and InGaAs LED's used for the experiment are laser diodes operated under threshold current level with output power less than 0.1 mW. The optical beam from the LED was transmitted through a transmitting multimode optical fiber to a remotely located gas sensor head, and then returned to the transmitter / receiver location via the receiving multimode optical fiber. The transmission loss of the silica optical fibers was lower than 1 dB/km in the 1.3 - 1.7 µm region. Absorption measurement was performed by using a grating monochromator to adjust the detecting spectral width.

In order to demonstrate the feasibility of the system depicted in Fig. 7, remote measurements of C_2H_4 and C_3H_8 gases were carried out using a 5-km long, low-loss silica optical fiber link. In Fig. 8, the result of real-time detection of C_2H_4 gas in 1 atm. C_2H_4-air mixture contained in a remotely located 50-cm long absorption cell is shown. The transmission of the evacuated cell was measured for comparison, before and after the absorbance measurements of the cell containing C_2H_4 gas. The

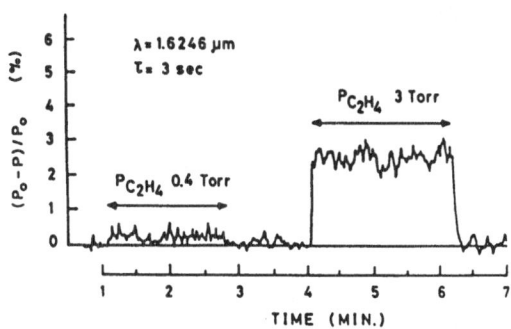

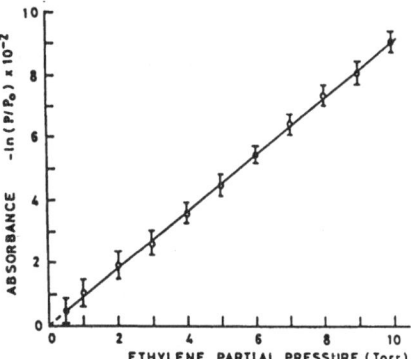

Fig. 8   Real-time detection of C_2H_4 absorption signals in 1 atm. C_2H_4-air mixture contained in a remotely located 50-cm long absorption cell, utilizing a 5-km long, low-loss silica optical fiber link and a 1.64 μm InGaAs LED.

Fig. 9   Measured dependence of the absorbance at 1.6246 μm on C_2H_4 partial pressure in 1 atm. C_2H_4-air mixture contained in a 50-cm long absorption cell using a 5-km long, low-loss silica optical fiber link in conjunction with an InGaAs LED. Spectral resolution was 0.8 nm.

wavelength was set at 1.6246 μm, and the spectral resolution was 0.8 nm. Detection of C_2H_4 gas was performed for two partial pressures; 0.4 Torr (∼530 ppm) and 3 Torr (∼ 3950 ppm). As is seen from Fig. 8, one could confirm that this kind of optical-fiber based remote measurement can be realized in situ, and the response is essentially instantaneous with respect to the gas pressure.

Figure 9 depicts the result of 5-km long, fiber-optic remote detection of C_2H_4, in which the dependence of absorbance at the wavelength of 1.6246 μm on C_2H_4 partial pressure in the sample cell was measured. P_0 and P represent the received optical powers, which passed through the evacuated and C_2H_4 gas cell, respectively. Spectral resolution of this detection system was also 0.8 nm. Based on the present preliminary experiments, it was confirmed that the system was able to detect C_2H_4 pressure in air lower than 0.4 Torr. This minimum detectable pressure in air was limited by the output power of the InGaAs LED and the sensitivity of the Ge detector, and could be improved. Since the lower explosion limit (LEL) of C_2H_4 gas in air is known to be 2.7 vol%, i.e., 20.5 Torr, this measured result corresponds to about 2% of the LEL. Thus, it demonstrates well the feasibility of the present system as an all-optical remote C_2H_4 gas sensor, and also indicates the potentiality for applying to remote vehicle exhaust monitoring.[18,22,23]

Remote detection of low-level C_3H_8 gas was also performed employing the experimental setup of Fig. 7 as well.[10-12,16] Curve A in Fig. 10 illustrates the measured dependence of absorption coefficient of C_3H_8 gas at 1.6837 μm on the spectral resolution of the receiving system, which was adjusted by changing the slit width of the monochromator. It is seen that the absorption coefficient decreases gradually as the resolution becomes lower. However, we note that lower resolution yields increasing received optical power. On the other hand, curve B in Fig. 10 shows the measured ratio of received power P_r to noise power P_n at 1.6837 μm through the

624

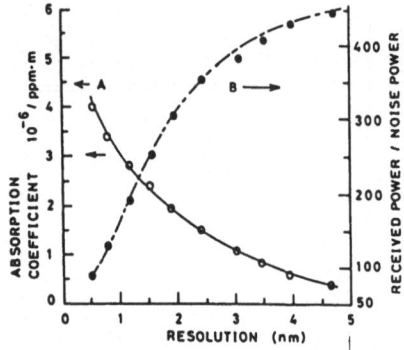

Fig. 10   Measured dependences of the absorption coefficient of C_3H_8 gas (curve A), in 1 atm. C_3H_8-air mixture contained in a 50-cm long absorption cell, at 1.6837 μm, and the signal-to-noise ratio (curve B) on the spectral resolution, using a 5-km long, low-loss silica optical fiber link and an InGaAs LED.   Both the received power and noise power were detected at 1.6837 μm through the evacuated cell.

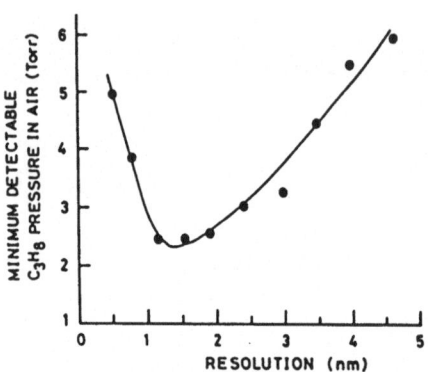

Fig. 11   Measured result of the minimum detectable pressure of C_3H_8 gas in air at 1.6837 μm as a function of the spectral resolution, employing a 5-km long, low-loss silica optical fiber link and an InGaAs LED.   The absorption cell was 50-cm long, and the detection time was 3 sec.

evacuated sample cell, using a 5-km long, low-loss silica optical fiber link.

This signal-to-noise ratio determines the detection sensitivity of the system and hence the minimum detectable density (or pressure) N_{min} which is approximately given by[4,11]

$$N_{min} = (1/\alpha L)/(P_r/P_n) \qquad (1)$$

where α is the absorption coefficient and L is the absorption path length, respectively.   Thus, it is evident from Fig. 10 that the maximum detection sensitivity of C_3H_8 gas can be achieved by suitably selecting the spectral resolution of the remote sensing system.

Figure 11 shows the result of 5-km long fiber-optic remote measurement of the minimum detectable C_3H_8 pressure in air as a function of the spectral resolution. As expected, the maximum sensitivity is realized with the resolution in the range of 1.2 - 1.5 nm for the present arrangement operated at 1.6837 μm.   Minimum detectable C_3H_8 pressure in air in this experiment was confirmed to be lower than 2.4 Torr (~3160 ppm), which corresponds to about 14 % of the LEL of C_3H_8 gas in air.   Hence, this result demonstrates that the present system can be effectively utilized for the remote monitoring and surveillance of C_3H_8 gas in air within the range of 5 km in diameter or more.

IV.   Optical Fiber Remote Sensing System for Near-Infrared Differential Absorption
      of Molecules

This section describes first a compact but highly reliable and sensitive remote sensing system employing differential absorption technique,[2-5, 13, 22, 23]   and then presents the experimental result applied to CH_4 gas utilizing its 1.66 μm absorption

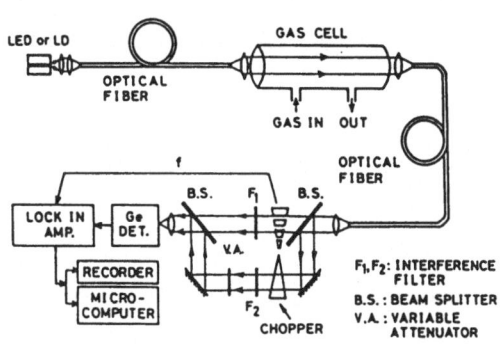

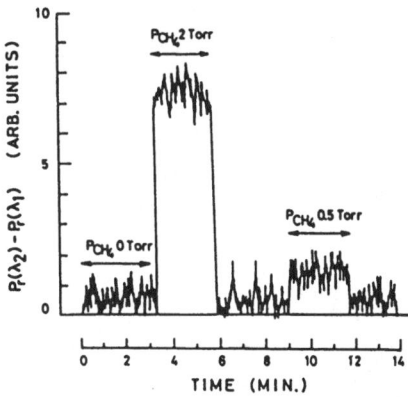

Fig. 12 Block diagram of the low-loss fiber-optic remote sensing system for differential absorption measurements of various molecular species in a remotely located absorption cell employing a LED or LD as the light source and dielectric interference filters.

Fig. 13 Detected differential absorption signals in the remote measurement of low-level CH_4 gas contained in a 50-cm long absorption cell, utilizing the system shown in Fig. 12, with a 2-km long, low-loss silica optical fiber link and a 1.61 μm InGaAsP LED. The time constant of the lock-in amplifier was 1 sec.

spectrum.[13]

The block diagram of the experimental setup is shown in Fig. 12. It is basically similar to the previous one depicted in Fig. 7 except the receiving system. The optical beam returned via the receiving optical fiber is divided into two beams by a beam splitter. These two optical beams are alternately chopped by a chopper with a frequency of about 100 Hz, and then pass through an on-resonance dielectric interference filter F_1 with the transmission center located at λ_1, and an off-resonance dielectric interference filter F_2 with transmission center at λ_2, respectively. Here, λ_1 is the wavelength at the peak absorption and λ_2 is at the valley or smaller absorption of the molecule. When the received optical powers $P_r(\lambda_1)$ and $P_r(\lambda_2)$ of the two beams after passing through the evacuated or empty cell are adjusted to be equal, the molecular concentration N in the absorption cell is approximately given by[13, 22]

$$N = \frac{1}{[\sigma(\lambda_1) - \sigma(\lambda_2)]L} \frac{P_r(\lambda_2) - P_r(\lambda_1)}{P_r(\lambda_2)} \tag{2}$$

where $\sigma(\lambda_i)$ is the absorption cross section at the wavelength λ_i(i=1, 2) and L is the absorption path length. This simple but useful expression is resulted from the technique called the power-balanced, two-wavelength differential absorption.[22]

For the remote differential absorption measurement of CH_4 gas, we selected a dielectric interference filter with transmission center at 1.666 μm, i.e., in the Q branch of the $2\nu_3$ band, and 3 nm FWHM as F_1,[9] and another dielectric interference filter with transmission center at 1.528 μm, and 5 nm FWHM as F_2. The differential absorption coefficient was then measured at 2.8×10^{-3}/Torr·m for this combination of interference filters. The light source was a 1.61 μm InGaAsP LED used previouly.

Figure 13 shows a chart recorder trace of the output of the lock-in amplifier,

626

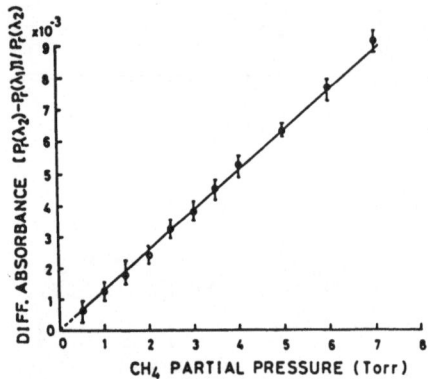

Fig. 14 Measured dependence of the differential absorbance on CH_4 partial pressure in 1 atm. CH_4-air mixture contained in a 50-cm long absorption cell at 1-km remote location, employing a low-loss silica optical fiber link and an InGaAsP LED operated in the 1.61 μm region.

which is proportional to the value of $P_r(\lambda_2)-P_r(\lambda_1)$, or the molecular concentration (pressure) according to Eq. (2), in the remote differential absorption measurement of CH_4 gas using a 2-km long, low-loss silica optical fiber link. As is seen from this result, differential absorption signals corresponding to CH_4 partial pressure of 2 and 0.5 Torrs mixed with 1 atm. air in the 50-cm long absorption cell were actually detected. The major advantages of employing the power-balanced, two-wavelength differential absorption method in the fiber-optic remote sensing system are high reliability and sensitivity resulting from the direct detection of differential absorption signals between the received optical powers at the two wavelengths through the same transmission trip in the low-loss optical fiber link.

Figure 14 illustrates the result of 2-km long fiber-optic remote measurement of CH_4 gas in which the dependence of the differential absorbance $[P_r(\lambda_2)-P_r(\lambda_1)]/P_r(\lambda_2)$ on CH_4 partial pressure in a 50-cm long absorption cell was depicted. A linear relation was observed within the low pressure range of 0.3 - 7 Torr. Hence, the minimum detectable CH_4 pressure in air was confirmed to be lower than 0.3 Torr, i.e., nearly 400 ppm. This value corresponds to about 0.8 % of the LEL of CH_4 gas in air, and thus demonstrates well the practical capability of the present system as an all-optical remote sensor for monitoring and continuous surveillance of low-level CH_4 gas in various environments and situations.

V. Optical-Fiber-Based Remote Gas Detection in the Near-Infrared Region Over a Wide Area

One of the attractive features of our all optical fiber-based, near-infrared remote sensing system is its wide area coverage, e.g., over several tens km.[10,12,14] To verify this practical advantage, we have carried out the remote absorption measurement of CH_4 gas using 5 - 20 km long, very low-loss silica optical fiber links. Experimental setup used in this measurement was basically similar to that described in Section III, besides the detection system which was arranged to utilize either a monochromator or an appropriate dielectric interference filter for the absorption signal.

Figures 15 and 16 show the results of remote detection of CH_4 gas employing a monochromator. The measured dependence of absorption coefficient of CH_4 gas in 1

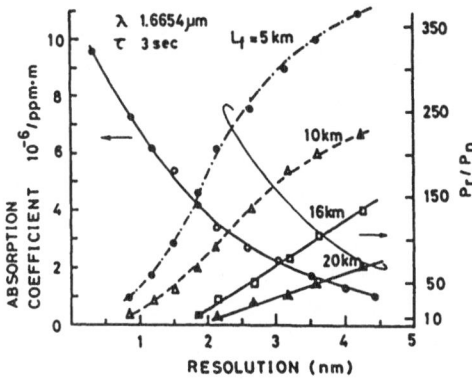

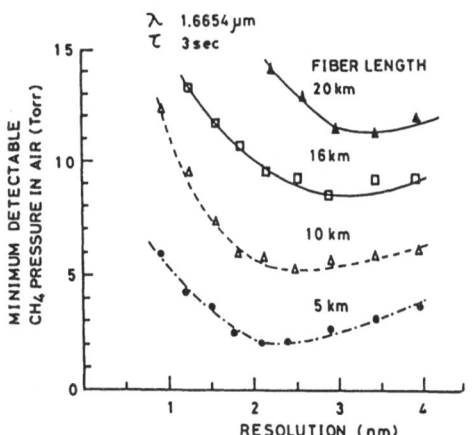

Fig. 15 Measured dependences of the absorption coefficient of CH_4 gas in 1 atm. CH_4-air mixture contained in a 50-cm long absorption cell at 1.6654 μm, and the signal-to-noise ratio on the spectral resolution, employing 5 - 20 km-long low-loss silica optical fiber links and an InGaAs LED. Both the received and noise powers were detected at 1.6654 μm through the evacuated gas cell.

Fig. 16 Measured results of the minimum detectable pressure of CH_4 gas in air at 1.6654 μm as a function of the spectral resolution and the length of low-loss silica optical fiber links in conjunction with an InGaAs LED and a 50-cm long absorption cell.

atm. CH_4-air mixture contained in a 50-cm long absorption cell at 1.6654 μm on the spectral resolution of the receiving system was depicted in Fig. 15. Also shown are the measured ratios between the received power and noise power P_r/P_n at 1.6654 μm through the evacuated sample cell, using four very low-loss silica optical fiber links with the length of 5, 10, 16 and 20 km, respectively. The transmission loss of the silica optical fiber was less than 1 dB/km in the wavelength region of 1.3 - 1.7 μm. The light source was a 1.64 μm InGaAs LED and the detection time was 3 sec. As mentioned in Section III, the maximum detection sensitivity of the system can be achieved by suitably selecting the spectral resolution of the receiving system, with reference to the operation wavelength and kind of gas to be monitored. Moreover, it is also understood from Fig. 15 that the optimum spectral resolution depends on the optical fiber length as well, because of the change of its transmission loss.

Figure 16 represents the measured results of the minimum detectable pressure of CH_4 gas in air at 1.6654 μm against the spectral resolution for four different optical fiber lengths. It is seen evidently that the optimum detection sensitivity is realized within the range of 2 - 4 nm resolution. More precisely, higher sensitivity can be generally achieved with higher spectral resolution in the case of shorter optical fiber link. Thus, if we use a LED with higher output power, e.g.,~ mW, or even a LED array, it will be possible to accomplish higher sensitivity detection of CH_4 gas, not only with an improved signal-to-noise ratio, but also with broader area coverage by longer optical fiber links. For a much higher sensitivity, if necessary, a LD tuned to the strongest absorption spectrum of CH_4 molecules at 1.6654 μm could be utilized in this remote sensing system as well.[2,3,5]

Figure 17 summarizes the measured results of the minimum detectable CH_4 pressure

628

in air at 1.33 and 1.66 μm for the low-loss silica optical fiber links of the length of 5, 10, 16 and 20 km, connected to a 50-cm long absorption cell. A dielectric interference filter with the transmission center at 1.331 μm and about 2.5 nm FWHM, and a similar filter around 1.666 μm as mentioned above were utilized, respectively. It was verified by these experiments that the detection sensitivity of 25% of the LEL of CH_4 gas density in air, which is normally required for any practical CH_4 gas sensor, can be realized up to a 20 km long optical fiber link with the detection time longer than 3 sec for 1.666 μm band, and also up to nearly 12 km with the detection time of 30 sec for 1.331 μm band. We should note that since the absorption of CH_4 molecules in 1.66 μm region is usually stronger than that in

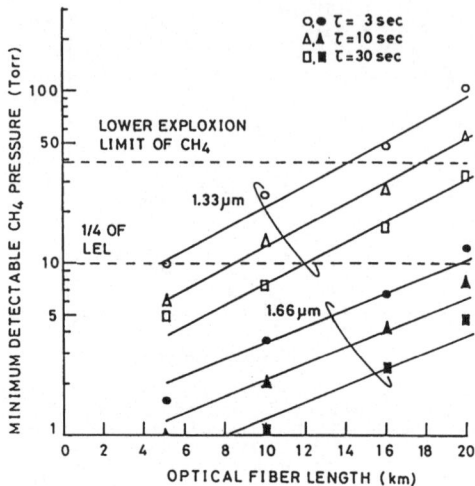

Fig. 17 Minimum detectable pressure of CH_4 gas in air at 1.331 and 1.666 μm as a function of the optical fiber length and the detection time in the all-optical remote measurement employ-ing 5 – 20 km-long, low-loss silica optical fiber links incorporating InGaAsP and InGaAs LED's and a 50 cm-long absorption cell as the gas sensor head.

1.33 μm region, higher sensitivity or broader detection coverage for the fiber-optic CH_4 gas remote sensing is achievable in 1.66 μm as seen in this result.

VI. Conclusion

We have measured and analyzed the near-infrared absorption bands of a number of polluting, combustible and/or explosive, and toxic molecules using high-radiant InGaAsP and InGaAs LED's. Based on these spectroscopic studies, we have developed a fully optical fiber-based remote sensing system for low-level ethylene, propane and methane gases over a wide area ranging up to 20 km. The detection sensitivity achieved for these gases completely satisfies the normal requirement for a practical gas sensor. Thus this all-optical system is demonstrated to be reliable, sensitive, feasible, nonhazardous, economical, and applicable to a great deal of various sub-stances in our environment and the atmosphere. Consequently, the similar remote sensing system is being developed for other species by suitably selecting the mea-suring wavelength in the near-infrared region as depicted in Fig.1.

On the basis of the present state-of-the-art of optical electronics technology, this kind of optical fiber network incorporating LED's or LD's and detectors in the near-infrared region should prove a very powerful scheme and provide a fully optical safe way for remote sensing of various dangerous and toxic gases/vapors at strategic points within the environment, such as industrial complexes, factories, mines, fuel storage yards, tunnels, undergrounds, ships, offices, hospitals, hotels, apartments, and so on. Finally, we are convinced that system level demonstrations of this technology, as presented in this and earlier papers, applied to a number of substances including not only gases but also liquids and solids will establish its validity and also secure the marketability in the near future.

Acknowledgement

The author is much indebted to Drs. S. Tanaka and K. Inada of Fujikura, Ltd. for kindly supplying the ultralow-loss silica optical fibers. He is also grateful to Drs. T. Ikegami and H. Kawaguchi of Atsugi Electrical Communication Laboratory, Nippon Telegraph and Telephone Corporation (NTT) for supplying the InGaAsP and InGaAs LD's. He wishes to thank to members of his laboratory who have been involved in the work decribed in this paper.

References

(1) E.g., T. G. Giallorenzi, J. A. Bucaro, A. Dandridge, G. H. Siegel, Jr., J. H. Cole, S. C. Rashleigh, R. G. Priest: IEEE J. Quantum Electron. QE-18, 626 (1982)
(2) H. Inaba: in Conf. Abstracts, 9th Int. Laser Radar Conf., Munich, West Germany, July 1979, Invited Paper 2-2, pp. 61-67
(3) H. Inaba, T. Kobayasi, M. Hirama, M. Hamza: Electron. Lett. 15, 749 (1979)
(4) T. Kobayasi, M. Hirama, H. Inaba: Appl. Opt. 20, 3279 (1981)
(5) H. Inaba: in "Optical and Laser Remote Sensing", D. K. Killinger and A. Mooradian, Eds. (Springer, Berlin, 1983), p. 288
(6) K. Chan, H. Ito, H. Inaba: Appl. Phys. Lett. 43, 634 (1983)
(7) K. Chan, H. Ito, H. Inaba: Appl. Opt. 22, 3802 (1983)
(8) H. Inaba, K. Chan, H. Ito: J. Opt. Soc. Am. 73, 1854 (1983)
(9) K. Chan, H. Ito, H. Inaba: IEEE / OSA J. Lightwave Technol. LT-2, 234 (1984)
(10) H. Inaba, K. Chan, H. Ito: in Digest of Tech. Papers, Conf. on Lasers and Electro-Optics (CLEO'84), Anaheim, Calif. U.S.A., July 1984, Invited Paper WN3, pp. 118-120
(11) K. Chan, H. Ito, H. Inaba: Appl. Phys. Lett. 45, 220 (1984)
(12) H. Inaba, K. Chan, H. Ito: in Conf. Proc., 2nd Int. Conf. on Opt. Fiber Sensors, Stuttgart, West Germany, Sept. 1984, P8, pp. 211-214
(13) H. Inaba, K. Chan, H. Ito: Appl. Opt. 23, 3415 (1984)
(14) H. Inaba, K. Chan, H. Ito: J. Opt. Soc. Am. A1, 1334 (1984)
(15) K. Chan, T. Furuya, H. Ito, H. Inaba: Opt. Quantum Electron. 17, 153 (1985)
(16) K. Chan, H. Ito, H. Inaba: Optics and Lasers in Eng. 6, 119 (1985)
(17) K. Chan, H. Ito, H. Inaba, T. Furuya: Appl. Phys. B38 (1985) to be published
(18) E. D. Hinkley, Ed., "Laser Monitoring of the Atmosphere" (Springer, Berlin, 1976)
(19) E.g., K. Inada, IEEE J. Quantum Electron. QE-18, 1424 (1982)
(20) E.g., G. Herzberg, "Molecular Spectra and Molecular Structure II, Infrared and Raman Spectra of Polyatomic Molecules " (Van Nostrand, New York, 1966)
(21) K. Chan, T. Furuya, H. Inaba: in Tech. Dig. 10th Nat. Laser Radar Symp. Japan, Awara, Fukui, May 1985, Paper E12, pp. 76-77 (in Japanese)
(22) M. Hamza, T. Kobayasi, H. Inaba: Opt. Quantum Electron. 13, 187 (1981)
(23) M. Hamza, T. Kobayasi, H. Inaba: Opt. Quantum Electron. 14, 339 (1982)

Laserfernmessung von Stickstoffdioxid und Schwefeldioxid

W. Staehr, W. Lahmann, C. Weitkamp und W. Michaelis

GKSS-Forschungszentrum, Institut für Physik

Max-Planck-Straße, D-2054 Geesthacht

Durch den Einsatz von gepulsten Lasern als Strahlungsquellen können atmosphärische Bestandteile ortsaufgelöst über größere Entfernungen nach dem Lidarprinzip nachgewiesen werden. Dabei wird das an Luftmolekülen und Aerosolpartikeln rückgestreute Licht zeitabhängig mit einem Empfänger registriert. Über die Laufzeit des Lichtes ist eine Bestimmung des Streuortes und damit eine ortsauflösende Messung möglich. Mit einem reinen Rückstreulidar können Bereiche erhöhter Aerosolkonzentration, wie z.B. Abgasfahnen von Schornsteinen, Dunstschichten, Wolken usw., nicht aber gasförmige Luftbestandteile festgestellt werden. Zum quantitativen Nachweis von Schadgasen in der Troposphäre erweist sich das Lidarverfahren nach dem Prinzip der differentiellen Absorption und Streuung (sog. DAS-Lidar) als am besten geeignet. Es beruht auf der unterschiedlich starken Absorption von Licht zweier verschiedener, jedoch eng benachbarter Wellenlängen durch das zu messende Schadgas.

Sendet man einen Laserpuls mit der Pulsdauer τ und der mittleren Leistung $\bar{P}_o$ aus, so wird das durch einen geeigneten Empfänger am Ort des Senders nachgewiesene atmosphärische Rückstreusignal $P(R,\lambda)$ aus der Entfernung R durch die Lidargleichung beschrieben:

$$P(R,\lambda) = \bar{P}_o \frac{c\tau}{2} \beta(R) \frac{A}{R^2} \eta \, O(R) \cdot \exp\left(-2 \int_o^R (\alpha(r) + N(r) \, \sigma(\lambda)) \, dr\right). \tag{1}$$

Dabei sind c die Lichtgeschwindigkeit, $\beta(R)$ der atmosphärische Rückstreukoeffizient, A die Fläche des Hauptspiegels und η die Effizienz des Empfangssystems. $O(R)$ beschreibt die Überlappung von Sendestrahl und Empfangsgesichtsfeld. $\alpha(r)$ ist der atmosphärische Extinktionskoeffizient in der Entfernung r, und $N(r)$ und $\sigma(\lambda)$ sind die Teilchenzahldichte bzw. der Absorptionsquerschnitt des zu messenden Schadgases. Aus den Rückstreusignalen $P(\lambda_o)$ und $P(\lambda_1)$ für zwei verschiedene Wellenlängen des Lasers, für die die Absorptionsquerschnitte $\sigma(\lambda_o)$ und $\sigma(\lambda_1)$ unterschiedlich sind, kann $N(R)$ bestimmt werden:

$$N(R) = \frac{1}{2 \left(\sigma(\lambda_1) - \sigma(\lambda_o)\right)} \frac{d}{dR} \ln \frac{P(R,\lambda_o)}{P(R,\lambda_1)} . \tag{2}$$

Die Wellenlänge mit dem geringeren Absorptionsquerschnitt wird dabei als die Referenzwellenlänge, die mit dem größeren als Meßwellenlänge bezeichnet. Atmosphärische und apparative Parameter in Gl. (1) kürzen sich heraus; das DAS-Lidarverfahren ist also selbsteichend, d.h. bei bekannten Absorptionsquerschnitten können Schadgaskonzentrationen absolut bestimmt werden.

Zum Nachweis von SO_2 und NO_2 mit dem Verfahren der differentiellen Absorption eignet sich der Wellenlängenbereich um 300 bzw. 450 nm besonders gut, weil diese Schadgase dort jeweils ausgeprägte Strukturen in ihrem Absorptionsspektrum aufweisen. Die Strahlung im Bereich um 450 nm wird direkt mit Farbstofflasern erzeugt, die UV-Strahlung um 300 nm durch Frequenzverdopplung von Farbstofflaserstrahlung mit 600 nm Wellenlänge.

Das im GKSS-Forschungszentrum entwickelte DAS-Lidarsystem für die SO_2- und NO_2-Messung ist in einen 20-Fuß-Standardcontainer eingebaut (Abb. 1). Der gesamte optische Teil des Meßsystems befindet sich in einem hydraulisch ausfahrbaren Meßkopf, der motorisch sowohl horizontal gedreht als auch vertikal geschwenkt werden kann. Außerdem ist der Container mit verschiedenen meteorologischen Sensoren sowie mit lokalen Meßgeräten für den NO_x- und SO_2-Nachweis ausgerüstet.

Abb. 1. GKSS-Lidarmeßstation für SO_2 und NO_2

Als Strahlungsquellen für das Lidarsystem werden vier durch koaxiale Blitzlampen angeregte Farbstofflaser verwendet, die sich durch kurze Pulsdauer und gute Strahlqualität auszeichnen [1]. Als aktives Medium wird für den NO_2-Nachweis der Farbstoff Coumarin 2, für den SO_2-Nachweis Rhodamin 6G verwendet. Der Laserresonator besteht aus einem Plan-Plan-Resonator mit einer Durchstimmeinheit aus drei Prismen zur Wellenlängenselektion und zur Einengung der Bandbreite. Für die UV-Erzeugung durch

Frequenzverdopplung wird als Kristallmaterial deuteriertes Kaliumdihydrogenphosphat (D-KDP) verwendet. Werden jeweils zwei Laser zum Nachweis eines Schadgases eingesetzt, so können bei einer Aussendung beider Wellenlängen innerhalb eines Zeitraumes von wenigen Millisekunden zeitliche Veränderungen der optischen Parameter der Atmosphäre als Fehlerquellen ausgeschlossen werden [2]. Im hier beschriebenen Lidarsystem werden die Laser für Meß- und Referenzwellenlänge nicht gleichzeitig, sondern mit einem zeitlichen Abstand von 50 µs gezündet, damit auf eine spektrale Trennung der Rückstreusignale verzichtet werden kann. Zur Kontrolle der Laserwellenlängen dient ein Spektrometer, das als dispersives Element ein Echelle-Gitter enthält und mit einer Diodenzeile zur Echtzeitausgabe der spektralen Verteilung auf einem Oszilloskop ausgestattet ist; die Wellenlänge läßt sich mit dem Spektrometer auf etwa 0,01 nm genau bestimmen [3]. Die Laserstrahlung wird über Flüssigkeitslichtleiter in das Spektrometer eingekoppelt, so daß eine simultane Kontrolle der Wellenlänge während der Lidarmessung möglich ist.

Die Laserstrahlen gelangen über jeweils zwei Strahlsteuerspiegel in Strahlaufweiter, die der Verringerung der Strahldivergenz dienen. Die rückgestreute Strahlung wird mit einem Newton-Teleskop aufgefangen und mit zwei Seitenfenster-Photomultipliern nachgewiesen. Die Strahlengänge für das Blaue und das UV werden durch einen dichroitischen Strahlteiler getrennt. Zur spektralen Filterung der nachzuweisenden Strahlung werden im Blauen ein Interferenzfilter und im UV eine Kombination aus Solarblind-Filter und Spike-Filter verwendet. Zur Einengung des Empfangsgesichtsfeldes und zur Verringerung der Signaldynamik des Rückstreusignals nach dem Prinzip der geometrischen Kompression [4] dient eine Irisblende in der Brennebene des Empfangsspiegels. Der Himmelslichtuntergrund wird durch die genannten Maßnahmen gegenüber dem Rückstreusignal soweit reduziert, daß Schadgasmessungen auch bei vollem Sonnenschein ohne Einbuße in der Nachweisempfindlichkeit durchgeführt werden können.

Zur Steuerung des Meßvorganges und zur Auswertung und Archivierung der Daten dient ein Prozeßrechnersystem auf der Basis eines Minicomputers PDP 11/34. Die Rückstreusignale werden von einem Transientenrekorder mit 10 bit Auflösung und 10 MHz maximaler Digitalisierungsrate in einem Zyklus digitalisiert und zwischengespeichert. Zur Bearbeitung der Rohdaten ist ein umfangreiches Programmpaket für so unterschiedliche Meßaufgaben wie die großflächige empfindliche Immissionskontrolle oder die erheblich höhere Ortsauflösung erfordernde Bestimmung der Emissionsrate einzelner Schornsteine entwickelt worden.

Durch zahlreiche konstruktive und prozedurale Maßnahmen konnte bei Immissionsmessungen eine Empfindlichkeit von 10 ppb für den NO_2-Nachweis und 15 ppb für den SO_2-Nachweis erreicht werden [5]. Diese Daten gelten für eine räumliche Mittelung über 300 m und eine zeitliche Mittelung über 120 Signalpaare, was etwa einer Meßdauer von

4 min entspricht. Die Reichweite des Meßsystems beträgt 6 km für die NO$_2$-Messung und 3 km für die SO$_2$-Messung. Durch die Zusammenfassung mehrerer Konzentrationsprofile, die in einem bestimmten Zeitraum in verschiedenen Richtungen gemessen wurden, erhält man eine synoptische Darstellung der Schadgasverteilung über dem betreffenden Gebiet, wie sie Abb. 2 für eine SO$_2$-Messung zeigt. Man erkennt deutlich die erhöhte SO$_2$-Konzentration über dem Gebiet einer Industrieanlage im Bereich von etwa 0,5 bis 1,5 km Entfernung vom Meßstandort.

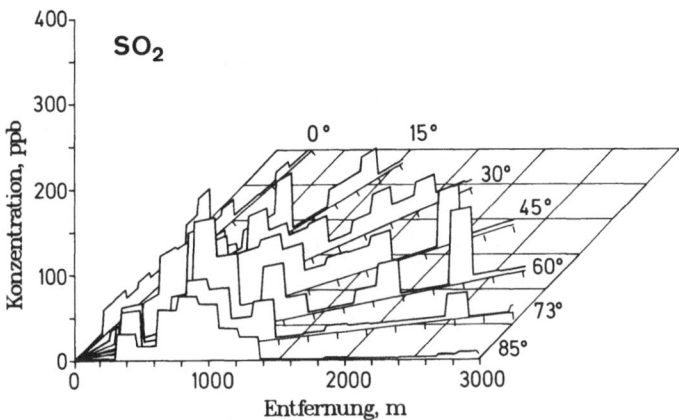

Abb. 2. Großflächige SO$_2$-Messung über Hamburg am 23. September 1983

Die Emissionsraten einzelner Schornsteine können mit dem Lidarsystem durch die Messung der Schadgasverteilung über einen Querschnitt der Abgasfahne bestimmt werden. Dazu wird das Produkt aus Teilchenzahldichte $N(\vec{r})$ und Teilchenmasse m, also die Schadstoffkonzentration am Orte $\vec{r}$, mit der Windgeschwindigkeit $\vec{u}(\vec{r})$ multipliziert, über die Querschnittsfläche F der Fahne integriert. Das Integral

$$\phi = \int_F m\, N(\vec{r})\, \vec{u}(\vec{r})\, d\vec{f} \qquad (3)$$

ist die gesuchte Emissionsrate. Zur Verifizierung der auf diese Weise bestimmten SO$_2$-Emissionsraten wurden Lidarmessungen am Schornstein einer Buntmetallhütte durchgeführt. Der Vergleich mit den vom Betreiber lokal gemessenen Werten zeigt gute Übereinstimmung.

Literatur
[1] T.B. LUCARORTO, T.J. MCILRATH, S. MAYO, and H.W. FURUMOTO, Appl. Opt. 19, 3178 (1980).
[2] N. MENYUK and D.K. KILLINGER, Opt. Lett. 6, 301 (1981).
[3] M.B. MORRIS and T.J. MCILRATH, Appl. Opt. 18, 4145 (1979).
[4] J. HARMS, Appl. Opt. 18, 1559 (1979).
[5] W. STAEHR, W. LAHMANN and C. WEITKAMP, Appl. Opt. 24, No 12 (1985), erscheint demnächst.

Einsatz eines Mini-Lidar bei lufthygienischen Überwachungsaufgaben

H.C. Salfeld, W.J. Müller
Institut für Arbeitsmedizin, Immissions- und Strahlenschutz
Davenstedter Str. 109, D - Hannover 91

Das Institut für Arbeitsmedizin, Immissions- und Strahlenschutz ist als niedersächsische Landesbehörde u.a. zuständig für die Analyse von Luftverunreinigungen. Es werden sowohl Emissionsmessungen z.B. an Schornsteinen oder diffusen Volumenquellen (Fabrikhallen) als auch Immissionsmessungen in direkter Nähe von industriellen Anlagen durchgeführt. Hierbei erlangt die Untersuchung des interregionalen grenzüberschreitenden Luftschadstofftransportes zunehmende Bedeutung. Mit dem hier beschriebenen Rückstreu-Lidar wird die räumliche Ausdehnung von Abgasfahnen (Staub, Aerosole) untersucht, ein in Kombination eingesetztes Doppler-Sodar erlaubt die Messung des höhenabhängigen Windfeldes und der Turbulenzbedingungen und bildet damit die Grundlage für Ausbreitungsmodellberechnungen.

Der Laser des Lidar-Systems strahlt einen kurzen Impuls in die Atmosphäre ab. Das Licht wird dort absorbiert und gestreut, ein Teil des zurückgestreuten Lichtes wird von einem Empfänger gesammelt. Zur Streuung tragen hier im wesentlichen Wassertröpfchen (Wolken, Nebel) sowie natürliche und anthropogene Aerosole bei (Mie-Streuung). Die Rückstreuung wird durch die sog. Lidar-Gleichung

$$P(r) = P_o \cdot G \cdot r^{-2} \; \beta(r) \cdot \tau^2 \tag{1}$$

beschrieben. $P(r)$ ist die vom Lidar empfangene Leistung, P_o die Ausgangsleistung des Lasers, G eine Systemkonstante (Spiegelfläche, Verluste) und r der Abstand vom Lidar. β ist der ortsabhängige Rückstreukoeffizient und τ^2 beschreibt die Absorptionsverluste. In den hier verwendeten Darstellungen wird das Produkt $\beta\tau^2$ als Meßergebnis dargestellt, τ kann bis zur Abgasfahnenvorderkante vernachlässigt werden, führt jedoch innerhalb einer dichten Abgasfahne zu einer stetigen Signalverringerung und muß bei der Berechnung von β entsprechend berücksichtigt werden. Bei der geometrischen Vermessung von Abgasfahnen ist $\beta\tau^2$ eine gute Näherung für die Abmessung, es ist in den folgenden Darstellungen in der Einheit $m^{-1}sr^{-1}$ angegeben. Tabelle 1 zeigt die technischen Daten.

Sender		Empfänger		
Laser	: Nd:YAG 1064 nm	Newton-Teleskop		
Hersteller	: ILS	Durchmesser	:	15 cm
Pulsenergie	: 100 mJ	Feld	:	2-10 mrad
Pulsbreite	: 18 ns	Detektor	:	EG&G YAG 444
Wiederholrate	: 10 oder 20 pps			
Divergenz	: 1.8 mrad			
Daten-Auswertung				
Transienten-Recorder Biomation 8100, 8 bit Auflösung, 2, Kanäle Computer cbm 8032 bzw. pdp11/34				

Tabelle 1. Technische Daten

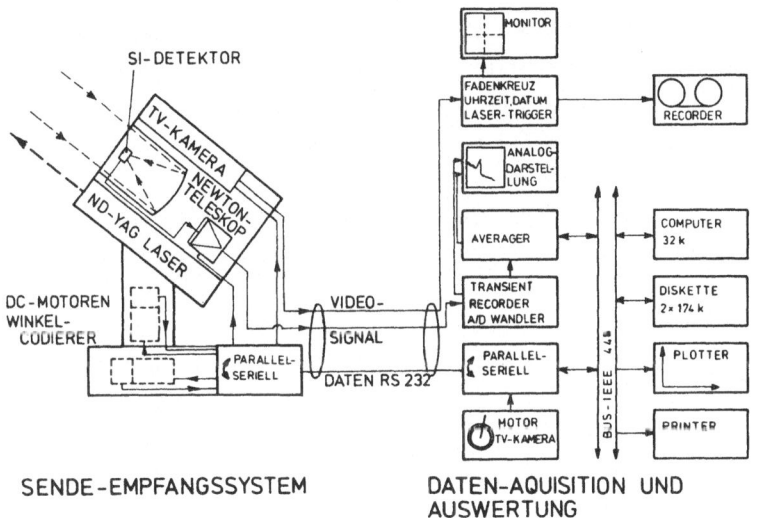

SENDE-EMPFANGSSYSTEM   DATEN-AQUISITION UND AUSWERTUNG

Bild 1. Blockdiagramm Lidar

Das Lidar wurde von der DFVLR (1) entwickelt und ist eine Sonderausführung des sog. Mini-Lidar. Die Anlage besteht aus einem Sende-Empfangssystem und davon räumlich getrennt die Auswerte- und Steuerelektronik. Das Lidar kann außerhalb eines Meßfahrzeugs auf einem Stativ montiert oder von einem Container aus einer Dachluke herausgefahren werden. Es besteht aus einem wasser/luftgekühlten Nd:YAG-Laser, Empfangsoptik, Si-Detektor mit umschaltbaren Verstärkungsstufen, TV-Kamera mit Zoom-Optik sowie der Schwenk-Neige-Mechanik. Die Positionierung erfolgt durch Gleichstrommotoren, die Position (Elevation, Azimut) kann über Winkelcodierer (Auflösung 0.1°) abgerufen werden. Sämtliche Funktionen können über eine serielle Datenleitung vom Meßfahrzeug aus fernbedient werden. Wegen der relativ geringen Pulsenergie des Lasers wird das Rückstreusignal mit einem fest verdrahteten Addierer schnell (10 - 20 Hz) gemittelt. Die Meßdaten werden über IEC-Bus ausgelesen und zusammen mit den sonstigen Meßparametern gespeichert, so daß bei der späteren Auswertung eine vollständige Dokumentation vorliegt. Dazu dient auch ein an die TV-Kamera angeschlossener Vieorecorder. Für Messungen bei Dunkelheit steht ein Nachtsichtgerät zur Verfügung.

Seit 1969 werden flüssige hochchlorierte Kohlenwasserstoffe von Verbrennungsschiffen auf der Nordsee verbrannt. Das Verbrennungsgebiet befindet sich südöstlich der Doggerbank, ca. 100 bzw. 180 km von den nächsten Küsten entfernt. Die entstehenden Abgase sind hauptsächlich Chlorwasserstoff, Kohlendioxyd und Wasserdampf. Im Rahmen von Untersuchungen zur Ausbreitung von Schadstoffen in der maritimen Grenzschicht (2) und spätere Verifikation von Ausbreitungsmodellen wurde eine Vermessung von Abgasfahnen dieser Verbrennungsschiffe in den Jahren 82 und 83 durchgeführt. Das Vermessungsschiff 'Gauss' des Deutschen Hydrographischen Instituts war eine ideale Plattform für den kombinierten Einsatz verschiedener Meßsysteme, neben konventionellen meteorologischen Gebern war ein Doppler-Sodar unseres Instituts installiert, neben dem hier beschriebenen Lidar war ein Lidar (694 nm, 2J) des Deutschen Wetterdienstes im Einsatz (3).

Bei der Vermessung der Abgasfahnen wurden 3 verschiedene Verfahren angewandt. Zur Ermittlung der Zugrichtung sowie Fahnenbreite wurde das Lidar jeweils auf die Mitte der Abgasfahne gerichtet und das Rückstreusignal über eine Zeit von typ. 15 min gemittelt. Dagegen wurde bei der Aufnahme eines Querschnittes jeweils über den Zeitraum von 1 min gemittelt, dann die Elevation verstellt usw. Bei beiden Verfahren war das Schiff gestoppt und befand sich 500 - 1000 m außerhalb der Rauchfahne. Bei der dritten Methode   'Traversierung' fuhr das Schiff quer zur Ausbreitungsrichtung unter der Fahne hindurch, das Lidar war mit einem konstanten Elevationswinkel geneigt. Der genaue Standort des Schiffes wurde von einem Navigationsrechner festgehalten, so daß später die Zugrichtung der Fahne sowie die Entfernung zur Quelle berechnet werden konnte. Bei allen Messungen wurde durch das Doppler-Sodar Windrichtung- und Geschwindigkeit sowie Turbulenzzustand ermittelt.

Ein Beispiel für das erste Verfahren zeigt Bild 2. Die Rauchfahne war hier bis zu einem Quellabstand von ca. 9 km zu verfolgen, in einem anderen Fall bis zu einer Entfernung von 20 km. Bemerkenswert ist die Drehung der Zugrichtung. Auf dem Bild

sind gleichzeitig Windrichtung und Geschwindigkeit in verschiedenen Höhen dargestellt. Die Richtungsdrehung der Abgasfahne stimmt gut mit der Windrichtungsdrehung
in der Höhe 45 m bis 155 m überein. Tabelle 2 zeigt die ermittelten Fahnenbreiten
im Vergleich der beiden Lidar-Systeme (4). Wird berücksichtigt, daß sich die beiden
Geräte in unterschiedlicher Höhe an Deck befanden, die Mittelungszeiten unterschiedlich waren und wegen der Schlingerbewegungen des Schiffes die Elevationsangabe unge-
nau war, ergibt sich doch eine recht gute Übereinstimmung.

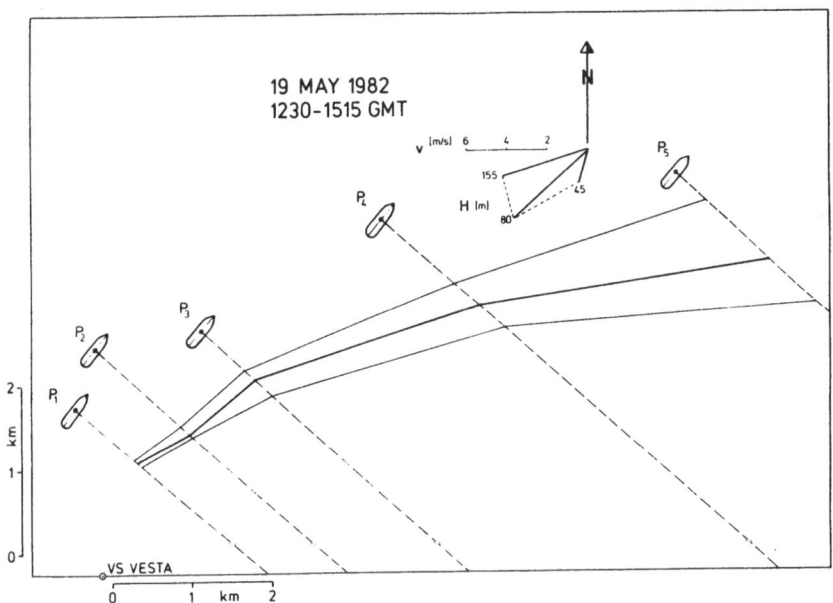

Bild 2 . Zugrichtung der Rauchfahne - Vergleich mit Doppler-Sodar
Messung

Profil	Quellentf.	Fahnenbreite(694nm)	Fahnenbreite(1064nm)
P1	900 m	375 m	400 m
P2	2200 m	450 m	475 m
P3	3000 m	750 m	500 m
P4	4000 m	800 m	800 m

Tabelle 2 . Dimension einer Abgasfahne im Vergleich

638

Bild 3 zeigt einen Vergleich mit der Ausbreitungsrechnung mit dem Ansatz eines
Gitterpunktmodelles (4). Abmessung, Form, und Neigung stimmen gut überein, die
stark asymmetrische Ausbildung der Querschnitts resultiert aus der starken höhenabhängigen Windrichtungsdrehung. Eingangsparameter für das Ausbreitungsmodell sind
die aus Sodar-Daten gewonnenen Windprofile sowie als Turbulenzmaß die Standardabweichung der Vertikalwindkomponente σ_w .

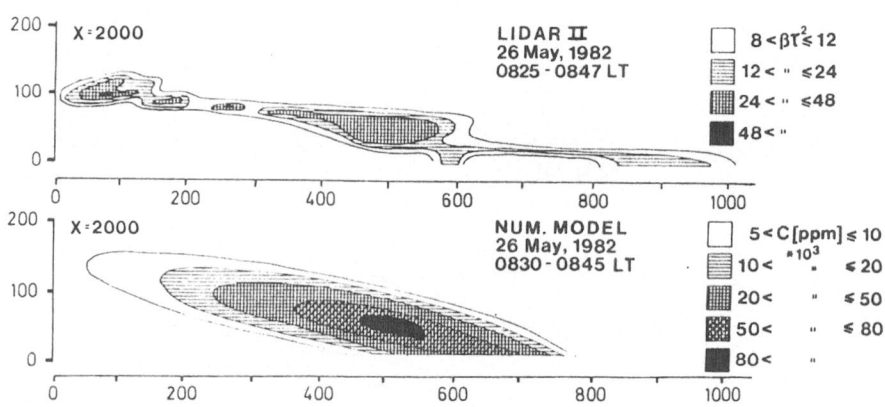

Bild 3 . Abgasfahnenquerschnitt Lidar oben, Modellrechnung unten

Die vorliegenden Ergebnisse zeigen, daß das hier beschriebene Lidar-System gut zur
Vermessung von Abgasfahnen geeignet ist. In diesem Sinn ist es ein wertvolles Instrument zur Verifizierung von Ausbreitungsberechnungen. Die Leistungsfähigkeit des
Mini-Lidar für diesen Zweck entspricht dem wesentliche aufwendigerer Lidar-Systeme
mit 20-facher Pulsenergie, wobei deutliche Vorteile in der leichteren Handhabbar-
keit liegen.

Literatur
(1) WERNER, CH. : Opt. Engeneering 20, 5 , 759 - 764 (1981)
(2) HOTZLER, I : UBA-FE Abschlußbericht 104 02 556/01(1984)
(3) KLAPHECK, K : Proc. of IGARSS'84 Symp. ESA SP - 215 (1984)
(4) WAMSER, C., HOTZLER, I., KLAPHECK, K., MÜLLER.W.J., SALFELD, CH.:
                UBA Bericht 10402 556/01 (1982)

Einsatz eines Minilidars für Umweltschutzmessungen

H. Herrmann und J. Streicher

DFVLR - Institut für Optoelektronik

D - 8031 Oberpfaffenhofen / Wessling

1. Einleitung

Die Fernmessung atmosphärischer Parameter mit Laser-Methoden beruht auf
dem Studium der Streuung und Absorption des ausgesandten Laserlichts
durch die Moleküle und Partikel(Aerosole) der Atmosphäre. Informationen
über das Aerosol erhält man mit einem Laser-Radar, auch Lidar genannt,
wenn die Wechselwirkung des Laserpulses mit den Partikeln untersucht
wird. Als Laser-Sender dienen Rubin-, Neodym- oder CO_2-Laser.

Die Abb.1 dient zur Erläuterung des Lidar-Prinzips. Die Öffnungswinkel
von Sender und Empfangsteleskop sind vergrößert dargestellt. Der Laser
sendet einen Impuls von etwa 20 Nanosekunden Dauer, entsprechend einem
Lichtpaket von 6 m Länge. Der Puls wird auf dem Weg durch die Atmosphäre gestreut und geschwächt. Das rückgestreute Licht kann durch ein
neben dem Sender angeordneten Teleskop mit Detektor empfangen werden.
Die Abbildung zeigt den typischen Verlauf des Empfangssignals U beim
Anmessen einer Aerosolwolke in etwa 1000 m Schrägentfernung.Die nachfolgend vereinfachte Lidar Gleichung beschreibt den Signalverlauf:

$$U\ (R)\ =\ G\ \frac{\beta\ \tau^2}{R^2}$$

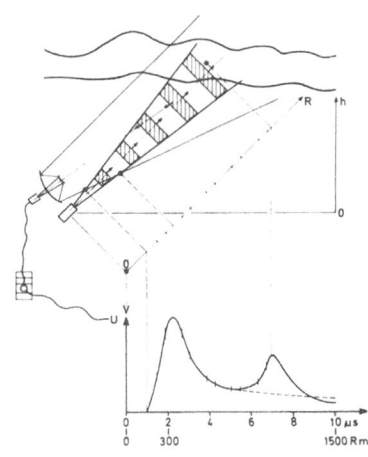

Abb. 1. Lidar-Prinzip

G ist eine Systemkonstante, die aus der
Laserausgangsleistung und optischen Parametern gebildet wird. τ^2 gibt die Extink-
tionsverluste des Laserpulses zum Meßort und zurück an ($\tau \leq 1$). Der Wert des
Rückstreukoeffizienten β (Einheit m^{-1})
ist ein Maß für die Aerosolkonzentration
in $\mu g/m^3$. Bei bekanntem G und einer betrachteten Aufpunktentfernung R wird das
Empfängersignal U damit proportional zur
Aerosolkonzentration.

2. Aerosolmessungen in Österreich und Süddeutschland

Die DFVLR führte im Herbst 1983 im Auftrag des Österreichischen Bundes-
instiuts für Gesundheitswesen Fernmessungen der Aerosolverteilung in In-
dustriegebieten Österreichs durch. Dazu wurde das auf der Laser 81 vorgestellte DFVLR-Minilidar in einem Meßbus betrieben. Die Abb.2 zeigt den
Einsatz an einer Papierfabrik in Hallein bei Salzburg. Mit dem in Azimut

und Elevation veränderbaren Lidar konnte an zwei Tagen eine
Übersicht der Aerosolmassenver-
teilung gewonnen werden, die in
Abb.3 dargestellt ist. Der Elevationswinkel blieb mit 10° konstant, während der Azimutwinkel
schrittweise variiert wurde.
Die Isolinien stellen gleiche
Massenkonzentrationen dar und
zeigen die Ausbreitung der
Rauchfahnen in Abhängigkeit vom
Wind und Geländeformation.

Abb.2. Lidareinsatz an Papierfabrik

Massenkonzentrationsänderungen von 25 - 500 $\mu g/m^3$ zeigen die Isolinien
A bis E vom Meßtag 19.9.83 und a bis g vom 24.9.83.

Das Ziel einer weiteren Meßkampagne mit demselben Lidarsystem war im
Frühjahr 1984 das Motorenwerk einer bekannten Automobilfirma im Süddeut-
schen Raum. Es sollten Hinweisen von Werksangehörigen nachgegangen werden,
deren auf dem Werksgelände geparkten Privatfahrzeuge durch Staubablage-
rungen beeinträchtigt wurden. Von einem Hallendach aus konnte durch Azi-
mutabtastung wieder die allgemeine Aerosolverteilung bestimmt werden. Die
Suche nach defekten Filteranlagen im Werk und Aerosoleintrag von außen
brachte keine auffälligen Resultate. Nur der Bereich um die Gießerei
zeigte erhöhte Aerosolkonzentrationen. Erst ein Standortwechsel in eine
der Gießerei nahegelegene Werkstraße lokalisierte die unbekannten Emmi-
tenten: Loren mit Gußstaub, die in regelmäßigen Abständen entleert wurden.
Diese stellten diffuse Aerosolquellen dar, deren unsichtbare Staubparti-
kel durch Windverfrachtung im Werksgelände verteilt wurden. Die Abb.4
zeigt einen Vertikalschnitt dieser Quelle. Bei der derzeit herrschenden
Westwindlage verblieben die Partikel etwa noch 15 Minuten am Meßort. In
der Abb.5 ist die zeitliche Änderung der Konzentration dargestellt.

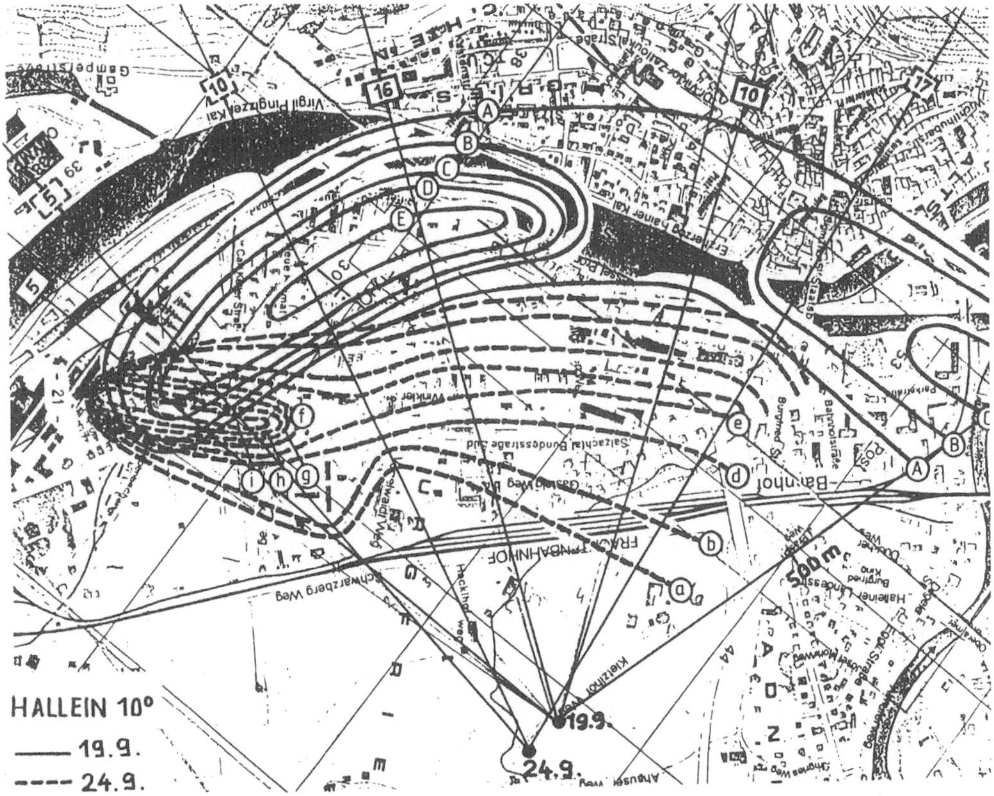

Abb. 3 Isolinien gleicher Massenkonzentration

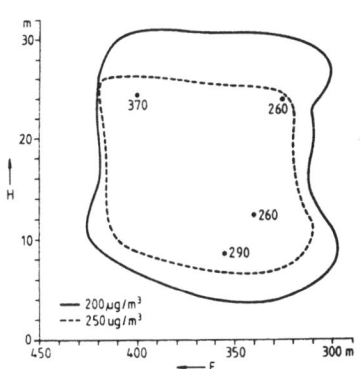

Abb. 4 Vertikalschnitt durch Aerosolquelle

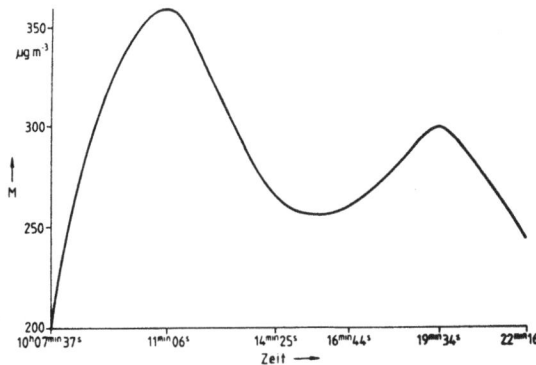

Abb. 5 Zeitliche Änderung der Konzentration nach Lorenschüttung

3. Ein koaxiales Minilidar kompakter Bauart

Basierend auf Erfahrungen mit dem Bau eigener Lidarsysteme und eines Lidars für das Niedersächsische Landesverwaltungsamt in Hannover, konnte
ein koaxiales Lidar für das österreichische Bundesinstitut für Gesundheitswesen Wien hergestellt werden. Abb.6 zeigt dieses System.

Links angeordnet ist das wetterfeste und thermostatisierte Lidar mit der
Stromversorgung und Laserpulsenergie-Anzeige im rechten Bildteil. Im unteren Teil des Lidars befindet sich ein Nd:YAG Laser als Sender mit etwa
60 mJ Pulsenergie und 20 ns Pulsdauer. Unter der weißen Schutzhaube montiert ist das justierbare Spiegelteleskop als Empfänger, eine Farbfern-
sehkamera mit der minimalen Ansprechschwelle von 7 Lux sowie ein Zielferr
rohr für die Peilung der Meßobjekte. Der Laserpuls wird über zwei 90° -
Umlenkprismen in die optische Achse des Empfangsteleskops geleitet, sodaß auch im Nahbereich Aerosolmessungen möglich sind.

Unterhalb des Rechners HP 85 mit dem Diskettenlaufwerk ist der für das
Lidarsignal erforderliche Transient Recorder erkennbar. Ein Transient Recorder arbeitet in der gleichen Weise wie ein Speicher-Oszillograph, d.h
einmalig vorkommende Signale können gespeichert werden. Das Lidar-Echo
als Analog-Signal wird mit einer digitalen Auflösung von 8 bit im Recorder Datalab DL912 digitalisiert und gespeichert. Dieses Gerät verfügt
über zwei Eingangskanäle mit der Speichertiefe 4 bzw. 8 k Byte. Die Ab-
tastrate 20 MHz (50 ns Meßintervall) ermöglicht eine Entfernungsauflösung von 7,5 m. 16 bzw. 32 Einzelsignale können seriell gespeichert
werden zu je 256 Worten. Mittelung dieser Signale ist damit möglich.

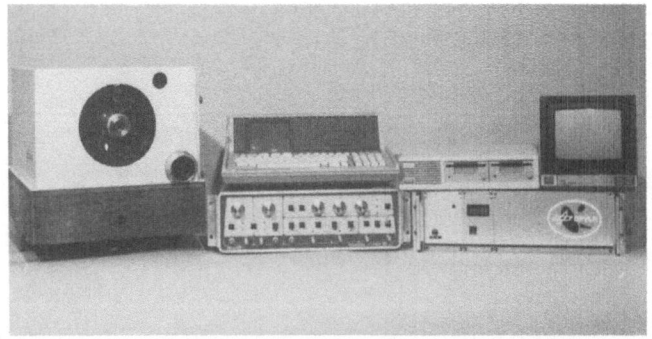

Abb. 6. Koaxiales Lidar-System

Literatur
H. HERRMANN, F. KÖPP und CH. WERNER: Laser 81 Conference Proceedings 269
Editor W. Waidelich, IPC Press 1981

Integrated FLIR/CO$_2$ Laser Rangefinder

C.O. Taylor, B.N. Berdanier & R.K. Powell, Texas Instruments Inc., P.O.Box 660246
Dallas, Texas 75266 / USA

Eyesafe CO$_2$ Laser Rangefinders have been fabricated and field tested utilizing
common module FLIRs as the receivers. This approach eliminated the necessity of
an additional set of cooler, detector, preamp, and receiver optics required with
a stand alone CO$_2$ laser rangefinder and allows total integration into some sights
such as the M6 0A3 tank thermal sight wherein the visible channel optics are shared
with the transmitter.

An Airborne Lidar System for Oceanographic Measurements

D. Diebel-Langohr, K.P. Günther, T. Hengstermann, K. Loquay, R. Reuter, R. Zimmermann

Universität Oldenburg, Fachbereich Physik

Postfach 2503, D 2900 Oldenburg

In the period 1979-82 a first version of an airborne fluorosensor was developed at the University of Oldenburg and utilized for oceanographic research. The system had been designed for monitoring the evolution of Rhodamine tracers /1,2/ and for near synoptic investigations of the biological productivity over extended areas of the sea.

Based on the experience achieved with this instrument a new sensor, the Oceanographic Lidar System OLS, was developed in 1983 which meets the requirement to pursue various tasks in the field of oceanography. The system allows airborne measurements of the water turbidity at different UV/VIS wavelengths, of dissolved organic matter (Gelb-stoff) and chlorophyll a concentrations, and of marine oil pollution. Lasers with pulse lengths in the nanosecond range and a fast signal detection system permit measurements of depth profiles of hydrographic parameters in the upper water column.

Installed in a DO 28 Skyservant of DFVLR Oberpfaffenhofen the sensor has been operated in various oceanographic experiments. Results obtained in the hydrographic depth profiling mode and measurements performed over marine oil spills are discussed in sep-arate papers given in this volume. In this paper we describe the OLS characteristics, and results obtained in the experiment ADRIA'84 are presented.

The Oceanographic Lidar System

An excimer laser with a peak power of 10 MW emitting at 308 nm serves as the main light source. Front and rear outputs of the laser are utilized as the lidar beam or as pumping beams for two dye lasers with emission wavelengths of 450 and 533 nm, respectively. The maximum pulse repetition rate is 20 Hz.

The signal receiver is a Schmidt-Cassegrain telescope with a 40 cm aperture. The field of view is set to 5 mrad which corresponds to the beam divergence of the excimer laser. Dichroic beam splitters deflect selected spectral ranges to filters and fast gated photomultipliers. Lasers and the detector system are mounted on an optical table to obtain a rigid alignment of the optical setup.

Signal digitization is done with a transient recorder at a sampling rate of 500 MHz. Since the digitizer is a one-channel instrument, 3 photomultiplier signals selected by the operator are sequentially combined on one signal line and fed to the transient re-corder. The system is controlled by a microcomputer, by which laser selection and triggering, selection of different detection wavelengths, quick-look data output and data storage is achieved (Fig. 1).

Principle of measurement

A spectral analysis of the laser-induced radiation from oceanic water yields structures as shown in Fig. 3 from which the following hydrographic data are derived:

- the sum of light attenuation coefficients $c_L + c_R$ at the laser excitation and the
  water Raman scattering wavelengths from the Raman scatter intensity,
- the concentration of Gelbstoff and of chlorophyll a from the intensity of their
  specific fluorescence with maximum values in the blue/green and at 685 nm.

Due to the partly overlapping signatures of these signal contributions 7 individual
detection channels are installed at wavelengths which permit a specific identification of the different spectral compounds.

A quantitative interpretation of the data is derived from the lidar equation /3/
giving the signal dP received from a depth interval dz at depth z:

$$dP = A\eta \frac{1}{(z+mH)^2} \exp\left(-\int_0^z c \, dz'\right) dz$$

where A represents instrumental factors, the quantum efficiency of fluorescence or
scattering, c the sum of light attenuation coefficients at the laser and the detect-
ion·wavelengths, H the aircraft flight height, and m the refractive index of water.
Assuming $c(z+mH) \gg 1$ which holds for all practical cases, the lidar equation can be
integrated over a finite homogeneous water layer at depth $z_1 < z < z_2$ /4/:

$$P_{z_1 \to z_2} = A \exp\left(-\int_0^{z_1} c \, dz\right) \frac{\eta}{c} \left(\frac{1}{(z_1 + mH)^2} - \frac{\exp(-c(z_2 - z_1))}{(z_2 + mH)^2}\right)$$

The data given in the following section are obtained by a total integration of the
signal peaks recorded versus time after subtraction of the sunlight-induced background. This corresponds to a depth integration between z=0 and $cz \to \infty$. A homogeneous
water column within this interval yields $P = A/(mH)^2 \, \eta/c$.

Experimental results

A number of OLS flights were performed in August 1984 during the international experiment ADRIA'84 in the northern Adriatic Sea. As an example, the distribution of
Gelbstoff and of chlorophyll a obtained on August 30 is shown in Fig. 3. The Po river
plume and the presence of different water masses with characteristic Gelbstoff and
chlorophyll a concentrations separated by hydrographic fronts with gradients of typically a factor of 10 over distances of 1-3 km are clearly seen. It is interesting to
note the high correlation of both parameters despite their very different nature.
Gelbstoff is mainly brought into the sea by river run-off and distributed by advective
processes; chlorophyll a productivity is dominated by local factors as light conditions, nutrients and stratification of the water column. A correlation up to this
degree was not found in other areas, e.g. the North Sea.

According to preliminary results of in situ investigations the detection limits of OLS
are 0.05 mg/l for Gelbstoff and 0.05 µg/l for chlorophyll a.

646

A detailed and sensitive delineation of the hydrographic situation in the northern Adriatic Sea is expected from the further data evaluation and from a comparison with in situ findings. Moreover, a combination of these data with measurements performed with optical radiometers along the same flight tracks by V. Amann, DFVLR Oberpfaffenhofen, will enable a step forward in the field of ocean remote sensing.

Acknowledgements

The development of the Oceanographic Lidar System was financed by a grant from the Bundesministerium für Forschung und Technologie, Bonn. The experiment ADRIA'84 was supported by the Commission of the European Communities, ISPRA Establishment, and by DFVLR Oberpfaffenhofen.

References

/1/ U. Gehlhaar et al.: Appl. Opt. 20, 3318 (1981)
/2/ H. Franz et al.: Deep-Sea Res. 29, 893 (1982)
/3/ E.V. Browell: NASA TN D-8447, 39 p. (1977)
/4/ D. Diebel-Langohr et al.: Archimedes I Final Report, Commission of the European Communities (in press)

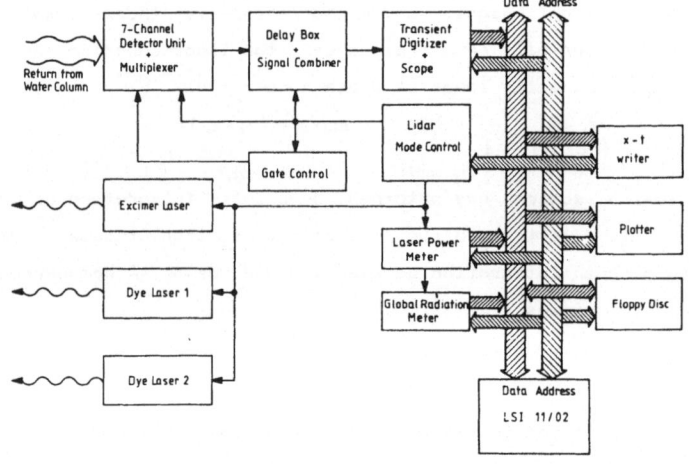

Fig. 1.
Schematic of the signal flow of the Oceanographic Lidar System

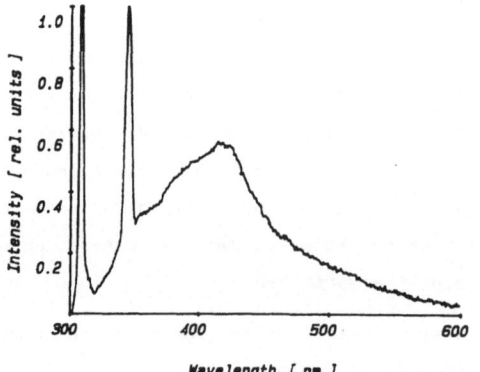

Fig. 2.
Emission spectrum of a natural water sample taken from the German Bight. Excitation wavelength is 308 nm. The peaks at 308 and 344 nm are due to elastic and water Raman scattering. The fluorescence centered at 420 nm is due to Gelbstoff. Chlorophyll a yields a fluorescence peak at 685 nm

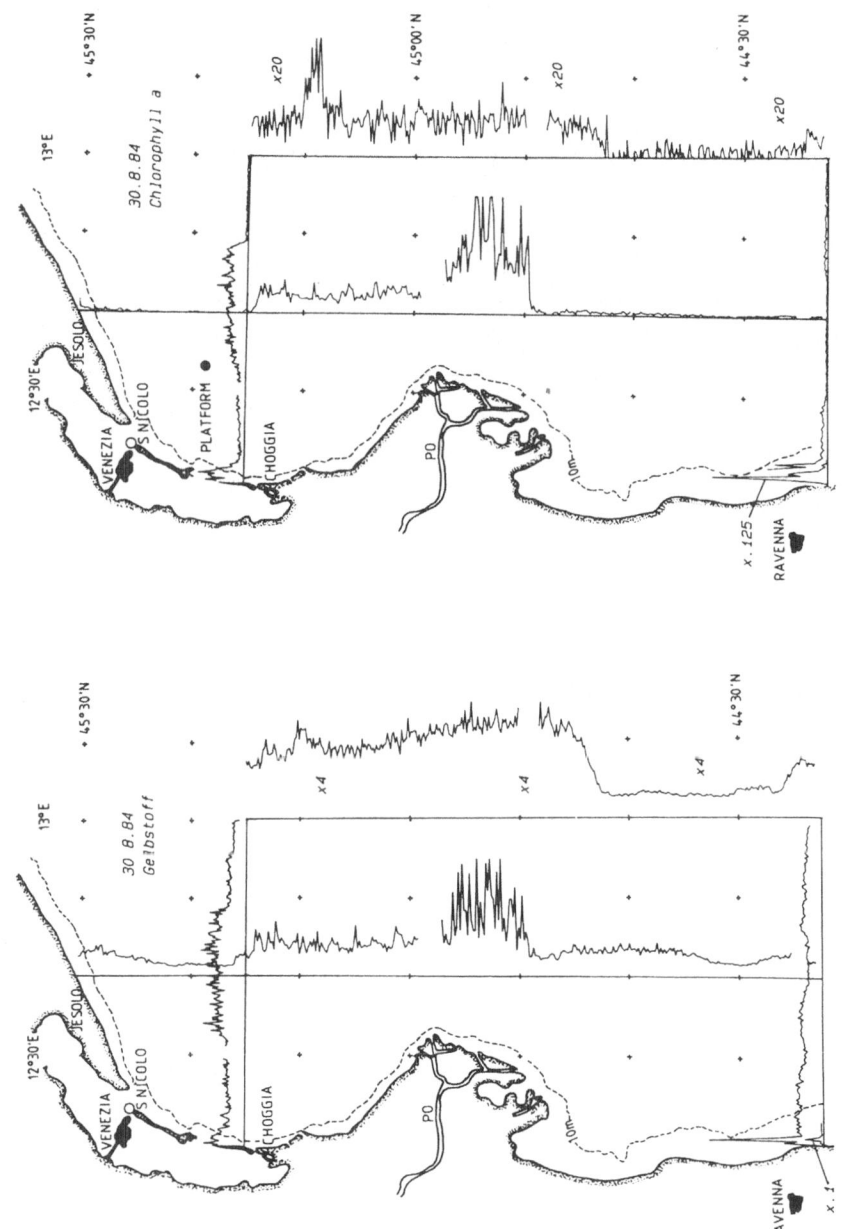

Fig. 3. Relative distribution of Gelbstoff and chlorophyll a in the northern Adriatic Sea measured with the Oceanographic Lidar System on 30.8.1984

Depth Profiles of Hydrographic Parameters – Measurement and Interpretation of Lidar Signals

D. Diebel-Langohr, T. Hengstermann, R. Reuter

Universität Oldenburg, Fachbereich Physik

Postfach 2503, D 2900 Oldenburg

The lidar method applied from low flying aircraft over the ocean has a potential for the depth resolved measurement of hydrographic parameters as the light attenuation coefficient and the concentration of water constituents (chlorophyll a, Gelbstoff, suspended minerals) in the upper water layer. This requires an excitation of fluorescence and scattering processes by laser pulses in the nanosecond range and a time resolved detection of the excited signals at time intervals small compared to the signal duration. Depth profiles can be derived by appropriate algorithms from lidar signals measured at detection wavelengths which corresponds to the scattering and fluorescence bands of the substances under investigation.

The following abbreviations are utilized :

$P_L(t)$ laser pulse shape

$P_\lambda(t)$ fluorescence or scattering signal measured at the detection wavelength λ

z water depth

$n(z)$ concentration of the substance fluorescent or scattering at wavelength λ

$c(z)$ sum of the attenuation coefficients at the laser and at the detection wavelength

v light velocity

m refractive index of water

A signal $S(z)$ originating from depth z depends on the local concentration of the fluorescent or scattering substance and on the attenuation of the laser light and of the light emitted at depth z in the water column between $0 \leq z' \leq z$. This is described by :

$$(1) \qquad S(z) \sim n(z) \cdot \exp(-\int_0^z c(z')dz') \qquad z \geq 0$$

A transformation to the time t with $t = 2zm/v$ yields :

$$(2) \qquad S(t) \sim n(t) \cdot \exp(-(v/2m)\int_0^t c(t')dt') \qquad t \geq 0$$

The signal $P_\lambda(t)$ received by the lidar system is a composition of signals from all water depths. Mathematically $P_\lambda(t)$ is a convolution of the depth function $S(t)$ and of the laser pulse $P_L(t)$ (with $S(t) = 0$ at $t < 0$) :

$$(3) \qquad P_\lambda(t) \sim \int_{-\infty}^{+\infty} P_L(t')S(t-t')dt'$$

Algorithms for the derivation of depth profiles from lidar signals

In order to obtain the depth function $S(t)$ from the measurable quantities $P_\lambda(t)$ and $P_L(t)$ the convolution integral (3) must be solved.

One possible way is to perform a <u>Fourier analysis</u>. This yields :

(4) $\qquad S(t) \sim \mathcal{F}^{-1}[\,\mathcal{F}(P_\lambda(t))\,/\,\mathcal{F}(P_L(t))\,]$

Alternatively the laser pulse shape is approximated by an infinitely short δ - pulse : $P_L(t) \sim \delta(0)$. Then the convolution integral (3) can be solved directly and it follows from this <u>Delta pulse analysis</u> :

(5) $\qquad S(t) \sim P_\lambda(t)$

The depth function $S_R(t)$ derived from a water Raman scattering signal with one of these methods is only a function of the depth dependent attenuation coefficient c since the concentration of the scattering water molecules n is constant with depth :

(6) $\qquad S_R(t) \sim \exp(-(v/2m)\int_0^t c(t')dt')$

If the lidar signals are recorded time resolved at intervals of length Δt, one can calculate the quotient

(7) $\qquad S_R(t)/S_R(t+\Delta t) = \exp((v/2m)\cdot c(t)\cdot \Delta t)$

assuming $c(t') = c(t)$ in the small time interval $t \leq t' \leq t+\Delta t$.
The solution for the attenuation coefficient c is :

(8) $\qquad c(t) = (2m/v\Delta t)\cdot \ln(S_R(t)/S_R(t+\Delta t))$

Signals originating from fluorescent water constituents yield the depth function :

(9) $\qquad S_F(t) \sim n(t)\cdot \exp(-(v/2m)\int_0^t c(t')dt')$

If the detection wavelength for the fluorescence signal is only slightly different from that of the water Raman signal, the attenuation coefficients at both wavelengths are nearly equal and the exponential part of (9) can be approximated by the depth function $S_R(t)$ derived from the water Raman signal. This yields for the concentration of the fluorescent substance :

(10) $\qquad n(t) \sim S_F(t)/S_R(t)$

Application of the algorithms to model signals

The algorithms given above were applied to model signals $P_\lambda(t)$ (3) of water Raman scattering and Gelbstoff fluorescence for testing their practicability. These model signals are calculated from model depth profiles of the attenuation coefficient and the concentration of Gelbstoff, and from a realistic laser pulse shape. The results shown in Fig. 1 demonstrate that in principle both algorithms allow a reconstruction of the given depth profiles. Using the Fourier analysis (4) slight errors occur when the Lidar signals have become very flat with increasing water depth. This is due to the limited bandwidth of the frequency range used. On the other hand the Delta pulse analysis (5) is only able to calculate values of the attenuation coefficient and the concentration representing averaged values over layers in the water column with a length corresponding to the length of the laser pulse. This is a consequence of the approximation of the laser pulse by a δ - pulse.

650

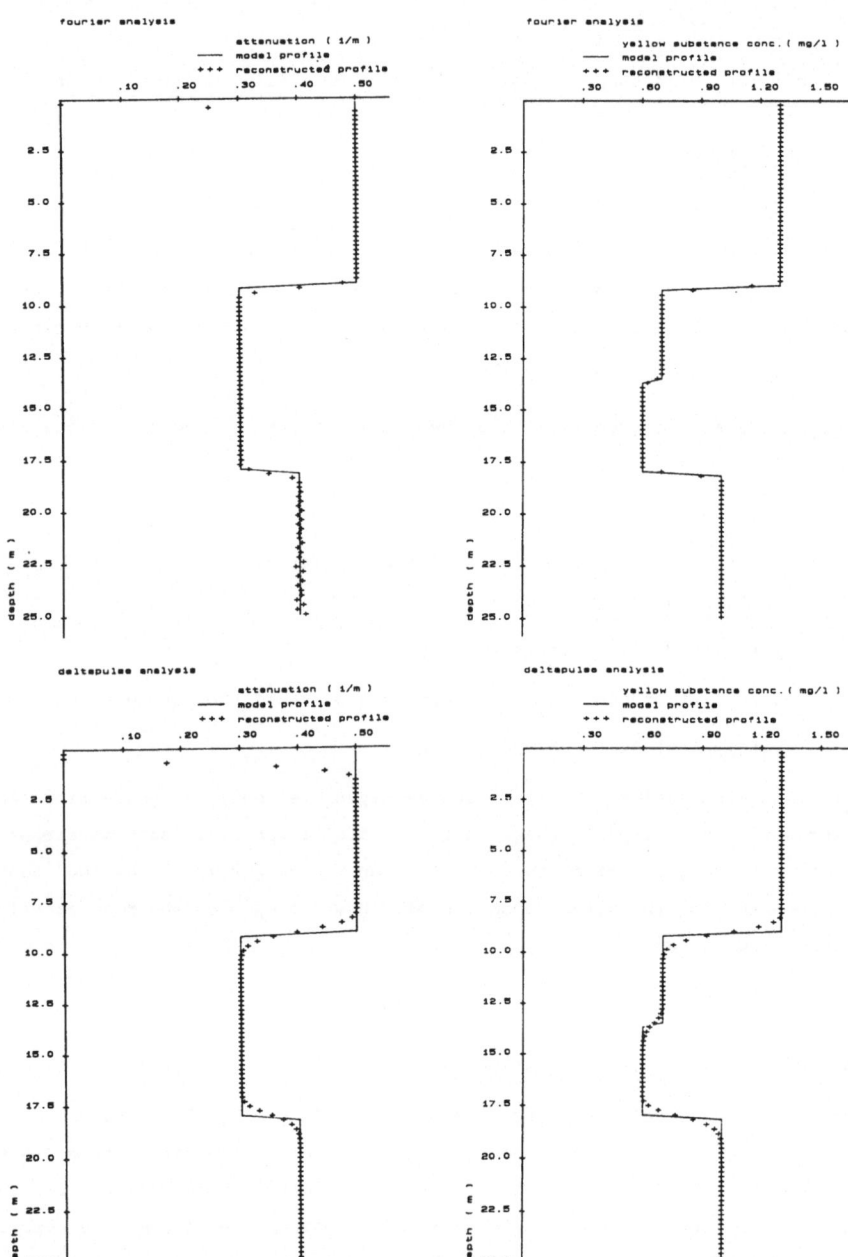

Fig.1 : Depth profiles of the attenuation coefficient and the Gelbstoff concentration
reconstructed with the Fourier analysis, and with the Delta pulse analysis

In reality signals are recorded with a finite amplitude resolution. The application of
the described algorithms to model signals with a simulated digitization shows that the
Fourier analysis is so sensitive to the digitization that no valuable profiles are reconstructed with an amplitude resolution of less than 12 bit. Using the Delta pulse
analysis the maximum error is less than 10% with 10 bit or higher amplitude resolution.

Measurement of depth profiles with the Oceanographic Lidar System (OLS)

In August 1984 measurements of the depth dependent attenuation coefficient were performed over the Adriatic Sea using the Oceanographic Lidar System of the University of
Oldenburg (OLS). The excitation wavelength was 450 nm, the laser pulse length 6 ns,
the peak power 1 MW, and the repetition rate 2 Hz. The water Raman signal was detected
at 533 nm. The signals were logarithmically amplified and recorded with a 2 ns time
resolution and a 6 bit amplitude resolution. To improve the insufficient 6 bit resolution an averaging over 20 signals was performed which corresponds to an averaging
over half a kilometer flight distance. With the equipment described depth profiles co-
vering 5-6 attenuation lengths could be measured. An east-west flight track of 25 km was chosen on the latitude of the Lagoon of Venice. The profiles of the attenuation coefficient were calculated with the Delta pulse analysis according to (5) and (8).

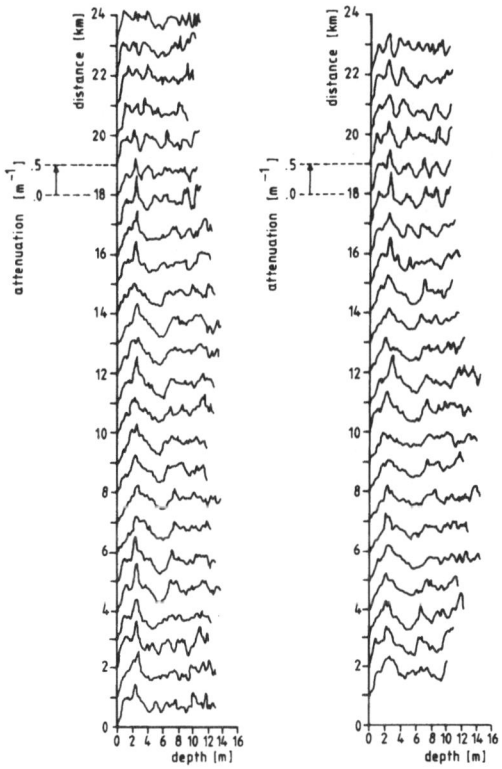

In Fig. 2 the derived profiles along the described track are shown for two flights which are identical except for a time dif-ference of 20 minutes and navigation un-certainties. Hence the fine structures of the calculated profiles are not equal for both data sets, but there exists a good corresponce for the main structures. A striking maximum appears at a depth of 2 m; in the middle part of the track a broad minimum around 6 m water depth is found; in the western part of the flight track a relatively homogeneous attenuation coefficient with depth is observed. As a consequence of the approximation of the 6 ns laser pulse by a δ-pulse all profiles start at a value of 0 and attain realistic values at a depth of about 1 m.

Fig.2 . Profiles of the attenuation coefficient (from 45°19'N, 12°39'E
to 45°19'N, 12°57'E) in the Adri-
atic Sea . λ_{ex}=450nm, λ_{em}=533nm

Conclusion

It is evident that the measurement of depth profiles of the attenuation coefficient
can be achieved with lidar down to water depths of at least 5 attenuation lengths.
For future flights a simultaneous detection of Gelbstoff fluorescence is planned to
obtain also profiles of the Gelbstoff concentration.
For a better examination of depth profiling lidar investigations laboratory measurements will be performed using a water column with well-defined structures of the attenuation coefficient and of the concentrations of different substances.

Fluorescence-Lidar Remote-Sensing of the Environment: Laboratory Experiments for the Characterization of Oil Spills and Vegetation

Giovanna Cecchi, Luca Pantani
IROE-CNR, Via Panciatichi 64, 50127 Firenze, ITALY

Piero Mazzinghi
IEQ-CNR, Via Panciatichi 56/30, 50127 Firenze, ITALY

Antongiulio Barbaro
Physics Student, University of Florence, ITALY

Introduction

Laboratory experiments and computer simulations has been carried out at IROE and IEQ Institutes in order to identify, in controlled conditions, the operational limits of fluorescence lidars in environmental remote-sensing. The attention was focused on the detection and characterization of oil spills (1,2,3) and on plants (4).

A particular care was devoted to the analysis of the potential of "differential fluorescence" techniques, i.e. techniques which use the comparison between the emitted fluorescence-radiation at two or more wavelengths in the remote-sensing of target characteristics.

The differential fluorescence lidar

When a laser beam at the wavelength λ_o illuminates a fluorescent target at a distance R, the ratio between the received signals at two different fluorescence wavelengths can be expressed as:

$$F(\lambda_o,\lambda_i,\lambda_j,R) = K\ E(\lambda_o,\lambda_i,\lambda_j)\exp\left[-D_{ij}(R)R\right] , \tag{1}$$

beeing:

$$E(\lambda_o,\lambda_i,\lambda_j) = \eta(\lambda_o,\lambda_i)/\eta(\lambda_o,\lambda_j) , \tag{2}$$

$$D_{i,j}(R) = \bar{\sigma}(\lambda_i,R)/\bar{\sigma}(\lambda_j,R) , \tag{3}$$

and $\eta(\lambda_o,\lambda_n)$ the fluorescence efficiency at λ_n with an excitation at λ_o, $\bar{\sigma}(\lambda_n,R)$ the average over R of the atmosphere extinction coefficient, and K a constant which depends on the lidar structure.

The target can be identified by means of a finite number of the signal
ratios (Eq.1), if the corresponding values of Eq.2 are characteristic
of the target itself and the atmospheric factors (Eq.3) are known or
unrelevant.

Characterization of oil films

As it was previously outlined by the authors (2), the oil-film
characterization can be achieved by means of the film thickness and
the identification of the oil type.

The accuracy of the film thickness measurement by means of both
fluorescence intensity and depression of the water Raman signal was
investigated using different oil samples; the data in Table I concern
samples of fuel oil and chocolat mousse taken during the Archimedes
experiment (5). As it is possible to see from Table I, the order of
magnitude of the film thickness can be evaluated by means of these
techniques for thicknesses ranging from 0 to 10 μm, while above these
values the film can be estimated optically thick for the excitation
wavelengths employed in the experiment (308 nm and 337.1 nm).

TABLE I

OIL SAMPLE	FILM THICKNESS (μm)	THICKNESS BY FLUORESCENCE (μm)	THICKNESS BY RAMAN (μm)
FUEL	6	1.3	-.-
FUEL	2	1.5	-.-
FUEL	1	0.7	0.4
FUEL	0.5	-.-	0.3
FUEL	0.1	0.4	0.2
FUEL	0.05	-.-	0.3
MOUSSE	6	4	-.-
MOUSSE	2	2.3	-.-
MOUSSE	1	0.9	1.2
MOUSSE	0.5	0.3	0.4
MOUSSE	0.1	-.-	0.3
MOUSSE	0.05	-.-	0.04

In order to analyze the possibility of an identification of the oil by
means of a differential fluorescence technique, the emission spectra

(with 308 nm and 337.1 nm excitations) of more than 60 different oils were stored on a computer and processed by means of a least square technique in order to find out the most suitable values of λ_i and λ_j . As a first step the analysis has been restricted to the use of one couple of wavelengths; in the future it will be extended to more wavelengths. The results have been compared with the results obtained by processing the same spectra by means of cross-correlation techniques (3,6). The comparison showed that the two techniques have about the same resolving power and give the discrimination between three different oil classes (crude, light, and heavy oil). Moreover there are some indications that the use of more than two wavelengths will allow the differentiation inside the class.

Vegetation analysis

The analysis of vegetation by means of the fluorescence lidar is based on the teledetection of chlorophyll-a fluorescence signal (4,7). Since the chlorophyll-a is involved in the photosynthetic process, its fluorescence spectrum is deeply influenced by the health state of the plant, and the exposure to the light. In a first approach, the experimental conditions have therefore to be settled carefully, in order to obtain reproducible results.

In laboratory experiments the fluorescence spectra of living plants were carried out with a HeNe cw and a Rhodamine 6G pulsed dye laser excitation. The spectra were made in different light conditions, and plants in different water stress state were investigated. The analysis of the spectra showed that the exposition to the cw excitation changes the spectral behaviour, while the phenomenon is practically not observable with the pulsed excitation. Moreover the spectral behaviour changes also with the background light level and with the water stress. For plants in different water stress conditions but with the same laser excitation and the same background light level an evident influence of the water stress on the value of eq. 1 was observed when $\lambda_i \sim 685$ nm and $\lambda_j \sim 720$ nm. In this situation $F(\lambda_c, \lambda_i, \lambda_j, R)$ increases with the water stress.

Conclusions

The laboratory experiments showed that:

- a measurement at least of the order of magnitude of the oil film thickness is possible at thickness lower than 10 µm by means of fluorescence and Raman techniques;

- a single wavelength couple differential fluorescence lidar can identify the class of the oil with about the same accuracy as cross-correlation techniques;

- the plant water stress can be detected by means of a single couple differential fluorescence lidar;

- the experiment conditions have to be carefully settled when the analysis is made on living plants in order to assure reproducible data.

References

(1) P. Burlamacchi et al., Appl. Opt. 22, 48 (1983)

(2) G. Cecchi et al., "Lidar Investigation of Oil Films on Natural Waters" in: Optoelectronic in Engineering - Springer Verlag p.517 (1984)

(3) G. Cecchi et al., ESA SP-233

(4) G. Cecchi et al., Proc. ECOOSA 84 SPIE, in print

(5) D. Diebel-Langohr et al., "Lidar Measurements Performed during Archimedes 83", in print

(6) L. Pantani et al., "Cross-correlation of Fluorescence Emission Spectra", Report IROE, in print

(7) K.P. Gunther, "Chlorophyll-a Detection by Fluorescence Lidar", in Laser 85 Opto-Elektronik, Munchen 1-5 juli 1985

Lidar Monitoring of Marine Oil Spills: Results of Archimedes '83

D. Diebel-Langohr, T. Hengstermann, R.Reuter

Universität Oldenburg, Fachbereich Physik

Postfach 2503, D-2900 Oldenburg

The pollution of the sea has increased steadily in recent years. Particularly in coastal waters extreme loads have been observed: e.g. for the German Bight one estimates a hydrocarbon input of 150 000 tons per year. The major source is assumed to be from the illegal release of mineral oil by ship traffic. This situation has led to increasing demands for airborne survey methods in order to reduce these permanent contributions to the marine oil pollution.

Archimedes'83 was a joint experiment of different European institutions involved in research on airborne oil spill monitoring, and in the application of these techniques for oil pollution survey and spill combatting operations. The experiment was initiated by the Commission of the European Communities with the intent to further improve the potential of existing remote sensing methods and to define combined sensor packages. Different airborne microwave radiometers and SLAR systems were utilized in the experiment /1/. In this paper we discuss measurements performed with the OLS of the University of Oldenburg; the sensor is described in a separate paper given in this volume.

Principle of measurement

The lidar method for the investigation of marine oil spills is based on the optical properties of mineral oil. Oil exhibits fluorescence at all visible wavelengths when irradiated with near UV light. Fluorescence spectra of different crude oils and of heavy fuel oil are shown in Fig. 1. Light oils are characterized by a strong fluorescence with maximum values at blue wavelengths; the fluorescence yield of heavy oils is markedly reduced and the maxima are shifted to green wavelengths. These spectral characteristics allow an identification of the oil type according to the main oil groups.

Measurements of the thickness of optically thin oil films on the water surface can be done in two different ways by analyzing the variation of the fluorescence yield along the flight track of the aircraft above the oil spill, or the depression of water Raman scattering originating from the water column below the oil. The reduction of water Raman scattering due to an oil film with varying film thickness is shown in Fig. 2.

With an excitation wavelength of 308 nm, Raman scattering from the water below the oil is sensitively damped at film thickness values lower than 10 μm. The fluorescence intensity reaches a constant value at a film thickness exceeding 50 μm which corresponds to optically thick oil films. The range of sensitivity for film thickness measurements

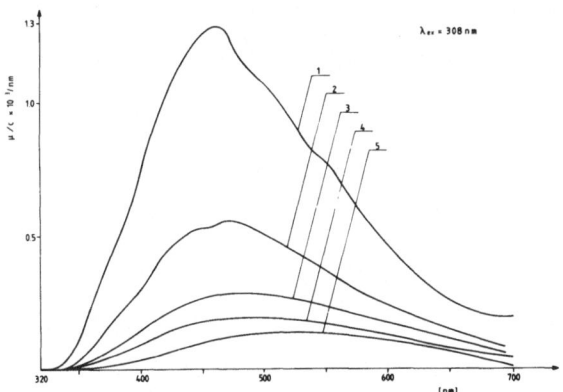

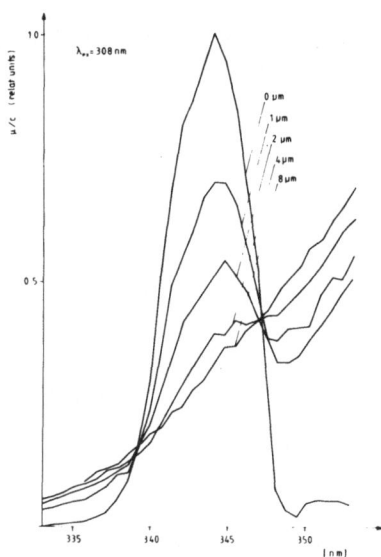

Fig. 1. Emission spectra of the normalized fluoreszence efficiency of (1) Statfjord crude oil, (2) Arabian light, (3) Arabian medium, (4) Arabian heavy crude oil, (5) fuel oil used for Archimedes'83. The predominance of heavy or light oil fractions determines the intensity and the shape of the spectrum, from which information on the oil type is deduced. Excitation wavelength 308 nm

Fig. 2. Fluorescence spectrum of Statfjord crude oil near the water Raman scattering wavelength 344 nm. Excitation wavelength 308 nm

is restricted to values of that order due to the strong light absorption within the oil, depending on the oil type, and on the excitation and detection wavelengths.

A quantitative formulation of the measuring process is derived from the lidar equation /2/. In the presence of an optically thin oil layer within a thickness d on the water surface, a depth integration of the lidar equation yields a detector signal /3,4/

$$P = A/(nH)^2 [\eta_o/c_o + (\eta_w/c_w - \eta_o/c_o) \exp(-c_o d)] \qquad (1)$$

where A describes instrumental factors, the signal loss in the atmosphere and effects of sea surface roughness; n is the refractive index, H the flight height, η_o, η_w the quantum efficiency of fluorescence or water Raman scattering corresponding to the ratio of emitted and absorbed photons, c_w and c_o the sum of light attenuation coefficients of water and oil at the excitation and detection wavelengths. In the case of a water surface free of oil or an optically thick oil film, equ. (1) reduces to

$$P = A/(nH)^2 \eta_w/c_w \qquad (2) \qquad\qquad P = A/(nH)^2 \eta_o/c_o \qquad (3)$$

Thus, lidar measurements of P_o performed over an optically thick portion of an oil slick yield a direct determination of the spectral fluorescence efficiency normalized to the oil attenuation coefficient η_o/c_o according to (3). A comparison of P_o with the characteristic signatures obtained in laboratory experiments enable an identification of the oil type and an estimation of its light attenuation coefficient. This information allows the evaluation of the oil film thickness from data taken at those parts of the spill where optically thin films are present. Algorithms are deririved from equ. (1), (2) and (3) and are discussed in /1/, /4/, /5/.

The experiment

Two oil spills were produced off the coast of Holland on October 21, 1983 consisting of fuel oil ($40m^3$) and of chocolat mousse ($50m^3$) produced from the same fuel oil. At the end of the experiment the oil was removed by recovery ships.

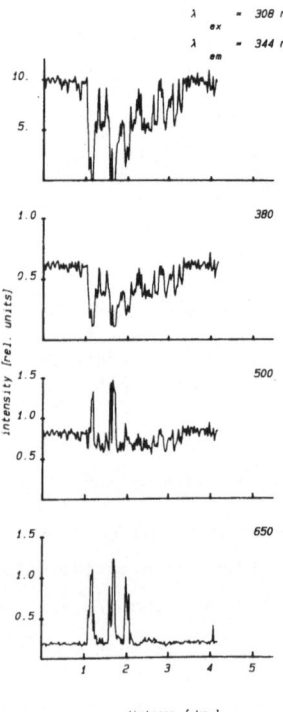

flight no.: 404   spill 2
date: 21.10.83   time: 16:34

Fig.3. Results obtained during flight no. 404 21.10.83 16:34 h, over the oil spill. Excitation wavelength 308 nm, detection wavelengths 344, 380, 500 and 650nm

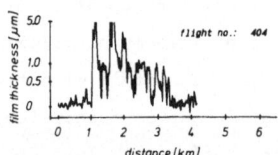

Fig. 4. Oil film thickness distribution obtained for flight no. 404

The data of one selected flight obtained over the fuel oil spill about 13 h after oil spilling with 308 nm excitation and 344, 380, 500, and 650 nm detection wavelengths are shown in Fig. 3. Aircraft flight heigth was 200m. The signal repetition rate was 5 Hz, the aircraft speed 50 m/s. The curves represent data of integrated signal peaks which have been corrected for flight height variations and for the spectral sensitivity of the detection channels.

The 344 nm signal represents water Raman scattering. It shows a drastic decrease in intensity at the edge of the spill when compared to the slightly varying signals received over clear water. With increased detection wavelengths, the signal change due to the presence of oil becomes restricted to small spots with typical diameters of 100–200 m. These spots with vanishing water Raman scattering are attributed to the bulk of the spilled oil, the extended areas surrounding these spots to films with a thickness of less than 10 μm.

A closer inspection of spectral signatures obtained at different locations of the spill reveals that the oil has been decomposed into different fractions with individual fluorescence characteristics. Particularly this is seen in the green and red detection channels, Fig. 3. The signal change over oil with respect to clear water is according to equ. (1) and (2) given by

$$\Delta P = A/(nH)^2 \ (\ \eta_w/c_w - \eta_o/c_o \ )(1-\exp(-c_o \ d)) \qquad (4)$$

the sign of the signal change thus depending on the relative values of the clear water and the oil efficiencies.

The film thickness over the spill, Fig. 4, is calculated from the variation of water Raman scattering, and with an attenuation coefficient $c_o = 0.7/\mu$m according to results of laboratory investigations. Due to the uncertain attenuation coefficient the accuracy will be about 50 %.

Conclusions

The results of OLS measurements achieved during Archimdes'83 demonstrate the capabilities of the lidar for airborne oil spill monitoring. The OLS is thus a sensor type
which allows an analysis of the oil type and of the film thickness up to 50 μm. For
operational applications, advances in the technical layout including an optimization
of the excitation and detection wavelengths are in progress as well as an expanded
catalogue of the optical properties of oils including weathering effects.

Acknowledgements

We are indebted to Dr. R.H. Gillot and Dr. F. Tosselli, Ispra Establishment for taking
the initiative for Archimedes'83. We are grateful to the colleagues of the Flight Department of the DFVLR Oberpfaffenhofen for the availability of the aircrafts and for
the support during Archimedes'83. Laboratory investigations are supported by the
Ministerium für Wissenschaft und Kunst, Hannover.

References

/1/ Archimedes I, Final Report: Commission of the European Communities (in press)
/2/ Browell E.V.: NASA TN D-8447 (1977)
/3/ Kung,R.T.V. and I.Itzkan: Appl. Opt. 15, 409, (1976)
/4/ D. Diebel-Langohr et al.: Lidar Measurements performed during ARCHIMEDES I,
    Archimedes I, Final Report, Commission of the European Communities (in press)
/5/ Hoge F.E. and R.N. Swift: Appl. Opt. 22, 3316-3317, (1983)

Passive Optical Remote Sensing of the Sea at the I.R.O.E. – C.N.R.

F.Castagnoli, L.Pantani, I.Pippi, B.Radicati

Istituto di Ricerca sulle Onde Elettromagnetiche

Consiglio Nazionale delle Ricerche

Via Panciatichi 64, I 50127 FIRENZE (Italy)

R.Bonsignori - Galileo S.p.A., Firenze (Italy)

Introduction

This paper describes two different activities carried out at the IROE-CNR on the remote sensing of the sea. The first activity deals with the analysis of the potential of infrared scanners in the real-time monitoring of oil spills over the sea surface and of thermal disconti-nuities of the sea, particularly in coastal waters. The IROE partici-pated to some field experiments in the last years in the North Sea (Archimedes 83), in the Marano lagoon (November 83, May 85),and in the Northern Adriatic Sea (Adria 84) successfuly testing the ability of small I.R. scanners in the real-time identification of oil spills and thermal discontinuities ([1, 2, 3]). The most suitable image processing techniques were also investigated. The second activity concerns the design and construction of an high-resolution spectrometer for the re-mote sensing of the sea surface in the visible region of the spectrum. The first test of this spectrometer was done during the Adria 84 ([3]) experiment.

Visible/Infrared Sensor

A sensor was assembled at the IROE-CNR in order to obtain a system for real-time monitoring and recording of environmental images in the visible and infrared. The sensor was thought as a light weight, low power system which can be carried also by small aircrafts for the control of coastal and internal waters.

The system is composed by an AGA Thermovision 782 SW infrared camera and a standard color TV camera. The infrared and visible images are mo

nitored in real time in front of the operator and recorded on video ta
pes. Two audio channels are allowable for comments, data recording,
etc. The principal characteristics are indicated on table I.

Table I. Characteristics of the visible/infrared sensor AIA-2

Infrared	Visible
AGA 782 SW Thermovision	Standar color TV camera
Bandwidth: 3 - 5.6 μm	Minimum sensitivity: 10 lux
field of view: 12°	Field of view: 12°
Observed scene from 100 m	
21 m x 21 m	Recording
	VHS videotape

During the Archimedes 83 only the IR camera was allowable and the IR
images were therefore recorded taking photographic pictures of the monitor screen. The system was installed in a D-28 of DFVLR and flown
over two oil spills one done with fuel-oil and the other with chocola-
te mousse (a stabilized mixture of fuel-oil and sea water). The two
spills were detected night and day and the processing of the images
showed the possibility of an indication of the different thicknesses
of the film at least from a qualitative point of view.

In the other experiments the complete sensor was installed in a Cessna
185 Skywagon and flown over the waters of the North Adriatic Sea and
of its coastal lagoons. During these experiments many thermal anomalies were observed in real time and recorded, these anomalies were cau
sed by natural events, like at river mouths, and by human actions, like discharges. Some of the observed anomalies were never observed before. Many oil slicks were also detected, and during the last experiment (Marano 1985) small quantities of seed-oil were spilled on the
sea in order to test the ability of the sensor in tracking a spilling
boat. The images recorded during the last two experiments, Adria 84
and Marano 85, are now in processing at the IROE.

In all the experiments the sensor showed a good potential in coastal
water sourveillance from small aircrafts. Oil spills and thermal anomalies, both natural and human made, can be detected in real-time and
recorded for legal use. The only limit of the system is in the require
ment of a transparent atmosphere.

The high resolution spectrometer

An high resolution spectrometer was built for the remote sensing of spectral radiances and spectral reflectances in the visible and near infrared part of the electromagnetic spectrum. This spectrometer was thought as a research tool in order to define the best characteristics of the bands of future remote sensors and scanners.

A 20 cm diameter Cassegrain-type telescope was employed which has in its focal plane the entrance slit of a fast grating spectrograph. At the spectrograph output the spectrum is intensified by means of an image intensifier and detected by a 1024 linear CCD-array. The system is controlled by an HP 86 desk-top computer which records the output signal on a floppy disk, this computer is also used for the processing of the recorded signals.

The electronic scanning and receiving circuitry, the spectrograph with the CCD array, and the telescope are enclosed in a waterproof metallic box connected to the computer by cables.

The main advantages of this spectrometer are its modular structure, the absence of mechanically moving parts and   the acquisition speed which is 30 ms for the whole spectrum with a resolution better than 1nm. The lowest detectable spectral radiance is less than $7 \ W \ cm^{-2} sr^{-1} nm^{-1}$.

The spectrometer has been installed on the CNR oceanographic ship "Bannock" during the "Adria 84" experiment and spectral measurements of water-leaving radiance have been performed. The recorded data are still in processing at the IROE.

References

1) F.Castagnoli, L.Pantani, I.Pippi:"The IROE participation to the ARCHIMEDES project, Report IROE, Firenze January 1984

2) L.Alberotanza et alii:"Un esperimento di controllo ambientale delle zone costiere dell'alto Adriatico" Atti del Convegno AITA-SITE, Bari Maggio 1984 (in print)

3) R.Bonsignori et alii:"First report on the "Adria 84" experiment" Report IROE, Firenze March 1985

Chlorophyll a Detection by Fluorescence Lidar – The Influence of Global Irradiation on the Fluorescence Efficiency

K.P.Günther

Universität Oldenburg, FB 8 Physik

Postfach 2503, D-2900 Oldenburg

The chl a detection with laser fluorosensors have shown their potential of monitoring phytoplankton distributions synoptically (KIM 1973; BRISTOW 1981; HOGE 1983). In coastal waters the lidar method gives more reliable results than optical radiometers due to the substance specific excitation and detection. The measurement with passive sensors is influenced by the high load of suspended matter and yellow substance disturbing the correlation of the green-blue ratio to chl a concentration found in oceanic waters. To deduce the chl a concentration from the laser induced chl a fluorescence at 685 nm, normalized to the water Raman scattering signal, one assumes a constant in vivo fluorescence efficiency for phytoplankton. Laboratory and in situ measurements of the chl a fluorescence show a variation of the in vivo chl a fluorescence efficiency with environmental parameters (KIEFER 1973a,b; BLASCO 1975; VINCENT 1979). Neglecting the influence of nutrients, temperature and phytoplankton composition the global irradiation has the most prominent impact on the fluorescence efficiency . At high light conditions a reduction of the fluorescence efficiency in the surface layer up to a factor 4 has been observed (VINCENT 1979).

Theoretical background

Based on the knowledge of the photochemical mechanism of photosynthesis BUTLER et al. (1975) developed the bipartite model of the photosynthetic units. Within this formalism, two photochemical pigment systems operate in series to drive the electrons from water to pyridine nucleotide, while the antenna pigments and the light harvesting chlorophyll proteins are the origin of the fluorescence. An increase of the fluorescence at the onset of continuous light is observed due to the strong interaction of the fluorescent pigments with the reaction center. The reaction centers are special chl a molecules in the photochemical pigment system where the charge separation takes place. This effect was first observed by KAUTSKY et al. (1931) and is discussed in terms of a variable fluorescence efficiency Φ_{FII} ,introduced by BUTLER et al. (1975).

$$\Phi_{FII} = \Psi_{FII} * \beta * f(A_{II})$$

where Ψ_{FII} represents a constant fluorescence efficiency, determined by the desacti-
vation processes at the antenna pigments, β an energy distribution parameter and A_{II}
the relative number of open reaction centers. The function $f(A_{II})$ describes the connection of the photosynthetic units in the cell. A reaction center is called open,
if it is in the ground state to accept excitation energy from the surrounding antenna. The reduction of the fluorescence due to high light, called photoinhibition, is
not included in the bipartite model. Introducing an intensity dependent energy distribution parameter β which describes the light dependent state of the thylakoid membrane, and an intensity dependent parameter A_{II}, it is possible to model the photoinhibition in a quantitative way. The results of HORTON et al. (1980) and ALLEN et al.
(1981) confirm that the membrane state is regulated by the redox state of a molecular
complex, called plastoquinone. The redox state of plastoquinone induces an enzymatic
reaction, the phosphorylation of the light harvesting chlorophyll proteins, changing
the structure of the membrane. In turn the redox state of plastoquinone is controlled
by the amount of light absorbed by the cell. $\beta(I)$ decreases from a maximum level
β_{MAX} to a minimum level β_{MIN} with increasing light. The decrease is determined by
an intensity parameter I_1, describing the state of the membrane. In contrast, the
light dependence of A_{II} is modeled by an exponential increase with increasing light
determined by a parameter I_0, indicating the adaption of the reaction centers according to the growth conditions, e.g. shade or sun adapted cells.
With this assumptions, confirmed by biochemical and physiological results, one can
calculate the relative variation of the chl a fluorescence efficiency due to global
irradiation. It is important to note that the light influencing phytoplankton is restricted to a wavelength band from 350 nm to 750 nm, called photosynthetic active
radiation. In figure 1, the relative decrease of the fluorescence efficiency with
increasing light is shown. The dark value of the fluorescence efficiency is set to 1.

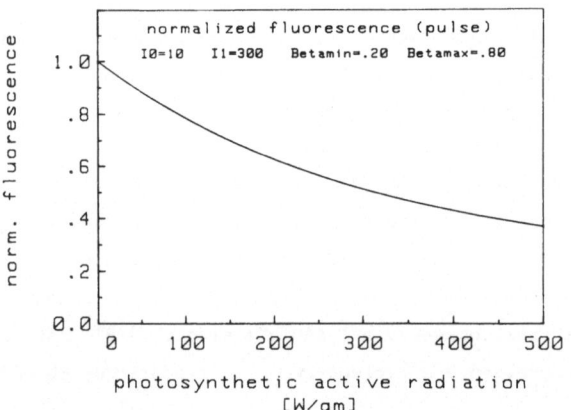

Fig. 1. Relative variation of the in vivo chl a fluorescence efficiency with the
photosynthetic active radiation

Experimental results

During the Fluorescence Remote 'Sensing Experiment FLUREX '82 measurements with a two-channel lidar system were performed at the research platform NORDSEE. The lidar system described by GEHLHAAR et al. (1981) was modified for the detection of chl a fluorescence. The excitation wavelength was set to 532 nm, the detection wavelenghts to 650 nm and 685 nm, the wavelengths of water Raman scattering and of chl a fluorescence, respectively. The lidar signals give the depth integrated chl a concentration and water turbidity at excitation and detection wavelength. In addition, continuous recording with an in situ fluorometer and an in situ attenuation meter, held at constant water depth, was performed. Water samples were taken at regular intervals for optical and biological analysis . The correlation of the fluorescence lidar and in situ data over the whole experimental period was high with a correlation coefficient of 0.97, indicating that the depth integrated chl a signals are influenced by the photosynthetic active radiation in the same way as the in situ fluorescence.

To analyse the influence of daylight on the fluorescence efficiency in detail , the continuously recorded in situ data were normalized to constant chl a concentration taking into account the results of the analysis of the water samples. The normalized fluorescence data represent the relative variations of the fluorescence efficiency. For the two days, April 20 and 21, figure 2 shows the daily cycle of the fluorescence efficiency together with the daily cycle of the photosynthetic active radiation given by the dotted lines . The solid line shows the result of the expanded photosynthetic model. The model parameters β_{MIN} , β_{MAX} , I_0 and I_1 were fitted by a computer program . For the next days , only the parameter β_{MIN} had to be changed to 0.3 to describe the observed photoinhibition with a high correlation. A detailed analysis of the proposed model shows that two parameters determine the reduction of fluorescence, while two parameters are insensitive.

To demonstrate the high correlation of depth integrated lidar and in situ fluorescence data the lidar equation (BROWELL 1977) was used. Assuming a constant chl a concentration with depth, as measured during FLUREX '82, the fluorescence efficiency increases with depth due to the decreasing light level according to the proposed model. To introduce a mean diffuse attenuation coefficient for the photosynthetic active radiation the data of BAKER et al. (1982) were used and integrated from 350 nm to 750 nm. The results of the calculation show the same decrease of the fluorescence lidar signals up to 55% at photosynthetic active radiation above the water surface of 500 W/m² compared to the situation during night where the fluorescence efficiency is constant with depth.

666

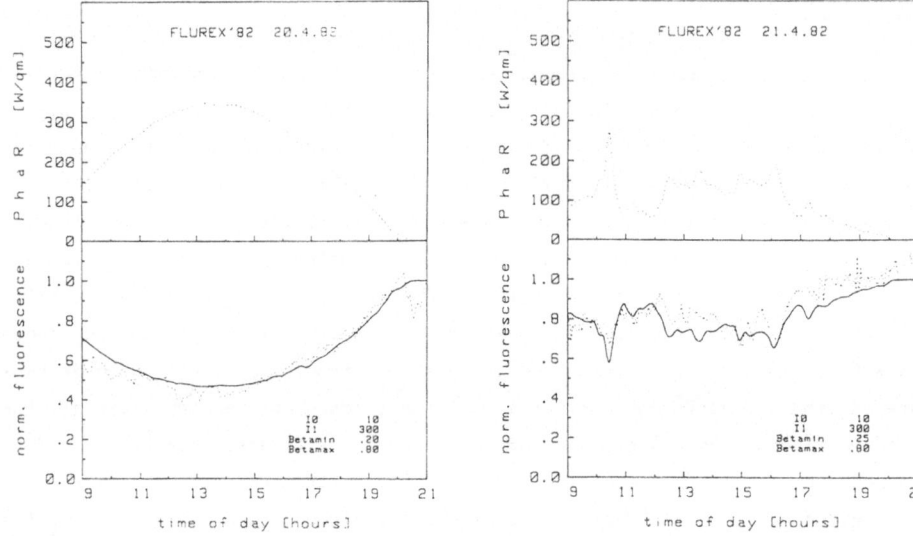

Fig. 2. Daily cycle of the relative chl a fluorescence efficiency and of the photosynthetic active radiation on April 20 and 21, 1982.   Dotted line: measured data;
solid line: results of the proposed model for the fluorescence efficiency taking
into account the influence of photosynthetic active radiation

Acknowledgements

The experiment at the research platform NORDSEE during FLUREX '82 was financed by a
grant from the Bundesministerium für Forschung und Technologie.

References

J.F. Allen, J.Bennett, K.E.Steinback, C.J.Arntzen: Nature 291,25 (1981)
K.S. Baker, R.C.Smith: Limnol.Oceanogr. 27,500 (1982)
D.Blasco: NASA TT F-16,317 (1975)
M.Bristow, D.Nielsen, D.Bundy, R.Furtek: Appl.Opt. 20,2889 (1981)
E.V. Browell: NASA TN D-8447,39 (1977)
W.L.Butler, M.Kitajima: BBA 396,72 (1975)
U.Gehlhaar, K.P.Günther, J.Luther: Appl.Opt. 20,3318 (1981)
F.E.Hoge, R.N.Swift: Appl.Opt. 22,2272 (1983)
P.Horton, M.T.Black: FEBS Letters 119,141 (1980)
H.Kautsky, A.Hirsch: Naturwissenschaften 19,964 (1931)
D.A.Kiefer: Mar.Biol. 22,263 (1973a)
D.A.Kiefer: Mar.Biol. 23,39 (1973b)
W.F.Vincent: J.Phycol. 15,429 (1979)

The Potential of Differential-Reflectance LIDAR in Environmental Monitoring

L.Pantani and I.Pippi

I.R.O.E. - C.N.R., Via Panciatichi 64, I 50127 Firenze, Italy

P.Vujković Cvijin and D.Ignjatijević

Institute of Physics, P.O.Box 57, 11001, Beograd, Yugoslavia

Introduction

Wavelength-tunable lasers allow the construction of multiple wavelength lidars with a carefull selection of the working wavelengths. This possibility allowed interesting applications to the remote-sensing of the environment the most popular of which is the "differential absorption" lidar widely used in the monitoring of atmospheric pollution ([1]).

As it was shown by different authors a multiwavelength lidar operating in the visible/NIR region ([2]) or in the 9÷10 μm region ([3,4]) of the electromagnetic spectrum may be used for the identification of target characteristics on the base of their spectral reflectance.

The IROE-CNR and the Institute of Physics carried out a cooperative research on the potential of differential reflectance lidars in the monitoring of the environment with a particular attention to vegetation monitoring and to the detection of soil surface. This paper is a first report on this activity.

The Differential-Reflectance LIDAR

When a multiwavelength lidar illuminates a target at a distance R the ratio between the echoes at two different wavelengths, λ_i and λ_j can be expressed as ([5]):

$$E_{i,j}(R) = K \, S_{i,j} \exp\{-2R \, D_{i,j}(R)\} \tag{1}$$

where:

$$S_{i,j} = \rho(\lambda_i)/\rho(\lambda_j) \tag{2}$$

$$D_{i,j} = \sigma(\lambda_i,R) - \sigma(\lambda_j, R) \tag{3}$$

K is a constant,$\rho(\lambda_n)$ the target spectral-reflectance at a wavelength
λ_n, $\sigma(\lambda_n, R)$ the average over the propagation path R of the atmospheric extinction coefficient at λ_n. The identification of the target by
means of one or more of the ratios (1) can be achieved if the corresponding set of (2) is characteristic of the target and the terms (3)
are known,or unrelevant, or can be measured through the inversion of equation (1) over reference targets with known spectral reflectance.

The experiments

In both the overmentioned wavelength-fields reflectance spectra were
recorded and stored on a computer. The spectra were than computer investigated in order to detect the wavelength couples which maximize the
influence of each particular phenomenon on the value of the ratio (1).

Reflectance spectra of living plants were recorded between 400 nm and
800 nm taking into account mainly two phenomena: leaves senescence and
plant water-stress. After some preliminary experiments the attention
was concentrated on three vegetables, tomato, endive, and bean. The existence of wavelength-couples for the identification of leaves senescence was demonstrated while more critical results were obtained for the
identification of water-stress. The possibility of detecting two olive
tree diseases was also shown. Some results are shown on table I.

The experiments in the 9÷10 µm wavelength range were carried out by
means of a lidar-simulator which uses a tunable CO_2 laser ([6]). The
first analysis was carried out on reflectance standards and the possibility of relyable reflectance standards was shown. Than 18 geological
samples, 4 construction materials, and 13 vegetatio samples(leaves,
grass, etc.) were taken into account. The spectral reflectance was measured for each CO_2 laser line and the data are now in processing for
the extraction of the more suitable wavelength-couples. The first results showed that the identification of the target material with a limited number of wavelength couples is feasible, an example is shown on
table II.

Wav. Coup.(nm)	tomato 1	2	3	endive 1	2	3	bean 1	2	4
740 - 700	3.9	3.0	2.3	3.0	2.0	1.4	3.1	2.0	1.5
760 - 700	3.9	3.0	2.3	3.2	2.1	1.5	3.3	2.0	1.5

	Fumaggine reference	1	Cycloconium reference	1	2
740 - 700	-	-	4.5	3.1	2.0
760 - 700	5.8	4.4	5.0	3.3	2.1

Table I. Influence of senescence on the reflectance ratio $S_{i,j}$ of leaves (top), the senescence increases from 1 to 3. Influence of two olive-tree diseases on the reflectance ratio of leaves (bottom), the disease increases from 1 to 2

Sample	Reflectance ratio values			
Granite (black)	A = 0.9	B = 1.1	C = 1.0	D = 1.3
Gabbro	A = 1.1	B = 0.9	C = 1.6	D = 1.3
Silica sand	A = 2.0	B = 0.9	C = 1.6	D = 1.3
Kaolin	A = 0.5	B = 1.8	C = 0.1	D = 0.4
Limestone	A = 1.1	B = 1.0	C = 1.0	D = 1.0

Table II. Identification of geological samples by means of four CO_2 laser wavelengts. The reflectance ratios correspond to the following wavelength couples: 9R28/9P10 (A), 10R30/10P20 (B), 10P20/9P10 (C), 10R30/9R28 (D)

Conclusions

The differential reflectance lidar shows a good potential in the monitoring of the environment both in the visible-NIR region, where it is possible to work with tunable dye-lasers, and in the 9÷10 µm region where the CO_2 laser operates.

The visible-NIR region of the electromagnetic spectrum seems to be particularly attractive for the remote-sensing of leaves senescence and and of some olive-tree disesases, other plant diseases having the same behaviour of the investigated ones may also probably be detected.

The experiment carried out in the wavelength range of CO_2 lasers showed the possibility of an identification of some geological species by means of two or more laser wavelengths.

References

1) KW.ROTHE, V.BRINKMAN, H.WALTHER: Appl. Phys., $\underline{4}$, 181 (1974)
2) L.PANTANI, I.PIPPI: Optica Acta, $\underline{30}$, 1473 (1983)
3) S.M.SHUMATE et alii: Appl. Opt., $\underline{21}$, 2386 (1982)
4) W.B.GRANT: Appl. Opt., $\underline{21}$, 2390 (1982)
5) L.PANTANI, I.PIPPI: Les Colloques de l'INRA, $\underline{23}$, 823, INRA, Paris 1984
6) P.VUJKOVIC CVIJIN et alii:"Spectral reflectance of topographic target surface matherials: implications on wavelength scanning CO_2 laser - based lidars" 12th Int. Laser Radar Conf., Aix en Provence, August 1984

This research work is supported by the CNR, Italy. Special Grant I.P.R.A. - Subproject 1.

Paper n. 511

Laser – Anwendung in der Chemie
Laser – Application in Chemistry

Laser und chemische Reaktionen

J. Wolfrum

Physikalisch-Chemisches Institut der Universität Heidelberg

Im Neuenheimer Feld 253, D - 6900 Heidelberg

Einstrahlung von Licht kann einen wesentlichen Einfluß auf den Ablauf chemischer und biologischer Prozesse haben. Ein bekanntes Beispiel hierfür ist die Photosynthese organischer Substanzen aus Wasser und Kohlendioxid durch das Sonnenlicht in Pflanzen. Daneben stellt elektromagnetische Strahlung ein wichtiges Hilfsmittel zur Bestimmung von Konzentration und Struktur chemischer und biologischer Substanzen dar. Mit dem Licht herkömmlicher Lichtquellen konnten jedoch viele interessante Fragen nicht beantwortet werden, weil Phase und Amplitude des Lichtes normaler Lichtquellen mit spontaner Emission statistisch schwanken und die spektrale Energiedichte meist gering ist. Durch die stürmliche Entwicklung der Lasertechnik in den letzten Jahren stehen nun in einem weiten Wellenlängenbereich Lichtquellen von großer Parallelität und Energiedichte zur Verfügung, deren Impulsdauer und Frequenzschärfe mit hoher Präzision abgestimmt werden können. Damit eröffnen sich eine Vielzahl neuer Möglichkeiten sowohl zur Untersuchung wie zur Beeinflussung chemischer und biologischer Prozesse.

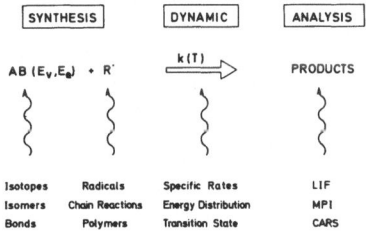

LASER AND CHEMICAL REACTIONS   Abb. 1.

Im vorliegenden Beitrag wird der Einsatz von Lasern zur Untersuchung der mikroskopischen Dynamik chemischer Reaktionen, zur empfindlichen Analyse chemischer Prozesse mit hoher räumlicher und zeitlicher Auflösung und zur gezielten Synthese chemischer Produkte an Hand verschiedener Beispiele diskutiert.

1. Untersuchung der mikroskopischen Dynamik chemischer Reaktionen

Laserlichtquellen mit ihrer hohen Lichtleistung innerhalb eines engen Spektralbereiches sind in der Lage, eine große Zahl chemisch reagierender Substanzen in bestimmte Energiefreiheitsgrade anzuregen und nach Ablauf der Reaktion die Anregung der Produkte zustandsselektiv zu untersuchen. Auf diese Weise kann man die Relativgeschwindigkeit von Reaktionspartnern sehr genau über einen weiten Bereich einstellen, Moleküle langsamer oder schneller rotieren lassen, Atome in einem Molekül verschieden weit auslenken und zu Schwingungen anregen, mit polarisierten Lasern die gegenseitige Orientierung der Teilchen während des Reaktionsablaufes festlegen und sogar die kurzlebigen "Übergangszustände" chemischer Reaktionen gezielt anregen und beobachten.

Die Abhängigkeit der Geschwindigkeit chemischer Reaktionen von der zugeführten Energie ist für den Chemiker tägliche Laborerfahrung. Dabei kann meistens die Energie der Reaktionspartner durch eine Temperatur und die Temperaturvariation der Reaktionsgeschwindigkeit durch eine einfache Arrhenius-Gleichung dargestellt werden. Die auf diese Weise erhaltenen Arrheniusparameter liefern jedoch keine direkte Information über den jeweiligen Beitrag der Energiefreiheitsgrade der beteiligten Reaktionspartner zur Überwindung der Potentialbarriere der chemischen Reaktion.

Wie in Abb. 2 schematisch gezeigt, können durch Photodissoziation mit Hilfe schmalbandiger Laser hoher Intensität und kurzer Pulslänge hohe Konzentrationen von reaktiven Atomen nahezu gleicher Geschwindigkeit erzeugt und primäre Produkt- und Energieverteilung chemischer Reaktionen zeitaufgelöst untersucht werden.

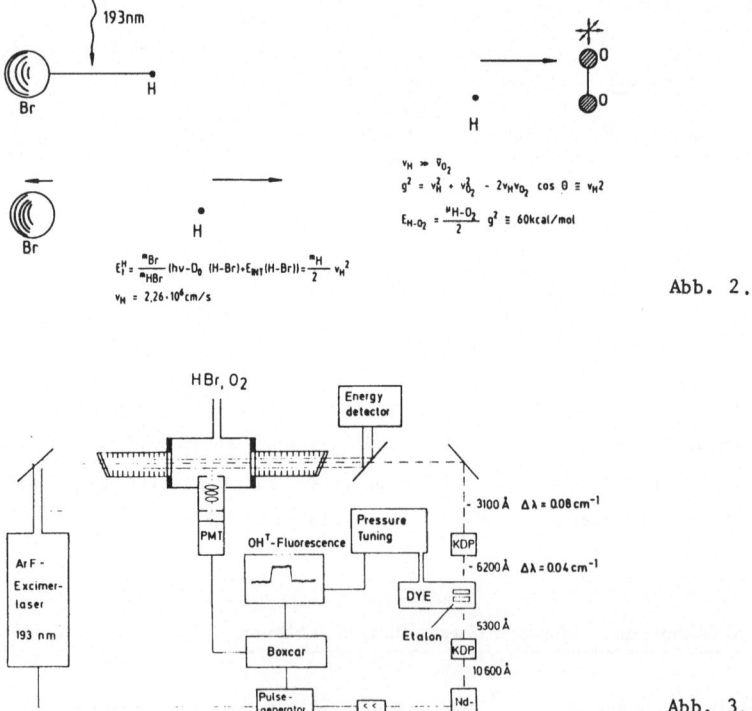

Abb. 2.

Abb. 3.

Eine experimentelle Anordnung, mit der man den bei zahlreichen Oxidations- und
Verbrennungsprozessen wichtigen Elementarschritt H + O_2 → OH + O und andere
Reaktionen schneller Atome mit Molekülen untersuchen kann, ist in Abb. 3 wiedergegeben. Sie besteht aus einem Strömungssystem in Verbindung mit einem Photoly-
se-(Excimer)Laser zur Erzeugung schneller Wasserstoffatome und einem frequenzverdoppelten Farbstofflaser zum zustandsspezifischen, zeitaufgelösten Nachweis des
OH-Produktradikals mit Hilfe der laserinduzierten Fluoreszenz (LIF).

Das Experiment zeigt, daß die in der Reaktion gebildeten OH-Radikale sehr hoch
rotationsangeregt werden. Die experimentelle Verteilung kann nun mit einer theoretisch berechneten Verteilung verglichen werden. Hierzu verfolgt man die Bewegung der
drei Atome auf einer quantenmechanisch ("ab initio") berechneten H-O_2-Potential-
hyperfläche durch schrittweise Integration der klassischen Bewegungsgleichungen bei
verschiedenen Anfangsbedingungen (Relativgeschwindigkeiten, Orientierung der Stoß-
partner, Rotations- und Schwingungszustände von O_2). In qualitativer Übereinstimmung mit den experimentellen Ergebnissen zeigen die Rechnungen, daß der größte Teil
der relativen Translationsenergie der Reaktanden in Rotationsenergie des Produkt-
moleküls OH umgewandelt wird. Bei einem genauen quantitativen Vergleich ergeben sich
jedoch deutliche Diskrepanzen. Insbesondere ist der theoretisch ermittelte Reak-
tionsquerschnitt um einen Faktor 3 kleiner als der experimentell bestimmte Wert.
Diese Abweichung ist insbesondere im Hinblick auf die Bedeutung der genauen Kenntnis
der absoluten Geschwindigkeitskonstanten dieser Reaktion bei hohen Temperaturen
wichtig, da dieser Wert empfindlich z. B. in die mathematische Modellierung von Verbrennungsprozessen eingeht.

Irgendwann im Verlauf einer chemischen Reaktion werden neue Bindungen geformt und
alte Bindungen gebrochen, so daß man weder "Reaktanden" noch "Produkte" hat. Die
direkte Untersuchung solcher "Übergangszustände" ist ein alter Traum der Chemiker.

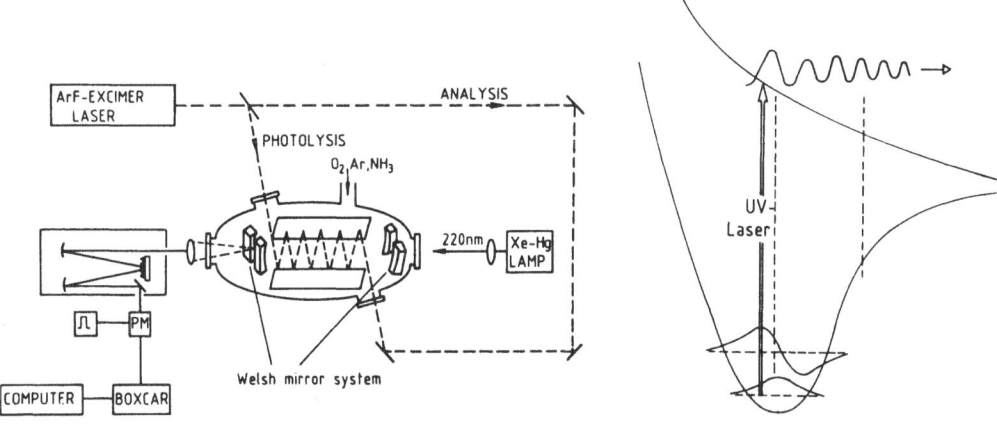

Abb. 4.   Abb. 5.

Abb. 4 zeigt eine experimentelle Anordnung zur Untersuchung des Übergangszustandes der Reaktion H + O_2 → OH + O, d.h. der Potentialhyperfläche des instabilen Radikals HO_2. Durch Laserphotolyse von NH_3 werden Wasserstoffatome erzeugt, die bei niedriger Temperatur in Gegenwart von Argon über die Reaktion H + O_2 + Ar → HO_2 + Ar HO_2-Radikale bilden. Durch einen verzögerten Laserimpuls wird anschließend HO_2 in einen höheren elektronischen Zustand angeregt. Die Photodissoziation von HO_2 ist ungefähr 6 Größenordnungen schneller als die spontane Emission, dennoch kann Fluoreszenz beobachtet werden. Aus der Frequenzverschiebung der Fluoreszenzlinien relativ zur Anregungswellenlänge erhält man die Schwingungsfrequenzen des HO_2 im elektronischen Grundzustand bis zur Dissoziationsgrenze. Die relativen Linienintensitäten charakterisieren die obere Potentialfläche und den zeitlichen Ablauf der Photodissoziation.

Weitere Beispiele über die selektive Anregung und Untersuchung bimolekularer chemischer Reaktionen durch Laser sind in Übersichtsarbeiten (1), (2), (3) dargestellt.

2. Chemische Analytik mit Lasern

Elektromagnetische Strahlung ist das wichtigste Hilfsmittel zur Bestimmung von Struktur, Eigenschaften und Verhalten chemisch reagierender Substanzen. Trotz großer Erfolge konnte die Spektroskopie mit herkömmlichen Lichtquellen viele interessante Fragen nicht beantworten. Insbesondere die Einführung abstimmbarer Laserquellen, wie z.B. des Farbstofflasers und der Diodenlaser und die Entwicklung nichtlinearer optischer Techniken ermöglichen praktisch jeden spektroskopischen Zustand eines Atoms oder Moleküls vom langwelligen Ultrarot bis zu Wellenlängen von 100 nm im Vakuum-ultravioletten Spektralbereich mit hoher Auflösung und Nachweisempfindlichkeit zu erfassen. Als neue Methoden haben zahlreiche Techniken wie die laserinduzierte Fluoreszenz (LIF), Einzel- und Mehrphotonen-Ionisation (MPI), Kohärente anti-Stokes-Ramanspektroskopie (CARS), Oberflächen Ramanspektroskopie (SERS), Photoakustische (PAS) und lasermagnetische Resonanzspektroskopie (LMR), Infrarot Diodenlaser Spektroskopie (IRLS), Dopplerfreie Absorptionsspektroskopie und andere Verfahren Eingang in die analytische Praxis finden können.

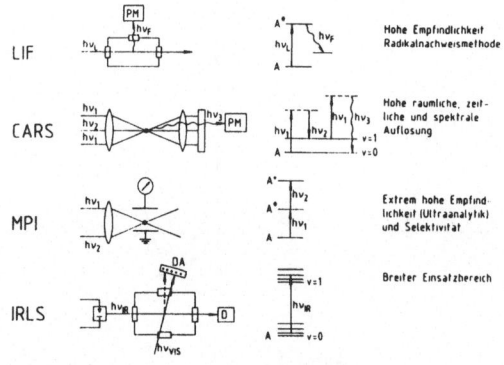

Laserspektroskopische Messung von Temperatur, Dichte und Konzentrationen

Abb. 6.

Schematische Darstellung verschiedener Verfahren der Laseranalytik

Neben dem sehr empfindlichen Nachweis, der in Form der Ultraanalytik bis zur Beobachtung einzelner Atome reicht, dem Nachweis von Spurenstoffen in der Atmosphäre, ist insbesondere die berührungsfreien Beobachtung rasch veränderlicher chemischer Prozesse, z. B. von Verbrennungsvorgängen, mit hoher zeitlicher, spektraler und räumlicher Auflösung ein wichtiges neues Anwendungsgebiet der Laseranalytik.

Abb. 7 zeigt als Beispiel den Einsatz der laserinduzierten Fluoreszenz zur zweidimensionalen Darstellung von Radikalkonzentrationen beim Zündprozeß im Automobilmotor. Ersetzt man die Zündkerze durch einen CO_2-Laser (s. Abb. 8) und beobachtet den Zündvorgang mit Hilfe eines Infrarotdiodenlasers mit hoher Repetitionsfrequenz, so ist ein quantitativer Vergleich der experimentellen Daten mit den Vorhersagen von mathematischen Modellen für den instationären chemischen Prozeß möglich. Ein detailliertes Verständnis der Zündprozesse ist nicht nur im Hinblick auf die Optimierung von Verbrennungskraftmaschinen sondern in gleicher Weise zur Vermeidung von Schadensfeuern wichtig.

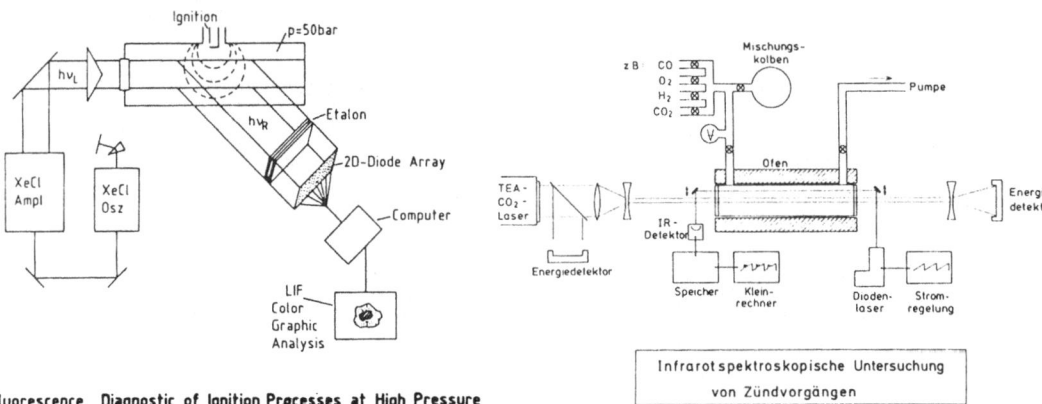

2D-Fluorescence Diagnostic of Ignition Processes at High Pressure

Abb. 7.

Infrarotspektroskopische Untersuchung von Zündvorgängen

Abb. 8.

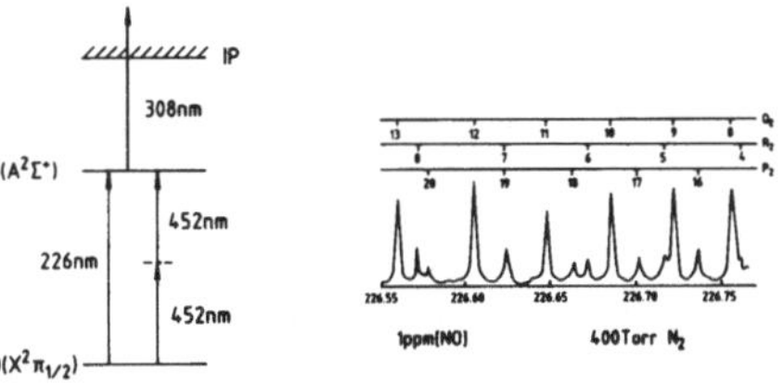

DETECTION OF NO BY RESONANCE IONIZATION SPECTROSCOPY (RIS)

Abb. 9.

Als Beispiel für den Nachweis von Schadstoffen sei hier das Stickoxid (NO) betrachtet. Eine Möglichkeit des empfindlichen Nachweises von NO über einen sehr weiten Konzentrationsbereich mit hoher zeitlicher und örtlicher Auflösung ist die resonante Mehrphotonenionisation. Wie in Abb. 9 gezeigt, wird dabei mit Hilfe eines Ein- oder Zweiphotonenprozesses zunächst ein resonanter Zwischenzustand besetzt, von dem aus

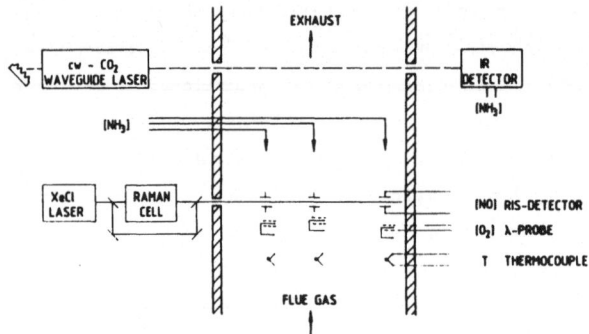

Abb. 10.

SELECTIVE CONTROLLED NO - REDUCTION

die selektive Ionisierung mit einem weiteren Laserphoton erfolgt. Wie in Abb. 10 schematisch angedeutet, können die durch resonante Laserphotoionisation entstehenden Ladungsimpulse an unterschiedlichen Stellen im Rauchgas registriert werden, so daß lokale durch das Strömugnsfeld beeinflußte Konzentrationen erfaßt werden. Eine berührungslose Messung der NH_3-Konzentration ist mit Hilfe eines Waveguide-CO_2-Lasers möglich, der durch Füllung mit einer $^{13}CO_2$-Isotopenmischung ein hochauflösendes spektroskopisches Verfahren mit Referenzsignal liefert. Die laserspektroskopisch gewonnenen Meßdaten erlauben unter Einsatz eines schnellen Rechners eine optimierte Regelung zur selektiven homogenen Reduktion von NO in Verbrennungsgasen durch Zusatz von NH_3.

MPI bietet auch eine Fülle neuer Möglichkeiten für die Massenspektroskopie größerer Moleküle. Wie in Abb. 11 anhand verschiedener Isomere des Buten schematisch angedeutet, ist es nicht möglich, mit Hilfe der klassischen Elektronenstoßionisation zwischen den verschiedenen Isomeren massenspektrometrisch unterscheiden zu können. Verwendet man Laser-MPI zur Erzeugung der Ionen im Massenspektrometer, so können Unterschiede im optischen Absorptionsspektrum der verschiedenen Isomeren ausgenutzt und eine Trennung der Isomeren erreicht werden. Da bei jeder verfügbaren Absorptionslinie ein getrenntes Massenspektrum erzeugt werden kann (Abb. 12) stehen wesentlich mehr Informationen zur massenspektrometrischen Analyse zur Verfügung.

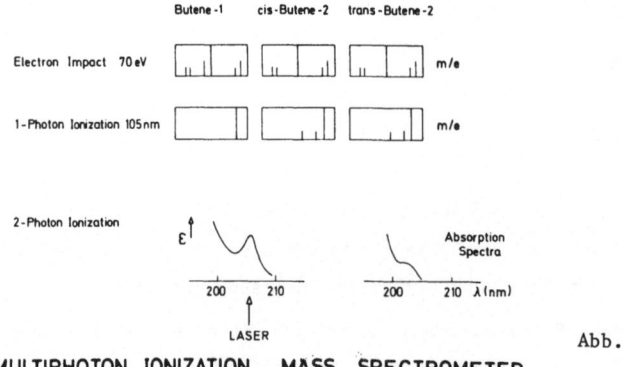

Abb. 11.

MULTIPHOTON IONIZATION   MASS   SPECTROMETER

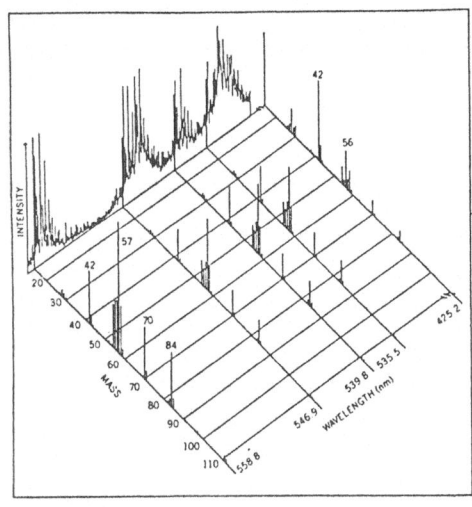

Abb. 12.

In ähnlicher Weise kann durch spektral und zeitlich aufgelöste Aufnahme der laserinduzierten Fluoreszenz (LIF) mit Hilfe einer Streak-Kamera ein zweidimensionaler
Fluoreszenz-"Fingerabdruck" erstellt werden. Wie in Abb. 13 beschrieben, kann die
Messung von Fluoreszenslebensdauern wesentlich zur Klärung der Rolle der Aminosäuren
Tryptophan und Tyrosin in Protein-Strukturen und bei der Protein-DNS-Wechselwirkung
beitragen. Insbesondere bei Erkennungsprozessen in DNS-Reparaturvorgängen und bei
der Bindung von Regulatorproteinen an DNS. So überträgt Tyrosin Fluoreszenzenergie
über weite Distanzen an die bei Zimmertemperatur in wässrigen Lösungen fast nicht
fluoreszierende DNS. Dies führt zu einer Verkürzung der Fluoreszenzlebensdauern.
Benutzt man zur Erzeugung der Fluoreszenz abstimmbare ultrakurze Laserimpulse im
ps-Bereich, so ist, wie in Abb. 14 dargestellt, eine schnelle Charakterisierung
biologischer Zellen unter Ausnutzung der zelleigenen Fluoreszenz möglich.

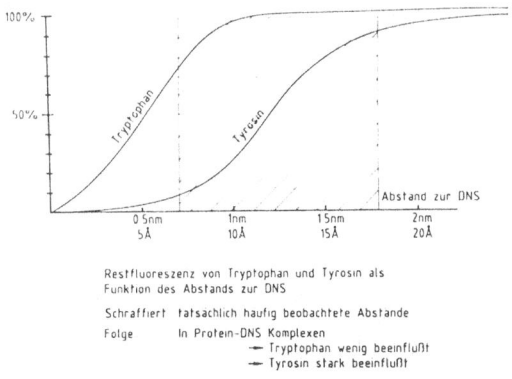

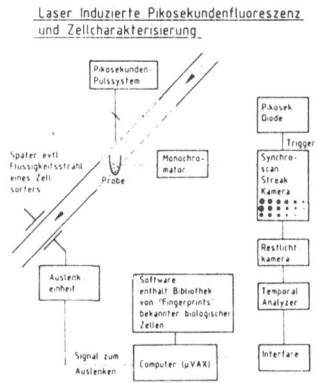

Abb. 13.   Abb. 14.

Die hier beschriebenen Beispiele können nur einen sehr kleinen Ausschnitt aus dem umfangreichen und sich sehr rasch weiterentwickelnden Gebiet der Laser-Analytik geben (4)-(8).

3. Einsatz von Lasern zur Erzeugung chemischer Produkte

Es liegt nahe, die beschriebenen Methoden der selektiven Steuerung von chemischen Reaktionen zur Herstellung neuer oder verbesserter chemischer Produkte zu nutzen. Unter den typischen Bedingungen chemisch-technischer Prozesse - hohe Drucke, Reaktionen in der Flüssigphase, große Moleküle - läßt sich jedoch ein selektierter Molekül- oder Übergangszustand in den meisten Fällen nicht bis zum Reaktionsschritt stabil erhalten. Der Energieaustausch zwischen den Freiheitsgraden innerhalb eines Moleküls und zwischen verschiedenen Molekülen findet in vielen Fällen in einem Zeitbereich zwischen 10^{-13} und 10^{-10} s statt und ist damit meist wesentlich schneller als die Reaktionszeit. Demgegenüber können nichtthermische Isomeren- und Isotopenverteilungen sowie nichtthermische Radikalkonzentrationen aus laserinduzierten Dissoziations- und Isomerisierungsprozessen durchaus unter chemisch-technischen Bedingungen aufrechterhalten und zur Herstellung neuer, besonders reiner oder mit geringerem Energieaufwand erzeugter Produkte verwendet werden. Dabei bietet die Verwendung des Lasers gegenüber konventionellen Lichtquellen eine Reihe von Vorteilen:

(1) Eine Erweiterung des Bereichs verfügbarer Wellenlängen

(2) Aufgrund der geringen spektralen Bandbreite des Laserlichtes die Möglichkeit der selektiveren Anregung nur einer Molekülart, z. B. zur Isotopentrennung oder Ultrareinigung

(3) Durch die Möglichkeit der starken Strahlbündelung eine räumlich und zeitlich kontrollierbare homogene Anregung des Reaktionsvolumens

(4) Durch die hohe Leistungsdichte die Möglichkeit der Benutzung von Mehrphotonenprozessen

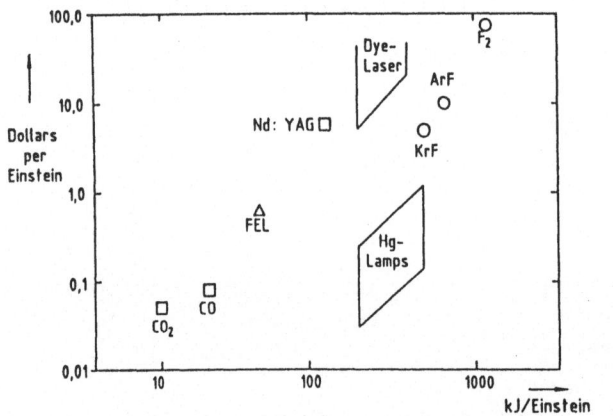

Abb. 15.

COST OF LASER PHOTONS AS FUNCTION OF ENERY

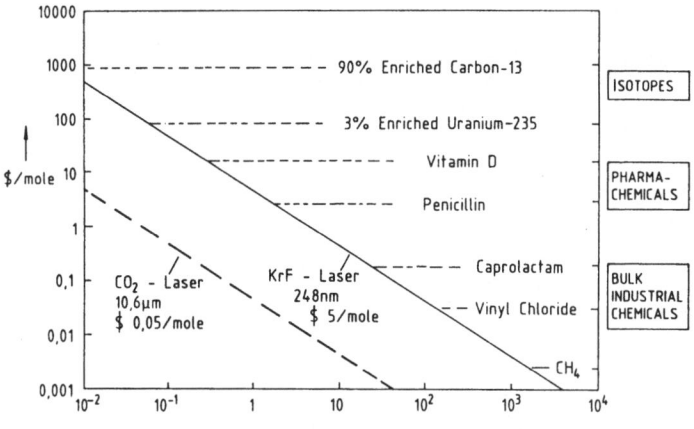

Abb. 16.

EFFECT OF QUANTUM YIELD IN LASER PHOTOCHEMISTRY

Wie an Hand der Kosten pro Photon in Abb. 15 dargestellt, sind zur Zeit als Lichtquellen für die industrielle Photochemie Hg- oder Xe-Lampen in den Investitions- und
Betriebskosten sowie im Wartungsaufwand, der Dauerleistung und Lebensdauer günstiger
als Laserlicht. Für den wirtschaftlichen Einsatz des Lasers in diesem Bereich müssen
jedoch die effektiven Kosten der erzeugten Photonen deutlich unterhalb der Kosten
für das gewünschte Produkt liegen (s. Abb. 16). Dabei ist zu bedenken, daß der
Produktpreis meist nur zu einem geringen Teil durch den photochemischen Verfahrens-
schritt bestimmt wird und photochemische Verfahren mit Lasereinsatz nicht nur mit
einem konventionellen photochemischen Prozeß, sondern auch mit verschiedenen anderen
Synthesewegen konkurrieren müssen. Wie generell bei der Einführung neuer Verfahren,
sollte der Einsatz des Lasers möglichst mehrere Verbesserungen gleichzeitig bewirken, wie etwa günstigeres Ausgangsmaterial, weniger oder höherwertige Nebenprodukte,
weniger oder billigere Verfahrensstufen, Verbesserung des Produktes u.a.

Wie Abb. 17 zeigt, wird bei Laserisotopentrennung eine Isotopenart in einer Mischung
mit anderen Isotopen durch schmalbandiges Laserlicht selektiv angeregt und in ein
Ion, Isomer oder Dissoziationsprodukt überführt, das sich leichter chemisch oder
physikalisch abtrennen läßt. Als Beispiel zeigt Abb. 18 die Darstellung von
^{13}C-Verbindungen durch Infrarotlaserinduzierte Multiphotonendissoziation (MPD)
ausgehend von Freon-22 (CF_2HCl), das isotopenselektiv in C_2F_4 überführt wird, aus
welchem durch Addition von HCl das Ausgangsprodukt CF_2HCl wieder gewonnen werden
kann, so daß eine mehrstufige Anreicherung möglich ist. Interessante Anwendungen für
^{13}C markierte Verbindungen ergeben sich in verschiedenen Bereichen der medizinischen
Diagnostik, so etwa bei Stoffwechseluntersuchungen als Ersatz für die risikoreichere
Markierung mit radioaktiven ^{14}C Isotopen und in der NMR-Tomographie. Es ist zu
erwarten, daß hierfür weltweit jährlich ein Bedarf von einigen hundert kg mol
vorhanden ist.

682

LASER ISOTOPE ENRICHMENT

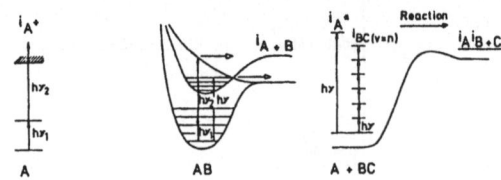

Abb. 17.

TPI : Two-Photon   TPD: Two-Photon Dissociation   LIR: Laser-Initiated Reaction
     Ionization   SPP: Single-Photon Predissociation   MPD: Multiphoton Dissociation

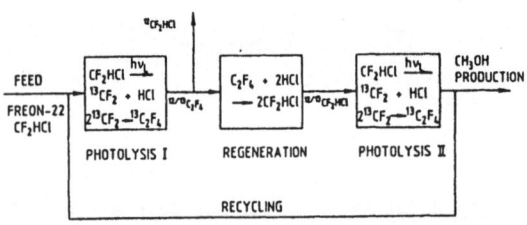

Abb. 18.

TWO STAGE INFRARED LASER MPD CARBON-13 ENRICHMENT PROCESS
(P.A. Hackett et. al 1984)

Neben der Darstellung isotopenreiner Verbindungen ist auch die photochemische
Entfernung von Verunreinigungen mit Lasern möglich. Interessante Anwendungen sind
hier etwa die Ultrareinigung von Gasen in der Halbleitertechnik (Entfernung von
PH_3, B_2H_6, AsH_3 aus Silan (SiH_4)) oder die Entfernung von H_2S-Verunreinigung aus
CO/H_2-Synthesegasmischungen.

Eine Reihe laserphotochemischer Untersuchungen liegen auch für die Darstellung
pharmazeutisch wirksamer Produkte vor. So für Vitamin D (9,10) Pheromone und Prosta-
glandine (11). Der Einsatz des Lasers bringt gegenüber der klassischen Photochemie
eine Reihe neuer Möglichkeiten. Einmal erlaubt die große Frequenzschärfe eine
gezieltere Anregung, wobei unerwünschte Absorptionen von Folge- und Zwischenproduk-
ten vermieden werden. Daneben kann, wie in Abb. 19 dargestellt, durch Mehrphotonen-
absorption von intensiven Laserimpulsen im ns- und ps-Bereich eine Anregung von
höheren Triplet- und Singulet-Zuständen in organischen Molekülen erreicht werden,
die zu neuen Reaktionsprodukten führen.

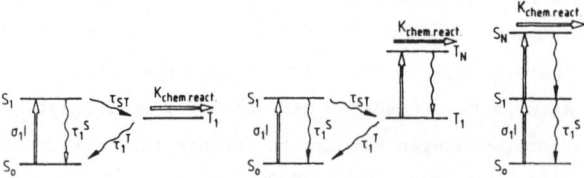

Abb. 19.

LASER EXCITATION OF POLYATOMIC MOLECULES

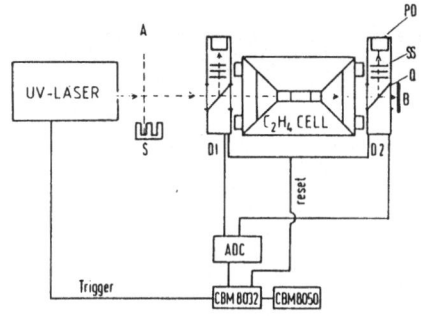

Abb. 20.

LASER-INDUCED HIGH-PRESSURE POLYMERIZATION OF ETHYLENE

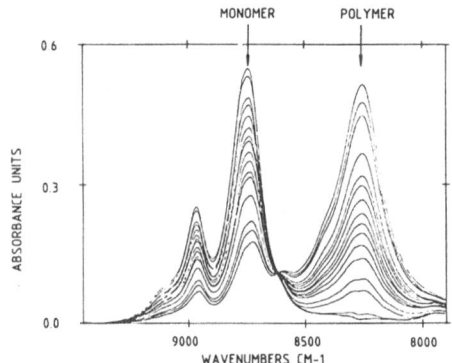

Abb. 21.

LASER-INDUCED HIGH-PRESSURE POLYMERIZATION OF ETHYLENE

Der Einsatz von Lasern zur Herstellung billiger Massenchemikalien lohnt sich nur bei sehr hohen Quantenausbeuten (s. Abb. 16). Als Beispiel für derartige Kettenreaktionen ist in Abb. 20 eine Apparatur zur Untersuchung der UV-laserinduzierten Polymerisation von Ethylen bei Drücken bis zu 3000 bar und Temperaturen bis 300°C in der fluiden Phase dargestellt. Durch Verwendung von quantitativer IR-Spektroskopie können die Quantenausbeuten sowie die zeitliche Entwicklung des Polymerisationsvorganges bei Variation von Druck, Temperatur und Laserintensität in einem weiten Umsatzbereich direkt beobachtet werden (12).

Abb. 21 zeigt die Entwicklung der Polymerbildung durch Aufnahme des Obertons der C-H-Streckschwingung beim Monomer-Polymer-Übergang nach Bestrahlung mit einem KrF-Laser. Die Zuführung der notwendigen Startenergie für die Kettenreaktion über Laserstrahlung erlaubt eine kontrollierte und homogene Auslösung des Polymerisationsprozesses. In den obigen Experimenten wurden Quantenausbeuten von über 10^4 erzielt. Sehr hohe Quantenausbeuten können auch bei der laserinduzierten Darstellung von Vinylchlorid erhalten werden. Vinylchlorid (VC), das Monomere des PVC, wird technisch vorwiegend durch thermische Abspaltung von HCl aus 1.2-Dichloräthan (DCE) durch eine Kettenreaktion hergestellt. Mit einem weltweiten Produktionsvolumen von über 3 x 10^7 Jahrestonnen gehört VC zu den mengenmäßig führenden Produkten der chemischen Industrie. Der Vorteil einer photolytischen gegenüber einer thermischen Auslösung der Kettenreaktion liegt in der Tatsache, daß dabei ein unimolekularer Prozeß mit geringer Barriere geschwindigkeitsbestimmend wird. Das führt zu einer niedrigeren Aktivierungsenergie und damit zu niedrigeren Reaktortemperaturen, höherem Umsatz, geringeren Energiekosten und weniger Nebenprodukten.

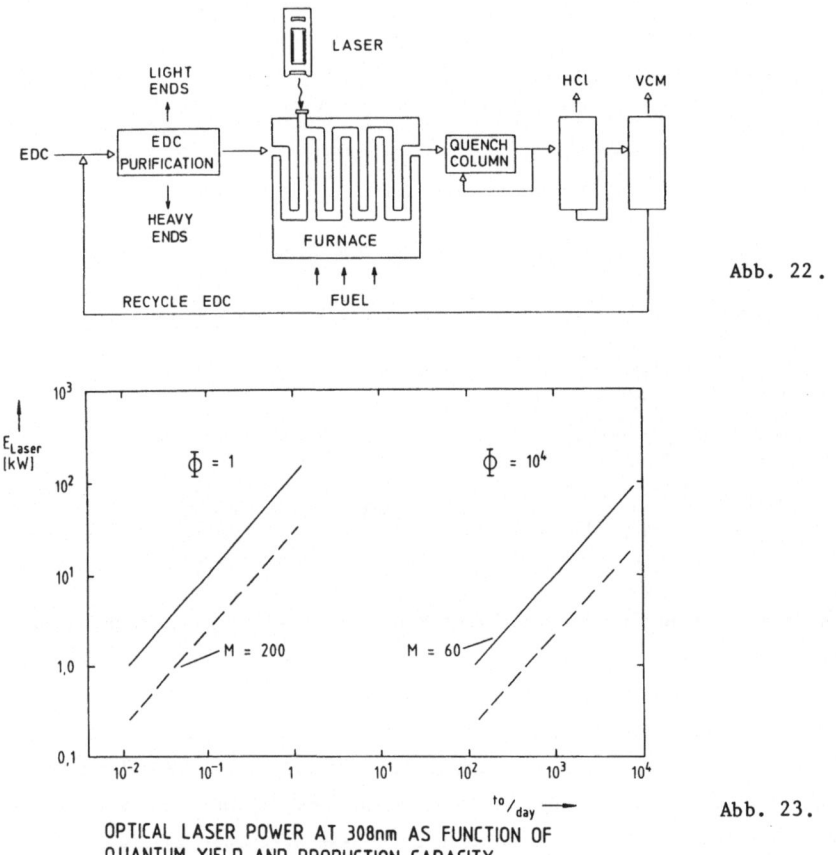

Abb. 22.

Abb. 23.

OPTICAL LASER POWER AT 308nm AS FUNCTION OF QUANTUM YIELD AND PRODUCTION CAPACITY

Damit ergibt sich die Möglichkeit, UV-Laserstrahlung zur Verbesserung eines großtechnischen Prozesses einsetzen zu können. Eine zur Zeit im Bau befindliche Pilotanlage ist in Abb. 22 schematisch dargestellt. Die thermische Spaltung des Dichlorethans erfolgt in einem Rohrreaktor. Nach Aufheizung des Dichlorethans kann mit Hilfe eines leistungsstarken Excimerlasers ein Segment des Reaktors bestrahlt werden. Der Laser wird dabei auf eine Wellenlänge eingestellt, bei der eine möglichst geringe Absorption im Medium eintritt. Durch die hohe Parallelität des Laserstrahls kann so ein sehr großes Volumen bestrahlt und gleichmäßig eine geringe Konzentration aktiver Chloratome und damit eine große Kettenlänge erzielt werden. Mit Hilfe der

Laserstrahlung kann gegenüber dem bisherigen Verfahren ein sehr viel rascherer Umsatz bei niedrigen Temperaturen erreicht werden. Dadurch sind höhere Gesamtumsätze bei gleichzeitig verringerter Bildung von Nebenprodukten möglich. Es besteht Aussicht, besonders diejenigen Nebenprodukte zu verringern, die langfristig als geringe Beimengungen im PVC zu einer verminderten thermischen oder photochemischen Stabilität dieses Kunststoffs beitragen. Abb. 23 zeigt die erforderliche optische Laserleistung bei Verwendung eines XeCl-Excimerlasers bei 308 nm in Abhängigkeit von der Quantenausbeute des photochemischen Prozesses (M), dem Molekulargewicht des Produktes (Φ) und der geforderten Anlagenleistung (13).

Durch den Einsatz von Gasmischungen, die durch Laserstrahlung pyrolysiert werden,

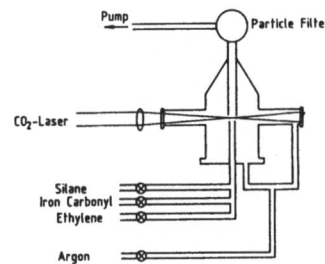

Abb. 24.

PRODUCTION OF Fe/Si/C-CATALYST BY CO_2-LASER PYROLYSIS
(J. Yardley, A. Gupta 1984)

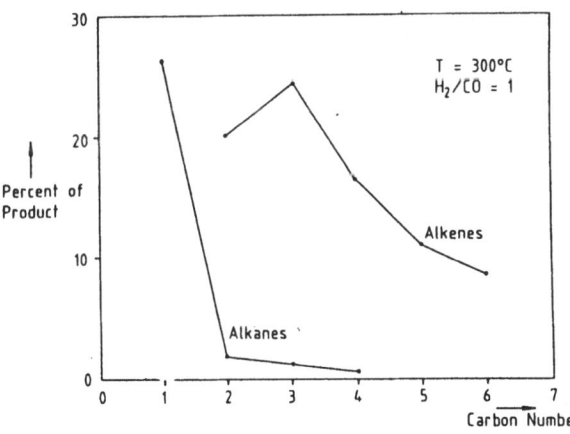

Abb. 25.

LASER PRODUCED Fe/Si/C-CATALYST IN FISCHER-TROPSCH SYNTHESIS

lassen sich katalytisch aktive Festkörper mit variabler Zusammensetzung herstellen (s. Abb. 24). Durch die vollständige Vermischung der gasförmigen Ausgangsstoffe und ihre schnelle Aufheizung im Laserstrahl werden sehr kleine Festkörperteilchen homogener Struktur und großer Oberfläche erhalten, deren Zusammensetzung sich in weiten Grenzen variieren läßt. Abb. 25 zeigt die Ausbeute an Alkanen und Alkenen aus der Fischer-Tropsch Synthese mit lasersynthetisierten Fe/Si/C-Katalysatoren. Bei

Optimierung der Zusammensetzung führen die durch Lasersynthese erhaltenen Katalysa-
toren zu höherer Selektivität und insbesondere zur bevorzugten Darstellung der
wertvollen Leichtolefine (C_2-C_4) (14). Ähnlich wie in Abb. 24 dargestellt, können
Pulver definierter Zusammensetzung z. B. für extrem hitzefeste Keramiken hergestellt
werden. CO_2-Laserbestrahlung von $(Me_3Si)_2NH$, aber auch von SiH_4-NH_3-CH_4-Gasmischun-
gen führt zu besonders feinen, reinen, kugelförmigen, wenig zusammenklumpenden und
nahezu gleich großen Si/C/N-Körnern zur Herstellung hochwertiger Keramik (15).

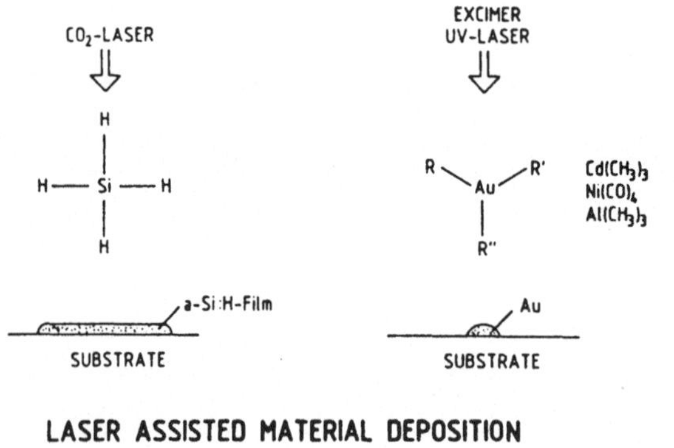

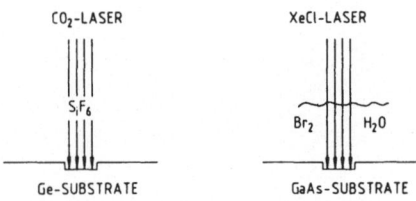

Abb. 26

LASER ASSISTED MATERIAL DEPOSITION

Abb. 27

LASER ASSISTED SURFACE ETCHING

Photochemische Abscheidung mit Hilfe fokussierter UV- und IR-Laser erlaubt auch die
Darstellung von Strukturen im µm-Maßstab auf Festkörperoberflächen. Feine Metall-
oder Halbleiterstreifen wurden durch Photodissoziationen von metallorganischen
Dämpfen und Ablagerung der entstehenden freien Metallatome auf ein Substrat (s. Abb.
26) erhalten. Durch Laserphotoablagerung eines geeigneten Elektronenmangelhalb-
leiters wie z. B. p-Cd auf einen lasererhitzten Elektronenüberschußträger wie n-InP
können Halbleiter im Mikromaßstab dotiert und Ohmsche Kontakte hergestellt werden.
Ziel dieser Experimente ist es, Metall-, Halbleiter-, Isolatoren- oder einfach
Ätzmuster mit hoher räumlicher Auflösung auf Oberflächen zu schreiben, um beschädig-
te oder fehlerhaft produzierte integrierte Schaltungen durch räumlich lokalisierte
Ablagerungen auszuheilen und letzlich "Chips" ohne Masken herzustellen (16).
Neben der Abscheidung können durch Laserzersetzung von an Oberflächen absorbierten
Halogenverbindungen auch feine Muster gezielt geätzt werden (s. Abb. 27).

Freie Halogenatome können durch Radikalfänger aus der Gasphase entfernt werden, so daß der Ätzprozeß auf die unmittelbare Nähe des Laserstrahls räumlich konzentriert wird. Das Verfahren ist auch an flüssig/fest Grenzflächen anwendbar. Benutzt man sehr kurzwellige UV-Laser wie z.B. ArF-Laser bei 193 nm, so lassen sich direkt die chemischen Bindungen in der Materialoberfläche spalten. Hierbei können hohe Ätzgeschwindigkeiten und große Flankensteilheiten der Strukturen erzielt werden (17). Dieses Verfahren erscheint besonders interessant im Zusammenhang mit einer kontrollierten Abtragung von biologischem Material, wie z.B. in der Kardiologie zur Beseitigung von arteriosklerotischen Plaques.

Literatur

(1) M. KNEBA, J. WOLFRUM: Ann. Rev. Phys. Chem. 31, 47 (1980)
(2) J. WOLFRUM: Chemical Kinetics in Combustion Systems: The Specific Effect of Energy, Collisions, and Transport Processes, 20th Symposium (International) on Combustion, The Combustion Institute, Pittsburgh, USA (1985)
(3) J. WOLFRUM: Laser Stimulation and Observation of Bimolecular Reactions, Int. Conference on Chemical Kinetics, Gaithersburg, USA (to be published in J. Phys. Chem. 1985)
(4) G. I. BEKOV, V. S. LETOKHOV: Appl. Phys. B, 30, 161 (1983)
(5) W. DEMTRÖDER (Ed.): Laser Spectroscopy, Springer, Berlin (1982)
(6) D. R. CROSLEY (Ed.): Laser Probes for Combustion Chemistry, Am. Chem. Soc., Washington, USA (1980)
(7) T. R. EVANS (Ed.): Applications of Lasers to Chemical Problems, J. Wiley, New York, USA (1982)
(8) K. L. KOMPA, J. WANNER: Laser Applications in Chemistry, Plenum Press, New York, USA (1984)
(9) P. A. HACKETT, C. WILLIS, M. GAUTHIER, A. J. ALCOCK: SPIE Vol. 458, Appl. of Lasers to Industrial Chemistry S. 65 (1984)
(10) N. GOTTFRIED, W. KAISER, M. BRAUN, W. FUSS, K. L. KOMPA: Chem. Phys. Lett. 110, 335 (1984)
(11) R. M. WILSON: SPIE Vol. 458, Appl. of Lasers to Industrial Chemistry S. 58 (1984)
(12) M. BUBACK, H.-P. VÖGELE: Laser-Induced High-Pressure Polymerization of Pure Ehtylene (submitted to "Die Makromolekulare Chemie" 1985)
(13) K. J. SCHMATJKO: Möglichkeiten des Einsatzes von Hochleistungslasern zur Erzeugung chemischer Produkte, Bericht Kraftwerk Union, Erlangen (1982)
(14) A. GUPTA, J. T. YARDLEY: SPIE Vol. 458, Appl. of Lasers to Industrial Chemistry 131 (1984).
(15) J. H. FLINT, J. S. HAGGERTY: SPIE Vol. 458, Appl. of Lasers to Industrial Chemistry S. 108 (1984)
(16) D. BÄUERLE (Ed.): Laser Processing and Diagnostics, Springer Series in Chemical Physics, 39 (1984)
(17) R. SRINIVASAN: in: Photophysics and Photochemistry above 6 eV, S. 595, Elsevier Science Publ., Amsterdam (1985)

Hochauflösende Mehr-Photonen-Massenspektrometrie

H. Kühlewind, A. Kiermeier, U. Boesl, H. J. Neusser und E. W. Schlag
Institut für Physikalische und Theoretische Chemie der Technischen Universität
München, Lichtenbergstraße 4, D-8046 Garching

Die Massenspektrometrie ist eine vielseitige und weit verbreitete physikalische Methode
in der modernen chemischen Analytik sowie bei der Strukturaufklärung. Eine neue Entwicklung auf diesem Gebiet ist die selektive Massenspektrometrie mit Lasern, bei der
die Probe mittels resonanzverstärkter Mehr-Photonen-Ionisation ohne thermisches Aufheizen ionisiert wird. Diese Art der Ionisation in einem Flugzeit-Massenspektrometer
mit reflektierendem elektrostatischem Feld führt zu einigen herausragenden Eigenschaften, die besonders für die chemische Analytik von Bedeutung sind.

I. Selektive Ionisation

Bei der Mehr-Photonen-Ionisation wird ein Laserlichtbündel mittels einer Linse in einen
effusiven Molekularstrahl fokussiert; dabei nimmt ein Molekül innerhalb kurzer Zeit
mehrere UV-Photonen auf und wird ionisiert. Die gebildeten Ionen werden in einem
Massenanalysator massenselektiert nachgewiesen. Der beobachtete Ionenstrom nimmt
drastisch zu, wenn reelle Zustände der Moleküle mit der Energie der Photonen des eingestrahlten Laserlichts in Resonanz sind /1/. Eine resonanzverstärkte Zwei-Photonen-
Ionisation wird zum Beispiel dann erzielt, wenn die Laser-Wellenlänge auf einen Ein-
Photonen-Übergang des Moleküls abgestimmt wird. Unter diesen Bedingungen läßt sich
bei relativ geringen Lichtintensitäten von einigen 10^6 W/cm^2 die Ionisation beinahe aller im Laserfokus befindlichen Moleküle erzielen. Die Resonanzverstärkung des Ionen-
stroms führt zu einer für die Spurenanalytik bedeutenden Eigenschaft der Mehr-Photonen-Ionisation: in einer Mischung verschiedener Verbindungen wird selektiv und empfindlich diejenige Spezies ionisiert und im Massenspektrum beobachtet, deren
Absorptionsbande sich mit der eingestrahlten Wellenlänge in Resonanz befindet. Anhand
der Massenspektren in Abb. 1 ist zu sehen, daß die spektrale Selektivität ausreicht, um
verschiedene ^{13}C-Isotopen-Moleküle geringer Konzentration in einem Gemisch natürli-
cher Isotopen-Zusammensetzung des Benzols nachweisen zu können. Dabei wird bei geeigneter Wahl der Wellenlänge eine optische Selektivität A $\approx$ 800 erreicht. Die
Verwendung eines Lasers in der Massenspektrometrie fügt auf diese Weise der Masse
mit der Wellenlänge eine zweite Dimension hinzu und macht so die Massenspektrometrie zu einer zweidimensionalen Methode - eine Eigenschaft, die für die Spurenanalytik
von großem Nutzen sein wird.

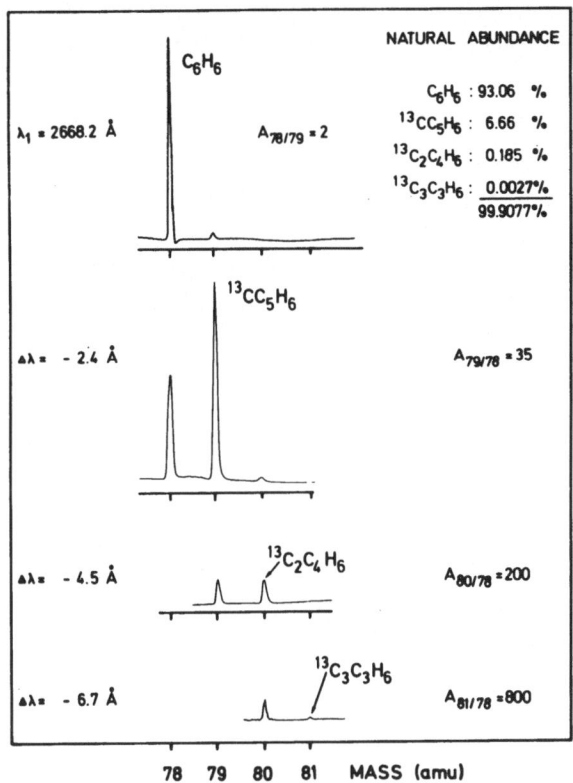

Abb. 1. Selektive Mehr-Photonen-Ionisation verschiedener seltener Benzolisotope im natürlichen Isotopenvorkommen

II. "Weiche" und "Harte" Ionisation

Für die Spurenanalytik ist eine "weiche" Ionisation ohne Fragmentierung bedeutsam, da die gewünschte Substanz in diesem Fall direkt nachgewiesen werden kann. Mittels resonanzverstärkter Mehr-Photonen-Ionisation kann bei geeigneter Intensität des Laserlichts eine "weiche" Ionisation erzielt werden. Dies wurde kürzlich von unserer Arbeitsgruppe an einigen aromatischen Verbindungen wie z. B. Benzol, Naphthalin und Thiophen demonstriert /2/. Auf der anderen Seite führt eine Erhöhung der Intensität des Laserlichts zu einer starken Fragmentierung. Eine graduelle und mittels der Laserintensität dosierbare Dissoziation der Molekülionen ist für eine Strukturaufklärung mittels Massenspektrometrie von großem Interesse.

III. "Leiterwechsel"-Mechanismus und Strukturspezifische Fragmentierung

Besondere Bedeutung kommt dem für das beobachtete Fragmentierungsmuster verantwortlichen Mechanismus der Mehr-Photonen-Anregung zu. In experimentellen /3/ und theoretischen /4/ Arbeiten konnten wir zeigen, daß die Absorption in einem "Leiterwechsel"-Mechanismus erfolgt. Die Photonenabsorption wechselt nämlich vom neutralen

Molekül zum Molekülion und dann vom Molekülion zu Fragmentionen usw., bis schließlich im Falle des Benzols oder anderer Kohlenwasserstoffe das Kohlenstoffkation C$^+$ gebildet wird. Die Mehr-Photonen-Ionisation führt daher anders als bei der Elektronenstoß-Anregung zu schmalen Energieverteilungen der Molekülionen. Dies ist für Ionen-kinetische Studien von Bedeutung und beeinflußt die Fragmentierungsmuster vielatomiger Molekülionen. Die Auswirkungen des Leiterwechselmechanismus lassen sich beispielsweise anhand der Mehr-Photonen-Fragmentierungsmuster von isomeren Butylkationen, die aus den entsprechenden Butyliodiden gebildet werden, demonstrieren /5/. Im Gegensatz zu den Elektronenstoß-Massenspektren, die im Bereich der Kohlenwasserstoff-Kationen ununterscheidbar sind, zeigen die entsprechenden Mehr-Photonen-Massenspektren sehr ausgeprägte und strukturspezifische Unterschiede (Abb. 2). Die beobachteten Unterschiede sind eine direkte Folge des Leiterwechsel-Mechanismus, der zu anderen Energieverteilungen führt, als die Elektronenstoß-Anregung. Alkylkationen spielen in den Massenspektren von Kohlenwasserstoffen und damit vor allem in der Petrochemie eine bedeutende Rolle. Stark unterschiedliche Fragmentierungsmuster von isomeren Alkylkationen lassen die Unterscheidung von Strukturisomeren zu und sind daher bei der qualitativen und quantitativen Analytik von Kohlenwasserstoffgemischen von großem Interesse.

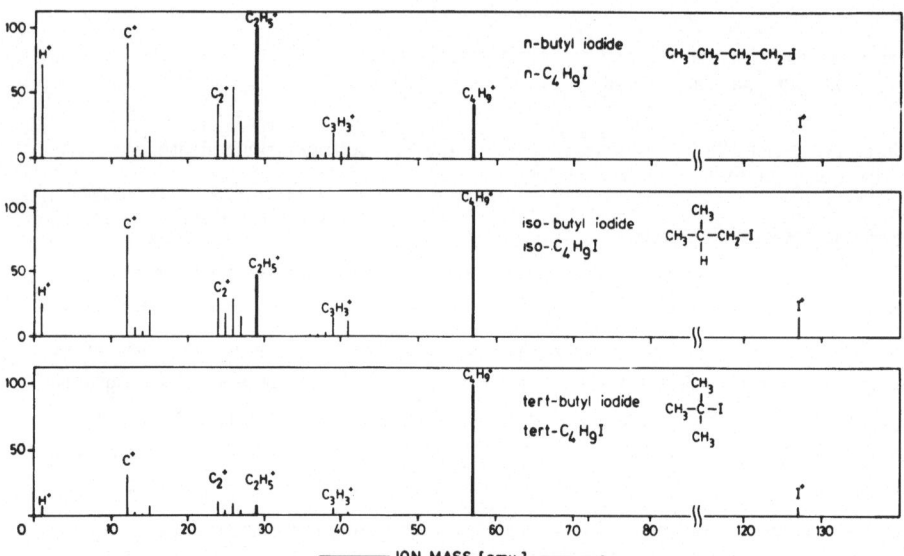

Abb. 2. Strukturspezifische Fragmentierung isomerer Butyl-Kationen in den Mehr-Photonen-Massenspektren von Butyliodiden

IV. Flugzeit-Massenspektrometer

Aufgrund der hohen örtlichen (≈ 50 μm) und zeitlichen (≈ 5 ns) Schärfe der Ionenquelle stellt die Mehr-Photonen-Ionisation eine ideale Ionisationsmethode für die Flugzeit-Massenspektrometrie dar. Die Flugzeit-Massenspektrometrie ihrerseits zeichnet sich durch hohe Transmission aus und die Kombination mit einem Laser führt ferner dazu, daß mit einem einzigen Laserimpuls das gesamtes Massenspektrum einer Probe aufgenommen werden kann. Auf diese Weise können chemische Reaktionen zeitaufgelöst verfolgt werden bzw. sehr kleine Mengen herab bis zum Pikogramm-Bereich nachgewiesen werden. Ein genereller Nachteil der klassischen Flugzeit-Massenspektrometrie ist ihre begrenzte Massenauflösung. Diese kann jedoch durch Kombination der Mehr-Photonen-Ionenquelle mit einem elektrostatischen Reflektorfeld, das Unschärfen der kinetischen Energie der Ionen kompensiert, deutlich verbessert werden. Die bisher in unseren Labors erzielte Massenauflösung $M/\Delta M \approx 5000$ stellt sicher noch nicht die Grenze der Leistungsfähigkeit dieses Gerätes dar /6/. Somit erfüllt die Mehr-Photonen-Massenspektrometrie eine wichtige Voraussetzung für die Analyse großer organischer Moleküle.

V. Metastabile Ionen

Ein wichtiger Schritt bei der Strukturanalyse unbekannter Verbindungen besteht in der Charakterisierung von unimolekularen Zerfällen und Fragmentierungswegen großer Molekülionen. Dies ist mit Hilfe von metastabilen Ionen möglich, die langsam auf dem Weg von der Ionenquelle zum Detektor zerfallen. Von diesen Zerfällen können sowohl die Tochter- als auch Muttermasse eindeutig bestimmt werden. Ein Vorzug der vorgestellten Laser-Massenspektrometrie ist die empfindliche und selektive Beobachtung von Massenspektren metastabiler Ionen. Am Beispiel des Anilins zeigten wir kürzlich, daß die Signale metastabiler Ionen bei der Mehr-Photonen Ionisation im Gegensatz zur Elektronenstoß-Ionisation beträchtliche Intensität besitzen und bei geeigneter Wellenlänge sogar zu den größten Signalen im Spektrum werden können /7/.

Die hier vorgestellten Eigenschaften der Laser Mehr-Photonen-Ionisation in einem energiekorrigierenden Reflektron-Flugzeit-Massenspektrometer führen zu einer Vielzahl von interessanten Anwendungsmöglichkeiten des Lasers in der Massenspektrometrie und chemischen Analytik. Besondere Bedeutung kommt dabei der Wellenlänge als einer neuen Dimension zu, da sie eine selektive Ionisation und selektive Fragmentierung ermöglicht. Dies ist eine Eigenschaft, die herkömmlichen Ionisationsmethoden nicht zu eigen ist.

Literatur
/1/ U. BOESL, H. J. NEUSSER, E. W. SCHLAG, Z. Naturforsch. 33a, 1546 (1978)
/2/ U. BOESL, H. J. NEUSSER, E. W. SCHLAG, Chem. Phys., 55, 193 (1981)
/3/ U. BOESL, H. J. NEUSSER, E. W. SCHLAG, J. Chem. Phys., 72, 4327 (1980)
/4/ W. DIETZ, H. J. NEUSSER, U. BOESL, E. W. SCHLAG, S. H. LIN,
     Chem. Phys., 66, 105 (1982)
/5/ H. KÜHLEWIND, H. J. NEUSSER, E. W. SCHLAG, J. Phys. Chem., im Druck
/6/ U. BOESL, H. J. NEUSSER, R. WEINKAUF, E. W. SCHLAG,
     J. Phys. Chem., 86, 4857 (1982)
/7/ H. KÜHLEWIND, H. J. NEUSSER, E. W. SCHLAG, J. Chem. Phys., im Druck

Der Laser als Werkzeug in der präparativen Photochemie

W. Adam, K. Hannemann, S. Grabowski, P. Hössel, U. Kliem & H. Platsch
Institut für Organische Chemie, Universität Würzburg, Am Hubland, 8700 Würzburg / D
R.M. Wilson, Department of Organic Chemistry, University of Cincinnati, Ohio / USA

Die Intensität und Monochromasie des Argonionenlasers wird genutzt, um durch direkte und triplettensensibilisierte n, n*-Anregung im Nah-UV-Bereich (334, 351, 364 nm) Azo- und Carbonylchromaphore anzuregen. Die durch charakteristische Photoreaktionen (Stickstoffabspaltung, Norrish-Typ II, Perno-Büchi, usw.) erzeugten Triplett-Diradikale werden mit molekularem Sauerstoff (Triplettfänger) in Form von Peroxiden (Endoperoxide, 1,2,4-Trioxane, Hydroperoxide, usw.) abgefangen. Repräsentative Beispiele von synthetischer Bedeutung sind die Strukturen 1 - 3. Das Potential dieses Werkzeugs für die Synthese von Naturstoffen (Prostaglandin, Plastochinon, Qinghaosu, Frontalin, usw.), die selbst Peroxide oder von Peroxiden abgeleitet sind, ist bisher noch kaum genutzt.

(1)   (2)   (3)

Laser Surface Chemistry

H. Schröder and K.L. Kompa
Max-Planck-Institut für Quantenoptik
8046 Garching, Germany

Abstract

Laser induced chemistry at surfaces is a new trend in laser chemistry which is likely to meet many needs in materials research and processing. In this contribution some background is given and one specific example - platinum film deposition on different substrate materials - is discussed.

I. Introduction and Background

Chemistry contains as one of its most prominent areas the <u>study and synthesis of materials</u>.

This suggests a stronger consideration of the use of lasers in materials research from a laser chemist's point of view. Let us briefly recall of what kind the questions were which were so far asked (and sometimes even answered) in laser chemistry. For a concise summary Figure 1 may be useful to consult. One sees in general that lasers provide very detailed insight into the dynamics and the controlling parameters of chemical reactions. This knowledge can subsequently be

Laser Source Property	Types of Applications	Typical Examples
Monochromaticity Tunability	State Selective Excitation / Probing	Basic Chemical Dynamics (Beams, Small Molecules) Laser Assisted Collisions
High Power / Energy	Multiphoton Excitation, High Excitation Rates	Separation, Purification (Isotopes) Controlled Radical Formation, Photoionization
Short Pulses	High Temporal Resolution	psec Phenomena (Condensed Phases, Biophysics)
Collimation Focusability	High Spatial Resolution, Remote Heat Transfer	High Temperature Chemistry - Thermal Process Control, Microchemistry

Fig. 1.   Systematics of laser chemistry in relation to the most important laser features.

694

transformed into better defined reaction conditions. These conditions may be realized either by conventional measures or again by lasers, using them e.g. for the preparation of reagents in desired chemical composition, desired state of excitation and desired concentration. But it is important to note that this second step is not necessarily following the first one and that the analytical application of lasers in chemistry is of prime importance. The general picture of a chemical reaction that most workers in the field had in mind, however, was that of a gas phase reaction between more or less isolated molecules. This picture changes and indeed becomes more colourful if we consider reactions at or near a surface.

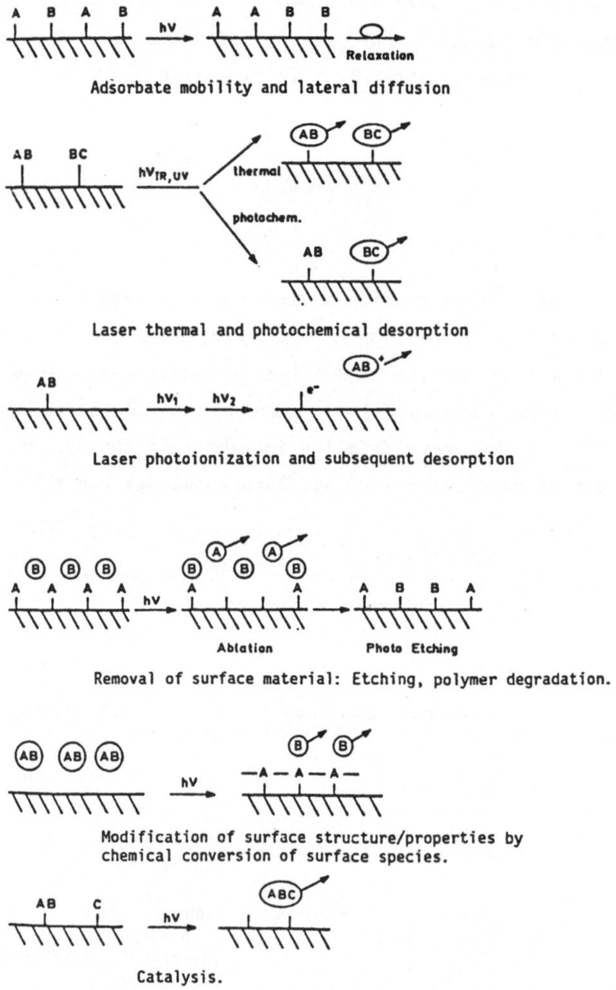

Fig. 2.   Schematic classification of laser-induced photochemical
and photopyhsical surface processes

The current literature on applied laser surface chemistry shows a long and diversified list of topics. Examples are: Photodeposition of metals, insulators and semiconductors, photoetching and structuring, diagnostics of semiconductor structures, photochemical doping, modification of surface properties, production of metastable (amorphous, glasseous) materails, production of very fine sinterable powders, production of new catalysts. There have been attempts for a systematic classification according to the details of the molecule/surface interaction. Such a classification scheme is reproduced in Fig. 2. A simpel guideline may be obtained by considering just three process parameters, namely:

1) The state of the molecular partner (radical, excited state) interacting with the surface

2) The state of the surface with regard to thermal or charge activation (photoemission).

3) The rate of mass transport and temperature change in comparison to the relevant times of relaxation, crystallization, phase transitions, impurity transport, cluster growth etc.

II. Discussion of specific results

We will try to illustrate these three parameters by considering as a specific example the metal film formation of platinum on a substrate starting with a volatile organometallic platinum compound and using pulsed ultraviolet (excimer) laser sources. Such an experiment is rather simple as Fig. 3 shows.

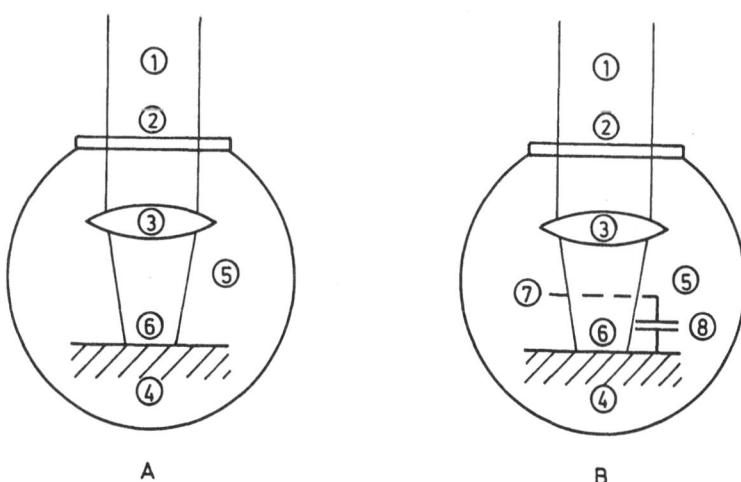

A   B

Fig. 3.   Experimental arrangement for UV laser-induced metal deposition with (B) and without (A) electric field. (1) Laser beam, (2) entrance window, (3) focussing lens, (4) substrate, (5) gas cell, (6) volatile organometallic complex, (7) grid electrode, (8) voltage supply

1) Preparation of rective species in the gas phase

In the special case of platinum film deposition the parent gaseous organometallic compound is tetrakis-trifluophosphine-platinum(o), $Pt(PF_3)_4$. This is a volatile molecule whose (1-photon) KrF-laser photolysis is reversible in the gas phase involving transient luminescent fragment species

$$Pt(PF_3)_4 \xrightleftharpoons{h\upsilon_{UV}} PT(PF_3)_3^* + PF_3$$

In contact with a suitable surface, however, photolysis becomes irreversible and platinum deposition is the end result.

$$Pt(PF_3)_4 \xrightarrow{h\upsilon_{UV}} Pt + 4\ PF_3\uparrow$$

This is a very simplified picture which ignores several important intermediates.

2) Activation of the surface

There is one key point to notice, however: In this case as in other cases the deposition of materials on surfaces shows a strong dependence on geometry (irradiation parallel or perpendicular to the surface) and on wavelength. Without a detailed discussion we want to mention here that this can be explained by photoelectric surface activation. It also is apparent from Fig. 3 (B) that once photoelectrons have been generated by the UV laser additional energy may be provided by an applied electric field, in this way enhancing the deposition according to the following sequence of steps.

$$MeL_4\ MeL_4 \atop MeL_4\ MeL_4 \xrightarrow{h\nu_{UV}} MeL_3^\ominus + L \atop MeL_4\ MeL_4 \quad \text{E Field} \longrightarrow$$

$$+nL\ MeL_3^\ominus \atop MeL_3\ MeL_3 \longrightarrow \longrightarrow Me\ Me\ Me\ Me + (n+3)L$$

The basic assumption here is that the work function of the substrate material be low enough that charge mobilisation can occur with the UV laser wavelengths available. Fig. 4 summarizes some of the relevant laser sources for this type of deposition studies.

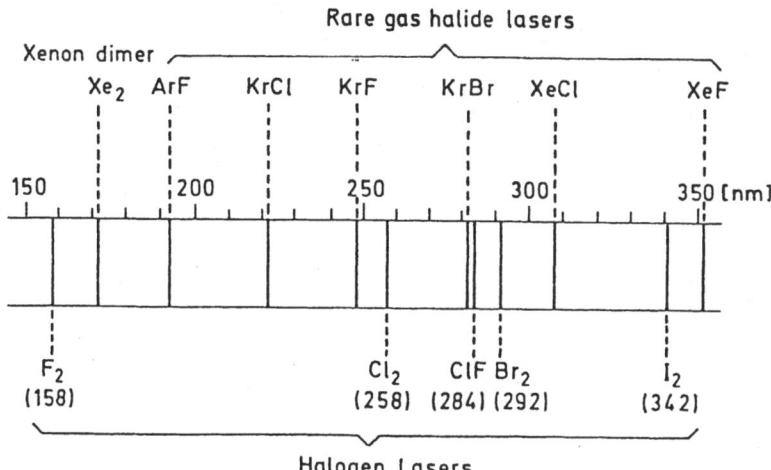

Fig. 4: Survey of the most developed rare gas excimer, rare gas halide and diatomic halogen lasers (potential tuning ranges not shown, only centre wavelengths [nm] of emission given)

3) Deposition rate and film characteristics

The deposition quantum yield (without additional electric field) differs for different substrates in the sequence Si > Al > C > SiO_2, Cu. For aluminum at 200 mJ/cm^2 of a KrF excimer laser the quantum yield is $\phi = 3 \cdot 7 \times 10^{-3}$ corresponding to 0.6 monolayer in one laser shot. Fig. 5 exemplifies that point for two substrate materials. The morphology of the deposit is shown in Fig. 6. The conclusions are the following. A nonporous film can be obtained and amorphous primary deposits are likely to be produced as a result of rather high deposition rates. Reorientation (crystallization) of the deposited material yielding good adhesion is observed under repetitive laser irradiation. This also implies that the pulse properties and irradiation conditions to some extent can be tailored to produce the desired result.

III. Conclusion

As a general conclusion one can say that several mechanisms may be operative in laser induced metal film deposition,

A) <u>Surface initiated</u>
   dissociation of adsorbates
   nucleation centers/barriers
   heat transfer to the vapor
   emission of electrons

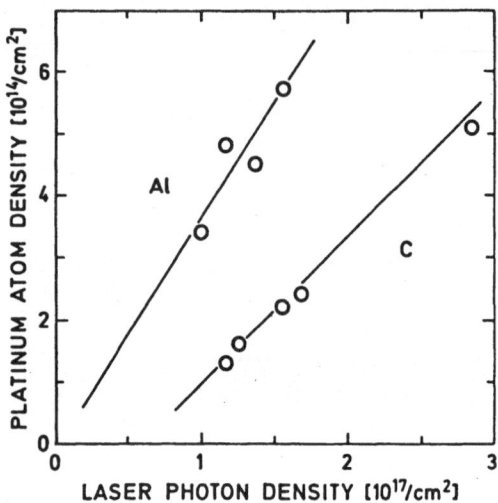

Fig. 5.   Platinum deposition density per laser shot as measured by
Rutherford backscattering

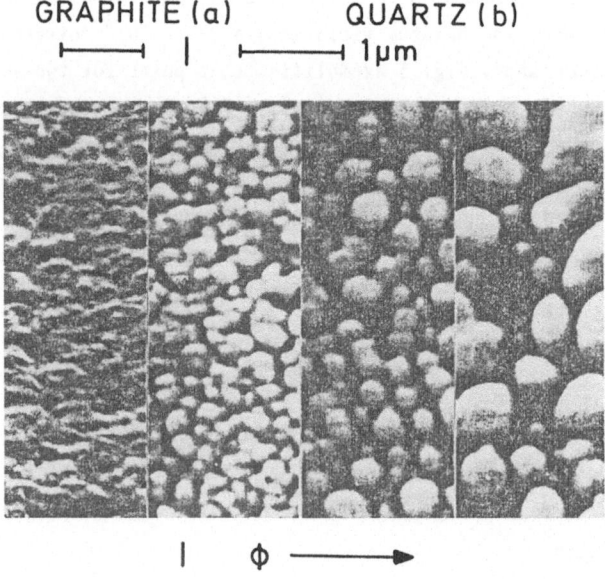

Fig. 6.   Typical development of Pt deposition. a) Primary anchoring
on surface, b) spheres form by remelting with increasing laser
energy density ϕ

Periodic Table of the Elements

Group Ia	Group IIa	Group IIIa	Group IVa	Group Va	Group VIa	Group VIIa	Group VIII			Group Ib	Group IIb	Group IIIb	Group IVb	Group Vb	Group VIb	Group VIIb	Group 0
1 [H]																1 H	2 He
3 Li	4 Be											5 [B]	6 C	7 [N]	8 [O]	9 [F]	10 Ne
11 Na	12 Mg											13 Al	14 Si	15 [P]	16 S	17 [Cl]	18 Ar
19 K	20 Ca	21 Sc	22 Ti	23 V	24 Cr	25 Mn	26 Fe	27 Co	28 Ni	29 Cu	30 Zn	31 Ga	32 Ge	33 [As]	34 Se	35 Br	36 Kr
37 Rb	38 Sr	39 Y	40 Zr	41 Nb	42 Mo	43 Tc	44 Ru	45 Rh	46 Pd	47 Ag	48 Cd	49 In	50 Sn	51 Sb	52 Te	53 I	54 Xe
55 Cs	56 Ba	57* La	72 Hf	73 Ta	74 W	75 Re	76 Os	77 Ir	78 Pt	79 Au	80 Hg	81 Tl	82 Pb	83 Bi	84 Po	85 At	86 Rn
87 Fr	88 Ra	89** Ac															

Fig. 7. Elements in the periodic table for which deposition (and doping) experiments are reported up to now

B) **Vapor initiated**

complete photolysis/pyrolysis

gaseous breakdown/plasma deposition

chemical chain reaction

C) Surface (auto)catalytic decomposition

For the one example discussed here the photoelectric deposition mechanism appears to be dominant. This is a mechanism with should be more studied and developed.

The range of applications of these types of laser chemistry is extremely wide and covers practically all areas of materials research from semiconductor processing out to catalyst formation. This wide spectrum manifests itself in the large number of chemical elements for which related reports exist already in the literature and which are marked in Fig. 7, which concludes this brief survey.

References
(1) Laser Processing and Diagnostics, ed.: D. Bäuerle, Springer Series Chemical Physics 39, Springer Verlag 1984
(2) Laser Diagnostics and Photochemical Processing for Semiconductor Devices, Eds.: R.M. Osgood, S.R.J. Brueck, H.R. Schlossberg, North Holland 1983
(3) H. Schröder, I. Gianinoni, D. Masci, K.L. Kompa, Appl. Phys. B. in print

Fern- und in situ-Analyse von Gasen in der Atmosphäre

W. Michaelis

Institut für Physik, GKSS-Forschungszentrum Geesthacht

D-2054 Geesthacht

Die Verfügbarkeit schmalbandiger Laser mit ausreichener Pulsenergie hat die Möglichkeit eröffnet, die Konzentrationen von Gasen in der Atmosphäre mit Reichweiten von
mehreren km zu bestimmen. Die Langpfadabsorption und die Methode des Lidar (= light
detection and ranging) auf der Basis der differentiellen Absorption und Streuung
(DAS) sind dabei besonders leistungsfähig. Während das erstgenannte Verfahren durch
eine sehr hohe Empfindlichkeit gekennzeichnet ist, aber nur integrale Messungen
längs des Laufweges des Laserlichts gestattet, vereinigt das DAS-Lidar die Vorteile
der Zusammenfassung von Sende- und Empfangsoptik in einem Gerät sowie der ortsauflö-
senden Bestimmung der Gaskonzentration. Die in situ-Messung von Gasen erhielt wesentliche Impulse durch die Entwicklung der durchstimmbaren Diodenlaser. Ein besonders interessantes Verfahren stellt die Derivativspektrometrie dar. Der vorliegende
Beitrag gibt an Hand einiger Beispiele einen kurzen Überblick über den heutigen
Stand der Fern- und in situ-Analyse.

Beim DAS-Lidar werden zwei Laserpulse engbenachbarter Wellenlänge in die Atmosphäre
ausgesandt. Ein kleiner Teil des Lichts wird durch Mie-Streuung an luftgetragenen
Partikeln und im kurzwelligen Bereich auch durch Rayleigh-Streuung an Molekülen
zurückgestreut. λ_1, die Signalwellenlänge, wird so gewählt, daß das Licht von dem
nachzuweisenden Gas stark absorbiert wird, während bei der Referenzwellenlänge λ_o
diese Absorption möglichst gering sein soll. Liegen λ_o und λ_1 hinreichend dicht
beieinander, so kann die Wellenlängenabhängigkeit apparativer und atmosphärischer
Parameter vernachlässigt werden. Sind $P(x,\lambda_i)$ die aus der Entfernung x empfangene
Leistung und $\sigma(\lambda_i)$ der Absorptionsquerschnitt bei der Wellenlänge λ_i, so gilt für
die ortsabhängige Teilchenzahldichte

$$N(x) = \frac{1}{2[\sigma(\lambda_1) - \sigma(\lambda_o)]} \frac{d}{dx} \ln \frac{P(x,\lambda_o)}{P(x,\lambda_1)}.$$

Diese Gleichung besagt, daß das Verfahren selbsteichend ist. Die Gaskonzentration
wird aus dem Verhältnis der empfangenen Lidarsignale abgeleitet ungeachtet der absoluten Amplituden. Eine allgemeine Bedingung für die Wahl der Wellenlängen ist neben
der Existenz von Absorptionslinien eine hohe Transmission der Atmosphäre. Diese
Forderung ist im sichtbaren Bereich, im nahen Ultraviolett und in einigen "Fenstern"
des nahen Infrarot erfüllt.

Für die Fernmessung von SO_2 und NO_2 eignen sich sehr gut Blitzlampen-gepumpte Farbstofflaser. Als Farbstoffe kommen in Betracht Rhodamin 6G für SO_2 und Coumarin 2 für

NO$_2$. Für das erste Gas ist eine Frequenzverdopplung erforderlich, die am günstigsten mit deuteriertem Kaliumdihydrogenphosphat (D-KDP) erfolgt (1). Die Signal- und Referenzwellenlängen betragen 296,17/297,35 nm bzw. 448,3/449,8 nm. Einzelheiten der vom GKSS-Forschungszentrum betriebenen Fernmeßstation sind an anderer Stelle beschrieben (1 - 4). Die Tabelle 1 gibt die Leistungsdaten dieser Station wieder.

Tabelle 1. Ortsauflösende Fernmessung von SO$_2$
und NO$_2$: Leistungsdaten

	SO$_2$		NO$_2$	
Anzahl Pulspaare	120		120	
Reichweite	3.0 km*		5,5 km*	
Ortsauflösung	75 m	300 m	75 m	300 m
Empfindlichkeit	30 ppb	15 ppb	20 ppb	10 ppb

* Tageszeitunabhängig

Zur Erzielung der aufgeführten Reichweiten und Empfindlichkeiten müssen verschiedene Voraussetzungen erfüllt sein: Geometrische Kompression der Signaldynamik (5), quasisimultaner Doppelpulsbetrieb (Pulsabstand 50 µs) (1), richtige Wahl der Datenreduktion (1) und Funktionswechsel der Laser für das nachzuweisende Gas bezüglich der beiden Wellenlängen (1, 3). Charakteristisch für das Fernmeßverfahren ist, daß keine Probennahme erfolgt, d.h. systematische Fehler durch Kontamination, Verluste oder Memory-Effekte entfallen. Gegenüber lokal messenden Geräten werden räumliche Beschränkungen in weitem Maße aufgehoben. Abb. 1 zeigt ein typisches Ergebnis einer Immissionsmessung für NO$_2$ im Stadtgebiet von Hamburg.

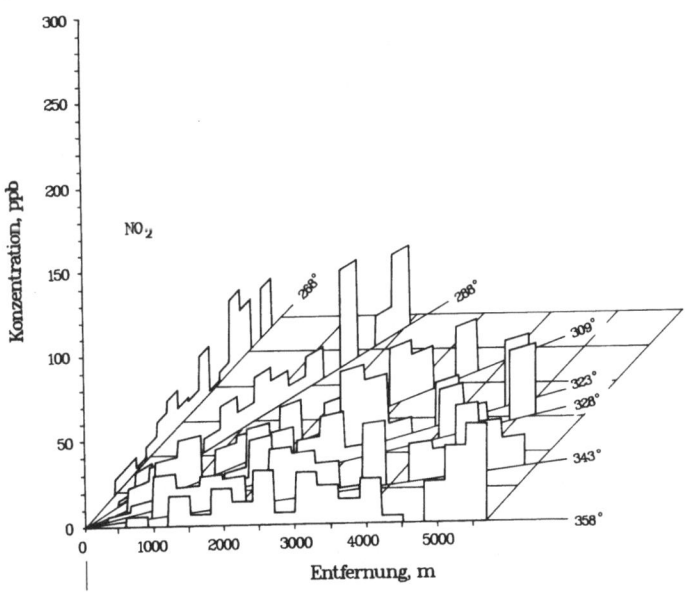

Abb. 1. Großflächige Immissionsmessung von NO$_2$

Während NO$_x$-Emissionen weitgehend aus einer Vielzahl verteilter Quellen erfolgen, wird SO$_2$ überwiegend von einzelnen, gut lokalisierbaren Emittenten abgegeben. Für dieses Gas ist daher auch die Fernmessung der Emission von Interesse, da sie ohne Kenntnis oder Mitwirkung des Betreibers einer Anlage erfolgen kann. Für die Messung von Emissionsmassenströmen aus Schornsteinen führt ein Verfahren zum Ziel, bei dem aus einer Position seitlich unter der Abgasfahne durch Variation des Elevationswinkels fächerförmig Konzentrationsprofile über den Fahnenquerschnitt bestimmt werden. Ist neben der Konzentrationsverteilung auch der Betrag der Windgeschwindigkeit in Hauptwindrichtung bekannt, so ergibt das über die Querschnittsfläche gebildete Integral des Produkts die durch diese Fläche pro Zeiteinheit hindurchtretende Gasmenge. Diese ist gleich dem Massenstrom am Ausgang des Schornsteins, wenn - wie im Falle des SO_2 - das Gas auf kurzen Wegstrecken keinem nennenswerten Abbau unterliegt.

Zum Nachweis der Richtigkeit des Verfahrens wurden mit Unterstützung der Hamburger Umweltbehörde Vergleichsmessungen durchgeführt, bei denen parallel aus den Volumenströmen und SO_2-Konzentrationen in einer Industrieanlage die Emissionen bestimmt wurden. Tabelle 2 gibt einen Auszug aus diesen Meßreihen. Die Windgeschwindigkeit wurde aus Messungen mit einem meteorologischen Mast am Lidarstandplatz abgeleitet.

Tabelle 2. Emissionen von SO_2 aus zwei Schornsteinen einer Buntmetallhütte gemessen nach dem DAS-Lidar-Verfahren bzw. aus Betreiberdaten bestimmt.

| Datum | Emission | |
	aus Betreiberdaten g/s	Lidar (GKSS) g/s
01.09.83	12,9 ± 2,6	13,8 ± 2,8
06.09.83	17,9 ± 3,6	19,9 ± 3,6
09.08.84	10,8 ± 2,1	14,3 ± 2,9
14.08.84	16,0 ± 3,2	15,1 ± 3,0
17.08.84	20,2 ± 4,0	19,8 ± 4,0

Gasförmiger Chlorwasserstoff ist der wichtigste Luftschadstoff, der bei der Verbrennung Chlorkohlenwasserstoff-haltiger Abfälle auf See entsteht. Die Untersuchung seiner Ausbreitung und seines Abbaus ist nicht nur von Bedeutung für die Beurteilung der Seeverbrennung unter dem Gesichtspunkt des Umweltschutzes, sie liefert auch grundlegende Informationen über Ausbreitungsprozesse in der marinen Atmosphäre. Es wurde daher ein schiffsgebundenes Lidarsystem entwickelt, das erfolgreich für solche Untersuchungen eingesetzt wurde (2).

Die Fernmessung von HCl gelingt mit einem DF-Laser, dessen $P_2(3)$-Linie bei 3,636 µm mit einem Wirkungsquerschnitt von (204 ± 10) 10^{-21} cm² von der $P_1(6)$-Linie des $H^{37}Cl$ absorbiert wird. Als Referenz eignet sich die $P_2(5)$-Linie des DF bei 3,698 µm, die nur eine geringe Überlappung mit der $P_1(8)$-Linie des $H^{35}Cl$ aufweist. Der Wirkungsquerschnitt beträgt $(0,56 \pm 0,03)$ 10^{-21} cm². Querempfindlichkeiten anderer Gase sind vernachlässigbar. Die Nachweisgrenze liegt bei 300 ppb, die maximale Reichweite ist 2 km.

HCl wird durch verschiedene Prozesse aus der Gasphase entfernt. Die wichtigsten sind die trockene Deposition auf die Wasseroberfläche, wo das HCl neutralisiert wird, Bildung von Aerosolen und Ausregnen, Aufnahme durch Seesalzaerosole und kondensierten Wasserdampf, Auswaschen durch Niederschlag sowie schließlich chemische Reaktionen. Der so definierte Abbau des gasförmigen HCl wurde nach dem oben am Beispiel der SO_2-Abgasfahnen erläuterten Verfahren in mehreren Meßkampagnen untersucht. In Abb. 2 ist das Ergebnis der Transportbestimmungen einer Meßfahrt wiedergegeben. Je nach meteorologischen Bedingungen liegen die ermittelten Lebensdauern des HCl zwischen 16 und 59 min.

Die im Infrarotbereich arbeitenden Diodenlaser gestatten die Durchstimmung der Wellenlänge auf molekulare Absorptionslinien mit großer Oszillatorstärke. Durch Auswahl geeigneter Linien und Reduzierung des Drucks in der Meßküvette können Querempfindlichkeiten eliminiert und Nachweisstärken im sub-ppb-Bereich erzielt werden. Um große optische Weglängen zu erreichen, werden Multireflexionsküvetten in 2- oder 3-Spiegelanordnung eingesetzt.

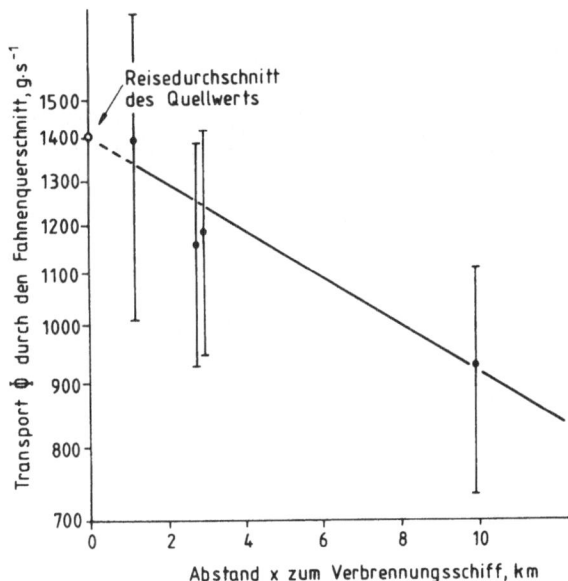

Abb. 2. Transport von HCl durch den Fahnenquerschnitt als Funktion des Abstandes vom Verbrennungsschiff

Bei sehr kleinen Gaskonzentrationen kann die geringe Schwächung des Lichtstrahls durch das Amplitudenrauschen der Strahlungsquelle überdeckt werden. Da dieses Rauschen aber mit wachsender Frequenz schnell abnimmt, läßt sich das Signal-zu-Rausch-Verhältnis dadurch verbessern, daß die Variation des Detektorsignals durch die Absorption im Gas bei einer höheren Frequenz registriert wird. Wellenlänge und damit auch die Schwächung des Lichts können beim Diodenlaser über den Strom leicht moduliert werden. Dieses Prinzip der Derivativspektrometrie wurde in verschiedenen Ordnungen bereits auf zahlreiche Gase angewandt (s. Lit. in (6)); Messungen an HCl wurden jedoch erstmals während der oben beschriebenen Meßkampagnen zur Seeverbrennung durchgeführt (2, 6).

Wird ein Lichtstrahl der Intensität I_0 und der Wellenzahl ν in einer Küvette mit der optischen Weglänge 1 durch ein Gas mit der Teilchendichte n und dem Absorptionsquerschnitt σ geschwächt, so gilt für die durchgelassene Intensität das Beersche Gesetz. Wird ν mit der Frequenz ω moduliert entsprechend

$$\nu = \nu_0 + a \sin \omega t,$$

wobei ν_0 auf das Maximum der Absorptionslinie abgestimmt ist, und gilt $nl\sigma \ll 1$, dann ergibt die Entwicklung in eine Taylor-Reihe bei Vernachlässigung der Glieder höherer Ordnung

$$I(\nu)/I_0(\nu) = 1 - nl\sigma(\nu_0) - nl \frac{d\sigma}{d\nu} (\nu_0) a \sin \omega t - \frac{1}{4} nl \frac{d^2\sigma}{d\nu^2} (\nu_0) a^2 (1 - \cos 2\omega t).$$

704

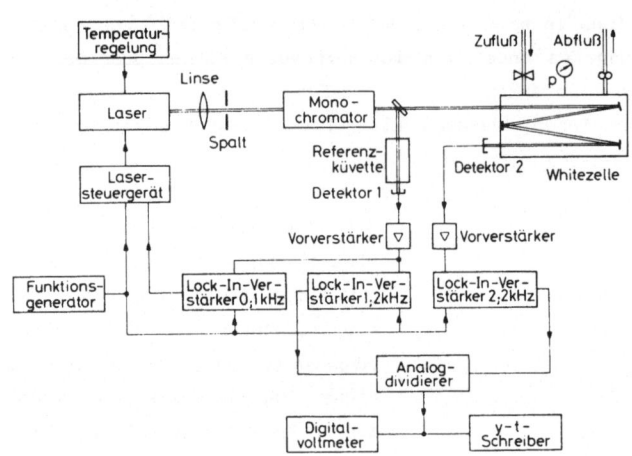

Abb. 3. Schematische Darstellung des Diodenlaser-Derivativspektrometers

Diese Gleichung zeigt, daß das Signal bei der Frequenz ω gut zur Stabilisierung der Laser-Wellenlänge geeignet ist (Nulldurchgang der ersten Ableitung). Das Signal bei 2 ω ist proportional zur Teilchendichte und zur zweiten Ableitung des Wirkungsquerschnitts und stellt ein empfindliches Maß für die Gaskonzentration dar. Das Blockdiagramm eines Spektrometers zweiter Ordnung ist in Abb. 3 wiedergegeben. Für den Nachweis von HCl eig-

net sich die $P_1(6)$-Linie des $H^{35}Cl$ bei 3,634 μm. Die Nachweisgrenze beträgt < 20 ppb $\sqrt{Hz}$. Fragen der Kalibrierung werden in (2, 7) ausführlich diskutiert.

Literatur
(1) W. STAEHR: Dissertation, Universität Hamburg, 1985
(2) W. MICHAELIS, C. WEITKAMP: Fresenius Z. Anal. Chem. 317, 286 (1984)
(3) W. LAHMANN, W. STAEHR, C. WEITKAMP, W. MICHAELIS: Proc. of IGARSS'84 Symposium, Strasbourg, 27 - 30 August 1984. ESA SP-215
(4) W. STAEHR, W. LAHMANN, C. WEITKAMP, W. MICHAELIS: Laser 85 Opto-Elektronik, Kongreßbeitrag 5.1
(5) J. HARMS: Appl. Optics 18, 1559 (1979)
(6) P. POKROWSKY, W. HERRMANN: SPIE Technical Meeting on Optics and Electro-Optics, April 20 - 24, 1981, Washington, DC, USA; GKSS 81/E/23
(7) C. WEITKAMP: Appl. Optics 23, 83 (1984)

Uran-Anreicherung mit Laser

H.L. Jetter, G. Meyer-Kretschmer, URANIT GmbH, Abt. TW, Postfach 1411
5170 Jülich / D

Die Anreicherung von Uran für Leichtwasserreaktoren der westlichen Welt stellt derzeit ein Geschäftsvolumen von ca. 8 Mrd DM pro Jahr dar. Da der Anreicherungsmarkt einerseits durch Überkapazitäten und andererseits durch hohe Produktionskosten gekennzeichnet ist, kommt den Bemühungen um kostengünstigere Anreicherungsverfahren große Bedeutung zu. Insbesondere von Laserverfahren verspricht man sich wegen ihrer hohen Isotopenselektivität eine deutliche Senkung der Anreicherungskosten.
Für zwei konkurrierende Verfahren wurde in den letzten 10 Jahren die Isotopentrennung im Labormaßstab demonstriert und die wesentlichen Verfahrensschritte erarbeitet sowie Konzepte für Lasersysteme gefunden, die das Potential für den großtechnischen Einsatz besitzen. Die weitere Entwicklung vom Laborexperiment zur Pilot- oder Demonstrationsanlage stellt eine große technologische Herausforderung dar - sowohl für die Materialhandhabung als auch für die Lasertechnik. Beispielsweise müssen große Massenströme bei z.T. extremen Temperatur- und Druckverhältnissen geführt werden. Für die Lasersysteme sind Wiederholraten und Durchschnittsleistungen im KHz und KW-Bereich nötig, wobei strenge Anforderungen an Strahlqualität, Wellenlänge und Bandbreite zu stellen sind. Erst der Ausgang dieser risikoreichen Entwicklungen kann Aufschluß über das wirtschaftliche Potential der Uran-Anreicherung mit Lasern liefern.

Laserentwicklungen für die Isotopentrennung

H.-J. Cirkel

Kraftwerk Union AG

Hammerbacherstr. 12 + 14, D-8520 Erlangen

Die einzelnen Prozeßschritte eines Laseranreicherungsverfahrens - insbesondere im großtechnischen Maßstab - stellen extreme Anforderungen an die Lasersysteme. Das UF_6-Trägergasgemisch strömt mit rund 500 m/s durch die Entspannungsdüse, in der das adiabatisch abgekühlte UF_6 auf einer Querschnittsfläche von ungefähr 10 cm² als Monomer existiert. Für eine lückenlose Bestrahlung muß das gepulste Lasersystem (Pulsdauer <100 ns) mit einer Wiederholrate von 10 kHz, die durch Multiplexen von 1 kHz-Lasern erreicht werden soll, emittieren.

Um das Laserlicht effektiv zu nutzen und die mittleren optischen Leistungen der Laser zu reduzieren, wird das Laserlicht mehrmals durch den "kalten" Gasstrahl reflektiert. Diese Art der Strahlführung, das Multiplexen einer Kette von Lasern und die Notwendigkeit, die abgekühlte Gaszone homogen auszuleuchten, bedingen eine hohe optische Qualität der Lasersysteme.

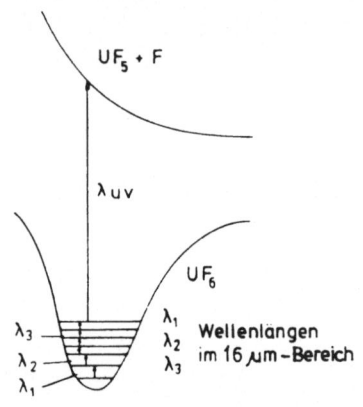

Fig. 1. Anregungsschema UF_6

Um aus Kostengründen bei einmaliger Belichtung einen hohen Anteil des 235U abzutrennen, muß eine Mindestenergiedichte auf den "kalten" Gasstrahl einwirken. Mit zusätzlichen Prozeßdaten errechnen sich aus diesem Wert die in einer industriellen Anlage zu installierenden mittleren Laserleistungen. Sie dürften im 16 µm-Bereich zwischen einigen 100 W und 10 kW liegen, während bei 308 nm bis 100 kW optische Leistung gefordert werden.

Die resonante Einstrahlung in den Q-Zweig des 235 UF_6 bei 16 µm mit der Wellenlänge λ_1 hebt bevorzugt die 235 UF_6-Molekühle in einen angeregten

Vibrationszustand. Ein weiterer 16 μm-Laser der Wellenlänge λ_2 überführt - auch durch resonante Einstrahlung- diese Moleküle in einen
noch höheren Vibrationszustand.

Bei λ_1, λ_2 sind Energiedichten von einigen mJ/cm² erforderlich. Die angeregten UF_6-Moleküle heizen sich bei Energiedichten von über 100 mJ/cm²
durch die Absorption mehrerer Quanten eines weiteren 16 μm-Lasers λ_3
noch stärker auf, dessen Wellenzahl durch den Prozeß auf einige cm⁻¹
genau bestimmt ist. Nur die "heißen" Moleküle des Gasstrahls sollen
durch UV-Strahlung von 308 nm bei Energiedichten von mehreren Joule/cm²
zu einem hohen Anteil dissoziiert werden.

Zur Erzeugung der 16 μm Strahlung hat sich als Verfahren mit techni-
schem Potential die Rotations-Raman Streuung von CO_2-Laser Strahlung
an para-Wasserstoff durchgesetzt (1, 2, 3).

Durch Abstimmung des CO_2-Pumplasers auf verschiedene Wellenlängen emittiert dieser Laser zwischen 13 μm und 18 μm. Jedoch erfordern die geringen Verstärkungsfaktoren für Raman-Streuung im mittleren Infrarot
Spitzenleistungen des Pumplasers bis zu mehreren 10 MW bei gleichzeitig großen Verstärkungslängen.

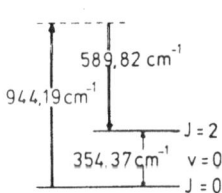

Ein zirkular polarisierter TEA CO_2-Laser im
TEM 00-Mode wird in den Resonator des Raman-
Lasers eingekoppelt, den eine Vielfach-Reflexionszelle mit einem Lichtweg von über 100 m
bildet. Schwingt der Laser an, so emittiert
er über 100 mJ/Puls bei 16 μm und erfüllt mit
dem linienabstimmbaren TEA-CO_2-Pumplaser die
Prozeßanforderungen für die Multiphotonenan-
regung des UF_6 bei λ_3.

Fig.2. Energieniveau-
schema für die Raman-
streuung an p-H_2

Die auf 0,01 cm⁻¹ vorgegebenen Wellenlängen λ_1, λ_2 werden bei gleichzei-
tigem Einkoppeln eines kontinuierlich abstimmbaren Hochdruck (HD)-CO_2-
Lasers in die Vielfach-Reflexionszelle durch 4-Wellenmischung erzeugt
(4). In einer nicht optimierten Anordnung konnten wir bei Energien des
HD-CO_2-Lasers von 400 mJ bei 16 μm kontinuierlich abstimmbares Laserlicht mit 10 mJ Energie pro Puls messen.

Für den Prozeß muß der HD-CO_2-Laser frequenzstabil mit einer Bandbreite kleiner 250 MHz und auf 10 ns genau synchronisiert mit dem star-

ken TEA-CO_2-Pumplaser emittieren. Diese harten Anforderungen konnten wir mit einem kontinuierlich abstimmbaren HD-CO_2-Laser (Druck 10 bar) erfüllen, der bei Ladespannungen von knapp über 40 kV mit einem Thyratron geschaltet wird.

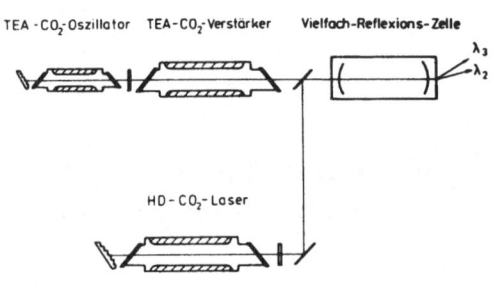

TEA-CO_2-Oszillator   TEA-CO_2-Verstärker   Vielfach-Reflexions-Zelle

λ_3
λ_2

HD-CO_2-Laser

Fig.3. 16μm-Erzeugung durch Vierwellenmischung

Die UV-Dissoziation stellt in dem technischen Anreicherungsverfahren extreme Anforderungen in der Pulsenergie und Wiederholrate an den XeCl-Excimer-Laser. Um die geforderte Energie von 10 J/Puls zu erreichen, haben wir eine neue Konzeption für das Anregungssystem von TE-Lasern - insbesondere Excimer-Lasern - entwickelt (5).

Das pulserzeugende Netzwerk hoher Kapazität und niedriger Eigeninduktivität, das die elektrische Anregungsenergie in einer großen Anzahl parallel geschalteter Kondensatoren mit Wasser als Dielektrikum speichert, speist - bei Ladespannungen von 30 kV - in weniger als 100 ns hohe Energien in das Lasergas ein. Als Schaltelement werden auf 1 ns genau synchronisierte Thyratrons eingesetzt. Das Lasergas wird mit Röntgenstrahlung vorionisiert. Die geometrisch günstige Anordnung aller Komponenten des sehr kompakt aufgebauten Lasers erlaubt es, ohne Schwierigkeiten das Lasergas transversal zur optischen Achse auszutauschen.

In einem Laserfleck von 48 x 54 mm² (54 mm Elektrodenabstand) emittiert der XeCl-Laser bei einem Druck von 4,5 bar und einer Verstärkungslänge von 45 cm in einem 80 ns langen Impuls 3,6 J (Fig.4a). Fig.4b zeigt den Strom-Spannungsverlauf im Entladekreis des Lasers, in dem die zunächst parallel geschalteten Kondensatoren von 275 nF auf knapp 33 kV aufgeladen werden. In Fig. 4c wurde die Spannung über den Laserelektroden U_L um den induktiven Spannungsabfall korrigiert. Die elektrische Energie wird bei einem Plasmawiderstand von 0,15 Ω und einer Brennspannung von 16,8 kV in den Laser eingekoppelt. In Abhängigkeit von der Gasmischung liegen die E/p-Werte für den XeCl-Laser zwischen 500 V/cm bar und 800 V/cm bar.

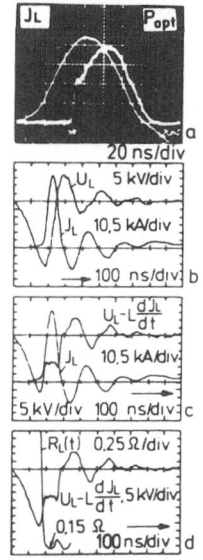

Fig.4. Impulsver-
läufe Excimer-Laser

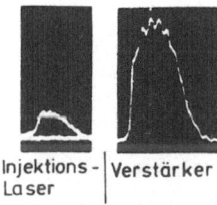

Injektions- | Verstärker
Laser

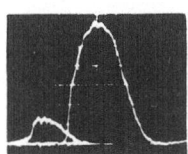

Injektions-Laser und Verstärker

40 ns Voreilung   20 ns/div

Die hohe Divergenz > 2 mrad dieses Lasers konnten wir durch Injizieren der Strahlung aus einem Injektionslaser hoher Strahlqualität mit einer Energie von rund 100 mJ auf < 100 µrad verringern.

Mit einem instabilen Resonator erreicht der Injektionslaser gegen Ende des Impulses eine Divergenz von rund 50 µrad. Bei einer Voreilung von 40 ns wird die Strahlung geringer Divergenz des Injektionslasers optimal verstärkt und der Pulsverlauf des Verstärkers - insbesondere in der Anstiegsflanke - stark beeinflußt (Fig. 5).

Durch Ramanstreuung an einer Vielzahl von Gasen kann mit einem Lasersystem dieser Strahlqualität das gesamte Spektrum bis in das nahe Infrarot nahezu dicht mit leistungsstarken Laserlinien überdeckt werden.

Bei Ausnutzung der physikalisch möglichen Verstärkungslänge von 1 m und der Steigerung der Wiederholrate auf 1 kHz dürfte unser System für Excimer-Laser mittlere optische Leistungen von 10 kW ermöglichen.

Ein Teil der Arbeiten für den Excimer-Laser wurde durch das Bundesministerium für Forschung und Technologie gefördert.

Fig.5. Kopplung von Injektionslaser und Verstärker

Literatur

(1) R.L. BYER: IEEE J.Quantum Electronics QE 12, 732 (1976)
(2) RABINOWITZ, STEIN, BRICKMAN, KALDOR: Appl.Phys.Lett. 35, 739 (1979)
(3) MIDORIKAWA, TASHIRO, TOYODA, NAMBA, AOKI, NAGASAKA: Cleo 85 WG 5
(4) R.L. BYER, R.L. HERBST: Topics Appl. Phys. Vol 16
(5) H-J. CIRKEL, W. BETTE: Deutsches Patent Nr. 2932781

Optoelektronische Sensor-Systeme
Optoelectronic Sensor-Systems

Optoelektronische Sensorsysteme in der Fertigungstechnik

J. Hesse
Fraunhofer-Institut für Physikalische Meßtechnik
Heidenhofstraße 8, D-7800 Freiburg

1. Einleitung

Die ökonomischen Randbedingungen zwingen die industrielle Fertigung zum Übergang auf flexible, automatisierte Herstellungsabläufe und Qualitätskontrollen unter größtmöglicher Integration der Material- und Informationsflüsse.

In einer so strukturierten Fertigung ist der Einsatz rechnergestützter Meß-, Steuer- und Regelsysteme (MSR-Systeme) zwingend (1,2). Die Verfügbarkeit preiswerter und zuverlässiger Mikroelektronik hat die kompatible Auslegung der MSR-Systeme in der Meßwertverarbeitung schnell vorangebracht und zur forcierten Nachentwicklung der daran angepaßten Meßwertaufnehmer (Sensoren) gezwungen (3). Im Sinne einer überwiegend durch mechanische Abläufe geprägten Produktion war dabei der Schwerpunkt auf die Bereitstellung von Sensoren zu legen, die geometriebezogene Kenngrößen für die Produktion (z.B. Abstand, Form, Position, Lage) und das Produkt (z.B. Oberflächenrauhigkeit, -welligkeit; Rißfreiheit) liefern.

Für derartige Aufgaben bietet die Optik interessante Lösungsansätze. Sie erhielten neue, richtungsweisende Impulse von den im Rahmen der Nachrichten- und Kommunikationstechnik entwickelten optoelektronischen Sender- und Empfängerbauelemente.

Als Senderbauelemente sind heute Lumineszenz (LED)- und Laserdioden (LD) in einem breiten Wellenlängenbereich von etwa 0,5 bis 1,6 µm kommerziell verfügbar, die bis 5 mW (LED) bzw. 50 mW (LD) Lichtleistung im Dauerstrichbetrieb abgeben, bis 200 MHz (LED) bzw. 2 GHz (LD) modulierbar sind und mindestens 10^4 bis 10^5 Betriebsstunden garantieren. Verbesserte Eigenschaften sind bei noch weiter fallenden Preisen zu erwarten. Die jetzt für 120 DM erhältliche HITACHI-LD (Typ HL 7801 E, 3mW, ca. 1m Kohärenzlänge) belegt beispielhaft diese Tendenz.

714

Als Empfänger stehen schnelle und empfindliche Fotodetektoren als Einzeldioden seit längerem zur Verfügung. Diodenzeilen und Diodenmatrixanordnungen werden zunehmend verfeinert. Tabelle 1 nennt beispielhaft einige kommerziell angebotene lineare Arrays unterschiedlicher Bildpunktzahl und -größe. Auch hier fällt der Preis (z.B. derzeit 800 DM für Texas Instruments RL 2048 H) stetig.

| Hersteller | Typ | Bildelemente | |
		Zahl	Größe(µm)
Texas Instruments	RL 2048 H	2048	12,7 x 12,7
Toshiba	TCD 103 C	2592	11 x 11
Texas Instruments	TC 104	3456	10,7 x 10,7
Fairchild	CCD 151	3456	7 x 7

Tabelle 1. Kommerziell erhältliche lineare CCD-Arrays (Beispiele)

Neben diesen optoelektronischen Bauelementen sind viele passive Bauelemente entstanden, die die Miniaturisierung und die Flexibilität optischer Systeme wesentlich fördern. Die dämpfungsarme Lichtleitfaser ist ein in diesem Sinne besonders herausragendes, systemtechnisch neues Element.

Dieser Aufsatz stellt einige Sensor-Systeme vor, die mit solchen modernen optoelektronischen Komponenten entstanden sind und einen Beitrag zur Bewältigung der anstehenden geometrierelevanten MSR-Probleme in der Fertigungsautomatisierung leisten können. Der Kompromiß zwischen Übersicht und Informationstiefe zwingt zur Beschränkung auf zwei Anwendungsgebiete: die automatisierte Geometrie-/ Oberflächenprüfung und das automatische Schweißen.

2. Sensor-Systeme für die automatisierte Geometrie-/Oberflächenprüfung

Die anzustrebende höhere Automatisierung der Fertigung setzt voraus, daß die visuelle Sichtprüfung durch den Einsatz rechnergestützter optischer Sensor-Systeme ersetzt wird. Problematik und prinzipielle Lösungswege dazu sind übersichtlich z.B. in (4,5) beschrieben.

Bei einem Teil der Entwicklung bestimmt der Lichtemitter die Eigenschaften im Sensor-System:

Zur Einrichtung von Maschinen sind Laserinterferometer mittlerweile eingeführte Geräte höchster Genauigkeit (10^{-7}), hierfür werden Kostennachteile in Kauf genommen (6).Für den routinemäßigen Einsatz in der

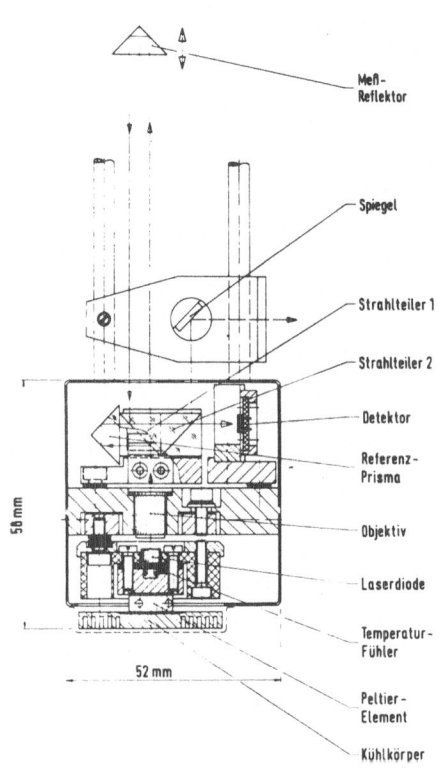

Fertigung, z.B. zur berührungslosen Abmessungskontrolle, würde man sie ebenfalls gern benutzten, allerdings besser in kompakter, stabiler Bauweise und zu geringerem Preis. Der Übergang vom He-Ne- zum Halbleiterlaser könnte dies leisten. Bild 1 zeigt eine auf LD-Basis entwickelte Version (7). Sie benutzt die eingangs erwähnte Hitachi-LD mit integrierter Fotodiode zur Emissionsstabilisierung und kommt damit zu einer Baugröße des Interferometerkopfes von nur 58 x 52 x 51 mm^3. Man verliert dabei zwar etwas an Meßgenauigkeit (1 µm/m=10^{-6}), deckt damit aber immer noch genügend interessante Einsatzfälle ab.

Labels in figure: Meß-Reflektor, Spiegel, Strahlteiler 1, Strahlteiler 2, Detektor, Referenz-Prisma, Objektiv, Laserdiode, Temperatur-Fühler, Peltier-Element, Kühlkörper, 58 mm, 52 mm

Bild 1. Konstruktionsprinzip eines miniaturisierten LD-Interferometers nach (7)

Optische Heterodyn-Doppler-Verfahren erweitern die Palette der laserinterferometrischen Verfahren zur Rauhigkeits- und Welligkeitsklassifizierung noch dahin, daß sie diese Messungen auch mehrachsig am rotierenden Werkstück ermöglichen. Bild 2 zeigt den prinzipiellen Aufbau eines kürzlich vorgeschlagenen Systems (8). Das optische Subsystem wirkt hier als selbstjustierender Sender und Empfänger für die Multifrequenz-Lasermeßstrahlen. Für gute Frequenzselektion und Signalaufbereitung sorgt eine Polarisationsoptik. Das Verfahren ist mit modernen Echtzeit- Datenverarbeitungssystemen (Tracking-Prozessoren, FFT-Spektralanalysatoren) in Verbindung mit Mikro- und Prozeßrechnern verwendbar.

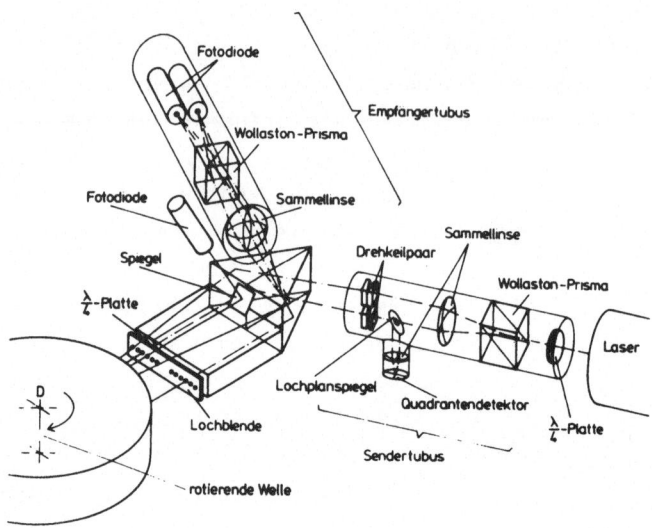

Bild 2. Prinzip eines optoelektronischen Sensor-Systems nach dem
Heterodyn-Doppler-Verfahren nach (8)

Zur berührungslosen mehrdimensionalen Vermessung bzw. Positionierung
von Werkstücken wurde ein optisches Lotungsverfahren auf LD-Basis vorgestellt (9). Gemäß Bild 3a wird das Licht einer mit 500 MHz hochfre-
quent modulierten LD über ein Dreh-/Schwingspiegelsystem auf das Werkstück rasterförmig abgebildet. Das diffus reflektierte optische Signal
wird in einer schnellen Fotolawinendiode in ein hochfrequentes
elektrisches Signal rückgewandelt. Die Phasendifferenz zwischen Emp-
fangs- und Modulationssignal ist ein Maß für den Laufweg des Lichts und
damit auch für die Höhe des reflektierten Oberflächenelements. Mit
diesem System läßt sich noch aus 2-3 m Abstand LD/Werkstück eine
dreidimensionale Profilbestimmung eines $0,5 \times 0,5\ m^2$ großen Werkstücks
mit 256 x 256 Bildpunkten und einer Auflösung ◀ lmm (lateral und
vertikal) innerhalb einiger 1/10 s erreichen. Bild 3b zeigt eine reale
Szene und Bild 3c das mit diesem optoelektronischen Sensor-System
gewonnene Höhenrasterbild (hier als S/W- Fotografie eines Fehlfarben-
bildes mit 16 Farb- bzw. Höhenstufen). Das System wird z.Zt. für den
Einsatz in der Handhabungstechnik in Verbindung mit Industrierobotern
adaptiert.

In einer Reihe anderer Entwicklungen bestimmt der optoelektronische
Empfänger die Systemeigenschaften:

In einfacher Kombination von konventioneller Lichtquelle und Einzeldioden sind mehrere Varianten auf dem Markt. Oft werden auch bereits

(a)

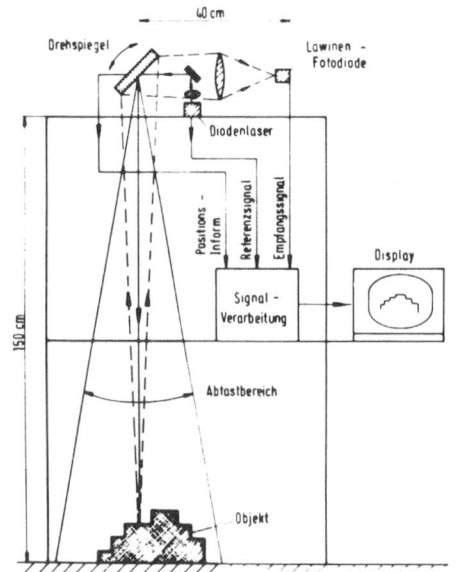

Bild 3. Optoelektronisches
Lotungsverfahren nach
(9): Prinzip (a),
Foto einer realen
Szene (b) und zuge-
höriges optoelektro-
nisches Rasterbild (c)

(b)

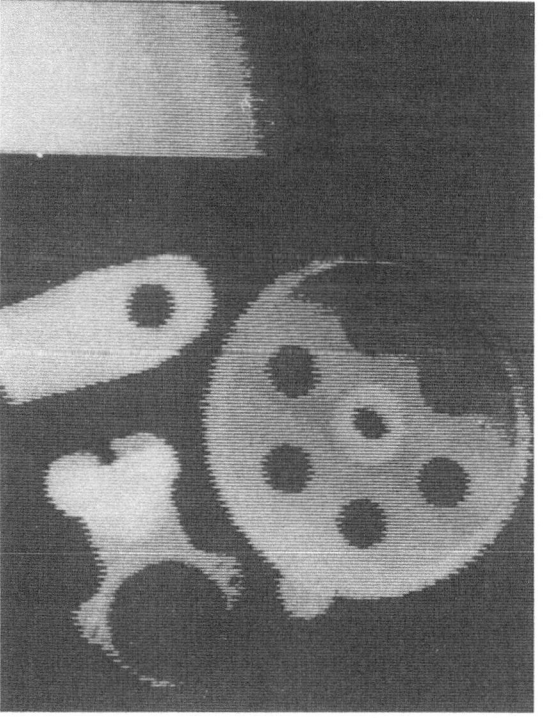

(c)

718

Lichtwellenleiter einbezogen, so z.B. bei dem in Bild 4 skizzierten Prinzip (10). Das Licht wird hier über eine Glasfaser zum Werkstück geleitet. Das von dessen Oberfläche reflektierte Licht gelangt über eine zweite Faser zum Empfänger. Eine Auswerteelektronik korreliert die Intensitätsänderung mit Fehlern wie Verfärbungen oder Oberflächenrissen noch in 0,05 mm^2 kleinen Bereichen.

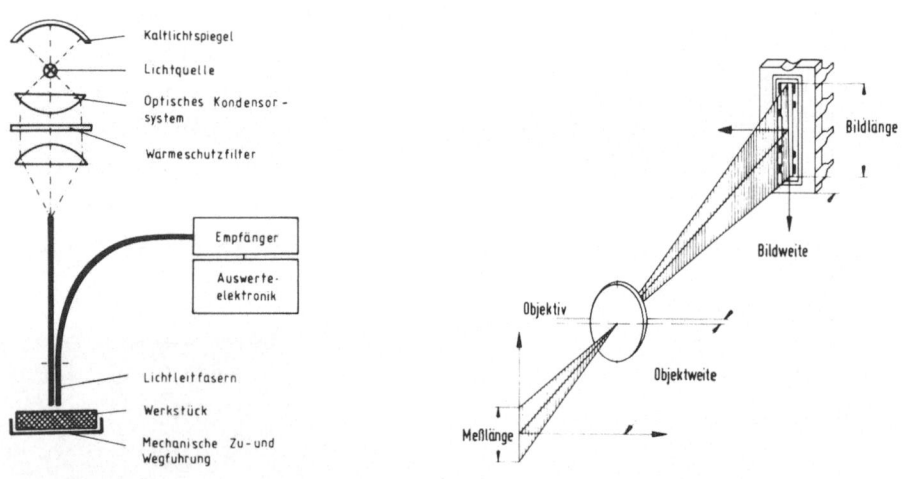

Bild 4. Prinzip des VOLPITRONIC   Bild 5. Meßanordnung mit CCD-Array

Geometrisch höher auflösende Systeme setzen CCD-Zeilen oder -Arrays ein, und mit verfeinerter Auswertung läßt sich diese Auflösung noch unter die Bildelementgröße (vgl. Tabelle 1) drücken (11). Sensor-Systeme für die flächenhafte Prüfung von Bauteilen lassen sich bereits mit zwei linearen Arrays in orthogonaler Anordnung aufbauen. In (11) wurde dieses Prinzip zur Kontrolle von Außen- bzw. Innendurchmesser und Konturverlauf von Stanzteilen eingesetzt. Über einen Meßbereich von 15 mm konnten Teile mit 0,05 mm Abweichung sicher erkannt werden. Die Bildabtastzeit betrug dabei 0,5 ms, entsprechend einer maximalen Durchlaufgeschwindigkeit der Stanzteile von 300 m/s.

Durch Hinzunahme von Verfahren der modernen Bildverarbeitung kann man diese Systeme dahingehend ausbauen, daß sie der Konzeption für eine voll automatisierte Fertigung schon recht nahe kommen. Einige grundsätzliche Überlegungen dazu sind z.B. in (12) zusammengefaßt. In (13) wird ein funktionsfähiges System vorgestellt, das seine große Flexibilität aus der nachgeschalteten Signalverarbeitungselektronik bezieht.

Seine Leistungsfähigkeit wurde bei der Kontur-Überprüfung von Ätzteilen mit ± 12 μm Genauigkeitstoleranz demonstriert und auf der Hannover-Messe 1985 vorgeführt.

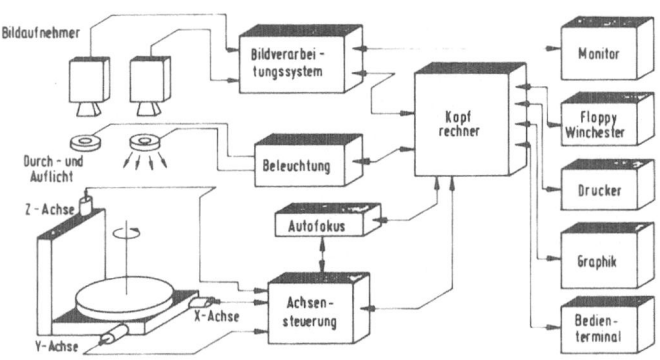

Bild 6. Systemkonzept des GEOPRÜF

3. Sensor-Systeme für das automatische Schweißen

Die Automatisierung des Schweißens als Fügeprozeß gewinnt besonders seit Einführung des Roboters in der Fertigung an Gewicht. Die Umstellung vom manuellen auf das automatisierte Schweißen erfordert in der Regel einen höheren Aufwand bei der Positionierung des Werkstücks. Zum Beispiel müssen schwankende Fertigungsmaße des Werkstücks und wechselnde Temperaturverzüge in situ ermittelt und dem Roboter als Führungsdaten übergeben werden. Meßtechnisch bedeutet dies die kontinuierliche Bestimmung von Breite und Lage bzw. Verlauf der Schweißnaht während des Schweißens in unmittelbarer Nähe des Brenners.

Dieses komplizierte sensorische Problem läßt sich optoelektronisch lösen.

Eine sensorisch einfache, in der Signalverarbeitung allerdings aufwendige Lösung ist die Beobachtung mit üblichen Bildwandlern (Fernseh-, Halbleiterkamera), deren Überstrahlungsempfindlichkeit durch besondere Belichtungstechniken zurückgesetzt wird. In Verbindung mit programmierbaren Bildverarbeitungseinheiten sind derartige Systeme spezifischen Anwendungsrandbedingungen (z.B. Nahtform) anpaßbar (14).

Einfacher wird die Signalverarbeitung bei Sensorsystemen, die Diodenzeilen verwenden. Bild 7a zeigt ein solches System, das die im Fugen-

bereich herabgesetzte Auflichtreflexion ortsaufgelöst (hier mit 256
Bildelementen) mißt (15). Die mikroprozessorgesteuerte Analyse der
nahtformspezifischen Sensorsignale nach Bild 7b stellt den Bezug zwischen detektierter, aktueller Position der Fugenmitte im Meßfeld zu
einer Referenzposition her und übergibt den ermittelten relativen
Nahtversatz nach Größe und Richtung der nachgeschalteten Steuerung zur
Schweißbrennerführung im on-line-Betrieb.

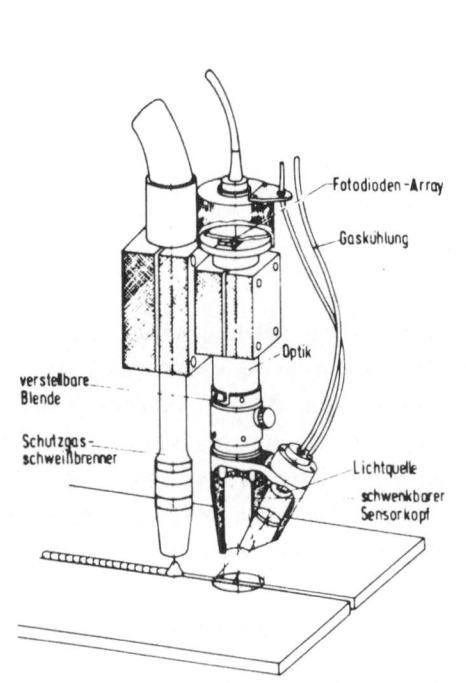

(a)

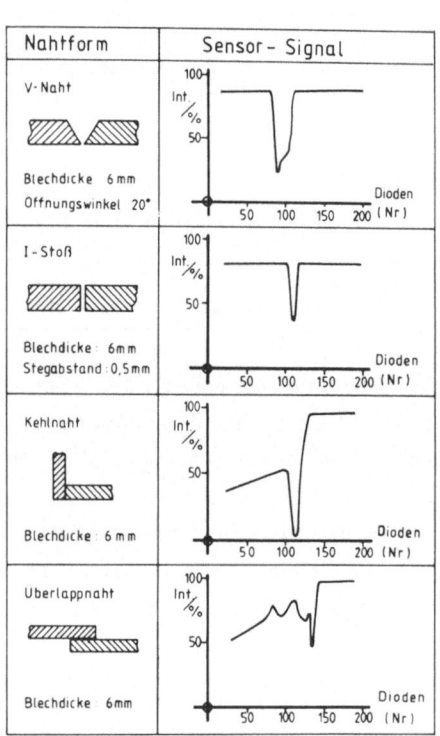

(b)

Bild 7. Prinzip (a) und Signalformen (b) eines optoelektronischen
Sensorsystems mit Dioden-Array zur Schweißnahtverfolgung

Geometrische Auflösung und Signalbildungsgeschwindigkeit lassen sich
noch steigern, wenn kombinierte Laser/Diodenzeilen-Anordnungen eingesetzt werden. Nach dem in (16) vorgestellten Prinzip (Bild 8) wird die

(diffus) reflektierte Laserstrahlung auf nur einige Detektorelemente konzentriert und damit die höhere Winkelauflösung erreicht. Zugleich sorgt die Intensitätkonzentration dafür, daß verwertbare Signalstärken bereits nach einigen 1/10 ms zur Verarbeitung anstehen. Das zu vermessende Nahtprofil wird über eine Entfernungsmessung durch schnelles Schwenken der Meßebene gewonnen. In einer typischen Auslegung für die Nachführung eines Schweißroboters wird ein Winkelbereich von 40 $^{\circ}$ mit 10 Hz Taktfrequenz abgetastet, wobei jedes Profil aus 200 Entfernungsmessungen aufgebaut wird. Ein derartiges System erreicht geometrische Auflösungen um 0,2 mm und wird von der Advanced Robotics Corp./USA seit Ende 1984 auch kommerziell angeboten.

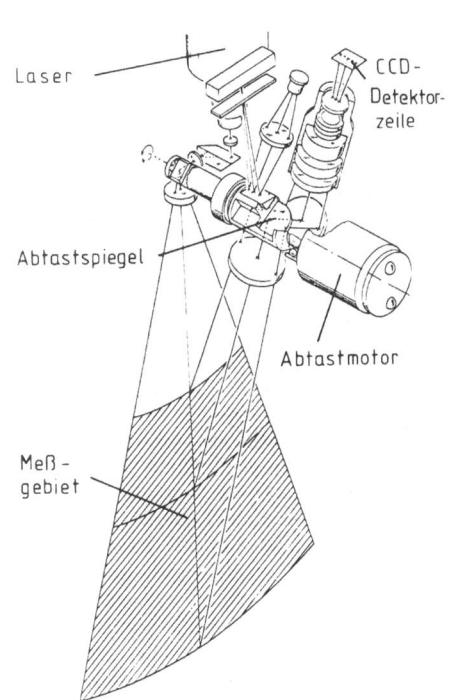

Laser

CCD-Detektorzeile

Abtastspiegel

Abtastmotor

Meß-gebiet

Bild 8. Optoelektronischer Profilsensor für Schweißroboter

5. Ausblick

Für die hier skizzierten repräsentativen Beispiele gibt es weltweit Parallelen. Vorsprünge des einen oder des anderen Landes gibt es auf Teilgebieten, insbesondere durch Japan und die USA bei Halbleiterlasern und -detektoren, aber m.E. nicht bei kompletten Systemen. Das liegt vor allem daran - und trifft für die Bundesrepublik Deutschland zu -, daß bei hoher Verfügbarkeit und geringem Preis der Komponenten einerseits und den kleinen Geräte-Stückzahlen in zudem anwendungsspezifischer Auslegung andererseits sich das Systemgeschäft geradezu als Domäne flexibler mittelständischer Unternehmen anbietet (17).

Die bekannte anwendungstechnische Problematik optischer Systeme bezüglich Verschmutzung und mechanischer Robustheit werden natürlich zu bedenken bleiben. Die optische Technik kann aber von sich aus zur Behebung ihrer Nachteile beitragen, indem der Einsatz störungsunempfindlicher Lichtwellenleiter und justierfester integriert-optischer Komponenten (Strahlteiler, Ablenker, Modulatoren etc.) konsequent betrieben wird. Zusätzlich werden auch günstigere Randbedingungen in der Fabrik der Zukunft und der automatisierten Fertigung technologisch hochwertiger Produkte von selbst entstehen, so daß sich optoelektronischen Sensor-Systemen eher noch bessere Marktchancen eröffnen werden.

Literatur

(1) W. DUTSCHKE: "Entwicklungstendenzen in der Fertigungsmeßtechnik". Technisches Messen 50, 171 (1983)

(2) H.-J. WARNECKE, W. DUTSCHKE (Hrsg.): "Fertigungsmeßtechnik". Springer-Verlag (1984)

(3) J. HESSE: "Sensoren als Grundelemente der Automatisierungstechnik".In: M. SYRBE, M. THOMA (Hrsg.): Messen-Steuern-Regeln 10, 1 (1983)

(4) K. MELCHIOR, G. PAVEL: "Automatisieren von Prüfvorgängen mit bildverarbeitenden Sensoren". Technisches Messen 50, 185 und 225 (1983)

(5) G. SEGER, U. SCHEIDING, H. SCHMALFUSS: "Qualitätskontrolle durch automatisierte Sichtprüfung". messen + prüfen, 312 (1984)

(6) H.-H. SCHÜSSLER: "Laserinterferometrische Längenmeßtechnik".Technisches Messen 52, 225 (1985)

(7) H.-J. BOEHNEL, H. HÖFLER: "Miniaturisiertes Halbleiter-Interferometer für die industrielle Wegmessung". Technisches Messen 52, Heft 9 (1985), in Druck

(8) O. WEGNER, M. HORSTMANN: "Optoelektronisches Sensorsystem zur mehrachsigen Geschwindigkeitsmessung rotierender fester Oberflächen". Technisches Messen 52, 106 (1985)

(9) G. GRABOWSKI, J. MOLNAR, W. SCHWEIZER, L. UNGER: "Optische Lotung für dreidimensionale Werkstückerkennung". Technisches Messen 51, 227 (1984)

(10) P. JOST: "Berührungslose optische Oberflächenprüfung in der Praxis". Feinwerktechnik & Meßtechnik 92, 375 (1984)

(11) G. JOBS: "Lineare Halbleiterbildsensoren zur berührungslosen Erfassung geometrischer Merkmale. In: M. SYRBE, M. THOMAS (Hrsg.): Messen-Steuern-Regeln 10, 37 (1983)

(12) G. STEIN: "Automatische Erkennung von Oberflächenfehlern mit Methoden der Bildverarbeitung". Technisches Messen 52, 67 (1985)

(13) H.-J. WARNECKE, C. KEFERSTEIN: "Optoelektronisches Sensorsystem mit automatisierter Bildverarbeitung". Technisches Messen 52, Heft 9 (1985), im Druck

(14) R. NIEPOLD, F. BRÜMMER: "Visuelles Sensorsystem zum Einsatz beim Lichtbogenschweißen". Technisches Messen 51, 264 (1984)

(15) P. DREWS, G. Starke: "Opto-elektronischer Nahtführungssensor zur Schweißnahtverfolgung". Technisches Messen 51, 270 (1984)

(16) L.H.J.F. BECKMANN: "Elektrooptischer Sensor für die Automatisierung des Lichtbogenschweißens". Technisches Messen 51, 259 (1984)

(17) "Sensoren für die Fertigungstechnik; Marktübersicht". Infratest Industria, München (1985)

Optoelektronische Bildanalyse, -verarbeitung
und -speicherung
Optoelectronic Image Analysis, – Processing
and – Storage

Automatische Bildanalyse im industriellen Bereich: Problemstellung, Lösungsmethoden, Anwendungen

Rainer Ott

AEG-TELEFUNKEN Forschungsinstitut

Sedanstr.10, 7900 Ulm

Einleitung

Die moderne Technologie ist dabei, wie die schon heute installierten, höchst automatisierten Fertigungsstraßen zeigen, in den Bereich industrieller Fertigungstechnik einzudringen. Kennzeichnend für den Stand dieser Technik ist, daß für den ordnungsgemäßen Ablauf eines Fertigungsprozesses weitgehend starre Gegebenheiten vorliegen müssen, was etwa durch die Notwendigkeit der präzisen, positionsgenauen Zuführung von Teilen bei der Montage verdeutlicht werden kann. Technische Systeme, die die Funktion des Sehens realisieren, so wie sie dem Menschen dabei hilft, sich in der Umwelt zu orientieren, und die eine weitaus fexiblere Fertigung ermöglichen würden, finden sich bis jetzt noch selten. Grund dafür ist, daß zur automatischen Bildanalyse außerordentlich große Datenmengen mit teilweise hochkomplexen Algorithmen innerhalb angemessener Zeit verarbeitet werden müssen. Erst die umwälzenden Fortschritte auf dem Gebiet der modernen Mikroelektronik haben die Voraussetzungen für die aufwandsgünstige technische Realisierung automatischer Bildanalysesysteme geschaffen.

Der Vortrag befaßt sich mit den Entwurfskonzepten für automatische Bildverarbeitungs- und Analysesysteme im industriellen Bereich. Die Systeme können eingesetzt werden zur automatischen Sichtsteuerung von Industrierobotern, bei Palettierungs-, Montage- und Greifaufgaben und zur Qualitätskontrolle in der Fertigung. Im Mittelpunkt der Darstellung steht die Methodenentwicklung, die Bereitstellung von Verfahren und Algorithmen, die imstande sein sollen, das von einem optischen Sensor aufgenommene Bild der industriellen Szene in eine gewissermaßen verbale Beschreibung umzusetzen, auf die dann Entscheidungen über den weiteren Ablauf eines automatischen Prozesses gestützt werden können.

Gesamtablauf einer automatischen Bildanalyse

Entprechend dem Typ der jeweils zu verarbeitenden Information wird zwischen dem ikonischen und dem symbolischen Teil der Bildverarbeitung unterschieden. Beim ikonischen Teil sind sowohl Eingabe wie Ergebnis bildhafte Informationen. Im symbolischen Teil wird ausschließlich auf symbolische Beschreibungen bereits analysierter Bildobjekte sowie auf Attribute, die diese Objekte näher beschreiben,

728

Bezug genommen. Dazwischen gibt es den Übergang von der ikonischen zur symbolischen Darstellung.

Die ikonische Bildverarbeitung besteht aus den Teilschritten Bildaufnahme, die etwa mit Laser-Scannern, Röhren- oder Halbleiterkameras realisiert werden kann, Bildvorverarbeitung, mit deren Hilfe unerwünschte Bildbestandteile weitestgehend unterdrückt und Strukturen hervorgehoben werden sollen, die für die weitere Verarbeitung von Interesse sind. Segmentierung, in der die Berandungen von Gebieten oder Objekten ermittelt wird, und Bildobjektvermessung, in der charakteristische Merkmale von Objekten bestimmt werden, schließen sich an. Der Übergang von der Ikonik zur Symbolik wird durch Prozeduren geschaffen, die die bildhafte Information in die Form von Bildgraphen umsetzen. Die Knoten der Graphen stellen die in dem Bild entdeckten Objekte dar, die Kanten repräsentieren jeweils die geometrische Beziehung "Gebiet liegt innerhalb von". Begleitparameter sind Attribute der Knoten und stellen die erfaßten charakteristischen Meßdaten an den Bildobjekten dar. Aus den im Bildgraphen enthaltenen Daten können jederzeit gezielt Objekte mit vorgegebenen Eigenschaften angesprochen und wieder bildhaft dargestellt werden.

Im symbolischen Teil der Bildverarbeitung geht es im wesentlichen um die Auswertung und Verarbeitung abstrakter Datentypen. Dabei spielen Modellbildung, Wissensakquisition, Wissensrepräsentation und eigentliche Wissensverarbeitung eine Rolle. Über diese Themenkreise wird heute unter der Überschrift künstliche Intelligenz viel diskutiert. Von besonderer Bedeutung für die Bildverarbeitung sind dabei die verschiedenen Graphsuchverfahren.

Methoden zur Realisierung wichtiger Bildanalyse-Verfahrensschritte

Eine wichtige Stellung im Bereich der ikonischen Bildverarbeitung nehmen die lokalen Operatoren ein. Mit ihnen können lineare und nichtlineare Filter realisiert werden. Das Prinzip beruht darin, die Funktionswerte eines Ergebnisbildelementes aus den Werten von Originalbildelement und Bildelementen der Nachbarschaft zu berechnen. Je nach Wahl der Abbildungsfunktion und des Nachbarschaftsraumes können so die verschiedensten Bildfilterprozesse realisiert werden. Häufig verwendete und wichtige Varianten sind Hochpassfilterung, zur Betonung von Bildfeinstrukturen, Tiefpassfilterung zur Betonung von Bildgrobstrukturen, Gradientenfilterung zur Hervorhebung von Bildobjektkanten, Erosion und Dilatation zur Beseitigung einzelner dunkler bzw. heller Flecken, Medianfilter zur Beseitigung von Störungen bei Erhaltung der Bildschärfe /1/.

Charakteristisch für die Realisierung derartiger Filterprozesse ist, daß sehr große Datenmengen verarbeitet werden müssen. Damit verbunden ist ein relativ hoher Rechenaufwand, der bei zeitkritischen Aufgaben häufig den Einsatz schneller, speziell zur Bearbeitung der ganz konkreten Aufgabe ausgelegter Prozessoren erfordert.

Die Bildanalyse im industriellen Anwendungsbereich zielt in vielen Fällen darauf ab, Objekte in der Bildszene zu erkennen, sie zu vermessen und deren Drehlage und Position zu bestimmen. In einfachen Fällen entspricht das Bildobjekt direkt einem Realweltobjekt - z.B. einem Werkstück. In einem solchen Fall ist die Objektkontur das wichtigste Kennzeichen zur Beschreibung, die Klassifikation kann unmittelbar auf der Basis der am Objekt gemessenen Merkmale und Merkmalskombinationen durchgeführt werden. Hierzu sind Objektmerkmale wie Fläche, Umfang und Formfaktor geeignet. Spezielle Konturmerkmale, die die Berandung des Objektes beschreiben, sind die Fourierkoeffizienten und daraus abgeleitete normierte Größen. Diese Merkmale haben die Eigenschaft, Lage und Gestaltparameter zu trennen. Objektkonturen, die durch Drehung, Verschiebung und Skalierung aus einer anderen Kontur hervorgehen, haben die gleichen Fourierdeskriptoren. Klassifikation und Drehlagenbestimmung sind damit ganz einfach möglich.

Bei komplexen Objekten, bei denen das Bildobjekt aus zahlreichen Teilobjekten besteht, ist die Methode der "modellgesteuerten Bildanalyse" der wichtigste Lösungsansatz. Ausgangspunkt ist bei diesem Verfahren das Modell des zu erkennenden Objektes. Es enthält eine Liste der Teilobjekte aus denen es zusammengesetzt ist. Für jedes Teilobjekt ist der gemessene Satz charakteristischer Merkmale sowie die Lagerelation zu den übrigen Teilobjekten abgespeichert. Im ersten Schritt der Analyse werden im auszuwertenden Bild auf der Basis der gemessenen Merkmale Kandidaten für jedes Teilobjekt des Modells ermittelt. Im weiteren Verlauf wird schrittweise eine Zuordnungskonfiguration von Kandidaten zu Modellteilobjekten aufgebaut. Für jede Konfiguration wird ein Bewertungsmaß berechnet, das dann besonders gut ist, wenn Merkmale sowie Lageparameter der jeweiligen Modellteilobjekte und zugehörigen Kandidaten weitestgehend übereinstimmen. Unter allen überhaupt möglichen Zuordnungskonfigurationen wird diejenige mit bestem Bewertungsmaß gesucht. Da die Zahl dieser möglichen Zuordnungskonfigurationen wegen der kombinatorischen Vielfalt sehr schnell anwächst, müssen Strategien angewandt werden, die den Suchaufwand klein halten. Verfahren, die dieses leisten, wurden auf dem Gebiet der künstlichen Intelligenz entwickelt /2/.

Anwendungen und Beispiele

Die Bilder geben Beispiele für industrielle Bildszenen, wie sie automatisch zu analysieren sind. Bild 1 stellt ein Grauwertbild eines Motorblockes dar. Die Aufgabe des Sichtsystems besteht in der Bestimmung der Lage und Orientierung von Motorblock und Zylinderbohrungen zur Steuerung eines Portalroboters, der die auf Paletten liegenden Blöcke entstapeln soll. Bild 2 zeigt das Ergebnis der Bildvorverarbeitung, die Grauwertänderungen deutlich hervorhebt. Dieses Beispiel stammt von einem sichtgesteuerten Robotersystem, das sich in einem großen Automobilwerk in praktischem Einsatz befindet.

Bild 3 zeigt ein durch geeignete Bildvorverarbeitung und Quantisierung entstandenes Binärbild eines Stanzteiles, in dem deutlich typische schwarze und weiße Zusammenhangsgebiete erkennbar sind. Davon wurden einige ausgesuchte zur Modellbildung des Teiles verwendet. Mit einem an unserem Forschungsinstitut aufgebauten Experimentiersystem kann jedes einzelne Teil der Bildszene von Bild 4, in der sich mehrere Teile ungeordnet überlappen, erkannt und Position und Drehlage bestimmt werden. Realisiert wird dies mit der modellgesteuerten Bildanalyse unter Anwendung des A-Stern Algorithmus /2/.

Literatur

(1) W.K. Pratt: Digital Image Processing (1978)
(2) Nils J.Nilsson: Principles of Artificial Intelligence (1982)

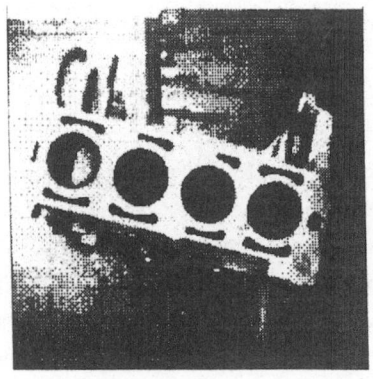

Bild 1.

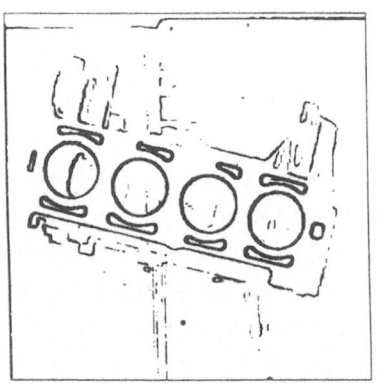

Bild 2.

Bild 3.

Bild 4.

Optical Methods of Information Storage

M. Hartmann
Philips GmbH Forschungslaboratorium Hamburg, D-2000 Hamburg 54, F.R.G.
A.M.L. Spruijt
Philips Optical Media Development SPG Eindhoven, The Netherlands
B.A.J. Jacobs
Philips Research Laboratory Eindhoven, The Netherlands

1. Introduction

The application of a focussed, scanning laser beam for data storage and retrieval is referred to as optical recording. The concept of this 'bit-oriented' optical storage has been under development since the early seventies. The Laser Vision video-disk system (1) as well as the more recently introduced digital audio-disks (2) offer areal information densities up to about 100 Mbit/cm^2, one to two orders of magnitude higher than in magnetic recording media. Both are playback (read-only) systems, offering combined video and audio and high-quality audio, respectively. One major advantage of the optical disk is its contactless read-out. A laser is focussed into a small spot (full width half maximum 0.8 µm) on the disk, thus reading the information. The distance between the objective lens and the information carrying layer is several millimeters; this allows for a transparent substrate of approximately 1 mm thickness to be situated between the storage layer and the reading head. The information layer is located inside a substrate sandwich. Small scratches or dust do not affect the reading quality, since the outer surface is out of focus.

Video and audio disks prepared by a replication process of a master recording are utilized for data distribution (3,4). The present optical recording research and development are directed towards storage systems, using information layers that can be used to 'write' or 'write and erase' optically readable effects, enabling data storage for any user. For this purpose a variety of optical readable effects can be thermally induced in a large number of materials, but only a few combinations are of practical interest. The restrictions on the storage materials are set by the typical system requirements that will be discussed later on.

After giving some general characteristics of optical recording in chapter 2, the material aspects of read-only, write-once and erasable storage will be reviewed in chapter 3. Since the optical video disk and the digital audio Compact Disk system are already commercially available products, their read-only media are only briefly discussed. Details of write-once and erasable disks are reported, with emphasis on the magneto-optical storage materials, these being the most promising candidates for erasable storage right now. Information about some applications in the market with respect to optical storage techniques is presented in chapter 4.

2. General principles

a) The disk

Optical read-out principles can be applied to any type of physical effect that changes the state of the incoming light upon reflection. The transmissive mode is generally not used in practice and thus will not be discussed here. In addition to the phase structures as used in Laser Vision and Compact Disk media, reflectivity and polarization changes are applied for write-once and erasable storage media. The effects used for the latter media are all thermally induced, due to a local raise in temperature as a result of the absorption of the power from the focussed laser spot in the storage layer. Just to give a rough idea: An optimum read-out signal must be created by physical effects in the medium within 50-100 ns with incident power of about 10 mW focussed into a 1 μm-sized spot with an pulse energy of 0.3-1.0 nJ and with a power density of about 10^6 W/cm^2. This energy is the laser power that can be delivered by the available semiconductor lasers in pulse operation, these being the most suitable light source for optical recording (5). This implies that the write-once and erasable storage media must have a reasonable optical absorption in the near infrared at emission wavelengths between 780 and 850 nm of available semiconductor lasers. By correcting the optical abberations of the emitted power in an optical laser pen set-up using aspheric lenses the focus dimensions can be reduced to the diffraction limit. The effectiveness determines the output power of the laser. An effective light pass is transmitting up to 25% of the output power to the disk. In order to create effects with a minimum amount of energy the volume to be heated must be as small as possible. Write-once and erasable media are used as thin films having a thickness range from 5-100 nm. The thin storage layers are deposited on a plastic or a lacquer covered glass substrate. To achieve an optimum sensitivity at the writing process, substrates of low thermal conductivity and very short write pulses are used to achieve adiabatic heating (6). At pulse times below 100 ns the radial heat flow in the plane of the thin storage films (e.g. metals, see chapter 3) is so small that the effective size is almost completely determined by the diameter of the spot. This implies that similar as with read-only applications the storage density is determined by the optical cut-off frequency 2 NA/λ (NA numerical aperture of the focussing objective lens, λ = wavelength).

Another very important requirement of information written on a storage disk is a sufficient signal to noise ratio (SNR) to enable error-free read-out of stored information. The field of application for a storage medium is strongly depending on the SNR value (and the bandwidth needed), which is mostly determined by the intrinsic material properties. For analogue (pit length modulated) video storage, a much higher SNR is required than for digital (pulse length or pulse-position-modulated) data storage. The read-out signal is determined by the optical contrast between a written spot and its surrounding and the amount of reflected light received by the detector. In most practical storage systems, the dominant contribution to noise is produced by the storage medium itself. Typical sources of disk noise are surface or layer/substrate interface roughness and small variations in the domain shape formed after writing processes.

Optical data storage is often performed for archival purposes in contrast to magnetic storage. The required and nowadays established media life ranges up to 10 years or even more. This aging is studied by accelerated life tests and depends on the system concept in which the medium is used. In digital data storage, error correction techniques are applied to restore erroneous data that are produced due to aging effects. The end of life of media is due to an increase of the error rate by a factor of 100 to 1000. Raw error rate are initial in the rate of 10^{-5} to 10^{-6}, corrected error rates are in the range of 10^{-12}-10^{-20} and the end of life raw error rate in the range of 10^{-3}.

b) The optical system

An optical disk player/recorder can be seen as a microscope which scans the pits or written bits on the rapidly rotating disk. The information is stored as a sequence of depressions or pits (Laser Vision, Compact Disk) or as written bits (data storage) in the form of a spiral or a set of concentric circles on the disk. During read-out the position of the pit or bit centers and their geometries reproduce the time dependent signal that was originally recorded on the disk.

The distance between the disk and objective lens has to be dynamically adjusted well within the focal depth of about 2 μm. The read-out beam must have a rather high numerical aperture (e.g. 0.4-0.6) in order to obtain sufficient resolution. The demands on the vertical and radial positioning of the light spot are severe; servosystems are indispensable. The tolerances for tracking and focussing have to be kept below a certain

signal. Different methods for focussing and tracking error signal generation have been developed (11,12).

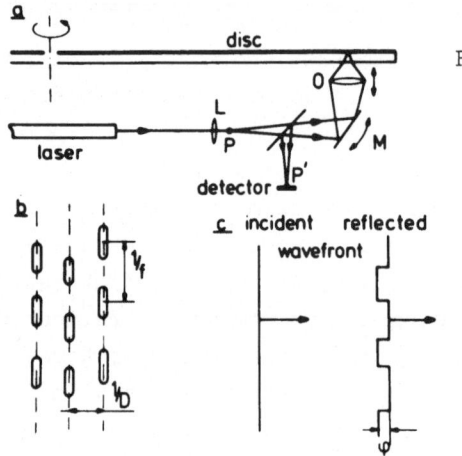

Fig. 1. a) Optical read-out scheme:
Upper side is the infor-
mation-carrying side of
the disk. Mirror M is par
of a servoloop for radial
tracking. The objective O
is positioned by the focu
servo.

b) Usual track configuration
1/D = track density; f =
spacial frequency.

c) Phasefront of an incident
plane wave and of the re-
flected wave. φ = phase-
difference in the reflect
ed light.

Fig. 1 shows a schematic representation of an optical read-out arrangement. An auxiliary lens L expands and applies optical correction on the
laser beam in order to fill the aperture of the objective O. The primar
focus P of the lens L is imaged by the objective O and the moving mirro
M to a spot on the information carrying surface of the rotating disk
which is placed on a turntable of the main drive-motor. The storage
medium is deposited on a pre-grooved substrate for tracking and focussing purposes. For some Philips products a method is employed based on
a photopolymerization process known as the '2p process'. The pre-groove
structure is replicated from a stamper made with a galvanic process fro
a master plate in a UV-curable 2p-lacquer deposited on a glass substrate (7). Pre-grooving is often combined with pre-formating. The pre-
recorded information is replicated in the same way as the pre-grooved
spiral 'tracks'. In plastic products the information is stored using a
high precision moulding technique (i.i.m.) using similar stampers.

3. Description of the three optical systems for information storage on disks

a) Read-only optical disks

With optical storage techniques 1 hour continuous video (100.000 video
frames $\approx 10^{11}$ digital bits) (8) can be stored on one side of a 30 cm
diameter Laser Vision disk (9,10). The light source of the optical head
is a HeNe laser, which emits a collimated monochromated beam of red
light (λ = 633 nm). The information on the disk is encoded in a spiral
track of small depressions (pits) of varying length and distance in ana
logue form. The amount of light that is reflected during read-out varie
due to the fraction of pits along a track representing the encoded vide

information (11). On the Compact Disk (2) using a diode laser as a light
source the signal is recorded in a similar manner, but the information
is present in the track in digital form. Only a limited number of defined pit lengths are used. Each pit and each land represent a series of
(channel-) bits. The density of the information on the Compact Disk is
also very high, 1 hour of digital audio is stored on one side of a 12 cm
disk (2-8 GB data). The channel bit of audio information covers an area
of 1 μm^2. In table 1 the specifications of the present Video and Audio
Disk are summarized. For details of the typical system aspects of the
mentioned read-only optical data storage applications, the reader is referred to the references of this chapter. Some product-oriented information will be discussed in chapter 4.

Table 1. Specifications of read-only optical disks

	Video Disk	Digital Audio Disk
Laser	HeNe	Semiconductor Diode
Wavelength	633.0 nm	780/850 nm
NA	0.4	0.45
Diameter	30.0 cm	12.0 cm
RPM	1500[a]	480-210[b]
Thickness	1.25 mm	1.2 mm
Track pitch	1.66 μm	1.6 um
Pit depth	0.12 μm	0.12 μm
Pit width	0.4 μm	0.6 μm
Pit length	0.5-2.0 μm	0.9-3.3 μm

[a] PAL video, constant angular velocity mode;
[b] Constant linear velocity mode, 1.25 m/s

3. b) Write-once and erasable optical disks

In all write-once and erasable storage materials of practical interest,
the effects are thermally induced by absorption of the power in the laser
spot that is focussed on a light absorbing material. Up to now, no practical implementation of any photon-induced effect in optical media like
photocromic   and photo-dichroic materials (13) exists. Therefore, this
chapter will be restricted to commonly applied, thermally-induced effect
and the related material combinations. Most widely used are ablative recording, bubble formation, phase-change and magneto-optical effects.

Ablative recording:

When archival aspects are important, the erasability of media is not essential, if may even be undesired. It is the typical application of write-once effects, of which hole burning in thin-films is probably the most studied phenomenon. Hole burning has been studied in materials like In, Bi, Te, various chalcogenides and some organic dyes (14,15). These are materials having a low melting point and a high optical absorption in the infrared. The hole formation process strongly depends on the thermal properties of the film, the film thickness and surface tension effects. It has been concluded from experiments and calculation that the temperature at which hole formation occurs e.g. in Te and In on a plastic substrate, can be slightly above the melting point of the material (6,14). In Te as the most sensitive recording material, the boiling point at 1263 K is almost reached in hole-opening times of less than 10 ns. A molten area in a thin film does not necessarily lead to a hole as it is shown in the SEM photograph in Fig. 2.

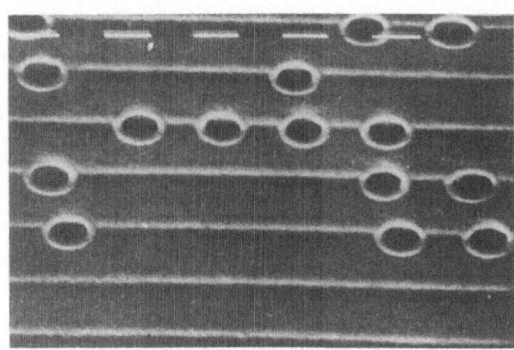

Fig. 2. SEM photograph of a DOR disk. The data a recorded in the continuous groove as micrometre sized holes in the reflective metallic layer. The white stripes are 1 μm long.

The reason for hole formation is a reduction of the total energy which is a function of the hole geometry and of the surface and interfacial energies of the thin film material and the substrate. Due to oxidation storage, layers of pure Te lack the stability required in archival storage applications. Alloying Te with e.g. Se and other chalcogenides lead to a stability that is sufficient for archival use of these media.

In order to optimize the sensitivity and stability, further elements like As, Sb can be added (16). Typical disk parameters are given in table 2 in comparison with the phase change and magneto-optical materials discussed at the end of this chapter.

Another important class of recording effects uses bubble formation (17) The sensitive layer consists of a polymer-metal bi-layer. The laser

light is absorbed in the metal, the heat is transferred to the polymer
which decomposes at elevated temperatures. Due to the resulting gaseous
emission the metallic layer separates from the polymer and deforms in-
elastically the metallic layer into a bubble with a diameter almost
equal to the spot size. The bubble has a permanent character and acts
similar as a pit, also introducing a phase shift of the reflected light.

Phase change recording:

Erasable-optical memory disks can be based on reversible amorphous-crystalline phase transitions in tellurium suboxide (18) and tellurium alloys (19). Usually information is stored by writing amorphous regions
in a crystalline structure when it is molten and cooled down rapidly.
Both states show a different index of reflectivity resulting in an optically detectable effect. The stability of an amorphous region is less
than that of a hole. It is determined by the amorphous-crystalline phase-
transition activation energy and kinetics, which depends strongly on the
composition. The requirements on a phase-change material is also included in Table 1. Erasure is achieved by recrystallization via thermal
annealing. Recently reported results described the possibility of single-
pass erasure by a laser beam in a chalcogenide system of Te in a TeO_2
matrix with Ge and Sn additives (18). Due to an optimized Sn concentration the dynamic of recrystallization is changed such that the erasure
after a 500 ns pulse of 0.8 mW/μm^2 incident power density can be achieved. Writing occurs in 50 ns with 6 mW/μm^2. Due to the difference in time
the rise in temperature differs. Single-pass erasure, though, can be
performed using separate laser spots, one for writing and reading and
another one for erasure; both are projected through the same objective
lens. Erasable optical memory media based on phase-change materials are
still in a research phase.

Magneto-optical recording:

During the last few years the suitability of amorphous magneto-optic
rare earth (RE)-transition metal (TM) alloys for erasable optical recording has been established (20-22). Magneto-optic recording combines
the advantages of optical storage with the reversible characteristic of
magnetic media. The magnetic structure strongly resembles that used in
'vertical' magnetic recording (recent trend in high-density magnetic
storage). Digital information is stored as a sequence of small circular
magnetic domains ($\leq$ 1 μm Ø). Writing and erasure is carried out by a

local temperature rise induced by an AlGaAs laser pulse in combination
with an oriented external magnetic field. The number of switching cycles
is almost unlimited; cycles have been performed over 10^6 times without
any measurable degradation of the storage layer (22). Non-destructive
read-out is achieved using the magneto-optical Kerr effect. The magneto-
optic medium consists of a thin film of a ferro- or ferrimagnetic material having a strong magnetic uniaxial anisotropy perpendicular to the
film plane (positive anisotropy constant K_u). The local temperature rise
by a laser pulse of e.g. 50 ns and about 5 mW focussed on the storage
layer introduces a local sensitivity of the material to the low external
magnetic field and can be written or erased depending on the field direction. The writing characteristics are determined by the temperature
dependence of the intrinsic magnetic material properties (Fig. 3). At
the end of a write pulse, the heated area cools down and keeps a permanent structure as a small magnetic domain. Fig. 4 presents the thermo-
magnetic process in principle. Fig. 5 demonstrates written domains on a
disk (with pre-grooves) observed with a polarization microscope. The sta
bility of stored information is mainly determined by the value of the
coercive energy that prevents domain wall motion. The switching properties and stability of the storage layers are strongly related to the
film composition (23,24) and preparation conditions (25).

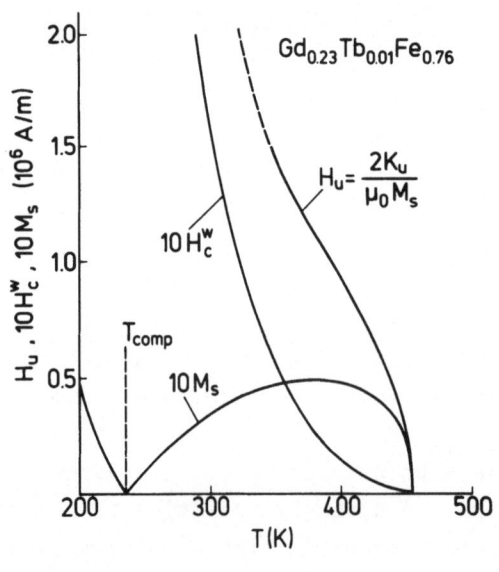

Fig. 3. Temperature dependence
anisotropy field H_u,
wall coercivity
H_c^ω and saturation
magnetization M_s.
At the compensation
temperature T_{comp} the
M_s-values is zero

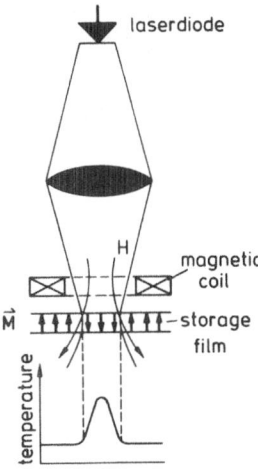

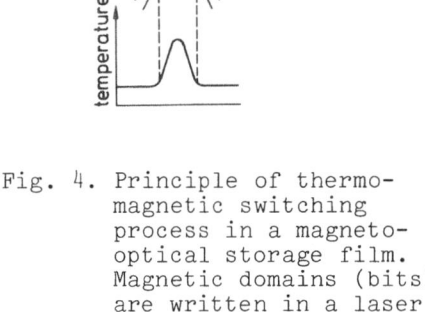

Fig. 4. Principle of thermo-
magnetic switching
process in a magneto-
optical storage film.
Magnetic domains (bits)
are written in a laser
heated spot with in-
fluence of an external
magnetic field

Fig. 5. Written domains (bits) on a
magneto-optic pre-grooved
disk, taken with a polarizing
microscope. Periodicity of
grooves is 1.6 μm

The corrosion sensitivity of RE-TM alloys has to be low in order to achieve a sufficient lifetime of the disk. The most promising way to improve the stability is to protect the magneto-optic layer in a sand-wich structure wich dielectric and/or metal coatings. This also gives the possibility of enhancing the read out contrast. The thickness and optical constants of the individual layers forms an optical interference structure (22). Basically, the magneto-optic read and write head has the same characteristics as other optical heads, but additionally a polari-zation sensitive (differential) detection and a magnetic coil have to be introduced. Because of the limited magnetic field modulation frequen-cies attainable, single-pass erasure is not applied in the recently re-ported prototype performance of magneto-optical storage systems (26,27).

Erasure is performed sector by sector using a single beam head. Only the writing is done at the high data rate given in the next chapter. Right now magneto-optical recording combines erasable recording with random access and high density on removeable disks. Table 2 presents a comparison of the thin film requirements for the mentioned optical write-once and erasable optical media.

Table 2. Comparison of thin film requirements for optical recording

	Ablative	Phase-Change	Magneto-optical
Substrate	Glass + 2P plastic i.i.m.	Glass + 2P plastic i.i.m.	Glass + 2P plastic i.i.m.
Deposition	magnetron sputtering	flash evaporation	E-beam evaporation, sputtering techniques
Film	TeSeSbS	TeSeSb	GdTbFe, TbFeCo
Thickness (nm)	30	100	80
Reflectivity (S.I)	35%	35%	43%
Film-Structure	Poly cryst.	Poly cryst.	Amorphous
Protection	Air Sandw.	Top coating	Sandwich, T.C.
Sensitivity	0.8 nJ	0.4 nJ	0.5 nJ
CNR; 1 MHz (BW 30 kHz)	58 dB	55 dB	53 dB
Data Storage	Holes	amorphous regions	Magnetic domains.

4. The optical disk applications

The current performance in magnetic and optical storage in general optical disk systems offer the capability of high density information stor age and retrieval including random access and removeability as another advantage over tape and film. The Laser Vision system (50.000 still pic tures per side) covers picture and sound applications for consumer entertainment, education and training, program and video data-base distri bution. The Compact Disk covers sound as well as digital data distribution.(CD-ROM). Digital optical recording provides digital data storage for documents (e.g. 500.000 typed DIN-A4 pages or 1.000 DIN-A4 digital images in dot representation) for archival data per disk side at 1 Mbyt image base. Systems are built for mass storage with fast retrieval up t 10^{11} bytes (juke-box). It is expected erasable optical storage being right now in the development phase, may enter the market soon and may b adapted for use in personal computers for home applications (BTX, gra phics) or even still picture systems as an ideal medium for storing and distributing computer software and to reduce storage cost per Mbyte e.g CAD/CAM applications. This may also be a competition for drives in some areas where capacity is more important than very fast access times (optical is typically 75-150 ms and Winchester drives 15-50 msec access time). Fig. 6 summarizes the current and realistically expected areal b densities in conventional and optical recording.

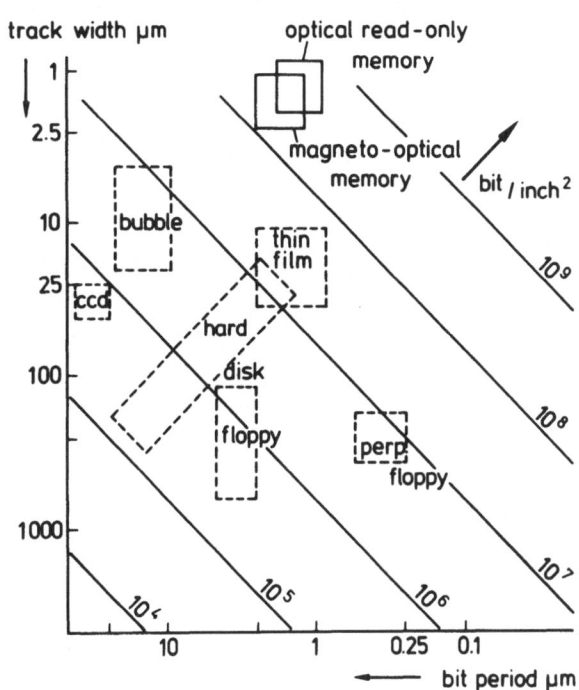

track width μm

optical read-only memory

magneto-optical memory

bit / inch²

bubble

thin film

ccd

hard

disk

floppy

perp floppy

bit period μm

Fig. 6. Areal bit density of proposed recording techniques

The market will ultimately determine the actual form of the products and its applications and areas of use. The next years will certainly show whether the technological advantage of optical media will be exploited in the sense mentioned by E.S. Rothchild in one of the recent Technology Reports: 'History has shown that no matter how much storage is offered, users find data to fill up the memories and then ask manufacturers for more'.

References

(1)   G. GOUWHUIS, J. BRAAT: Appl. Opt. and Opt. Engineering IX;
      R. SHANNON, J. LOYANT: eds. Academic Press 3, N.Y. (1983)
(2)   M. CARRASSO, J. PEEK, J. SINJOU: Philips Techn. Rev. 40, 151 (1982)
(3)   B. JACOBS: Appl. Optics 17, 2001 (1978)
(4)   H. HAVERKORN, V. RIJSEWIJK, P. LEGIERSE, G. THOMAS: Philips Techn.
      Rev. 40, 297 (1982)
(5)   K. BULTHUIS, M. CARRASSO, J. HEEMSKERK, P. KIVITS, W. KLEUTERS,
      P. ZALM: IEEE Spectrum, August issue (1979)
(6)   P. KIVITS, R. de BONT, P. ZALM: Appl. Phys. 24, 273 (1981)
(7)   A. v.d. BROEK, H. HAVERKORN, V. RIJSEWIJK, P. LEGIERSE, G. LIPPITS,
      G. THOMAS: J. Rad. Curing 11, 1 (1984)
(8)   K. COMPAAN, P. KRAMER: Philips Techn. Rev. 33, 178 (1973)
(9)   G. BOUWHUIS, J.J.M. BRAAT: Appl. Optics and Optical Engineering
      IX, 73 (   )
(10) B. HOEKSTRA: Phys. Technol. 14, 241 (1983)
(11) J.J.M. BRAAT, G. BOUWHUIS: Appl. Optics 13, 2013 (1978)
(12) G. BOUWHUIS, J.J.M. BRAAT: Appl. Optics 17, 1993 (1978)
(13) L. RALSTON: SPIE Proc. 420, 186 (1983)
(14) P. KIVITS, R. de BOUT: Appl. Phys. 24, 307 (1981)
(15) J. WROBEL, A. MARCHANT, D. HOWE: Appl. Phys. Lett. 40, 928 (1983)
(16) D.J. GRAVENSTEIJN, C. STEENBERGEN, J. van der VEEN: SPIE Proc.
      420, 327 (1983)
      A. HUIJSER: SPIE Proc. 382, 270 (1983)
(17) R. FREESE, R. WILLSON, L. WALD, W. ROBBINS, T. SMITH: SPIE Proc.
      329, 174 (1982)

(18) M. TAKENAGA, N. YAMADA, S. OHARA, K. NISHINCHI, M. NAGASHIMA,
      T. KASHIHARA, S. NAKAMURA, T. YAMASHITA: SPIE Proc. 420, 173
      (1983)
(19) C.M.J. van UIJEN: Opt. Mass Data Storage Conf., Los Angeles (1985)
(20) T. DEGUCHI, H. KATAYAMA, A. TAKAHASHI, K. OUTA, S. KOBAYASHI,
      T. OKAMOTO: Appl. Opt. 23, 3972 (1984)
(21) F. TANAKA, Y. NAGAO, N. IMAMURA: IEEE Trans. Magn. 30, 1033
      (1984)
(22) M. HARTMANN, J.J.M. BRAAT, B. JACOBS: IEEE Trans. Magn. 20,
      1013 (1984)
(23) J. BIESTERBOS: Journal de Physique 40, C5-274 (1979)
(24) M. URNER-WILLE, P. HANSEN, K. WITTER: IEEE Trans. Mag. MAG-16,
      1188 (1980)
(25) H. HEITMANN, M. HARTMANN, M. ROSENKRANZ, H.J. TOLLE:
      M.M.A. '85 Conference, Grenoble, paper A3-I (1985), to be
      published in J. Physique

High Resolution CCD-Line-Sensor-Camera for Multispectral Environmental Imaging

H. Seidlitz, S. Berber, P. Hutzler, Department Applied Optics, Gesellschaft für Strahlen- und Umweltforschung mbH München, Ingolstädter Landstr. 1, 8042 Neuherberg

The effect of air pollutants on plant growth is presently the subject of a comprehensive experimental program. Many new results are expected from trees grown in so called exposition boxes under controlled fumigation.

These studies require continous optical monitoring of plant growth both for quick analysis and for documentation purposes. The amount of data to be handled makes digital data processing desirable and electronic imaging is therefore advantageous. The requirements for an adequate imaging system are

- high spatial resolution
- small geometric distortions
- linear response
- spectral range 400 - 1000 nm.

We found that the most economic approach would be a linear charge coupled device (CCD) sensor which is scanned mechanically over the entire field.

The camera developed to meet the requirements stated above is shown in Fig. 1.

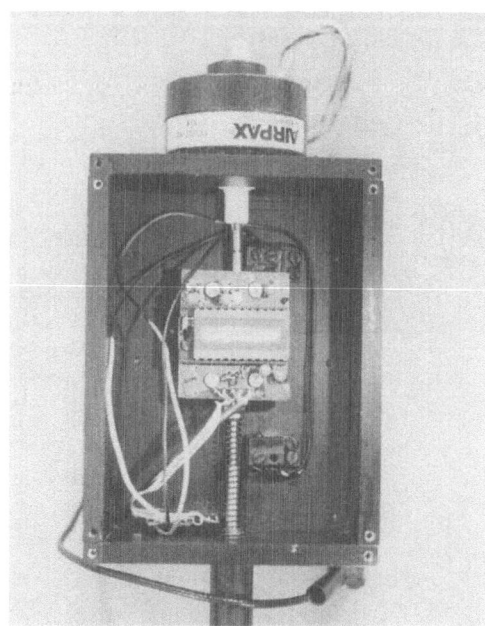

fig. 1. CCD camera front view

The CCD sensor with 1024 elements at 13 µ distance is in the center. Fig. 2 presents the side view. Stepper motor, driving mechanism and camera electronics can be seen. The sensor is advanced by 25 µm each step. The dimensions of the camera head are 12 x 9 x 9 cm³ (without lens).

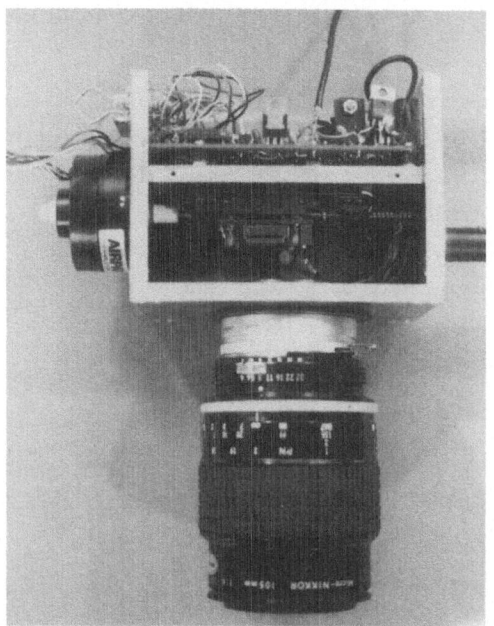

fig. 2. CCD camera front view

The schematic circuitry of the complete system is shown in Fig. 3.

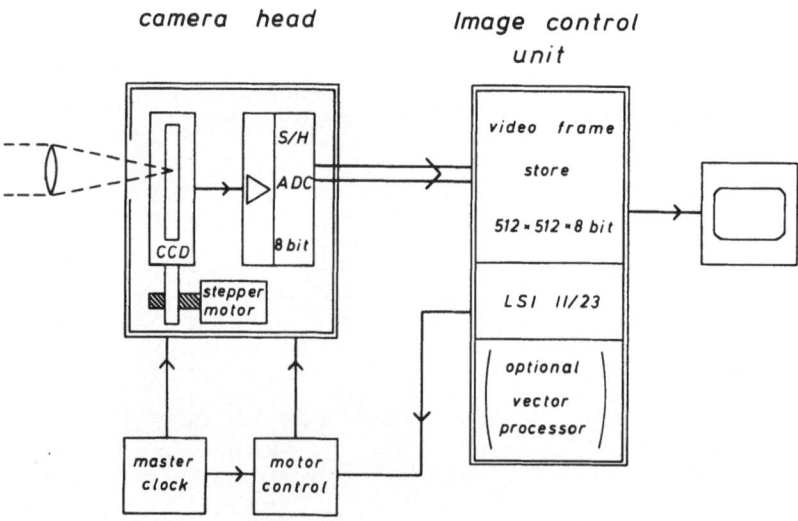

fig. 3. CCD imaging system schematic diagram

The camera head contains CCD chip, stepper motor and analogue components as amplifiers, sample hold circuit (SH), and an 8 bit analogue digital converter (ADC). The control box houses power supply (not shown) and a master clock wich provides pixel transport and line transfer pulses and clock pulses for the motor control unit which also resides in the box.

Due to slow scanning (several seconds per frame) real time display is not possible and a image memory is necessary. The memory size is 512 x 512 x 8 bit therefore only every second pixel can be used. The whole system is controlled by an LSI 11/23 lab computer. An additional vector processor allows fast image operations e.g. shading correction segmentation, contrast enhancement etc.

Optical and electrical data are listed in table 1 and 2 respectively.

optical specifications		electrical specifications	
picture elements	512/line	SNR (detector)	3500
lines	512	dynamic range	8 bit
pixel size	13 x 25 μm^2	integration time	20 – 80 ms
dimension of image plane	approx. 13x13 mm^2	video frame time	10 – 40 s
resolution (measured)	17 lp/mm		
spectral range	400 – 1000 nm		

table 1.   table 2.

The number of picture elements and number of lines are only limited by memory size. The fastest video frame time is limited by the stepper motor.

The analysis of a spoke target image proves that resolution is equal both in horizontal and vertical direction.

Color separations required for plant analysis can be obtained by inserting appropriate filters at 480, 550, 680 and 800 nm. These particular wavelengths have been obtained from reflexion spectra of various conifer needles.

Fig. 4 shows the four multispectral separations of a small spruce tree as it is grown in the exposition boxes. The vertical bars in the center of each picture belong to a grey chart. Typical for healthy vegetation is the pronounced reflexion in the near infrared range (700 - 900 nm) which is between 30 and 60% whereas the plant reflects only 15% to 20% of green light. The green separation shows some brighter branches at the top. These are fresh shoots.

746

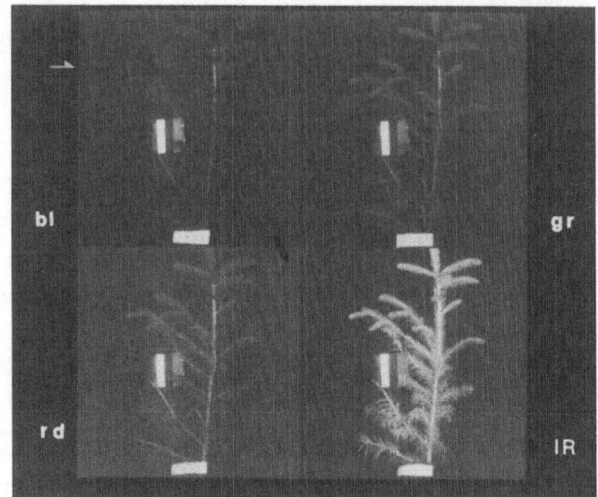

fig. 4. blue. green, red and infrared images of a spruce tree

A histogram of brightness taken from that area (fig. 5) yields the distribution
of grey levels of these fresh shoots. Converting these particular grey level into
white produces density slice emphasizing all fresh shoots. The total "shoot area"
can also be obtained using simple counting operations. Other characteristics e.g.
diseased needles or branches can be analysed in a similar fashion.
The images from each of the four bands can also be combined to produce true or
false color composites.

fig. 5. green separation with histogram and density slice

Einzelthemen
Individual Papers

Verzeichnungsarme Aufnahme einmaliger lichtschwacher Vorgänge

B. Lieberoth-Leden; W. Pfeiffer; H. Reinhard; D. Wittmer

Fachgebiet Elektrische Meßtechnik, Technische Hochschule Darmstadt

Schloßgraben 1, D - 6100 Darmstadt

Einleitung

Unterstützt durch das starke Vordringen der Digitalrechner und deren Einsatz zur Bildverarbeitung nimmt die Bedeutung der Videomeßtechnik ständig zu. Die mit der digitalen Bildverarbeitung mögliche Genauigkeit läßt auch verstärkt den Wunsch nach entsprechend leistungsfähigen Bildaufnehmern aufkommen.

Für viele Anwendungsfälle ist die geometrische Verzeichnungsfreiheit von großer Bedeutung, wie zum Beispiel im wissenschaftlichen Bereich bei der Aufzeichnung von Spektren. Die bei normaler Beleuchtungsstärke verwendeten Bildaufnehmer (z.B. Vidikons, Newvicons, CCDs) haben im allgemeinen gute geometrische Linearität, im Gegensatz zu SIT-Röhren, die bei geringen Beleuchtungsstärken oder kurzen Belichtungszeiten häufig verwendet werden. Als Alternative wird hierzu eine Kamera mit einem Newvicon und vorgesetztem Nahfokus-Bildverstärker vorgestellt. Für eine anschließende Einzelbildauswertung ist die Kombination der Kamera mit einem Video-Bildspeicher zweckmäßig. Zur Erfassung kurzzeitiger, nicht mit der Kamera synchronisierbarer Leuchterscheinungen, wurde ein Halbleiter-Bildspeicher mit Bildpunkttriggerung entwickelt.

1. SIT-Röhre und Newvicon mit Nahfokus-Bildverstärker im Vergleich

Untersucht wurden zwei ähnliche Kameras, Bosch, TYC 9A mit 1"-SIT-Röhre (RCA) und TYK 9A mit 1"-Newvicon und glasfasergekoppeltem zweistufigem Bildverstärker (Nahfokus-Prinzip). Die SIT-Röhre besteht aus einem Multidiodenvidikon, dessen Target ca. 0,6 Millionen Dioden enthält. Davor befindet sich ein elektrostatischer Bildverstärker (Bild 1).

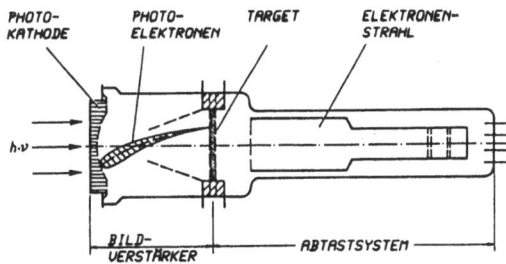

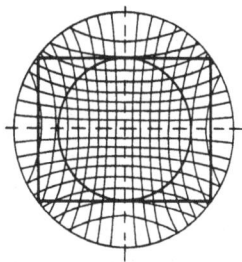

Bild 1. Prinzipieller Aufbau eines SIT-Vidikons

Bild 2. Verzeichnung eines elektrostatischen Bildverstärkers

Dem hohen Empfindlichkeitsgewinn von ca. 1000 der SIT-Röhre im Vergleich zu einem
Vidikon stehen allerdings erhebliche kissenförmige Verzeichnungen gegenüber, wie sie
in Bild 2 schematisch dargestellt sind /1/. Diese Fehler und zusätzliche Unschärfe
im Randbereich entstehen durch die unterschiedlich langen Wege der Elektronen zwischen der Photokathode und dem Target. Während die Photokathode zur Kompensation der
Wegdifferenzen konkav gewölbt ist und über Glasfasern in ein planes Bildfenster angekoppelt wird, muß zur Anpassung an das Abtastsystem des Multidiodenvidikon-Teils
das Target eben ausgebildet sein. Bild 3 zeigt die gemessenen geometrischen Verzer-
rungen der untersuchten SIT-Kamera in horizontaler Richtung und in Abhängigkeit vom
Ort. Man sieht deutlich, daß in den oberen und unteren Randbereichen die Streckung
des Bildes am größten ist. Aber auch in der mittleren Zeile beträgt die Verzeichnung
am Bildrand noch mehr als 10 %. Wie in /2/ ausführlich beschrieben wurde, sind deshalb mit dem Einsatz einer SIT-Kamera, z.B. in der Spektroskopie, gravierende Nachteile verbunden, die sich auch mit digitaler Bildverarbeitung nur schlecht kompensieren lassen.

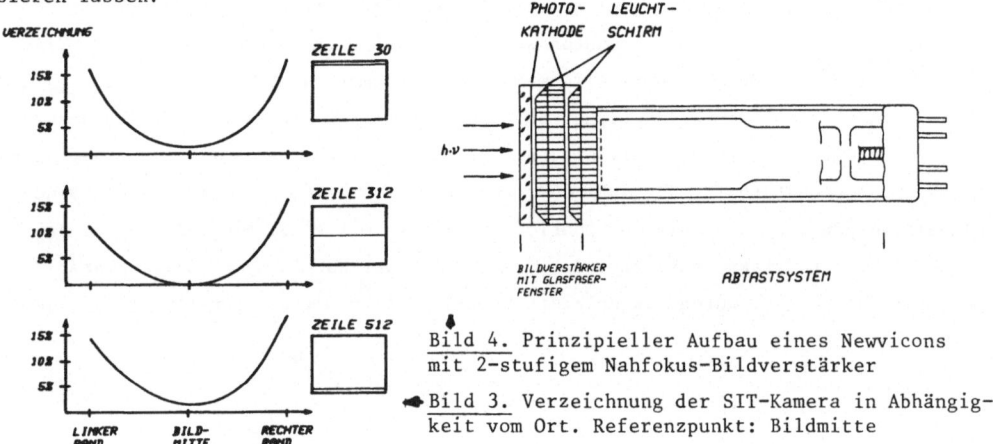

Bild 4. Prinzipieller Aufbau eines Newvicons
mit 2-stufigem Nahfokus-Bildverstärker

Bild 3. Verzeichnung der SIT-Kamera in Abhängigkeit vom Ort. Referenzpunkt: Bildmitte

Die zweite untersuchte Kamera besitzt ein 1"-Newvicon mit Glasfasereingang. Über diesen wird ein zweistufiger Nahfokus-Bildverstärker angekoppelt (Bild 4). Beim Nahfokus-Bildverstärker befindet sich die als Leuchtschirm ausgebildete Anode planparallel
in wenigen Millimetern Abstand zur Kathode. Dadurch erhält man eine praktisch ver-
zeichnungsfreie Abbildung. Bei der untersuchten Kamera lag sie unter 1 %, begrenzt
durch die Linearität des Ablenksystems des Newvicons.

Die Auflösung der Kamera wird im wesentlichen durch die Faseroptiken und insbesondere deren Übergänge, die Nähefokussierung, die Körnigkeit des Phosphors und durch
Kontrastminderung aufgrund der Streuung im Bildverstärker bestimmt /3/. Für das Ge-
samtsystem wurde eine Grenzauflösung von 18 Lp/mm gemessen (Herstellerangaben für das
Newvicon allein: 34 Lp/mm, /4/). Bei der SIT-Röhre ist die Auflösung im wesentlichen
durch die elektrostatische Elektronenoptik bestimmt. Dadurch ergibt sich im Randbereich eine deutlich schlechtere Auflösung verglichen mit der Röhrenmitte. Als Auf-
lösungsgrenze wurde in der Röhrenmitte ein Wert von 22 Lp/mm ermittelt.

Häufig werden Restlichtkameras zusammen mit anderen Bildwandlern (z.B. Streak-Kameras, getastete Bildverstärker) verwendet, die in der Regel einen P20-Phosphorschirm am Ausgang haben. Deshalb wurde die Empfindlichkeit der Kameras beim Maximum der relativen spektralen Strahlungsverteilung vom P-20 Phosphor gemessen. Die SIT-Kamera benötigt zur Vollaussteuerung 0,67 nW/cm^2 (= 4,7 mLux) und damit ein Drittel weniger Bestrahlungsstärke als das Newvicon mit zweistufigem Bildverstärker. Der höheren Empfindlichkeit der SIT-Kamera steht dafür allerdings ein um den gleichen Faktor kleinerer Dynamikbereich gegenüber (Bild 5). Der Dynamikbereich hat aber gerade bei meßtechnischen Problemen entscheidende Bedeutung.

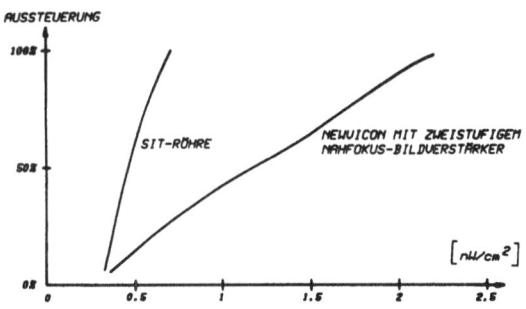

<u>Bild 5.</u> Aussteuerung in Abhängigkeit von der Bestrahlungsstärke mit monochromatischem Licht (λ = 546 nm)

2. Aufzeichnung einmaliger transienter Vorgänge mit Videokameras

Die Fernsehbilder einmaliger transienter Leuchterscheinungen müssen zur Betrachtung und Auswertung zwischengespeichert werden. Hierfür kommen analoge (z.B. Videorekorder) oder digitale Speicher (Halbleiterspeicher) zum Einsatz. Bei triggerbaren Vorgängen kann vom Aufzeichnungssystem der zu untersuchende Vorgang während der vertikalen Austastlücke gestartet werden. Beim nächsten Auslesezyklus wird dann ein vollständiges Fernsehbild ausgelesen. Schwieriger ist die Erfassung nicht triggerbarer Ereignisse. Hier befindet sich der Abtaststrahl zum Triggerzeitpunkt an einer beliebigen Stelle auf dem Target und die Auslesung beginnt an einer undefinierten Stelle. Damit enthält das gerade ausgelesene Halbbild nur die Information, die sich unterhalb der Lage des Abtaststrahls zum Triggerzeitpunkt befindet. Das folgende Halbbild enthält dann den darüberliegenden Bildteil und die beim ersten Halbbild noch nicht vollständig ausgelesene Restinformation (Bild 6a). Um in solchen Fällen ein vollständiges Fernsehbild aufzuzeichnen, kann man zum Triggerzeitpunkt den Abtaststrahl dunkel tasten und erst mit dem nächsten Bildsynchronimpuls wieder einschalten. In dem ausgetasteten Bildbereich wird aber der Dunkelstrom und das Rauschen der Bildverstärker doppelt so lange aufintegriert und macht sich gerade bei SIT-Röhren deutlich bemerkbar (Bild 6b). Außerdem ist ein Eingriff in die Kamerasteuerung erforderlich. Diese Nachteile lassen sich mit einem auf wenige Bildpunkte genau triggerbaren Bildspeicher vermeiden. Bild 7 zeigt das Blockschaltbild eines solchen Gerätes.

Aus dem Videosignal der Kamera werden in einer Synchronabtrennstufe die Horizontal- und Vertikalimpulse gewonnen. Das eigentliche Bildsignal gelangt nach Klemmung auf einen definierten Gleichspannungspegel an den 8-Bit-A/D-Wandler. Sofern im Videosignal Frequenzanteile oberhalb von 5 MHz, der halben Abtastfrequenz, vorhanden sind, müssen diese von einem Tiefpaß abgeschnitten werden, um Aliasing zu verhindern.

Bild 6. Halbbilder einer gepulsten Leuchtdiode bei verschiedenen Ausleseverfahren. Der Abtaststrahl befand sich zum Triggerzeitpunkt (→) zufällig in der Bildmitte.
a) 1. Halbbild nach dem Triggerereignis;
b) 1. Halbbild nach dem Triggerereignis bei zuvor ausgeschaltetem Abtaststrahl;
c) Vom Bildspeicher zusammengesetztes Bild. Die Informationen unterhalb des Triggerpunktes stammen aus dem laufenden, die darüberliegenden aus dem folgenden Halbbild

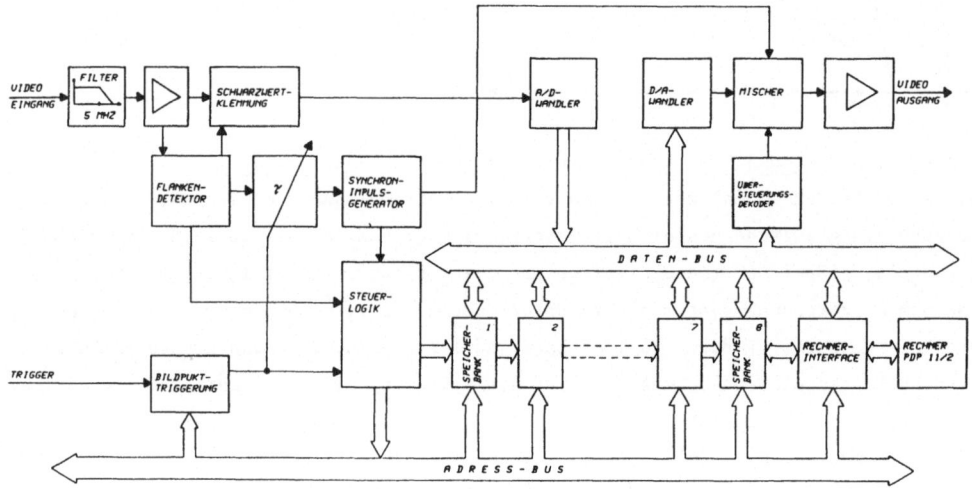

Bild 7. Blockschaltbild des auf einen Bildpunkt triggerbaren Bildspeichers

Das Gerät liest im Ruhezustand zyklisch die Speicherinhalte in den 8-Bit-A/D-Wandler. Beim Eintreffen des Triggerimpulses zu einem beliebigen Zeitpunkt speichert das Gerät die Adresse des momentan gerade ausgelesenen Bildpunktes, und das Gerät beginnt mit dem Abspeichern der digitalisierten Grauwerte. Der Einlesevorgang ist beendet, wenn die Ausgangsadresse wieder erreicht ist.

Das Speicherformat ergibt sich aus der Forderung nach einer Auflösung von ca. 5 MHz bei gleichzeitiger Ausnutzung der Formate handelsüblicher Speicher-ICs. Bei 512 Bildpunkten/Zeile ergibt sich eine Abtastfrequenz von f_o = 9,8 MHz, so daß die höchste auflösbare Frequenz nach Shannon f_g = 4,9 MHz bei einer theoretischen Modulationstiefe von 64 % gemäß der Modulations-Transfer-Funktion

$$MTF = \frac{|\sin(\pi f_g/f_o)|}{\pi f_g/f_o}$$

beträgt. Das Gerät kann im Zwischenzeilenverfahren 2 Vollbilder à 512 Zeilen oder im Nicht-Zwischenzeilenverfahren 4 Halbbilder à 256 Zeilen speichern.

In Bild 6c sieht man deutlich, daß die Information der unteren und der oberen Bildhälfte, die wie anfangs beschrieben, in verschiedenen Bildern vorhanden sind, wieder richtig rekonstruiert worden sind. Der auf einen beliebigen Bildpunkt triggerbare Bildspeicher ermöglicht auf einfache Weise die Untersuchung transienter Leuchterscheinungen mit Standard-Fernsehkameras ohne Eingriff in die Kamerasteuerung und ohne zusätzliches Dunkelstromrauschen.

Ein zuschaltbarer Aussteuerungsdekoder überprüft das eingelesene Bild auf Übersteuerung. Die übersteuerten Bildbereiche werden dann schwarz gezeichnet, so daß sie sich deutlich von den umliegenden Bildpunkten abheben (Bild 8).

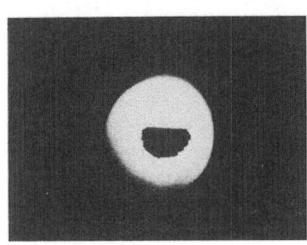

Bild 8. Übersteuertes Fernsehbild. Der Bildspeicher zeichnet die übersteuerten Bereiche schwarz

Literatur

/1/ Bildverstärkerröhren. Technische Information für die Industrie,
Valvo Firmenzeitschrift, Nr.: 780626, 1978
/2/ B. AULBACH: Entwicklung neuer elektrooptischer Meßverfahren zur Untersuchung des Verhaltens gasförmiger Isolierstoffe
Dissertation TH Darmstadt, 1984
/3/ D. WITTMER: Das Verhalten nähefokussierter Bildverstärker bei Tastung im Nanosekundenbereich. Dissertation TH Darmstadt 1985
/4/ Kameraröhren Datenbuch, Valvo, 1982

Phase Shifting TV Moire Interferometry to Investigate Quick Phase Fluctuations

Ervin Tanos, András Pilinyi
Research Institute of the Electrical Industry
Budapest, Hungary, 1601 Pf.45

Introduction

Our task was to investigate quick phase changes inside a fluctuating medium. For this purpose, the holographic interferometry is a good tool but it was not fast enough in our case. We have solved the question using a modified Mach-Zender interferometer, which contained a phase modulator. The interferometric fringe system was fixed on a TV recorder thus we could take some 50 pictures in a second, which is about 1500 times faster than holography.

The basic arrengement

The scheme of the interferometer is shown on Fig.1.

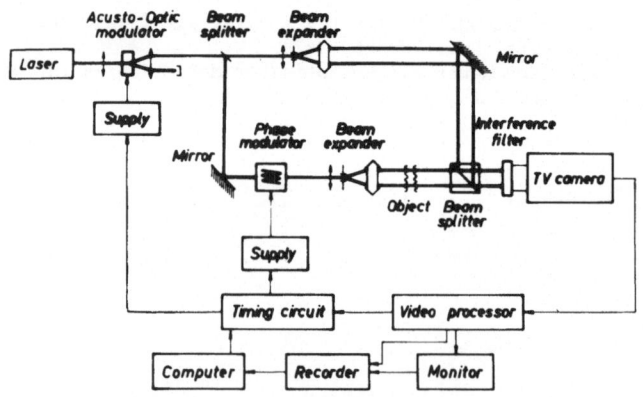

1. Figure
The scheme of the interferometer

It is a usual interferometer of Mach-Zender type as mentioned above.
The TV camera gives a synchronous signal at the beginning of   each

frame. Then the amplitude modulator switches on the light and the first exposure is fixed on the vidicon. After a short time, the phase modulator is turned on to shift the phase of the object beam by π, and the light is switched on for the second exposure. For one frame the vidicon tube stores the pictures, that is the sum of the intensity distributions, until the electron beam erases the information. Practically, both exposures have to be taken in the 1. quarter of the frame, i.e. the time between the exposures has to be less than 5 ms.

The theory

The theoretical principle of the interferometer is very simple. The basic equations are shown on Fig.2.

Original

$$I_1 = I_0 \cdot \cos^2 \phi(x)$$

Shifted

$$I_2 = I_0 \cdot \cos^2 \left[\phi(x) + \frac{\pi}{2} \right]$$

$$I = I_1 + I_2 = I_0$$

_____ × _____

Plane waves

$$\phi(x) = k \cdot x$$

Pert. plane waves

$$\phi(x) = k \cdot x + \psi(x)$$

$$I = I_0 \cdot \left[\cos^2(kx) + \sin^2(k \cdot x + \psi(x)) \right]$$

2. Figure
The basic equations

Obviously, if there is no change between the exposures, then the distribution of the summarized intensities is uniform. In case of phase changes this uniform picture is modulated and some fringes appear in the place of changes, there and only there. Fig.3. and Fig. 4. show some theoretical examples for this.

The experiment

In order to demonstrate the working ability of the interferometer some experiments were carried out. A part of the experimental setup

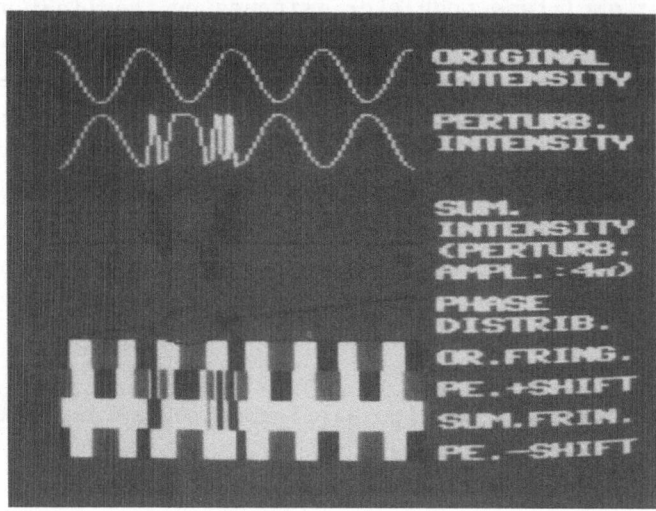

3. Figure
The theoretical intensity distributions and fringes

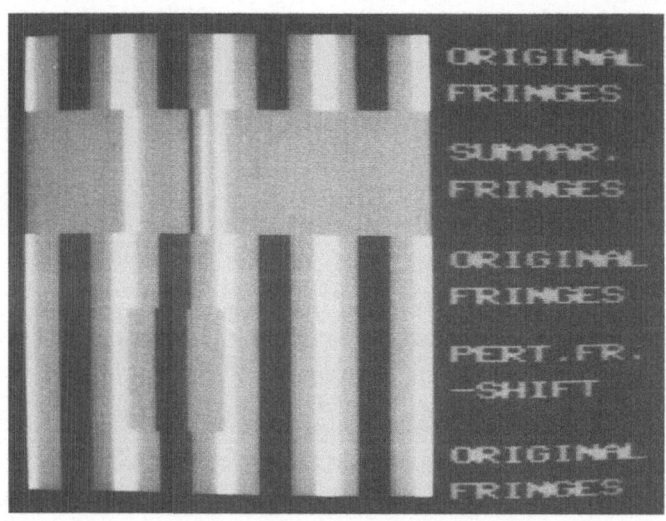

4. Figure
The theoretical fringes

is shown on Fig. 5. The model medium was free air convection surrounding a hot metal peak. The basic fringe systems are shown on Fig.6.,
without and with phase shift. Phase change occured when applying cold
air flow. These fringe systems are shown on Fig.7., without and with

phase shift.

5. Figure
A part of the experimental setup

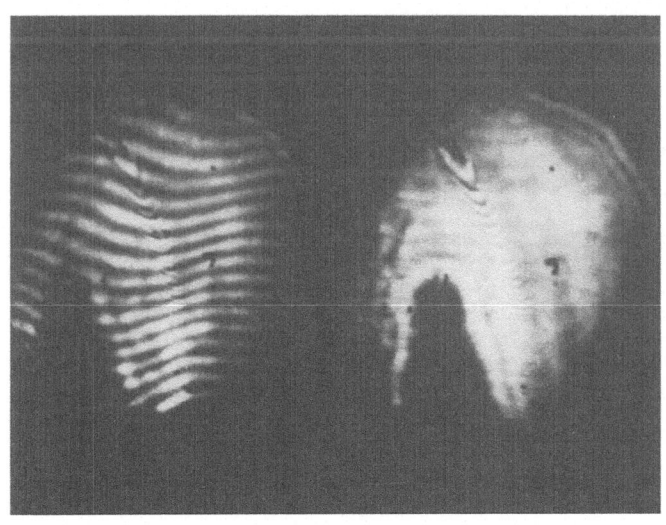

6. Figure
The basic experimental fringes
without phase shift   with phase shift

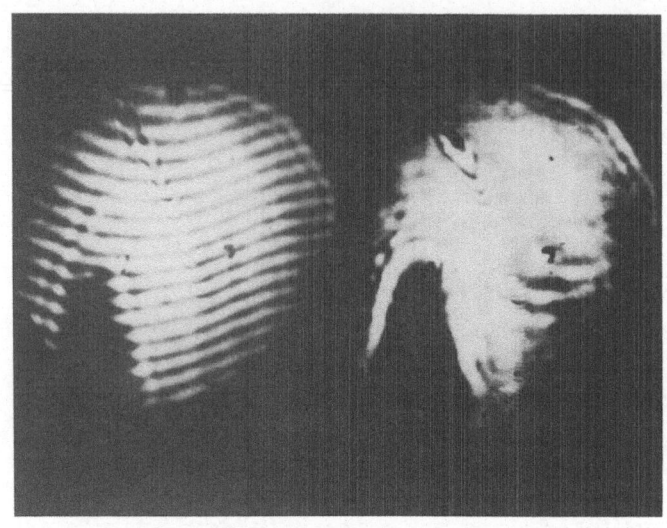

7. Figure
The perturbated experimental fringes
without phase shift   with phase shift

Summary

This interferometer is to be used if
   -one has to collect a lot of data,
   -the basic fringe system is unknown,
   -the characteristic time of phase changes to be investigated is less
     than that of the fluctuations and less than 5 ms.

Neue Entwicklungstendenzen in der Technologie der Reinigung von Laser-Komponenten und anderer optischer Elemente

H.R. Suter,CH„ und H. Knödler,D

Anhand eines Beispieles werde ich aufzeigen, mit welchen Entwicklungstendenzen in der Reinigungstechnologie für anspruchsvolle optische Komponenten in der Zukunft zu rechnen sein wird. Die wesentlichen Grundlagen meiner Aussagen basieren auf den Erkenntnissen, die in einem gemeinsamen Entwicklungsprojekt der Firmen Litton Systems, Canada, und ROAG, Fabrikationstechnik, Schweiz, erarbeitet wurden.

Meine Aussagen gliedern sich in:

- Aufgabenstellung
- Vorgehen und Prüfmethoden
- Ergebnisse
- Entwicklungstendenzen

Aufgabenstellung

Bei dem zu reinigenden Objekt handelt es sich um einen Ring-Laser-Gyro.
Die wichtigsten Zielsetzungen der Arbeiten waren:

- die konventionelle Handreinigung zu ersetzen
- ein stets gleich hohes Qualitätsniveau zu erzielen
- keine gesundheitsgefährdenden Chemikalien einzusetzen
- die Oberflächenqualität für das Ansprengen der Laserspiegel zu gewährleisten
- auf/im Zerodur keine Fremdelemente zu hinterlassen, insbesondere keine Fluoride oder Chloride
- gleichzeitig drei Gyros in einem Prozessablauf bearbeiten zu können

Vorgehen und Prüfmethoden

In einer ersten Phase wurden

- verschiedene Detergenzien auf ihre Reinigungswirkung und Verträglichkeit mit Zerodur geprüft
- die optimale Ultraschallkonfiguration für die in Frage kommende Badgröße in Abstimmung mit den drei zu reinigenden Gyros festgelegt

In einer zweiten Phase wurde

- die Reinigungswirkung in den Pathholes der Gyros geprüft
- das Reinigungsverfahren hinsichtlich Badabfolge,Zeiten, Konzentrationen festgelegt

In einer weiteren Phase wurde

- das allgemein bekannte und angewandte Trocknungsverfahren mit FKW durch ein neues, industriell anwendbares Verfahren mit Stickstoff ersetzt und die entsprechenden Konstruktionsarbeiten ausgeführt.

Die Prüfung der Teststücke erfolgte durch das WIW-Institut (Knödler) in Strasslach mit Hilfe der Analysemethoden REM, SIMS und XPS

Ergebnisse

Die gesetzten Ziele wurden erreicht und in einer Reinigungsanlage realisiert.
Hinsichtlich der einzelnen Projektzielsetzungen bedeutet dies konkret:

- die Handarbeit wird ersetzt durch eine Anlage mit automatischem Prozessablauf
- drei Gyros werden gleichzeitig bearbeitet
- die eingesetzten Detergenzien sind nicht gesundheitsgefährdend und erfordern keine speziellen Entsorgungsmassnahmen
- die durchgeführten SIMS Analysen zeigen eindeutig keine Restverschmutzungen in/auf dem Material, insbesondere keine Fluoride und Chloride
- das Ansprengen der Spiegel erfolgt ohne Schwierigkeiten
- die Trocknung aller drei Gyros mit Stickstoff erfolgt in weniger als 120 Sekunden

Entwicklungstendenzen

Die in diesem Beispiel angesprochenen Arbeitsfelder zeigen in Übereinstimmung mit den allgemein erkennbaren Bemühungen in der optischen Industrie die folgenden Entwicklungstendenzen:

1. Die manuelle Handreinigung wird durch die maschinelle ersetzt
2. Die maschinelle Reinigung erfolgt prozessgesteuert und ist in den Produktionsprozess integriert
3. Trocknungsverfahren auf der Basis von FKW oder CKW werden ersetzt werden durch kostengünstigere und umweltfreundlichere
4. Mit Präferenz werden umweltfreundliche Detergenzien eingesetzt werden. CKW und FKW als Reinigungsmedium werden an Bedeutung verlieren.

Reaction Kinetics of Nitrogen-Ion Laser in a Fast Electric Discharge

R. Sadighi-Bonabi   C.B. Collins
Atomic Energy Organization of Iran   University of Texas at Dallas (U.S.A)

ABSTRACT

In this research work a unique kind of preionized fast discharge system was constructed. Uniform operation of the system at repetition rates up to 30 Hz and for pressures up to 10 atm. showed the potential of this system for supporting kinetic measurements at new regions of pressure (exceeding one atmosphere). This new instrument was optimized for system integrety and dependable preionization while maintaining an exceptionally low inductive loading of the discharge circuit.

The light due to fluorescence was pectrally dispersed with a grating manochrometer and dectected by photomultiplier capable of 1.5 n.sec resolution. The transient intensity was then recorded with a biomation 8100 transient digitizer placed in an electrostatically screened room and conected to a computer. In this arrangement the rate coefficients of bimolecular and termolecular reactions of He-N_2 mixtures was measured which were $(1.1 \pm 0.02) \times 10^{-9}$ and $(1.44 \pm 0.12) \times 10^{-29}$ respectively. The accuracy obtained in this experiment was about an order of magnitude better than the previous results at much lower pressures.[21]

INTRODUCTION

The nitrogen ion laser, successfully operated by Collins, et al[1] in 1974, was the first member of the class of charge transfer lasers. Through experimentation with an e-beam generator they were able to obtain stimulated emission from the afterglow of a high pressure mixture of helium and nitrogen. Coincidentally an electric discharge version of nitrogen ion laser was discovered by the Soviet group of Ischenko, et al.[2] in the same year, 1974. Since then, the resulting N_2^+ laser has shown the promise of yielding high efficiencies at high pressures.[3] In fact it has been so successful that a growing number of new lasers have developed which use the same basic principle of charge transfer. This finally motivated the development of a model for ion molecule reactions at high pressures.[4]

Since its discovery, the N_2^+ laser has proven to have many advantages over other gas lasers. The combination of visible wavelengths, narrow linewidth, relative freedom from run-away superfluorescence, operation as a four level system, and simultaneously high output power densities and efficiencies give the nitrogen ion laser considerable promise as a practical device.

When N_2^+ laser was excited by an intense electron beam with currents of the order of $1 KAcm^{-2}$ quasi-cw operation of the laser was achieved, at 427.8 nm and output power was found to accurately follow input power after the onset of threshold. Data obtained with e-beam excitation showed power conversion efficiencies of 3%.[5]

Experimental results from an electric discharge arrangement demonstrated power conversion efficiences of 2% and overall efficiencies ranging from .2 to 0.4%.[6] These results varied from the kinetic model, which projected an overall efficiency of 0.9% in an electric discharge system. This apparent reduction of power density and overall efficiency led Ischenko et al.[2] to misidentify the principal kinetic step as being direct collisional excitation of neutral nitrogen by hot electrons in the tail of the energy distribution present in an wall estabilized discharge. However, the studies of the instantaneous power transfer efficiencies that have been reported[6] showed that lowered output efficiencies generally occured as the result of the loss of coupling between the electrical load and driving circuit caused by the time-waving impedance of the laser tube. This suggested that the nitrogen ion laser would perform best when coupled to the field in an oscillator-amplifier configuration. This arrangement demonstrated for the first time in 1978.[7]

KINETIC BACKGROUND

It is the combination of Helium ions such as He_2^+ reacting with some other gas that introduces the features peculiar to this scalable charge transfer laser, giving it both unique advantages and difficulties. Different reports have given the rate at which the charge transfer reactions of the Helium ions with molecular gases such as those mentioned earlier occur. The reactions of interest were mostly binary reactions which serve to transfer the stored energy into excitation of product states capable of supporting the extraction of stimulated emission (for example, ref. 8). A typical example is reactions of the type :

$$He_n^+ + X \longrightarrow nHe + (X^+)^* . \qquad\qquad 1$$

where n = 1 or 2 or 3, X is the reacting additive and (X^+) is the product state.

To form appreciable concentrations of He_2^+, as opposed to He+, requires either high pressures of helium or long-lived plasma since the conversion of the ions from atomic to molecular form proceeds most favorably by three-body collisions which will be discussed in following section in detail. Both conditions pose particular problems. The latter is generally precluded by the time scale set by the spontaneous lifetimes of the excited states of the reaction products considered as laser candidates. The former causes significant pressure broadening of the molecular transitions because of the relative frequency of collision with the background helium atoms. The result is that the cross sections for stimulated emission are typically peaked at 0.01 to 0.1 nm^2,[9] values which ions through the well known three body irreversible reaction

$$He^+ + He + He \longrightarrow He_2^+ + He . \qquad\qquad 2$$

Values reported for the rate coefficient for this three-body association process range between[10] 6.3×10^{-32} and[11] $1.4 \times 10^{-31} cm^6 S^{-1}$ at room temperature. In other words the characteristic time for converting the initial atomic He+ population at a concentration n_1 into molecular He_2^+ ions at a concentration n_2 would range between 12 and 27ns at atmospheric pressure.

Since the vibrational ground state of He_2^+ lies about 2eV below the energy of H^+ ions (see Fig. 1) , the He_2^+ molecular ions are formed in a highly excited vibrational state (v=15) which quickly relaxes to the ground vibrational state.[10]

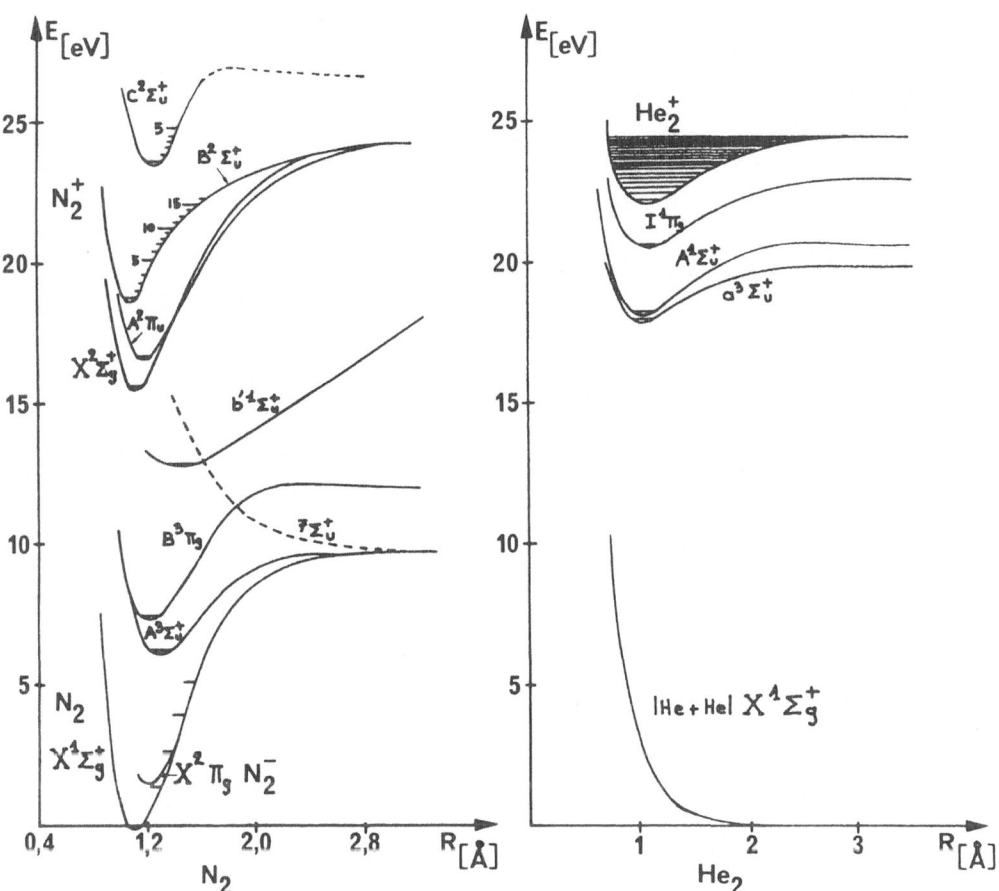

Figure 1 . Energy levels of He and potential energy curves of He_2^+ , N_2, and N_2^+. This figure illustrates the pumping of N_2^+ by charge transfer from He_2^+

Molecular ions He_2^+ are also formed by collisions between excited atoms and ground state atoms, as had first been proposed by Hornbeck and Molnar :[12]

$$He^*(i) + He \longrightarrow He_2^+ + e. \qquad\qquad 3$$

The corresponding cross sections have been measured for various energy levels i.[13,14] Theoretical and experimental investigations have indicated that this is a nearly resonant reaction, resulting in molecular ions in an excited vibrational state (v=4 for the 3^1D level).[10] Because of the very low concentration of the excited atomic species, associative ionization by (3) is negligible with respect to the ion conversion (2) in the production of molecular ions.

The He_2^+ ions then maintain a dynamic equilibrium with the He_3^+ ions and in the same way the He_3^+ ions maintain a dynamic equilibrium with the He_4^+ ions as :

$$He_2^+ + 2\,He \rightleftharpoons He_3^+ + He, \qquad\qquad 4$$

$$He_3^+ + 2\,He \rightleftharpoons He_4^+ + He, \qquad\qquad 5$$

These reactions are considered reversible because of the very low binding energies of additional helium atoms.[15]

Nitrogen ion laser pumped not only by bimolecular reactions of pike veation 1, but it is pumped by temolecular analog which is dominant at high pressures.

$$He_2^+ + N_2 + He \longrightarrow N_2^+ + 3\,He \qquad\qquad 6$$

Before the discovery of termolecular reactions it was generally assumed that the extensive literature[16] on charge and excitation transfer reactions derived from low pressure experiments could be used to adequately model the high pressure gas systems. Since the analog termolecular reactions of bimolecular reactions dominate collisions at atmospheric pressures, the introduction of such 3-body reactive collisions,[17,18] provided significant guidance in estimating new reaction channels. This could be very important to a variety of lasers operating at high pressures.

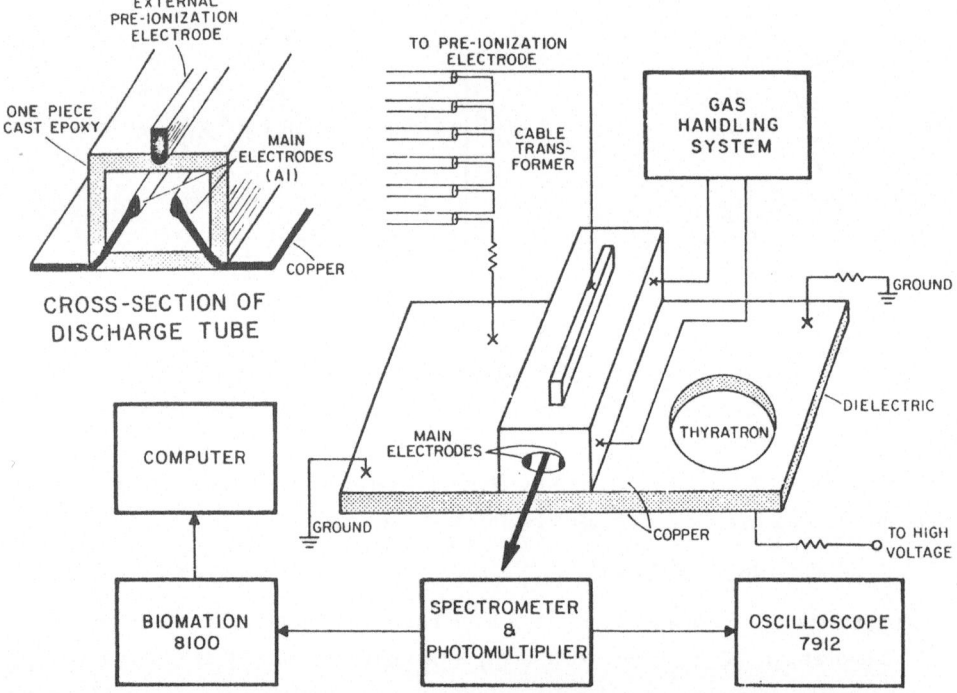

Figure 2. Schematic diagram of the experimental arrangement used in this experiment. Cross section of the cast tube is represented

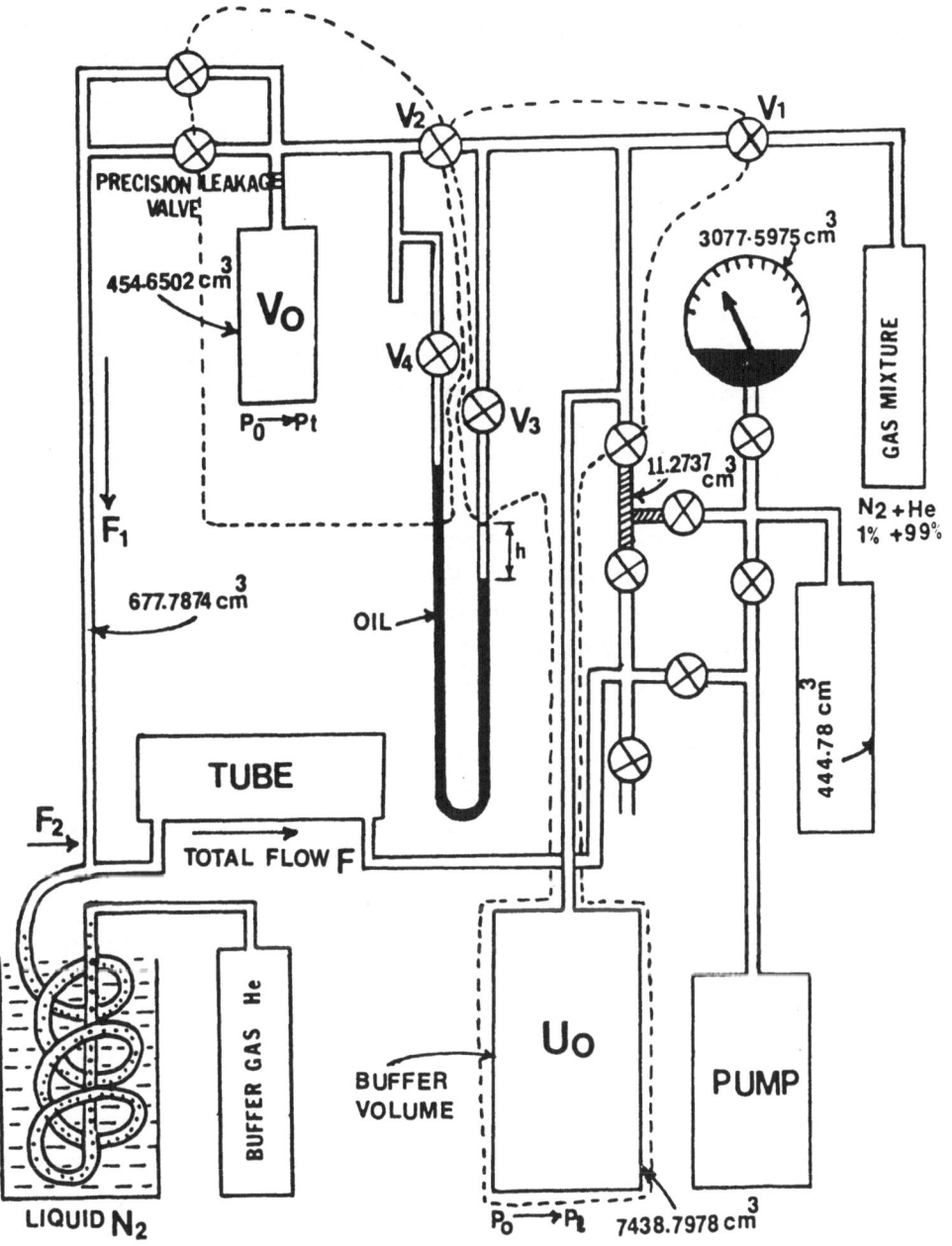

Figure 3 . Schematic representation of gas handling device which was used when extremely slow flow rates were desired

Three major lines have been excited in mixtures of helium and nitrogen pumped by charge transfer. These lines correspond to transitions from the same upper vibrational state the v=o of the $B^2\Sigma_u^+$ electronic state, to three different vibrational states of the $X^2\Sigma_g^+$ electronic state of the N_2^+ molecular ion. The output at 427. 8nm (v=0, ^{g}v' = 1) was reported to be the strongest and to have a pulse duration of approximately 25 nsec. This was the transition which was selected to monitor the decay of the H_2^+ ion population. For mixtures of He-N_2, a three-body rate coefficient (K_{13}) of $(1.44 + 0.02) \times 10^{-29} Cm^6$ sec^{-1} was obtained. The two-body rate coefficient (K_{12}) was found to be $(1.10 + 0.02) \times 10^{-9} Cm^3$ sec^{-I}.

EXPERIMENTAL ARRANGEMENT

Recently an electric discharge system was used for kinetic measurements at pressures below one atmosphere.[21] These systems have several advantages over electron beams including : ease of operation, compactness, long discharge length, repetition rate of up to several KHz and finally, their attainable high power and high efficiency. Extension of these measurements to atmospheric pressures may have an important imact on the development of a new class of high pressure gas lasers namely, high power fast pulse lasers.

In this work a unique version of an electric discharge system was designed and constructed. A schematic reperesentation of this apparatus is shown in Figure 2.

The light due to fluorescence was spectrally dispersed with a grating monochrometer and detected by a photomultiplier capable of 1.5 nsec resolution. The transient intensity was then recorded with a Biomation 8100 transient digitizer placed in an electrostatically screened room and connected to a computer. A fast transient digitizer (Tektronix 7912) was used to monitor the pulses visually.

Figure 3 is the schematic reperesentation of the gas handling device which was constrcted in this research work. In this arrangement flow of He+N_2 (F_1) and total flow (F) was measured, from which F_2, the flow of buffer gas (F_2=F-F_1) was calculated.

Therefore, we know the gaseous composition of the total flow. The precision of this system is great because of the extremely low flow through a precision leakage valve (theoretical flow rate from 10^{-6} to 2.5 Cm^3/sec) and because of the accuracy of the large manometer (10^{-2} Torr). This device was very useful especially for kinetic measurements, where precise determination of flow rates was necessary.

The discharge tubes were cast from various epoxy materials. Although the construction of such a refined cell required a great deal of laboratory time, the experience gained warranted the effort. The positive results finally obtained, after testing various materials and various techniques of construction, may have great future impact on the improvement of laser technology. Three of the best materials were found to be (stycast 1269A) crystal clear epoxy, (Stycast 1267) clear epoxy and (Stycast 3180M) black epoxy.

Figure 4 shows the cross-section of two different discharge tubes. These discharge tubes were tested at pressures of up to 10 atmospheres. By using this system the rate coefficients of bimolecular and termolecular reactions of He-N_2 mixtures at high pressures were measured.

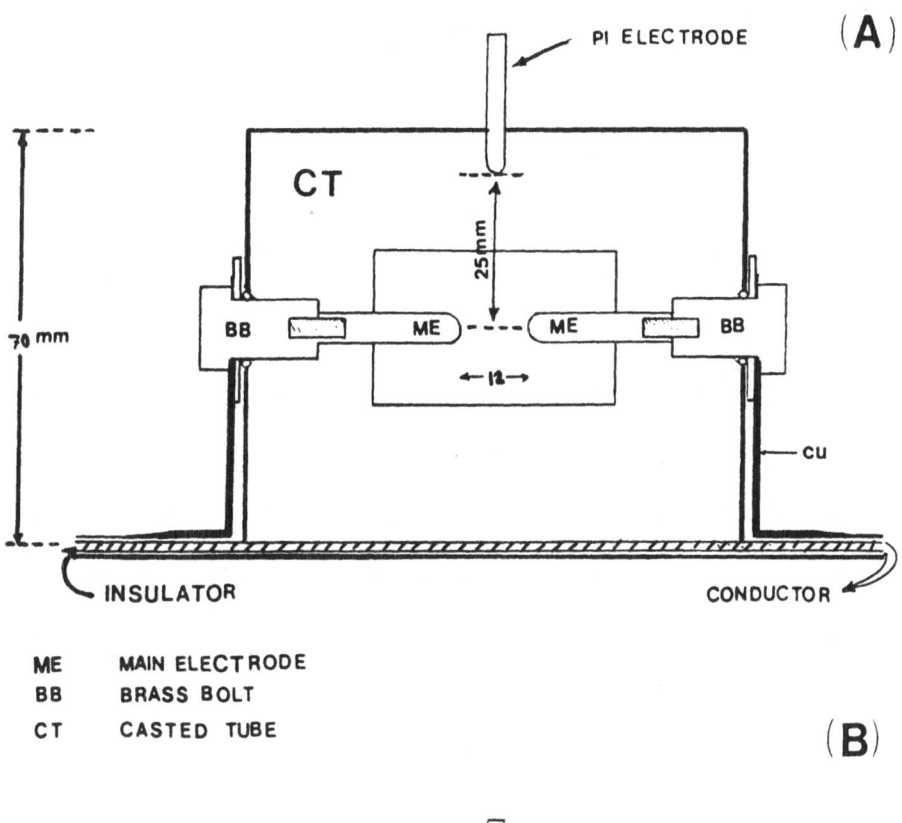

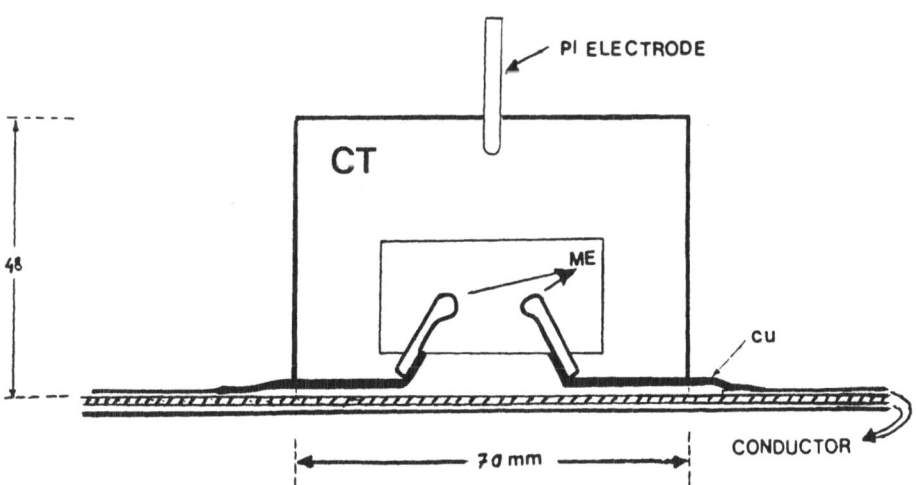

Figure 4 . Reperesention of cross section of discharge tubes casted by two different techniques. A, The cross section of a tube which casted in the mold. B, The cross section of a tube casted by partial pouring

RESULTS AND CONCLUSION

    The intensity of the light at 427.8nm produced by the system is directly proportional to the concentration of He_2^+ ions.

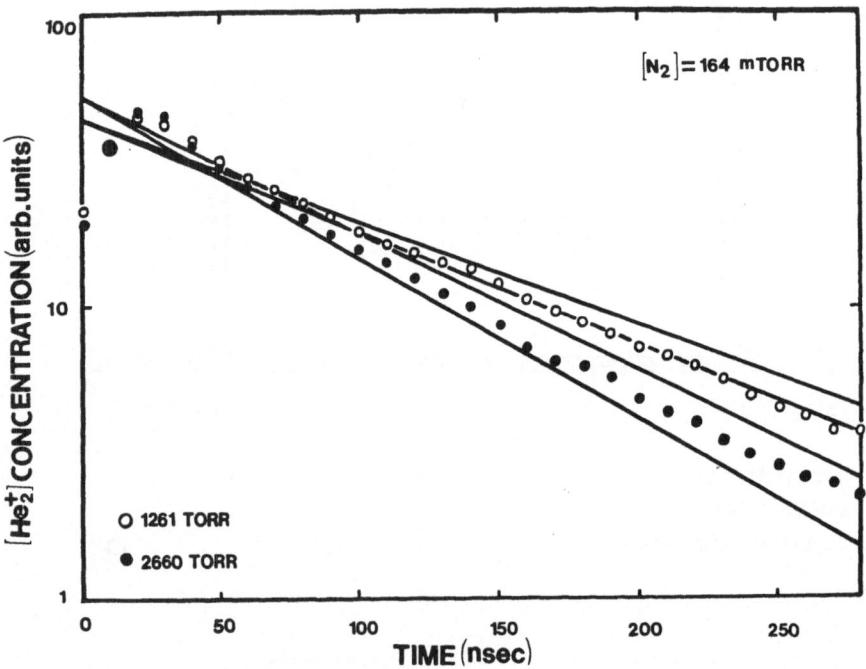

Figure 5 . Graph of the transient dependence at 427.8 nm (B-X) of N_2^+ fluorescence at a fixed nitrogen concentraiton for two different helium pressures

    One can see evidence of a termolecular reaction channel by studying Figure 5 which shows the transient dependance at 427.8 nm (B(v=o)-x(v=1)) of N_2^+ fluorescence at a fixed nitrogen concentration (164 m torr) for two differenct pressures. This difference can only be explained by considering the presence of a three body reaction. At 2260 torr of Helium pressure the termolecular rate coefficient (1.15×10^{-9}) and bimolecular rate coefficient (1.1×10^{-9}) are of the same order. The solid lines plot the time evolution of the He_2^+ concentration computed from literature[19,20,21] values of rate coefficients. Data points record values inferred from the intensity of the 427.8 nm transition in N_2^+. The advantage of this measurement is that as we can see from Fig. (5) the three body component of the reaction rate appearing in the usual decomposition, (written for $He_2^+ + N_2 \longrightarrow$).

$$v = vo + K_2 \; [N_2] + K_3 \; [N_2] \; [He] \qquad\qquad 7$$

can be made directly perceptable. Where K_2 and K_3 are the two body and three body rate coefficients respectively.

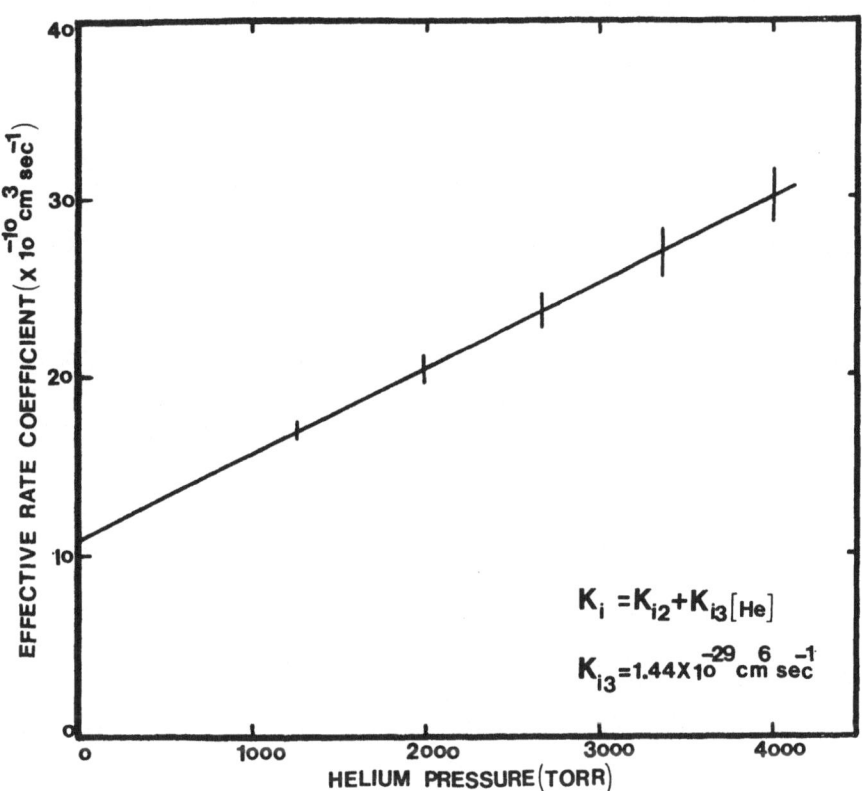

Figure 6 . Graph of the data showing the effective rate coefficient obtained from stern-volmer plot, for various values of helium pressures (up to 5.5 Atm.)

From the data similar to Figure 5 the destruction frequency can be obtained and plotted as functions of nitrogen pressure for the various helium pressures (Stern-Volmer plot) .[21] Then the effective rate coefficient (K_i) which is the combination of both two body and three body rate coefficients can be measured. It is the slope of the Stern-Volmer plot. Finally by plotting the calculated effective rate coefficient versus the helium pressure both two body and three body rate coefficients can be measured. Figure 6 shows the effective rate coefficient as a function of helium pressure up to 4000 torr which is more than 5 atmospheres. These measurements can be extended up to 10 atm very easily. From Figure 6 one can see the three body rate coefficient (K_{13}) is $(1.44 \pm 0.12) \times 10^{-29}$

$Cm^6 sec^{-1}$ and two body rate coefficient of (K_{i2}= the effective rate coefficient at zero helium pressure) $(1.1\pm0.02) \times 10^{-9}$. These results are in very good agreement with earlier measurements at much lower pressures[19,20,21]. More interesting, is the accuracy obtained in this experiment which is about an order of magnitude better than the previous results from He-N_2 system.[21]

REFERENCES

1. C.B. Collins, A.J. Cunningham, S.M. Curry, B.W. Johnson, and M. Stockton, Appl. Phys. Lett. 24, pp. 477-478, 1974.

2. V.N. Ishchehko, V.N. Lisitsyn, A.M. Razhev, and V.N. Starinskii, JETP Lett., vol. 19, pp. 233-234, 1974.

3. R. Sadighi-Bonabi, High Power Fast Pulse Lasers, The University of Texas At Dallas Dec. 1983.

4. C.B. Collins, F.W. Lee, W.M. Tephenhart, and J. Stevefelt, J. Chem. Phys. 78, 6079 (1983).

5. C.B. Collins, J.M. Carroll, F.W. Lee, and A.J. Cunningham, Appl. Phys. Lett. 28, 535 (1976).

6. C.B. Collins, J.M. Carroll, and K.N. Taylor, J. Appl. Phys. 49, 5093 (1978).

7. C.B. Collins, J.M. Carroll and K.N. Taylor, Appl. Phys. Lett., 33, 175 (1978).

8. C.B. Collins, seventh Semi-Annual Technical Report UTDP-ML-04 (1975).

9. J.V. Peterson, Theoretical Modelling of Charge Transfer Laser, Ph.D. Disertation, the University of Texas at Dallas, 1978.

10. R.Deloche, P. Manchcourt, M. Cheret, and F. Lambert, Phys. Rev. A, 13, 11440 (1976).

11. J.M. Pouvesle, thesis, universite' d' Orleans, France, 1981 (unpublished).

12. J.A. Hornbeck and J.P. Molnar, Phys Rev. 84, 621 (1951).

13. H.F. Wellenstein, and W.W. Robertson, J. Chem, Phys. 56, 1077 (1972).

14. C.B. Collins, B.W. Johnson, and J.J. Shaw, J. Chem. Phys. 57, 5310 (1972).

15. F.W. Lee, C.B. Collins, and R.A. Waller, J. Chem. Phys, 65, 1605 (1976).

16. D.L. Albritton, Data and Tables, 22, 1(1978).

17. C.B. Collins and F.W. Lee, J. Chem. Phys. 68, 1391 (1978).

18. C.B. Collins and F.W. Lee, J. Chem. Phys. 70, 1275 (1979).

19. F.W. Lee, C.B. Collins and R.A. Waller, J. Chem. Phys. 65, 1605 (1976).

20. C.H. Chen, J.P. Judish, and M.C. Payne, J. Chem, Phys. 67, 3376 (1977).

21. J.M. Pouvesle, A. Bouchoule and J. Stevefelt, J. Chem. Phys. 77, 817 (1982).

Pumping Mechanism for Single Line Fluorine Laser

R. Sadighi - Bonabi
Industrial university of sharif
Azadi st. Tehran Iran

ABSTRACT

For most of the period over which the atomic fluorine laser has been studied, the principal mechanism of excitation has been in doubt. Originally, it was assumed that the upper level was populated either by direct dissociative excitation of fluorine containing molecules or by dissociation of excimer HeF^* Through collison with a metastable helium atom. However theoretical models and experiments showed that these channals can not populate the uper level of this system.

In this report different pumping sequences of atomic fluorine laser are discussed. It was found that two different charge transfer reactions are responsible for pumping this system.

At low pressures, the charge transfer from He^+ excites the fluorine, Where at high pressures the upper radiotive levels pumped by charge transfer between He_2^+ and fluorine ions. The main reason of uncertainty was because of these different nature of pumping mechanism, at lower pressures only the doblet lines are seen while at atmospheric press-ures, The quartet lines are dominant. This new pumping mechanisms sugg-ests the single line operation of this laser which partially was achieved in this research work. The normal excitation of a superradiant fluorine laser results in the uncontroled generation of a group of output lines. Two principal outputs having wavelengths of 745 and 635n.m were selected and amplified. The extraction of all available energy from the excited level through a single line molivated this experiment.

INTRODUCTION

The atomic fluorine laser was discovered in 1970 by Kovacs and Ultee[1] using a pulse discharge arrangement in a mixture of He and CF_4, they measured a total peak power of 150 Watts with a pulse dura-tion of 1 μs. The resulting wavelengths (703.7 nm, 712.8 nm and 720.2 nm), were due to the doublet $3p^2p^o-3sp$ multiplet of atomic fluorine. A short time later, Jeffers and Wiswall reported lasing action on the same transitions from a discharge in a He-HF mixture.[2] Their work, together with other early observations,[2-4] showed that several

components of the $3p^2p^0-3s^2p$ transition manifold of F could be stimulated in pulsed electrical discharges through various fluorine donors
diluted in helium. Outputs were of unifromly low intensity at wavelengths lying in the red region of the spectrum. Single line operation
was achieved only through the use of a dispersive element to "filter"
the intensity within the laser cavity. The introduction of fast -
pulsed discharge devices by Bigio and Begley in 1976 led to significant improvements in scaling the outputs. A value of 70 KW on five doblet lines near 710 nm with a pulse duration of 25-30 nm was achieved.[5-6]
Later,a peak power of 300 KW was reported by a Soviet group of Chapovsky et al,[7] in a similar system.

However,still less than half of the total intensity was found in
any single line. Later attempts to scale the output power by increasing the pressure succeeded only in increasing the number of components
of the 3p-3S complex into which it was distributed.[6-10]

Since its discovery,the atomic fluorine laser has been investigated very optimistically,[1-12] because it showed substantial potential
as a source for pumping numerous stable dyes having laser bands in the
near infrared (ir).[16] It is a pump source similar to the nitrogen
laser. Special attention has been paid to the nonlinear mixing of the
lines of the laser with a pumped dye, which produces an output in the
range of tens of u m.[6] This laser is also considered to be suitable
for photo chemical applications. More recently it has been observed
that laser output from atomic fluorine can be readily obtained from
mixtures of helium and fluorine donors excited in many of the commercial or standard excimer devices. Nevertheless, the effective operation
of this system has not yet been reported. particular difficulties
occur from the superradiant excitation of many lines with little control of their relative intensities.

Figure 1 shows the energy levels of atomic fluorine and all observed lasing transitions. Because the power is distributed among all
these transitions the extracted power through single line is low in
comparison to energy pumped into the 3p manifold.

In 1980 a new kinetic channel was reported for excitation of the
quartet lines of this laser.[14] partly based on those results,extrac-
tion of all available energy from the upper level through a single
line motivated the experiments reported here concerning the atomic
fluorine laser. In the following sections different proposed pumping
sequences of this laser will be discussed and proper kinetic reactions
will be introduced. Transitions of the F laser and the possiblity of
operating this laser as a single line laser will be investigated.Fina-
lly,operation of a fluorine amplifier will be explained.

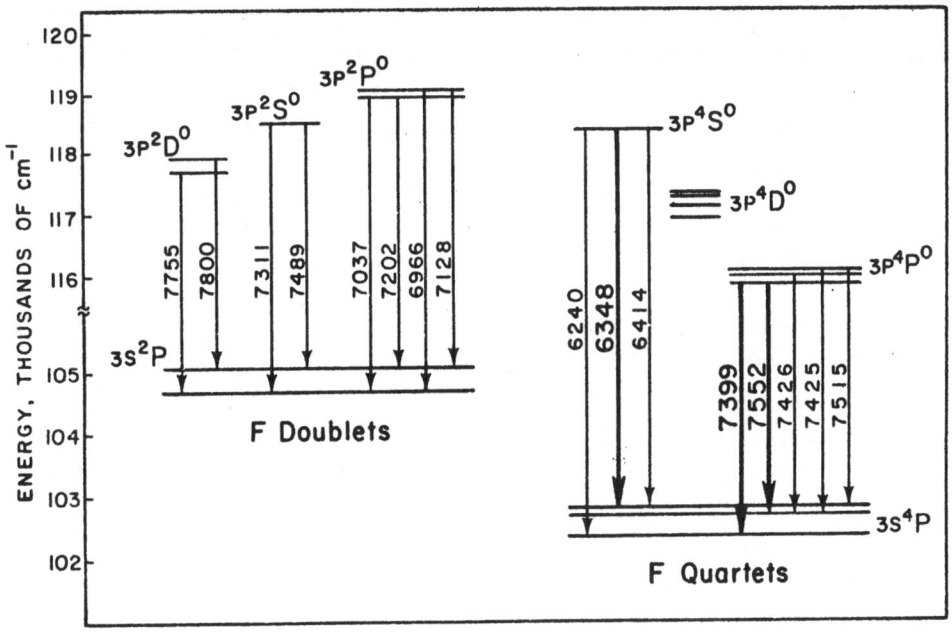

Figure 1 . Energy levels of atomic fluorine showing all observed lasing transitions. Doublet and quartet manifolds are indicated. The three strongest transitions examined in this experiment are shown by heavier arrows

PUMPING MECHANISMS OF F LASER

Jeffers and Wiswall,[2] in 1970 proposed the excitation and de-excitation mechanisms which make possible laser emission from atomic fluorine. They suggested that the pumping mechanism for the He+HF laser was due to excitation of fluorine in dissociative collisions between He(2^1S) metastable and HF. They based their conclusions on an analysis of the energy defects in the proposed collisional pumping mechanism. Jeffers and Wiswall also observed that the addition of H_2 to an He-SF_6 mixture increased laser power and that He is always a necessary component of the system. They concluded that the pumping mechanism of a neutral atomic fluorine laser is the dissociative excitation of an F donor by He metastable state. In 1976,[6] Hocker and Phi hypothesized that population inversion in the fluorine laser resulted from the dissociation of HeF into ground state helium and 3p fluorine atoms. Hocker later suggested that the HeF molecules are formed from ground state fluorine atoms and 2^1S or 2^1p helium atoms.[9]

In addition to the two kinetic sequences just mentioned,many other
sequences have been proposed. Through a careful examination of all
conceivable mechanisms, a proper pumping channel can be selected.
In the following sections XF designates fluorine containing molecules
such as F_2, NF_3, SF_6 etc. He designates either the $2s^1S$ or $2p^1p^o$ helium
metastables. Important reactions which have been suggested are the
following:

$$He^+ + XF \longrightarrow He + X^+ + F^* \ (3P). \qquad\qquad 1$$

$$e + XF \longrightarrow e + X + F^* \ (3P). \qquad\qquad 2$$

$$e + XF \longrightarrow e + X + F, \qquad\qquad 3a$$

$$e + F \longrightarrow e + F^* \ (3P). \qquad\qquad 3b$$

$$e + XF \longrightarrow X + F^-, \qquad\qquad 4a$$

$$F^- + He^+ \longrightarrow He + F^* \ (3P). \qquad\qquad 4b$$

$$XF + H_2 \longrightarrow HF(^1\Sigma^+, v=0) + H + X, \qquad\qquad 5a$$

$$HF(^1\Sigma^+, v=0) + He(2s^1S) \longrightarrow He(1s^2 1S) + H(1s^2 S) + F^* (3P) + \ E. \ 5b$$

$$He^* + XF \longrightarrow He + X + F^* \ (3P). \qquad\qquad 6$$

$$e + XF \longrightarrow e + X + F, \qquad\qquad 7a$$

$$F + He^* \longrightarrow HeF, \qquad\qquad 7b$$

$$HeF \longrightarrow He(1s^2 \ 1S) + F^* (3P). \qquad\qquad 7c$$

Since mechanism (1) is endothermic, the available energy to excite
the fluorine atom in this reaction is much less than 14.75 ev which is
the energy required to excite the F atom. For example if NF_3 is assumed
to be the fluorine containing molecule in this reaction, only 10.2 ±
0.1 ev would be available to excite fluorine atom.[11] In reactions (2)
and (3) helium is not a necessary element. Lasing does not occur on
the 3P-3S transition when helium is substituted with any other gas.
This has been investigated throroughly (see reference 2 and 11). Thus
clearly these reactions can also be eliminated.

Charge transfer from He^+ described by reaction (4) could be a
possible candidate for pumpng the F laser at lower pressures. The He^+
are dominated ions at lower pressures. At high pressures these ions
will produce He_2^+ ions. (See Fig.2)

Mechanism (5) which was proposed by Jeffers and Wiswall,is not a
general one. Because this pumping sequence can occur only in presence
of H_2 which is not a required element in the operation of the fluorine
laser. However, there have been claims that addition of H_2 will enhance the output.[2] This may be explained by the reduction of loss during
the pumping sequence.

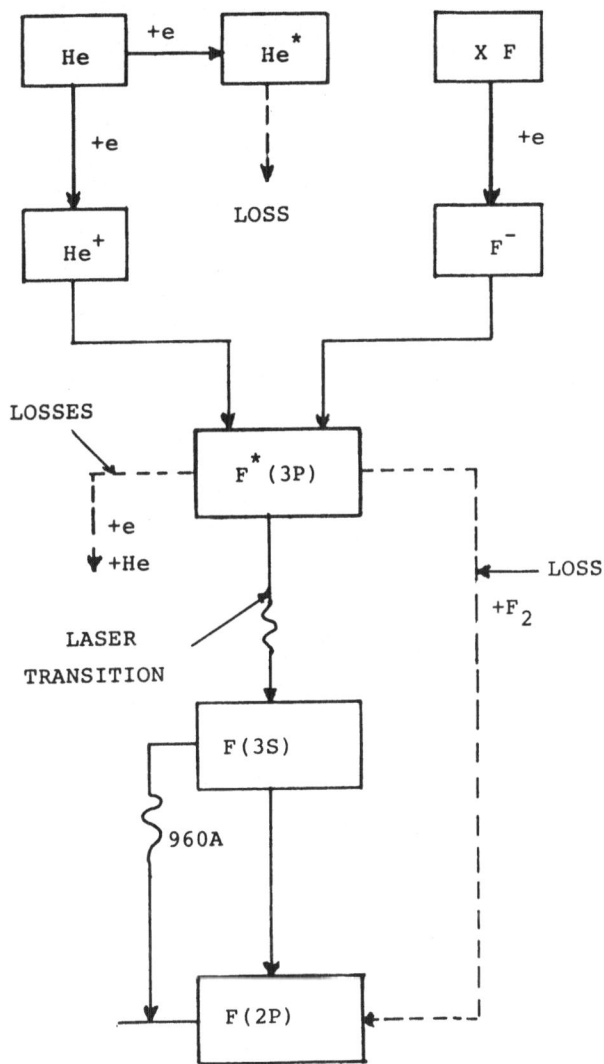

Figure 2. Flow chart of the kinetic process relevent to the atomic
fluorine laser at low pressures. Solid arrows denote the dominant
kinetic channels dashed arrows represent the competing channels
within pumping sequence

For most of the period over which the atomic fluorine laser has been
studied,the principal mechanism of excitation has been in doubt. Originally,[1-10] it was assumed that the upper laser level was populated
either be direct dissociative excitation of the fluorine containing
molecules or by dissociation of excimer HeF* through collison with a
metastable helium atom, He*. These two mechanisms are represented by
reactions (6) and (7)respectively. However,theroetical modeling of

Miller and Morgner,[13] showed that the attractive part of the potential curve for HeF* could not be effectively coupled to the input channel, He* and F. This conclusion was confirmed experimentally,[11] and that essentially precluded any significant contribution from He* to the population of the upper laser levels at the very early times character-istic of the fast discharge devices. As a consequence, the only accep-table mechanism among all conceivable pumping sequences for the F laser is reaction (4) which has been assumed with some doubt one of the possible mechanisms in earlier reports,[11,13] We conclude that this cou-ld be the pumping channel for F laser, but only at low pressures.Thus we have to search for another process or processes for pumping of this laser at high pressures.

ATOMIC FLUORINE LASER PUMPED BY CHARGE TRANSFER FROM He$_2^+$ AT HIGH PRESSURES

In 1977 Collins et al,[15] reported strong correlation between the population of He$_2^+$ ions excited by an e-beam and spontaneous emission of components of the quartet system of the atomic fluorine laser. They noticed the presence of a charge neutralization step in the pumping sequence. Both qualitatively and quantitively they concluded that, the upper radiative level is mostly pumped by charge transfer between He$_2^+$ ions and F ions. They accepted the possiblity that doublet and quartet lines also were excited weakly by helium metastables. In this report we will show that such contributions,in fact,do not exist. Some insight into the kinetic mechanism of this laser was provided by two recent reports,[14,17] together with earlier discussions in this section. These reports confirm that the performance of the atomic fluorine laser exci-ted at high pressures by intense, fast pulsed discharges to be consist-ent with the pumping sequence.

$$F_2 + e \longrightarrow F^- + F, \qquad\qquad\qquad 8a$$

$$He_2^+ + F^- \longrightarrow F^*(3P) + 2He. \qquad\qquad\qquad 8b$$

This is a two-electron jump, charge transfer process. In step (8b) one 2P electron of F$^-$ ion is transfered to the He$_2^+$ and another electron is promoted to the 3P orbital by the excess energy. The 3P electron can combine with the remaining $2P^4(3P)$ core to give a S, P, or D configura-tion with total spin corresponding to either a doublet or quartet state. The coupling scheme is sufficiently near L.S. that the fine structure splitting of these six basic levels is relatively small compared to their separation in energy,but it is still easily resolved (see figure 1).

Figure (3) summarizes all of the process which populate the upper radiative level at high pressures. At low pressures the mechanism is different, charge transfer between He^+ ion and F^- ion appears to be the most suitable one (Figure 2).

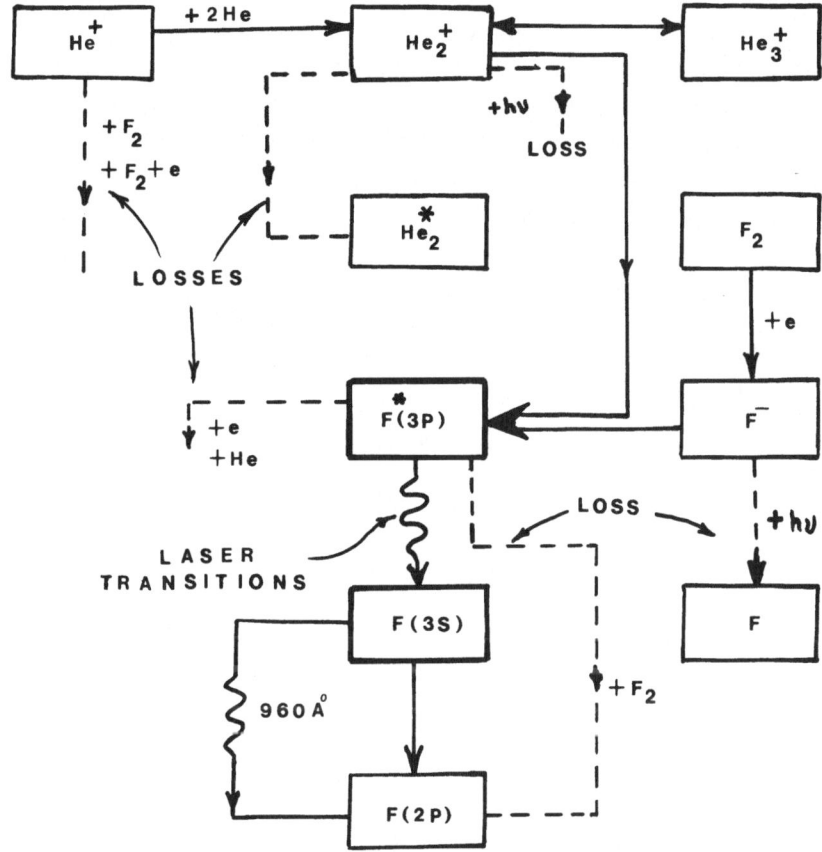

Figure 3 . Flow chart of the Kinetic process relevant to the atomic fluorine laser at high pressures

TRANSITIONS OF ATOMIC FLUORINE LASER AND THEIR PRESSURE DEPENDANCE

Selection rules allow 27 different transitions from the various fine structure components to members of the $2P^4 3S$ manifold. These potential laser transitions span the wavelength range from 623.9 to 780.0 nm. Over half of them have been observed (see figure 1). Several papers report the pressure dependence of these laser transitions.[6,8,10,12] These studies indicate that the atomic fluorine laser is pressure tunable over a rather large spectral range.In general, as the pressure

778

was increased, it was found that the output tended to switch from the doublet lines to the quartet transitions belonging to the same manifold.[19] The 731.1 nm and 712.8 nm doblet transitions were dominant at pressures below one atmosphere, while the 634.9 nm, 739.9 nm and 755.2 nm quartet transitions contained about 97% of the total intensity,[14] at atmospheric pressures. These three quartet lines are the focus of the experiments reported earlier.[17]

The spectral variation for various total pressures and for various concentrations of fluorine donor shows that the pumping mechanisms at high pressures are different from those at low pressures. This also supports the idea discussed earlier that two different charge transfer mechanisms exist in the atomic fluorine laser. Figure 4 shows the results obtained in this laboratory by exciting atmospheric pressures of helium containing 0.2% F_2 in a preionized,traveling-wave,discharge device.[14] No optics were used in obtaining these output, and the powers measured were divided between at least five major lines of the quartet system.

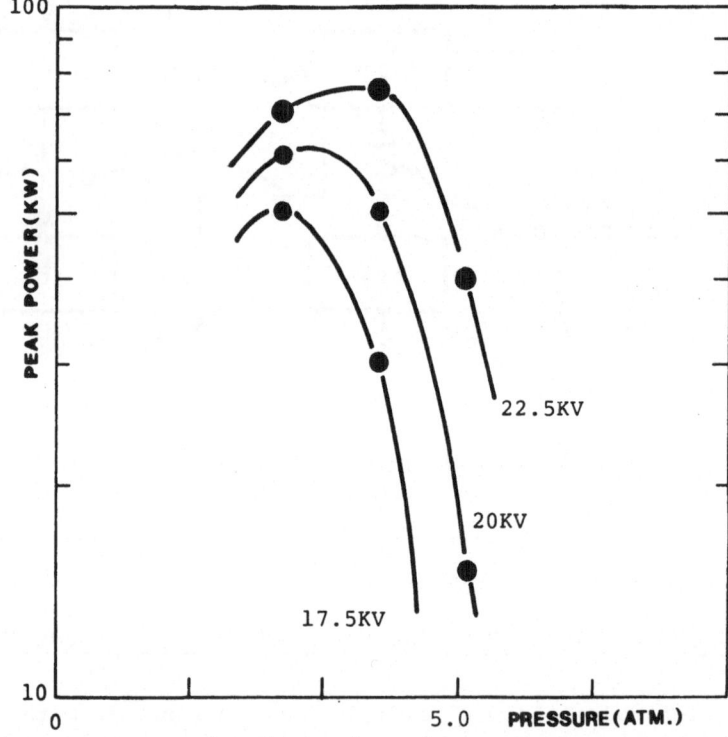

Figure 4   pressure and voltage dependance of the peak power outputs emitted from a 1.5-m-long traveling-wave laser pumped by charge transfer from He_2^+ containging 0.2%F_2

TABLE 1. Total peak power and their durations have been obtained for various manifolds are represented. The ratio of gas compositions are indicated.

Year	Manifold	Reference	Gas Composition	Total Power (KW)	FWHM (n sec)
1970	Doublet	1	CF_4,C_2,F_6,SF_6:He	0.15	1.000
1970	Doublet	2	HF:He(1:6)	0.042	15,000
1976	Doublet	5	NF_3:He(1:100)	70	25-30
1977	Quartet	8	F_2:He(2:100)	75	8
1977	Both	7	NF_3:He ——	300	2
1979	Both	12	NF_3:He(1:400)	150	10
1979	Both	10	F_2:He(0.1 to 2%)	19	30
1980	Quartet	14	F_2:He(0.2:100)	80	8

Table 1 shows the total peak power or energy obtained from the indicated systems of the atomic fluorine laser. Using a high power fast pulse laser system,a power of 300 KW was achieved. This clearly suggested the benefits of employing such system. By studying the kinetics it appears that the quartet atomic fluorine laser pumped by charge transfer from He_2^+ holds an even greater potential for the development of a scalable high power high efficiency laser than the analagus N_2^+ laser. If extraction of all the invested population through a single transition is achieved, it could present a magnificant advantage over similar systems, because of possibility of choosing between its numerous transitions.

SINGLE LINE ATOMIC FLUORINE LASER

The 27 laser transitions of fluorine laser can be usefully classified by three criteria: 1) wavelength, 2) quantum numbers S, L, J for the upper level, and 3) quantum numbers S and L for the lower level, where S,L. and J are the total spin, orbital angular momentum, and total angular momentum numbers, respectively. While there could be 27 transitions that differ in wavelengths, criterion(1); there are only 13 originating form different upper laser levels, in the 3P manifold, criterion(2); and only 6 upper levels that differ in some quantum number other than J. The significance of this type of classification is that it can guide the analysis necessary to develop a strategy for channeling the excitation produced by reaction (8b) into a single output line.

The normal excitation of a superradiant fluorine laser results in the uncontrolled generation of a group of output lines that are distinguished only by the first criterion, the wavelength. It is clear that selective saturation of transitions having different upper states could reduce the number of lines to 13, one for each upper state. However, the general experience with electronic transition lasers excited at an atmospheric pressures has indicated that a rapid collisional relaxational is to be expected for small scale excitation such as the fine-structure splitting in atoms and the vibration in molecules. It might be reasonably expected that at high pressures, reaction(8) would be followed by this type of relaxation.

$$F_J^* + He \longrightarrow F_{J+1}^* + He \qquad\qquad 9$$

where it is to be recalled that F^* forms inverted multiplets for which higher values of total angular momentum, J have lower values of energy. Experimental confirmation of this point has been indicated by the reports that the principal lines at high pressure originated on the lowest energy multiplet.[8,14] Since higher pressures favor the formation of the precursive He_2^+ from He^+, the observations that high pressures strongly favor quartet states[8,10,14] imply that reaction(8) produces principally quartet state populations.

These general considerations suggested that operation of the atomic fluorine laser with high pressures of helium diluent should concentrate the excitation principally into three upper levels, $3P\ ^4S_{3/2}$, $3P\ ^4D_{7/2}$ and $3P\ ^4P_{5/2}$. Although laser emission has been reported from all three, it cannot be concluded simply that reaction(8) produces each directly, It might populate only one directly and the other two could result from collisional mixing or collisional relaxation. If this occurred at a sufficiently rapid rate to produce a kinetic equilibrium, the selective saturation of a single laser transition from one of the levels would serve to extract in a single line all of the energy pumped into the 3P manifold. The possible implementation of this concept motivated the work reported here.

ATOMIC FLUORINE AS AN OPTICAL AMPLIFIER

In prior studies of the atomic fluorine laser it had been shown that[14] if this laser could be operated in an oscillator-amplifier arrangement it would show its best performance. In this experiments a dilute fluorine plasma pumped by the ion-ion recombination of He_2^+ and F^- has been successfully operated as a pulsed amplifier. Two synchronously excited plasmas were produced by a electric discharge system at atmospheric pressure. Two principal transitions from $F^*(3P^4S_{3/2})$ and F^* $(3P^4P_{5/2})$ were obtained from the first plasma and amplified in the second. Figure 4, showes the amplication of varous inpout pulses. to avoid the posibility of super fluorescence from the amplifier, the physical length of the amplifier plasma had to be made quite small. The discharge voltage were found to be optimized at 17 Kv.

Two basic lines at 745 and 635 nm was isolated and amplified. from these experiments effective saturation intensities.Which were 1.2 and 4.5 Kw/cm^2 for 745 and 635 nm respectively . Corresponding small signal gains were found to be 0.44 and 0.31 respectively.

CONCLUSIONS

In this report various proposed pumping sequences of atomic fluorine laser were discussed and concluded that this system pumped by two different charge transfer reactions. Charge transfer from He^+ Which populates the doublet lines is dominant at low pressure charge transfer between He_2^+ and F^- ions pumps the upper radiati e level at high pressure is the pumping mechanism for quartet lines.In this work also the possiblity of extraction of all available energy from the upper level through single line was discussed. Two principal outputs at 745 635 nm were seperated and amplified, from which effective saturation intensity and small signal gain were measured.

ACKNOWLEDGMENT

I would like to thank Mr.A. Behjat for this cooperation. Through out this report.

REFERENCES
1. M.A. Kovacs and C.J.Ultee, Appl.phys.Lett.17,39 (1970).
2. W.Q. Jeffers and C.E. Wiswall, Appl.phys.Lett.17, 444 (1970).
3. A.E. Florin and R.J. Jensen, IEEE J. Quant. Electron.QE-7,472(1971).
4. D.G. Sutton,L. Galvan,P.R. Valensuela,and S.N. Suchard, IEEE J. Quant. Electron. QE-11, 54 (1975).
5. I.J. Bigio and R.F.Begley, Appl. phys.Lett.28, 263 (1976).
6. L.O. Hocker and T.B. phi,Appl.phys.Lett. 29, 493 (1976).
7. P.L. Chapovsky, S.A. Kochubei, V.N. Lisitsyn, and A.M. Raxhev,Appl. phys. 14, 231 (1977).
8. T.R. Loree and R.C. Sze, Opt. Commun. 21, 255 (1977).
9. L.O. Hocker, J. Opt. Soc. Am. 68, 262 (1978).
10. S.Sumida,M. Obara,and T.Fujioka,J. Appl.phys. 50,3884(1976).
11. J.E. Lawler,J.W.Parker,L.W. Anderson and W.A. Fitzsimmons,IEEE J. Quant. Electron. QE-15,609 (1976).
12. A. Rothem and S. Rosewaks, Opt. Commun. 30, 227 (1976).
13. W.H. Miller and H. Morgner,J. Chem. phys. 67, 4923 (1977).
14. C.B. Collins, F.W. Lee and J.M. Carroll, Appl.phys.Lett.37,857(1980)
15. C.B. Collins, Ni, trogen Ion Laser,Final Technical Report,UTDP-ML (1977).

16. See. for example, Y. Miyazoe and M. Maeda, Appl. phys. Lett. $\underline{12}$, 206 (1968); J.P. Webb, F.G. Webster, and B.E. Plourde, IEEE J. Quantum Electron. $\underline{QE-11}$, 114 (1975).
17. R. Sadighi-Bonabi, F.W. Lee, and C.B. Collins, J. APPl. phys. $\underline{33}$, (1982).
18. Bengtson, M.H. Miller, D.W. Koopman, and T.D. Wilkerson, phys. Rev. A. $\underline{3}$, 16 (1971).
19. R. Sadighi-Bonabi High Power Fast Pulse Lasers University of Texas at Dallas Dec.(1983).

Thermal Stability Condition of Optical Resonator with Internal Lens

Wei Guang Hui

Dept. of Engi. Optics, Beijing Institute of Tech., Beijing

Abstract

The thermal stability condition of lens resonator against thermal disturbance, $2G_1G_2 = 1 - \frac{G_2}{G_1}(\frac{L_2}{L_1})^2 - 2G_2(\frac{L_2}{L_1})$, derived by J. Steffen et al is a necessary but not a sufficient one. The necessary and sufficient condition is given in the paper.

Resonator of high power stability and low output divergence can be achieved by carefully choosing its parameters. In 1972, Steffen et al (Ref.1) analyzed the condition of stability for resonators against thermal disturbance and suggested that a resonator shown in Fig.1 would be thermal stable provided that

$$2G_1G_2 = 1 - \frac{G_2}{G_1}(\frac{L_2}{L_1})^2 - 2G_2(\frac{L_2}{L_1}) \tag{1}$$

where G_1 and G_2 are the G-parameters of the resonator and L_1, L_2 the distances of the saying thermal induced lens of focal length f to the end mirrors R_1 and R_2 respectively.

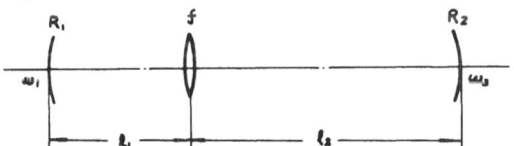

Fig.1. Resonator with one internal lens

In what following it is shown that Eq.(1) is not sufficient to be used to determine wether a resonator is of stability against thermal disturbance, i.e., resonators satisfying Eq(1) may not be,as in general case, thermal stable. For a resonator with a thermal induced lens of focal length f, the optical parameters R_1, R_2, L_1, L_2 and f should satisfy, in addition to Eq.(1), a certain relation. Unless the relation derived below is satisfied, it is not confident to judge wether the resonator is thermal stable.

The G-parameters of the resonator shown in Fig.1 can be written as following (Ref.2 and 3)

784

$$G_1 = -\frac{l_2(l_1-R_1)}{R_1}\left[\frac{1}{l_2} + \frac{1}{l_1-R_1} - \frac{1}{f}\right] \qquad (2a)$$

$$G_2 = -\frac{l_1(l_2-R_2)}{R_2}\left[\frac{1}{l_1} + \frac{1}{l_2-R_2} - \frac{1}{f}\right] \qquad (2b)$$

If the thermal induced focal length f is, as often the case, treated as a variable, the curve $G_2 = F(G_1)$ will be a straight line in the stability diagram, of which the equation is read as

$$G_2 + \frac{l_1}{l_2} = \frac{l_1(l_2-R_2)R_1}{l_2(l_1-R_1)R_2}\left[G_1 + \frac{l_2}{l_1}\right] \qquad (3a)$$

or $\qquad G_2 - \frac{(l_2-R_2)R_1}{(l_1-R_1)R_2} = \frac{l_1(l_2-R_2)R_1}{l_2(l_1-R_1)R_2}\left[G_1 - \frac{(l_1-R_1)R_2}{(l_2-R_2)R_1}\right] \qquad (3b)$

Eqs.(3) show that resonators with fixed R , R , L , and L will be located on this line when the thermal induced focal length f is changing as shown in Fig.2.

The normalized spot sizes on surfaces of the two end mirror R_1 and R_2 are

$$\left(\frac{\pi W_1^2}{\lambda}\right)^2 = \left(\frac{\pi \omega_1^2}{\lambda}\right)^2 \left(\frac{1}{R_1} - \frac{1}{l_1}\right)^2 = -\frac{(D-1)(D-1-A)}{D(D-A)} \qquad (4a)$$

$$\left(\frac{\pi W_2^2}{\lambda}\right)^2 = \left(\frac{\pi \omega_2^2}{\lambda}\right)^2 \left(\frac{1}{R_2} - \frac{1}{l_2}\right)^2 = -\frac{(D-A)(D-1-A)}{D(D-1)} \qquad (4b)$$

where the normalized refractive power

$$D = \frac{l_1(l_1-R_1)}{R_1}\left(\frac{1}{l_1-R_1} - \frac{1}{l_2-R_2} - \frac{1}{f}\right) \qquad (5)$$

and

$$A = \frac{l_1(l_1-R_1)R_2}{l_2(l_2-R_2)R_1} \qquad \text{with} \qquad |A| > 1 \qquad (6)$$

It should be noted that in the case that $|A| < 1$, the above results re-

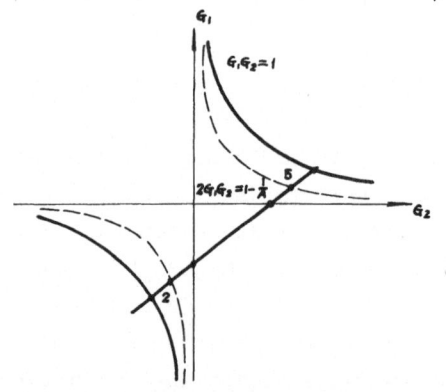

Fig.2. Stability diagram with the line of variable refractive power 1/f and the hyperbola of thermal stability.

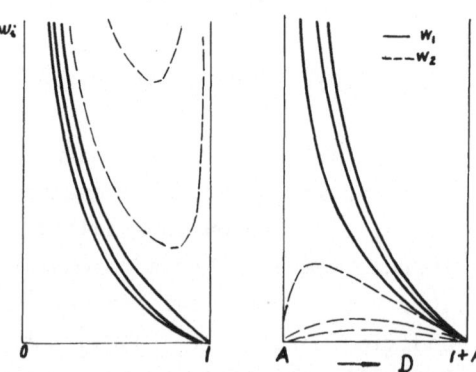

Fig.3. Normalized spot sizes as a function with respect to the normalized refractive power D in the case that

main valid provided that the subscripts of all quantities involved are interchanged , i.e., with 1 changing to 2 and 2 to 1. The two normalized spot sizes W_1 and W_2 as function of D are shown in Fig.3, where W_2 reaches its extrem values for certain values of D.

$$D = \frac{A+1}{2} \pm \frac{1}{2}\sqrt{A^2-1}$$

According to the definition of thermal stability condition proposed by Steffen et al, the resonator is thermal stable when

$$\frac{d\omega_2}{d(1/f)} = 0 \tag{7}$$

At this moment

$$\frac{1}{f} = \frac{1}{2}\left[\frac{1}{l_1} + \frac{1}{l_1-R_1} + \frac{1}{l_2} + \frac{1}{l_2-R_2} \mp \sqrt{\left(\frac{1}{l_2-R_2} - \frac{1}{l_2}\right)^2 - \left(\frac{1}{l_1-R_1} - \frac{1}{l_1}\right)^2} \right] \tag{8a}$$

or

$$G_1 = -\frac{l_2(l_1-R_1)}{2R_1}\left[\frac{R_1}{l_1(l_1-R_1)} - \frac{R_2}{l_2(l_2-R_2)} \pm \sqrt{\frac{R_2^2}{l_2^2(l_2-R_2)^2} - \frac{R_1^2}{l_1^2(l_1-R_1)^2}} \right] \tag{8b}$$

and

$$G_2 = \frac{l_1(l_2-R_2)}{2R_2}\left[\frac{R_1}{l_1(l_1-R_1)} - \frac{R_2}{l_2(l_2-R_2)} \mp \sqrt{\frac{R_2^2}{l_2^2(l_2-R_2)^2} - \frac{R_1^2}{l_1^2(l_1-R_1)^2}} \right] \quad \begin{matrix}(8c)\\ \\(8d)\end{matrix}$$

From Eqs.(8), it follows that

$$2 G_1 G_2 = 1 - \frac{l_2(l_2-R_2)R_1}{l_1(l_1-R_1)R_2} \tag{9}$$

And by Eqs.(3) and (8), we know that Eq.(9) is another form of Eq.(1).

Hence, resonators satisfying Eq.(1) are denoted by the hyperbola, but only those resonators satisfying Eqs.(8) , i.e., denoted by point 2 and 5, in Fig.2, are thermal stable for fixed parameters R_1, R_2, L_1 and L_2.

Furthermore, it is obvious that the resonator denoted by point 5 has larger mode volume than point 2.

From the above discussion, there are something evident to us. First, optical resonators satisfying Eq.(1) are not always thermal stable, or, Eq.(1) is not the sufficient condition for a resonator of thermal stability. Second, Eq.(8) is, with regard to the very end mirror,the sufficient condition for fabricating a thermal stable resonator. Third, as a consequence of the previous discussion, the equations of determining the thermal stability condition for the resonators with more than one internal lens have the similar form.

References

(1) J. Steffen and J. Lortscher, IEEE J. QE-8,239(1972)
(2) H.P. Kortz, R. Iffänder and H. Weber, App. Opt., 20, 4124(1981)
(3) Wei Guang Hui, Laser 83 Opto-Elektronik Abstracts,102

Laser als Wirtschaftsfaktor
Lasers – An Economic Factor

VDI-Technologiezentrum Düsseldorf

Laser Technology in the United States*

David A. Belforte

Now entering its 20th year in industrial materials processing, the laser has experienced slow but steady acceptance as an alternative to conventional manufacturing operations.

Industry Size

In the past few years industrial laser device sales have kept pace with the growth of the entire commercial device market. Table 1 is a compilation of data from the annual surveys as reported in Laser Report.

The data presented in Table 1 is considered to be conservative, since the compilers of the annual survey do not receive accurate information from Japan, a major laser manufacturer.

Table 2 is historical data on lasers sold exclusively for materials processing. Compiled from the Lasers & Applications magazine annual reports, the data is, as in the previous table, conservative due to unreported sales by Japanese laser manufacturers.

Table 3 shows the estimates of the number of annual installations of materials processing lasers. Worldwide data is extracted from Lasers & Applications magazine reports, U.S. share is estimated using percentages reported in Laser Reports annual survey and 1985 estimates are made by Belforte Associates.

The U.S. share of materials processing lasers appears to be quite conservative and Belforte Associates estimates that the U.S. now consumes about 55% of the annual output.

Industrial lasers are enjoying increasing usage in manufacturing operations . As shown in Table 4, the laser as a percentage of machine tool sales will reach 4.5% in 1987, with an expected 6.4% ten years from now.

---

*Presented at Laser '85 Opto Elektronik, Munich, July 1, 1985

Identifying the major U.S. users of industrial lasers is a diffi-
cult task. However, using information reported in the 1984 Electrical
Power Research Institute report, Materials Processing Lasers, Belforte
Associates has been able to identify eight major industry segments.

Table 5 is an analysis of all lasers installed in the U.S. which
are currently being used in a manufacturing operation. Not surprisingly
the top three industry segments employ 84% of all laser types and 66% of
all CO_2 lasers.

In 1984 the distribution of laser applications was 42% metalworking,
40% non-metal, and the remainder in marking applications.

Figure 1 represents Belforte Associates estimates of laser applications for 1984.

In their report Lasers in Metalworking, Tech Tran Corporation analyzed the generic laser applications for growth potential. Table 6 is a
summary of their findings as modified by Belforte Associates.

In the short term; welding, cutting (metals and non-metals) and
marking are expected to show high growth rates. In the long term; welding, non-metal cutting, and marking offer the highest growth potential.

The reasons behind the choice of lasers in manufacturing operations
are illustrated in Table 7. Here, cost benefits are judged against the
current cost of the laser in a matrix which should give indications as
to application growth.

It is obvious from this analysis that laser selling prices will have
to decrease significantly in order to sustain growth rates.

In the U.S. four industry segments are expected to spend moderate
to high amounts of money for new plant and equipment in the next five
years, according to Predicasts Inc.

In their 1985 capital equipment spending survey Production magazine
learned that these four segments will invest, in varying amounts, in laser processing.

These estimates are summarized in Table 8 and compared to current
laser usage as estimated by Belforte Associates.

Current Technology

In the U.S. fast axial flow lasers, rated in the 500 - 1500 watt range, enjoyed the highest growth in 1984.

Product introductions by Raytheon and Photon Sources expanded the choice of vendors. Coherent General and Spectra Physics continued to enjoy customer acceptance of their fast axial flow units.

Laser Corporation of America introduced and recently delivered the first units of their new Falcon 800, an 800 watt, compact, transverse flow CO_2 unit.

Three and six axis machining centers and three and five axis metal cutting systems are being installed in increasing numbers in U.S. manufacturing plants. General Motors will take delivery of the first five axis robotically manipulated laser spot welder for auto assembly operations.

Both YAG and CO_2 lasers are being used by aerospace companies for cutting applications.

Welding with multi-kilowatt CO_2 lasers is growing in auto manufacturing operations. Both 5 and 6 kilowatt systems are employed by General Motors and Ford.

The first industrial cladding system was installed in a plant manufacturing gate valves for the petroleum industry.

Both Q-switched YAG and TEA CO_2 lasers are being installed in large numbers in a cross section of U.S. manufacturing operations for laser marking.

Current Technology Trends

Recently introduced by Laakmann Electro Optics is a flexible metal CO_2 laser beam delivery system. Some investigators believe that flexible waveguide type delivery may be a better method than multi-mirror articulated arm.

Anticipated heavy funding for the U.S. Strategic Defense Initiatve
(Star Wars) is expected to lead to commercialization of high energy excimer lasers. If true, then we can expect a faster growth for excimers
in industrial applications.

Spectran Corporation (Sturbridge, MA) recently announced the availability of a new fiber material said to be able to pass higher CO_2 laser
powers. If these materials can transmit several hundred watts of power,
then multi-dimensional cutting will be facilitated.

Cincinnati Milacron announced the development of a prototype through
the robot beam delivery system. This approach reduces the number of mirrors in the systems, aiding in system alignment.

As stated earlier, General Motors has purchased a multi-axis laser
spot welding system from Prima Progetti. This unit will be the initial
assembly line laser welding system.

After lagging in usage, U.S. companies are now using pulsed CO_2
for precision metal cutting applications. Better edge quality, easier
programming, and reduced dross are the benefits experienced.

Future Technology Trends

If the cost of industrial lasers can not be significantly reduced
more power for less money is an alternative to sustain industry growth.

High average power YAG slab (TIR) lasers are now in the prototype
stages. Developers expect 1 kilowatt units to be commercialized in
three years.

Higher average power TEA or excimer lasers will be products within
five years, opening new applications opportunities.

High power CO lasers, currently under development in Japan, are expected to be ready for industrial applications in three years. This laser, with a wavelength more compatible with fiber optic materials, could
become the next metalworking standard.

Investigation into the results from multi-kilowatt pulsed CO_2 welding and cutting could lead to the development of high average power
pulsed CO_2 units.

REFERENCES

Laser Report Annual Economic Reviews, 1984 and 1985, Published by Penn-Well Publications, Littleton, MA, USA

The Laser Marketplace - 1985, January 1985, Lasers & Applications, p47-56

The Laser Marketplace - 1984, January 1984, Lasers & Applications, p51-58

U.S. Machine Tool Markets, January 1984, Published by Predicasts Inc., Cleveland, OH, USA

Lasers In Metalworking, 1983, Published by Tech Tran Corporation, Naperville, IL, USA

Assessment of Materials Processing Lasers, EPRI EM-3465, May 1984, Published by Electric Power Research Institute, Palo Alto, CA, USA

Capital Spending 1985-1986, Production Magazine, March 1985, p53-59

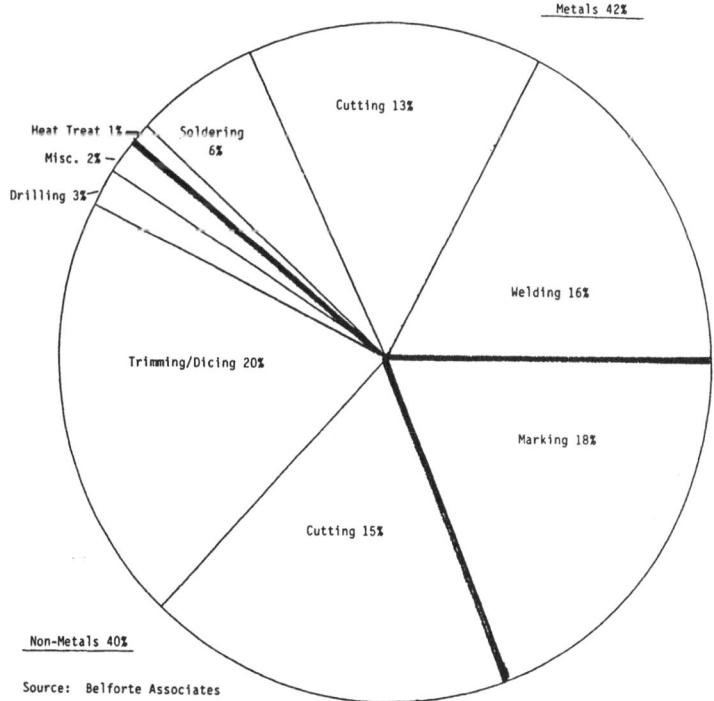

FIGURE 1, DISTRIBUTION OF INDUSTRIAL APPLICATIONS
(1984 Installations)

TABLE 1. COMMERCIAL LASER DEVICE SALES (WORLDWIDE)

($ million - % growth in brackets)

Laser Type	1982	1983	1984	1985
CO_2	60.3	74 (23)	89 (20)	116 (30)
Solid State	57.3	69 (20)	82 (18)	100 (22)
Excimer	8	10.7 (34)	14 (30)	19 (36)
Total	125.6	153.7 (22)	185 (20)	235 (27)

Source: Laser Report

TABLE 2. LASER SALES - MATERIALS PROCESSING (WORLDWIDE)

($ million)

Laser Type	1982		1983		1984	
	Units	Unit $	Units	Unit $	Units	Unit $
Solid State	600	23.0	630	24.0	733	27.6
CO_2	575	41.8	725	46.0	942	55.7
Excimer	4	1.4	3	7.0	6	8.3
Total	1179	66.2	1358	77.0	1681	91.6
System Sales ($ million)		145		186.4		232.9

Source: Lasers & Applications

TABLE 3. INSTALLATIONS - MATERIALS PROCESSING LASERS

(Solid State and CO_2)

	1982	1983	1984	1985
Worldwide	1175	1355	1675	2010
United States	588 (50)	610 (45)	804 (48)	965 (48)

Source: 1. Lasers & Applications Annual Reports (1/85)

2. Laser Report Annual Economic Reviews (1/85)

3. Average Industry Estimates

TABLE 4. MATERIALS PROCESSING EQUIPMENT TRENDS - UNITED STATES

	1967	1982	1987	1995	% Annual Growth Historic 67/82	Projected 82/87	87/95
Machine Tool Shipments (mil. $)	1816	3726	7580	16430	4.9	15.3	10.2
Industrial Laser Shipments (mil. $)	<1	120	345	1055	10.0	20.0	15.0
Laser as % Machine Tool	0	3	4.5	6.4			

Sources: Lasers & Applications
          Predicasts
          Belforte Associates

TABLE 5. INDUSTRIAL LASER APPLICATIONS BY INDUSTRY

(% of Total Installations)

Industry Segment	All Lasers	CO_2 Lasers
Electrical/Electronic Machinery	78	26
Fabricated Metal Products	6	27
Transportation Equipment	5	13
Machinery	4	10
Tobacco	1	6
Rubber & Plastic Products	1	5
Paper & Allied Products	1	3
Textiles & Apparel	1	2
Misc.	3	8
	100	100

Source: EPRI Report EM-3465

         Belforte Associates

TABLE 6. APPLICATIONS GROWTH RATE (U.S.)

	1984-85	1987-89	1990-
Welding	H	H	H
Cutting (Metal)	H	H	L
Cutting (Non-Metal)	H	H	M
Heat Treating	L	L	M
Cladding	L	L	M
Alloying	L	L	M
Melting	L	L	M
Marking	H	H	M
Soldering	M	H	H
Drilling (Non-Metals)	M	M	M
Drilling (Metals)	M	L	L

Key: H = High annual rate of increase (over 20% per year)
      M = Medium annual rate of increase (10-20% per year)
      L = Low annual rate of increase (0-10% per year)

Sources: Belforte Associates
          Tech Tran

TABLE 7. MANUFACTURING COST BENEFITS — LASER VS. CONVENTIONAL PROCESSING

	Rate	Secondary Operations	Quality	Scrap	Automation	Laser Cost
Welding (Thin section)	H	L	L	M	M	H
Welding (Thick section)	H	M	L	L	M	H
Cutting (Metal)	H	H	H	H	M	M
Cutting (Non-metals)	H	L	M	M	H	H
Heat Treat (On-line)	L	M	L	M	M	H
Cladding	H	H	M	M	M	H
Alloying	M	H	M	M	H	H
Marking	H	-	M	-	H	M
Soldering	M	-	H	H	H	M
Drilling	H	M	L	M	H	M

Source: Belforte Associates

Key:

H = Significant cost advantage
M = Marginal cost advantage
L = Insignificant cost advantage

TABLE 8. MARKETS OF OPPORTUNITY FOR CO_2 LASERS

Industry Segment	Projected[1] P & E Exp.	Planned Spending[2] Metal Cutting	Heat Proc.	Weld	Current Laser[3] Use
Fabricated Metal Products	Moderate	High	Low	Low	High
Machinery, Except Electrical	High	High	Moderate	Moderate	Moderate
Elec. & Electr. Machinery	High	Low	Low	Low	High
Transportation	High	High	Low	Moderate	High

Sources:

1. Predicasts Inc., U.S. Machine Tool Markets 1984
2. Production Magazine, 16th Capital Spending Survey 1985
3. Belforte Associates

Laser Technology in Japan

T. Maruyama, Marubun Corporation, 8-1 Nihombashi, Tokyo 103 / JAPAN

More than 20 years ago, during my soul searching days I have encountered with this magical and mysterious word "holography" that experts tell me that this may eventually replace all mannequins in all show windows and also televisions with three dimensional images.

A heart of this holographic system would be a laser, some light source which was known as a death ray.

My natural instinct tells me that this laser would someday lead me to a success and fulfilment of my desires and ambitions.

In the beginning I have consulted with many well-known researchers and recent returnee from abroad and found that these researchers were looking for a person or an organization which would handle importation of laser components and accessories such as ruby rods, flash lamps, detectors and others. Of course I jumped to this opportunity.

This was the beginning of my involvement and association with researchers and professors in this ever-expanding industry of optoelectronics. Through past 23 years in this field, I was very fortunate to meet many researchers and organizations not only in Japan, but also in the world. In my talk I would like to present the history and future trends of Japanese laser industry and its market through my experience and association with domestic researchers and organizations.

* more information is provided by VDI Technologiezentrum, Physikalische Technologien, Graf-Recke-Strasse 84, 4000 Düsseldorf 1 / D

Lasertechnik in der Bundesrepublik Deutschland, Bemerkungen zu Markt und Wirtschaftsfaktor

G. Rauscher, München

Die Lasertechnik berührt und integriert eine Vielzahl von Industrie- und Wissenschaftsgebieten, wie

- Maschinen- und Werkzeugmaschinenbau
- Elektrotechnik und Elektronik
- Datenverarbeitung und numerische Steuerungen
- Optik, Feinmechanik und Meßtechnik
- Werkstoffwissenschaften und Fertigungstechnik
- Nachrichtentechnik

Die Lasertechnik gibt diesen Gebieten neue Impulse für die Entwicklung neuer Produkte und rationeller Fertigungslinien.

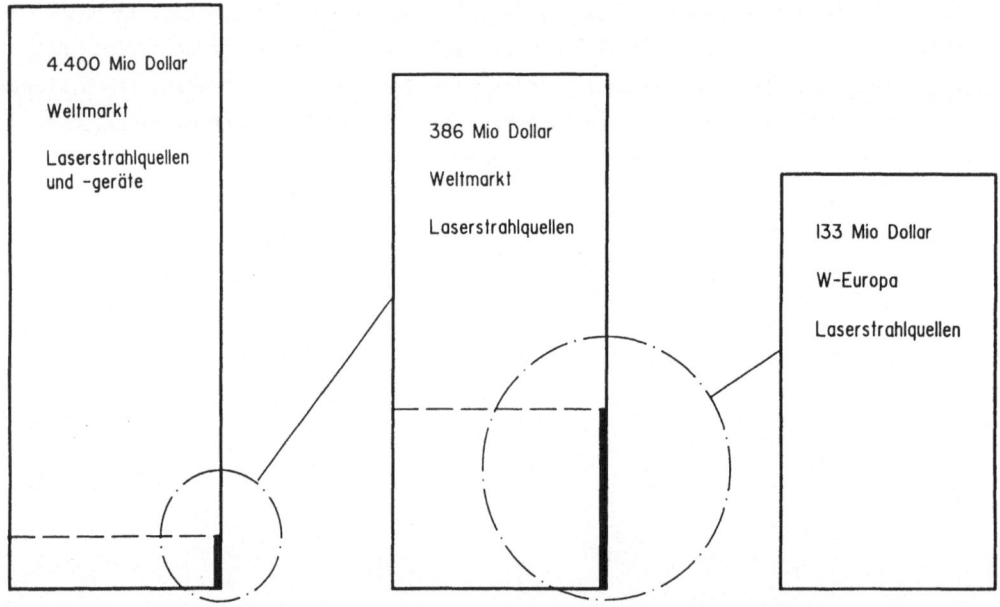

4.400 Mio Dollar

Weltmarkt

Laserstrahlquellen und -geräte

386 Mio Dollar

Weltmarkt

Laserstrahlquellen

133 Mio Dollar

W-Europa

Laserstrahlquellen

LASERMARKT 1984

Markt

Der Lasermarkt in der Bundesrepublik Deutschland für Laserstrahlquellen und Lasersysteme betrug 1984 ca. 380 Millionen DM, davon
entfielen etwa 1/3 auf die Laserstrahlquellen und etwa 2/3 auf
die Lasersysteme. Den größten Marktanteil hatte - ähnlich wie in
W-Europa - das Gebiet der Materialbearbeitung, das am schnellsten
wachsende Gebiet ist das der Nachrichtentechnik.

Faßt man den Begriff "Lasermarkt" etwas weiter und zählt auch die
Umsätze an Systemen und Geräten mit hinzu, bei denen der eingebaute
Laser zwar wesentlich, aber von den Kosten nicht bedeutend ist,
so ergibt sich für 1984 in der Bundesrepublik Deutschland ein Markt
für "Geräte mit Laser" von ca. 500 Millionen DM.

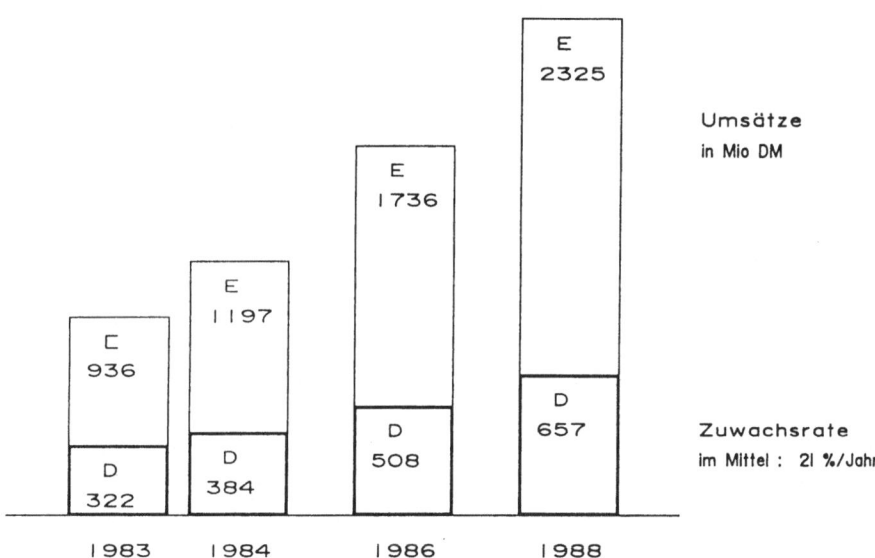

Laserfirmen :   Deutsche Firmen : 21
                  Vertr. ausländ. Firmen : 29

LASERMARKT BR DEUTSCHLAND

Markt Laserstrahlquellen: 1984

ca. 120 Mio DM

	Zuwachs %	Umsatz Laser-strahlquellen	Multiplikations-faktor Syst.: Laser	Umsatz Systeme mit Laser
Materialbearbeitung	+30	40	3	120
Information processing	+20	10	11	110
Messen/Ausrichten (Bau)	-	5	5	25
Medizin	+30	15	5	75
Meßtechnik	+15	10	4	40
Opt. Nachrichtentechnik	+40	5	10	50
Forschung u. Entwicklg.	+20	30	3	90

Markt Systeme mit Laser: 1984   ca. 500 Mio DM

Umsätze in Mio DM

MARKT "LASER" (BR DEUTSCHLAND)

Dieser Markt hat Zuwachsraten von 20% bis 40%, die Innovationslokomotive ist immer der Laser als kohärente Strahlquelle.

Der Markt von Laserstrahlquellen in der Bundesrepublik wird je
etwa zur Hälfte von einheimischen bzw. von ausländischen Firmen
getragen. Bei leistungsstarken CO_2-Lasern sind das insbesondere
Firmen aus den USA, bei den Festkörper-Impulslasern Firmen aus
der Schweiz und Großbritannien.

Laser als Wirtschaftsfaktor

Lasertechnik ist eine Schlüsseltechnologie mit großer Breitenwir-
kung, sie ist wie die Mikroelektronik ein Innovationsmotor für
viele Branchen.

An einigen, fast willkürlich herausgegriffenen Beispielen soll
verdeutlicht werden, in welchem Umfang der Markt mit Laserstrahlquellen einen wesentlich größeren von Geräten mit Lasern trägt:

Fertigungstechnik (Laser-Materialbearbeitung)

* Schneiden, Ritzen
* Schweißen
* Bohren
* Oberflächenveredelung

* Beschriften
* Materialbearbeitung elektronischer
  Komponenten

Medizintechnik (Laser-Medizin)

* Chirurgie, Endoskopie
* Bestrahlungen, Dermatologie

* Augenheilkunde

Kommunikation und Informationstechnik

* Nachrichtenübertragung
* Audio- und Videodisc-Plattenspieler,
  Massenspeicher

* Reprographie
* Farbscanner
* Laserdrucker
* Kassenterminals

Laser-Meßtechnik und -Analytik

* Entfernungsmeßtechnik, Justieren
* Oberflächenrauhheit
* Kurzzeitmeßtechnik

* Mikroanalyse
* Massenspektrometrie
* Prozeßkontrolle

Energieforschung mit Lasern

* Laser-Isotopentrennung

* (Laserfusion)

Grundlagenforschung mit Lasern

* Spektroskopie
* Nicht-lineare Optik
* Integrierte Optik

* Laser-Chemie
* Kurzzeitphysik und -chemie

Quelle: BMFT-Broschüre 1985

LASERANWENDUNG, wesentliche Gebiete

a) Materialbearbeitung mit CO_2- und Nd:YAG-Laser.

1985 wird in der Bundesrepublik Deutschland mit CO_2-Laser
im Wert von ca. 20 Millionen DM ein Umsatz an Werkzeugmaschinen
mit ca. 80 Millionen DM erzeugt. Bei den Nd:YAG-Lasern ist die
Umsatzvervielfachung etwas geringer; mit ca. 10 Millionen DM
an Strahlquellen, werden ca. 30 Millionen DM an Systemen umgesetzt.

Die Zuwachsraten mit 30% - 50% sind beachtlich. Der Laser als
flexibles Werkzeug, die Mehrfachnutzung (Strahlteiler) und die
Automatisierungsmöglichkeiten sind wesentliche Triebfedern des
Lasereinsatzes.

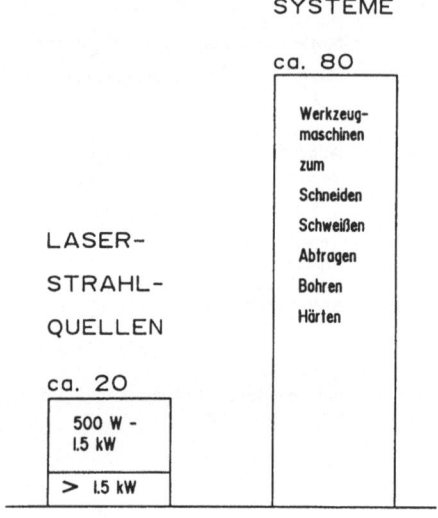

SYSTEME

ca. 80

Werkzeug-
maschinen
zum
Schneiden
Schweißen
Abtragen
Bohren
Härten

LASER-
STRAHL-
QUELLEN

ca. 20

500 W -
1.5 kW

> 1.5 kW

Umsatz 1985 in Mio DM   (D)

- Zuwachsraten: 30 % – 50 % p.a.

- Automatisierung (Roboter)

- Verbesserung der Verfügbarkeit von
  Laser und Komponenten
  Ziel: > 90 %

- Job-shops (Schweißen, Schneiden,
                        evtl. Härten)

- Ausstrahlung auf Maschinenbau
  Führungsmaschinen
  CNC-Systeme

- Innovation über
  Konstruktionslabors

MATERIALBEARBEITUNG CO_2-LASER

Feinwerktechnik, elektrische Bauteile

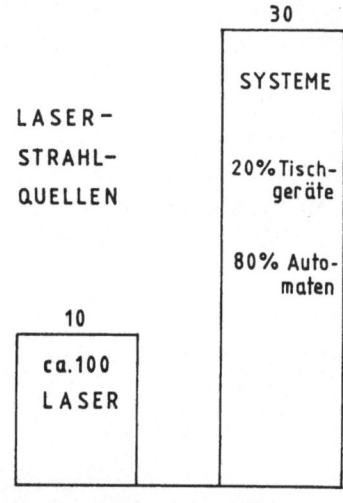

30

SYSTEME

20% Tisch-
geräte

80% Auto-
maten

LASER-
STRAHL-
QUELLEN

10

ca.100
LASER

Umsatz 1985 in Mio DM
(D)

— Werkzeug zur Rationalisierung
   und Automatisierung

— Justieren - Fixieren
   dynamisch   justieren / fixieren

— Vielfachpunkte durch Strahlaufspaltung

— Verfügbarkeit > 90%

Trends :

— Kostensenkung durch Mehrfachnutzung
   (Lichtleitfasern)

— Höhere mittlere Leistung
   (Evweiterung der Einsatzfelder)

— µP- ansteuerbar

MATERIALBEARBEITUNG   Nd :YAG/Nd:GLAS-LASER (gepulst)

b) Bauelementeherstellung

Für den Abgleich von Schichtschaltungen und für die Ausbeute-
erhöhung bei der Herstellung elektronischer Speicher (durch
Einführung sogenannter Laser-Redundanzkonzepte) sind Laserverfahren heute die Standardverfahren geworden. Die Weiterentwicklung der Lasersysteme zu rechnergesteuerten, vollautoma-
tischen Produktionsanlagen wurde hier konsequent durchgeführt.
Mit Nd:YAG-Lasern im Wert von ca. 5 Millionen DM wird in der
Schweiz, Österreich und Deutschland ein Volumen von 30 Millionen DM an Geräten umgesetzt. Die Systemkosten dieser komplexen, rechnergesteuerten Automaten liegen um die 700.000,--
bis 800.000,-- DM. Als längerfristige Zuwachsraten werden auf
diesem Markt 10% bis 15% erwartet, kurzfristig können es durch
den zunehmenden Einbau elektronischer Schaltungen in die Produkte
der KFZ-Industrie auch wesentlich mehr sein.

Abgleich von Schichtschaltungen − Redundanzkonzepte für Speicher

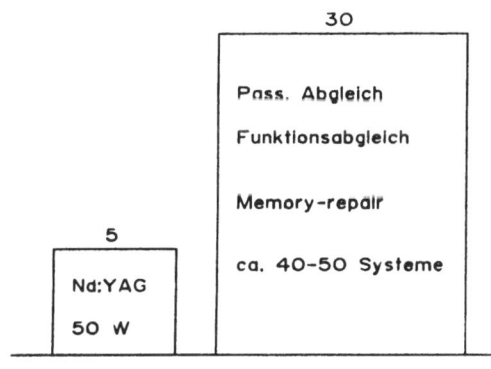

Umsatz 1985 in Mio DM
(D / CH / A)

- Zuwachsrate: 10 % - 15 %
- Zunehmend: Funktionsabgleich

- Entwicklungstrends:
  * Memory-repair:
    Strahldurchmesser ($< 3\ \mu$m)
    Positioniergenauigkeit ($< 1\ \mu$m)

  * Schichtschaltungen:
    Produktivitätserhöhung

- Laser Standardverfahren

MATERIALBEARBEITUNG Elektron. Komponenten

c) Drucktechnik

Die Drucktechnik ist ein typisches Beispiel einer Branche, bei
der die Laser in den letzten 10 Jahren große Innovationsschübe
ermöglichte.
Die in der Drucktechnik führende Fa. Dr. Hell, Kiel, hatte im
Geschäftsjahr 83/84 einen Umsatz weltweit von 714 Millionen
DM, ca. 50% davon waren Geräte mit Laser. In Deutschland setzte
die gleiche Firma ca. 60 bis 70 Millionen DM an Geräten mit
Laser um. Die Systemkosten dieser drucktechnischen Geräte liegen
im Bereich von 250.000,-- bis 800.000,-- DM; die Laserstrahlquelle hat davon nur einen Anteil von 2,5% bis 10%. Das Problem
der Laser in den drucktechnischen Geräten ist die noch nicht
ausreichende Zuverlässigkeit, verbesserten Systemen steht hier
ein großer Markt offen.

Beispiele : Farbscanner, Fotosatz, Flachbettgeräte (Zeitungssatz)

—Systemkosten :   250000,- bis 800000,- DM

davon Laserstrahlquelle ca. 2,5% -10%

bzw. Laserstrahlquelle komplett

mit opt. Komponenten ca. 20% – 35%

Opt. Komponenten : Modulatoren

Polygonspiegel

Objektive. Spiegel etc.

—Problem : Laserzuverlässigkeit

125

ca. 50%
Systeme
mit
Laser

Umsatz Fa. Hell in D
1984, in Mio DM

—Laser als Innovationslokomotive

DRUCKTECHNIK (information processing)

Auch die Schnelldrucker der Großrechner sind erst durch den Einsatz von Lasern möglich geworden. Von diesen sogenannten Laserdruckern im Wert von 300.000,-- DM bis 500.000,-- DM werden in der Bundesrepublik pro Jahr etwa 1.500 Stück hergestellt, ca. 500 werden jährlich in Deutschland installiert. Das wesentliche Teil dieser Drucker sind schnell modulierbare HeNe-Laser im Wert von 3.000,-- bis 10.000,-- DM.

c) Meßtechnik - Optische Prüftechnik

Die holographische Qualitätsprüfung durch lasergestützte, bildverarbeitende Prüfsysteme ist die Innovation der letzten Jahre.

In der Bundesrepublik werden für diese Sparte ca. für 10 Millionen DM Systeme umgesetzt; der Wert der installierten Laser liegt bei ca. 25% bis 33% der Systemkosten.

z. B. Holographische Qualitätsprüfung
   Lasergestützte und bildverarbeitende Prüfsysteme

10^7

Optische Prüftechnik: vorwiegend analog

Trend zu Digitaltechnik
   d.h. zu Systemen mit Laser
      und elektron. Datenverarbeitung

Erwartung: steigender Umsatz,
      mit gegenüber heute
      sinkendem Laseranteil ($\leq 10\%$)

25% — 33%
Laser -
strahlquellen

Umsatz 1985 Mio DM

(D)

Opt. Prüftechnik

Probleme: Laserkosten
      Zuverlässigkeit der Laser

MESSTECHNIK — OPTISCHE PRÜFTECHNIK

Der Trend der optischen Prüftechnik in der Bundesrepublik geht mit zunehmenden Maße von der Analog- zur Digitaltechnik. Durch diesen Trend nimmt der Umfang pro System an elektronischer Datenverarbeitung stark zu, relativ dazu der Laseranteil ab. Experten schätzen, daß der Laseranteil bei lasergestützten, bildverarbeitenden Systemen auf unter 10% absinken wird, allerdings wird gleichzeitig eine starke Ausweitung des Geräteabsatzes erwartet.

Markenpotential (Materialbearbeitung)

Für das Gebiet der Materialbearbeitung werden die großen Zuwachsraten des Lasergerätemarktes in den nächsten Jahren anhalten. Das Anwendungspotential ist groß. In der metallverarbeitenden Industrie wird die Firmenanzahl der potentiellen Laseranwender auf ca. 30.000 geschätzt, in der nicht-metallverarbeitenden Industrie sind es ca. 5.000 Firmen. Die heutige Durchdringtiefe dürfte noch unter 10% liegen.

In diesem Potential von 35.000 möglichen Laseranwendern liegt sicherlich ein Markt, der 3 bis 5mal größer als der heutige ist. Derzeit sind in der Bundesrepublik vermutlich kaum mehr als 700 bis 1.500 Lasergeräte im technischen Einsatz, davon in einigen Großfirmen schon mehr als 50 Systeme.

A) Metallverarbeitende Industrie

	Firmenanzahl	
	D	Europa
Maschinenbau	15 600	79 600
Mechan. Bauteile	8 300	44 700
Elektr. Bauteile	5 200	22 600
Fahrzeugbau	2 100	16 400
sa.	31 200	163 300

B) Nicht-Metallverarb. Industrie

	Firmenanzahl	
	D	Europa
Bekleidung	1 760	7 950
Schuhe	320	2 230
Holzwaren	560	7 530
Kartonagen	370	2 650
Lederwaren	340	2 300
Plastik / Gummi	1 300	10 340
Uhren / Schmuck	310	2 910
Glas / Keramik	490	2 110
sa.	5 450	38 020

Markt für Lasersysteme (Potentielle Laseranwender)

Arbeitsplätze

In der Bundesrepublik gibt es etwa 20 einheimische Laserfirmen
und ca. 30 Vertretungen bzw. Tochtergesellschaften ausländischer
Firmen. Über die Zahl der Mitarbeiter in diesen Laserfirmen liegen
keine Werte vor. Schätzt man aber aus den bekannt gewordenen europäischen Zahlen entsprechend dem Umsatz, so ergeben sich für die
Bundesrepublik für 1983 ca. 1.100 Mitarbeiter in Laserfirmen; diese
werden bis 1988 auf ca. 2.700 ansteigen. Die Zahl der Arbeitsplätze,
die durch den Bau und Einsatz von Geräten mit Laser in Firmen der
unterschiedlichsten Branchen geschaffen bzw. erhalten wurde, ist
wesentlich höher; sie wird für 1983 auf knapp 6.000 und für 1988
auf ca. 19.000 geschätzt.

Ausblick

Von der Lasertechnik geht ein Innovationsschub, ein Technologie-
schub für eine Vielzahl von Branchen aus. Mit der Entwicklung von
Lasern und deren Komponenten sowie durch deren Einbau in Systeme
ist ein Partizipieren an stark wachsenden Märkten möglich. Die
Bundesrepublik Deutschland mit den traditionell starken Branchen
Maschinenbau, Feinwerktechnik, Optik und Elektrotechnik hat gute
Voraussetzungen auf dem Lasermarkt und auf dem Markt der Geräte
mit Laser eine führende Position einzunehmen.

810

Quellenangaben:

- BMFT (Broschüre zur Hannover-Messe 1985)

- Angaben der Firmen:

  BIAS, Bremen
  ESI GmbH, München
  Haas, Schramberg
  Dr. Hell GmbH, Kiel
  W.C. Heraeus GmbH, Hanau
  MAN NT, München
  Meditec, Heroldsberg
  Messer-Griesheim Strahltechnik, Puchheim
  Rofin-Sinar, Hamburg
  Rottenkoeber Holo-System, München
  Siemens AG, München

- Studien:

  Frost and Sullivan, "The commercial and industrial
                       Lasermarket in W. Europe (1984)"

  IRD, "Laser Markets Opportunities (1984)"

- Zeitschriften

  Laser-Focus (1.85)
  Laser-Report (07.01.85)
  Laser and Applications (1.85)

Wieviel Laser braucht die Produktionstechnik?

Prof. Dr.-Ing. Dr.h.c. W. König; Dr.-Ing. Cl. Schmitz - Justen
Fraunhofer-Institut für Produktionstechnologie, Aachen

Will man diese Frage beantworten, d.h. die Stellung und Bedeutung der
Lasertechnik innerhalb der industriellen Fertigungstechnik - d.h. dem
gesamten Wertschöpfungsprozeß vom Rohteil bis zum Fertigprodukt - näher
untersuchen, so sind als Prämisse zunächst die Zielsetzungen eines heutigen Produktionsbetriebes zu betrachten. Diese können im wesentlichen
unter den Begriffen "Flexibilität", "Qualität" und "Wirtschaftlichkeit"
zusammengefaßt werden und konzentrieren sich darauf, mit minimalem Aufwand eine marktgerechte Produktqualität bei sinkenden Losgrößen und
steigender Variantenzahl zu erreichen.
In der Verfolgung dieser Zielsetzung ist das Unternehmen verschiedenen,
zeitlich veränderlichen äußeren Einflußgrößen ausgesetzt, wie sie beispielhaft in Abbildung 1 aufgezeigt werden. Neben dem Produktspektrum,
dem Personaleinsatz und auch der Fertigungstiefe ist hierbei das technologische und wirtschaftliche Potential der eingesetzten Fertigungsprozesse ein wesentlicher Faktor für den Erfolg des Unternehmens.
Wenn nun der Laser - wie Stanley Ream es formuliert - zumindest in den
USA als industrielles Werkzeug die Phase von "promotion and pessimism"
durchschritten hat und sich gegenwärtig in einem Stadium von "acceptance and application" befindet, so müssen die zentralen Fragen an dieser Stelle für den Fertigungstechniker lauten: Welchen Beitrag leistet
der Laser heute in der Bundesrepublik Deutschland, um die erwähnten
Zielsetzungen der Produktionstechnik zu erfüllen? Auf welche Art und
Weise geschieht dies, und welche Entwicklungen aus der jüngeren Vergangenheit weisen den Weg für entscheidende zukünftige Trends?
Das Ziel dieser Überlegungen soll jedoch auch sein, aus fertigungstech-
nischem Blickwinkel hier auf der Lasermesse, die ja eigentlich das Fo-

rum der Ausrüster und Anbieter ist, die Sichtweise und das Problemver-
ständnis aktiver und potentieller Anwender der Lasertechnik in der Entwicklung über die letzten Jahre hinweg darzustellen.

Eine erste Antwort auf die gestellten Fragen ergibt sich, wenn man die
Anwendungsfelder des Lasers innerhalb der Produktionstechnik betrachtet.
Diese sind zum einen in der Materialbearbeitung mit dem Leistungslaser
zu sehen, zum anderen aber auch in der fertigungsbegleitenden Lasermeß-
technik.

In der Materialbearbeitung mit CO_2-Lasern im Leistungsbereich zwischen
250 W und 10 kW kann die Gesamtzahl der in der Bundesrepublik Deutschland gegenwärtig installierten Anlagen auf über 200 Stück beziffert
werden. Auch heute noch konzentriert sich ein Großteil dieser Anwendungen auf das Schneiden von ebenen Bauteilen geringer Materialstärke
im Apparatebau und in der Elektrogerätetechnik. Eine herausragende Rolle
nehmen die Automobilindustrie und ihre Zulieferer ein, die nicht nur
den größten Einzelanteil im Anwenderspektrum ausmachen, sondern das
Innovationspotential der Lasertechnik bisher in der größten Fertigungs-
breite und -tiefe nutzen. Durchweg zurückhaltend verhält sich hingegen
bisher der klassische Maschinenbau. Als Ursache ist die geringere Bedeutung der bislang dominierenden Schneidtechnik in dieser Branche anzusehen. Hier ist jedoch ein Anstieg der Nutzung durch die Ausdehnung
der Lasertechnik auf weitere Fertigungsverfahren durchaus zu erwarten.
Dabei ist in erster Linie das Laserschweißen zu nennen, das im Multi-
kW-Bereich ja derzeit erste Anwendungen unter Produktionsbedingungen
erfährt, und darüber hinaus auch die Oberflächenwärmebehandlung.

Ein Beispiel dafür zeigt <u>Abbildung 2</u> mit der martensitischen Randzonen-
härtung des Lagersitzes auf einer Getriebewelle aus Vergütungsstahl.

Nicht zu vernachlässigen ist auch der Einsatz von Festkörperlasern in
der Produktionstechnik, wobei hier nur die Materialbearbeitung mit dem
Ziel der Bauteilfertigung berücksichtigt werden soll. Allein hier sind
über die letzten Jahre Installationsquoten von über 20 Anlagen / Jahr
in der Bundesrepublik Deutschland zu verzeichnen, wobei die Anwendungen
hauptsächlich im Triebwerks- und Feingerätebau mit den Mikrobearbei-
tungsverfahren Bohren, Schweißen und Schneiden zu finden sind.

Die Nutzung der Laserfertigungsmeßtechnik betrifft inzwischen ebenfalls
eine wachsende Anzahl von Industriebereichen. Die berührungslose "in-
process-Vermessung" von Bauteilen begleitend zum Bearbeitungsvorgang
sollte im zunehmenden Automatisierungstrend von großem Interesse sowohl
für die Hersteller von Werkzeugmaschinen als auch für deren Nutzer sein.
Schon etabliert ist die Lasergradheits-, -positions- und -geschwindig-
keitsmeßtechnik als Abnahmeverfahren für Maschinenstrukturen. Eine in-

teressante Erweiterung dieses Meßprinzips zeigt <u>Abbildung 3</u> mit einem
Laser-Ebenheitsmeßsystem, das zur Flächenprüfung, beispielsweise an
Granitplatten oder Maschinenbetten, herangezogen wird.

Hochpräzise Werkzeugmaschinen sind bereits heute mit interferometrischen
Laserwegmeßsystemen ausgestattet. Die Sensorentechnik auf Laserbasis
für die berührungslose fertigungsbegleitende Bauteilvermessung steht
gegenwärtig erst am Beginn einer stark expansiven Entwicklung, die
auch in enger Verflechtung mit Arbeiten auf dem Gebiet der Nachrichtentechnik zu sehen ist.

Wie selbst dieser knappe Abriß der aktuellen Laseranwendungen in der
Produktionstechnik zeigt, hat der Laser heute durchaus den Status eines
unübersehbaren Wirtschaftsfaktors erreicht. Dieser Sachverhalt wirft
die Frage auf, wer insgesamt an der industriellen Bereitstellung dieser
Technik nicht nur für, sondern in der Bundesrepublik Deutschland betei-
ligt ist. Und hier hat in der Tat in der jüngeren Vergangenheit - etwa
über die letzten 5 Jahre - ein gewaltiger Umbruch stattgefunden, der am
Beispiel der Materialbearbeitung mit Leistungslasern dargestellt sein
soll (<u>Abbildung 4</u>).

So existieren inzwischen mehrere inländische Laserquellenhersteller mit
Produkten bis in die Multi-kW-Leistungsklasse hinein, die auch international bereits Marktgeltung besitzen, weitere werden gewiß noch hin-
zukommen. Noch wesentlicher ist eigentlich in diesem Zusammenhang die
Tatsache, daß substantielle Komponenten, wie etwa reflektierende Optik-
bauteile, in hoher und verläßlicher Qualität aus inländischer Herstellung zum Einbau kommen. Dies löst zumindest hier das Problem der teilweise unkalkulierbaren Handelsrestriktionen bei Abhängigkeit von ausländischen Anbietern. Ein ähnliches Aufholen des Technologievorsprunges
deutet sich für die nächste Zukunft auch auf dem Sektor der Transmissionsoptik an.

Deutlich verbessert hat sich über die letzten Jahre auch das Angebot
und die Qualität der Handhabungseinrichtungen, die unter realen Produktionsbedingungen einsetzbar sind. Ebenso gelten die Steuerungssysteme
mehrerer Hersteller inzwischen als industriereif in Bezug auf spezifische Anforderungen und Voraussetzungen der Lasermaterialbearbeitung.

Applikationslabors, von denen derzeit bereits über 1o in der Bundesrepublik Deutschland existieren, bieten eine wesentliche Hilfe bei der
Absicherung technologischer Parameter im Vorfeld produktionsmäßiger Anwendungen. Darüber hinaus hat sich in den letzten Jahren eine Reihe von
reinen Lohnfertigungsbetrieben etabliert, die ein weitgespanntes Fer-
tigungsspektrum offerieren. F + E - Institutionen schließlich haben mit
grundlagen- und anwendungsorientierten Arbeiten eine Basis für die wei-

tere Laserquellenentwicklung und für zahlreiche Anwendungen gelegt, die
permanent weiter ausgebaut wird.

Dieser Abriß, dessen Tendenz sicher ebenso - wenn auch unter anderen
Randbedingungen - Gültigkeit für die Laseranwendung in der Fertigungsmeßtechnik besitzt, belegt deutlich das Wachstum der Lasertechnik auch
in wirtschaftlicher Hinsicht. Selbst wenn man die weitläufig prognosti-
zierten Zuwachsraten mit Skepsis betrachtet, so ist die wachsende Bedeutung dieses Industriebereiches doch unübersehbar.

Galten die bis hierher ausgeführten Überlegungen eher global und über-
greifend der Stellung des Lasers in der Produktionstechnik, so läßt
sich eine Reihe von Fragen und Herausforderungen bei der Einführung
dieses Werkzeuges besser in mikroskopischer Sicht eines einzelnen Anwenders erörtern. Auch hierbei soll als Beispiel wieder die Lasermaterialbearbeitung gewählt werden. Von welchen Voraussetzungen und Randbedingungen muß also etwa der Fertigungsleiter eines Produktionsbetriebes ausgehen, um eine fundierte Entscheidung zur Einführung der Lasertechnik zu treffen? Wo liegt das Einsatzspektrum und welche Faktoren
wirken sich fördernd bzw. hemmend auf die Anwendung aus? (Abbildung 5)

Im Zusammenhang mit diesen Fragestellungen ist es äußerst interessant,
die Struktur der Unternehmen zu betrachten, die sich mit der Anwendung
des Lasers auseinandersetzen oder sie aktiv betreiben. Hierbei sind
zahlreich solche Klein- oder Kleinstbetriebe vertreten, die sich ausschließlich und spezialisiert mit "exotischen" Fertigungsverfahren befassen. Daneben finden sich die Großunternehmen, wobei die Firmenorga-
nisation fast immer eine Kleingruppe von Spezialisten als "Nucleus" in
bezug auf die Lasertechnik vorsieht. Hieran zeigt sich, daß der Laser -
abgesehen von den bereits weit ausgereiften Anwendungen - derzeit noch
nicht als Werkzeugmaschine in der Produktionstechnik ebenso akzeptiert
wird wie beispielsweise ein Bearbeitungszentrum. Dies begründet auch
die Zurückhaltung der vielen mittelgroßen Firmen, sich intensiv mit der
Lasertechnik zu befassen: der anscheinend zur Produktionseinführung un-
erläßliche Spezialistenapparat werde das Unternehmen - auch wirtschaftlich - überfordern. Die Konsequenzen dieser Situation sind bilateral:
einerseits müssen sich die Anbieter aufgefordert sehen, solche Laserquellen zu entwickeln, die in bezug auf Bedienung, Zuverlässigkeit und
Wartungsfreundlichkeit den Ansprüchen des Produktionsbetriebes genügen,
diese in eine hochwertige Werkzeugmaschinentechnik zu integrieren und
eine abgesicherte Prozeßtechnologie zu gewährleisten.

Andererseits ist es an den Anwendern, das vielfach bestehende Wissens-
defizit im Hinblick auf den Laser abzubauen, und die oft mit dem Stich-

wort "Laser" verknüpfte konturlose Fortschrittshoffnung durch eine un-
ternehmerisch mutige, differenzierte, sachliche Auseinandersetzung mit
dieser Technik zu ersetzen. Den Hochschulen und Universitäten gilt dabei die Aufforderung, in der Ingenieurausbildung das hierfür notwendige
Rüstzeug zu vermitteln.

Ein weiterer Entscheidungsfaktor für die Lasereinführung in den Pro-
duktionsbetrieb liegt in den potentiellen Anwendungsfeldern. Hier ist
eine interessante Entwicklung festzustellen: Bei Anwendungen der Schneidtechnik war und ist hauptsächlich die Flexibilität und Arbeitsgeschwin-
digkeit im Verfahrensvergleich ausschlaggebend für die Entscheidung zu-
gunsten des Lasers und bestimmt damit auch meist den Einsatz in der
Kleinserien- und Einzelfertigung. Bei den gegenwärtig aufkommenden
Schweißanwendungen hingegen sind hohe Anforderungen hinsichtlich der
Bauteilqualität und die Automatisierbarkeit des Verfahrens maßgeblich:
der Einsatz verschiebt sich damit zur Großserie, was die aktuellen Pro-
duktionsanwendungen des Schweißens ja auch durchweg belegen. Hiermit
ist auch eine Verlagerung des wirtschaftlichen Effektes der Lasernutzung
verbunden, die nicht übersehen werden sollte: Die übergeordnete Wertschöpfung liegt im ersten Fall neben geringeren Bearbeitungskosten in
einer Verkürzung der Bauteildurchlaufzeit in der Fertigung und der damit verbundenen niedrigeren Kapitalbindung. Im zweiten Fall ist es die
Möglichkeit einer Optimierung des Fertigungsflusses und damit günstigeren Abstimmung von Fertigungsprozessen und -folgen oder etwa einer
energieärmeren Fertigung.

Hieran zeigt sich deutlich, daß bei Entscheidungen zum Lasereinsatz in
der Produktionstechnik über die reine Fertigung hinaus weitere Bereiche
des Unternehmens einbezogen und berücksichtigt werden müssen. Dies beginnt bei der "lasergerechten" Konstruktion von Bauteilen, d.h. einer
solchen Auslegung, die eine konsequente Nutzung technologischer Vorteile der Laserbearbeitung ermöglicht. In der Arbeitsvorbereitung muß
darüber hinaus die Fertigungsfolge so geplant und strukturiert werden,
daß die durch Laserbearbeitung eröffneten Produktivitätsreserven er-
schlossen werden. Das beinhaltet beispielsweise die Laserstrahlhärtung
im fertigbearbeiteten Zustand oder die Möglichkeit der Integration von
Laserschweiß- und -wärmebehandlungsanlagen in Fertigungslinien oder
flexible Fertigungssysteme.

Bereits zu Beginn der Auseinandersetzung mit der Lasertechnik begegnet
der potentielle Anwender aus der Produktionstechnik einem ihm unbekannten und größtenteils nicht vertrauten Anbietermarkt. Hinzu kommt noch,
daß ein junger, wachsender Industriezweig, wie die Lasertechnik ihn
gegenwärtig darstellt, typischerweise nicht gerade übersichtlich struk-

turiert ist. Dennoch sind auch hier bemerkenswerte Entwicklungen über
die letzten Jahre hinweg zu verzeichnen. So hat sich ein potentes An-
bieterspektrum konstituiert, die eigentlichen Laserquellenhersteller
haben sich in mehreren Fällen zu echten Systemanbietern konsolidiert.
Es sind dort inzwischen doch breite Erfahrungen in produktionstech-
nischen Anwendungen gereift. Viel Lehrgeld ist bezahlt worden, dafür
existieren jetzt aber auch erfolgreich produzierende Anlagen, die die
"Referenzen" der Anbieter ausmachen und dem potentiellen Nutzer eine
Vertrauensbasis vermitteln können. Dies zeigt sich auch in der Umstrukturierung der beteiligten Firmen, die aus bescheidenen Anfängen teilweise zu veritablen und renommierten Häusern gewachsen sind, womit in
der Marktentwicklung eine gewisse Segmentaufteilung und Bereinigung
verbunden ist. Insgesamt betrachtet, läßt sich feststellen, daß die
fachliche Kompetenz für produktonstechnische Anwendungen, die noch vor
4 - 5 Jahren fast ausschließlich auf seiten des interessierten und deshalb engagierten Nutzers lag, heute auch beim Anbieter vorhanden ist.
Dennoch muß die Frage gestellt werden, ob wirklich alles getan ist, um
das gesamte technische Potential der Industrie bei der Einführung des
Lasers in die Produktionstechnik bestmöglich auszuschöpfen. Diese Pro-
blemstellung ist im internationalen Vergleich der Bundesrepublik Deutschland besonders akut, wo es darum geht, eine gesicherte Marktstellung zu
erreichen und möglicherweise auszubauen.
In erster Linie muß hierbei der Werkzeugmaschinenbau angesprochen werden, der sich - abgesehen von einem frühen und äußerst erfolgreichen
Einstieg in die Schneidtechnik - bisher recht zurückhaltend in bezug
auf die Lasereinführung verhält. Äußerst bedeutsam erscheint in diesem
Zusammenhang der Kommentar von John Eckersley in der April-Ausgabe des
"Laser-Focus", der von einem aktuellen, gravierenden Umbruch im Laser-
systemgeschäft der USA durch das massive Kapital- und, wesentlicher
noch, knowhow-Engagement der bedeutenden Werkzeugmaschinenhersteller
berichtet. Die hierdurch entstehende wirtschaftliche und technische
Potenz muß auch den hiesigen Unternehmen zu denken geben: Die Werkzeug-
maschinenbauer haben mit dieser Angebotserweiterung ihrer amerikanischen Konkurrenten zu rechnen und darauf einzugehen, die Laserbauer
müssen ihren Status als reine Komponenten- oder Systemanbieter ebenfalls sorgfältig überdenken, um dieser zukünftigen Herausforderung gewachsen zu sein. Um auf den wirtschaftlichen Aspekt zurückzukommen: Es
geht einerseits um die Wertschöpfung durch Erschließung neuer Kunden-
kreise und andererseits um die Konsolidierung durch Aktualisierung der
Unternehmensziele.
Resümierend sind noch einmal die fördernden und hemmenden Faktoren bei

der Einführung des Lasers in der Produktionstechnik zu betrachten (Abbildung 6). Positiv wiegen die Flexibilität, hohe erzielbare Qualität und die Steuerbarkeit der Laserverfahren, die teilweise völlig neue Fertigungsperspektiven eröffnen. Hinzu kommt der wachsende Reifegrad der Systeme und schließlich die zunehmende Anwendungserfahrung.

Negativ wirkt sich die undifferenzierte Betrachtungsweise gegenüber der Lasertechnik aus, die durch ein Informationsdefizit bezüglich der technisch und wirtschaftlich vorteilhaften Anwendungsmöglichkeiten hervorgerufen wird. Neben dem hohen notwendigen Kapitalaufwand schreckt auch die Unsicherheit im unbekannten Anbietermarkt ab sowie die Tatsache, daß die vollständige Ausschöpfung der Vorteile oft erst durch eine Umstrukturierung der Fertigung erzielbar ist.

Betrachtet man den Wirtschaftsfaktor "Laser" in der Produktionstechnik, so muß dies Anwender und Anbieter gleichermaßen umfassen. Hierbei geht es zum einen um die Erschließung bedeutender Produktivitätsreserven durch den Lasereinsatz, zum anderen um den Weg vom ehemals exotischen Randbereich zur Schlüsselindustrie. Wird durch ein engagiertes Zusammenwirken aller Beteiligten die vielversprechende Entwicklung der letzten Jahre weitergeführt, so scheinen diese hochgesteckten Ziele - auch im internationalen Vergleich - durchaus erreichbar.

Hiermit schließlich wird auch die eingangs gestellte Frage beantwortet: "Wieviel Laser braucht die Produktionstechnik?"

Die industrielle Fertigungstechnik braucht gegenwärtig und zukünftig genau soviel Laser, wie diese Technik zur Erfüllung der Unternehmensziele beiträgt. Wieviele Laser dies sind und sein werden, hängt einerseits vom Vermögen der Anwender ab, diese Technik vorteilhaft einzusetzen, und andererseits von der Fähigkeit der Entwickler, hierfür das geeignete Werkzeug bereitzustellen.

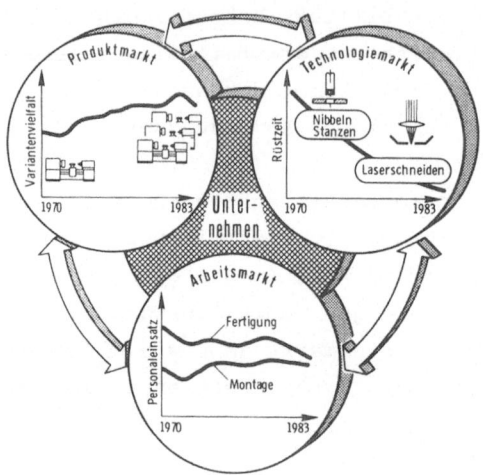

Abbildung 1.

Einflußfaktoren auf

das Unternehmen

Abbildung 2.

Laserstrahlhärten einer

Getriebewelle

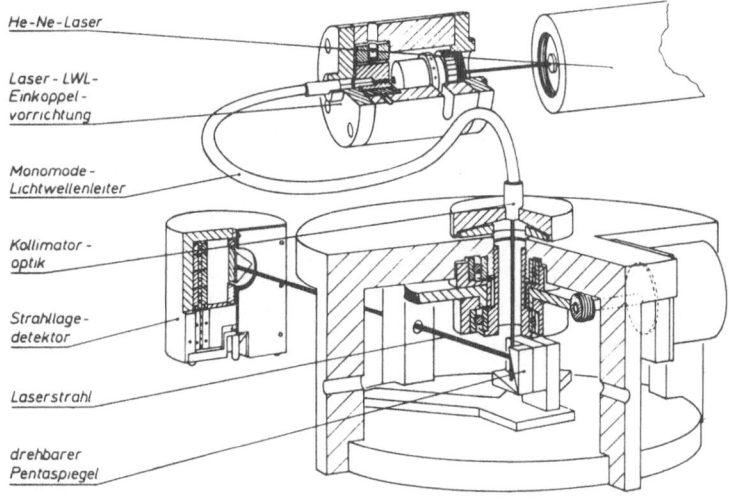

He-Ne-Laser

Laser-LWL-
Einkoppel-
vorrichtung

Monomode-
Lichtwellenleiter

Kollimator-
optik

Strahllage-
detektor

Laserstrahl

drehbarer
Pentaspiegel

Abbildung 3.   Laserebenheitsmeßsystem

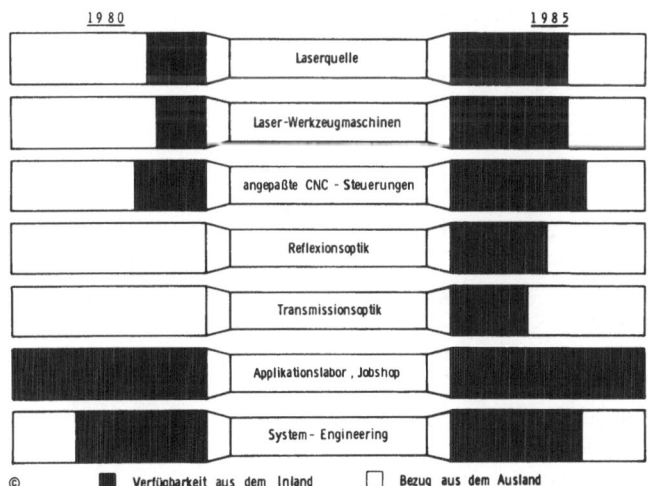

1980   1985

Laserquelle

Laser-Werkzeugmaschinen

angepaßte CNC - Steuerungen

Reflexionsoptik

Transmissionsoptik

Applikationslabor , Jobshop

System- Engineering

© ■ Verfügbarkeit aus dem Inland   □ Bezug aus dem Ausland

Abbildung 4.   Verfügbarkeit der Lasertechnik

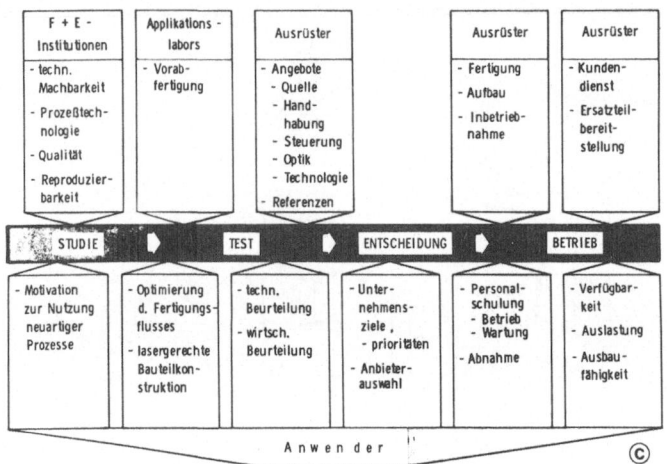

Abbildung 5.   Problemstellungen bei der Lasereinführung

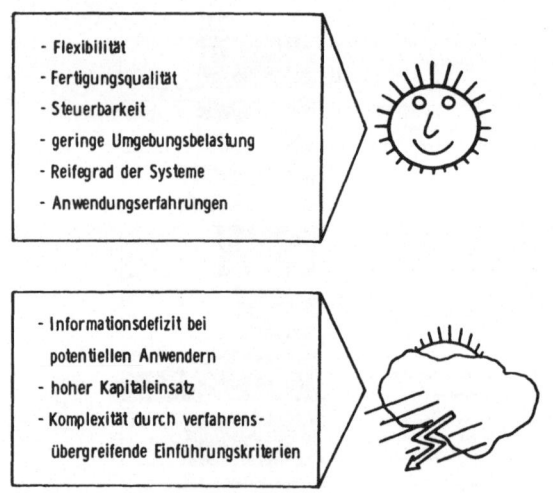

Abbildung 6.   Fördernde und hemmende Faktoren bei der
Einführung des Lasers in die Produktionstechnik

Entwicklungen und Entwicklungstrends des Lasers in der Medizin

G.J. Müller, Laser-Medizin-Zentrum GmbH, Kramerstr. 6-10, 1000 Berlin / D

·Es werden kurz die wichtigsten biophysikalischen Wechselwirkungen des Lasers aufgezeigt und dabei anhand von einigen Beispielen der Stand der heutigen Technologie demonstriert.

Daran anschließend wird anhand einiger statistisch aufgearbeiteter Marktdaten die wirtschaftliche Bedeutung dieser neuen Geräteklasse in der Medizin verdeutlicht und zum Schluß werden einige zukünftige Trends aufgezeigt, die insbesondere deutlich machen, daß es nicht allein der bzw. die Laser als Instrument in der Hand des Chirurgen sind, die dieses Geschäftsfeld beeinflussen, sondern mit wachsender Tendenz in praktisch gleichem Umfange das für diese spezielle Art der Therapie notwendige Zubehör.

---

* nähere Informationen erhalten Sie über das VDI Technologiezentrum, Physikalische Technologien, Graf-Recke-Strasse 84, 4000 Düsseldorf 1 / D

Der Excimerlaser – Ein kurzer Überblick über die Technik, seine Anwendungen und die Marktaussichten

D. Basting

Lambda Physik GmbH

Hans-Böckler-Straße 12, 3400 Göttingen

Der Excimerlaser ist einer der wenig "wirklich guten" Laser. Was sind nun gute Laser? Als Definition gilt, daß sie kommerziell erfolgreich sind und Eigenschaften besitzen, die von größerer als wissenschaftlicher Bedeutung sind, die, wenn es um industrielle Anwendungen geht, eine gewisse elektrische Effizienz aufweisen und die in der Leistung skalierbar sind.

In den 25 Jahren des Lasers sind eine Vielzahl von Lasertypen erfunden worden. Die Zahl der tatsächlichen aber eingesetzten Laser ist recht gering.

Bild 1 zeigt die Wellenlängenbereiche, in denen einige der wichtigsten Laser arbeiten.

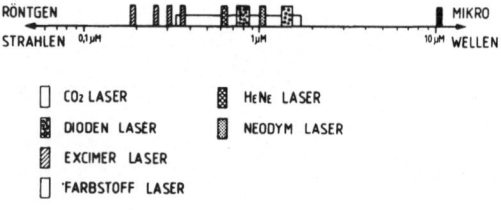

Bild 1. Wellenlängenbereiche einiger wichtiger Lasertypen

Das Paradebeispiel für einen wissenschaftlichen Laser ist der Farbstofflaser, der die schmalste Brandbreite und die kürzeste Pulsbreite und kontinuierliche Abstimmbarkeit von 320 nm bis 1.800 nm bietet und das Gebiet der optischen Spektroskopie revolutioniert hat.

Diodenlaser finden aufgrund ihrer Nahinfrarotwellenlängen und miniaturisierten Formen Anwendung in der optischen Kommunikation. Während Helium-Neon-Laser meßtechnisch und zum optischen Lesen von Daten verwendet werden. Die verbleibenden 3 Laser: der CO_2-Laser, der Nd:YAG-Laser und der Excimerlaser können im weiteren Sinn in der Materialbearbeitung angewendet werden. Ein äußerst wichtiger Punkt ist hierbei, daß sich mit der Verwendung von Excimerlasern völlig neue Aspekte in der Materialbearbeitung eröffnen.

Der CO_2-Laser ist auf der Wellenlängenskala im Infraroten angeordnet, seine nieder-
frequente-niederenergetische Strahlung führt zum Aufheizen des Materials. Dies kann
Erwärmen, Schmelzen und Verdampfen sein und bei hohen Intensitäten Plasmen erzeugen. Im Gegensatz dazu werden bei Excimerlasern mit Wellenlängen im UV Photonen-
energien erreicht, die zu elektronischen Anregungen zur Ionisation und zum Auf-
brechen chemischer Bindungen führen kann.

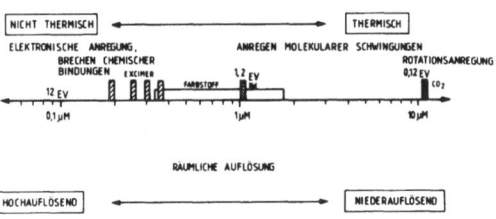

Bild 2. Typische Wechselwirkungen von CO_2-Laser, Neodymlaser und Excimerlaser mit
Materie

Damit ist die Möglichkeit gegeben, substanzielle Energien in absolut nichtthermische Verteilungen zu übertragen und strukturelle Veränderungen in kalten Medien
herbeizuführen. Natürlich sind auch thermische Heizprozesse möglich, indem man die
elektronischen Anregungen relaxieren läßt. Man kann erwarten, daß bei der Anwendung
von Excimerlasern völlig neue und interessante Effekte auftreten werden. Die
Wellenlängenskala enthält noch eine weitere wichtige Information, die eine Besonderheit des Excimerlasers zeigt, nämlich dann, wenn man sich um die Herstellung und
Erkennung kleiner Strukturen kümmert.

Prinzipiell läßt sich der Excimerlaserstrahl feiner bündeln als der eines CO_2-
-Lasers, nämlich entsprechend dem Verhältnis der Wellenlängen, d.h., der Excimerlaser läßt sich etwa 50 mal feiner bündeln als der Strahl des CO_2-Lasers. Damit
sind Excimerlaser zur Erzeugung besonders kleiner Strukturen geradezu prädestiniert.

Unter Excimerlasern versteht man eine Gruppe von gepulsten Hochdruckgaslasern, die
alle bei verschiedenen UV-Wellenlängen emittieren und die in der Lage sind, sowohl
sehr hohe Spitzenleistungen im Megawattbereich als auch beträchtliche mittlere
Leistungen im 100 Wattbereich abgeben. Gemeinsam ist diesen Lasern, daß ihre oberen
Laserzustände Excimers sind. Im Falle der kommerziellen Excimerlaser sind das
zweiatomige elektronisch angeregte Moleküle, die aus einem Edelgas und einem
Halogenatom entstehen.

Excimer haben die Besonderheit, daß ihr elektronischer Grundzustand entweder anti-
bindend oder nur äußerst schwach bindende Potentialkurven aufweist und das Molekül
damit im Grundzustand praktisch nicht existiert. Dieses Thermdiagramm erlaubt

besonders effizienten Betrieb (Bild 3).

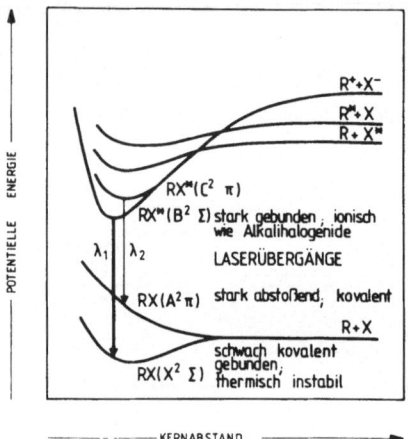

Bild 3. Potentialkurven eines zweiatomigen Edelgashalogenid-Excimers. R: Edelgas-atom, X: Halogenatom; der Stern deutet die elektronische Anregung an

Kommerziell verwendet werden Edelgashalogenide, wie ArF, KrF, XeF und XeCl. Die Emissionswellen liegen im Vakuum-UV bei 157 nm (F2) und gehen bis 350 nm im nahen UV (XeF) (Bild 4).

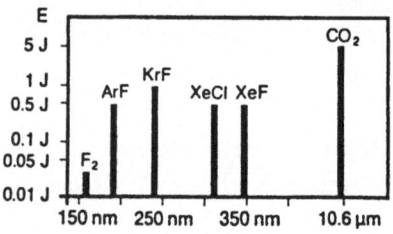

Bild 4. Emissionswellenlängen von Excimerlasern.

Der erste Excimerlaser wurde 1975 entwickelt. Ein Jahr darauf brachte Tachisto die leicht modifizierte Version eines CO_2-Lasers auf den Markt, der auch als Excimer-laser verwendbar war. 1977 führte Lambda Physik den ersten Laser, der bezüglich seiner Designparameter und der Halogenkompatibilität ganz als Excimerlaser ausge-legt war, auf den Markt. Das Bild (5) zeigt wie sich die Leistungsfähigkeit des Excimerlasers seit seinen Anfängen in den letzten Jahren vergrößert hat.

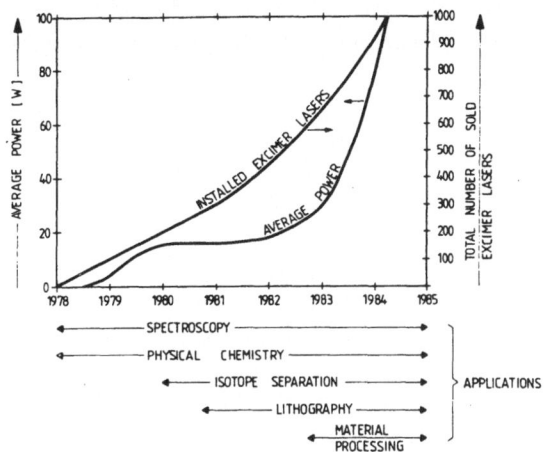

Bild 5. Entwicklung der Leistungsfähigkeit der Excimerlaser

Mit steigender Leistungsfähigkeit des Lasers und zunehmenden Bekanntheitsgrad hat sich auch die Zahl der verkauften Lasergeräte erhöht. Bis im Jahre '84 hat sie schon weltweit mehr als 1.000 Stück betragen. Wie bei allen neuen Lasergeräten war der Einsatz am Anfang im wesentlichen in der universitären Forschung in der Spektroskopie und in der physikalischen Chemie. Man sieht, daß in den letzten Jahren zu den rein wissenschaftlichen Anwendungen immer stärker industriell ausgerichtete Anwendungen dazukommen.

Viele Anwendungen erfordern aber höhere Leistungen. Wir erwarten, daß noch vor dem Jahre 1990 Excimerlaser mit Ausgangsleistungen über 1 kW zur Verfügung stehen werden (Bild 6).

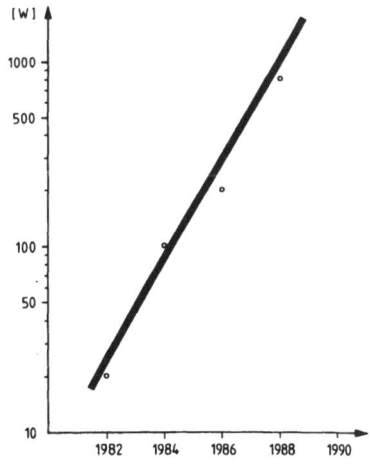

Bild 6. Erwartete Entwicklung der Ausgangsleistung von Excimerlasern

Die Vielzahl der wissenschaftlichen Einsatzmöglichkeiten des Excimerlasers ist in Publikationen eindrucksvoll dargelegt worden. Aus einigen besonders praxisnahen wissenschaftlichen Anwendungen haben sich in letzter Zeit aber auch die ersten industriellen Anwendungen für den Excimerlaser ergeben.

Industrielle Anwendungen

Im Industriebereich sind bereits der CO_2- und der Nd:YAG-Laser fest etabliert. Beide bedienen im wesentlichen den Materialbearbeitungssektor im weitesten Sinne (Bild 7).

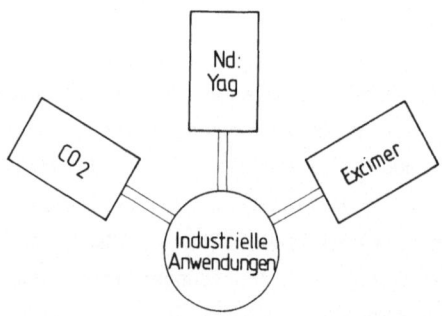

Bild 7. Laser in der Materialbearbeitung

Der Excimerlaser kommt nun als dritter Laser hinzu und schafft sich aufgrund seiner ganz andersartigen physikalischen Eigenschaften einen ganz neuen Markt, der - so die Prognose - wesentlich vielfältiger als der der herkömmlichen Industrielaser sein wird. Die hohe Pulsenergie bzw. Spitzenleistung des Excimerlasers ermöglicht auch, durch Plasmaerzeugung und Ionisation Lichtemission im Vakuum-UV bis ins Röntgengebiet zu erzeugen. Solche Lichtquellen werden in einigen Jahren für die Lithographie hochintegrierter Schaltungen zunehmend interessant. Für viele Anwendungen steht die Frage der Effizienz, d.h. der Umwandlung der elektrischen Leistung in Laserleistung im Vordergrund. Hier ist der Excimerlaser von den üblichen kommerziell eingesetzten und gepulsten Systemen in der Effizienz an zweiter Stelle, wobei er gleichzeitig die - physikalisch gesehen - wesentlich schwieriger zu erzielenden UV-Wellenlängen emittiert (Tabelle 1).

CO_2	10,6 u	5%
Excimer	248 nm	2,9%
Nd:YAG	1,06 u	0,6%
Nd:YAG	532 nm	0,2%
Nd:YAG	355 nm	0,1%
Ar^+	Vis	0,006%

*Effizienz = Steckdoseneffizienz = $\dfrac{\text{Laserleistung [W]}}{\text{elektrische Leistung [W]}}$ 100%

Tabelle 1. Effizienz kommerzieller Pulslaser

In Tabelle 2 sind einige industrielle Anwendungen aufgezählt. Sie sind zum Teil unterschiedlich weit vorangeschritten. Während sich einige noch in der Laborerprobung befinden, sind andere bereits im industriellen Einsatz.

- Isotopentrennung
- Annealing
- CVD (Chemical vapour deposition)
- Markierung
- Photochemisches Stanzen
- Lithographie
- Augenmedizin
- Materialbearbeitung.

Tabelle 2. Industrielle Anwendungen von Excimerlasern

Eine weit fortgeschrittene Anwendung des Excimerlasers ist die Ausnutzung seiner kurzen Wellenlänge bei der Herstellung von integrierten Schaltkreisen mit höchster Integrationsdichte. Bild 8 zeigt, wie sich die gewünschte Auflösung in den nächsten Jahren im Bereich der Produktion weiter entwickeln wird.

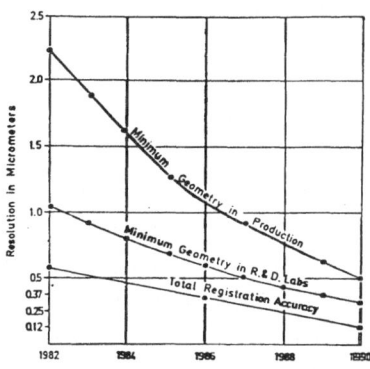

Bild 8. Entwicklung der gewünschten Auflösung von integrierten Schaltungen

Im Jahre '84 ist es der Firma Carl Süss mit einem Lambda Physik Excimerlaser gelungen, lithographische Strukturen im Bereich von 0.3 bis 0.5 um zu erzielen. Bild 9 zeigt die Präzision und Auflösung dieses Verfahrens.

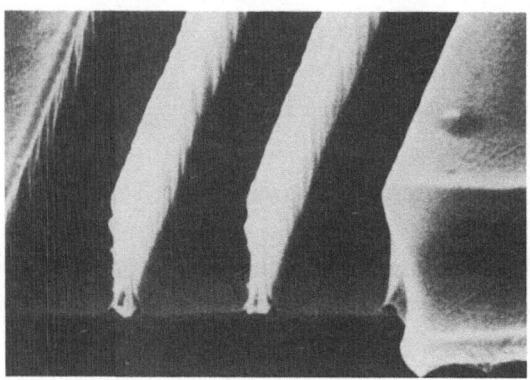

Bild 9. Halbleiterstrukturen im Submikrometer-Bereich in einem Photoresist (Firma Süss, München)

Verwendet wird dabei das Kontaktdruckverfahren (Bild 10). Aber auch bei modernen Projektionsverfahren sind Mikrostrukturen herstellbar. Auch hier sind amerikanische Gerätehersteller führend in der Erprobung von Excimerlaser-Projektionsverfahren. Das Potential für den Excimerlaser liegt in diesem Bereich zwischen einigen 100 bis 1000 Geräten in 2 Jahren.

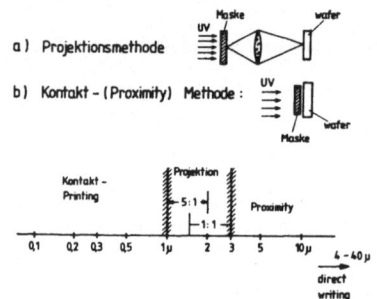

Bild 10. Prinzip des Kontaktdruckverfahrens

Eine weitere und vielleicht langfristig noch interessantere Anwendung ist das photochemische Stanzen, der sogenannte Srinivasan-Effekt, bei dem man mit Hilfe des Excimerlaser UV-Lichtes bei Kunststoffen wie Mylar und Kapton mikroskopisch scharf definierte Bereiche bearbeiten kann. Diese Mikromaterialbearbeitung unterscheidet sich von der herkömmlichen Materialbearbeitung dadurch, daß sie auf einem photo-chemischen Effekt beruht und daher völlig hitze- und zerstörungsfrei für den umgehenden Bereich abläuft. Der Effekt tritt bei geringen Leistungsdichten von etwa 100 mJ pro cm^2 bereits auf. Wie fein mit diesem Effekt Bearbeitungen durchgeführt werden können, zeigt einmal ein menschliches Haar (Bild 11), auf dem Einschnitte durchgeführt wurden, als auch eine Kaptonfolie, in der Löcher und Borhungen unterschiedlicher Größe ausgeführt wurden (Bild 12)

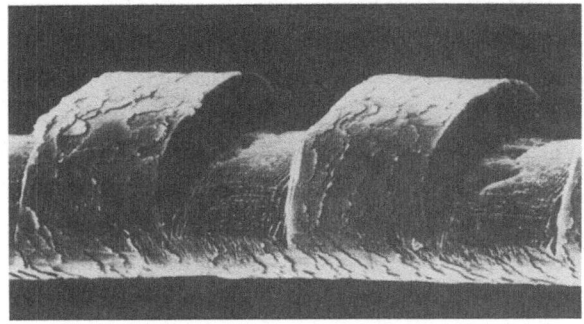

Bild 11. Einschnitte in einem menschlichen Haar

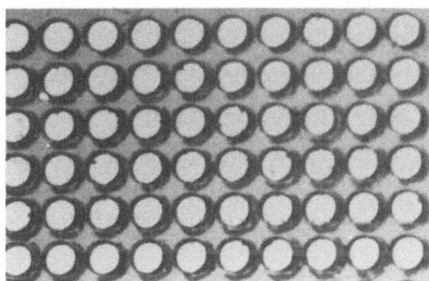

Bild 12. Bohrungen in hitzebeständiger Kaptonfolie

Das Verfahren wird bereits im kleineren Maßstab zur Herstellung von intraokularen Plastiklinsen, zur Herstellung eines flexiblen Solarzellensegels für die amerikanische Raumfähre "Discovery", Perforierung von Folien und vielen anderen - zum Teil noch firmengeschützten-Anwendungen - verwendet.

Medizinische Anwendungen

Eine spektakuläre und in ihrem Potential noch nicht absehbare Anwendung wurde in Zusammenarbeit zwischen der Columbia-Universität, Prof. Trocel, Dr. Srinivasan - von IBM und Lambda Physik gefunden. Es ist mit dem Excimerlaser möglich, die Hornhaut des Auges so einzuschneiden, daß sich eine Änderung der Brechkraft der Linse ergibt. Man hofft, dadurch extreme Kurzsichtigkeit behandeln zu können (Bild 13).

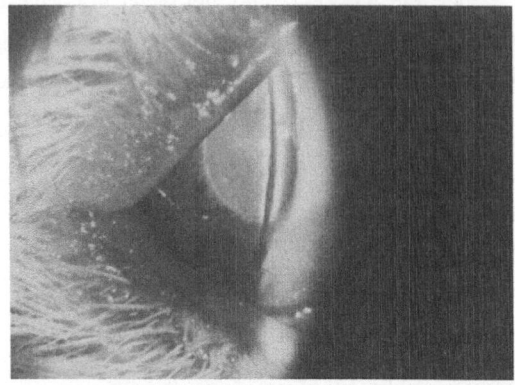

Bild 13. Hornhautschnitte mit einem Excimerlaser

Wachstumschancen

Das Wachstum des Excimerlasermarktes betrug in den letzten Jahren ca. 30% p.a.

Bei den rein wissenschaftlichen Anwendungen ist mit einer kontinuierlichen Steigerung bis zum Ende der 80iger Jahre zu rechnen.

Die in diesem Artikel beschriebenen Anwendungen lassen aber darüberhinaus gute Wachstumschancen in den Bereichen Mikroelektronik, Materialbearbeitung und Medizin erwarten.

Vorsichtige Prognosen gehen davon aus, daß bis 1990 der Excimerlaser neben dem CO_2- und dem Nd:YAG-Laser der Hauptumsatzträger auf dem zivilen Lasermarkt sein wird (Bild 14).

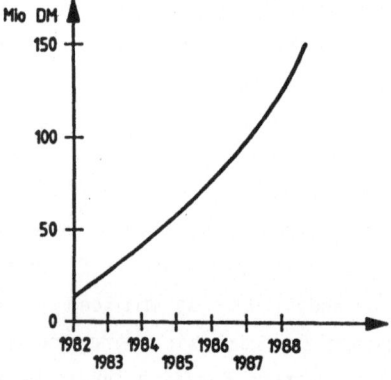

Bild 14. Wachstumschancen des Excimerlaserumsatzes in den 80iger Jahren

Die Bedeutung staatlicher Fördermaßnahmen auf dem Gebiet der Lasertechnik im Vorfeld wirtschaftlicher Nutzung

R. Röhrig

Bundesministerium für Forschung und Technologie
Heinemannstr. 2, D-5300 Bonn 2

25 Jahre nach Erfindung des Lasers kann kein Industrieland auf den industriellen Einsatz des Lasers verzichten. Wenn heute die Lasertechnik mit Wachstumsraten von 30 % und höher in einen schnell wachsenden Markt vordringt, wenn die Bedeutung des Lasers verglichen wird mit anderen Schlüsseltechnologien, so ist das auch ein Ergebnis intensiver staatlicher Förderung von Forschung und Entwicklung in nahezu allen Industriestaaten in der Vergangenheit (1)(2).

Vor der Schilderung der Situation des eigenen Landes soll kurz auf staatliche Maßnahmen in USA und Japan eingegangen werden, also jener Staaten mit denen sich die Bundesrepublik Deutschland in der Erschließung von Hochtechnologien messen lassen muß.

USA: Die Bundesregierung der Vereinigten Staaten, insbesondere die Ministerien für Verteidigung (DOD) und Energie (DOE) gibt im Jahr 1985 rd. 936 Mio$ für die Entwicklung von Laserlichtquellen, Lasersystemen und der zugehörigen Anwendungstechniken aus. Nicht eingerechnet sind hierbei die Ausgaben der National Science Foundation und der NASA. Zur Erfüllung der Verteidigungsaufgaben werden eine Reihe von strategischen und taktischen Lasersystemen entwickelt wie z. B. Festkörperlaser, chemische Laser, Excimer- und Freie Elektronen-Laser. Für die Energietechnik werden große Laserprojekte durchgeführt insbesondere in den Bereichen Trägheitsfusion und Isotopentrennung (z. B. für Uran).

Die Steigerung der Haushalte von DOD und DOE vom Fiskaljahr 1985 auf 1986 wird entsprechend den Anforderungen der US-Bundesregierung an den Kongreß bei über 50 % liegen. Diese erhebliche Steigerung auf rd. 1.415 Mio$ ist in erster Linie auf die "Strategische Verteidigungs-Initiative "(SDI) des Präsidenten der Vereinigten Staaten zurückzuführen. Zum Vergleich: Mit diesem Betrag allein für die Lasertechnik wird in den USA eine finanzielle Größenordnung erreicht, die über dem Betrag liegt, der dem Bundesminister für Forschung und Technologie für die gesamte Förderung industrieller FuE zur Verfügung steht.

Diese von der amerikanischen Bundesregierung geförderten Entwicklungen auf dem militärischen Sektor werfen in aller Regel nur indirekten Nutzen für die zivile Laserentwicklung ab. Die Wirkung dieser hohen staatlichen Förderung in den USA liegt vielmehr darin, daß

- FuE-Kapazitäten auch in der Industrie aufgebaut und vorgehalten werden können,

- eine hohe Zahl hochqualifizierter Wissenschaftler und Techniker auch für zivile Anwendungen verfügbar wird,

- ansonsten unwirtschaftliche Entwicklungen und Produktionen vorfinanziert und damit möglich werden, z. B. in der Kristallzucht für Festkörperlaser oder bei optischen Komponenten.

Der zivile Spin-off der in den USA eingesetzten staatlichen Mittel des DOD und DOE wird mit rd. 10 bis 15 % beziffert. Die aus dieser Abschätzung folgende Größenordnung von 300 bis 450 Mio DM jährlich zeigt, wie groß die internationale Wettbewerbsverzerrung anzusetzen ist.

Japan: Schon Mitte der 70er Jahre hat man in Japan die Bedeutung des Lasers für die flexible Fertigung erkannt. Dabei sollten die Eigenschaften des Lasers genutzt werden, als ein verschleißfrei und berührungslos arbeitendes Werkzeug eingesetzt zu werden und in seinem Strahlungs-output vom Rechner steuerbar zu sein. Nach einem rd. zweijährigen Abstimmungs- und Planungsprozeß förderte das japanische Ministerium für internationalen Handel und Industrie (MITI) von 1977 bis 1985 ein nationales Großprojekt mit dem Thema "Flexible Manufacturing System Complex provided with Lasers" (FMS). Hierfür wurden umgerechnet 150 Mio DM ausgegeben (3).

Im Rahmen des FMS-Projekts wurden u.a. ein 26 kW CO_2-Laser, ein 600 W Nd:YAG-Laser (einschließlich Kristallzucht) und zahlreiche optische Komponenten (KCl; ZnSe) entwickelt. Darüber hinaus wurde die Einsatzfähigkeit des Lasers in der Materialbearbeitung demonstriert. Typisch für die japanische Vorgehensweise war, daß ausschließlich für dieses Großprojekt ein Konsortium, bestehend aus verschiedenen Firmen und staatlichen Forschungsinstituten gegründet wurde. Weitere Laser-Fördergebiete in Japan waren in den letzten 10 Jahren die Anwendung des Lasers in der Medizin (Laser-Skalpel), die optische Meßtechnik (OMCS-Projekt; rd. 220 Mio DM) und Laser-Isotopentrennung für Uran.

Das Ergebnis der staatlichen Förderung durch das MITI wurde von einer BMFT-Delegation von Laserexperten in Augenschein genommen, die im Juni 1985 Japan bereiste. Danach hat die japanische Industrie einen Vorsprung von 2 bis 3 Jahren bei der Entwicklung von Laserstrahlquellen (Ausnahme: Excimerlaser). Demgegenüber ist der Stand in der Materialbearbeitung vergleichbar dem in der Bundesrepublik Deutschland (4). Bemerkenswert ist, daß mit Unterstützung des MITI in den letzten Jahren ein Wirtschaftsverband(OITDA)entstand, der sich ausschließlich auf dem Gebiet der Optoelektronik betätigt (siehe Vortrag von Dr. Maruyama auf der Laser '85).

Insgesamt plant Japan, im 21. Jahrhundert die weltweite technologische Führerschaft einzunehmen. Um dieses Ziel in der Lasertechnik zu erreichen, werden derzeit weitere nationale Großprojekte vorbereitet wie:

- Entwicklung von Excimer-Lasern der 1 kW-Klasse (Fernziel: 10 kW).
- Anwendung von Leistungslasern zusammen mit fortgeschrittenen Robotertechnologien.

Notwendigkeit staatlicher Maßnahmen in der Bundesrepublik Deutschland: Angesichts massiver staatlicher Hilfen vor allem in den USA und Japan, sowie des Nachholbedarfs auf deutscher Seite hält der BMFT eine verstärkte Förderung der angewandten Laserforschung, der industriellen Laserentwicklung und der Anwendungstechniken für erforderlich. Denn die Beherrschung der Lasertechnik im eigenen Land und eine höhere Abdeckung des Inlandbedarfs an Laserquellen und Laserkomponenten ist entscheidend für die künftige Wettbewerbsfähigkeit von erheblichen Teilen deutscher Traditionsbranchen wie dem Maschinen- und Fahrzeugbau, der Nachrichten- und Elektrotechnik. Es gilt:

- Die deutsche Industrie darf in der Lasertechnik nicht vom Ausland abhängig bleiben, z. B. wegen langer Lieferfristen bis hin zu Embargoproblemen.
- Die Industrie muß sich auf qualitativ hochstehende Zulieferung aus eigener Produktion verlassen können (Versorgungssicherheit).
- Der ungewollte Know-how-Abfluß zu Laserherstellern im Ausland sollte reduziert werden.
- Erforderlich ist eine hochentwickelte Anwendungstechnik in einer Vielzahl von Industriezweigen.

Förderung der Lasertechnik durch den BMFT: Als Ziele der ab 1986 verstärkten Förderung werden gesetzt:

- Schaffung einer Basis für eine eigenständige und international wettbewerbsfähige Laserindustrie
- Aufbau einer FuE-Infrastruktur in der angewandten Laserforschung.

Hierzu ist für 1985 ein Haushaltsansatz in Höhe von 27 Mio DM (1985: 15 Mio DM) vorgesehen. Für den Zeitraum 1986 bis 1989 ist

eine Fördersumme von rd. 140 Mio DM eingeplant (bisher rd. 59 Mio
DM). Gefördert werden sollen im Rahmen von Projekten:
- Laserstrahlquellen und Laserkomponenten
- Lasersysteme bis zum Funktionsnachweis
= Verfahrensentwicklung und Anwendungstechniken.

Der Schwerpunkt der Maßnahmen wird im Bereich Laser und Laseranwendungen für die Materialbearbeitung, Meßtechnik und Analytik sowie Medizin
liegen. (Darüber hinaus werden Projekte im Rahmen des EUREKA-Proramms
geplant). Die oben genannten Maßnahmen umfassen nicht die
Gebiete "Integrierte Optik", die Entwicklung von Halbleiterlasern
für die Nachrichtentechnik, die Laserfusion und die Urananreiche-
rung, die anderen Förderschwerpunkten des BMFT zugeordnet sind.

Die Maßnahmen auf dem Gebiet der Lasertechnik werden vom VDI-
Technologiezentrum "Physikalische Technologien", Düsseldorf
getragen. Sie werden ergänzt durch Technologie-Transfer-Akti-
vitäten wie Informationsvermittlung, Beratung, Teilnahme an
Messen. Die vom VDI-Tz durchgeführte kontinuierliche Anfinanzierung
des Lasergebiets in den letzten Jahren hat wesentlich dazu beigetra-
gen, daß die nun verstärkt einsetzende BMFT-Förderung auf fruchtbaren
Boden fällt. Die Chancen für den Erfolg der Maßnahmen sind gut, wenn
es weiterhin verstärkt gelingt, in der Lasertechnik das bisherige
gemeinsame Vorgehen von Wirtschaft, Wissenschaft und Staat auszubauen.

Literatur

(1) Roger Main: Lasers & Applications $\underline{6}$, 73 (1985)
(2) Jeff Hecht: Lasers & Applications $\underline{6}$, 65 (1985)
(3) T. Maruyama, T. Satoh: Lasers & Applications $\underline{2}$, 71 (1984)
(4) L. Cleemann: LASER MAGAZIN $\underline{3}$, 6 (1985)

Laser/Optoelektronik in der Medizin 1985

Laser/Optoelectronics in Medicine 1985

Vorträge des 7. Internationalen Kongresses
Proceedings of the 7th International Congress
Laser 85 Optoelektronik

Herausgeber/Editor: **W. Waidelich, P. Kiefhaber**

1986. Etwa 500 Seiten
Broschiert DM 108,-. ISBN 3-540-16018-3

Inhaltsübersicht: Sitzungsleiter. – Referenten. – Laser-
Dermatologie. – Laser-Chirurgie. – Laser-Gynäkologie. –
Photodynamische Therapie. – Laser-Biostimulation. –
Laser-Sicherheit. – Optoelektronische Meßverfahren und
Laser für medizinische Anwendungen. – Laser-Photobio-
logie. – 2nd International Nd: YAG Laser Conference
(New Developments; Gastroenterology; Neurosurgery;
Ophthalmology; Otolaryngology; Pulmonary Cardiology;
Dermatology; Oral Surgery; Gynaecology; Urology).

In der Optoelektronik werden neue Forschungsergeb-
nisse in rasantem Tempo in technische Entwicklungen
und Anwendungen umgesetzt. Die seit 1973 alle zwei
Jahre veranstaltete Kongressmesse LASER OPTO-
ELEKTRONIK ermöglicht einen überblick über den aktuellen Stand.

Das Programm von LASER 85 OPTO-ELEKTRONIK
umfaßte – wie seine Vorgänger – wieder ein breites
Spektrum der Bereiche Technik und Medizin. In Basis-
Seminaren wurden physikalische Grundlagen erläutert,
Plenarvorträge vermittelten eine Übersicht über wichtige
Fachgebiete, in Einzelvorträgen wurden neue Ergebnisse
mitgeteilt.

Springer-Verlag
Berlin Heidelberg
New York Tokyo

Der vorliegende Band „Laser/Optoelektronik in der
Medizin 1985" enthält die Vorträge des 7. Internationalen
Kongresses mit medizinischen Themenschwerpunkten.

Springer

Optoelektronik in der Medizin 1983
Optoelectronics in Medicine 1983

Vorträge des 6. Internationalen Kongresses
Proceedings of the 6th International Congress
Laser 83 Optoelektronik

Herausgeber/Editor: **W. Waidelich**

1984. 181 Abbildungen. XV, 273 Seiten
Broschiert DM 78,-. ISBN 3-540-12778-X

Inhaltsübersicht: Highlights in der Lasermedizin. – Laser in der Neurochirurgie. – Laser in der Chirurgie. – Laser und Infrarot Koagulator in der Gastroenterologie. – Laser in der Urologie. – Laser in der Gynäkologie. – Laser in der Dermatologie. – Laser Photomedizin. – Laser Photobiologie. – Laser Optoelektronische Diagnosesysteme. – Laser Sicherheit.

Optoelektronik in der Technik 1983
Optoelectronics in Engineering 1983

Vorträge des 6. Internationalen Kongresses
Proceedings of the 6th International Congress
Laser 83 Optoelektronik

Herausgeber/Editor: **W. Waidelich**

1984. 537 Abbildungen. XXII, 680 Seiten
(320 Seiten in Englisch)
Broschiert DM 118,-. ISBN 3-540-12779-8

Inhaltsübersicht: Laser-Systeme für Forschung. – Optoelektronische Komponenten und Sensoren. – Optronisches und lasertechnisches Messen und Prüfen. – Laser in der Materialbearbeitung. – Optoelektronische Signalübertragung. – Laser und Optoelektronik in der Weltraumtechnik. – Laser in der Umweltmeßtechnik. – Optoelektronische Displays. – Laser-Chemie. – Laser Sicherheit. – Photovoltaische Solartechnik. – VDI-Technologiezentrum: Industrielle Umsetzung der Lasertechnologie.

Springer-Verlag
Berlin Heidelberg New York Tokyo

Optoelectronics in Medicine

Proceedings of the 5th International Congress
Laser 81

Editor: **W. Waidelich**

1982. 150 figures (11 figures in colour).
XI, 239 pages
Soft cover DM 62,-. ISBN 3-540-10968-4

Contents: Milestones in Laser Medicine. – Laser in Surgery. – Laser in Urology. – Laser in Dermatology. – Laser in Gynaecology. – Laser in Otorhinopharyncology. – Photobiology and Laser Photomedicine. – Laser and Optoelectronics in Medical Diagnosis. – Laser in Dental Technique.

Optoelektronik in der Technik
Optoelectronics in Engineering

Vorträge des 5. Internationalen Kongresses
Proceedings of the 5th International Congress
Laser 81

Herausgeber/Editor: **W. Waidelich**

1982. 504 Abbildungen. XXII, 580 Seiten
(266 Seiten in Englisch)
Broschiert DM 88,-. ISBN 3-540-10969-2

Contents: Laser Systeme/Laser Systems. – Laser-Spectroskopie und Laser-Chemie/Laser Spectroscopy and Laser-Chemistry. – Lasertechnisches Messen und Prüfen/Laser Measurement and Testing. – Laser in der Materialbearbeitung/Lasers in Material Processing. – Laser in der Umweltmesstechnik/Lasers in Environmental Measuring Techniques. – Laser und Optoelektronik in der Weltraumtechnik/Lasers and Optoelectronics in Space Techniques. – Optoelektronische Komponenten/Optoelectronic Components. – Optoelektronische Signalübertragung/Optoelectronic Signal Transmission. – Optoelektronische Bildaufnahme/Optoelectronic Image Pickup. – Optoelektronische Bild- und Datenaufzeichnungen/Optoelectronic Image and Data Recording. – Optoelektronische Bildverarbeitung/Optoelectronic Image Processing. – Optoelektronische Solartechnik/Optoelectronic Solar Technique.

Springer